Geocryology

Geocryology

Characteristics and Use of Frozen Ground and Permafrost Landforms

Stuart A. Harris
Department of Geography, University of Calgary, Calgary, Alberta, Canada

Anatoli Brouchkov
Geocryology Department, Faculty of Geology, Moscow State University, Moscow, Russia

Cheng Guodong
Cold and Arid Regions Environmental and Engineering Research Institute, Chinese Academy of Sciences, Lanzhou, China

CRC Press
Taylor & Francis Group
Boca Raton London New York

CRC Press is an imprint of the
Taylor & Francis Group, an **informa** business

A BALKEMA BOOK

CRC Press
Taylor & Francis Group
6000 Broken Sound Parkway NW, Suite 300
Boca Raton, FL 33487-2742

First issued in paperback 2020

ISBN-13: 978-1-138-05416-5 (hbk)
ISBN-13: 978-0-367-52895-9 (pbk)

Typeset by MPS Limited, Chennai, India

Library of Congress Cataloging-in-Publication Data

Names: Harris, Stuart A., author.
Title: Geocryology : Characteristics and Use of Frozen Ground and Permafrost Landforms / Stuart A. Harris, Department of Geography, University of Calgary, Calgary Alberta, Canada, Anatoli Brouchkov, Geocryology Department, Faculty of Geology, Moscow State University, Moscow, Russia, Cheng Guodong, Cold and Arid Regions Environmental and Engineering Research Institute, Chinese Academy of Sciences, Lanzhou, China.
Description: First edition. | Boca Raton, Florida : CRC Press/Balkema is an imprint of the Taylor & Francis Group, [2017] | Includes bibliographical references and index.
Identifiers: LCCN 2017024668 (print) | LCCN 2017034592 (ebook) | ISBN 9781315166988 (ebook) | ISBN 9781138054165 (hardcover : alk. paper)
Subjects: LCSH: Frozen ground.
Classification: LCC GB641 (ebook) | LCC GB641 .H37 2017 (print) | DDC 551.3/8–dc23
LC record available at https://lccn.loc.gov/2017024668

Published by: CRC Press/Balkema
                 Schipholweg 107C, 2316 XC Leiden, The Netherlands
                 e-mail: Pub.NL@taylorandfrancis.com
                 www.crcpress.com – www.taylorandfrancis.com

**Visit the Taylor & Francis Web site at
http://www.taylorandfrancis.com**

**and the CRC Press Web site at
http://www.crcpress.com**

Table of contents

Preface   xv
About the authors   xvii
Acknowledgements   xix
Dedication   xxi
List of figures   xxiii
List of tables   xxxix
List of symbols   xli

Part I Introduction and characteristics of permafrost   I

I   Definition and description   3
   1.1   Introduction   3
   1.2   Additional terms originating in Russia   7
   1.3   History of permafrost research   8
   1.4   Measurement of ground temperature   9
   1.5   Conduction, convection and advection   9
   1.6   Thermal regimes in regions based on heat conduction   10
   1.7   Continentality index   15
   1.8   Moisture movement in the active layer during freezing and thawing   16
   1.9   Moisture conditions in permafrost ground   18
   1.10 Results of freezing moisture   20
   1.11 Strength of ice   22
   1.12 Cryosols, gelisols, and leptosols   22
   1.13 Fragipans   22
   1.14 Salinity in permafrost regions   24
   1.15 Organic matter   28
   1.16 Micro-organisms in permafrost   30
      1.16.1 Antarctic permafrost   31
      1.16.2 High-latitude permafrost   31
      1.16.3 High altitude permafrost in China   32
      1.16.4 Phenotypic traits   32
      1.16.5 Relation to climate change on the Tibetan plateau   34
   1.17 Gas and gas hydrates   35

1.18	Thermokarst areas	37
1.19	Offshore permafrost	38

2 Cryogenic processes where temperatures dip below 0°C — **43**

2.1	Introduction	43
2.2	The nature of ice and water	43
2.3	Effects of oil pollution on freezing	50
2.4	Freezing and thawing of the active layer in permafrost in equilibrium with a stable climate	51
2.5	Relation of clay mineralogy to the average position of the permafrost table	53
2.6	Ground temperature envelopes in profiles affected by changes in mean annual ground surface temperature (MASGT)	54
2.7	Needle ice	58
2.8	Frost heaving	59
2.9	Densification and thaw settlement	60
2.10	Cryostratigraphy, cryostructures, cryotextures and cryofacies	60
2.11	Ground cracking	61
2.12	Dilation cracking	63
2.13	Frost susceptibility	64
2.14	Cryoturbation, gravity processes and injection structures	65
	2.14.1 Cryoturbation	65
	2.14.2 Upward injection of sediments from below	69
	2.14.3 Load-casting	69
2.15	Upheaving of objects	71
2.16	Upturning of objects	72
2.17	Sorting	73
2.18	Weathering and frost comminution	74
2.19	Karst in areas with permafrost	78
2.20	Seawater density and salinity	80

3 Factors affecting permafrost distribution — **85**

3.1	Introduction	85
3.2	Climatic factors	85
	3.2.1 Heat balance on the surface of the Earth and its effect on the climate	85
	3.2.2 Relationship between air and ground temperatures	91
	3.2.3 Thermal offset	93
	3.2.4 Relation to air masses	96
	3.2.5 Precipitation	99
	3.2.6 Latitude and longitude	103
	3.2.7 Topography and altitude	105
	3.2.8 Cold air drainage	107
	3.2.9 Buffering of temperatures against change in mountain ranges	108
3.3	Terrain factors	109
	3.3.1 Vegetation	109

3.3.2   Hydrology   111
3.3.3   Lakes and water bodies   115
3.3.4   Nature of the soil and rock   117
3.3.5   Fire   118
3.3.6   Glaciers   119
3.3.7   The effects of Man   122

4   Permafrost distribution   123
4.1   Introduction   123
4.2   Zonation of permafrost   126
4.3   Permafrost mapping   127
4.4   Examples of mapping units used   129
4.5   Modeling permafrost distribution   130
4.6   Advances in geophysical methods   131
4.7   Causes of variability reducing the reliability of small-scale maps 131
4.8   Maps of permafrost-related properties based on field observations   135
        4.8.1   Permafrost thickness   135
        4.8.2   Maps of ice content   135
        4.8.3   Water resources locked up in perennially frozen ground   136
        4.8.4   Total carbon content   139
4.9   Use of remote sensing and airborne platforms in monitoring
        environmental conditions and disturbances   140
4.10   Sensitivity to climate change: Hazard zonation   140
4.11   Classification of permafrost stability based on mean annual ground
        temperature   142

Part II Permafrost landforms   145
II.1   Introduction   145

5   Frost cracking, ice-wedges, sand, loess and rock tessellons   149
5.1   Introduction   149
5.2   Primary and secondary wedges   153
        5.2.1   Primary wedges   153
                 5.2.1.1   Ice-wedges   153
                 5.2.1.2   Sand tessellons   167
                 5.2.1.3   Loess tessellons   170
                 5.2.1.4   Rock tessellons   170
        5.2.2   Secondary wedges   170
                 5.2.2.1   Ice-wedge casts   171
                 5.2.2.2   Soil wedges   173

6   Massive ground ice in lowlands   179
6.1   Introduction   179
6.2   Distribution of massive icy beds in surface sediments   181
6.3   Sources of the sediments   182
6.4   Deglaciation of the Laurentide ice sheet   183
6.5   Methods used to determine the origin of the massive icy beds   186

6.6   Massive icy beds interpreted as being formed by cryosuction   186
6.7   Massive icy beds that may represent stagnant glacial ice   187
6.8   Other origins of massive icy beds   189
6.9   Ice complexes including yedoma deposits   189
6.10   Conditions for growth of thick ice-wedges   190
6.11   The mechanical condition of the growth of ice-wedges and its
       connection to the properties of the surrounding sediments   192
6.12   Buoyancy of ice-wedges   193
6.13   Summary of the ideas explaining yedoma evolution   195
6.14   Aufeis   195
6.15   Perennial ice caves   198
6.16   Types of ice found in perennial ice caves   200
6.17   Processes involved in the formation of perennial ice caves   202
6.18   Cycles of perennial cave evolution   204
       6.18.1 Perennial ice caves in deep hollows   204
       6.18.2 Sloping caves with two entrances   205
       6.18.3 Perennial ice caves with only one main entrance but air
               entering through cracks and joints in the bedrock walls   206
       6.18.4 Perennial ice caves with only one main entrance and no
               other sources of cooling   206
6.19   Ice caves in subtropical climates   207
6.20   Massive blocks of ice in bedrock or soil   210

7   Permafrost mounds   213
    7.1   Introduction   213
    7.2   Mounds over 2.5 m diameter   214
          7.2.1   Mounds formed predominantly of injection ice   215
                  7.2.1.1   Pingo mounds   215
                  7.2.1.2   Hydrostatic or closed system pingos   216
                  7.2.1.3   Hydraulic or open system pingos   218
                  7.2.1.4   Pingo plateaus   222
                  7.2.1.5   Seasonal frost mounds   225
                  7.2.1.6   Icing blisters   226
                  7.2.1.7   Perennial mounds of uncertain origin   228
                  7.2.1.8   Similar mounds that can be confused with
                            injection phenomena   228
          7.2.2   Mounds formed dominantly by cryosuction   229
                  7.2.2.1   Palsas   230
                             7.2.2.1.1   Palsas in maritime climates   231
                             7.2.2.1.2   Palsas in cold, continental climates 234
                             7.2.2.1.3   Lithalsas   239
                             7.2.2.1.4   Palsa/Lithalsa look-alikes   243
          7.2.3   Mounds formed by the accumulation of ice in the thawing
                  fringe: Peat plateaus   244
    7.3   Cryogenic mounds less than 2.5 m in diameter   249
          7.3.1   Oscillating hummocks   252

	7.3.2	Thufurs	256
	7.3.3	Silt-cycling hummocks	260
	7.3.4	Niveo-aeolian hummocks	261
	7.3.5	Similar-looking mounds of uncertain origin	263
	7.3.6	String bogs	264
	7.3.7	Pounus	265

8 Mass wasting of fine-grained materials in cold climates   267

8.1	Introduction		267
8.2	Classification of mass wasting		267
8.3	Slow flows		269
	8.3.1	Cryogenic creep	269
		8.3.1.1 Needle ice creep	270
		8.3.1.2 Frost heave and frost creep	272
		8.3.1.3 Gelifluction	274
		8.3.1.4 Other creep-type contributions to downslope movement of soil	276
	8.3.2	Landforms produced by cryogenic slow flows in humid areas	279
	8.3.3	Landforms developed by cryogenic flows in more arid regions	284
8.4	Cryogenic fast flows		287
	8.4.1	Cryogenic debris flows	287
	8.4.2	Cryogenic slides and slumps	296
	8.4.3	Cryogenic composite slope failures	297
		8.4.3.1 Active-layer detachment slides	298
		8.4.3.2 Retrogressive thaw failures	300
		8.4.3.3 Snow avalanches and slushflows	304
		8.4.3.3.1 Snow avalanches	306
		8.4.3.3.2 Slush avalanches	310
8.5	Relative effect in moving debris downslope in the mountains		313

9 Landforms consisting of blocky materials in cold climates   315

9.1	Introduction		315
9.2	Source of the blocks		315
9.3	Influence of rock type		317
9.4	Weathering products		318
9.5	Biogenic weathering		319
9.6	Fate of the soluble salts produced by chemical and biogenic weathering		320
9.7	Rate of cliff retreat		321
9.8	Landforms resulting from the accumulation of predominantly blocky materials in cryogenic climates		322
	9.8.1	Cryogenic block fields	322
		9.8.1.1 Measurement of rates of release of blocks on slopes	326

	9.8.2	Cryogenic block slopes and fans		326
	9.8.3	Classification of cryogenic talus slopes		329
		9.8.3.1	Coarse blocky talus slopes	331
	9.8.4	Protection of infrastructure from falling rock		332
9.9	Talus containing significant amounts of finer material			333
	9.9.1	Rock glaciers		334
		9.9.1.1	Sedimentary composition and structure of active rock glaciers	337
		9.9.1.2	Origin of the ice in active rock glaciers	338
		9.9.1.3	Relationship to vegetation	339
	9.9.2	Movement of active rock glaciers		340
		9.9.2.1	Horizontal movement	340
		9.9.2.2	Movement of the front	341
	9.9.3	Distribution of active rock glaciers		345
	9.9.4	Inactive and fossil rock glaciers		347
	9.9.5	Streams flowing from under rock glaciers		348
9.10	Cryogenic block streams			349
	9.10.1	Characteristics		351
	9.10.2	Classification		354
		9.10.2.1	Siberian active dynamic block streams – kurums	355
		9.10.2.2	The Tibetan type of active dynamic block streams	357
		9.10.2.3	Active cryogenic lag block streams	359
		9.10.2.4	Inactive, relict block streams	359
9.11	Surface appearance of blocky landforms			365
10	**Cryogenic patterned ground**			**367**
10.1	Introduction			367
10.2	Forms of cryogenic patterned ground			368
10.3	Factors affecting the development of cryogenic patterned ground			369
10.4	Macroforms of cryogenic patterned ground			373
	10.4.1	Cryogenic nonsorted circles		374
		10.4.1.1	Cryogenic mudboils	375
			10.4.1.1.1 Arctic mudboils	376
			10.4.1.1.2 Subarctic mudboils	379
		10.4.1.2	Xeric nonsorted circles	381
		10.4.1.3	Nonsorted circles in maritime climates	384
		10.4.1.4	Frost boils	385
		10.4.1.5	Plug circles	386
10.5	Cryogenic sorted patterned ground			387
	10.5.1	Cryogenic sorted circles		388
	10.5.2	Cryogenic sorted polygons, and nets		391
		10.5.2.1	Sorted stripes	391
		10.5.2.2	Stone pits	392

10.6   Identification of active versus inactive forms of
        macro-sorted patterns   393
10.7   Microforms of cryogenic patterned ground   394

11   Thermokarst and thermal erosion   **397**
11.1   Introduction   397
11.2   Causes of thermokarst   400
11.3   Cavity development in permafrost   402
11.4   Effect of thermokarst on soil   403
11.5   Thermokarst landforms   405
        11.5.1   Thermokarst pits   406
        11.5.2   Thermokarst mounds   407
        11.5.3   Pingo, palsa and lithalsa scars   409
        11.5.4   Beaded streams   411
        11.5.5   Thermokarst lakes   412
        11.5.6   Oriented lakes   415
        11.5.7   Alases   417
        11.5.8   Cycle of alas formation   418
11.6   Thermokarst and thermal erosion along river banks   424
        11.6.1   Ice jams   425
11.7   Thermal erosion and thermokarst processes along sea coasts   429
        11.7.1   Effects of seasonal sea ice   430
        11.7.2   Effects of geology   433
        11.7.3   Topographic effects   433
        11.7.4   Sea conditions   434
        11.7.5   Deposition of sediments   435
11.8   Processes involved in the erosion of ice-rich arctic
        coastal sediments   435
11.9   Importance of coastal erosion of sediments containing
        permafrost   439

Part III Use of permafrost areas   **441**
III.1   Introduction   441

12   The mechanics of frozen soils   **445**
12.1   Introduction   445
12.2   Strains and stresses in the freezing and thawing of soils
        resulting in frost heaving   445
12.3   Rheological processes   456
12.4   Frost susceptibility   460

13   Foundations in permafrost regions: building stability   **465**
13.1   Introduction   465
13.2   The effect of construction on permafrost stability   468
13.3   Choice of method of construction   470
13.4   Building materials   471

13.5	Timing of construction	472
13.6	Types of foundations	473
	13.6.1 Pads	473
	13.6.2 Slabs and rafts	474
	13.6.3 Sills	475
	13.6.4 Spread footings	476
	13.6.5 Piles	477
	13.6.6 Thermosiphons	482
	13.6.7 Artificial refrigeration	489
	13.6.8 Ventilation ducts	490
	13.6.9 Angle of slope of the embankment sides	492
	13.6.10 Snow removal	492
	13.6.11 The diode effect: use of rocks	493
	13.6.12 Shading	496
	13.6.13 Insulation	497
	13.6.14 Use of geotextiles and waterproof plastics	498

14 Roads, railways and airfields — **501**

14.1	Introduction	501
14.2	The problems	501
14.3	Types of roads	502
14.4	Experimental embankments	504
14.5	Winter roads	505
14.6	Environmental effects of winter roads	507
14.7	Embankment heights	508
14.8	Unpaved embankments	509
14.9	Main problems with embankment stability	518
14.10	Concrete versus ballast railway tracks	524
14.11	Paving of road and airfield runways	527
14.12	Use of white paint	529
14.13	Bridges	530
14.14	Icings	532
14.15	Cut slopes	538
14.16	Airfield construction	538

15 Oil and gas industry — **543**

15.1	Introduction	543
15.2	Oil and gas exploration	543
15.3	Drilling rigs	546
15.4	Production and keeper wells	547
15.5	Sump problems	549
15.6	Pipelines	550
	15.6.1 Buried mode	551
	15.6.2 Pipelines on piles	559
	15.6.2.1 Design parameters	563

		15.6.2.2	Construction methods	564
		15.6.2.3	Failures in the buried section	564
	15.7	Monitoring		566
	15.8	Compressor stations		566
	15.9	Pipeline crossings		569
	15.10	Effects of heat advection from producing wells		571
	15.11	Gas hydrates in permafrost ice		571

16   Mining in permafrost areas — **577**

	16.1	Introduction		577
	16.2	Placer mining		577
	16.3	Open cast/pit mining		580
		16.3.1	Exploration	582
		16.3.2	Extraction of the ore	582
	16.4	Underground mining		587
		16.4.1	Transport of the ore around the mine	589
		16.4.2	Support facilities	590
	16.5	Waste materials and tailings ponds		590
		16.5.1	Toxic wastes	592

17   Provision of utilities — **597**

	17.1	Introduction		597
	17.2	Water supply		598
		17.2.1	Sources of water	598
		17.2.2	Dams to impound water on permafrost	600
		17.2.3	Municipal water storage	602
		17.2.4	Water treatment	602
		17.2.5	Water requirements	603
		17.2.6	Transportation methods for water and waste water	604
	17.3	Waste disposal		607
		17.3.1	Wastewater treatment and disposal	607
			17.3.1.1 Undiluted wastes	608
			17.3.1.2 Moderately diluted wastes	608
			17.3.1.3 Conventional strength wastewater	609
			17.3.1.4 Very dilute waste water	609
		17.3.2	Solid waste disposal	610
	17.4	Electric transmission lines		610
		17.4.1	Foundation problems for transmission lines built on permafrost	611
		17.4.2	Transmission tower foundation types	614

18   Agriculture and forestry — **617**

	18.1	Introduction		617
	18.2	Zonation of natural vegetation across Siberia		621
	18.3	Zonation of natural vegetation in North America		623

18.4   Southern and Eastern Kazakhstan, Mongolia and the
        Qinghai-Tibet Plateau   625
18.5   The Eichfeld zones   626
        18.5.1   Eichfeld zone I   627
        18.5.2   Eichfeld zone II   627
        18.5.3   Eichfeld zone III   629
                  18.5.3.1   The northern Taiga   629
18.6   Asian steppe grasslands and deserts   630
18.7   The development of modern agriculture in permafrost areas   632
18.8   Forestry   633
18.9   Potential effects of climate changes   634

References   637
Subject index   755

Preface

This book is intended to be a general survey of the young science of Geocryology, which is the study of permafrost, its nature, characteristics, processes and distribution. Permafrost is the product of a combination of a number of climatic and environmental factors that produce frozen ground and icy layers. It has an enormous impact on the use of unglaciated cold regions and environments on the Earth. These regions differ from warmer areas by the presence of seasonally and/or perennially frozen ground. Movement of moisture to centres of freezing produces heaving, while contraction of the ground in cold weather can cause cracking. A unique group of landforms and processes occur in these areas that are not found elsewhere.

Permafrost occurs on all continents except perhaps Australia. Its distribution is closely related to climate and the local environment, and any changes in the microenvironment will result in either expansion or loss of ground ice. Once water appears at the surface, thawing of the ground ice is difficult to stop, and results in the development of a group of landforms called thermokarst. The hydrology of permafrost areas is much more complex than in other environments, and water itself is a source of heat that can destabilize the ground ice. Areas with saline soils are common in arid lands, as well as those regions which have been under the sea. Any soluble salts modify the properties of the ground producing reduced bearing strength and lowered freezing point of water.

These cold conditions tend to cause breakup of materials such as concrete, bricks and mortar, as well as variations in thermal expansion that put equipment and structures at risk. The growth of ice causes heaving, while thawing of ground ice leads to subsidence and soil flowage. Coastal areas are subjected to serious erosion of different kinds to those in warmer regions, while the spring thaw produces substantial concentrations of water in the surface layers of the ground that result in flows and slides. Bearing capacity of the ground varies with season and is often significantly lower than elsewhere. Total incoming solar radiation varies with aspect, resulting in gradual tilting of structures. These structures can destabilize the permafrost resulting in failure of foundations. Thus these regions exhibit numerous problems for humans which are unique to cold climates.

Increasingly Mankind is extracting resources from these regions, and this requires a good knowledge of the field of Geocryology in order to carry out projects that will be successful, economic ventures. Initially, there has to be exploration which can be carried out using winter roads, often along cut lines through the vegetation. When a suitable deposit is found, it is necessary to prove its quality and extent, which involves building better quality roads to allow access for drilling equipment. When an adequate venture has been discovered, further upgrading of the access is necessary to bring in buildings, equipment, workers, and supplies, and to construct suitable waste disposal facilities such as sumps. Power sources must be obtained, water supplies found and buildings constructed as necessary. In order to get the products to market, it is essential to have good, reliable, linear transportation routes such as roads and railways. People must live in these areas to carry out the necessary work.

Indigenous people usually inhabit these remote areas, and also require modern facilities and infrastructure. Where possible, they should be given employment in return for the disturbance of their local surroundings. Furthermore, the natural environment and wildlife must be protected as far as possible. These landscapes are very fragile and it is essential to avoid serious damage, particularly of the type that makes the ways of life of the indigenous people impossible. Pollution of the environment must be minimized, and the waterways treated with respect. At the end of the life of a project, suitable funds should have been set aside to clean up and rehabilitate the area. Unfortunately, this is often not done.

Some permafrost areas are important grazing lands at low latitudes as well as at high elevations such as on the Qinghai-Tibet Plateau and in Mongolia. Large areas of tiaga represent some of the largest contiguous forest resources. Any disturbance of the vegetation can produce changes in ground temperature conditions, so that sustainable use of these landscapes can be problematic.

This book is written by three scientists representing three countries with extensive areas of permafrost and three different language groups. Together, they have had over 120 years of experience in dealing with permafrost problems around the world, and they try to present a world view of what can be found in these areas. Similar problems are found in each part of the permafrost realm, and a comparison of the knowledge gained in different parts of the world shows considerable similarities. However, climate and other factors produce significant local modifications, so that no one continent or country exhibits the whole range of variability. For this reason, a global view seems justified.

This book is organized into three parts. Part one consists of an introduction to the characteristics of permafrost. It is organized in four chapters dealing with the definition and characteristics of permafrost, the unique processes operating in areas of frozen ground, the factors affecting it, and an explanation of its distribution. Part two consists of seven chapters describing the characteristic landforms unique to permafrost areas. Finally, the third part describes the extra problems encountered by engineers in construction projects in permafrost areas, as well as by foresters and agriculturalists. A good bibliography is provided, along with more than 350 illustrations to make the book more user-friendly.

About the authors

Stuart Arthur Harris was born on January 14th, 1931, in Cheltenham, Gloucestershire, England. He earned the degrees of B. Sc. (Honours), M.Sc. and Ph.D. in Geology and D.Sc. in Geography from Queen Mary University, University of London. During his National Service, he advised the Chief Engineers Branch, British Troops Egypt and the Arab Legion Engineers in Jordan, solving problems in geology, water supply and engineering. Subsequently, he was a soil surveyor for the consulting firm, Hunting Technical Services, before becoming Government Soil Surveyor in Guyana. He taught in the geography Departments of the University of Chicago, Wilfred Laurier University, and the University of Kansas before joining the University of Calgary in 1969. The National Research Council of Canada asked him to study the relationship of climate to permafrost in 1973, and he mapped the permafrost distribution from Northern New Mexico to Inuvik, Northwest Territories. Subsequently, he carried out detailed studies of the permafrost landforms and processes in northwest Canada, as well as on the Tibetan Plateau, China. He has carried out field work in Iceland, the Alps, Poland, Russia, China, Mongolia, New Zealand and Kazakhstan, publishing over 200 papers, books and reports. The Russian Geographical Society awarded him the Nikolai Mihailovich Prjevalsky Medal for his research on Alpine permafrost in 1996. He has also organised three International Field Trips in the Rocky Mountains for overseas scientists in connection with International meetings in Canada.

Professor Anatoli Brouchkov was born April 18, 1957 and raised in Khatanga of Arctic Siberia. He obtained his Ph.D and D.Sc degrees from the Geocryology Department of Geology Faculty of Lomonosov Moscow State University, studying under the tutelage of some of the recent famous Russian permafrost scientists such as V.A. Kudryavtsev, S.S. Vyalov, E. D. Yershov and N. N. Romanovski. Over the years, he has run a geocryological laboratory for the Russian Academy of Sciences as well as an underground permafrost laboratory in Amderma, involving research all over the Russian Arctic. He has specialized in the study of the effects of salinity on the properties of frozen ground and the effects of climate

change on permafrost. He has also carried out a research on the survival of microorganisms in permafrost. In addition, he has acted as a geocryological consultant to
Gazprom and other Russian and international companies, and a permafrost expert
for World Meteorological Organization. He was a professor at Hokkaido University
(Japan, 2001–2004) and Tyumen State University (since 2005), publishing over 150
papers and books. In 2010, he succeeded the late Edward Yershov as Professor and
Head of the Geocryology Department of Moscow State University.

Academician Cheng Guodong was born on July 11th, 1943
in Shanghai, China. He earned his B.Sc. at Beijing Geology
College (China University of Geosciences), and carried out
fundamental research in the CREEL laboratories at Hanover,
New Hampshire, before returning to China. Guodong became
an Academician of the Chinese Academy of Sciences in 1993.
He was responsible for reorganizing the Institute of Glaciology
and Geocryology of the Chinese Academy of Sciences to form
the Cold and Arid Regions Environmental and Engineering
Research Institute. He has led the research very successfully,
thus enabling the modernization of the construction techniques
in the permafrost areas of China during the last 15 years. One of his most important
achievements was the development of the most effective methods of cooling the beds of
linear transportation routes using blocks of rock. He has written eight books, and his
name appears below the title of numerous papers dealing with the use of permafrost
lands. He has received many awards in China, as well as being the recipient of one
of the first three IPA Lifetime Achievement awards by the International Permafrost
Association.

Acknowledgements

A number of living and deceased colleagues and organizations have kindly permitted the use of their photographs in this book. Thanks are due to Vasily Bogoyaavlensky (Figure 7.9), B. Burton (Figure 10.3), Mike Chambers (Figure 10.2), Lee and Barbara Clayton (Figure 6.6), Hanna H. Christiansen (Figure 7.16), R. V. Desyatkin and A. V. Desyatkin (Figure 11.17), R. O. van Everdingen (Figure 11.8), S. Fomin (Figure 5.20), D. Froese (Figure 5.22), Aldar Gorbunov (Figure 9.13), M. Grigoriev (Figures 1.31, 6.4 & 6.8A), Gennady Griva (Figures 15.6 & 15.9), A. Gubarkov (Figures 5.3, 8.5 & 11.2), Bernard Hallet (Figure 10.1), Owen L. Hughes (Figure 11.6), B. M. Jones (Figure 11.37), V. Kondratiev (Figures 13.35, 14.18 & 14.19), A.G. Kostyaev, (Figure 6.8B), I. W. Lee (Figure 14.25), A. G. Lewkowicz (Figure 7.43), J. Ross Mackay (Figures 7.3 & 11.13), V. Melnikov (Figure 13.7), A. Osokin (Figures 14.1, 14.12 & 14.22), E. Pike (5.16), L. and S. Rollinson (2.8), M. Rosen (Figure 13.37), Vera Samsonova (Figure 5.31, 6.2 & 13.25), M. K. Seguin (Figure 7.21), W. W. Shilts (Figure 11.10), C. Scapozza (Figure 9.11), V. Singhroy (Figure 8.33), Tourismusverband Werfen (Figures 6.12 & 6.13), A. Cheng (Figure 11.8), J.-S. St. Vincent (Figure 11.35), G.-S. Wang (Figure 17.9), Sizhong Wang (Figure 5.9), Yakutic Reindeer Tours (Figure 18.4), Y.-H. You (Figure 17.5) and M. Zheleznyak (Figure 4.12). The rest of the photographs were taken by one or other of the authors as indicated in the captions.

Several Journals have permitted the use of illustrations previously published in their volumes including *Arctic* (Figures 1.8, 1.10, 3.17, 3.5, 3.6, 4.2, and iii.1), *Arctic, Antarctic and Alpine Research*, formerly *Arctic and Alpine Research*, courtesy of the Regents of the University of Colorado (Figures 7.22, 8.6, 8.7A, 10.10 and Table 1.1). In addition, Elsevier (Table 3.1), Matti Seppälä (Figure 7.18) and V. Romanovsky (Figure 2.7) have permitted the use of previously published figures. Figures 1.9, 1.13, 1.14, 2.23, 4.17, 13.8, 16.5, 18.12, 18.13 and Tables ii.2 are reproduced from the book, *The Permafrost Environment* written by Stuart A. Harris and published in 1986. In most cases, the diagrams have been redrawn by Robin Poitras of the Department of Geography, University of Calgary, in order to bring the drawings to a consistent style throughout this book. Pamela Harris kindly proof-read the manuscript, along with Anatoli Brouchkov and Stuart Harris.

Dedication

This book is dedicated to our wives, Pamela Rosemary Harris, Marina Brouchkova and Zhang Youfen, in appreciation of their endless patience, support and companionship over the last several decades while we were carrying out research in Geocryology. Without them, this book would not have been possible.

List of figures

1.1	Typical profile showing icy permafrost	2
1.2	Names of the parts of the upper layers in a permafrost profile	4
1.3	Rotten porous ice	5
1.4	Block diagram of an area with permafrost showing the names of the constituent parts	5
1.5	Change in active layer thickness in bedrock at Plateau Mountain, Alberta	6
1.6	Geothermal gradients in boreholes at Prudhoe Bay, and Barrow, Alaska, showing the effect of different thermal conductivities of rock	11
1.7	Effect of differences in texture on active layer thickness	11
1.8	Ground temperatures measured at 4–10 m depth in Shargin's well	13
1.9	Temperature distribution versus time for wet tundra and alpine tundra	13
1.10	Variation in duration of the zero curtain effect in clay and shattered rock as a function of depth	14
1.11	Thermal temperature envelope for a site with permafrost	15
1.12	Change in ground temperature during the year in relation to moisture movement	17
1.13	Diagram showing the normal distribution of H_2O in permafrost	18
1.14	Changes in the perched water table after drilling	19
1.15	Frost action on sandstone and ice-wedge casts	21
1.16	Sorted stone circles on Plateau Mountain, Alberta	21
1.17	Distribution of fragipans in post-Wisconsin sediments in N. America	23
1.18	Changes in salinity along a traverse across the Slims River Delta	25
1.19	Warm spring along the Golmud-Lhasa road, Tibet	26
1.20	An open system (East Greenland type) pingo at a spring in the Kunlun Pass along the Golmud-Lhasa road	27
1.21	Distribution of the main areas of the two types of salinity	28
1.22	Distribution of peatlands in the northern Hemisphere	29
1.23	Relative abundance of micro-organisms in the four main permafrost areas	30
1.24	Variation in community composition of micro-organisms in permafrost between various Chinese mountain ranges	33
1.25	Micro-organisms, decomposition of organic matter in soils and release of greenhouse gases	34
1.26	Zonation of permafrost and gas hydrates with latitude in Russia	35
1.27	Stability of the methane-water-gas hydrate system	36
1.28	Gas hydrates burning in Chinese permafrost	37

1.29	Distribution of subsea permafrost in the Arctic	39
1.30	Horsts and graben in the Laptev Sea	40
1.31	Ice-wedges in eroding cliffs on Big Lyakovsky Island, Arctic coast, Siberia	40
2.1	Vertical ice crystals perpendicular to the freezing plane	44
2.2	Structure of a water molecule in the gaseous state	45
2.3	Phase diagram for pure water in the range of temperature and pressure encountered on Earth	46
2.4	Ground temperature changes at the onset of freezing	47
2.5	Comparison of thermograms for nonsaline and saline soil containing sodium chloride	49
2.6	Terminology for the parts of the ground temperature envelope proposed by van Everdingen	50
2.7	Unfrozen water contents in nonsaline and saline oil-polluted soils	51
2.8	Multiple ice lenses beneath a moss layer	52
2.9	Geotherms for site #2, Plateau Mountain, Alberta, in bedrock	52
2.10	Variation in timing completion of refreezing of the active layer at Marmot #2 borehole, Jasper	53
2.11	Theoretical effects of climatic changes on mean annual ground surface temperature	55
2.12	Ground temperature profile at Plateau Mountain #2 borehole	55
2.13	Changes in mean annual air temperature at Yakutsk	56
2.14	Thermal disequilibrium conditions beneath water bodies	56
2.15	Modeled evolution of the ground temperatures beneath a thaw lake	57
2.16	Needle ice along the Denali Highway, Alaska	58
2.17	Fossil crack picked out by iron-staining	63
2.18	Dilation cracks on an icing mound	64
2.19	Comparison of load casting, upward injection and formation of ball-and pillow structures	66
2.20	Cryoturbation in a soil	66
2.21	Telescoping nested rods used to measure differential heave	68
2.22	Load casting in China	70
2.23	Section showing the sharp boundary between the sandy silt infilling forming a cast of a former ice block and the undisturbed outwash gravels in the lower part of a load-cast structure	71
2.24	Diagram showing the mechanism of upturning of stones	73
2.25	Relationship between mean annual ground temperature and weathering index	75
2.26	Karst features in northern Canada	79
2.27	Seawater density as a function of temperature	81
2.28	Changes in seawater density as a function of temperature during the Holocene at Amundson Bay	82
2.29	Salinity and temperature profiles along a traverse perpendicular to the coast in the shallow (<6 m) Laptev Sea	83
3.1	Global net radiation measured by space satellites	87
3.2	Comparison of the carbon dioxide content of the atmosphere over the past 4000 million years with mean annual air temperature	88
3.3	Mean annual ground and surface air temperatures overlying or close to permafrost	90

3.4 Variation in mean soil heat flux and diurnal range of heat flux at 5 cm   91
depth on silt loams under various slope conditions

3.5 Distribution of continuous permafrost (>70%), discontinuous   92
permafrost and sporadic permafrost (<30%) using annual freeze-thaw
indices

3.6 Comparison of the degree of continentality of selected permafrost   93
areas

3.7 Thermal offsets in winter in the boreal forest of Alaska   94
3.8 Evolution of Rossby Waves with time as they move east   97
3.9 North-south section through the lower atmosphere showing the climatic   98
conditions occurring in 1980 A.D. in North America

3.10 The effect of positive (warming) snow pack thermal offset on the depth   100
of the active layer

3.11 Diagram showing the effect of various depths of snow cover on the   100
underlying ground temperatures

3.12 Effect of mean snow depth in winter on ground temperature at 150 cm   101
soil depth on Plateau Mountain

3.13 The lower limit of permafrost in the Andes compared with the average   103
in the Northern Hemisphere

3.14 Altitude of the 0°C isotherm and some other geographical boundaries   104
plotted against latitude

3.15 Comparison of modeled temperature field on the Matterhorn with that   105
for Plateau Mountain (a former nunatak)

3.16 Section north-south across Alaska showing the position and thickness of   106
permafrost

3.17 Anatomy of cold air drainage west of Fort Nelson, B.C.   107
3.18 Temperatures of cold still cP air mass plotted against presence or   108
absence of cold air drainage at Fox Lake, Yukon

3.19 Results of placing a road culvert too low in an area of ice-wedges   113
3.20 Ground temperatures in the upper 12 m between 1996 and 1999 at   114
Marmot Basin #2 borehole, Jasper National Park

3.21 Temperature regimes of cold, freshwater lakes in North America   116
3.22 Energy budget and thermal offsets in lakes   117
3.23 The influence of small lakes on the ground temperatures conditions   117
3.24 Relationship of the distribution of glaciers to the mean annual freezing   120
and thawing indices and permafrost zones

3.25 Permafrost distribution in relation to the Columbia Icefield in Banff   121
National Park

3.26 Ground temperatures in the borehole through the Antarctic ice cap   121
beneath the Vostock station in Antarctica

4.1 Approximate distribution of permafrost in the Northern Hemisphere   124
4.2 Distribution of permafrost zones along the Eastern Cordillera of western   124
North America

4.3 Diagrammatic north-south transect of permafrost in Central Siberia and   125
in North America

4.4 Distribution and relationship of permafrost with non-frozen ground as   126
traced from south to north

4.5 Nomenclature for permafrost in China   127

4.6	The current geocryological map of Russia (1:2,500.000) showing mean annual ground temperatures	128
4.7	The "Probability of Permafrost map" of Bonnaventure *et al.*, 2012	132
4.8	Airborne electrical resistivity traverse at the junction of the Yukon and Porcupine rivers	133
4.9	Detail of the variability in ground temperature around Twelvemile Lake, Alaska, based on airborne resistivity	133
4.10	Detailed changes in permafrost temperatures over short distances	134
4.11	Permafrost in Russia	136
4.12	Permafrost thickness under the Russian Platform underlain by the Yakutian Shield	137
4.13	Distribution of ice content in permafrost areas	138
4.14	Stratigraphy of three typical boreholes on the Qinghai-Tibet Plateau	139
4.15	Oxygen isotope palaeotemperature record and geomagnetic polarity timescale	141
4.16	Sketch showing various types of permafrost islands	143
4.17	Stability of permafrost in North America based on the distribution of mean annual ground temperature	144
4.18	Vertical zones of permafrost in various climatic regions	144
5.1	Changes in dominant slope processes on slopes underlain by permafrost in relation to mean annual air temperature and mean annual precipitation	146
5.2	Distribution of individual cryogenic landforms with permafrost zonation	147
5.3	An open frost crack on the Yamal Peninsula, Western Siberia	150
5.4	Polygons with raised marginal ridges and a lower centre, Prudhoe Bay, Alaska	151
5.5	Distribution of thermal contraction cracking in peat and mineral sediments with freezing and thawing indices	152
5.6	Diagram showing the development of ice-wedges	154
5.7	Ice-wedges of different ages cutting through part of the massive ice on Herschel Island	155
5.8	The southern-most inactive ice wedge found at Yitulihe, China	156
5.9	"Shoulders" (upfolding of host) and "belts" (banding in ice) of the left upper side of an ice-wedge	156
5.10	Diagram showing the difference in shape between epigenetic, syngenetic and multi-stage ice-wedges	157
5.11	Multistage ice-wedge forming in peat at Yitulihe	158
5.12	Details of an active ice-wedge in Siberia	158
5.13	Relationship between soil and ice-wedges and mean annual ground temperature for A, clay substrates and B, sand and gravel	159
5.14	Relationship between the distribution of active ice-wedges polygons and freezing and thawing indices	160
5.15	Stresses and resultant changes that appear to have occurred in ice-wedges on the Aldan River terraces shown diagrammatically	162
5.16	Irregular shapes of ice-wedges	162
5.17	Block diagram of ice-wedges forming a tessellated pattern	163

5.18	An area of ice-wedges on a pingo	163
5.19	Low centre polygons, Lena delta, Siberia	164
5.20	Ice-wedges over and around a peat bog	165
5.21	Typical arrangement of first, second order and third order ice-wedges	165
5.22	The oldest known ice-wedge in North America	166
5.23	Pleistocene sand tessellons west of Laramie, Wyoming	167
5.24	Intersecting sand sheets in Pleistocene sand wedges	168
5.25	Loess tessellon in bedrock at Yellow River village	171
5.26	Vertical structure in loess tessellons at Sandar, China	172
5.27	Rock tessellon in frost-shattered fissile sandstone	173
5.28	High centre polygons south of Prudhoe Bay, Alaska	174
5.29	Thermocast mounds in cleared larch forest in Siberia	174
5.30	Ice-wedge cast of younger Dryas age in Holland	175
5.31	Cracks filled with ice in Yamal peninsula, Western Siberia	176
5.32	Cross section of a soil wedge in Iceland	177
6.1	Active thawing icy cliff of a retrogressive thaw flow slide in the sea cliffs on the Yamal peninsula, Russia	179
6.2	Aufeis, river Artuk, in summer	180
6.3	Distribution of buried massive icy beds in the Northern Hemisphere	181
6.4	Massive ice in the cliffs of Big Lyakhovsky Island, Siberia	182
6.5	Distribution of buried glacier ice and ice stagnation deposits in relation to the ice margins and two retreat stages of the former Laurentide Ice Sheet in North America	184
6.6	Dead stagnant ice at the terminus of the Martin River Glacier in southern Alaska	185
6.7	Massive ice along the shoreline at Tuktoyaktuk, Northwest Territories	188
6.8	Ice-wedge structures featuring vertical flow	194
6.9	Aufeis along a river floodplain in August in the North Fork Pass, Dempster Highway, Yukon Territory	196
6.10	Relationship of aufeis occurrences to freezing and thawing indices	197
6.11	Relationship between accessible ice caves and freezing and thawing indices	199
6.12	Speleothems and icy slopes where water has entered the roof of the cold Werfen Ice Cave, south of Salzburg, Austria	201
6.13	Banded ice coating the walls of Werfen Cave, Austria	201
6.14	Plate-like ice crystals from the coating of the wall of Plateau Mountain Ice Cave in 1978	202
6.15	Diagram showing the pattern of air movement and ice accumulation in Candelaria Ice Cave, New Mexico	205
6.16	A. Temperatures measured at monthly intervals during 1974 at a series of 10 stations in the Canyon Creek Ice Cave, Alberta, and B., comparison of the average air temperatures in the ice cave on specific days with the mean monthly air temperature outside the cave	206
6.17	Plate-like crystals coating the walls of Plateau Mountain Ice Cave	208
6.18	Stalagmites developed on the floor of Plateau Mountain Ice Cave	209
6.19	Thaw pits left after the melting of large blocks of ice at Fairbanks, Alaska	210

6.20   Massive ice blocks in the ground along the Haul Road in Happy Valley,   211
       Sagavairtok Quadrangle, Alaska
7.1   Spilt pingo, Tuktoyaktuk Peninsula   213
7.2   Solid pingo ice with occasional soil inclusions   216
7.3   Developing pingo scar, Mackenzie Delta   217
7.4   Hydrostatic (closed system) pingos growing on the floor of an alas near   218
       Yakutsk, Siberia
7.5   Relationship of hydraulic and hydrostatic pingos to freezing and   219
       thawing indices
7.6   A hydraulic pingo at Harigqiong, Qinghai-Tibet Plateau   220
7.7   Hydraulic pingo with a depression in the top   221
7.8   Section through the Kunlun Pass pingo   222
7.9   The Yarmal crater   223
7.10   A flat-surfaced, open system pingo, Qinghai-Tibet Plateau   223
7.11   Section in the pingo plateau shown in Figure 7.10   224
7.12   Massive injection ice in the pingo plateau   225
7.13   Seasonal frost mounds, North Fork Pass, Dempster Highway   226
7.14   Cross section of a seasonal frost mound   227
7.15   Icing blister and icing, Siberia   227
7.16   Drilling in a continental palsa and resulting core   228
7.17   Segregated ice in a lithalsa, Qinghai-Tibet Plateau   228
7.18   General model of the evolution of a maritime palsa   232
7.19   Distribution of maritime and continental palsas with freezing and   234
       thawing indices
7.20   Distribution of active lithalsas and continental palsas with freezing and   236
       thawing indices
7.21   Palsas on a flood plain at Sheldrake River, Québec   237
7.22   Water content (% by volume) in a Manitoba palsa   238
7.23   Distribution of peaty mounds and lithalsas with mean annual air   239
       temperature and precipitation in the Yukon Territory
7.24   A lithalsa beside Marsh Lake, Yukon Territory   241
7.25   Lithalsa at Kangqiang, Qinghai-Tibet Plateau   243
7.26   Margin of a peat plateau, Robert Campbell Highway   245
7.27   Comparison of the average height of floating palsas with the thickness   245
       of the icy core
7.28   Discrepancy between the expansion of water on freezing and actual   246
       elevation of peat plateaus
7.29   Comparison of moisture contents and dry density for fen and peat   246
       plateau samples
7.30   Thickness of the thawing fringe in peat plateaus   247
7.31   Relationship between stable icy peat plateaus and freezing and thawing   249
       indices
7.32   Cryogenic earth hummocks, Lake Hosvgul   250
7.33   Pika mounds, Tibetan Plateau   251
7.34   Textures of mineral sediment in oscillating earth hummocks   253
7.35   Cross section of an oscillating earth hummock, Mongolia   254
7.36   Radiocarbon ages of organic matter in oscillating earth hummocks   255

7.37	Distribution of oscillating and niveo-aeolian earth hummocks with freezing and thawing indices	257
7.38	Thufurs in southern Iceland	258
7.39	Cross section of a thufur	258
7.40	Posulated mechanism of the formation of thufurs	259
7.41	Grain sizes of sediments in Icelandic thufurs	259
7.42	Distribution of thufurs with freezing and thawing indices	260
7.43	Niveo-aeolian hummocks on the Fosheim Peninsula	262
7.44	Temperatures in a hummock at 49° North	263
7.45	String bog in Norway	264
7.46	Anastomosing string fen in Sweden	265
7.47	Distribution of string bogs with freezing and thawing indices	266
8.1	Turf-banked terraces near Nome, Alaska	268
8.2	Distribution of needle ice with latitude	271
8.3	The mechanism of frost creep	273
8.4	Gelifluction on frozen ground at 4700 m, Kunlun Shan	275
8.5	Fast gelifluction in ice-rich silty sediments dissecting the tundra vegetation on the Yamal Peninsula.	277
8.6	Thermal creep caused by diurnal temperature changes	277
8.7	Relation of soil creep and slope	278
8.8	Turf-banked terrace, Rat Pass, Dempster Highway	279
8.9	Gelifluction deposit, Kunlun Shan	281
8.10	Lower front of a gelifluction lobe, Kunlun Shan	281
8.11	Braking block, Kunlun Shan	282
8.12	Elongation of the buried stem of a plant on the gelifluction slope, Kunlun Shan	282
8.13	Ploughing block in melting snow, Marmot Ski Area, Jasper	283
8.14	Development of multiple buried organic layers in gelifluction lobes	283
8.15	Comparison of pediments and bajadas with and without permafrost	285
8.16	Tors, North Fork Pass, Yukon	286
8.17	Altiplanation terraces at Boundary, Yukon Territory	286
8.18	Mudflow in the Central Yakutia, Russia	288
8.19	Embryonic cryogenic debris flows, Summit Lake, B.C.	288
8.20	Debris flows along the Slims River valley, Yukon Territory	289
8.21	Oblique aerial photograph of the Vulcan debris flow fan	290
8.22	Photographs of parts of the Vulcan debris flow fan	291
8.23	Timing of the debris flows on the Vulcan debris flow fan compared with precipitation and air temperature	292
8.24	Processes operating along the thalweg, Vulcan debris flow fan	293
8.25	Density of the fan sediments with age from deposition	293
8.26	Debris flows after the 1988 fire, Marshall Creek, Yukon	294
8.27	Filter dams as defences against debris flows, Bormio, Italy	295
8.28	Several types of defences against frequent debris flows	296
8.29	Elongate active layer detachment slides, Kluane Lake, Yukon	298
8.30	Surface remaining after the passage of an active layer detachment slide	299
8.31	Diagrams showing the parts of retrogressive thaw slumps	300

8.32   Headwall of retrogressive thaw slump along the Aldan river of the   301
       Central Yakutia, Russia
8.33   Polycyclic retrogressive thaw slump, Northwest Territories   302
8.34   Headwall height relative to rate of regression   303
8.35   Retrogressive thaw slumps along the Mackenzie River   303
8.36   Cycle of slump initiation, stabilization and re-initiation   305
8.37   Snow avalanche distribution before and after fire compared with that   307
       during the last Neoglacial period, Vermilion Pass, Kootenay National
       Park
8.38   Lee cliff avalanches in winter and summer, Vermilion Pass   307
8.39   Results of wet gulley avalanches, Ischgl, Silvretta Alps, Austria   308
8.40. Relationship between mean angle of slope and avalanche frequency,   309
       Rogers Pass, British Columbia
8.41   Avalanche shed, Rogers Pass   311
8.42   Slush avalanches in the Austrian Tirol   311
8.43   Close-up of the terminus of a slush avalanche   312
9.1   Talus slopes and Murtèl I rock glacier, Switzerland   316
9.2   Weathering of rock by *Lecidea auriculata*   320
9.3   Distribution of active cryogenic block fields with mean annual freezing   323
       and thawing indices
9.4   Grain size of the sedimentary cover of Plateau Mountain, Alberta   324
9.5   15 m ground temperatures in forest and alpine meadow with Altitude at   325
       Plateau Mountain, Alberta
9.6   Talus fans and a rock glacier, North Slope of the Brooks Range   327
9.7   Talus slopes on the south-facing side of Mt. Yamnuska, Alberta   328
9.8   Debris flow levees on two adjacent talus fans, Bow Summit Banff   329
       National Park
9.9   Longitudinal ERT resistivity profile of the Petit Mont Rouge #1   330
       talus cone
9.10   An active spatulate rock glacier, Kluane National Park   331
9.11   Distribution of permafrost in the Petit Mont Rouge talus fan   332
9.12   Stratified scree deposits near Kluane Lake, Yukon Territory   334
9.13   Active tongue-shaped rock glaciers, Zailijskiy Alatau, Kazahkstan   335
9.14   Typical complex rock glacier with overlapping and over-riding lobes   336
9.15   GPR profile of the Hiortjellet rock glacier, Svalbard   339
9.16   East Slims rock glacier in Kluane National Park with a split tree   340
9.17   Comparison of the fronts of the Gorodetsky and East Slims rock glaciers   342
9.18   Front of a rock glacier advancing across a former road in the Tian Shan   343
       mountains, China
9.19   Possible mechanisms for the frontal movement of rock glaciers   344
9.20   Actual measured movements of the Suvretta and Gruben rock glaciers   345
9.21   Relation of active lobate rock glaciers to mean annual freezing and   346
       thawing indices
9.22   View down the relict Roc Noir rock glacier, France   348
9.23   Measuring water flow from a spring, East Slims rock glacier   349
9.24   Discharge from two springs in front of the East Slims rock glacier   350
       compared with mean daily temperature and precipitation events

9.25   Cryogenic block streams, Maqú, China   351
9.26   Relation of active Tibetan and Siberian dynamic rock streams to mean   352
        annual air temperature and mean depth of winter snow pack
9.27   Typical location of active block streams in Yakutia   355
9.28   Plan and cross-section of a Tibetan dynamic block stream, Kumlun   358
        Shan
9.29   Movement of painted stones in one year, Kunlun Shan   358
9.30   An active lag block stream, Chinese Tien Shan   360
9.31   The active lag block stream below the Empty Cirque, Chinese Tien   361
        Shan
9.32   Lichen covered subrounded to subangular boulders on the surface of   362
        the active lag block stream in Figure 9.31
9.33   Inactive block stream at Odenwald, near Felsen, Germany   363
9.34   Relict block stream above the Middle Chalet, Marmot Ski Area, Jasper   364
9.35   Close-up of the margin of the relict block stream emerging from   364
        beneath glacial till, Marmot Ski Area, Jasper National Park
9.36   Relatively flat surface of the blocks on block fields in the Falkland   365
        Islands
9.37   A boulder pavement along a river bed, Macmillan Pass, Yukon   366
10.1   Sorted stone circles near Svalbard   367
10.2   Change in shape of sorted polygons down a slope in the Falkland   368
        Islands
10.3   Small scale (12 cm width) patterned ground in Katmai National Park,   370
        Alaska
10.4   Mature sorted circles formed in an ephemeral lake, Alaska   371
10.5   Distribution of inactive sorted patterns with slope angle, Plateau   372
        Mountain
10.6   Relationship of arctic, subarctic mudboils (A), and nonsorted circles   377
        (B) to freezing and thawing indices
10.7   Formation of arctic mudboils (nonsorted circles) according to the   378
        equilibrium model of Mackay (1980b)
10.8   Subarctic mudboils at the Arctic Circle, Dempster Highway   379
10.9   Diagrammatic cross-section showing the structure of subarctic   380
        mudboils in winter
10.10   Effect of modification of the surface vegetation on subarctic mudboils   381
        in Alaska
10.11   Change in shape of the mud flowing from embryonic xeric nonsorted   382
        circles, Plateau Mountain, during 1980 to 1983
10.12   Average rates of circulatory movement (mm/a^{-1}) over 3 years in a xeric   383
        nonsorted circle, Plateau Mountain, Alberta
10.13   Increasing age of Rhizocarpon geographicum thalli towards the margin   385
        of relict nonsorted circles in Sweden
10.14   Frost boils developed in soil on the summit of Heart Mountain,   386
        Alberta, and breaking up pavement in downtown Yakutsk
10.15   Relationship between active cryogenic sorted circles and freezing and   387
        thawing indices
10.16   Change in shape of patterning with slope   388

10.17   Apparent movement of sorted circles on raised beaches in Spitzbergen   389
10.18   Clean separation of fines and stones in a sorted polygon   390
10.19   Cross-section and appearance of sorted micro-stripes   392
10.20   Intermediate-sized stone stripes in Iceland   392
10.21   A stone pit in the Swiss Alps   393
11.1   Thermokarst produced by thawing of ice wedges in the Central Yakutia. The icy cliff retreats at about a few metres per year, with the resulting sediment flowing downslope to the river (retrogressive thaw slump)   397
11.2   Ground ice content of the permafrost areas of Siberia   398
11.3   Subsidence resulting from the thawing of the upper surface of the permafrost   404
11.4   Thermokarst features developed on an Arctic landscape   405
11.5   Thermal conductivity of the soil at 5 cm depth before and after an experimental fire in the larch forest near Yakutsk   406
11.6   The result of removing the surface soil over ice-wedges along the upper Blackstone River valley, Dempster Highway   407
11.7   Early stages of thawing of ice-wedges in a Siberian larch forest   408
11.8   Pond in the thawing summit of a hydraulic pingo, Dawson City   409
11.9   Sequence of decay of plateau palsas in Iceland   411
11.10   Beaded stream in the Northwest Territories   412
11.11   Thermokarst lakes on the Mackenzie Delta   413
11.12   Typical pingo (hydrolaccolithe) on the floor of a former thaw lake in Western Siberia   414
11.13   Oriented lake in the Northwest Territories   416
11.14   Depth of active layer and mean annual air temperature, Central Yakutia   418
11.15   Sequence of development of an alas in Yakutia in the larch forest   419
11.16   Alas formation at the Yukechi site, right bank of the Lena River, near Yakutsk: A, the first *bilar* stage; B, the second *dujoda* stage; C, the third *timpi* stage; D, the last, stable *alas* phase   419
11.17   Changes in soil stratigraphy during one alas cycle   421
11.18   Changes in thermokarst intensity during the last 13 ka on Kurunnakh Island, central Lena River, Siberia   422
11.19   Sequence of alas development on the grassy steppes of Mongolia   423
11.20   Processes and features resulting from thermokarst along streams incised into permafrost   424
11.21   Thermal erosionl notch in river bank   425
11.22   The three different kinds of slope failure along river cliffs in icy permafrost   426
11.23   Diagrammatic cross section of a valley affected by ice jams   427
11.24   Longitudinal profile of an ice jam and inshore shear lines   428
11.25   An inundated settlement resulting from an ice jam on the Yukon River   428
11.26   Stage frequency of ice jams versus summer floods on the Yukon River at Dawson City between 1896 and 1999   429
11.27   Circum-Arctic map of coastal erosion rates   430

11.28   Mean annual erosion rates for a 60 km segment of the Alaskan   431
Beaufort Sea between1955 and 2007

11.29   Cross-section of an ice pile-up along the shore at Cape Kellett,   432
Southern Banks Island, Northwest Territories

11.30   Conditions for rafting boulders   432

11.31   Ice-push ridges on the Sabine peninsula of Melville Island   433

11.32   Stages in coastal erosion of yedoma-type sediments with different ice   434
contents

11.33   Mean sea surface temperature (2000–2009) versus day of year in the   434
Beaufort Sea

11.34   Annual erosion rates for three distinct shorelines along the Alaska   435
Beaufort Sea shore

11.35   Retrogressive thaw slumps along the northwest coast of the Yukon   436
Territory

11.36   Wave-cut notch on the shores of the Laptev Sea   437

11.37   Block erosion due to failure along ice-wedges in coast erosion   437

11.38   An active layer detachment slide descending on to a beach   438

11.39   Results of thermal erosion along the northeast coast of Alaska   438

12.1   Results of frost heaving on a concrete apartment building in Vorkuta,   442
European part of Russia

12.2   Results of freezing a fine-grained soil   447

12.3   Mechanical interaction of soil underlain by a rigid horizon   450

12.4   Stresses developed in freezing glacial silt at −2°C   451

12.5   Stresses in water-saturated glacial silt at −2°C   452

12.6   Frost heaving stresses measured by sensors of frigidity   457

12.7   Creep stages of frozen sand at one-axis deformation (ε)   458

12.8   Curves for creep during the shear test including the third stage   458
(above) and the curve for a change in the frozen soil strength with
time (below)

12.9   Grain sizes of different classes of frost susceptibility according to the   462
Norwegian classification

12.10   Conditions of temperature, active ice lens formation, frost front,   462
freezing front and suction profile during the formation of ice lenses

13.1   Changes in ground temperatures from 1960 to 2000 A.D. at Norilsk,   466
Siberia: A and B, with depth, while C shows the lithology with depth –
(Norilsk permafrost station data).

13.2   Variation in permafrost temperatures in and around Yakutsk in 2015   467
on the Aldan river terraces.

13.3   Differential settlement and heave at the four corners of an unheated   468
building with time

13.4   Cross-section of the foundation of a heated building after 3 years   469

13.5   Result of the development of a thaw bulb beneath a building   469

13.6   Deterioration of a concrete wall in the dry permafrost climate of the   472
Qinghai-Tibet Plateau after 7 years

13.7   Result of the development of thermokarst on a parking in Yakutsk   473

13.8   Progressive degradation of permafrost beneath a gravel pad   474

13.9   Isotherms beneath a 75 cm air space and sill foundation overlying a   475
       90 cm gravel pad in Northeast China

13.10 Ground temperatures at maximum thaw depths beneath a   476
       pier-supported building on spread footings in Northeast China

13.11 Maximum ground temperatures beneath buildings constructed on piles   477
       in Northeast China

13.12 A building on piles on continuous permafrost at Harasavei   478

13.13 Piles with hydraulic jacking capability, Prudhoe Bay   479

13.14 Effects of slurry and pile spacing on temperature rise in the adjacent 480
       soil

13.15 Diagram of the forces acting on a pile in permafrost   481

13.16 Estimated short-term and sustained adfreeze strengths for wood and   482
       steel piles in icy frozen clays or silts

13.17 Failure of a geological survey building in Yakutsk due to loss of bearing 483
       capacity of its piles in permafrost

13.18 Diagram of a typical vertical support member (VSM) used on the   484
       TransAlaska Pipeline

13.19 Thermosiphons cooling a part of the foundation that is failing due to   485
       thawing of the underlying permafrost in Vorkuta, European part of
       Russia

13.20 Thermosiphons cooling the piles supporting the TransAlaska Pipeline   485
       north of the Yukon River

13.21 Sloping thermosiphons (thermoprobes) used to stabilize the   486
       Golmud-Lhasa road in an area of warm permafrost

13.22 Comparison of two phase and one phase thermosiphons   487

13.23 Three thermosiphon designs beneath buildings   487

13.24 Sloping thermosiphons beneath a sports complex in Yakutsk   488

13.25 A flat loop seasonal cooling system at the base of oil storage tanks in 489
       Western Siberia

13.26 Ventilation ducts in the embankment of the Golmud-Lhasa railroad   490
       near Beihu'he, Qinghai-Tibet Plateau

13.27 The results of closing or leaving open the air doors on vents in summer 491
       in a composite embankment

13.28 Diagram of an experimental tubular heat drain used to cool the   491
       shoulder of a road or airfield embankment

13.29 The four types of configuration of rock layers tested for use in building 494
       the Golmud-Lhasa railway

13.30 New design of the embankment using waterproof geotextiles   494

13.31 Temperature fields in summer and winter on the embankment using   495
       geotextiles as in Figure 13.39

13.32 Sod covering the impermeable membrane on the side slopes of a   495
       freeway being constructed north of Madoi, China

13.33 Shading of piles along the Golmud-Lhasa railroad   496

13.34 Detail of the support beneath the shaded rail bed   497

13.35 Experimental shading awnings along the BAM railroad   498

13.36 Cross-section of the road embankment showing the location of the   499
       moisture wicking material

13.37	Wicking fabric being laid down along the Dalton Highway, Alaska	499
14.1	Longitudinal cracking of the road embankment, Yamal Peninsula	503
14.2	Subsidence alongside the Golmud-Lhasa road	503
14.3	Relationship of permafrost problems to embankment height on the Original Golmud-Lhasa road during the first 30 years	504
14.4	The experimental embankment at the Beih'he Research Station	505
14.5	Trucks crossing the Albany River, Manitoba on an ice road	507
14.6	Relationship between minimum embankment height and mean annual air temperatures for different roadway surfaces	508
14.7	Determination of the critical MAAT and embankment heights to protect the underlying permafrost	509
14.8	Lateral and vertical movements measured on a road embankment in Alaska	510
14.9	Diagrammatic cross-section of the Canol Road at Heart Lake showing the talik that had developed after the first 20 years	511
14.10	Models of the long-term changes in ground temperature following removal of the vegetation along the Norman Wells pipeline	512
14.11	Erosion at a culvert under a road at Bovanenko gas field, Yamal Peninsula	513
14.12	Consolidation of the embankment on a local road, Yamal	513
14.13	The route of the Qinghai-Tibet Railway across the permafrost areas of the Qinghai-Tibet Plateau with mean annual ground temperatures	515
14.14	A train travelling along the rail line near Fenghoushan	516
14.15	Typical embankment on the north slopes of the Fenghoushan Mountains	516
14.16	The north portal of the Fenghoushan Tunnel	517
14.17	Sand dunes and fences to reduce problems with blowing sand	517
14.18	Problems with the rail lines in Siberia	519
14.19	Differential subsidence and lateral creep of the embankment along part of the railway in Siberia	519
14.20	Suggested methods of stabilizing railway embankments in Siberia	521
14.21	Suggested use of thermosiphons to counteract embankment flowage in Siberia	522
14.22	Development of thermokarst and thermal erosion on the Yamal railroad	523
14.23	The locomotive equipped to measure differential heave or settlement along the Qinghai-Tibet Railway	523
14.24	Change in bearing capacity of a non-cohesive soil from the frozen state, through thawing, to the thawed state with and without geogrid	524
14.25	Concrete tracks for a high speed railway in South Korea	525
14.26	Methods of correcting the level of rails on concrete railbeds	525
14.27	Concrete track rails before and after grouting	526
14.28	Predicted maximum thaw below the centre line of embankments using 10 cm thick EPS sheets on Highway G214, China	527
14.29	Maximum thickness of the active layer in the embankment of the Qinghai-Golmud road during 30 years with either asphaltic concrete or a flexible asphaltic surface	528

14.30   Cumulative incident and reflected radiation along highway test sections   529
comparing white-painted asphaltic concrete with a control section

14.31   Cross-section of the Eagle River at the Dempster Highway bridge   531

14.32   The old bridge at Goldstream Creek, Fairbanks   531

14.33   Part of the new bridge across the Yukon River along the Haul road,   532
Alaska

14.34   Buckling of the causeway across a dry stream bed along the old   533
Golmud-Lhasa road

14.35   Tilting of a pile support resulting in the collapse of a section of bridge,   533
Golmud-Lhasa Highway

14.36   Diagram of a typical icing affecting a road embankment   535

14.37   Ice blocking the outlet of a culvert along the Alaska Highway   535

14.38   Embankment on the Dempster Highway with multiple large culverts   536
intended to slow the onset of icings crossing the road

14.39   A "moose warmer" keeping a space open in a culvert   537

14.40   Some possible methods of stabilizing cut slopes   539

14.41   Snow cover and thaw depth below the runway centre line at the west   540
end of the Inuvik airstrip

15.1   Main producing oil and gas fields in the Arctic basin relative to the   544
population centres in the Northern Hemisphere

15.2   Pipelines carrying gas to Western Europe from the gas fields north of   545
Nadym, Western Siberia

15.3   Schematic diagram showing the special precautions taken in casing   548
production wells in permafrost

15.4   Parts of a successful sump   549

15.5   Pipeline routes proposed for construction in Canada by 2011   550

15.6   Deformation of the uninsulated, high-pressure gas pipeline in the   552
Medveje gas field, Western Siberia

15.7   Forces acting on a warm pipe buried in discontinuous or continuous   553
permafrost

15.8   Forces acting on a cold pipeline crossing patches of unfrozen ground or   554
buried in unfrozen sediments

15.9   Gas pipeline that was once buried and now is floating   555

15.10   Modelled excess pore-water pressure generated as a result of thawing   556
of ice-rich permafrost at a location along the Norman Wells pipeline

15.11   Modelling of geothermal changes along the Norman Wells pipeline   557
right-of-way showing the effects of different insulation values at an
initial ground temperature of $-2°C$ and a climate warming rate of
$0.08°C/a^{-1}$

15.12   Modelled variation in temperature of the oil at Mo'he pumping station   559
based on the Russian data

15.13   Modelled variation in temperature of the oil along the pipeline with an   560
annual flow of 15×10^7 tons

15.14   Modelled results of thaw and freezing depths along the pipeline route   561
in the warm permafrost along the China-Russian Pipeline

15.15   Probable thaw depths below the TransAlaska Pipeline as a function of   561
time, oil temperature and thermal conductivity of the ground

15.16   Types of vertical support members used for elevated sections of the   562
         TransAlaska Pipeline
15.17   The bends in the TransAlaska Pipeline to allow for expansion and   563
         contraction
15.18   A typical example of the pipeline being carried across a river on piles in   564
         Alaska
15.19   Conditions at the time of failure of the TransAlaska Pipeline at Mile   565
         Post 166 on the north slope of Atigun Pass, Alaska
15.20   Maximum subsidence designed for along the TransAlaska Pipeline   566
         compared with the conditions of failure at mile posts 166 and 734
15.21   The successfully repaired section of pipeline in Atigun Pass using   567
         closely spaced thermosiphons
15.22   The major compressor station and gas processing plant, Yamburg,   568
         Western Siberia
15.23   Pipeline temperature profile proposed for the Mackenzie gas pipeline   569
15.24   Thermosiphons in piles used to cool the ground on either side of the   570
         TransAlaska Pipeline when in buried mode
15.25   Production pipelines going through larger culverts to avoid thawing of   570
         permafrost at vehicle crossings, Prudoe Bay
15.26   Stability of the methane-water-gas hydrate system with pressure   572
         plotted as an equivalent depth assuming a hydrostatic gradient of
         $10^4 \, \mathrm{Pa\,m^{-1}}$
15.27   Prediction of the *in situ* hydrate zone using formation temperatures   573
         and pore pressure gradients
15.28   Locations of some known and inferred gas hydrates   573
15.29   Comparison of the three structures of ice that can contain gas hydrates   574
15.30   Effect of sodium chloride, methanol and polyethylene oxide (PEO) on   574
         the hydrate-aqueous liquid-vapour phase equilibria
16.1   An old gold mining dredge at Nome, Alaska   578
16.2   Comparison of the shape of the thawed holes in permafrost at Nome   579
         by the three methods in use in 1920
16.3   View of the Faro mine developed on the Pelly Mountain slopes,   581
         Yukon Territory
16.4   Aerial view of the Inter Diamond Mine in Central Siberia   581
16.5   Diagram of the way blasting is used to cut back a rock face   583
16.6   Typical blast pattern in ice-cemented ground containing excess ice and   584
         pore ice
16.7   Drill holes in the lake at Ekati Diamond Mine to establish the   585
         distribution of kimberlite beneath the water
16.8   Retaining dam to hold back the water from the mine operation, Panda   586
         Mine, Ekati
16.9   North-south section of the Panda mine kimberlite pipe showing the   586
         position of the proposed mining ramps for ore trucks
16.10   General view of the Ekati workings   587
16.11   Predicted extent of thaw around a 3 m drift or decline with depth as   588
         related to measured ground temperature and permafrost stability at
         Asbestos Hill, Quebec

16.12	Problems that may occur around tunnel entrances	589
16.13	Aerial view of the abandoned Giant Mine, Yellowknife	593
16.14	Aerial view of the position of the chambers and stopes in the old gold workings in permafrost, Giant Mine	594
17.1	Utilidors on piles at Inuvik, Northwest Territories	598
17.2	Aerial view of the Vilyuy Dam, 75 m high, 600 m long, with a total capacity of 35.9 km^3, generating 2,700 GWh of electricity	600
17.3	Underpasses or overpasses where roads and utilidors meet in Inuvik	605
17.4	Diagram of a typical above-ground utilidor in North America	606
17.5	Icings along the Tuotuohe valley adjacent to the Qinghai-Tibet High Voltage Transmission Line	612
17.6	Groundwater welling up in an excavation for a pile, Tuotuohe River valley	612
17.7	Diagram showing the evolution of the rupture resulting from penetrating the permafrost base when drilling	613
17.8	Foundation types selected by Qinghai-Tibet DC Interconnection Project for the bases of towers	615
17.9	Placement of thermosiphons used to cool the ground around the bases of towers	615
18.1	Distribution of the Taiga and Boreal Forest in the Northern Hemisphere	617
18.2	Typical former winter home of the Sakha people	618
18.3	Reindeer herding summer camp at the mouth of the River Ob, Western Siberia	619
18.4	Reindeer pulling a sleigh in winter in Siberia	620
18.5	Present-day land use in Asiatic Russia	620
18.6	The natural vegetation zone of Russia compared with the southern limit of permafrost	622
18.7	The boundaries of the four main regions of Russia	622
18.8	The distribution of Boreal Forest across North America	623
18.9	Distribution of lowlands outside of the Canadian Shield that are currently underlain by permafrost	624
18.10	Herd of Tibetan gazelle on meadow steppe, QTP	625
18.11	Use of stones to prevent wind erosion	626
18.12	The Eichfeld agricultural zones in Eurasia compared with permafrost zonation	627
18.13	Distribution of Eichfeld zones as applied to North America relative to permafrost zones and cultivable land	628
18.14	Typical pasture in the grassy steppes west of Ulan Bator, Mongolia	630
18.15	A yak pulling a cart carrying part of a yurt in Northwest Mongolia	631
18.16	Mongolian horses with their traditional saddles	632

List of tables

1.1	The continentality Index classification for mountains by Gorbunov	16
2.1	Results of experimental weathering of common minerals	76
3.1	Typical albedos of various common surfaces	86
3.2	Some suggested controls of climate change	87
4.1	Differences between the terminology used in China for permafrost zones in publications between 1988 and 2006	129
4.2	Details of typical boreholes in permafrost along the Qinghai-Tibet Highway	139
4.3	Classification of the stability of mountain permafrost based on mean annual temperature	142
5.1	The diagnostic characteristics of primary and secondary wedge structures, infilled thawed massive ice cavities and load-casting involutions in China	169
7.1	Summary of the different kinds of cryogenic earth hummocks currently known, developed in mineral soils	215
7.2	Stages of development and demise of lithalsas at Fox Lake, Yukon	240
7.3	Summary of the different kinds of earth hummocks	251
8.1	Some results of measurement of mean annual needle ice creep from different environments	272
8.2	Results of actual measurements of frost heave in different environments in North America	274
8.3	Characteristics of the five debris flows observed on the Vulcan debris flow fan in 1994–5	292
9.1	Typical results of measurements of average rate of rockwall retreat in northern environments since deglaciation	318
9.2	Characteristics of cryogenic active block streams over permafrost	354
10.1	Measured ranges of limiting slope angle for sorted patterned ground	373
10.2	Measured rates of downslope movement of the coarse and fine sections of sorted patterned ground as related to type of slope processes involved	373
10.3	Characteristics of the four main types of cryogenic nonsorted circles	375
11.1	Characteristics of the transient layer on the Abalah plain on the right bank of the Lena river, Yakutia	401

11.2   Calculated mean annual temperatures of saline and nonsaline soils and the resulting thickness of the active layer   406

11.3   Depth of alases and thickness of the ice-complex deposits in the Lena-Amga region   420

11.4   Accumulations of carbon and nitrogen in humus in various soils, together with carbonate content in the alas, meadows and forests around Yakutsk, Siberia   421

12.1   The U.S. Corps of Engineers frost design classification   461

12.2   U.S. Corps of Engineers classification of frost susceptibility   461

13.1   Percentages of buildings showing structural damage in Russian towns and cities located on permafrost   465

13.2   Numbers of different types of thermosiphons and thermopiles used in Canada between 1994 and 2008   488

14.1   Proposed counter-measures for reducing plastic flow on Russian railway embankments   520

14.2   Methods of adjusting the rails for settlement on concrete tracks   526

14.3   Advantages and disadvantages of the methods of dealing with track subsidence   526

14.4   Typical minimum thicknesses of asphaltic concrete, base and sub-base courses used in Canada for aircraft with various tire pressures   540

15.1   Summary of the advantages and some disadvantages of the main trenching methods   551

15.2   Comparison of different insulation materials tested for thermal resistance for the Norman Wells pipeline   556

16.1   The results of the Miles experiments comparing the different methods of thawing the ground   579

16.2   Diagram showing the arsenic trioxide alternatives preferred by the community of Yellowknife   595

16.3   Risk assessment of some of the options in Table 16.2   595

17.1   Classification and characteristics of groundwater in permafrost areas   599

17.2   Examples of residential and community water consumption   603

List of symbols

α albedo (decimal fraction or %)

E reradiated energy back into the atmosphere (W/m^2)

R_n Net solar radiation (W/m^2)

S Solar constant (roughly 1.362 kW/m^2)

H Sensible heat flux (W/m^2)

LE Latent heat flux (W/m^2)

G Ground heat flux (W/m^2)

Δt_{snow} is the thermal offset of annual mean temperature due to the warming influence of the snow (°C)

ΔA_{snow} is the decrease of the amplitude (physical) of the annual fluctuations of temperature under the snow (°C)

A_θ is the meteorological amplitude of the annual fluctuations of the temperature of air (°C)

z_s is the depth of snow cover (m)

C is the volumetric heat capacity of snow (J/m^3 * °C)

λ is the thermal conductivity of snow (W/(m * °C))

$\overline{R}_{snow}$ is the average thermal resistance of snow in winter ((m^2 * °C)/W)

Q is the heat of phase transitions (latent heat) in the ground (J/kg)

λ_f is the thermal conductivity of the frozen ground (W/(m * °C))

Ω_w a parameter equal to $|t_w| * \tau_w$, where t_w – average winter temperature

τ_w the duration of the winter (s)

t_ξ average annual ground temperature on bottom of the layer of the seasonal thawing (°C)

Δt_{toc} an increase in the average annual temperature on the bottom of the layer of the seasonal thawing (t_ξ) in comparison with the average annual temperature on the ground surface (t_s) due to the infiltration of atmospheric precipitation (°C)

ξ the thawing (freezing) depth (m)

T is the period of year (s)

Q_{pr} heat of atmospheric precipitation coming to the layer of the seasonal thawing (J/m^2)

λ_g effective thermal conductivity of the layer of the seasonal thawing (W/(m * °C))

$V_{pr,i}$ the monthly (or decade) sum of summer atmospheric precipitation (kg/m^2)

$t_{pr,i}$ the average monthly (or decade average) temperature of precipitation, approximately taken as the temperature of air (in the absence of data) (°C)

C_w the specific heat capacity of water (J/°C) (J/(kg * °C))

σ_z horizontal stresses (MPa)

ε_f heaving strain of about 9% of volume of freezing water (decimal fraction)

l_{fr} size of the frozen part of the soil (m)

l_{ice} size of the ice part (m)

E_{fr} strain (Young's or compression) modulus of frozen soil (MPa)

E_{ice} strain (Young's or compression) modulus of ice (MPa)

E_t strain (Young's or compression) modulus of the unfrozen part of soil (MPa)

H depth from the surface of the reservoir (m)

H_i the maximum for this region thickness of ice (m)

t_H annual mean temperature at the depth of H (°C)

t_{min} minimum average monthly temperature on the surface of ice under the snow (°C)

t_{max} maximum average monthly temperature of the water (°C)

σ stress (MPa)

ε strain (decimal fraction)

P external pressure (MPa)

σ_0 is the initial stress below which viscoplastic steady flow has not yet started, which is sometimes known as the creep threshold (MPa)

h height of ice-wedge (m)

η_s viscosity of the surrounding soil (Pa * s) or (Pl)

η_i viscosity of ice (Pa * s) or (Pl)

z vertical movement of ice wedge (m)

τ time (s)

l width of ice-wedge (m)

t temperature (°C)

W volumetric liquid moisture content (decimal fraction)

W_i volumetric ice content (decimal fraction)

W_{shr} shrinkage limit (decimal fraction)

p equal pressure of ice and water (MPa)

k_g a factor of frigidity (or directed to massive ground, or to engineering construction) (decimal fraction)

q the volume of migrating water (m³)

ε_s value of suppressed strains (decimal fraction)

ε_a value of allowed strains (decimal fraction)

dp_h dependence of stresses on the value of the "suppressed" strain in the linear form (MPa)

W_ξ the volumetric moisture content at the interface from the side of the thawed zone, or in the desiccated zone (decimal fraction)

W_{uf} the volumetric unfrozen (or thawed) water content (decimal fraction)

ε_h the value of potential free frost heaving strain in the case of an absence of the strain suppression (decimal fraction)

$d\varepsilon_{sh}$ shrinkage of the unfrozen part due to dehumidification of the soil (decimal fraction)

k_r rigidity of sensor (or object to affect, for example, pipe) (decimal fraction)

k_{gr} factor of decreasing of freezing pressure in soil, if the soil is allowed to expand (decimal fraction)

p_h pressure (stress) acting in the system (MPa)

l_t size of the unfrozen part of the soil (m)

K_{il} the coefficient approximately equal to the cosine of an angle of migratory ice lenses (decimal fraction)

$\partial W_{uf}/\partial t$ the thermogradient coefficient ($^\circ C^{-1}$)

J_W migratory moisture flux (kg/(s $*$ m^2))

λ_W the hydraulic conductivity of soil in case of migratory moisture flux (s/m^4)

K_W the moisture diffusivity (kg/(s $*$ m^2))

β the coefficient of the volumetric shrinkage (decimal fraction)

W_{in} the initial volumetric moisture content (decimal fraction)

ξ_{sh} the depth of the shrinkage zone being equal to the freezing depth plus the zone of desiccation under the freezing front (m)

Part I

Introduction and characteristics of permafrost

Geocryology is the science studying the effects of ground temperatures below 0°C on the surface layers of the crust of the Earth. In areas with tropical or subtropical climates, the ground remains above freezing throughout the year. In these regions, the normal geomorphological processes take place, excluding glaciations and processes unique to areas of permanently or seasonally frozen ground.

As used in this book, the term soil refers to either the upper surface layers of the ground that have been modified by weathering and soil formation, or unweathered rock, unless otherwise indicated. When the land areas are traced towards the poles, the climate becomes sufficiently cool for the surface layers of the ground to experience temperatures below 0°C. This cooling may be temporary or short-term, lasting one or more hours, or for a few days in a year. These soils grade into seasonally-frozen soils, where the sub-zero temperatures may last for several months, and the length of the frost-free period becomes a limiting factor for the biota. Yet further towards the Poles, areas of soil are present which remain below 0°C for more than two consecutive years. These cold layers are referred to as permafrost, and the soils are said to be cryotic. The surface layers of permafrost are in quasi-equilibrium with the present-day climate (see below), but the deeper layers may have formed partly under past, colder climates, and are called relict permafrost. Glaciers are not included, and are treated as separate landforms with their own characteristic processes and resulting topography.

In Russia, the layers of the ground undergoing temporal, seasonal, or long-term freezing are regarded as making up the cryolithozone, though this terminology is usually not used elsewhere. The depth of the frozen soil depends on the microenvironment and thermal history, and ranges from a few centimetres to hundreds of metres. Permafrost thickness increases with a decrease in the annual temperature, other factors being equal, while the depth of summer surface thaw called the active layer decreases. Temporary and seasonally frozen surface soils are usually continuously frozen in winter at the surface of the ground but they thaw in summer and are sometimes also incorrectly described as being the active layer, although they have a different ice distribution and content. Permafrost refers to the layers with temperatures below the active layer that are continuously below 0°C for more than two years. Note that it applies regardless of moisture content, or lack thereof.

The key property of permafrost is the tendency to accumulate H_2O in the form of ice in its upper layers, resulting in a unique moisture distribution in the ground (Figure 1.1). The quantity of moisture accumulated can reach up to 90% by volume below the surface of the perennially frozen ground. This results in severe modifications in the moisture regimes in these soils, together with an important dependence on there

Figure 1.1 Typical profile showing icy permafrost below the 72 cm thick active layer developed in Mohella (wind-blown volcanic ash) in Iceland. Note the concentration of ice in the surface layers of the permafrost, with ice content decreasing with depth. © S. A. Harris.

being a stable climate. Without it, the ground ice will tend to partly or wholly melt during warmer periods. The weather and climate are constantly fluctuating, while other parts of the microenvironment such as the vegetation and groundwater regime are also subject to changes with time.

The groundwater, ice and cold temperatures combine to cause special landforms unique to these areas, while even seasonal frost can result in the occurrence of processes not seen in warmer climates. In Part 1 of this book, chapter 1 summarizes the basic characteristics of permafrost, while chapter 2 examines the soil processes essentially unique to the cryolithozone. Chapter 3 examines the factors affecting the development of permafrost, while chapter 4 describes our present state of knowledge about its distribution. In Part 2, the unique landforms of these regions are described, while in Part 3, the limitations and problems affecting development in these areas are briefly examined.

Chapter 1

Definition and description

1.1 INTRODUCTION

Permafrost is defined as ground that remains below 0°C for more than two years (ACGR, 1988). It includes dry rock and soil, as well as all icy substrates including rock glaciers, but excluding true, icy glaciers. Icy glaciers are dealt with in science as a separate subject, although there are intergrades (Dobiński, 2012). Saline water bodies may remain unfrozen as part of permafrost. Ground that is colder than 0°C is said to be *cryotic*. The study of permafrost is called *geocryology* and is part of *cryology* which is the study of all frozen matter.

   Dry permafrost (0°C) is currently only known from Antarctica. Lacking moisture, it has different thermal properties to moist soil, since it takes extra energy to warm the water, and conversely, the water holds more heat energy than air, so more energy must be lost to change the temperature of the wet soil by 1°C. Elsewhere, cold, massive rock is not regarded as dry permafrost since ice is found in the cracks and joints. This is important since warm summer rainwater can partially thaw this ice, permitting meteoric water to descend to the intrapermafrost taliks or the ground water table beneath the permafrost.

   The nature of the ground affects the type of profile encountered and its thermal regime, but a typical profile in porous sediment is shown in Figures 1.1 and 1.2. At the surface is a layer that freezes and thaws with the seasons and is therefore called the *active layer*. In sediments that lack many through pores, it shows up as an ice-free zone or zone of rotten ice overlying the icy surface of the permafrost which is called the *permafrost table*. In more porous sediments such as peat, it includes a zone in which the upper layer of ice is porous and allows the ground water in the perched water table to move through it above the permafrost table (Figure 1.3). As the water moves, it slowly enlarges the pores by melting the outermost layer of the enclosing icy sediment, and this layer is called the *thawing fringe*. The ice is partly thawed, hence the inclusion of this layer in the active layer when it is present. At its upper surface is the *thawing front*. Its most frequent occurrence is in peat profiles where it can make up over 50% of the active layer. In porous mineral soils, it is much thinner, and is absent in clays and in bedrock. Since the actual position of the permafrost table varies from year to year, the zone of variation is called the *transient layer,* and in northwest North America, is characterized by an accumulation of ice. This is thin, except in mountainous areas with a highly variable continental climate. In Western Siberia, a thin dry zone

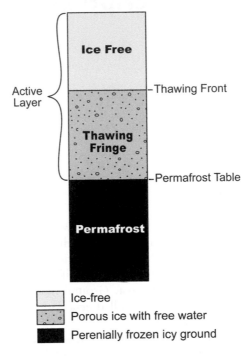

Figure 1.2   Names of parts of the upper layers in a permafrost profile in peat that is in equilibrium with the present-day climate (translated from Harris 2005).

has been described from above it in dense sediments (Shur, 1988a; 1988b). This dry layer is called the *shield layer* which protects the icy permafrost from thawing. Note that the *seasonally frozen layer* and the *seasonally thawed layer* are not necessarily identical. In any given year, the depth of the seasonally thawed layer is independent of the seasonally frozen layer which results from the maximum depth of thawing, though this ignores any results of climate changes. The structure and ice content in these layers is different, so the use of these terms in Russia is unique.

There is always a maximum concentration of ice in the sediments immediately below the permafrost table, and the ice content usually decreases with depth (Figure 1.1). The lower limit of permafrost is called the *permafrost base*, but the main mass of permafrost may include unfrozen zones called *taliks*. *Through taliks* may occur under rivers and the larger lakes, but under smaller water bodies, the taliks are shallow and overly the main permafrost body (the *blind* or *false taliks* in the Russian literature). In this case, these *supra-permafrost taliks* occur because water is a source of heat and needs to lose large quantities of energy ($152.855 \, \text{J/cm}^3$ or $640 \, \text{g.cals/cm}^3$) when changing its state from liquid to ice at $0°C$. The usual distribution of taliks and their terminology is shown in Figure 1.4. Sometimes the *closed, intra-permafrost taliks* or *sub- permafrost waters* are occupied by saline water that cannot freeze due to its salt content. The salts represent the brine left when the surrounding water froze, concentrating the salts in the remaining liquor, and these liquors are called *cryopegs*.

Figure 1.3 Rotten porous ice at the surface of an ice body in alluvium, Kluane, Yukon Territory, Canada. © S. A. Harris.

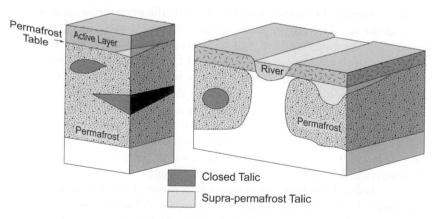

Figure 1.4 Block diagram of an area with permafrost showing the names of the constituent parts. The closed taliks can also form tubes parallel to the ground surface through which groundwater moves downslope.

This includes undersea water near shores. Because of the high salt concentrations, they can remain liquid at very low temperatures. These can result in saline water moving through the permafrost from the unfrozen ground below. Artificial cryopegs may also result from the injection of saline waters during fracking or drilling.

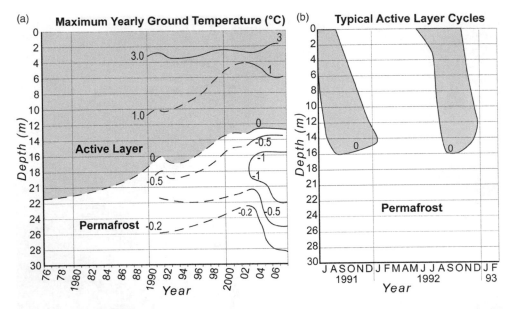

Figure 1.5   Change in active layer thickness a, during 30 years at Plateau Mountain #2 borehole drilled in massive limestone and dolomite, southwest Alberta, and b, the shape of the warm summer wave during a two year period from 1991–1993 (from Harris, 2008, Figure 4). In adjacent clastic sediments, the active later is only 1.5–2.5 m thick.

In North America, the unfrozen material below the permafrost base is sometimes referred to as the *sub-permafrost talik*. It develops where there has been substantial climate warming (see Chapter 3) so that the active layer does not refreeze completely and the relict permafrost has become separated from the upper layers with present-day freezing and thawing. This usually occurs where the permafrost is starting to degrade (Jin *et al.*, 2006). Most soils consist of clastic sediments with grain-to-grain contacts and pore spaces in between. These spaces may be partly or wholly filled with ice in the frozen layers. When this ice melts, there can be significant subsidence, while the resulting water can result in actual liquefaction of the sediment. The thermal properties including thermal conductivity depend on the abundance of the contacts between grains, the porosity, and the water/ice content.

In massive bedrock profiles, the permafrost is rather different, though it has rarely been examined. At Plateau Mountain in southwest Alberta, Canada, the bedrock consists of massive limestone and dolomite, some parts being covered by a thin layer of sediment, with some surfaces consisting of frost-fractured, thin-bedded limestone, while others consist of massive limestone with master joints (Harris & Brown, 1978). In the areas with a covering of clastic sediment, the thickness of the active layer is typical of profiles elsewhere in sedimentary soils, *i.e.*, about 1.5–2.5 m. Below the fissile, fractured rock, the active layer proved to be about 5–8 m thick, while in the massive bedrock, the active layer thickness measured 13–21 m (Figure 1.5). Similar deep active layers in bedrock have been recorded in Norway (Christiansen *et al.*, 2010; Farbrot *et al.*, 2011). This is due to the high thermal conductivity of the bedrock that was measured as being between 4.92 and 6.1 W/mK for the cores from the boreholes on Plateau Mountain (Harris & Brown, 1978). Since there is negligible pore space with trapped air in the massive rock, heat flow is very efficient compared with rocks or sediment containing air in the pores and fissures. Seasonal heating and cooling affects a much greater thickness of the surface material in this case. Air is a poor conductor of heat, while ice is a fairly efficient conductor, although it has a lower thermal conductivity than the bedrock. The boreholes were monitored for many years, and the subsequent cooling of the mean annual air temperature has resulted in the deepest active layer being reduced by several metres, though it is still deeper than 13 m in most years (Figure 1.5a).

The joints in the bedrock complicate measuring the thickness of the active layer in massive bedrock and also provide temperature anomalies in deciding the actual depth of the permafrost table. When the ice melts in spring, the resulting water flows down the joints and cracks in the rock, warming the surrounding rock. Similarly, the water from the warm summer rains descends through the cracks, further warming the walls of the joints, if only for a short period of time. Monitoring of boreholes in permafrost adjacent to joints demonstrates sudden peaks of warming above 0°C for a few hours or even one or two days before the ground temperature drops back to the normal permafrost temperatures recorded in the main mass of rock. These excursions are ignored when mapping permafrost since they only affect thin zones around major joints. Thus permafrost in these rocks is far from impermeable, and water can move freely downwards in areas of relatively warm permafrost. This does not occur in areas of colder, continuous permafrost, *e.g.*, in Nahanni National Park.

1.2 ADDITIONAL TERMS ORIGINATING IN RUSSIA

Russia was the first country where the study of permafrost was carried out. Until recently, it had more research scientists working on permafrost than any other country, and additional terms will be found in the Russian literature as discussed below.

Where the top of the permafrost coincides with the bottom of the active layer, it is called *confluent permafrost*, but if its top is located deeper, it is referred to as *non-confluent permafrost* (Dostovalov & Kudryavtsev, 1967). Sometimes the permafrost may consist of two or more layers, divided by thawed interlayers. In this case, it is called *layered or multilayered permafrost*. The deep layer of ancient, or *relict permafrost*, formed in the late Pleistocene, was first discovered by Zemtsov (1957) in Western Siberia. The relict permafrost extends considerably south of the *southern boundary* of the modern permafrost in Siberia (Berdnikov, 1970; Ananjeva *et al.*, 2003). Its existence, as well as islands of permafrost in areas of non-frozen soils, raises problems in the determination of the southern boundary of permafrost. Continuous permafrost is observed only in very northern regions or on the polewards-facing slopes of high mountains. However, there are still areas without permafrost under the large reservoirs and lakes, and in places of intensive circulation of underground waters, which are called *through taliks* and *blind*, or *false taliks*.

Seasonally frozen soils which do not melt for the summer and exist for several years are called *pereletok* in Russia. The difference between the pereletok and relatively "young" permafrost is based on the typical thermal mode for a region and the frequency of their appearance. Permafrost appears regularly and tends to persist, but pereletok appears occasionally and tends to disappear after a short, usually one-year time-period. The term is not used in North America or in Europe.

1.3 HISTORY OF PERMAFROST RESEARCH

Tsytovich *et al.* (1959), Harris (1986a), Yershov (1990; 1998a; 1998b) and French (2003) provide reviews of the development of permafrost research up to the late 20th century. The earliest inhabitants of these areas learned how to survive using earth walls for houses in winter in Siberia and Iceland where there was not enough snow, and igloos made from blocks of snow or ice in the Arctic. Tunnels in the permafrost are still used for storing meat, *etc.*, in Tuktoyuktuk, Canada, Khatanga in Russia and Ulan Bator in Mongolia. In winter, water was obtained by cutting blocks of ice from the frozen rivers or collecting snow and then melting it. The first ground temperature profile was obtained when a Russian trader (Fedor Shargin) of the Russian-Alaskan Trading Company arrived in Yakutsk and decided to dig a well. After digging down to 105 m., all he encountered was frozen ground. Academician A. F. von Middendorf examined the well in 1836, measuring the ground temperatures, and although the well was deepened, the permafrost base was not found (Middendorf, 1867–1878). The seasonal fluctuations in temperature cease below about 20 m depth.

Sumgin *et al.* (1940) developed the concept of zonation of permafrost based on the intensity of cooling of the ground at 10–15 m depth. Further work has resulted in percentage area being substituted for temperature for the definition of Sumgin's classes. It is based on the percentage of the ground surface underlain by permafrost, independently of its thickness. Although this is a useful concept for studying distribution and for planning, there is limited agreement regarding the class limits (see Chapter 4).

Shortly after the middle of the last century, the International Permafrost Association was formed and holds periodic international meetings. As a result, many countries that do not have permafrost have joined and carry out research in this subject. More recently, the European permafrost scientists have carried out joint research projects to determine the extent, characteristics and properties of permafrost on that continent and elsewhere. They meet regularly for a continental conference (EUCOP). Since the late 1970s, China has been carrying out a great amount of research on the use of permafrost areas within its borders, particularly in Northeast China and in the very high mountains and plateaus of central, south and west China. In about 2013, the Chinese were spending considerable sums of money on training specialists and have subsequently been carrying out research in support of the enormous development projects, many of which are in permafrost regions. Now the key development projects are completed and with the current economic downturn, permafrost research there has been greatly cut back.

Since the last summary of our knowledge on permafrost (Washburn, 1979), tremendous progress has been made in understanding where permafrost occurs, what it is, how it behaves, what are its controls, and how it affects the people who live and work in these areas. We also now know much more about the unique landforms in these regions, how they form and change with time, and what the consequences are if they are disturbed either by natural forces or by Mankind.

1.4 MEASUREMENT OF GROUND TEMPERATURE

Harris (1986a) discussed the methods available at that time for measuring ground temperatures. Of those, the *thermistors* (tiny bead resistors that change resistance with temperature changes) are generally the preferred method, but have to be protected against moisture. *Thermocouples* (two wires of different metals or alloys whose ends are fused together) are used in situations where this is impossible, *e.g.*, where they are buried in ice as on ice islands floating in the Arctic Ocean. Both can provide an accuracy on the order of $+/-0.01°C$ when the readings are made using *data loggers*. The latter can be for one-time use with a built-in thermistor, or they can be multichannel, reusable, and capable of measuring several different properties at the same time.

The latest innovation is in the experimental use of *distributed temperature sensing* (Rogers *et al.*, 2015), which uses a fibre optic cable to measure the temperatures at distances down to less than one metre apart along a right of way up to 22 km long. It may also be possible to measure moisture content using the same DTS, although so far, it has only been used for determining stream dynamics (Selker *et al.*, 2006). Strains can also be measured at the same time. This could permit careful monitoring of the temperature, stress fields and moisture regimes along linear rights of way, though the cost is likely to be high. However the cost of repairs when the embankment fails is also significant.

1.5 CONDUCTION, CONVECTION AND ADVECTION

Engineers and scientists have generally assumed that heat is moved in the ground by *conduction*, *i.e.*, the transfer of heat from grain to grain at the points of contact in sediments. The calculations of the necessary depths for the footings of engineering structures in clastic sediments are based on this (see Part 3). In massive rock, the thermal conductivity is far greater since the movement is not interrupted by voids. This is why the massive rocks and the soils around concrete or steel foundations exhibit much deeper active layers than the nearby sediments.

Convection is a term referring to circulatory movement of fluids (air or water) conveying heat from one place to another as a result of the development of thermal/density gradients. This is an important cause of the cooling of the ground beneath large rocks and boulders, and is also involved in the development of many ice caves (see Chapter 6).

Advection is the name used for heat moving through the ground by the movement of water or any other fluid, either vertically from the surface, or sub-horizontally downslope. In the past, this has generally been ignored, but lateral advection of heat in the form of the movement of water in the surface soil layers within taliks parallel to the ground surface, or surface meteoric waters descending down the through-pores in the sediments have produced serious problems. Liquid water has a temperature above 0°C and therefore adds heat to the ground. Veuille *et al.* (2015) have carried out experiments confirming the heating effect on the soil profile, and have shown that the effects depend on the rate of water flow for a given temperature difference. This has caused road embankment failure of the Alaska Highway near Beaver Creek as well as failure of the BAM railway embankment in Russia (Chapter 16). In China, it has created problems with the tower foundations of power lines on the Qinghai-Tibet

Plateau (see Chapter 17). The oil and gas flowing from producing wells also carries heat that must be dissipated into the surrounding environment, *e.g.*, at Prudhoe Bay and in Western Siberia (Harris, 2016b).

1.6 THERMAL REGIMES IN REGIONS BASED ON HEAT CONDUCTION

Permafrost is the result of a heat balance at the surface of the Earth, resulting in the mean annual ground surface temperature (MAST) being below 0°C. The Earth approximates a sphere in shape and the maximum solar radiation is received within the Tropics. The Polar Regions receive limited solar heat, but 30% of the heat absorbed in the Tropics is currently transported polewards in the Northern Hemisphere by surface ocean currents, thermohaline currents, and by movements of air masses (Harris, 2002a). In the Southern Hemisphere, the presence of the cold, dense Antarctic air mass and the shape of the continent minimize this transfer. That is why the Northern Hemisphere is more hospitable for Mankind (Budyko, 1980).

The heat balance at any one place depends on the balance between the inputs of heat from insolation, precipitation and geothermal heat flow, and the losses through reradiation from the surface of the Earth. The *geothermal heat flow* is regarded as being fairly constant at *c.* $1.6736 \times 10^6\ \mathrm{Jm^{-2}/a^{-1}}$ while the *incoming solar radiation* is potentially 6000 times greater in its effect (Judge, 1973, p. 37). The latter is modified by clouds, albedo, snow cover, soil moisture, vegetation cover, latitude, thermal conductivity and soil latent heat. Thermal conductivity of rock varies considerably, and sites with higher heat conductivity will have a thicker active layer as well as a greater thickness of permafrost, other things being equal (Lachenbruch, 1970). This is demonstrated in Figure 1.6. Thus permafrost is 650 m thick at Prudhoe Bay compared with 450 m thick at Barrow, even though both have similar air temperature regimes.

As already discussed, active layer thickness usually varies from 30 cm to 2 m in sedimentary deposits, depending partly on latitude and the mean annual air temperature. However, the nature of the substrate is also an important factor (Figure 1.7). As pointed out by Mackay (1970), the grain size of the sediment has an enormous effect on the thickness of the active layer, other things being equal. The greater the number of contacts of grains per unit volume, the greater the thermal conductivity of the sediment. In actual practice, the mineralogy, pore space in sediments, their grain shape and their density modify this relationship. The thickness of the active layer in rocks is merely an extension of this effect.

Cheng (2004) has used the term *thermal diode* to describe the important effect of two very different substrates that develop permafrost long before frozen ground develops in other materials. The first material is peat, which has enormous pore space if not compacted. In the fall, the first snows are melted by the warm ground surface and the peat becomes wet. As the temperatures drop, the water changes to ice which is a good conductor of heat. This effect is called the *temperature shift* in Russia (Dostavalov & Kudryavtsev, 1967). Additional water migrates to the cold surface layer from above and below, and heat readily flows from the ground into the atmosphere during the long, cold winter nights. In the spring, this ice has to change state to liquid water before the ground can warm up, but the water in the peat rapidly evaporates. The pore spaces

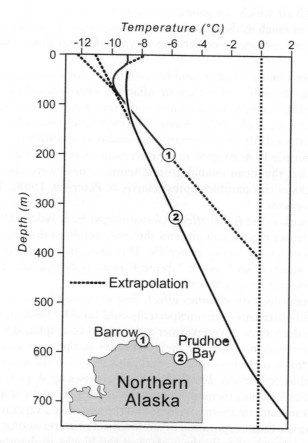

Figure 1.6 **Geothermal gradients** in boreholes at Prudhoe Bay and Barrow, Alaska, showing the effect of different thermal conductivities of rock (modified from Lachenbruch & Marshall, 1969, Figure 2, and Gold & Lachenbruch, 1973, Figure 1).

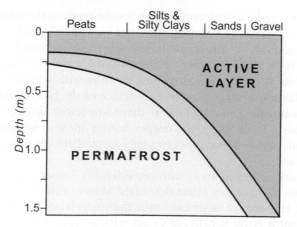

Figure 1.7 Effect of differences in texture on active layer thickness in a given Arctic environment (modified from Mackay, 1970, Figure 13).

become filled with air which is a poor conductor of heat, and so the underlying layers do not warm up as much as those in the surrounding soils, hence the thin active layer. If the peat becomes compacted or decomposes, the thermal diode effect is decreased or lost.

It has long been known that ground beneath large blocks in areas with low winter snow cover is significantly cooler than in adjacent, finer-grained mineral substrates, *e.g.*, in block fields, screes/talus, kurums, and even mine tailings (see for example, Cheng, 2004; Delaloye *et al.*, 2003). A year-long experiment was conducted at Plateau Mountain in Southwest Alberta, comparing ground temperatures in a block field with those in the soil profile 10 m away, with measurements every 10 minutes for a year. The results showed that the mean annual ground temperatures were about 5–7°C cooler beneath the blocks at comparable depths (Harris & Petersen, 1998). This was caused by four main processes.

The first process is the **Balch effect**. Climatologist E. S. Balch (1900) pointed out that cold air is denser than warm air, and therefore tends to displace the warmer air in the interstices of coarse blocky materials. This process is most effective in regions with low winter snowfall such as the Tibetan Plateau (Cheng *et al.*, 2008) and where there are large connecting spaces between the blocks.

The second process is the **chimney effect**, first suggested by von Wakonigg (1996). It is based on field observations in unexpectedly cold boulder fields in the eastern Alps. Where there is a deep snow cover, warmer air tends to be displaced from between the blocks by cold air entering wherever there are holes in the snow cover. This appears to be rather common at the base of scree and talus slopes in the maritime parts of the Swiss Alps (Lambiel & Pieracci, 2008). The warm air rises up slope through the spaces between the blocks, escaping through holes in the upper part of the slope. Morard *et al.* (2008) describe the resultant thermal regime and the associated vegetation distribution.

Thirdly, summer-time evaporation or sublimation of water and/or ice in the blocky deposits absorbs latent heat from the surface of the blocks in the upper layers of the block field, resulting in cooling them (von Wakonigg, 1996). This process is most effective in regions with dry summer air such as southern Alberta and the Tibetan Plateau.

Fourthly, there can be a continuous air exchange between the interstitial air and the overlying air mass (Harris & Petersen, 1998). This is particularly common in block slopes on the mountains of southern Alberta where strong winds and low snowfall aid the process. Since there are about eight months in the year with mean daily air temperatures below 0°C as opposed to four months with warmer air temperatures, this process results in overall cooling of the surface of the block field. Although there are strong winds on the Tibetan Plateau, there are fewer slopes, so this process is less important there. In the Alps, the deeper winter snow cover largely inhibits this process. In any one area, all four processes may act at different times of the year when the weather permits.

The same result is seen in areas with considerably higher winter snow cover, and this can be due to the chimney effect described above, although Gruber & Hoetlzle (2008) ascribe it to conduction of heat into the underlying blocks. This conduction from the winter snow cover is believed to act without thermally driven convection to move heat into the lower layers of the blocks which have low thermal conductivity. A layer of blocks with a porosity of 0.4 has a thermal conductivity of about an order

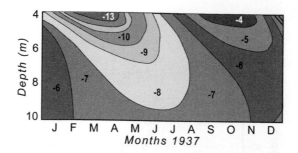

Figure 1.8 Ground temperatures measured at 4–10 m depth in Shargin's well in Yakutsk during 1937 (Melnikov, 1962). Note the gradual penetration of the warming (summer) and cooling (winter) waves into the ground, becoming weaker with depth.

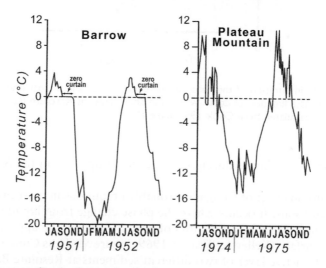

Figure 1.9 Temperature distribution versus time for wet tundra (25 cm depth) at Barrow, Alaska (after Brewer, 1958, Fig. 4) and Plateau Mountain #2 (20 cm depth), Alberta (from Harris, 1986a).

of magnitude less than solid rock. This reduces the warming effect of the snow cover, resulting in a cooling of about one or more degrees Celsius.

When ground temperatures at different depths in a permafrost area are plotted against time, the result shows a progressive lag in the arrival of the seasonal heating and cooling waves with increasing depth (Figure 1.4b and 1.8). The waves descend at a rate of about 23 m/a^{-1} through bedrock at Plateau Mountain, but only 13 m/a^{-1} at Yakutsk due to the much lower thermal conductivity in the icy silts at Shargin's well. Figure 1.5b also shows that the active layer freezes from above and below at the end of the thaw season. Note also that the amplitude of the waves become progressively weaker with depth. A *zero curtain effect* is seen in the trace of the cooling curve due to the change in liquid water to ice in the active layer (Figure 1.9), *e.g.*, on the wet tundra at Barrow, Alaska (Brewer, 1958). This is due to the output of 152.855 J/cm^3 of heat resulting from changing one gram of water to ice at 0°C. Note that this is miniscule in the case of the well-drained soils at Plateau Mountain. A similar zero curtain is seen

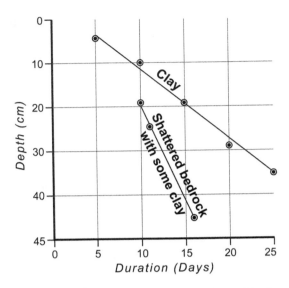

Figure 1.10 Variation in duration of the zero curtain effect in clay and in shattered rock with a clay coating as a function of depth within the active layer at Resolute Bay, Cornwallis Island (data recalculated from Cook, 1955, Table 1).

in profiles of icy active layers during thawing in spring, *e.g.*, in the Alps (Mutter & Phillips, 2012).

The zero curtain effect of Sumgin is probably ubiquitous in most situations except for essentially dry soils. It occurs where the phase change from liquid water to ice or *vice versa* produces enough gain or loss of heat to balance the applied heat moving down from the surface (Kelley & Weaver, 1969, Figure 5). Thus Cook (1955) demonstrated this in the active layer of two different sediments at Resolute Bay, Cornwallis Island, where all depths examined exhibited the zero curtain effect (Figure 1.10). The actual duration of the zero curtain effect increased linearly with depth as the seasonal wave of heat gain attenuated with depth. The soils at that site were essentially saturated by water from the melting snow, so that soil water content was more or less constant with depth. If there are layers with different thermal properties or moisture contents in the upper part of the active layer, these will modify the progression of ground temperature changes into the ground. The value was based on the earlier limit of accuracy of measurement of ground temperature before the development of data loggers, but we can now measure the temperature with considerably more precision. Harris & Brown (1982) suggested using the name *zone of minimal amplitude*, but it should be noted that these concepts assume stability in ground temperatures. We now know that differences in climate from year to year, shallow permafrost in the warmer permafrost areas, and climatic changes restrict its use. The ground temperature readings can also be plotted against depth to show the range of variation in a year or period of years (Figure 1.11). The decrease in amplitude of the ground temperature fluctuations with depth is obvious, as is the trend to warming with depth. The permafrost table is located where the warmer side of the inverted bell transects the 0°C isotherm. Where

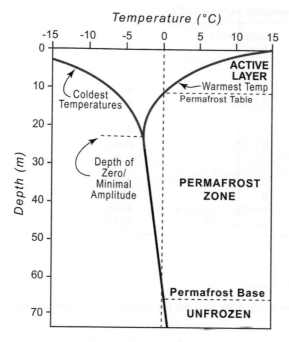

Figure 1.11 Thermal temperature envelope for a site with nonsaline permafrost, showing the gradual increase in temperature with depth. The geothermal gradient depends on the thermal conductivity of the ground as well as the local geothermal heat flow.

the yearly variation in ground temperature is less than 0.1°C, the upper limit of this depth has been referred to as the *depth of zero amplitude*. This terminology mainly applies to areas of relict permafrost and the colder, continental permafrost climates. In areas with warm permafrost, the temperature profile is not sufficiently stable from year to year, though close to zero.

1.7 CONTINENTALITY INDEX

The interiors of continents tend to exhibit a much greater annual range of air temperatures (up to c. 105°C) than maritime areas (<60°C). This is due to the low specific heat and poor conductivity of land vis-à-vis water (Duckson, 1967). Geographers have used this in two ways, namely as part of the regional climate classification and also as indices (Currey, 1974). In permafrost research, this can be demonstrated by plotting the *mean annual freezing index* (the sum of the negative mean daily air temperatures) against the *mean annual thawing index* (the sum of the positive mean daily air temperatures) for a given station (see Chapter 2). This can be used for all land areas, and is related to the Stephan formula, named after the Slovene physicist Josef Stefan in 1889 after studying ice formation and melting in the Polar oceans.

Gorbunov (1978) proposed a *continentality index* for mountainous areas based on the difference in elevation of snow line, the presence or absence of seasonal and

Table 1.1 The **Continentality Index** Classification of Gorbunov (1978).

Class	Continentality Index	Arrangement of Geocryological zones	American member
Continental Types			
Verkhoyansk	0.2	A	Brooks Range
Tien Shan	0.2–1–8	A/B	Mackenzie Mtns. North Rockies
Tibetan Mtns.	1.5	A/B/C	–
Central Andean	0.1–1.5	A/B/C/D	–
Maritime Types			
Ecuadorian	–	A/C/D	–
Chugach	0	A/B	Chugach
New Zealand	Possible −ve values	A/B/C	S. British Columbia
Himalayan	−0.1 to −0.3	A/B/C/D	–

Geocryological Zones:
A. Permafrost zone.
B. Seasonally frozen ground.
C Shot-term (nocturnal) freezing.
D. Frost-free zone.
Negative values of Continentality Index imply that the equilibrium line is lower than the lower limit of permafrost.

diurnal frost, and the lower limit of permafrost. This system produced eight classes when applied on a world-wide scale (see Table 1.1). Cui (1985) and Corte (1985) tested the system in China and Argentina and it worked for the mountains in the Himalayas and Andes. However, it has the obvious drawback of only being applicable to areas with all the diagnostic features, and cannot be used for most lowlands and plateaus. Harris (1989) tried to use it in western Canada and found that the index depended on whether the lower boundary of *continuous* permafrost, *discontinuous* permafrost or *sporadic* permafrost was used. The best results were obtained by using the lower limit of continuous permafrost, and this revealed a strong, systematic increase in continentality towards the higher latitudes. This correlates with increased moisture availability and the dominant vegetation. Edaphically dry areas have higher values than edaphically wet sites at a given latitude, but this variability is far less than the latitudinal gradient.

1.8 MOISTURE MOVEMENT IN THE ACTIVE LAYER DURING FREEZING AND THAWING

Except for some soils with dry permafrost in Antarctica and perhaps on the Qinghai-Tibet Plateau, the permafrost and the overlying active layer always contain moisture. The seasonal fluctuations in the thermal regime in the active layer play an important role in its movement (Figure 1.12). Prior to spring/summer thawing, the moisture profile shows a single upward movement towards the surface to the coldest part of the active layer (Figure 1.12b). Once the surface part of the active layer thaws, moisture moves up from below and down from above into the remaining cold zone below the thawed layer (Figure 1.12c), *i.e.*, producing two zones in the active layer with different

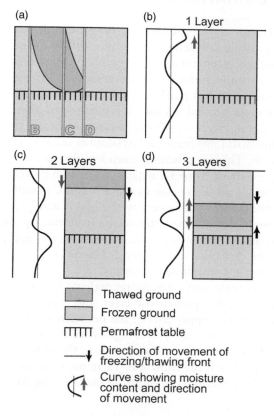

Figure 1.12 The change in ground temperature is shown diagrammatically in a, with the locations at different times of year of the three different patterns of movement of moisture (black lines) in the active layer (b, c, and d) shown by yellow vertical lines.

moisture regimes. When the active layer is refreezing (Figure 1.12d), a third zone develops with moisture moving towards the colder surface and down towards the permafrost table in the part that is refreezing from below. On completion of the refreezing, the moisture profile returns to that in Figure 1.13b as soon as the temperature profile in the active layer is coldest at the surface.

As noted earlier, the weather conditions fluctuate from year to year, the actual position of the permafrost table responds by moving up and down over time, the amount of variation depending on the variability of the climate. This zone of fluctuation (the *transient layer*) accumulates large amounts of ice beneath a dry *shield layer* (Figure 1.12b, c, and d), and these two layers together act as a buffer to thawing of the underlying icy permafrost in years when the permafrost table is lower than usual during the summer. The thickness of the ice increases with increasing climatic variability in continental climates such as central Siberia, where it acts as a buffer to protect the main upper surface of the permafrost beneath it. Elsewhere, it is relatively unimportant due to the thin active layer and the absence of appreciable fluctuations in depth of summer thaw.

1.9 MOISTURE CONDITIONS IN PERMAFROST GROND

Although the temperature of the ground is a fundamental part of the definition of permafrost, it is the presence and activity of the moisture in the ground that creates the special landforms and processes that are characteristic of these regions. A typical profile showing the moisture regime at the end of freeze-back is shown diagrammatically in Figure 1.13. The moisture at A is partly due to snowmelt infiltration into the surface layer of the soil prior to the ground freezing. It is augmented by water moving upwards to the freezing plane during freeze-up. The zone C represents the remains of the perched water above the permafrost table from the last summer season, while zone B represents the drier zone above it. There is always a considerable maximum of moisture in the upper layer of the permafrost (zone D) with an obvious high concentration of visible ice (Figure 1.1). Ice lenses will also occur in zones A and C in winter (Mackay, 1971a), but only occur in zone B if the site is very wet. Zone B will have lost water to zones A and C during freeze-back since the latter occurs from the top and the bottom of the active layer simultaneously. Water always tends to move through the soil towards the coldest place as will be explained in Chapter 2. Harris (1988a) provides examples of actual moisture profiles in permafrost landforms.

Water movement near the surface of the permafrost follows thermal and hydraulic gradients, the zone C, upper permafrost layer and basal active layer having the maximum H_2O content. The upper part of the permafrost (zone D) always contains the largest quantities of ice, and the percentage of ice by volume can exceed 90% in silty sediments, *e.g.*, in the lacustrine sediments along the Mackenzie River valley. This represents a serious problem if the climate is warming, or if the micro-environment is disturbed so as to produce thermal instability. In places in the Arctic, buried glacier ice

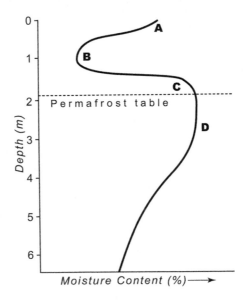

Figure 1.13 Diagram showing the normal distribution of H_2O in a profile through permafrost (after Harris, 1986a, Figure 2.14). The letters mark the parts of the profile referred to in the text.

may also occur (*e.g.*, Lorrain & Demeur, 1985; Dalimore & Wolfe, 1988; French & Harry, 1988; 1990; Astakhov & Isayeva, 1988; Astakhov, 1992; Kaplanskaya & Tarnogradskiy, 1986; Solomatin, 1986). As discussed in Chapter 6, differentiating this from segregated ice is often a challenge, both in Arctic Canada and in northern Russia (Mackay, 1971b; 1989; Rampton & Walcott, 1974; Mackay & Dallimore, 1992; Moorman & Michel, 2000; Vasil'chuk & Vasil'chuk, 2012). There are also other possible origins for the massive ice (*e.g.*, Fujino *et al.*, 1983; 1988; French & Pollard, 1986).

It is important to note that ice content can be described as a percentage by dry weight of the sediment, so that 90% by volume in silts translates into about 1000% by weight. Ice contents in peat can reach 2000% by weight, even in the absence of ice lenses (Kinosita *et al.*, 1979). Crystal size of the segregated ice in the upper layers of the permafrost may be 14–25 times the size in the seasonally frozen layers (Tsytovich, 1975).

Figure 1.14 shows the effects of a heavy rainstorm in summer on the moisture profile. Water perches on the relatively impermeable permafrost table, but also migrates

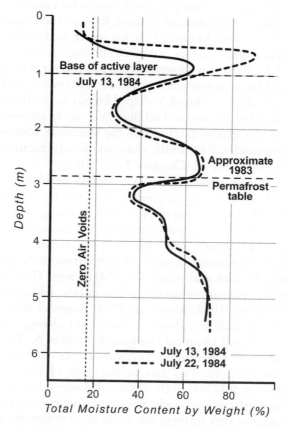

Figure 1.14 Perched water table at the base of the thawed layer measured by neutron probe in the borehole at Summit Lake A, two and a half years after thawing during drilling (Harris, 1986a, Figure 2.16).

into the surface layer of the permafrost, increasing the H_2O content (Harris, 1988a). Note also the substantial addition of H_2O in the profile in the two years after drilling in winter, 1992. Use of water as a lubricant during drilling had resulted in complete thawing of the surrounding soil at that time. This opens the question of the stability and value of cryostructures (Yershov, 1979).

1.10   RESULTS OF FREEZING MOISTURE

When liquid water changes to ice, it gives out $3.347 \times 10^8 \, J\,m^{-3}$ (640 calories) of latent heat of fusion for each gram or ml of water and its volume increases by 9% at $0°C$. This expansion is relieved partly by uplift of the overlying soil and partly by compaction of the sediment. It is this expansion that creates so many problems for man-made structures in permafrost areas, and there is similar contraction when the ice melts. This contraction results in subsidence of the ground, but the compacted substrate merely cracks, if it changes dry density at all. Repeated freezing and thawing results in the development of *over-compacted soil,* which is characteristic of former permafrost areas. The same freezing and thawing can break up rock, then rotate and move the resultant blocks to the surface of the ground (see Chapter 2). The ground contracts when cooled and develops polygonal contraction cracks which can become filled with water to form *ice wedges,* or by sand, loess or rock fragments to form *tessellations* (see Chapter 5). On thawing of the ice, an ice-wedge cast may be produced by wind-blown sand or other surface sediment infilling the space left by the former ice wedge. Figure 1.15 shows an example from south-central China on the eastern slopes of the Tibetan Plateau where the rock has been shattered, and the resulting sediment contorted. Ice wedges produced V-shaped breaks in the fractured rock, and as the wedge ice melted, wind-blown sand infilled the space to produce *ice-wedge casts.* Often there is a slow convection developed in the sediments forming the active layer producing *non-sorted circles, frost boils,* and convoluted structures including sorted patterned ground (Figure 1.16; see Chapters 2 and 10).

In the case of inclined, massive rock with cracks or clefts, water entering the fissures adds heat and can thaw any ice present (Hasler *et al.*, 2011b). This can lead to climate-dependent rockfall, consisting of a period of widening of cracks in summer and cryogenic widening in winter. This slowly modifies the geometry, eventually resulting in rockfalls (Gruber and Haeberli, 2007; Hasler *et al.*, 2012).

Another effect of frost action in areas of permafrost is the production of large quantities of silt particles with grain sizes of 0.05–0.01 mm. There are several theories as to why this occurs in primary minerals such as quartz, feldspars, amphiboles, and pyroxenes (see Konishchev, 1982; Minervin, 1982; Rogov, 1987), and these are discussed in Chapter 2. In addition, there are significant changes in the clay fraction involving both physical and chemical changes (Koinishev & Rogov, 1993). There is also evidence that small euhedral gold crystals are forming in the tailing piles at Dawson City, Yukon Territory. This may be the result of electrochemical reactions causing crystallization of gold due to the electrical charges that are developed during freezing of the ground affecting groundwater containing gold salts. In Argentina, ore mineral fragments tend to sink through the active layer sediments when they become saturated with water, so that they accumulate at the surface of the permafrost table at the base of the active layer, due to their greater density (Ahumada, 1986).

Figure 1.15 Frost-fractured sandstone exhibiting churning of the rock fragments and ice-wedge casts filled with wind-blown sand in a road cut on the eastern Tibetan slopes, China. The section is 3.5 m high. © S. A. Harris.

Figure 1.16 Professor Dr. Peter Höllerman inspecting sorted stone circles on top of Plateau Mountain, Alberta, with active non-sorted circles developed in the finer sediment in the centre. © S.A. Harris.

I.II   STRENGTH OF ICE

Ice is one of those substances which can deform or creep in response to pressure (Duval *et al.*, 1983). It is this property that results in the movement of glaciers. When sufficient ice is present to hold the individual mineral grains apart, the frozen material can move slowly down-slope under the influence of gravity, resulting in the formation of *rock glaciers* (see Chapter 9). If a heavy weight is applied to permafrost, it will slowly deform, even though the ice content is less than 40% by volume. It is also rather elastic, so that it reduces the effects of earthquakes or man-made explosions (Bauer *et al.*, 1965; 1973; USSR, 1972; Morgenstern *et al.*, 1978). Warm permafrost temperatures, moisture and the presence of salts decrease the strength of the ice.

I.I2   CRYOSOLS, GELISOLS, AND LEPTOSOLS

Cryosol is the name given by soil scientists for all soils with permafrost within the upper metre of the surface, or within 2 m if the pedon is strongly cryoturbated (FAO, 1988; Kimble, 2004; Canadian Soil Classification, 1998). They occupy about 30% of areas mountain permafrost, and about 12% of all the for permafrost areas. While the Americans use the term *Gelisols* for permafrost soils (USDA, 2014), the IUSS Working Group World Reference Base for Soil Resources (WRB, 2014) classifies those cryosols/gelisols with continuous hard bedrock within 0.25 m of the surface or soils with <10% of fine earth as *Leptosols*. In short, the classification is still under discussion.

I.I3   FRAGIPANS

Fragipans are soils with a non-calcareous horizon that is extremely hard, thus largely preventing the movement of air or water through it because of its lack of pores. Most plant roots cannot penetrate it, although it may exhibit a coarse prismatic or columnar structure. As used in the US Department of Agriculture soil classification, they include at least two different types of dense horizon. In one, the hardness is regarded as being due to cementation by silica in soils under forest vegetation where silica in the soil water is not absorbed into the trees (Franzenmeier *et al.*, 1989). This results in silica becoming deposited in the dry season as a thin layer on the soil particles. The small peds in these horizons do not slake readily in water, and the silica cement can be observed in thin sections of the material making up the layer. These are therefore a type of *duricrust*, often called *silcrete*. Because they have similar physical properties to the other type in the field, they have been classified with the original type in the U.S. soil classification system.

   The second, original type of fragipan is uncemented, but exhibits a very high density compared with the other soil layers, *e.g.*, Bryant (1989) and Payton (1992). It is an extreme form of over-compaction of loamy sediments of various origins, and is usually found in areas where ice wedge casts are present. This prompted FitzPatrick (1956) to suggest that these brittle, hard layers were produced by permafrost processes, and this has been further discussed by subsequent workers (Crampton, 1965; FitzPatrick, 1976). Grasses and sedges on the tundra preferentially absorb dissolved silica so that

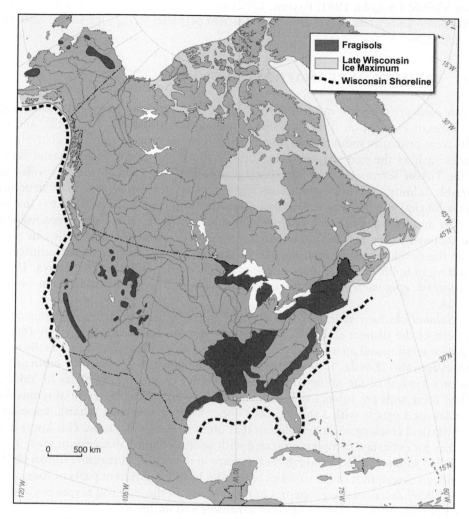

Figure 1.17 Distribution of fragipans in the post-Wisconsin sediments in North America.

the soil water beneath them lacks excess silica, and therefore silica cementation cannot occur. These fragipans are developed in loess or other loamy soils with moderate clay content and a range of grades. Recent research on the freezing of fine-grained soils suggests that contraction of the soil accompanies the development of a hexagonal pattern of vertical ice veins above the freezing front (Arenson *et al.*, 2008; Azmatch *et al.*, 2012). In fragipans, the loamy layer has been compacted so much that it has become hard enough for a crowbar to become bent if put in a vertical crack under sufficient force. The hardness tends to remain regardless of moisture condition, unlike the hard layers in *solonetzic soils*, which change to a jelly-like consistency when wet (Harris, 1964). These fragipans grade into hard, dense layers that tend to slake as the clay content either decreases or increases beyond the optimum composition. Fragipans of this type are therefore potentially useful indicators of former permafrost distribution (Figure 1.17), their distribution having been mapped in detail in Western Europe

(Van Vleit & Langohr, 1981; Payton, 1992) and found to be similar in distribution to other *paleo-indicators* of the former existence of permafrost, *e.g.*, ice-wedge casts and rock glaciers.

1.14 SALINITY IN PERMAFROST REGIONS

Saline soils are common in areas with limited precipitation in permafrost areas, *e.g.*, the Tibetan Plateau, on drier floors of valleys in Yakutia (Dolenko, 1913), and the salts were primarily sodium chloride originally. They also occur in other relatively dry places such as the rain shadow area north of the Wrangell-St. Elias Coastal Range in the Yukon Territory and along coastal plains. They are characterized by soils with variable salinity and often characteristic soil profiles consisting mainly of structure-less solonchaks, as well as desalinized solonetzic soils where the sediment is fine silt or clay and the salts are primarily sodium chloride. If the salts were deposited with the sediment, they are called *syngenetic*. Syngenetic salinization of soils can result from the development of thermokarst (Desyatkin, 1993), or by isostatic adjustment resulting in uplift of the sediments close to the shoreline (Dubikov & Ivanova, 1994). In contrast, *epigenetic* salinization can be found in most coastal sediments in western Russia.

Solonchaks have pale, pink, or grey soil profiles, usually lacking obvious development of the distinct soil horizons found in semi-arid or arid areas where there is seasonal wetting and an accumulation of salts near the surface left behind as the moisture evaporates (Kovda, 1946). Brouchkov (1998) calls this *continental salinization*, and it is typical of the saline soils found in the dry continental climates of Yakutia. When these soils are subject to leaching called desalinization, differential removal of salts leaves a profile with a thin, leached upper horizon overlying a hard, dense layer with vertical cracking forming columns that often have cap-like tops. This lower horizon has an exchange complex saturated with sodium and/or magnesium ions. These are called *solonetz soils*, although they are rare in the northern regions. Glinka (1931) provides a description of the varieties of profiles found in different parts of Asia, while Joffe (1939) describes the properties of the profiles. In the U.S. soil taxonomy system, they fall under various subdivisions of *aridosols* and *mollisols*.

In theory, when water freezes, the dissolved salts contained in it will become concentrated in the remaining fluid. This is regarded as the mechanism for the formation of cryopegs. However, this is not the entire story in the case of the saline soils on the Slims River Delta (Harris, 1990a). Chemical analysis shows the salts originate from springs coming up through taliks in the permafrost at the base of Sheep Mountain (Figure 1.18). They slowly migrate down the slight gradient across the delta to the Slims River, *i.e.*, they are *epigenetic*. Figure 1.18 shows the progression of thawing of the seasonal frost in the sands and coarse silts making up the delta sediments, together with the salinity of the thawed surface layers. As the soil commences thawing in early May, the upper layers are non-saline. As the thawed layer deepens, the trapped salts in the sediment become mobile and are carried to the surface of the soil with the soil water, where they accumulate as soil water at the soil surface evaporates. The concentrations are particularly marked on any slight ridges or mounds.

Periodic rain and migration of soil water perched above the frozen layers carries some of the salts towards the river bank where they eventually accumulate during the

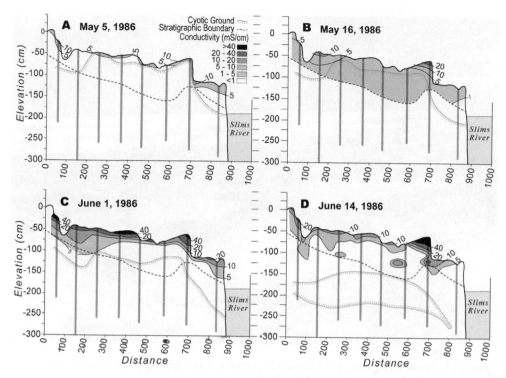

Figure 1.18 Changes in salinity along a traverse across the Slims River Delta from Sheep Mountain to the Slims River, Kluane, Yukon Territory, as thawing progressed in 1986 (modified from Harris, 1990a, Figure 7).

summer period. Continued seepage of saline water from the springs replenishes the salts in the soil during the summer in low concentrations throughout the unfrozen sediment, where they crystallize out as euhedral crystals as the sediment dries. These become entombed in the icy sediments as the soil freezes in the fall. There can be a four-fold change in salinity from year to year, depending on the timing and amounts of summer rains. Consequently the soil salinity is always changing.

Since the salts in the Slims River delta are predominantly magnesium sulphate, the saline soils are *solonchaks*. These are structureless, saline soils, usually with a hard crust underlain by a layer with loose crumb structure. If the crust is disturbed or destroyed, the underlying soil is readily eroded and transported by wind (Williams, 1996; Sneath, 1998). Grazing animals in Tibet have caused considerable desertification of solonchaks in the last 50 years (Williams, 1996; Miller, 1998; Sneath, 1998; Ho, 2000; Harris, 2013b), and probably the camel trains plying the various alignments of the Silk Road in the last few centuries damaged the soils and aided the desertification process in Central Asia.

A wide range of salts can occur on and around the Tibetan Plateau including borax, but the salt crusts of commercial importance there are confined to seasonally frozen playa lakes at lower elevations and currently lacking permafrost. These normally have

Figure 1.19 Warm spring along the Golmud-Lhasa road in Tibet, feeding water to saline soils on the valley floor to the east. © S. A. Harris.

a crust of salts that can readily be mined. Solonchaks are also common in permafrost areas in China around the numerous lakes dotting the plateau along the Golmud-Lhasa road (Figure 1.19), between the active sand dunes to the east, as well as around isolated open system (East Greenland) pingos and lithalsas in the mountains, *e.g.*, Figure 1.20. There is evidence that where continuously moist soils are present in the dry, cold climate on the Qinghai-Tibetan Plateau and surrounding mountains, the evaporation of water cools the ground, producing permafrost on the damp valley floors when it is absent elsewhere on the landscape. These locations are marked by a salt efflorescence around the vegetation.

Two types of saline soils are described from Russia (Brouchkov, 1998). The *"continental type"* occurs scattered throughout the main permafrost area in Central Yakutia (Figure 1.21). The saline soils in the forested taiga of northern Siberia tend to occur in the grassy alas depressions in both larch and poplar forests (Lopez *et al.*, 2006). The salts are regarded as epigenetic, and appear to be accumulated by evaporation of precipitation, possibly with additions washed in by moisture movement down-slope from the adjacent forest and pingo mounds, since the ions present have a similar composition to those in rainfall. Higher evaporation in the open areas brings salts to the surface in a zone around the water body. These saline soils are characterized by high concentrations of calcium carbonate and sulphate, as found in North

Figure 1.20 An open system (East Greenland type) pingo at a spring in the Kunlun Pass above 4600 m along the Golmud-Lhasa road, which was blown apart by the Highway Maintenance Squad to obtain water. © G. D. Cheng.

America in permafrost regions. Gradients of both temperature and salt content can cause mobility of salts in the ground (Brouchkov, 2000).

The second type is the *"marine type"*, found along the Arctic coast in the areas of former marine sedimentation (Velli, 1973; 1977; 1980; Hivon & Sego, 1993; Gigarev, 1997; Brouchkov, 2002; 2003). Isostatic uplift and tectonism have resulted in substantial areas of marine clayey sediments, especially in the area east of the Ob estuary. The former marine sediments uplifted by isostatic rebound to 250 m, east of Hudson Bay are also potentially saline for the same reason. The dominant salt is sodium chloride but the active layer is usually nonsaline. However, the quantity of salt normally increases with depth. The sodium-rich salinity confers special properties on the soils which cause considerable problems for the stability of foundations.

1.15 ORGANIC MATTER

Organic matter in the form of peat acts as a thermal diode, and can also oxidize to produce carbon dioxide. Accordingly, the *total organic carbon (TOC)* is of considerable interest. Unfortunately, its measurement is difficult because organic carbon

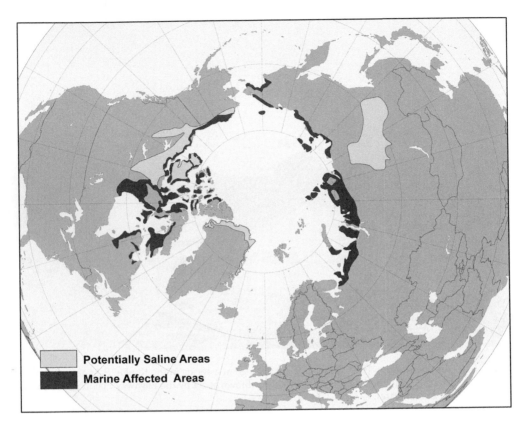

Figure 1.21   Distribution of the main areas of the two types of frozen saline soils in the Arctic. The continental, potentially saline areas are shown in yellow and the marine-affected areas in red.

can be elemental (charcoal), lignin (wood), cellulose, bone, chitin (horns) or humus (Schumacher, 2002). It varies in size from massive tree trunks to fine particles, so there is a sampling problem. In both the United States and Russia, it is measured by *loss on ignition (LOI)*, but the best results are obtained by keeping the temperature below 400°C to minimize decomposition of hydrated clay minerals and limestone particles. Typical values for LOI in frozen sediments in European Russia are 3.81% for alluvium, 11.44% for alluvial marine sediment, 60.29% for swamps, 3.92% for marine deposits and 29.1% for lacustrine-alluvial material. Glacial-marine sediments lack organic matter (Brouchkov, 1998). The errors are small from that area due to the lack of limestone bedrock and hydrated clay minerals. However in western North America, around Hudson Bay, in Greenland and parts of Scandinavia, the Palaeozoic limestone bedrock causes the LOI values to be too high due to decomposition of hydrated clay minerals and limestone.

Permafrost sediments and soils tend to have more undecomposed organic matter than the soils in similar but warmer environments. The total daily quantity of insolation received by the surface of the ground in summer at 54°N is similar to that being received

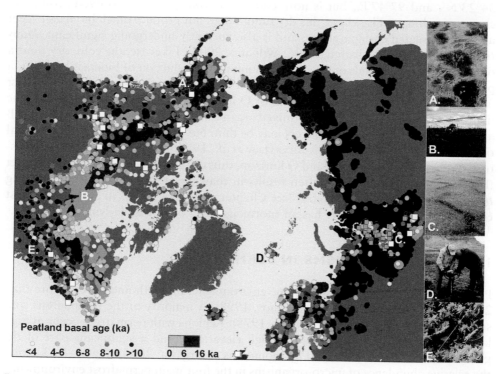

Figure 1.22 Distribution of peatlands in the Northern Hemisphere in relation to vegetation zones, partly after Beilman *et al.* (2010). Note the area of late expansion of peat formation in parts of Canada.

at 66°N, although it is spread out over more hours. As a result, plants grow well and produce a substantial quantity of biomass in northern Canada, Alaska and Siberia. The long winters and cold ground temperatures reduce decomposition of organic matter, particularly in wet situations such as in the Mackenzie River valley, the west Siberian lowlands and the former glacial lake beds in northwest Canada (Figure 1.22). Although most of Canada was covered by ice sheets until about 12,000 years ago, peat up to about one metre thick blankets considerable areas. In parts of Canada, the ice sheets persisted until about 7,000 years ago, so there has been relatively little time for the peat deposits to accumulate. In the Yukon Territory, peats blanket the lower parts of the mountain valleys in the MacMillan Pass at 64°N. Rates of peat formation in North America are about 100–200 cm/1000 years while the rates elsewhere range from 60–80 cm/1000 years. This peaty layer is extremely important due to the thermal diode effect, resulting in the development of permafrost in otherwise permafrost-free areas. Abundant *peat plateaus* and *palsas* are associated with these (see Chapter 7).

These areas contrast with the arid parts of the main Tibetan permafrost areas at c. 33°N. that have a lack organic matter in the soil or on its surface, except locally under the steppe grasslands of the northern-facing slopes of the high terrace at about 3700 m elevation, as well as on the eastern slopes of the Plateau. Blanket peat produces a smooth-appearing upland above about 4500 m at Qala Q'u Hue Q'u at about

34°23′N., and 97°47′E., but is now undergoing oxidation due to overgrazing and changes in climate. Elsewhere, this dry tundra does not produce much biomass except under the isolated cushion plants, and is also currently undergoing significant *desertification* (Williams & Balling, 1996; Williams, 2000). Likewise, the cold, dry tundra of the High Arctic Islands in Canada has a minimal production of biomass by cushion plants, so there is negligible organic matter present. In Central Yakutia, the larch forest is typically underlain by about 1–2 cm of O horizon (decaying accumulated organic matter), and 5–10 cm of A horizon (mixed organic and mineral matter), overlying the B and C horizons. These organic layers may be thin, but produce the significant thermal offsets discussed in Chapter 2 (Kudryavtsev *et al.*, 1974; 1977). Fire alters this by disturbing the surface vegetation and O horizon, thus changing the surface energy budget (Hinzman *et al.*, 2001). This often results in thawing of the permafrost, producing *thermokarst lakes* and depressions (see Chapters 2 and 11). Thus in all but the driest environments, organic matter has an enormous effect on any underlying permafrost.

1.16 MICRO-ORGANISMS IN PERMAFROST

Permafrost is one of the most extreme environments on earth and covers more than 20% of the earth's land surface (Yershov, 1998). A majority of the biosphere is constantly below +5°C (Baross & Morita, 1978). Regions with permafrost occur at high latitudes, in Antarctica, and also at high elevations, and a significant part of the global permafrost is represented by mountains (Margesin, 2009). Figure 1.23 shows the relative abundance of micro-organisms in the four main permafrost environments.

1.16.1 Antarctic permafrost

The Antarctic continent is covered by an ice cap with ice dating back to Miocene times, *i.e.*, the permafrost is over 20 Ma old. The micro-organisms in the permafrost have had to survive the coldest temperatures on Earth. Investigations of glacier ice at the Vostok

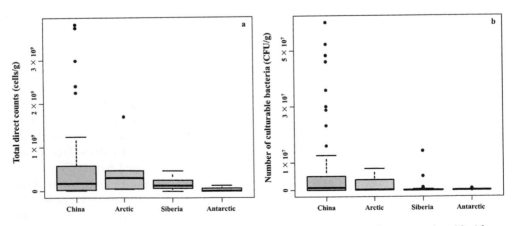

Figure 1.23 Relative abundance of micro-organisms in the four main permafrost areas (modified from Hu *et al.*, 2015).

station in Antarctica (Abyzov, 1993) revealed the presence of bacteria, fungi, diatoms and other micro-organisms which are believed to have been carried there by wind. They presumably became trapped in the snow and then remained in the ice for thousands of years. Microbial activity has recently been found in ice-sediment communities in the surface layers of perennial and permanent lake ice (Psenner & Sattler, 1998) as well as in Antarctic sea ice (Bowman *et al.*, 1997). Populations of bacteria are also present in surface snow at the South Pole (Carpenter *et al.*, 2000). Ashcroft (2000) discusses the physiological adaptation of organisms that live at great depths in Antarctic glaciers.

1.16.2   High-latitude permafrost

The permafrost in Siberia and northwest North America dates back to around 3.5 Ma (Harris, 1994; 2001a). Anaerobic psychrophilic micro-organisms have been isolated from a high Arctic glacier (Skidmore *et al.*, 2000) and a glacier in Greenland (Sheridan *et al.*, 2003). Although the microbial long-term survival in permafrost has been questioned, there is evidence that bacteria have been able to survive in 500,000-year-old permafrost (Johnson *et al.*, 2007). There is considerable diversity of micro-organisms, including bacteria, archaea, phototrophic cyanobacteria and green algae, fungi and protozoa that are present in considerable numbers in permafrost (Steven *et al.*, 2006, 2009; Gilichinsky *et al.*, 2008). The characteristics of these micro-organisms reflect the unique and extreme conditions of the permafrost environment. Permafrost soils may contain up to 20% or more of unfrozen water in the form of salt solutions with a low water activity ($a_w = 0.8$–0.85) (Gilichinsky, 2002). micro-organisms in this environment have to thrive under permanently frozen, oligotrophic conditions, complete darkness, constant gamma radiation and extremely low rates of nutrient and metabolite transfer (Steven *et al.*, 2006; Gilichinsky *et al.*, 2008). Substantial growth and metabolic activity (respiration and biosynthesis) of permafrost micro-organisms have been demonstrated at temperatures down to $-35°C$ (Panikov & Sizova, 2007; Bakermans, 2008; Amato *et al.*, 2010).

Recent studies of relict micro-organisms from ancient permafrost indicate their potential significance as objects of gerontology. *Bacillus* sp. was isolated from permafrost sands of the Mammoth Mountain in Central Yakutia (Brouchkov *et al.*, 2009) from an ancient Neogene deposit that was probably permanently frozen for 3.5 million years (Markov, 1973; Baranova *et al.*, 1976; Bakulina & Spector, 2000). This strain of *Bacillus* sp. (probably *Bacillus cereus*) demonstrated enhanced longevity as well as immunity and resistance to heat shock and UV irradiation in *Drosophila melanogaster* and mice (Brouchkov *et al.*, 2012; Kalenova *et al.*, 2011). Probiotic activity by a *Bacillus* sp. strain isolated from the same sample has recently been reported (Fursova *et al.*, 2012). Spores of *Bacillus* spp. are known to be one of the most resistant to decay (Nicholson et al., 2000). One of the spore-formers was isolated from a spacecraft-assembly facility, belonging to the genus *Bacillus* (Venkateswaran *et al.*, 2003). Diversity of micro-organisms isolated from multimillion-year-old amber includes *B. thuringiensis* and *B. mycoides* (Greenblatt *et al.*, 1999).

Bacterial isolates were able to grow at subzero temperatures and some were halotolerant. In spite of the ligninolytic activity of some strains, no biodegradation activity was detected. In general, sensitivity to rich media, antibiotics, heavy metals, and salt increased when temperature decreased ($20°C > 10°C > 1°C$). This could be explained as the reaction to an increased stress situation at low temperatures. However, further studies are needed to elucidate the mechanisms behind this process.

1.16.3   High altitude permafrost in China

The present-day permafrost on the high mountains of the Qinghai-Tibet Plateau mainly formed in the last 35 ka (Jin *et al.*, 2007). The permafrost is warm, usually with a thickness of under 50 m and temperatures warmer than $-2°C$. The active layer is relatively thick due to the high solar insolation by day, offset by marked cooling by long-wave re-radiation at night. The result is that there are diverse and novel microbial lineages, although the microbial community composition varies greatly across geographically separated areas (Figure 1.24). This is presumably due to variations in the available colonizers and in their ability to adapt to the colder conditions. The number of viable and total cells decreases along with the ability to recover viable cells with increasing depth (Hu *et al.*, 2014) and age, but the diversity of the bacterial isolates seems independent of permafrost depth (Feng *et al.*, 2004; Zhang *et al.*, 2007a; Wang *et al.*, 2011b; Ollivier *et al.*, 2013; Tai *et al.*, 2014). On Mount Everest, Zhang *et al.*, 2009) found that archeal *amo*A abundance decreased significantly below 5400 m a.m.s.l., although the situation was reversed at higher altitudes. Viable microbial numbers also varied from month to month at a given station (Chen *et al.*, 2011), and between vegetation types (Yu and Shi, 2011; Li *et al.*, 2012). Soil moisture, pH, organic carbon and total nitrogen content were closely correlated with numbers of microbes. The populations in permafrost in China tend to be halotolerant in lab tests (Zhang *et al.*, 2007b). Salts tend to be expelled and be concentrated in the remaining fluid during freezing, so this adaptation to high salinity is often combined with adaptation to cold (D'Amico *et al.*, 2006). Production of cold-active enzymes is frequently detected from the Chinese microbial permafrost communities (Bai *et al.*, 2005; Zhang *et al.*, 2007b). These tend to increase the flexibility of their structure to compensate for the freezing effects of their cold habitats (Feller, 2007). There are also genes encoding several features required for surviving in permafrost (Jansson & Taş, 2004).

Figure 1.24 also shows the functional roles of the micro-organisms in the chemical breakdown of organic matter and in fixing nitrates the soil. A significant number of micro-organisms are also involved in oxidizing methane and reducing sulphur in the soil to make these elements available to plants.

1.16.4   Phenotypic traits

Permafrost is an extreme environment and indigenous microrganisms must withstand long periods at geological time scales in a habitat with subzero temperatures, background radiation, low water availability and resultant rates of nutrient and metabolic transfer (Steven *et al.*, 2006; Margesin and Minerva, 2011). Live micro-organisms exist in frozen soils consisting of mineral particles and ice of different ages (Friedmann, 1994). Even microbial cells showing features of aging (Stewart *et al.*, 2005; Johnson & Mangel, 2006) are able to live or stay viable for a long time. While it is unknown whether these cells are individually surviving or growing, *Bacillus anthracis* can remain viable for more than 100 years (Nicholson *et al.*, 2000). Reliable reports

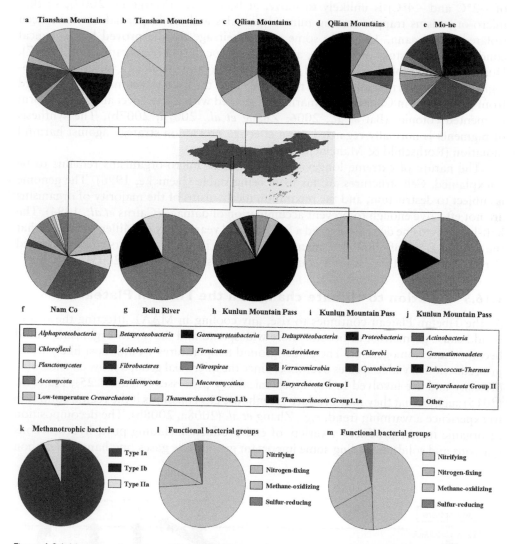

Figure 1.24 Variation in community composition of micro-organisms in permafrost between various mountain ranges in China (Hu *et al.*, 2015). They include bacteria, archaea, yeasts, filamentous fungi and microalgae, in 83 genera.

exist of the recovery and revival of spores from environmental samples as old as 105 years (Puskeppeleit *et al.*, 1992).

The viability of bacteria below 0°C has been investigated (Katayama *et al.*, 2007). Unfrozen water held tightly by electrochemical forces on the surfaces of mineral particles occurs even in very cold permafrost. Bacterial cells are not frozen at temperatures of −2°C and −4°C (Clein & Schimel, 1995; Ashcroft, 2000). The thin liquid layers provide a route for water flow, carrying solutes and small particles, possibly nutrients or metabolites, but movement is extremely slow. A bacterium of greater size (0.3–1.4 μm) than the thickness of the water layer (0.01–0.1 μm at temperatures of $-2°C$ and $-4°C$) is unlikely to move, at least in ice (Margesin, 2009), so that micro-organisms trapped among mineral particles and ice in permafrost have been isolated (Friedmann, 1994). In some cases, their age can be proved by geological conditions, the history of freezing, and radioisotope dating (Katayama *et al.*, 2007). During the long evolutionary period, the surviving organisms adopt a variety of physiological adaptations that allow them to survive. Thus the bacterial communities isolated from permafrost in China are primarily rod-shaped with a few cocci and always form pigmented colonies (Bai *et al.*, 2006; Zhang *et al.*, 2007a; 2007b). The synthesis of pigments is normally regarded as an effective protection strategy against harmful radiation (Rothschild & Mancinelli, 2001).

The nature of extreme longevity of permafrost micro-organisms remains to be unexplained. Cell structures are far from being stable (Jaenicke, 1996). The genome is subject to destruction, and the reparation mechanisms of the majority of organisms are not effective enough to prevent accumulation of damage (Cairns *et al.*, 1994). The half-life of cytosine does not exceed a few hundred years (Levy & Miller, 1998), so that the ancient DNA of mummies, mammoths, insects in amber appears to be destroyed (Greenblatt *et al.*, 1999; Rauser *et al.*, 2005; Willerslev & Cooper, 2005).

1.16.5   Relation to climate change on the Tibetan Plateau

As the Tibetan Plateau continues to rise, any cooling in MAAT affecting the region would cause permafrost to become more widespread and the micro-organisms would become more homogenous. The colder ground temperatures would most likely result in a reduction in viable species over time. Since the microbial groups show considerable diversity and are involved in key biochemical key processes (Figure 1.25), Hu *et al.* (2015) suggest that they may produce greenhouse gas emissions if the region continues to experience a warming trend, *e.g.*, Zhang *et al.* (2008a, 2008b). The decomposition of organic matter produces a variety of end products including peat, hydrocarbons (gaseous and solid), including some important greenhouse gases (methane and carbon

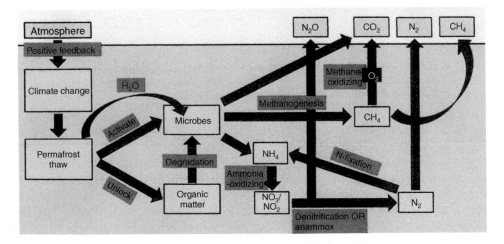

Figure 1.25 Position of micro-organisms and their role in decomposition of organic matter in soils resulting in release of greenhouse gases (from Hu *et al.*, 2015).

dioxide), as well as nutrients for plants. The quantity of methane production can be very high in wet, permafrost environments such as swamps, hence its common name, marsh gas (Brouchkov & Fukuda, 2002).

1.17 GAS AND GAS HYDRATES

Permafrost usually contains gases besides air bubbles which are located in pore spaces and within the crystal lattice of ice. The gases come from a combination of air, plus the gases produced by the decomposition of organic matter and exudates from living roots. These gases can move about by diffusion, and are often associated with oil/gas fields. These gases include significant amounts of carbon dioxide, hydrogen, and aliphatic and aromatic gases such as methane, ethane, propane and butane. These may gradually seep back into the atmosphere, or may also explode in the ground at low temperatures when oxygen is present in suitable quantities (see K. P. Harris, 1990).

Gas hydrates are where molecules of these gases become enclosed within the ice crystals. The crystal structure of ice is hexagonal, and the molecular lattice has large spaces in it. Small molecules such as methane, ethane, propane and isobutane can be trapped within this lattice if they are present in the ground (Davidson, 1973). The ice looks the same as ordinary ice. Under cold temperatures and increased pressure, these small gaseous molecules condense into solid crystals, so that the ice lattice can hold considerable quantities of the solid gases called *clathrate hydrates*. If decomposition occurs, the solid gaseous material changes to gas, resulting in a substantial change in volume such as occurred in the Markhinskaya well in northwest Yakutia in 1963. Makogon (1982) showed diagrammatically the approximate zone of occurrence of both permafrost and hydrates in the Russian Arctic (Figure 1.26) and similar relationships are found in North America (Harris, 1986a, pp. 142–145). Gas hydrates do not normally occur closer to the surface than about 140 m, nor deeper than about 1860 m in the ground (Judge, 1982), but they can also occur beneath oceans if the weight of the overlying water is great enough. Figure 1.27 shows typical pressure-temperature relationships for the methane-water-gas hydrate system after

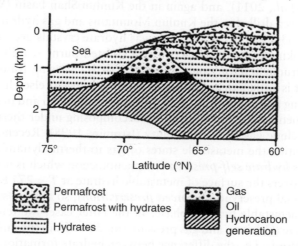

Figure 1.26 Zonation of permafrost and gas hydrates with latitude in the Russian Arctic (redrawn from Makogon, 1982).

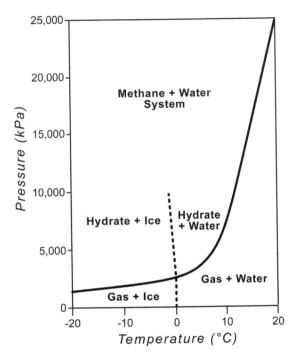

Figure 1.27 Stability of the methane-water-gas hydrate system (after Davidson *et al.*, 1978, page 938). Reproduced with the permission of the National Research Council of Canada.

Davidson *et al.* (1978, p. 938). Gas hydrates have recently been discovered in the Qilian Mountains, Qinghai, North-west China (38°06′N, 99°10′E at 4057 m elevation; Xu *et al.*, 1999; Lu *et al.*, 2011), and again in the Kunlun Shan basin (Wu *et al.*, 2015). Four wells have been drilled in the Kunlun Mountains and gas hydrates were found at depths between 133 and 396 m. The white gas hydrate crystals decompose producing a honeycomb-looking ice-core, while the gas emitted burns readily (Figure 1.28A). The typical spectrum curve of gas hydrates was detected using Raman spectrometry (Figure 1.28B). It is anticipated that gas hydrates may occur elsewhere in permafrost profiles, both along the north side of the Qinghai-Tibet Plateau and in northeast China. Gas hydrates sometimes stay stable without decomposing under thermodynamic conditions which exclude gas hydrate stability (Istomin, 1998). Recently, most authors have been attributing the metastable states of gas to thermodynamically stable solid phases. In the *gas hydrate self-preservation* phenomenon which is not fully explained yet, the ice film covers the surface of metastable hydrate at $T < 271$ K, and there is the possibility of special preservation (*forced preservation*), so a more stable gas hydrate film covers the metastable hydrate phase. In the latter case, the temperature can be negative or positive (Celsius), and gas pressure can differ from the ambient conditions. Gas hydrate hysteresis, *i.e.*, the difference between hydrate formation and dissociation conditions, can sometimes be thermodynamically stable solid phase.

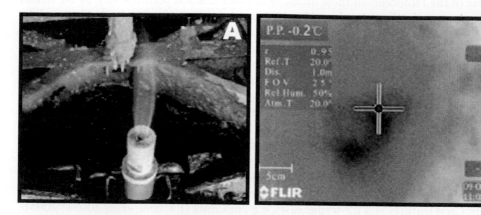

Figure 1.28 Gas hydrates in Chinese permafrost, A, burning at the exit of the drill stem, and B, the Raman infrared spectrum of the flame showing that the temperature in the flame was below 0°C (from Lu *et al.*, 2011).

It is unclear whether the ice cover is responsible for the metastability, however, it may play a key role in it (Istomin 1998; Istomin *et al.*, 2006). Gas hydrates are of widespread occurrence in northern Siberia and along the Arctic coastal plain of Alaska and in the Mackenzie Delta area.

1.18 THERMOKARST AREAS

This refers to all topographic features and structures produced by melting of ground ice. They are particularly common under the present climate due partly to the warming of the climate since the last glaciation about 25,000–12,000 years ago, and partly to warming in many areas such as the Qinghai-Tibet Plateau in the last half century coupled with changes in land use (Wang *et al.*, 2011; Jin *et al.*, 2011). During the last major glacial event, glaciers covered most of Canada. As a result, the permafrost found there today is mainly developed in equilibrium with the Holocene climate and is post glacial in age. Even so, climatic warming, culminating in the Climatic Optimum/Hysithermal/Altithermal warm event about 6000 years ago, followed by the Little Ice Age and subsequent warming, have produced some significant small thermokarst features in the area once occupied by permafrost during major glaciations (*e.g.*, Rockie, 1942; Hopkins, 1949; Harris, 1968; Morgan, 1972; Péwé, 1973a; 1983a; Rampton, 1974; Mackay, 1975; Murton & French, 1993a; 1993b).

Large areas of Siberia were not glaciated but endured very cold climatic conditions during each of the main glacial episodes of the last 3.5 Ma (Aubekerov & Gorbunov, 1999; Harris, 2013a). As a result, very icy permafrost has developed to thicknesses as great as 1100 m in some areas, and repeated cold events followed by warm Interglacials undoubtedly meant many periods of thermokarst development followed by redevelopment of vast areas of permafrost. In southern and central Siberia and Northeast China, most of the permafrost is relict, and still adjusting to the warmer climate of the Holocene. In these areas, major thermokarst features such as the *alases* (depressions and valleys developed by thawing of the ground ice and subsidence of the land) were produced primarily during the Climatic Optimum/Hysithermal/Altithermal warm event about 6000 years ago (Kachurin, 1961; Soloviev, 1973b; Fukuda *et al.*, 1995). During the Late Wisconsin Glaciation, or Sartan Glaciation in Western Siberia (29,000–12,000 years ago), very cold conditions prevailed as far south as the Central Qinghai-Tibet Plateau, as indicated by the *rock tessellons* (polygonal cracks filled by rock fragments)) that formed there during the last millennia of the Late Wisconsin Glaciation (Harris & Jin, 2012), but at that low latitude, the thickness of permafrost was not very great. However, the Plateau and the Himalayas are rising, and prior to a million years ago, the climate was wetter and probably warmer (Gerasimov & Zimina, 1968), so permafrost is a relatively recent phenomenon there.

In other parts of the world, the former presence of permafrost is indicated by thermokarst features and ice-wedge casts, *e.g.*, in China (J. Cheng *et al*, 2005a; 2005b; 2006), Europe (Poser, 1948; Karte, 1987; Vandenberghe, 1983; 1988; Pissart, 2003), New Zealand (Harris, 1983a), South America (Corte, 1967; Gonzalez and Corte, 1976; Graf, 1986; Corte, 1988), and Australia and New Guinea (Galloway *et al.*, 1973). The different thermokarst features are discussed in the chapters in Part 2, while features produced by water actively eroding icy permafrost (*thermoerosion* and *thermal abrasion*) are described in chapter 11.

1.19 OFFSHORE PERMAFROST

Only 29% of the surface of the Earth consists of land, and of this, about 22% is currently underlain by permafrost. In addition, around the Arctic Ocean are extensive areas of relatively shallow water underlain by *subsea permafrost* (Figure 1.29). Offshore permafrost is formed either as a response to negative mean annual sea-bottom temperatures, or by submergence of terrestrial permafrost. This area is particularly wide in the vicinity of the Mackenzie Delta and again along the shores of the Laptev Sea (Gigarev, 1997). The Mackenzie Delta is actively aggrading, developing permafrost in the new alluvial sediments where the adjacent sea water is cold enough and there is insufficient salinity to prevent freezing. Permafrost growth is primarily syngenetic with cryostructures and ice wedges (Mackay, 1973a). The silty sediments allow ice segregation (Mackay, 1972; Hollingshead *et al.*, 1978), and the resulting ice may underlie the sea floor or occur as a shallow layer underneath a thin layer of unfrozen sediment. Submarine pingos also occur, though their origin is uncertain.

Changing sea levels since the Late Wisconsin maximum are not usually reflected in the permafrost stratigraphy of the delta, but conditions are such that *in-situ* gas hydrates occur in boreholes in the Beaufort Sea (Weaver and Stewart, 1982). Taylor *et al.* (1996) provide a model suggesting the Late Quaternary history of the onshore-offshore transition zone.

Eastwards, an old mountain range has been partially submerged so that the former river valleys are now beneath the sea and the main upstanding mountain ranges currently form the Arctic Islands along the north coast of Canada. The former continental shelf extends northwards beyond these islands. Deep grabens separate Baffin Island from Greenland. However, the Laptev Sea is a different matter (Figure 1.30). Several

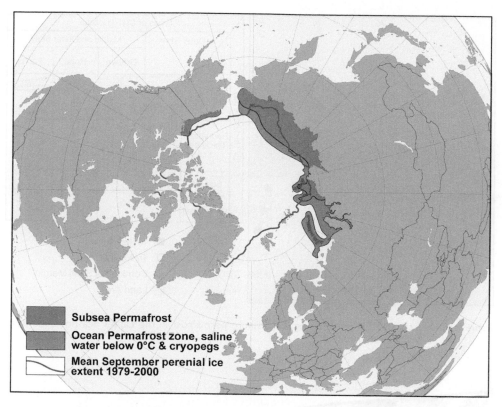

Subsea Permafrost

Ocean Permafrost zone, saline water below 0°C & cryopegs

Mean September perenial ice extent 1979-2000

Figure 1.29 Distribution of subsea permafrost along the shores of the Arctic Ocean.

large rivers supply sediment to the area and changing sea levels over time are certainly involved in the rather complex permafrost (Bondarev, *et al.*, 1999; Romanovsky & Hubberton, 2001a; 2001b). The situation is further complicated by the obvious results of tectonic movements of unknown age that have resulted in a series of horsts and grabens in the shallow coastal waters (Dobrovolskiy, 1974; Drachev *et al.*, 1995; 1998; Biryukov & Sovershaev, 1998; Sekretov, 1999; Franke *et al.*, 2001). The tops of the horsts are at a significantly different level in the ocean to the floors of the grabens, and there is considerable variation in the thermal regime with weather and the seasons (Berezovskaya *et al.*, 2002). The ice complexes displayed in the coastal sections (Figure 1.31) indicate a complex history of ice segregation and growth of spectacular ice wedges (Schirrmeister *et al.*, 2002), while the sea bed exhibits both frozen and partially frozen sediments (Gavrilov *et al.*, 2001). These are believed to have been formed during the cold permafrost events corresponding to the past glaciations, although the exact ages are not known (Bykov, 1938). The area is vast and drilling is expensive and challenging. As a result, modeling has been employed to suggest the history, *e.g.*, Gavrilov & Tumskoy (2003) and Gavrilov *et al.* (2006), although there are obviously numerous unknowns in the database which the models have to disregard. In general, the history assumes periods of low sea levels during very cold events, resulting

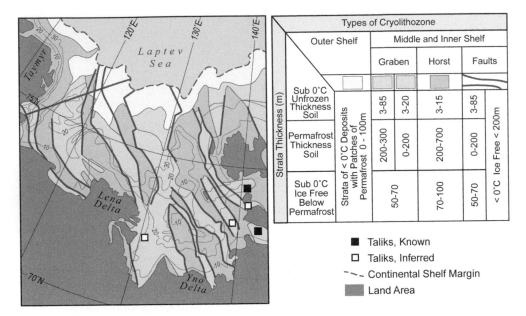

Types of Cryolithozone					
	Outer Shelf	Middle and Inner Shelf			
		Graben	Horst	Faults	
Strata Thickness (m)	Strata of < 0°C Deposits with Patches of Permafrost 0 - 100m				< 0°C Ice Free < 200m
Sub 0°C Unfrozen Thickness Soil		3-85 / 3-20	3-15	3-85	
Permafrost Thickness Soil		200-300 / 0-200	200-700	0-200	
Sub 0°C Ice Free Below Permafrost		50-70	70-100	50-70	

■ Taliks, Known
□ Taliks, Inferred
~-~ Continental Shelf Margin
▧ Land Area

Figure 1.30   Horsts and Graben affecting the offshore permafrost in the Laptev Sea (modified from Gavrilov *et al.*, 2006).

Figure 1.31   Ice-wedges in eroding cliffs on Big Lyakovsky Island, Arctic coast, Siberia. © M. Grigoriev.

in extensive permafrost development in a wide zone above the 100 m isobath, which is currently found well to the north, alternating with warmer events as at present with a retreating shoreline, degrading permafrost and venting of methane to varying degrees in the sea over the former extent of the permafrost (Shakhova *et al.*, 2010).

Biriyukov & Ogorodov (2002) summarize the bottom relief of the Pechora Sea, west of the Ural Mountains. They are the result of a series of land-ocean interactions during both the Pleistocene and the Holocene. Tectonics are also involved in the genesis of the permafrost on the surface deposits of the plains to the south (Baulin *et al.*, 1978a), where relict permafrost also occurs (Baulin *et al.*, 1978b). Belopukhova (1966) and Belopukhova & Sukhov (1986) describe the landforms of that area. Spesivtsev (2001) discusses the effects of economic development on the shelves of the Kara and Barents Seas.

Less is known about the processes taking place below sea level in the sediments on the continental shelf. However subsea permafrost has proved to be widespread in polar regions where there is often a wide, shallow sea offshore. These sediments are saline and their distribution and temperature are poorly known (Brouchkov, 1998).

Chapter 2

Cryogenic processes where temperatures dip below 0°C

2.1 INTRODUCTION

It will be obvious that regions with permafrost have a number of special processes that produce the unique landforms found in these areas. Many of the same processes are also involved in producing characteristic features in regions with seasonal frost in the ground. This chapter examines these processes in detail, so as to provide the reader with an understanding of what is occurring.

2.2 THE NATURE OF ICE AND WATER

Ice in permafrost on Earth is type 1 of the 14/15 known types (Militzer & Wilson, 2010), except at temperatures below −70°C in Antarctica (Murray *et al.*, 2005). When the temperature there rises, it changes back to type 1. The molecule consists of an oxygen atom attached by electrical bonds to two hydrogen atoms, forming an isosceles triangle (Figure 2.2). In the solid state, these molecules form a hexagonal crystal consisting of stacks of layers piled one on top of another. Each layer consists of three molecules, and these are held in place in the stack by relatively weak hydrogen bonds. Since freezing of the active layer develops from both the top and bottom of the active layer (Mackay, 1973b), the ice crystals in that layer appear to grow with the stacks perpendicular to the freezing plane, *i.e.*, the crystals have their long axes aligned vertically (Figure 2.1). Actually, the crystals start to grow parallel to the freezing front (by optical analysis), but because they all tend to grow at the same time, the growth becomes perpendicular. Even so, some crystals still retain their optical axis parallel to the freezing front. The weak hydrogen bonds provide an important mechanism for the creep of ice, since the individual hexagonal plates can glide laterally, deforming the ice crystal (Ollier, 2010).

Frozen ground in sediments normally consists of at least five constituents, *viz.*, organic matter, mineral grains, air, ice and water. The water has the unique property of being able to move over the surface of the solid constituents. It can also move salts and tiny mineral grains with it. Water is held in soil or rock under conditions ranging from the infilling of the available pores (*gravitational water*), through a film of water weakly bonded to the surface of solid particles by capillary forces (*capillary water*), to a thin film of strongly bonded water (*hygroscopic water*) held by electrical forces (Anderson, 1971). Water can also be more weakly bonded by cations which

Figure 2.1 Vertical ice crystals in the upper part of the active layer perpendicular to the freezing plane, Slims River alluvium, Kluane, Yukon Territory. The ice forms at the surface of the moist soil beneath the dry surface of the ground. It was fed by moisture moving to the freezing plane from the wet soil layers below. © S. A. Harris.

may be bonded by hydrogen bonds acting between water molecules. If however, they are bonded with cations or minerals instead of hydrogen bonds, they will add extra layers of water. As used in agriculture, the hygroscopic water may be structured with hydrogen bonds (more strongly bonded) or by molecular forces known as osmotic forces (weaker). The gravitational water slowly drains out of the soil or rock after

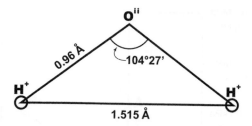

Figure 2.2 Structure of a water molecule in the gaseous state resulting in positive and negative sides producing a dipole effect (modified from RIGCDR, 1981).

rainfall to join the groundwater or streams. The capillary water is partly available to plants until it is depleted to the wilting point of a given species. The more strongly bonded hygroscopic water remains as a liquid even in dry soil and is unavailable to most micro-organisms and plants, whereas the weakly bonded, osmotic water is available. It is this hygroscopic water that can move over the surface of solid particles in otherwise frozen ground. Grechishev (1980) discusses the thermodynamics involved.

The movement of hygroscopic water depends on several factors. Firstly, the shape of the water molecule explains its potential movement (Figure 2.2). Obviously this molecule can become attached to other positively or negatively charged surfaces. Secondly, the solid phase in the soil also has electrical charges at its surface. These are usually negatively charged, so that positively charged atoms will be attracted to its surface (Yershov, 1998a). Thirdly, the available surface area of the solid phase determines the quantity of surface charges potentially available in a given mass of the soil, *e.g.*, silts will have far more available sites than sands. Amounts of hygroscopic water are minimal in cryotic gravels, but are greater than 20% in clays down to −20°C (Tsytovich, 1957). Even at −78°C, clays still hold some hygroscopic water. As a result, cryotic clays often have a consistency similar to chewing gum. Fourthly, the mineralogy of the sediment also affects the hygroscopic water content, *e.g.*, kaolinitic clays hold less water at a given temperature than clays consisting of montmorillonite (Williams & Smith, 1989). The *cryosuction* causes the development of ice lenses resulting from capillarity, increasing by approximately 1.3 MPa per °C below 0°C in an open system. There, water can move relatively freely to the freezing plane, or away from it if the pressure is raised. However in a closed system, the cryosuction increases by 13.4 MPa per °C below 0°C. This is not entirely due to capillarity, but may be due to the freezing itself. In addition, salts dissolved in soil water may largely occupy the exchange sites on the surface of the mineral grains, while concentrations cause shrinkage of the volume of the layer of bonded ions and a smaller amount of total bonded water. Hygroscopic and capillary water move towards the coldest part of the ground, or towards a zone with higher salinity (Nixon, 1982; Nixon & Lem, 1984). Significant quantities of water can move into the upper, frozen, cold part of the permafrost from above (Mackay, 1983) causing frost heaving in winter. However, the details of the processes occurring are rather complicated and not necessarily fully understood.

Figure 2.3 shows the phase diagram for pure water in the range of temperature and pressure that are normally encountered. In the gaseous phase, the molecules are

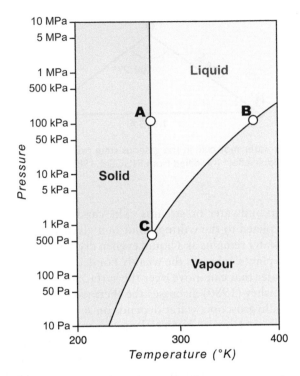

Figure 2.3 Phase diagram for pure water in the range of temperature and pressure encountered on Earth in permafrost. A is the freezing point of water at 1 atmosphere pressure (273.15°K; 101.325 kPa), B is the boiling point of liquid water (373.15°K; 101.325 kPa) and C is the triple point below which solid ice starts to sublime to water vapour (273.16°K; 611.73 Pa).

flying around in the atmosphere, avoiding collisions because of the electrical charges. Water vapour has a molecular weight of 18, which is sufficiently great for the pull of Earth's gravity to prevent all the molecules escaping into space, though on Mars, they have escaped. As the gas cools, the movement decreases and eventually the molecules collapse into a gyrating, pulsating mass with a distinct upper surface, thus becoming the liquid phase. Further cooling results in decreased movement and increased density as the pile becomes more compact. Maximum density is achieved at 4°C, after which the molecules start to become arranged in the typical, open, hexagonal structure of the solid type 1 crystals with a small change in the angles of the molecular structure, and though different to ice, continue to vibrate. This requires expansion and reduction in density, and is the reason for the annual turnover of water in lakes as the surface water cools below 4°C in the autumn. However, if the water is saline, the maximum density is at lower temperatures. This is fairly common in areas of permafrost. For marine water with a concentration of salt of about 36 g/l, the maximum density is below 0°C, which is why the sea water tends to sink during surface cooling, so that the sea bottom temperatures are lowest.

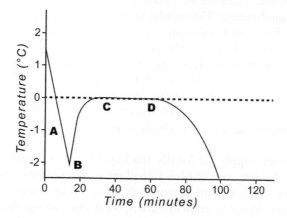

Figure 2.4 Ground temperature changes at the onset of freezing (modified from RIGCDR, 1981). The letters refer to the processes described in the text.

Pure liquid water changes to solid ice at 0°C (273.15°K) and 1 atmosphere pressure (101.325 kPa), provided suitable nuclei are available. This involves the water molecules of the gravitational water crystallizing out on the surface of mineral particles. In actual practice, this can only happen if one or more of the electrical charges on the surface of the mineral particles is not occupied by ions from the dissolved solids in water, or a suitable nucleus (mineral, organic or ice) is available. If the liquid water does not immediately find a suitable charge, the liquid phase continues in that form at temperatures below 0°C producing *super cooling* (Figure 2.4 A and B). As soon as a suitable electrically charged site is found, the molecules begin to accumulate in the form of hexagonal crystals. This involves more expansion, together with the loss of $3.347 \times 10^8 \, \mathrm{J \, m^{-3}}$ of latent heat of fusion. This heat is usually enough to neutralize the cooling from above (Figure 2.4 C), and the cooling plane will remain stationary until the supply of liquid water moving to the freezing plane in the form of both gravitational and capillary, mainly bonded water no longer produces enough heat to match the cooling being applied to the soil (Figure 2.4 D). Maximum super cooling of gravitational water in permafrost soils is about 3°C. The water that moves to the freezing plane is largely the bonded water.

The soil is essentially an open system, and much of the water migrates to the freezing plane from elsewhere, resulting in the temperature remaining at about 0°C, *i.e.,* the layer is *isothermal*. The period of constant soil temperature may extend for several days or even weeks in soils with high moisture contents, producing the *zero curtain effect* shown in Figure 1.9. The greater the pore-water pressure, the longer the freezing plane remains stationary at a given depth and the thicker the layer of segregation ice (Sumgin *et al.*, 1940; Mackay, 1971b). It typically occurs in the seasonally thawed layer between the upper and lower freezing fronts. Water transfer can temporarily stabilize the freezing front though the thawed layer is not isothermal. Ice segregation is favored by materials with grain sizes of 0.01 mm or less (Taber, 1929, 1930a, 1930b), and clay-rich sediment exhibits a longer zero curtain than shattered rock or gravel

(Cook, 1955). Volcanic ashes are particularly frost susceptible and good examples of water transfer during freezing. That is why frost heave is a particularly severe problem in the Russian Far East, the Kuril Islands and the Japanese Island of Hokkaido.

After this, the cooling front can then commence descending lower into the soil. In soils with a high content in the active layer, the zero curtain effect can also result in retarded warming in the spring (Harris, 2013b), thus shortening the growing season for plants in spite of an overall warming of mean annual air temperatures, *e.g.*, on the Tibetan Plateau (Yu *et al.*, 2010). However this is relatively rare. The zero curtain effect is most frequent in the autumn in both arctic and mountain permafrost (Kelley & Weaver, 1969).

Should the water supply be locally inadequate for the pore-water pressure to exceed the overburden pressure, the freezing plane would tend to move down unevenly, so that wide tongues of ice extend downwards. This can produce large blocks of ice in the ground with considerable vertical extent, as it does along the Dalton Highway (Figure 6.19) and in deep outwash gravels near Fairbanks, Alaska. The water in the rest of the soil freezes as pore ice. Fine-grained soils liable to ice segregation are said to be *frost-susceptible*.

The phase diagram (Figure 2.3) also indicates that ice can change to water vapour directly under certain conditions, *i.e.*, it *sublimes*. Although the diagram suggests that this might be rare, studies of the fate of the snowfall at Kananaskis (c. 1400 m elevation) in the Chinook belt in the foothills east of the Eastern Cordillera in southern Alberta indicate that 40% of the total snowfall sublimes from the snowpack between November and May (Harris, 1972; Peng *et al.*, 2007). This probably happens elsewhere in cold dry climates, especially at high altitudes, so that the apparent precipitation in these areas over-estimates the effective precipitation available for processes occurring in the ground.

Normally, permafrost is found at the higher elevations in the landscape due to the decrease in air temperature with altitude (the *lapse rate*). This is not always the case due to *cold air drainage* (see below), or in very dry climates such as on the Qinghai-Tibet Plateau where the intense radiation at low latitudes and high altitude combine to cause evaporation to exceed the amount of precipitation (Liu *et al.*, 2012). In this case, the wet low areas can be cooled sufficiently by the evaporation of water from the soil surface so that permafrost develops (An *et al.*, 2006). This process may be the cause of low, flat-topped pingos called *pingo plateaus*, see chapter 7.

The expansion of the water molecules during crystallization produces extra pressure in the direction parallel to the freezing plane. If there is sufficient lateral pressure, individual molecules move in the vertical direction from the sides of the growing crystals to sites that have lower pressure, *i.e.*, the top of the growing hexagonal crystals. This causes uplift of the material above, whether it consists of soil or stones, in the form of *frost heaving* or the formation of *needle ice* respectively. These will be discussed further below. The energy required for lifting up the overburden comes from the latent heat being released during freezing. Any inclusions of air being confined by the growth of ice are also put under lateral pressure and become elongated in the same direction at right angles to the cooling plane by movement of individual water molecules to sites with less pressure. This is important in demonstrating the direction of movement of the cooling plane in permafrost. In glacial till and ground that has flowed laterally, the air bubbles will be elongated in the direction of the last movement. There is no

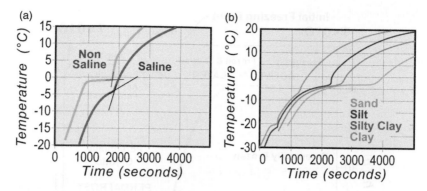

Figure 2.5 Comparison of thermograms for a nonsaline soil (a) with that for a saline soil containing sodium chloride, and (b) effect of texture. Modified from Grechishcheva & Motenko, 2015, Figures 1 and 5 respectively.

significant correlation between active layer thickness and annual thawing degree days (Christiansen *et al.*, 2008).

The presence of salts in solution in the soil water reduces the exact temperature at which freezing takes place (Wan *et al.*, 2015). Deliquescent salts like calcium chloride give the greatest depression of the freezing point, followed by hygroscopic salts such as sodium chloride and potassium chloride. Calcium sulphate and calcium carbonate have only minor effects. This depression of the freezing point is best measured using warming thermograms to avoid the effects of supercooling (Sergeev & Batyuk, 1978). Figure 2.5 shows the results for warming frozen non-saline soil compared with saline soil containing sodium chloride. The limit for saline soils is shifted lower by about 2°C. In this study of saline soil temperatures during thawing, the pore solution freezing temperature decreases with the decreasing of pore solution concentration as more ice melts, resulting in the curvature of the lines (Motenko *et al.*, 1997).

Pressure is also very important, since at 100 atmospheres, the ice melts at −0.9°C, while at 1500 atmospheres, the ice melts at −14.1°C (13.4 MPa/degree C). There is a reduction of 1°C for every 13.4 MPa increase in pressure. Within the soil, the points of contact between the individual grains take all the pressure, so that even at shallow depths, the ice will melt on the contact surfaces between grains (Tsytovich, 1973). This may partly explain the lack of massive icy beds and thick lenses of segregation ice at depth. Other possible causes are the higher density of the sediment decreasing ease of movement of water, and the increased overburden pressure that makes creating space for crystal growth more difficult.

This depression of the freezing point prompted van Everdingen (1976) to suggest a modified set of terms for the parts of the temperature-depth envelope (Figure 1.10) to allow for this (see Figure 2.6). However, that terminology is not usually used, partly because the freezing point depression varies with the seasonal changes in salinity of the active layer (see Figure 1.16), as well as with the increasing overburden pressure with depth. The depression is usually very small.

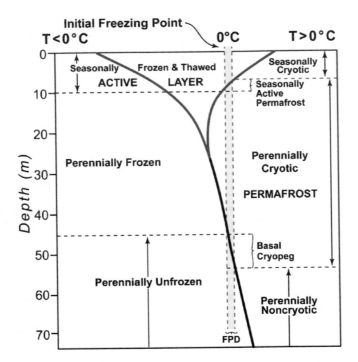

Figure 2.6 Terminology for the parts of the ground temperature envelope proposed by van Everdingen (1986) to allow for the *freezing point depression* (FPD).

2.3   EFFECTS OF OIL POLLUTION ON FREEZING

There are several sources of pollution of soil water including crude oil, commercial waste water, drainage from barns and feedlots, cleaning fluids, chemicals used in desalting oil (Solntsava, 1998) and the chemicals used in agriculture and in fracking. Oil spills, reservoir fluids and waste water usually result in salinization. These combine together to change the physical properties of the soil (Nebogina, 2009). The result is lower freezing temperatures and reduced bearing capacities that can result in subsidence and failure of engineering structures.

In cold climates, the rate of decomposition of hydrocarbons is low (Solomonov *et al.*, 2001), with chemicals with very aggressive properties (Chuvilin & Miklyaeva, 2005), and the associated water is highly mineralized. It represents a serious form of pollution that does not affect the rheological properties of the soil, but does alter the thermal properties (Sheskin *et al.*, 1992; Shevchenko & Shirshova, 2008). However, oil on its own, does not alter the freezing temperature.

Figure 2.7 shows the unfrozen water content in soils of various textures in the nonpoluted, nonsaline state (Figure 2.7a), contrasted with the saline (1%) oil polluted (8%) state (Figure 2.7b). The change in slope in the latter at about −20°C is due to a change in hydration of 1% sodium chloride (Grechishcheva & Motenko, 2015). The higher unfrozen moisture content is obvious, particularly in the sample of clay. It is this that causes the lower bearing capacity of the ground.

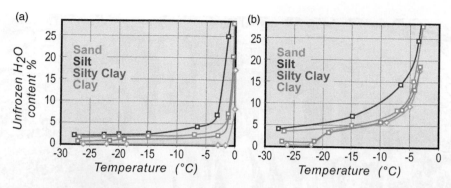

Figure 2.7 Unfrozen water contents in nonsaline oil polluted soils (a), contrasted with the addition of 1% sodium chloride and 8% oil to the same soils (b). Modified from Figures 7 and 8 of Grechishcheva and Motenko (2015).

2.4 FREEZING AND THAWING OF THE ACTIVE LAYER IN PERMAFROST IN EQUILIBRIUM WITH A STABLE CLIMATE

Initially, the freezing front tends to stop as it moves into the surface of the active layer due to the movement of soil water towards the freezing plane from below, producing a lens of ice crystals (Figure 2.1). After the amount of available gravitational and bonded water decreases sufficiently, the freezing front descends downwards until it meets the freezing front moving upwards from the permafrost table. During the descent, any increased supply of gravitational and capillary water will result in increased release of latent heat of fusion, so the rate of descent will slow or stop until the supply of latent heat of fusion no longer neutralizes the degree of downward cooling. The ice that is formed during the stationary freezing event will take the form of an ice lens, and multiple lenses can form if the sediment has alternating silty layers that favour water movement to the freezing plane (Figure 2.8). These are very common in the upper layers of the permafrost, but become rare at depth. The result of this movement of soil water is a drying out of the remaining part of the active layer. Ground temperature diagrams demonstrate that the cooling air temperatures in cold, continental climates usually compensate for the cooling losses to the soil, thus maintaining a fairly rapid descent of the freezing front (Figure 2.9). Part way through its descent, gradual up-freezing from the permafrost table develops until the entire active layer is refrozen. Once again, water migrates to this second freezing front. The total amount of upward freezing observed on the swampy Mackenzie Delta is only about 13 cm (Mackay, 1973b, 1974a). The timing of the completion of refreezing varied from August to January at Marmot Basin #2 borehole, Jasper National Park (Figure 2.10), measured for a 23-year period. This was caused by variations in the timing of the onset of the cold weather and snowfalls (Harris, 2005, Figure 8).

After the active layer is refrozen, the freezing front descends slowly into the permafrost at a rate proportional to the thermal conductivity of the material through which it is passing and the applied cooling from above. As demonstrated in Figures 1.8 and 2.9, the amplitudes of the warming and cooling waves decrease with depth until

Figure 2.8 Multiple ice lenses beneath a moss layer on the Appalachian Trail in Virginia, USA. Photo by Lon and Susan Rollinson (http://my.ilstu.edu~jrcartier/ice/diurnal/).

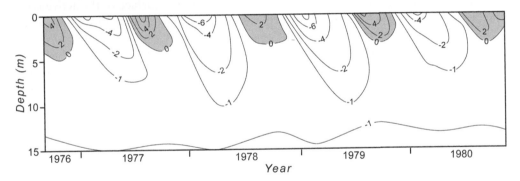

Figure 2.9 Geotherms for site #2, Plateau Mountain, Alberta between 1976 and 1980, showing the variable shape of the lower limit of the freezing front from year to year and the variation in upfreezing from below from year to year. Maximum amount of upfreezing is about 1.5 m in the dry, limestone-rich soil.

the seasonal range in temperature is less than 0.1°C. This is called the ***depth of zero amplitude*** or minimal amplitude, or also a depth of seasonal change in temperature, and is based on the former limit of accuracy of measurement of temperatures with thermocouples (0.1°C). When thermistors or thermocouples are used with data loggers, temperatures can be measured to better than 0.01°C and show that slight variations can be detected at greater depths. Figure 1.10 shows the typical ground temperature envelope of a deep borehole and the names given to the constituent parts. Thawing of the active layer commences from the surface. In areas with a substantial snowpack, the snowpack becomes isothermal at about 0°C. Since the snow is translucent, melting occurs throughout the snowpack during daylight hours, producing a layer of water at

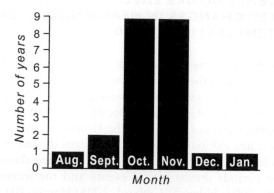

Figure 2.10 Variation in timing of completion of refreezing of the active layer at Marmot Basin #2 borehole in a 23-year period (translated from Harris, 2005).

its base. This water will partly drain away down any gradient, but it also carries heat that aids in producing a relatively fast descent of the thawing front, unless the active layer contains a layer with a lower thermal conductivity or one that contains more soil water. When this water is draining over the ground, it can result in significant erosion of the surface layers of the soil (Harris, 1998c).

2.5 RELATION OF CLAY MINERALOGY TO THE AVERAGE POSITION OF THE PERMAFROST TABLE

Zenin Xing *et al.* (1980, 1984) examined the clay mineralogy in Quaternary sediments between Wudaoliang and Fenghou Shan on the Qinghai-Tibet Plateau. They used the <2.0 mm fraction and treated it with glycol. X-ray diffraction and differential thermal analysis were used to obtain the 7 Å and 10 Å peaks. They found that the B value for the relative content of hydromica divided by the relative content of the chlorite and kaolinite was markedly lower at the average position of the permafrost table. Kaolinite is rare in these sediments.

Above the permafrost table, water passes through the sediments, dissolving and removing the soluble weathering products such as potassium (from sericite and muscovite), thus helping the weathering of the feldspars and micas. Since many of these chemical reactions are exothermic, they can modify the ground temperature to some extent. Below the permafrost table, there is only closely bonded water that can move but cannot readily transport cations. Oxygen is also available above the water table and increases the valency of iron and manganese, stabilizing them. It also helps break down chlorite, while hydration by liquid water results in the production of hydromicas. The result is considerably more intense weathering above the permafrost table than below it. This was demonstrated for three sites, and a deep borehole near Fenghou Shan indicated the presence of two earlier positions of the permafrost table, one at 3.8 m and one at 4.5 m. They were interpreted as corresponding to the permafrost table in the warm periods at 8 ka (2°C warmer than present) and 11 ka (4°C warmer than now).

2.6   GROUND TEMPERATURE ENVELOPES IN PROFILES AFFECTED BY CHANGES IN MEAN ANNUAL GROUND SURFACE TEMPERATURE (MASGT)

It is now becoming clear that the main climatic parameters are constantly fluctuating from year to year. Sometimes, these fluctuations vary to and fro around a reasonably constant mean value, *e.g.*, *the mean annual air temperature (MAAT)* for Watson Lake in the Yukon Territory from 1939 to 2002 (see Harris, 2009, Figure 10). However, examination of inferred values of MAAT for the Holocene from different parts of the world indicate that it has exhibited substantial changes during the last 10,000 years (Harris, 2010), while even larger fluctuations have occurred during the late Pliocene and the Pleistocene periods during glacial events and the intervening interglacials (Isarin, 1997; Harris, 2001; Ehlers & Gibbard, 2008; Harris, 2013a).

Any change in *mean annual ground surface temperature (MAGST)* will alter the heat flow into the surface of the ground, and will result in a bending of upper portion of the geothermal gradient in the upper layers (Figure 2.11). This can be accomplished either by a change in MAAT, a change in vegetation cover, soil erosion or deposition, or by a change in precipitation, especially variations in the winter snow cover. The geotherms are the result of the balance between the geothermal heat flow from below and the input of heat from the surface. The geothermal heat flow is usually reasonably constant below the depth of zero amplitude except in earthquake prone areas, and can be disregarded for most areas (Duchkov *et al.*, 1995). The resulting change in ground temperature with depth is called the **geothermal** gradient. The increase in temperature shown by the geothermal gradient is normally about 3°C/100 m, though it is inevitably affected by the thermal conductivity of the bedrock (see Figure 1.6).

In the case of warming (Figure 2.11A), an increase in MAGST results in an immediate increase in the MAGST of the surface, but with declining effect with depth. If the thermal condition is stabilized at the new MAGST, the warming slowly spreads downwards, eventually resulting in an upward movement of the permafrost table. Eventually, a new equilibrium is established with a constant geothermal gradient in the ground, provided the new MAAT remains stable. For a soil with 30% ice by weight subjected to a 2°C rise in MAGST, 15 m would be thawed from the top of the permafrost table in 100 years, but it would take 10,000 years to raise the permafrost base by 200 m (Terzaghi, 1952). The small amount of change in the permafrost base has been used by Balobayev (1978) to calculate the minimum air temperatures for the Wisconsin Glaciation in Russia using the thickness of permafrost in the contemporary outwash terraces. There would also be a decrease in surface elevation due to the melting of some ice.

Relict permafrost is sometimes encountered (Figure 2.12). It can readily be recognized by the decreasing ground temperatures with depth, which makes calculation of the thickness of permafrost impossible without drilling through the permafrost base. At that site on Plateau Mountain (Southwest Alberta), the ground temperature at 152 m is about −1.5°C and the relict permafrost dates back to the major glaciations. In the Tanggula Shan (Qinghai-Tibet Plateau) the relict permafrost is probably from the last Neoglacial (Little Ice Age) event. The thickness of the relict permafrost is related to its age and the length and degree of the cold period in which it developed.

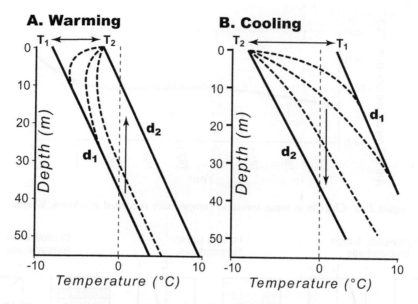

Figure 2.11 Theoretical effects of climatic changes involving warming (A) or cooling (B) of the mean annual ground surface temperature (after Lachenbruch, 1968).

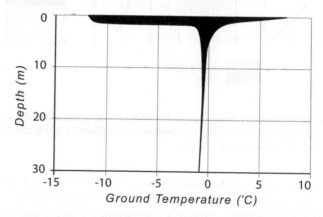

Figure 2.12 Ground temperature at Plateau Mountain #2 borehole, S. W. Alberta (Harris & Brown, 1978).

In actual practice, multiple increases in MAAT often occur (Figure 2.13) and thus the geothermal gradient at a given site can become rather complicated. At Yakutsk, there was a slow warming of about 1°C between 1830 and 1910, corresponding to the end of the last Neoglacial cold event. The MAAT stabilized and even cooled slightly until about 1975, when a further increase in MAAT began and is still continuing.

In the case of cooling on land (Figure 2.11B), a similar sequence of changes occurs, with the surface of the ground cooling first. Gradually the geothermal gradient adjusts from the surface downwards and the permafrost base slowly descends until a final equilibrium is reached. Taking the geothermal curves in Figure 1.6, Barrow

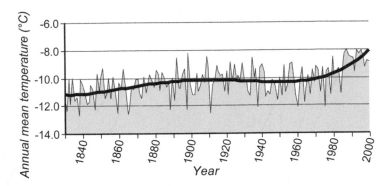

Figure 2.13 Changes in mean annual air temperature recorded at Yakutsk, Siberia.

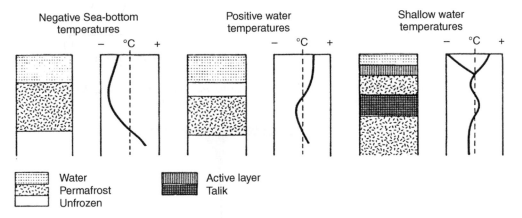

▦ Water			▮ Active layer	
⣿ Permafrost			▦ Talik	
☐ Unfrozen				

Figure 2.14 Thermal conditions of disequilibrium beneath various types of water bodies (after Mackay, 1972; Hollingshead *et al.*, 1978). The shallow water differs from the other cases by becoming frozen to the bottom each winter.

and Prudhoe Bay were obviously subjected to an MAGST of −12°C for a very long time during the Wisconsin Glaciation, followed by a warming of about 4°C. Then a cooling of about 1°C took place, probably corresponding to the last Neoglacial event. Similar disequilibrium conditions have been reported from Spitzbergen (Liestøl, 1977), with a warm period from 1920 to 1960 A.D.

Disequilibrium temperature profiles are also found beneath water bodies, *e.g.*, Figure 2.14, with extreme examples occurring in offshore permafrost and beneath thaw lakes. An example of disequilibrium conditions beneath a 60 m wide thaw lake in the Beiluhe Basin on the Tibet Plateau is described in Ling *et al.* (2012). It is less than 2 m deep and modeling suggests that it will be 733 years before an open talik becomes established (see Figure 2.15). In this case, the model indicates that the open talik will continue to expand, though at a steadily decreasing rate for over 1500 years. Similar models have been used in Alaska and Canada (*e.g.*, Mackay, 1979a), but they all necessarily assume that the present-day climatic conditions will continue unchanged for the entire length of the period studied in the model, in this case, over 1500 years.

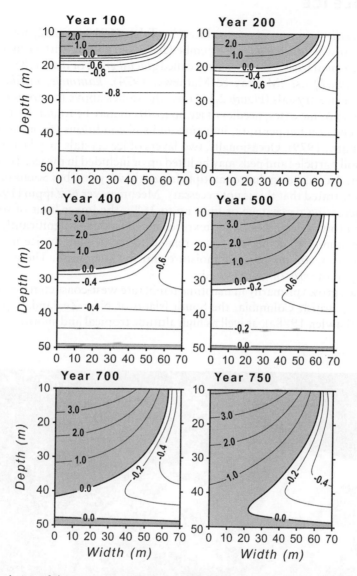

Figure 2.15 Evolution of the ground temperature profile beneath a thaw lake on the Tibetan Plateau by modeling (Ling *et al.*, 2012). The unfrozen ground is in orange.

Given the recent frequent climatic fluctuations in the region, this is unlikely to occur, but these models give some idea of the effect of the greater absorption of solar energy by water, which is approximately five times that of soil (Pavlov, 1999; Harris, 2002). In summer, there is an additional warming effect of water on the underlying ground due to mixing of the water so that its temperature does not change with depth. For this reason, thermokarst is very difficult to stop, once it begins (see Chapter 10). In winter, the water freezes downwards from the surface.

2.7 NEEDLE ICE

The term, *needle ice*, was first introduced by Taber (1918, p. 262) for elongate ice crystals that grow beneath stones lying embedded in the ground surface in maritime climates with seasonal frost. It has also been called *glacial grass* or *druza* (Baranov, 1949), *pipkrake* or *mushfrost* (Mackay & Mathews, 1974), *kammeis, eisfilamente*, and *hoar frost*. The ice crystals (Figure 2.16) raise the stones above the ground, usually by 0.5–3 mm. In extreme cases, needles of ice up to 40 cm may be found (Krumme, 1935). The size depends on temperature, moisture and soil conditions (Soons & Greenland, 1970; Washburn, 1979). Occasionally, two layers of ice crystals may be formed (Troll, 1944), and soil particles and peds may be lifted up or included in the ice. Troll explained the multi-tiered ice as being due to multiple freeze-thaw cycles, but Soons & Greenland (1970) demonstrated that this is not necessary. Meentemeyer & Zippin (1981) showed that with increasing percentage of fines up to 16%, the lower limit of soil moisture required for growth decreases. Finer textured soils produced significantly thicker ice crystals. The quantity of soil lifted was related to ice height, the ice mass to atmospheric void ratio, soil texture, soil moisture and soil roughness. There seemed to be no selectivity in the size of grains lifted.

Needle ice grows primarily in maritime temperate west-coast climates, *e.g.*, Alaska (Figure 2.16), British Columbia, the South Island of New Zealand, Japan, and the British Isles (Lawler, 1988a), as well as high altitude tropical and subtropical mountain

Figure 2.16 Needle ice about 1.5 cm long that had lifted a stone weighing about 1 kg out of the soil on tundra during an overnight frost in the discontinuous permafrost zone along the Denali Highway, Alaska. © S. A. Harris.

ranges such as the Andes, the east African Rift Mountains, the Drakensberg (Lawler, 1988b; Grab, 1999), Colorado (Fahey, 1973, 1974), and Tennessee (Matthews III, 1999). They are less frequent in areas of permafrost, though they can develop there during spring, summer and early fall if climatic conditions are right (Figure 2.16).

Substrates include stony loams and volcanic ejectamenta, *e.g.*, Mount Garibaldi (British Columbia) and in Iceland, and they develop in a few hours in a single night. The stones have significantly greater thermal conductivity than the dry, surrounding surface soil, so the freezing plane develops first beneath the stones. Moisture then moves to the freezing plane from the adjacent soil. Maximum crystal development occurs on frosty nights when the stones lie on wet soils with high silt content to provide the abundant water supply. Cobbles up to 15 kg in weight can be lifted (Mackay & Mathews, 1974). Next day, thawing of the ice on the sunny side of the ice pillar causes the stone to fall off. This process can cause significant sorting of material through differential heaving (Hay, 1936; Troll, 1944, 1958; Gradwell, 1957; Fahey, 1973), as well as contributing to down-slope movement and erosion (see Chapter 9). Outcalt (1971a, 1971b) and others have modeled the needle ice growth.

2.8 FROST HEAVING

The increase in volume of 9% when liquid water changes to ice has inevitable consequences, even without additional movement of water to the freezing plane. Since the latter can be considerable, an obvious space problem is developed. Initially, the ice tends to cement the grains together, but the greater volume of ice compared to water causes the sediment to become pressed together, *i.e.*, it becomes more dense. When more ice is formed, the pressure eventually equals the overburden pressure, and any further formation of ice results in heaving of the overlying layers in proportion to the excess ice (Taber, 1929, 1930a, 1930b) using the energy provided by the latent heat of fusion. This uplift is called *primary heaving*. The amount of heave increases as more water moves to the freezing plane. The direction of heave is parallel to the direction of crystal growth, which is indicated by the direction of elongation of inclusions and air bubbles (Gell, 1974). The rate and degree of cooling of the adjacent air do not correlate well with the amount of heaving, and the prediction of the amount of heave to be produced has proven to be very difficult (Abzhalimov, 1982; Anderson *et al.*, 1984; Grechischev, 1984; Chan, 1984), though it can be calculated (Orlov, 1962). The available water moving to the freezing plane and the amount of heat flow to the surface of the soil are the complicating factors. Unfortunately, these cannot readily be predicted.

Telescoping tubes (heave-meters) installed at different levels can show the amount of primary heave at different depths. The movements can also be matched with ground temperatures and read frequently during the year. Primary heave has been found to occur throughout the winter when the ground is frozen (Mackay *et al.*, 1979; Smith, 1985b). Likewise, magnetic sensors on a probe at 40 and 58 cm depth have confirmed an increase in distance of separation of 1.25 cm between 8th April, 1984 and 30th May, 1984 at Illisarvik (Mackay & Leslie, 1987).

Primary heaving usually amounts to about 60–70% of the eventual total heave. Continued study of the material after the primary heaving has been relieved shows that the material usually undergoes *secondary heaving* (Miller, 1972). This is believed to be caused by redistribution of water within the frozen ground resulting in accumulation of more ice in the profile. These changes alter the physical and mechanical properties of the soil, usually adversely. When the soil thaws, the ice melts and the sediment consolidates into a more dense deposit.

Saline soils adjacent to the ocean exhibit only about half the heaving of non-saline soils (Chamberlain, 1983). There are also far fewer ice lenses. Sea water causes the formation of a thick active layer with many growth sites for ice lenses and produces patches of unfrozen water with high concentrations of brine. The freezing point is lower and the brine-rich zones are a potential zone of low shear strength. However, the quantity of salts present varies with time (Figure 1.18),

Frost heaving also occurs in massive, jointed rock (Dyke, 1984). Yearly movements can be as great as 5 cm both vertically and horizontally. The movement is slow but continuous, unlike the primary and secondary heaving discussed above. Disruption of blocks and rupturing have been described from Baffin Island, affecting various rock types in a number of different types of unweathered rock outcrops (Dyke, 1978). These seem to be a special form of physical weathering of bedrock and heaving up of blocks by the growth of ice. Where the exposed rock has undergone weathering for 20,000–40,000 years, the heaving during a single winter may be up to 1 m. Frost wedging can also dislodge joint blocks from rock cliffs (Matsuoka *et al.*, 1997; Koštak *et al.*, 1998; Matsuoka, 2001).

2.9   DENSIFICATION AND THAW SETTLEMENT

Freshly deposited alluvial and deltaic sediments can contain up to 90% pore space (Pettijohn, 1949). When heaving of this sediment takes place in a closed system, followed by thawing, the amount of settlement may exceed the amount of heaving by 20% in the first few cycles (McRoberts & Nixon, 1975, p. 162). This is due to densification as a result of compaction of the sediment. Repeated freezing and thawing will continue to compact the soil up to a stage where there are few, if any free pore spaces in the material. A similar consolidation takes place during drying (Williams, 1967, pp. 1–10). Inevitably this alters the thermal and geophysical properties of the soil.

Hollingshead *et al.* (1978) have described over-compacted sediments from the Mackenzie Delta. They are found in the active layer of sediments on the older land surfaces and their compact, brittle structure is in marked contrast to the new sediments currently being deposited on the Delta. The compaction parallels the surface and is an example of the type 2 fragipan, lacking cementation (see Chapter 1). Over-compaction is also present in varying degrees in the surface layers of the sediments around Calgary, Alberta, Canada.

2.10   CRYOSTRATIGRAPHY, CRYOSTRUCTURES, CRYOTEXTURES AND CRYOFACIES

Cryostratigraphy involves the study of the layers of frozen ground in the upper layer of the Earth. It involves understanding how certain geometric relationships between rock layers arise and what these geometries mean in terms of their depositional environment. *Syngenetic permafrost* refers to permafrost that is formed essentially at the same time during the deposition of the host material (ACGR, 1988), as opposed to *epigenetic permafrost*, which is where the permafrost developed some time after the material was deposited. *Cryolithology* refers to the lithology, distribution and ice amounts in the ground (Popov *et al.*, 1985; Melnikov & Spesivtsev, 2000).

Cryostructure refers to the shape and distribution of ice within the host material making up frozen ground (Murton & French, 1994, p. 738). These authors use *cryotexture* to refer to the size and shape of the ice crystals. Unfortunately, there is always a mixture. Size, shape and roughness are structural parameters, while particle distribution in space and the relationship between components constitutes the texture (Shumski, 1964, p. 6). Therefore *cryostructire* is the size, shape and roughness of ice inclusions, and their distribution in space and the relationship between them is *cryotexture. Cryofacial analysis* was suggested by Katasonov (1969) following the well-established descriptive terminology and concepts developed early in the 20th century for sedimentary rocks in Western Europe and North America (Pettijohn, 1949; Bates & Jackson, 1987, p. 681). This terminology is important for producing standardized descriptions of permafrost profiles and of the ice within them. Given the fact that there is almost continuous movement of water in the cold ground under the influence of temperature gradients, it is unclear how stable cryotextures really are.

Murton & French (1994) recognized seven cryostructures in the frozen sediments of the Tuktoyaktuk lowlands, *viz.*, stuctureless (Sl), lenticular (Le), layered (La), regular reticulate (Rr), irregular lenticulate (Ri), crustal (Cr), and suspended (Su). These cryostructures can be *transitional* (meaning they are partly one type and partly another, *e.g.*, irregular reticulate and lenticular), or *composite* (meaning two structures occurring together, *e.g.*, structureless and crustal in a pebbly sand). The layered and lenticular categories were further subdivided into planar, wavy or curved bedding, that was further divided into parallel or non-parallel bedding, adopted from Collinson & Thompson (1989, Figure 2.6). They also used five cryofacies types (Murton & French, 1994, Table 1) that were arbitrarily distinguished based on apparent volumetric ice content. These can be further subdivided on the basis of the sediment in which the ice occurs. In practice, these could be grouped into *cryofacies assemblages* that were used to describe individual sedimentary formations. However, the relationship of cryostructures to the obvious frequent movements of large quantities of soil moisture as well as the age of the deposits remains to be determined.

2.11 GROUND CRACKING

When any material is cooled, it tries to contract. When a sheet of brittle, frozen sediment is cooled, the build-up of strain by contraction eventually exceeds the tensile strength of the material, resulting in the development of vertical fractures (*frost cracks*). These tend to form polygonal patterns if the contraction is large enough. The same pattern can be produced by drying sheets of swelling clays on a river bed, or by cooling of a lava flow. Sometimes the cracking in the Arctic results in loud, booming noises (Leffingwell, 1915; Lachenbruch, 1962), though this loud sound has been found to be rather unusual. In permafrost, Harris (1982a) found that it requires a minimum of 1000°C days of seasonal freezing index (the sum of the negative mean daily air temperatures during a given winter), but cracking will not occur if there is sufficient insulation by the snow pack (Mackay, 1978). Cracking can also occur in areas lacking permafrost during very cold winters, *e.g.*, Amsterdam, Holland (Washburn *et al.*, 1963). They are found in warmer climates in peats than in mineral soils, and are the first stage in the development of *ice, sand, loess and rock tessellons* (*polygons*, Chapter 5). Frost cracking can be found more often on artificial islands, gravel roads and building pads as well as on the tundra.

In the Arctic and Antarctic, frost cracking occurs each year in at least some of the polygons, depending on the severity of the winter cooling. In the autumn, the cold weather and ice produce a thick brittle, frozen mass. When this is subjected to sudden strong cooling without an adequate insulating layer of snow, cracking takes place. On Garry Island with a Mean Annual Ground Surface Temperature (MAGST) of $-7°C$, 40% of the ice wedges cracked by mid-March (Mackay, 1974a; Mackay & Mackay, 1974), whereas at Barrow, Alaska (MAGST of $-9°C$), about 50% of the ice wedges cracked (Black, 1963, 1974). One metre of snow cover is the critical thickness to prevent cracking on Garry Island. Many of the large wedges in the boreal forest and taiga in unglaciated Alaska and Siberia are relicts from the Wisconsin Glaciation (Péwé, 1966). In the Mackenzie Delta, the maximum depth of cracking is 5 m, but in northern Siberia, cracking may have extended down to much greater depths, based on the depth of penetration of fossil ice wedges (see Figure 1.24), possibly during major glaciations (Popov, 1962; Shumskiy & Vtyurin, 1966; Dostovalov & Popov, 1966). The latter paper reported that many of the wedges were syngenetic, *i.e.*, the cracking zone moved upwards as more and more sediment was deposited. If that is the case, then the depth of frost cracking during major glaciations would be similar to the present-day features. Dostovalov & Popov also interpreted the diameter of the polygons as indicating the severity of the climate, the largest diameters indicating a less cold climate. Under more severe climates, they suggested that secondary and tertiary cracking occurred inside the primary polygon. Cracking in the larger diameter polygons commences at about 3 m depth and spreads upwards and downwards. The average width of the cracking of the ground in subsequent years usually occurs in the same place. Sometimes the cracks close up in the spring without becoming filled with extraneous material. The average width of the cracks at the surface on the Mackenzie Delta is about 1 cm. If the cracks become filled with extraneous material such as ice or sand (Figure 2.17), subsequent cracking does not necessarily occur in the same place. What is unknown is whether there are additional smaller cracks with no obvious surface expression. It is now realized that these crack patterns represent planes of weakness in sediments that can act as convenient pathways for the upward injection of super-saturated, relatively low density sediment where there is a suitable frost susceptibility gradient in the active layer, as on Cornwallis Island (Washburn, 1997). It is possible that these may be involved in most load casting hexagonal-type patterns such as those described by Vandenberghe (1992) from the loose Pleistocene sediments from the Netherlands and northern Belgium, as well as those described from N.W. Svalbard by Van Vliet-Lanöe (1988). Hexagonal-type patterns occur preferentially in homogenic sediments in extensive areas, whereas orthogonal-type patterns tend to be associated with the natural orientation of river valleys, terraces and lakes.

Figure 2.17 Fossil crack picked out by iron staining, Richardson Roadhouse Gravel pit, Alaska. © S. A. Harris.

2.12 DILATION CRACKING

Dilation cracking refers to the development of straight or curved cracks in the soil over permafrost or seasonal frost mounds or icing blisters (Figure 2.18). On the latter, they can widen to about 10 cm at the surface and relieve the stress built up when the growth of an ice body or bodies underground up-domes the upper ground surface. The ground cracks from the surface downwards when the strain of trying to stretch the brittle, frozen ground over the expanding ice body exceeds its tensile strength. Curved cracks commonly occur on the crest of pingos (Chapter 7), radiating from the centre. In that case, they continue to widen during pingo growth and are often eroded by water. Dilation cracks are also abundant on palsas and lithalsas, where they are the locus of slides during decay of the mounds. Peat on palsas is split into blocks by dilation cracking and the individual surface blocks of peat tend to slide down the sides of the mounds.

Figure 2.18 Dilation cracks (black lines) on the surface of an icing mound, Alaska Highway east of the Donjek. River. © S. A. Harris.

2.13 FROST SUSCEPTIBILITY

This is a term that is used by engineers to indicate which soil materials will be subject to movement of soil water to the freezing plane, thus causing the development of ice lenses, heaving, upward movement of stones and other objects, and subsidence during thaw. There is no generally accepted criterion to characterize a frost-stable material, but minimum grain size, amount of fines, and degree of mixing (called grading) are key indicators. Casagrande (1932) suggested that it occurred in soils containing more than 10% finer than 0.02 mm. No ice segregation was observed in soils containing less than 1% of grains finer than 0.02 mm, even in waterlogged soils. Linell & Kaplar (1959) concluded that this was the best guideline, and the frost design classification system developed by the U.S. Corps of Engineers is based on it.

The problem is that some gravelly soils with 1% of material finer than 0.02 mm do heave, while some sandy materials with 20% of material finer than 0.02 mm do not. Accordingly it is necessary to carry out laboratory testing in order to arrive at a reasonably reliable conclusion. This is vital for successful design engineering in permafrost regions (see Chapter 12). Frost susceptibility also is involved in producing a variety of structures and landforms in permafrost areas (see Nichols, 1953; Mackay, 1953; Van Vliet-Lanöe, 1988; and Washburn, 1997). The presence of soil layers with different frost susceptibilities produces the variety of features discussed below under cryoturbation, gravity processes and injection structures that are so common in the profiles of present-day and past active layers. Coupled with frost cracking which provides a suitable pattern of planes of weakness, they are thought to play an important role in producing some types of patterned ground (see Chapter 9).

2.14   CRYOTURBATION, GRAVITY PROCESSES AND INJECTION STRUCTURES

When seen in a vertical exposure, the seasonally thawed active layer often exhibits structures (Figure 2.19) indicating a churning of the sediments, upturning of sediments due to injection of one layer into another, or blocks of sediment that have sunk into the layer below due to higher density (load-casting), *e.g.*, Washburn (1979), Heyse (1983), Van Vliet-Lanöe (1985, 1988, 1991), Vandenberghe (1988, 1992), Washburn (1997). As noted above, the active layer is usually very wet after the melting of the snowpack, and it appears that often these structures develop in this sodden state or during refreezing in the autumn.

Van Vliet-Lanöe (1985) introduced the concept of the frost susceptibility gradient in vertical sections. This has been further refined in subsequent papers (Van Vliet-Lanöe, 1988, Figure 11, 1991; Vandenberghe, 1992, Figure 6), with the addition of the effects of prior frost cracking as well as the effects of slope processes to explain the different types of structures and many forms of patterned ground (see Chapter 9).

2.14.1   Cryoturbation

Cryoturbation (Edelman *et al.*, 1936), *involutions* (Denny, 1936; Sharp, 1942b), or *brodelböden* (Gripp, 1926, in Jahn, 1975; Troll, 1944) all refer to distorted and deformed sediments that occur in the active layers of many permafrost soils (Figure 2.20). Deformations can occur in either the frozen or thawed state of soils, and these produce different results. The sources of distortions or deformations are of two different kinds, *viz.*, ice movement under gravity, and freezing resulting in volume

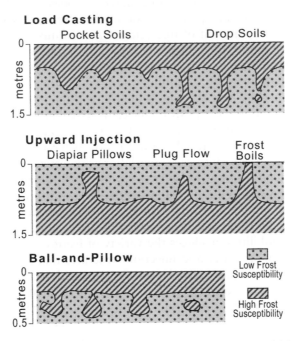

Figure 2.19 Schematic diagram showing the results of load casting, upward injection and formation of ball-and-pillow structures in soils with layers of sediments with high or low frost susceptibility (modified from Van Vliet-Lanöe, 1985, 1988).

Figure 2.20 Results of cryoturbation in a soil section.

expansion. These two major forces cause other mechanisms, for example, convection cells or deformation when the active layer is squeezed between a rigid, frozen surface layer and the underlying permafrost. The results of these are described below.

The term, "cryoturbation" is used here since it best conveys the idea that these features are developed under freezing conditions. They are developed in sediments with a uniform frost susceptibility throughout the active layer. As a result, load-casting, plug flow and injection of sediment from below will not normally occur. This is found in many Arctic soil profiles (Bockheim & Tarnocai, 1998). They are abundant in surficial Pleistocene deposits, particularly in non-glaciated areas that have been exposed to conditions under which permafrost probably developed (Heyse, 1983; French, 1996, p. 240; Owen *et al.*, 1998). In these situations, they are often associated with ice-wedge casts, and can then be used as one indicator of past permafrost conditions. However they can also develop by mass movements or flowage by solifluction-type processes on slopes (see Chapter 8), so they cannot safely be used as indicators of former permafrost unless there are other reliable indicators present, *e.g.*, ice-wedge casts or rock glaciers (Van Vliet-Lanoë, 1988). In areas with only a short period of time for permafrost to be present following deglaciation, cryoturbation features are present but very subdued (Everett *et al.*, 1971).

Cryoturbations can be developed by two different mechanisms:

(a)   "Cryostatic pressures" created when the active layer is squeezed between a rigid, frozen surface layer and the underlying permafrost during refreezing of the active layer (Kesseler, 1925). It may explain cryoturbations in poorly to imperfectly drained soils underlain by permafrost, but does not explain cryoturbations in non-permafrost soils or in well-drained soils, unless there has been a change in climate.

(b)   The convection theory (Nordenskjold, 1909), suggesting that there are convection cells that develop in the active layer. It has been invoked by Low (1925), Gripp (1926), Romanowsky & Cailleux (1942), Ray *et al.* (1983), Hallet & Prestrud (1986); Harris (1988) and Van Vieit Lanöe (1991). Rate of movement is slow with a typical cycle taking of the order of 300–400 years, but it is supported by field measurements in some well-drained sites such as Plateau Mountain and Svalbard.

They can also occur together with the other processes discussed below.

Cryoturbations are most obvious in stratified fine-grained sediments, but they can be found below the former soil surface developed in a wide variety of materials including frost-shattered sandstone (see Figure 1.13). Differential vertical movements in the active layer can be demonstrated by emplacing telescoping nested rods of different lengths (Figure 2.21) in the ground at right angles to the surface. Differential uplift of the sections of rods exposed to the soil at different depths indicates differential movement with depth. By making repeated measurements of the changes at each level over time, a picture can be obtained of the location of these movements, together with the time at which they occur (Harris, 1998a). If certain depths show greater upward movement than others, this provides an indication of where material is moving laterally as opposed to a vertical direction.

Figure 2.21 Telescoping nested rods that can be used to measure differential heave in the surface layers of the soil. The rods are emplaced in a vertical hole with the black tape at the soil surface. The rods move vertically with the soil layer with which they are in contact.

Lateral movement can also be demonstrated by emplacing plastic tubes or columns or pegs or dowels in holes drilled vertically into the active layer down to the permafrost table. Mackay (1981a) successfully used this technique in studies of earth hummocks on slopes in the Mackenzie Delta. Results of these measurements indicate that active movements occur in the spring or early summer when the ground thaws from the surface downwards. The water is provided by downward percolating snow-melt and by thawing of the ice in the previously frozen soil. Moisture contents in the active layer at this time often exceed the liquid limit and there is usually a very high plasticity index, although low plasticity indices have been reported from the Swiss Alps (Furrer *et al.*, 1971). Movements in this case are down-slope and vary with depth.

The main time of churning together with down-slope movement is in the autumn during freeze-up. In the case of gentle slopes with earth hummocks (Chapter 7), this autumnal down-slope movement tends to be identical throughout the upper part of the active layer and is therefore called *plug flow* (Mackay, 1981). The plane of movement is the rising surface of the frozen ground at the base of the active layer. Rates of movement averaged over 10–13 years were about 0.2 cm/year on very low slopes and 1.0 cm/year on steeper slopes. The earth hummocks move down-slope, progressively burying the inter-hummock peat to produce buried organic material in the hummocks. Radiocarbon dates on these buried organic materials indicated that the mean residence time of the organic matter ranged from approximately 1100 radiocarbon years B.P. about 4 m upslope to around 2000 radiocarbon years B.P. on the flat area below the slope. On the flat upper surface of Plateau Mountain, differential movements in the soil below undisturbed alpine-tundra vegetation were up to 12 mm/a, averaged over three years. Only circular movements within the upper part of the active layer could explain the results (Harris, 1998a). These movements were closely associated with the development of non-sorted circles (Chapter 9). Similar results were obtained by Hallet & Prestrud (1986) in sorted patterned ground on marine terraces in Spitzbergen.

Bockheim & Tarnocai (1998) have argued that all soils overlying permafrost are affected by cryoturbation in some way. One important effect of this is that it mixes up the layers, redistributing organic carbon through the active layer (Sokolov, 1980; Rieger, 1983; Bockheim, 2007). The distribution of buried organic matter that has been radiocarbon dated indicates that cryoturbation was at a maximum during the Altithermal/Hypsithermal warm period about 6000 radiocarbon years B.P. Experiments using radionuclides show that vertical mixing rates vary from insignificant to a few centimetres per year in a very short distance (Klaminder *et al.*, 2011). The vertical movement of carbon containing silt greatly affects the fate of the carbon in permafrost soils, and cryoturbation will also redistribute nutrients throughout the soil.

2.14.2   Upward injection of sediments from below

These are sometimes found in the active layer of present-day arctic soils, but are common in Pleistocene sediments. They consist of hexagonal polygons (Van Vleit-Lanöe, 1988; Vandenburghe, 1992), or else pillars or ridges of material squeezed up from an underlying sediment with higher frost susceptibility, *e.g.*, Washburn (1997) from Resolute Bay, Cornwallis Island. Washburn concluded that it was the result of differential frost heave of adjacent sediments of different textures and frost susceptibility in the active layer, as first suggested by Sharp (1942). The larger expansion of the highly frost-susceptible lower layer causes compression of that sediment, and the pressure is relieved by upward or lateral injections of sediment into an upper layer with low frost susceptibility (Van Vleit Lanöe, 1965, 1988a). It sometimes involves liquefaction during the thawing of the lower later. Dylikowa (1961), Corte (1972), Pissart (1976; 1982), and Van Vleit-Lanoë (1985, 1988) have invoked this origin or some variation of it. Some examples are shown in Figure 2.16.

The polygonal pattern or dyke-like form is interpreted as evidence of movement along planes of weakness developed by earlier frost cracking. These features are formed entirely within the active layer, but may or may not reach the surface of the ground. They are the rarest of the three deformation structures found in permafrost regions, but can also occur in poorly drained, non-permafrost areas. Good, present-day examples are known from Svalbard (Van Vleit-Lanöe, 1988), Cornwallis Island (Washburn, 1997), and the Mackenzie Delta (Mackay, 1963). They are also described from Pleistocene/Early Holocene sediments from Illinois (Sharp, 1942).

2.14.3   Load-casting

Load-casting (Figure 2.22) is associated with thawing (Mortenson, 1932; Kuenen, 1958), and has been demonstrated in small-scale experiments by Jahn & Cerwinski (1965) and Cegla & Dzulinski (1970). It occurs where an upper layer with high frost susceptibility overlies another layer with low frost susceptibility within the active layer. The volume of both sediments taking part in the movement should be equal

Figure 2.22 Load casting at Mengyuan, China (37°38′N, 101°10′E), at 3148 m. in the 5 m terrace. The measuring rod is 5 m long. © S. A. Harris.

unless the load-casting is being affected by additional processes. Vandenburghe (1992) regards load-casting as being more common than upward injection of sediments from below. It produces spectacular features that French (1996, Figure 14.6) following Eissmann (1994) has called **drop-soils, pocket soils**, and **ball-and pillow structures** (Figure 2.17).

Maximum thickness of the load cast appears to be about 3 m. They are known from Pleistocene/Early Holocene sediments in northern Belgium and Holland (Vandenburghe and Van den Broek, 1982; Gullentops and Paulessen, 1983; Huesse, 1983; Vandenberghe, 1983), central Europe (Strunk, 1983), the Mackenzie Delta (French, 1986), Mongolia (Owen *et al.*, 1998) and China (Figure 2.22). Present-day examples are described from Spitzbergen.

Where the underlying, less frost susceptible sediment consists of outwash gravels, these may include buried blocks of ice (Harris *et al.*, 2017). At the section at Mengyuan which lies along a valley on the eastern slopes of the Qilian Mountains, some of the load-casting features exhibit a sandy silt infilling extending down almost 5 m into gravels that were deposited during the early part of the last glaciation in that area. The upper part of the infilling exhibits the characteristic equal movement of the two sediments, but this relationship grades down into a zone where the sandy silts form a cast of the former ice blocks, exhibiting a sharp boundary between them and the enclosing, undisturbed gravels. The shape of the infilling is no longer wedge-shaped,

Figure 2.23 Section showing the sharp boundary between the sandy silt infilling that is forming a cast of a former ice block and the undisturbed outwash gravels in the lower part of a load-cast structure at Mengyuan, China. © S. A. Harris.

but has the angularity and shape of the former ice blocks (Figure 2.23). In this case, the structures can extend downwards to at least 4.5 m. It is postulated that as the climate ameliorated, the active layer extended downwards but the water from the melting ice in the sandy silts could not drain away due to the underlying permafrost. Eventually, the thawing reached the top of the ice blocks and the overlying soupy sandy silts descended into the spaces produced by the thawing ice, thus preventing collapse of the gravels surrounding the former blocks of buried ice.

2.15   UPHEAVING OF OBJECTS

Up-heaving or *frost jacking* of objects is commonly found in the active layer of permafrost regions, and occurs in sediments of all textures. It can occur at any depth in the active layer (Corte, 1962a, 1963a; Mackay, 1973b, 1984; Mackay & Burrows, 1979; Washburn, 1947). The first Russians that visited the Arctic in the 17th century described the area in a letter to the Tzar as "the land of such cold where even dead people jump from graves". As a result, coffins used to be placed in a pile of stones in Yakutia to avoid them being ejected from the ground. It disturbs and mixes up the layers at archaeological sites in permafrost areas (Johnson & Hanson, 1974; Johnson *et al.*, 1977), and is the reason for the northern limit of growing winter wheat in Canada. In areas of appreciable seasonal frost, it causes the fields on glacial till to "grow stones", which have to be removed periodically so that the fields can be ploughed in the Canadian Prairie Provinces. In more humid areas, it results in crushing and ejection of tile drains, and the forces involved can cause heaving of piles, culverts and concrete foundations (Harris, 1986a, pp. 44–45).

Washburn (1979, p. 86) regarded the heaving as being primarily autumnal in Greenland, and the up-heaving during a single freezing cycle is often proportional to the "effective height" of the object, *i.e.*, the vertical height of the stone above its greatest horizontal diameter (Hamberg, 1915, p. 609). This would correlate with the stone being pulled from above, called the *frost-pull theory*. First proposed by Högbom (1910, pp. 53–54), it suggests that the slowly descending freezing front causes ice lenses to develop in the soil around the top of the stones, freezing to them. As the ice lenses enlarge, they lift the stone upwards leaving a cavity beneath. The uplifting ends when the freezing front passes below the widest part of the stone. During thawing, the soil slumps into the space beneath the stone, thus completing the upward movement. Time-lapse photography confirms its effectiveness (Kaplar, 1965), and he produced a formula for calculating the maximum distance of heave in a given cycle (Kaplar, 1969, p. 36).

The second mechanism is called the *frost-push theory* (Högbom, 1914, p. 305). Since the thermal conductivity of stones is greater than that of the enclosing soils, the freezing front descends faster through the stone. Accordingly, ice will start to form a lens under the stone as water migrates to the freezing plane, and the stone will be lifted up while the surrounding finer material is still unfrozen. Bowley & Burghardt (1971), Corte (1962b, 1963a), and Mackay & Burrows (1979) have demonstrated the process using freezing from above and below in laboratory simulations.

Probably both processes occur together during a freezing cycle, and 5 cm uplift would be the typical total movement in a given year (Price, 1970). In areas of cold, seasonal frost, repeated freeze-thaw cycles during a single winter could produce even greater upward movement of objects, *e.g.*, Vorndrang (1972). However, upwards movement of objects can also occur during wetting and drying of soils (Springer, 1958; Jessup, 1960).

2.16 UPTURNING OF OBJECTS

Another characteristic of some permafrost environments is the rotation of stones so that their long axes become vertical. Vorndrang (1972) thoroughly mixed up the surface soils containing a mixture of grain sizes which previously exhibited 41% of the stone fraction with vertical axes. After the next winter, during which there were 46 freeze-thaw cycles, 38% of the stones had re-developed their vertical orientation. Harris (1969) found that the intermediate axes of the upturned stones faced in the same direction as the long axes of stones in the underlying till unaffected by the upturning. Thus the mechanism is not random.

The mechanism appears to be a modification of the frost-pull theory (Figure 2.24). The inclined stone is lifted and rotated due to the resistance of the soil around the lower

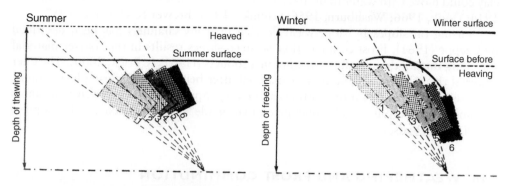

Figure 2.24 Diagram showing the process of upturning of stones (from Harris, 1986a).

part of the stone (Pissart, 1969, 1973; Schmid, 1955; Washburn, 1979). In the High Arctic, only one freeze-thaw cycle occurs in the active layer each year (Cook, 1966, p. 29) whereas at Plateau Mountain, there was an average of 16 cycles at 20 cm depth and 5 cycles at 50 cm depth each year. Thus the freeze-thaw cycles appear to increase in frequency during the winter at a given depth at lower latitudes in permafrost areas, though this could also be partly due to intensity and variability of the winter cold weather.

2.17 SORTING

The early work on *sorting* was extensively reviewed by Washburn (1956, 1973, 1979). Both needle ice and up-freezing of stones have already been discussed, as has the evidence for churning movements. Lötschert (1972, p. 7) described plants from the margins of sorted circles with tap roots that could be traced horizontally to the centre of the circle. Schunke (1975, p. 63) described stones sliding down the slope of circles towards the stony borders during thawing of the ground. Miller (1984) discusses evidence for thermally induced regelation which can permit movement of grains through ice. Ahumada (1986) reported ore minerals of high density accumulating at the base of the active layer after apparently sinking through the saturated soil during the spring thaw in Argentina. More recently, Van Vliet-Lanöe (1988) has demonstrated that the presence or absence of layers with different frost susceptibilities in the active layer greatly affects the nature of movement of material when the active layer is saturated (see Cryoturbations, *etc.*, discussed above).

In numerous studies, Corte (1961a, 1962a, 1962b, 1962c, 1962d, 1962e, 1963a, 1963b, 1966a, 1966b) and others have concluded that under the right conditions, finer material migrates ahead of the freezing front, while coarser material is more readily anchored by the freezing plane. Critical factors include orientation of the freezing front, rate of freezing, moisture content, together with the shape, size, and density of the mineral grains. Both horizontal and vertical sorting may occur. Ice can also transport small particles upwards under a freezing gradient towards the warm side (Hoekstra & Miller, 1965; Radd & Oertle, 1973; Römkens, 1969; Römkens & Miller, 1973), while clay could move with water to the freezing front (Thoroddsen, 1913, 1914; Johansson, 1914; Cook, 1966; Washburn, 1956; Schunke, 1975; Brewer & Haldane, 1957).

Sorting on slopes by slope-wash flowing in stony channels has been described by Czeppe (1961). Frost comminution on steep slopes results in the coarser material travelling further than the finer weathering products (Harris & Prick, 2000). The finer material then flows over the coarser material after being soaked during heavy rains. Whatever the cause or causes, sorting can be very rapid under the right circumstances (Strömquist, 1973). It can also operate on man-made structures such as tailings piles, gravel pads, roads, *etc.*

2.18 WEATHERING AND FROST COMMINUTION

Weathering is the breakdown of material *in situ* by physical, chemical or biological means (see Chapter 9). *Frost comminution* refers to a reduction in grain size of particles. It would produce very fine dust unless the molecular forces on the surface of the fragments cause the particles to accumulate as aggregates. This is why northern areas have large quantities of dust particles.

Both weathering and frost comminution occur in the active layer, and it has also been argued that rocks can be broken up within the permafrost itself (Harris, 1986a, p. 51) based on the depth of fractured rock exceeding the maximum depth of the active layer, *e.g.*, at Plateau Mountain, southern Alberta (~40 m broken rock versus ~18 m maximum active layer thickness). Ice-rich layers have been described in bedrock within the permafrost by Mackay (1999), and they obviously break apart the pre-existing mass of rock.

To prove that the broken-up surface material comes from the underlying bedrock, it is necessary to demonstrate that the end product has the same mineralogy, heavy mineral or clay mineral content as the bedrock. The weathering sequence of Jackson *et al.* (1948) is critical in the case of using clay minerals, and Willman *et al.* (1963) provide examples of clay mineral studies to determine the origin of Quaternary sediments in Illinois, while and Vogt & Larqué (1998) describe similar studies in Patagonia and the TransBaical area of southern Siberia. Such studies can indicate the type of climate during their formation.

There are several types of physical weathering that take place in polar and alpine areas, namely expansion of ice, hydration shattering, colloid plucking, ice segregation in micro-cracks, and growth of crystals in micro-cracks. In addition, thermal stresses have been suggested as a possible cause for break-up of rock (Hall & André, 2001; Hall *et al.*, 2002), although this has been disputed in the literature. However, Zhu *et al.* (2003) found evidence that this process can occur at low latitudes where the solar radiation is very intense and has acted on very cold ground during previous cold events (glaciations/permafrost extension southwards).

The material being subjected to weathering varies from massive igneous rocks to sedimentary materials, and the weathering occurs on the surfaces of both rocks and individual mineral grains. Chemical weathering can act on any exposed surface on the ground or in the active layer (Betelev, 1974). Although it is commonly assumed that the weathering processes in arctic climates are unique, Hall *et al.* (2002) have challenged this idea.

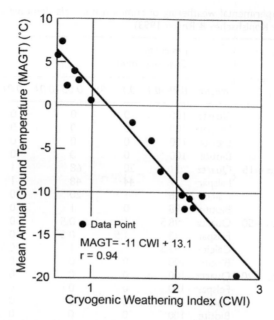

Figure 2.25 The relationship between mean annual round temperature (°C) and the cryogenic weathering index for soils collected from across central and Western Siberia (modified from Konishchev, 1998, p. 591).

Frost wedging of individual blocks can occur in the case of outcrops of massive igneous rock as discussed under the section on heaving (Dyke, 1978, 1984). The physical weathering of unconsolidated material has been studied in considerable detail, but even so, the details of the processes involved are still being debated. The original idea of water entering cracks in solid rock, expanding on freezing and breaking the rock apart was regarded as unlikely to be effective in most cases (Hallet *et al.*, 1991). The problem is to produce the pressure necessary to break up the rock when the moisture tends to move away to the freezing front, the water saturation level is less than 91%, and the system is not closed so that the necessary pressure cannot be developed. However Matsuoka (2008) demonstrated in field studies that crack widening occurred due to volumetric expansion, especially during refreezing of water derived from melting snow that had infiltrated cracks.

Konishchev (1998, p. 591) developed a *cryogenic weathering index (CWI)* based on the percentage of quartz and feldspar (%) in the 0.05 to 0.01 mm fraction. Using samples from across western and central Siberia, he found the relationship in Figure 2.25.

There is new evidence to indicate that the freezing of liquid water can produce frost cracking at all temperatures below 0°C (Girard *et al.*, 2013), based on acoustic emission measurements in steep bedrock (Girard *et al.*, 2012). Sustained freezing causes more damage than repeated freeze-thaw cycles, probably due to moisture moving to the coldest part of the rock near or at the surface. *Thermomechanical forcing* (rock thermal expansion and contraction) can occur through differences or irregularities in

Table 2.1 Results of experimental weathering of common minerals demonstrating the instability of Quartz (after Konishschev & Rogov, 1993).

Experimental conditions	Temperature Range (°C)	Mineral	Percentage Grain size (mm)				
			0.25–0.1	0.1–0.05	0.05–0.01	0.01–0.005	0.005–0.001
Before the experiment		Quartz	100	0	0	0	0
		Felspar	100	0	0	0	0
		Calcite	100	0	0	0	0
		Biotite	100	0	0	0	0
Freezing and thawing, Water saturated	−10 to +15	Quartz	11	20	68	1	0
		Felspar	7	44	48	1	1
		Calcite	6	1	20	30	29
		Biotite	98	0	1	0	0
Freezing and heating, dry	−10 to +50	Quartz	98.5	1	0.5	0	0
		Felspar	98	1.5	0.5	0	0
		Calcite	93.5	5	1	0.5	0
		Biotite	100	0	0	0	0
Wetting and drying at laboratory temperatures	+18 to +20	Quartz	100	0	0	0	0
		Felspar	100	0	0	0	0
		Calcite	100	0	0	0	0
		Biotite	100	0	0	0	0

the temperature field, and is known to contribute to fracture propagation at a wide range of depths (Gischig *et al.*, 2011). Acoustic emissions under freezing conditions are about two orders of magnitude greater than under thawing conditions, *i.e.*, they are due to freezing-induced processes. The statistical properties of acoustic emission events correspond to *microfracturing* (Amitrano *et al.*, 2012). However, the details of the silt-sized sediments are extremely common in areas that have been subjected to intense frost action, *e.g.*, in loess. Konishchev (1973, 1978) argued that these silt particles were the product of frost comminution. Konishchev *et al.* (1975), Konishchev (1982) and Konishchev & Rogov (1993) have analysed the silt sized particles (between 0.05 and 0.01 mm diameter) in loess-like sediments and have shown that they are made up of fractured primary minerals such as quartz and feldspar. However feldspar crystals mainly break down to 0.1 to 0.05 mm in size (Table 2.1). The surface of individual quartz grains often exhibit a conchoidal fracture. The products produce material with a grain size distribution obeying Rosin's Law of Crushing, as do those of hydration shattering.

Hydration shattering is the ordering and disordering of water molecules in sorptive interactions with water vapour and liquid water. It can produce shattering in shales, siltstones and argillaceous carbonate rocks (Dunn & Hudec, 1965, 1966, 1972; Hudec, 1974; Fahey & Dagesse, 1984). The presence of even small quantities of montmorillonite, which can expand to 20 times its dry volume when wetted, is an important factor to be considered. Repeated sorption and desorption of unfrozen water cause expansion and contraction, resulting in fatigue in sorption-sensitive rocks. This ultimately results in fracturing of the particles, but Potts (1970) and

Guillien & Lautridou (1974) have provided examples where silt and clay are not the end products of hydration shattering. This may be due to swelling clays becoming closely attached to the surfaces of mineral grains when dry, resulting in plucking of fragments of the underlying material when they are detached (*colloid plucking*). White (1976a, 1976b) has suggested that hydration shattering operates on block slopes and in most areas of mass wasting phenomena.

Walder & Hallet (1985) suggested an ice segregation model, in which the expansion is due to water moving to the freezing plane in the pores of the rock to provide extra pressure in the microcracks and relatively large pores within the body of the individual rock fragment. Fukuda (1983) has shown that water does migrate to these freezing planes, Prick *et al.* (1993) have demonstrated the dilation of rocks during freeze-thaw experiments, while other freeze-thaw experiments in the laboratory have resulted in frost-induced spalling (Letavernier & Ozouf, 1987; Matsuoka, 1990). However the details as to the stage and timing of the breakup have not been established.

Chemical weathering of the particles or rocks in the active layer tends to produce clay minerals (B. Meyer in Semmel, 1969; Leshehkov & Ryashchenko, 1973; 1978). It must be remembered that during weathering of the sediments in the active layer and seasonally frozen ground in permafrost regions, removal of significant amounts of solute can occur (Rapp, 1960a; Lewkowicz & French, 1982a; Thorn *et al.*, 2001) and can amount to 8 to 30 times that of the suspended sediment being removed. The presence of vegetation can increase this loss of soluble chemicals (Moulton & Berner, 1998), and even the growth of scattered lichens can greatly enhance weathering of minerals to produce additional soluble salts. McCarroll & Viles (1995) showed that the lichens penetrate the rock surface causing flaking of the rock, and increase the rate of surface lowering by up to 50 times compared with the rates for bare surfaces. They can also exude chemicals that react with the minerals in the rock to form crusts (Walton, 1985; Weed & Norton, 1991).

Ping *et al.* (1998) have reported that arctic soil profiles along the Haul Road to Prudhoe Bay are characterized by medium texture, poor drainage, and high organic matter content. The base saturation in the active layer decreases southwards as exchangeable aluminium increases. Where there is negligible cryoturbation, the carbonate and clay tend to be leached down the soil profile, where they accumulate above the permafrost table.

In Karkevagge, Sweden, the amount of chemical weathering depends considerably on the drainage. The soils on ridge-crests show minimal removal of dissolved solids in solution, but more secondary minerals (Thorn *et al.*, 2001). Total dissolved solids increase rapidly in the soil water and runoff downslope. Meanwhile, the valley floors show multiple thin buried soils due to the active slope processes. There were also coatings on rocks of several types including iron films, gypsum and calcium carbonate. On the North Slope of the Brooks Range in Alaska, the sediments in the mountains contain 58% calcite, but the carbonate decreases steadily down the slope towards the sea in the alluvial deposits, suggesting progressive leaching of soluble chemicals during transport (Robinson & Johnsson, 1997). However, it may also be the result of attrition of the softer calcite during transport.

All these solutions of chemicals can react with the minerals through which they pass, producing new minerals which have a different density and volume, thus subjecting the weathering rock or mineral grains to stresses that can lead to physical breakup by spalling of the outer layers. Thus plagioclase feldspar (found in most igneous rocks) reacts with water and carbon dioxide to produce calcium carbonate and silicic acid. Furthermore, the salts in solution can crystallize out in the pores of the rock during desiccation and the resulting stress can break up the rock fragments more efficiently than pure water (Fahey, 1985; Jerwood *et al.*, 1990a, 1990b). Examples of spalling, granular disintegration and honeycomb weathering have been described by Calkin & Cailleux (1962), Cailleux & Calkin (1963); Czeppe (1964); Selby (1971) and Watts (1983). These are particularly common in the Dry Valleys of Antarctica. The processes involved have been described by Williams & Robinson (1991).

These chemical reactions also occur on the rocks lying on the surface of the ground. Weathering of the rock surface produces a rind which then builds up a quantity of cosmic-ray-produced isotopes such as ^{36}Cl, ^{10}Be, and ^{26}Al which can be used to measure the duration of exposure of the rock surface to the atmosphere (Cerling & Craig, 1994). This permits dating of rock surfaces in glaciated areas which has been used to estimate ages of glaciations, *e.g.*, in Norway (Stone *et al.*, 1998), New Zealand (Chinn, 1981), and in Antarctica (Schaefer *et al.*, 1995).

The process of aggregation of the fractured material into dust of specific grain sizes as in the production of soil peds is not well studied.

2.19   KARST IN AREAS WITH PERMAFROST

Since calcium carbonate increases in solubility with decreased temperatures, one might expect karst topography to be exceptionally well developed in areas with permafrost. The precipitation interacts with the active layer and any taliks, but the solubility of limestone is always quite low, and once the water is saturated with the chemical, no further solution occurs. Increased solubility of carbon dioxide at low temperatures results in increased acidity of both the surface and underground waters. This results in increased sequestering of carbon dioxide in the Arctic Ocean, and potentially increases the chemical reactions in the soils.

At Sunshine Meadows, Banff National Park, Alberta, the surface water is fully saturated with calcium carbonate within the first kilometre, and there is little change in load along the Saskatchewan drainage system from there to Hudson Bay. Snow banks may not contain much carbon dioxide (Smith, 1972), so that solution as bicarbonate is similar to that elsewhere. A further complication is the ice-rich permafrost that can effectively seal out the surface water from the underlying rock in areas of continuous permafrost. In such areas, the solution of calcium carbonate is limited to the active layer or to tors (Bird, 1967, pp. 257–270). Likewise areas with gypsum deposits in areas of continuous permafrost lack evidence of appreciable solution, *e.g.*, on Ellef Ringes Island (St. Onge, 1959), in contrast to deposits further south in warmer climates.

Ford (1984, 1987) described four different landscape types from the extensive limestone bedrock of northern Canada. In the Arctic Islands, the continuous permafrost ensured that the karst is post-glacial in age and confined to the active layer. It is best developed along escarpments where drainage in the active layer is deeper. The surface of the limestone shows minor solutional features (karren). The southern Arctic Islands and northern mainland exhibit reticulate patterns of large "corridor" troughs, which Ford (1984) suggested were the result of sub-glacial drainage flowing beneath weakly

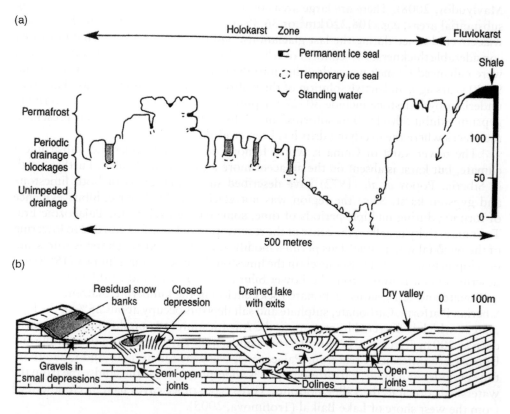

Figure 2.26   Karst features (a) in Nahanni, Northwest Territories (after Ford, 1987), and the Great Bear Lake region (after van Everdingen, 1981), and (b) Akpatok Island, Northwest Territories (from French, 1996). Large-scale karst features have been described at Vardeborgsletta, outer Isfjorden, Svalbard (Salvgsen & Elgersma, 1985).

erosive basal ice of the interior of the Laurentide ice cap. The glacier was probably "cold-based" in part, and frozen to the underlying rock. The third type occurs in the subarctic zone with discontinuous permafrost (Figure 2.20). This has resulted in the development of large fluvio-karst systems that have been described from the Great Bear Lake region (van Everdingen, 1981), the central Yukon (Harris *et al.*, 1983) and Akpatok Island (Lauriol & Gray, 1990). These exhibit closed depressions, depressions with open joints and dolines that fill with water at the end of the spring thaw, only to drain completely before the freeze-up in the fall. Well-developed karst springs are widespread, while karren and tors are common on bedrock outcrops. They may have developed over a long period of time and survived more than one glaciation. Figure 2.26 shows the type of features present in this environment.

Warm water coming to the surface from great depths has produced open taliks that have permitted the water to circulate through limestone rocks producing large-scale karst features such as dolines and ponds. The karst processes are active today aided by the cold climate in spite of the adjacent permafrost. Karst also occurs in European Russia (Shamanova & Uvarkin, 1973; Shavrina & Guk, 2005; Shavrina *et al.*, 2005;

Mavlyudov, 2008). There are large areas of limestone and gypsum bedrock covering substantial areas, *e.g.*, 106,220 km^2 or 36.3% of the Arkhangelsk region. The region was covered in ice during the last glaciations, and the glaciers left varying but often considerable thicknesses of sediment overlying a former karst plain. The karst processes were enhanced by melt-water draining into the underlying rock between 13,000 and 9,800 years ago, and active solution of the bedrock is still continuing today. The areas underlain by limestone include 30–270 depressions/km^2, while landscapes overlying gypsum exhibit 500–1500 mesoforms/km^2. A full range of karst features is present in the places where the overlying drift has been eroded away.

The tower karst of China is world famous near Guilin under the Sub-tropical climate, but karst is absent on the adjacent more arid Tibetan Plateau. Further north in Siberia, Popov *et al.* (1972) have described substantial areas of both limestone and gypsum karst. Since the region was not glaciated, this karst exhibits evidence of forming during multiple periods of time, some dating back to the Palaeozoic Era. Tectonic activity usually is the cause of the reactivation of karstification due to lowering of the regional water table. This permits modification of the existing karst features and development of karst at lower levels in the limestone bedrock. Thus Filippov (1998) has described karst features from the Lower Silurian and Lower and Middle Ordovician carbonate and terrigenous carbonate rocks enclosing the kimberlite diatremes on the Siberian Platform. Carbonate, sulphate and salt deposits occupy about 25% of the high plateau which is dissected by streams in the Angara source region flowing in valleys as deep as 200 m (Trzcinski, 1996). The rocks underwent karst erosion in the Middle and Upper Carboniferous, Cretaceous-Palaeogene, Neogene-Quaternary and Quaternary time periods. As a result, karst topography is widespread across the landscape from the watersheds down to the valley bottoms. Three types of ice caves have been described from the west shore of Lake Baikal (Trofimova, 2005).

Human-induced gypsum karst has been reported from southern Priangaria in Eastern Siberia (Trzhtsinsky, 2002). This is in an area of relic permafrost from the cold Pleistocene events. There were three stages in karst development in this region, namely, Middle Cambrian, pre-Jurassic and Quaternary (Vologodsky, 1975). Practically the whole thickness of the gypsiferous Angara Member has been subjected to karstification. Current chemical denudation rates are estimated to be between 0.02 to 0.08 mm/year, while the overall rate of surface denudation over the last 10,000 to 12,000 years is estimated to be 1 m. Typical landforms consist of sinkholes, caverns, blind valleys, karst trenches, and dissolution troughs (Red *et al.*, 1996). Construction of the Bratsk Reservoir, and the resulting fluctuating water levels has resulted in the development of evaporite karst.

Spector (2002) and Spector & Spector (2009) provide a good description of carbonate karst in the area of continuous permafrost along the Middle Lena River.

2.20   SEAWATER DENSITY AND SALINITY

Mean annual air temperatures in the Arctic usually fall in the range of −10 to −15°C. Pure water has its maximum density at about 4°C, causing the temperature in deep lakes to be close to this value. As a result, freshwater Arctic lakes can freeze to considerable depths.

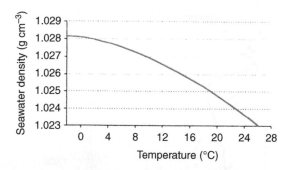

Figure 2.27 Variation in seawater density as a function of temperature for a salinity of 35⁰/₀₀ (modified from François, 2002).

In the case of sea water, the density of the water continues increasing as the temperature decreases (Figure 2.27, after François, 2004). This results in the entire column of sea water cooling to its freezing temperature since the lower part often has a freezing temperature below 0°C (Figure 2.28). Any ice that is formed in the water column rises to the surface due to its lower density, and protects the underlying sea water from further cooling. If this was not the case, the Arctic Ocean would have a temperature similar to the mean annual air temperature. As it is, much of the Arctic Ocean probably has sea bottom temperatures around 0°C. Near the surface, the oceanic water is influenced by solar radiation, turbulent heat exchange with the air, and phase transfers (evaporation, precipitation, freezing and thawing). This layer near the surface is called the *mixed layer*, and is also affected by waves and ocean currents. The depth of the mixed layer is usually about 10–20 m in summer, but is greater in winter due to cold surface water sinking due to its higher density.

Below the mixed layer where the temperature and salinity change abruptly are the *thermocline* and *halocline*. These occur at different depths in the Arctic Ocean, unlike elsewhere. A complication is that in summer, a low-salinity water mass spreads over the continental shelf being supplied from the rivers as well as the melting surface. In winter, brine of high salinity is formed as this water freezes. The halocline separates the relatively fresh water from the warmer and more salty Atlantic sea water. The water from the Atlantic Ocean entering the Arctic flows underneath the halocline, transferring heat to the low-temperature, low salinity water layers in the Arctic Ocean (Carmack, 1986).

Bottom sea temperatures are affected by not only the present temperature regime of the water, but by the past history. Sub-sea permafrost in the Laptev Sea was formed under subaerial conditions during the last few glacial periods (Romanovsky *et al.*, 2005). Subsequently, it became submerged due to a post-glacial rise in sea level (c. 120 m in the case of the Last Glacial Maximum), causing a warming of about 17°C (Soloviev *et al.*, 1987; Kim *et al.*, 1999; Romanovsky *et al.*, 2005). Temperatures may have reached 4–5°C at the seabed (Miller *et al.*, 2010). Current evaluation of the present thermal state and stability of subsea permafrost is mainly based on modelling results (Soloviev *et al.*, 1987; Kim *et al.*, 1999; Delisle, 2000; Romanovsky *et al.*, 2005). Permafrost could be destroyed after inundation by upward degradation

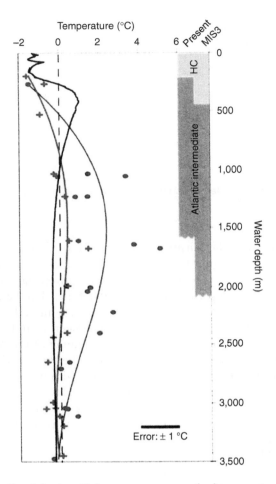

Figure 2.28 Depth profile of the late Holocene temperature (red temperature profile) and the MIS3 drillhole profile (blue line) from Amundsen Bay (Cronin *et al.*, 2012). The black line represents the modern temperature profile. The elevated borehole temperatures suggest a deeper halocline and an Atlantic Intermediate Water mass at depths of 1,000–2,500 m in the past.

under geothermal heat flux in the areas underlain by fault zones (Romanovsky & Hubberton, 2001b), or without excessive geothermal heat flux in border conditions. The dates of the the movements along the faults are unknown, as is their actual motion over time. Degradation could also be the result of the warming effect of large rivers (Delsile, 2000), or by changes in salinity and/or temperature at the sea floor. There were open taliks under large lakes and rivers even during the glaciations which could have remained unfrozen until today, or become frozen, depending on the salinization and temperature conditions after inundation. Similar attempts at modelling have been made for the permafrost dynamics of the Siberian Shelf (Romanovsky *et al.*, 2000) and on the North American Beaufort Sea (Mackay, 1972a; Osterkamp & Harrison, 1985; Nixon, 1986; Taylor *et al.*, 1996).

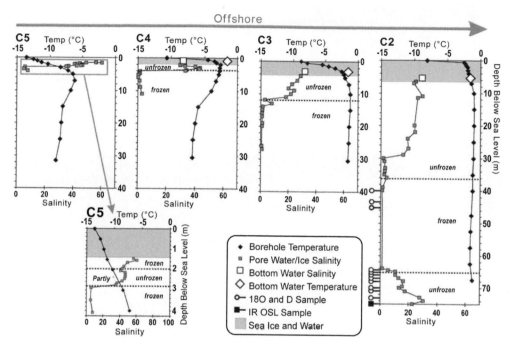

Figure 2.29 Salinity and temperature profiles of the borehole C2–C5 along a traverse of 12 km running perpendicular to the coastline at a water depth of under 6 m in the Laptev Sea (redrawn from Rachold *et al.*, 2007).

Taliks existing under the sea are important since they can also serve as conduits for methane to escape from the ground (Shakhova *et al.*, 2005; Shakova & Semiletov, 2007). This is believed to contribute to ***climate warming***. However these taliks may also indicate less likelihood of encountering gas hydrates when drilling, and will certainly eliminate the problems with subsidence due to thawing of the permafrost in the surface layers around production wells.

While there is new data on sea bottom temperatures during the Late Pleistocene (Miller *et al.*, 2010), there is limited data on the ground temperatures to help in predicting permafrost distribution and thickness. The fact that these are tectonic zones along plate boundaries adds to the unknowns. The air temperature data from the Vostok ice core refers to global air temperature variations, not those occurring in specific areas. What is needed is data on the changes in air temperature, air mass circulation, and precipitation in a given area during the Late Pleistocene.

The temperature of marine permafrost varies in short distances (Figure 2.29) and is significantly different to that found on adjacent land areas. Data from high-resolution seismic work in the Kara (Rokos *et al.*, 2001; Rekant *et al.*, 2005; Portnov *et al.*, 2013) and Pechora Seas (Bondarev *et al.*, 2002) suggests permafrost extends into areas below the 60 m isobaths, over 100 km from the present shoreline. Bondarev (1994) reported up to 115 m of permafrost below 4.5–29 m of unfrozen Cenozoic sands, silts and clays with negative temperatures.

Chapter 3

Factors affecting permafrost distribution

3.1 INTRODUCTION

As already noted, permafrost is defined as ground that remains below 0°C for more than two years. Determining this is difficult without actual temperature measurements since by merely looking at the ground surface, the temperature regime cannot be determined. Again, fluctuations in weather from year to year can change the status of frozen ground. Using actual temperature measurements and relevant observations, Brown & Péwé (1973) divided the factors affecting the presence or absence of permafrost into climatic and terrain factors which interact together to determine the actual distribution of permafrost. The climatic factors control the level of heat (temperature) and amount of heat (duration) applied to the surface of the Earth. Altitude, latitude, longitude and type, timing, and amount of precipitation are involved. Their combined effect is modified by the terrain factors, *e.g.*, local relief, aspect, vegetation, hydrology, nature of the substrate and fire. In addition, environmental changes over time and the activities of humans further affect the distribution of permafrost. The climatic factors are dominant in areas of continuous permafrost, whereas the terrain factors increase in importance as the percentage of ground underlain by permafrost decreases. Each of these factors will be discussed in turn in this chapter.

3.2 CLIMATIC FACTORS

3.2.1 Heat balance on the surface of the Earth and its effect on the climate

The actual ground temperature is a result of the heat balance on the surface of the Earth. This is dependent on two main factors, *viz.*, the geothermal heat flow and the solar radiation reaching the surface of the Earth. As noted in Chapter 1, the geothermal heat flow is small compared to the potential solar radiation, but can vary depending on the presence or absence of hot spots beneath the surface layers of the crust or the presence of plate margins.

Solar radiation comes to the surface directly from the sun (long-wave radiation) and as radiation from the atmosphere. A part of that radiation is reflected, and the fraction of solar energy (short-wave radiation) reflected from the Earth is called the *albedo* (α). Snow has a very high albedo, whereas dark coloured soil has a low albedo,

Table 3.1 Typical albedos of various surfaces (Dostovalov, B. N. & Kudryavtsev, 1967).

Surface	Albedo	Surface	Albedo
Limestone	56	Clay dry	16–23
Basalt	6	Clay wet	5–14
Granite	12–18	Grass green	16–27
Clay desert surface	29–31	Water (depending on sun position and condition of water surface)	2–78
Snow	30–85	Earth in total	40 (according to Budyko, 1980)

resulting in the surface of the soil absorbing most of the incoming energy arriving on its surface (see Table 3.1). Part of this absorbed energy is reradiated back into the atmosphere (*E*), the actual amount depending on the temperature (*T*) of the ground surface according to the ***Stefan–Boltzmann law***:

$$E = \sigma T^4 \tag{3.1}$$

The constant of proportionality σ is called the ***Stefan–Boltzmann constant***:

$$\sigma = 5.6703 * 10^{-8} \frac{\text{Watt}}{\text{m}^2 \text{K}^2} \tag{3.2}$$

The difference between the energy reaching the surface and that emitted from the surface radiation by reradiation is called the ***net solar radiation*** (***Rn***). It varies depending on the time of day, season, geographical position and micro-environment. Averaged over the year, net solar radiation (measured in the upper atmosphere) is negative for northern areas (Figure 3.1). On the surface, the negative net radiation only occurs in high latitudes or altitudes (Budyko, 1971). In the Tropics, there is an excessive amount of net solar radiation, and this energy imbalance is the driver of both the atmospheric and oceanic circulations. To maintain the mean annual air temperature in the Northern Hemisphere at the present-day levels, 30% of the heat absorbed in the Tropics must be moved polewards by air mass movements, surface ocean currents and the thermohaline circulation (Christofferson, 1994; Harris, 2002a). Any fluctuation in any of these processes can alter the mean annual air and ground temperatures which will result in a tendency to increase or decrease the active layer thickness and ultimately the temperature and thickness of the permafrost. Since energy and matter cannot be created or destroyed, one place will become warmer whereas another place will become proportionately colder, *i.e.*, it produces ***regional climate changes***. Cohen *et al.* (2012) showed that regional changes can also be produced by changes in the patterns of winter precipitation modifying the winter cooling on a regional basis.

One of the main causes of precipitation is the interaction between cold, dense, polar or sub-polar air and warm, humid temperate or tropical air, so the movements of the air masses are critical in maintaining the precipitation at any given location. Harris (2013a) discusses some of the main causes that are involved in producing climate change, listed in Table 3.2. The first order causes produce the largest temperature changes, reinforced by feedbacks from the second order causes.

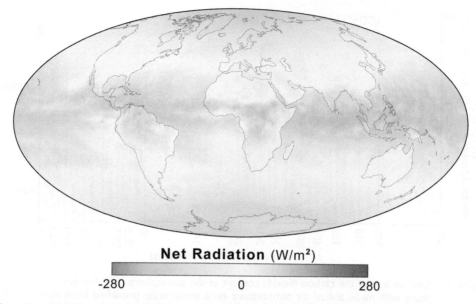

Net Radiation (W/m^2)

-280   0   280

Figure 3.1 Global net solar radiation as measured from space satellites. The measurements were made by the Clouds and the Earth's Radiant Energy System (CERES) sensors on NASA's Terra and Aqua satellites (Wielicki *et al.*, 1996).

Table 3.2 Some suggested controls of climatic change arranged into four orders based on the potential temperature change that they can cause (Harris, 2013a, with permission).

Order	Potential change in Temperature	Control
1st	20–30°C	1. Difference in heat absorption by sea and land as controlled by the position of continents and oceans.
		2. Changes in surface ocean currents and thermohaline circulations.
		3. Changes in Ocean Gateways.
		4. Plate Tectonics.
2nd	15–20°C	5. Eustatic sea level changes.
		6. Snow covering ice bodies.
		7. Changes in air masses.
3rd	5–10°C	8. Milankovich Cycles.
4th	<5°C	9. Fluctuations in CO_2 and greenhouse gases.
		10. Cosmic rays and fluctuations in the total solar irradiance.
		11. Variations in solar output; the Maunder cycle.
		12. Large-scale volcanic eruptions.
		13. Elevation of large tracts of land, *e.g.*, Tibet.
		14. Oscillatory ocean currents, *e.g.*, El Niño and La Niña.
		15. Other short-term cycles, *e.g.*, 2 and 7 years.
		16. Agriculture, deforestation and urbanization.

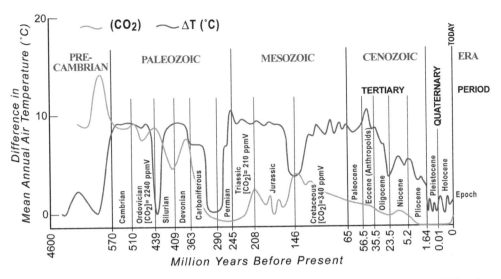

Figure 3.2 Comparison of the carbon dioxide content of the atmosphere over the past 4000 million years with mean annual air temperature on a linear scale (modified from Berner, 1990; Berner & Kothavala, 2001; Retallack., 2001; Ruddiman, 2001; Royer *et al.*, 2004; Pagani *et al.*, 2005).

The Milankovitch cycles are always present so long as the Earth and sun maintain their current trajectories. The fourth order causes produce minor fluctuations. These causes are all interacting at any one time, and the weather and climate merely show the product of all these factors. With so many processes oscillating with differing periodicities or gradually changing with time, it is no wonder that the weather is so unpredictable.

It has been proposed that the decreasing level of CO_2 was a primary driver of Phanerozoic climate (Royer *et al.*, 2004). Carbon dioxide is an unusual gas in that its solubility in water increases with decreasing water temperatures. The Earth has oceans and seas covering approximately 70% of its surface, so a small lowering in temperature of the Earth will result in considerable quantities of carbon dioxide leaving the atmosphere to become sequestered in the water bodies. This reduces the carbon dioxide content of the atmosphere. Conversely, a slight increase in the temperature of the Earth will result in release of some carbon dioxide from the water bodies, increasing the CO_2 content of the atmosphere as demonstrated by Landschützer *et al.* (2015). Thus mean annual air temperature and CO_2 content of the atmosphere will normally be closely correlated, unless some other factors are changing the situation. Examination of the long-term record of mean annual air temperature measured using proxies with the measured CO_2 content of the air shows that there have been spectacularly different independent fluctuations in air temperature and carbon dioxide content of the atmosphere in the past (Figure 3.2), so that other causes besides changes in carbon dioxide content of the atmosphere must have been involved in causing the changes in air temperature of the Earth.

The actual timing of the development of geocryological conditions depended on the geographical location and local history of plate tectonics. Antarctica was the first land area to be affected (c. 30 Ma – Tripati *et al.*, 2008), followed by South America (8.9 Ma – Rabassa *et al.*, 2004; Rabassa, 2008), northwest North America and eastern Siberia (3.5 Ma – Froese, 1998; Harris, 2001a), Kazakhstan (3.5 Ma – Aubekerova & Gorbunov, 1999), Southern Siberia (3.0 Ma – Fortiev, 2009), Europe and western Russia (1.8 Ma – Ehlers & Gibbard, 2008), and Tibet (c. 1.0 Ma). When the South American plate moved north to collide with the North American plate between 2.8 and 2.4 Ma, it closed off the Panamanian Seaway to El Niño, which resulted in reduced warming of Western Europe by the Gulf Stream.

The distribution of land and sea are also very important in producing either continental or maritime conditions, each of which result in different combinations of permafrost landforms being present or absent. The reduction in area of the Tethyan Sea during the Cenozoic Era correlates with a drastic drop in mean annual air temperature of the Earth (Figure 3.2). The presence of mountain ranges over which moist air masses must pass is also involved in controlling the amount of moisture that is available to produce ground ice.

In general, energy on the Earth surface should be balanced so that the amount of energy which comes in also goes out over the span of a year or else the average temperature will rise or fall. Solar radiation comes to the surface of the Earth in the amount of about 1368 Watts per square metre, which called *solar constant (S)*. The Earth is a globe, which receives energy depending on the Sun's position and angle of surface, $\pi r^2 S(1 - \alpha)$ in total, but reradiates from all over its surface, $4 \pi r^2 S T^4$ in total, where T is the Earth temperature. Then the result is:

$$\frac{\frac{S}{4}(1 - \alpha)}{\sigma} = T^4 \tag{3.3}$$

The average albedo of the Earth is about 0.3, so the incident solar radiation reflected back to space will be $1368/4 = 342$ and $342 \times 0.7 = 239.4$ Watts per square metre. Emitting radiation is the only way the Earth can get rid of energy. Most of the energy emitted is infrared radiation which is not warm enough to emit visible radiation. The equation above can be used to estimate the average temperature of the Earth, which is called the *effective radiating temperature*. It is the temperature that would be measured by an infrared radiometer in space, pointed at the Earth.

It follows from the equation above that the effective radiating temperature of the Earth is 255°K (−18°C). However, the average temperature of the surface of the Earth is about 287°K (+15°C). Accordingly, there is 32°C difference between the calculated and the actual temperature of the Earth surface, which can be explained by the *greenhouse effect*. The short wavelengths of visible light from the sun can pass relatively easily through the atmosphere, but the longer wavelengths of the infrared re-radiation from the Earth surface are absorbed by water vapour, carbon dioxide and other *greenhouse gases*. The trapping of the long wavelength radiation by the atmosphere causes heating of the atmosphere, and eventually, the Earth surface. Increase of carbon dioxide and methane concentration in the atmosphere by human activity can result in limited *global climatic change*, as well as *regional climate changes*. However, the ratio of water

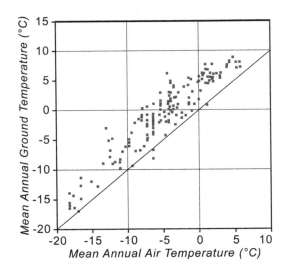

Figure 3.3 Mean annual ground surface and air temperatures for sites in areas overlying or close to permafrost in Canada and Norway (from Harris, 2015).

vapour to carbon dioxide is 500–750 to 3, so that any changes in relative humidity of the atmosphere can produce a far more significant change in the greenhouse effect (Harris, 2010b, p. 18).

The Earth surface and the atmosphere also exchange heat through convection between the surface and the air (the *sensible heat flux, H*) as well as through evaporation and transpiration (**latent heat, LE**). That happens due to the latent heat required for any phase transitions of water on the surface of the Earth. The rest of the heat on the surface goes into the lithosphere as the *ground heat flux* (*G*). Thus, the *heat balance equation on the Earth* surface is:

$$Rn = H + LE + G \qquad (3.4)$$

where: *Rn* – net solar radiation, *H* – sensible heat flux, *LE* – latent heat, *G* – ground heat flux.

In spite of the fact that in the full annual cycle (day, year etc.) the net solar radiation tends to be zero, at any given moment this may not be the case for the sensible heat flux, latent heat and ground heat flux. Usually the value of the ground heat flux is substantially (usually 0–7°C) warmer than the net solar radiation, sensible heat flux or latent heat (Figure 3.3). However, the ground heat flux is the reason for seasonal freezing and thawing of the ground and other thermal processes. Topography and slope aspect have a significant effect on the net diurnal heat flux, even when the slopes are a mere 5° (Figure 3.4). Study of the exact values for energy flows on the Earth surface is an ongoing research, supplemented by new estimates coming from satellite observations, ground-based observations, and numerical modeling.

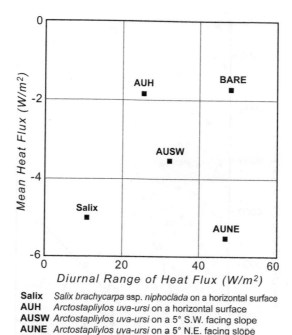

Salix   *Salix brachycarpa* ssp. *niphoclada* on a horizontal surface
AUH   *Arctostapliylos uva-ursi* on a horizontal surface
AUSW   *Arctostapliylos uva-ursi* on a 5° S.W. facing slope
AUNE   *Arctostapliylos uva-ursi* on a 5° N.E. facing slope
BARE   Bare silt loam on a horizontal surface

Figure 3.4 Variation in mean soil heat flux and diurnal range of heat flux at 5 cm depth on silt loam soils under various slope conditions (zero and 5°), aspects and vegetation covers on lithalsa #4, Fox Lake, Yukon Territory, on August 17th, 1992 (redrawn from Harris, 1998b, Figure 5).

3.2.2 Relationship between air and ground temperatures

A simple plot of mean annual air temperatures (MAAT) against *mean annual ground temperatures (MAGT)* indicates a considerable discrepancy and variability, with the ground temperature usually being noticeably colder than the air temperature (Figure 3.3). This difference is called the *thermal offset*, and will be discussed below. However in areas of low winter snow cover (under 50 cm), there is a strong relationship between the permafrost distribution and the *mean annual thawing index (MATI)* calculated by adding up the total positive mean annual daily air temperatures and the *mean annual freezing index (MAFI)* (see Figure 3.5). The relationship still shows some variability, but the relationship can be used to obtain a reasonable probability of the potential for the existence of permafrost around the weather station (Harris, 1981a, Figure 5). The same plot can also be used to show the relative continentality of permafrost areas (Figure 3.6).

In areas with higher winter snow covers, the *bottom snow temperature (BTS)* at the end of winter has been shown to indicate the presence or absence of permafrost (Haeberli, 1973) as well as the probable thickness of the active layer (Haeberli & Patzelt, 1983). Once again, the results are not always correct, but they have been used successfully to indicate areas of permafrost in the Alps (Haeberli, 1973), Colorado (Greenstein, 1983), Scandinavia (King, 1983), and in the mountain ranges of the southern Yukon Territory (Lewkowicz *et al.*, 2011).

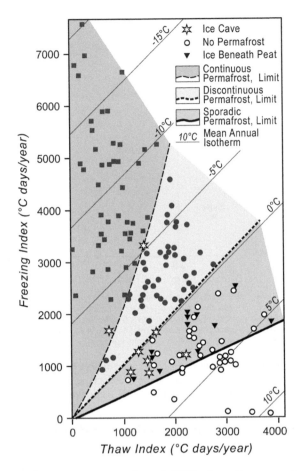

Figure 3.5 Distribution of continuous permafrost (>70%), discontinuous permafrost and sporadic permafrost (>30%) using annual freeze – thaw indices (from Harris, 1981a, Figure 5).

Nelson & Outcalt (1987) introduced the **Nelson frost number** for predicting the regional distribution of permafrost, following an earlier test in central Canada (Nelson, 1986). While this predicted permafrost distribution satisfactorily on the Qinghai-Tibet Plateau under cold, arid conditions, others have found that it does not always work. Additional models have been suggested by various authors, but a recent detailed test of these methods and suggested modifications produced a less than satisfactory result using ten-year data from ten sites in the Alps (Mutter & Phillips, 2012). Later algorithms are still under discussion, especially in regard to the effects of the necessary assumptions, *e.g.*, Xie & Gough (2013), and Kurylyk (2015).

The reasons for the rather unpredictable effect of the snow cover were explored by Goodrich (1983), who concluded that the effect of the snow cover is difficult to model accurately due to the timing of the snowfalls with variations in air temperatures, redistribution of snow, and latent heat effects. Subsequently, it has been found that the diode effect of peat and fractured rock, together with the processes operating within

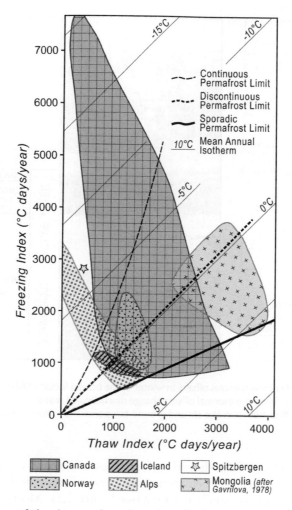

Figure 3.6 Comparison of the degree of continentality of selected permafrost areas (after Harris, 1981a, Figure 6). Continentality increases with increasing freezing and thawing indices.

them seriously interfere with the modeling (see Harris and Pedersen, 1998). In addition, sublimation of snow alters the balance of latent heat, which creates further problems in areas of low winter humidity (Harris, 1972; Peng *et al.*, 2007).

3.2.3 Thermal offset

This term was introduced by Kudryavtsev *et al.* (1974) to explain the change in observed temperature between the air and the upper surface layers of the soil or underlying rock (Figure 3.3; see also, Smith & Riseborough, 2002). Unfortunately, the term "thermal offset" is not usually used in a form signifying exactly what is being discussed, nor is the time period always provided for which the value is measured.

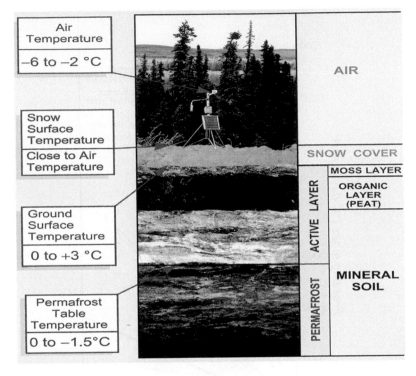

Figure 3.7 An example of the thermal offsets in winter in the boreal forest of Alaska (from Romanovsky *et al.*, 2011). Since the thermal offsets change diurnally and seasonally, the overall mean annual thermal offset between the air and ground temperatures is much less than shown here.

It is actually a product of the sum of several types of thermal offsets across distinct boundaries within a given microenvironment. These offsets may be positive or negative, but the total effect is usually negative (see Figure 3.2). Most thermal offsets are continually changing with the seasons, *e.g.*, snow cover, so specifying the time period being considered is essential.

Figure 3.7 shows the general concept of some of the layers involved in producing the various thermal offsets. The ***total thermal offset*** represents the sum of multiple parts. Within the atmosphere, there is usually a lapse rate approximating 0.4–0.6°C/100 m rise in altitude, assuming there are no temperature inversions present. In a coniferous forest in winter, there is an offset at the top of the forest canopy (the ***tree offset***), one at the surface of the snow pack (***snow pack offset***), one at the surface of the O horizon which is usually 1–2 cm thick (called the *organic layer offset*), one at the upper boundary of the A horizon (the top of the true mineral soil) which is only 5–15 cm deep (called the *A horizon offset*), and then the one at the surface of the B and C horizons (called the *mineral soil offset*). Each of these layers has a different thermal conductivity which means that significant temperature offsets will occur. The tree offset is usually positive by day in winter, whereas the summer tree offset together with the other offsets are usually negative. If ground or shrub vegetation is present, this too, will result in temperature offsets that will change with the time of year (the ***ground*** *vegetation offset* and the *shrub offset*). To add to these complications, the snow pack comes and goes with the seasons, and the thermal conductivity of the layers changes with moisture content as well as between frozen and unfrozen states. Meltwater will add to the moisture content in the soil during late spring-early summer, and summer rains will bring extra heat to the surface layers of the ground. Soil textures can vary considerably in a small distance. Inevitably, the thermal offsets are constantly changing and even fluctuate from year to year, and it is always necessary to specify the time period (day, season, annual) being discussed. Any interference with any one of the offsets will alter the overall thermal condition of the soil. Usually, the variations in vegetation and winter snow cover produce the largest offsets that can amount to several degrees C. Similarly, poor drainage and standing water have enormous effects.

Kudryavtsev *et al.* (1974) also considered the effect of slope and aspect in affecting the total thermal offset. The *exposure and slope thermal offsets* alter both the MAAT (t_{av}) as well as the amplitude of the ground surface temperature (A_s) by modifying the quantity of incoming and absorbed solar radiation. Aspect decreases both t_{av} and A_s when the maximum insolation comes from the south, through southwest to northeast to north in the northern hemisphere. Since winter cooling differs little on all slopes, it is the summer insolation that controls the effect. The combined effect of both exposure and slope angle can be estimated as follows (Kudryavtsev *et al.*, 1974):

$$H = k\Delta t_R = R_n - LE - G \quad \text{and} \quad \Delta t_R = (R_n - LE - G)/k \quad (3.5)$$

where: R_n is the net solar radiation, H is the sensible heat flux, LE is the latent heat, G is the ground heat flux, k is the coefficient of turbulent heat transfer between the ground and atmosphere minus the sensible heat flux.

One of the least complicated situations in the field is that for nearly vertical rock walls, but even there, thermal offsets have been reported (Hasler *et al.*, 2011a, 2011b), some being positive while others are negative. Lacelle *et al.* (2016) show that a similar simple case occurs in the McMurdo Dry Valleys of Antarctica.

Another simple case is found in the Prairies of North America, well south of the permafrost region. The main offsets there are the snow pack offset, the ground vegetation offset, the A horizon offset and the mineral horizon offset. Since the composition and structure of the vegetation cover and soil horizonation is similar throughout the region, the main variable is the snow pack offset. Todhunter & Popham (2008) examined the relationship of soil and air temperatures at sites throughout the Great Plains of the United States, using screen air temperatures and soil temperatures at 10.5 cm depth (within the A horizon). The results indicated that the total annual thermal offset was less than 1°C in the south, but increased to >4°C by the border with Canada. The southern stations showed no sensitivity to snow cover, so their offset is mainly due to variations in the ground vegetation and A horizon offsets. In the north, the annual thermal offsets were very variable and were obviously strongly affected by variations in snow cover, timing of snowfalls and duration of the snow cover. In an earlier study of the annual thermal offset at Fargo, North Dakota, Grundstein *et al.* (2005) showed that the mean annual thermal offset at Fargo varied wildly from year to year, ranging from slightly under 1.5°C to over 4.0°C. These authors discussed the variability in detail, but found no model that has been proposed so far which satisfactorily explains all the variations. Even wider discrepancies between MAAT and apparently stable permafrost can occur in ice caves (Harris, 1979).

Burn & Smith (1988) examined the total annual thermal offset at several sites near Mayo, Yukon Territory. Some sites were in hay fields but others were in the Boreal Forest. The air temperatures were up to 1.7°C warmer than the ground surface temperature, and the permafrost appeared to be in thermal equilibrium with the present-day climate, or actually aggrading at sites where the MAAT was between 0°C and 1.7°C. In a more detailed study of a transect across two hills near Fairbanks, Alaska, Romanovsky *et al.* (2011) found a range of total annual thermal offset for 2010–2011 ranging from −0.3 to 2.2°C. Brouchkov *et al.* (2005) working in central Yakutia (MAAT c. −10°C), found that the mean annual ground temperature was −3.0 to −3.5°C in larch forests, −2.2 to −2.7°C in birch forests, amd −1.8 to −2.3°C in disturbed areas. These results also suggest that there may be a reduction in the magnitude of the annual thermal offset at the latitudes above the 60° parallel. Seasonal variability will remain, but the long winter condition dominates the average. For comparison with these mean annual thermal offsets, those for the adjacent profiles in loessic soil and a block slope at the base of Plateau Mountain for 1995 were −0.8 and −4.5°C respectively at c. 5 cm depth (recalculated from Harris & Pedersen, 1998). There, the offset became positive at greater depths in the loessic soil, but not in the block field, at least not in the upper 90 cm.

Hasler *et al.* (2015) found that topoclimatic factors strongly influence the thermal offset over permafrost in the mountains of British Columbia. Excluding wetlands and peatlands, the seven sites studied exhibited only small total offsets between the air and ground temperatures. However there was a 4°C difference between steep north and south facing slopes on vertical rocks as opposed to a 1.5–3°C difference on exposed gentle slopes with a similar aspect. Small macroclimatic differences were ascribed to either differences in snow cover or variation in the magnitude of amplitude of the air temperature. There was no discussion of the possible role of variations in precipitation which are highly variable in that Province.

Further complications take place if the environment is disturbed. Forest fires, logging, meandering of rivers, and other types of modification by humans will upset the balance, resulting in significant temperature changes that can cause thermokarst landforms such as alas formation.

Forest fires are one of the triggers of thermal regime change and the formation of thermokarst lakes and depressions, which develop in the continuous permafrost area of Central Yakutia (Hinzman *et al.*, 2001). These fires modify ground surface conditions, causing deepening of the active layer. Mackay (1977) demonstrated a deepening of the active layer between 1968 and 1976 as a result of the Inuvik fire. The removal of the surface vegetation and organic layer by fire changes the energy budget of soil significantly. Even on the steppes of Mongolia, overgrazing can result in the development of thermokarst (Harris, 2002b), while the agricultural reforms on the Tibetan Plateau after the Chinese started administering the area, resulted in the beginnings of desertification of vast areas, resulting in widespread permafrost degradation (Harris, 2013a). Once started these processes are very difficult to stop, let alone reverse.

3.2.4 Relation to air masses

Permafrost distribution is closely related to the distribution of the Arctic and Antarctic air masses as well as the polar air masses. As indicated above, the subtropical air masses are critical in moving heat polewards to compensate for the imbalance in solar radiation

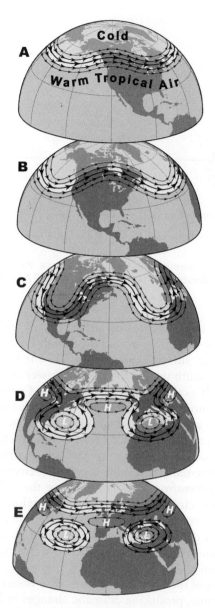

Figure 3.8 The evolution of the Rossby Waves with time as they move east (stages A to E). The exact rate of movement, shape of the lobes and rate of change depends on the air pressure differences in the air masses. Cyclones and associated fronts are omitted to simplify the diagram.

at the higher latitudes, particularly in continental areas such as Antarctica, interior Canada and Siberia. In the polar regions, the Arctic and Antarctic Polar air masses are very cold in winter, and the dense cold air spreads out towards the equator until it collides with the subtropical air moving polewards or the maritime Temperate air mass moving eastwards on-shore from the two major oceans. Wherever the cold air masses go for substantial periods in winter, permafrost and glaciers are commonly found.

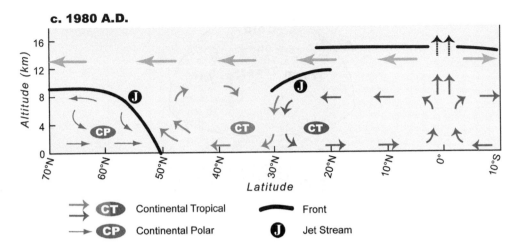

Figure 3.9 North-south section through the lower atmosphere showing the circulation of the air masses under the climatic conditions occurring in 1980 A.D. (from Harris, 2010a).

The rotation of the Earth causes the air masses to move eastwards, with the polar and arctic air masses rotating around the North Pole in a series of four waves (Figure 3.8), both at the surface and at the 500 mb level (the *Rossby waves*). Figure 3.9 shows a north-south section through the lower atmosphere showing the air mass movements currently occurring along the Eastern Cordillera of North America. The *polar jet streams* are normally found along the margins of the Rossby waves. The waves go through a cycle as they move (Figure 3.8), and the cyclical variations result in there being a zone with alternating warmer and colder fluctuations through much of the winter months. This zone only exhibits sporadic permafrost at locations where the microenvironment is particularly favorable for the development of permafrost. Low elevational, *latitudinal permafrost* usually occurs on the poleward side of this zone, unless there are mitigating circumstances, *e.g.*, a deep winter snow cover in Québec. A similar set of Rossby waves are found in the southern hemisphere, but are located primarily over the ocean due to the strong Antarctic high pressure air mass. These waves only affect the South Island of New Zealand, Tasmania, the West Antarctic archipelago, and Patagonia.

The average position of the boundary between the warm air and the colder, dense polar air can vary over time, producing dramatic changes in MAAT along the boundaries. Thus the average position moved north of Smithers and Fort Nelson (British Columbia) between 1982 and 1994, resulting in a change of MAAT of +5.23°C in 14 years, after which it stabilized (Harris, 2009, Figure 9). This was due to a strengthening of the maritime Pacific air mass and the continental tropical air mass relative to the continental polar air mass, resulting in the average position of the Arctic Front moving north past them.

Since the air masses can move freely towards areas of lower air pressure, they are constantly fluctuating. These boundaries between air masses are also the locus of considerable precipitation. In summer, it falls as rain, but in winter, the associated fronts bring snowfall. The other major sources of precipitation are the maritime temperate air masses coming onshore from the North Pacific or Atlantic Oceans. These produce maritime permafrost conditions on the cold west coasts of Alaska, Greenland, Iceland and Scandinavia. As the moisture-laden air moves east, it causes *orographic precipitation* on any mountain chains over which it must climb, *e.g.*, the Alps, the Rocky Mountains, Caucasus, *etc.* This will be discussed below.

3.2.5   Precipitation

In most permafrost areas, at least half of the annual precipitation falls as snow, and it accumulates as the snow pack. The snow pack offset is often regarded as the critical one in areas with permafrost by modifying the radiation balance. Snow changes the albedo of the ground surface, increasing its reflectivity to 0.7 to 0.85 compared with 0.1 to 0.3 for the vegetation during summer. As a result, under a thin snow cover, the quantity of absorbed radiation is reduced, and a cooling effect of the snow occurs. A large portion of the incoming radiation that is absorbed is used in the reduction of the snow pack through thawing or sublimation. On the other hand, a thick winter snow pack insulates the ground from heat loss into the atmosphere (Brown, 1966b; Mackay, 1978; Nicholson, 1978). It only operates for the cold part of the year, whereas most other thermal offsets affect the microenvironments on a given landscape throughout the year. Mackay (1978) introduced the idea of using snow fences to modify the thickness of the winter snow pack. He was able to demonstrate that cracking of the ground could be reduced or even halted by causing artificial accumulations of deep snow, resulting in an increase in the MAGT.

The snow influence as an insulator can be estimated as follows (Kudryavtsev *et al.*, 1974):

$$\Delta t_{\text{snow}} = \Delta A_{\text{snow}} = \frac{A_\theta}{2}\left(1 - e^{-z_s\sqrt{\frac{C\pi}{\lambda T}}}\right) \tag{3.6}$$

where: Δt_{snow} is the thermal offset of annual mean temperature due to the warming influence of the snow, °C; ΔA_{snow} is the decrease of the amplitude (physical) of the annual fluctuations of temperature under the snow in °C; A_θ is the meteorological amplitude of the annual fluctuations of the temperature of air, °C; z_s is the depth of snow cover, m; C is the volumetric heat capacity of snow (J/m^3 * °C); λ is the thermal conductivity of snow, W/(m * °C); T is the period, equal to the year, s.

Another formula for an estimate of the thermal offset – warming snow influence Δt_{snow} taking into account the annual heat exchange of grounds with the atmosphere was proposed by S.N. Buldovich (2001):

$$\Delta t_{\text{snow}} = \frac{\overline{R}_{\text{snow}}}{T}\sqrt{2\lambda_f\Omega_w Q} \tag{3.7}$$

where: $\overline{R}_{\text{snow}}$ – the average thermal resistance of snow in winter; Q – the heat of phase transitions in the ground; λ_f – the thermal conductivity of the frozen ground; Ω_w – a parameter equal to $|t_w| * \tau_w$, where t_w – average winter temperature, τ_w – the duration of the winter.

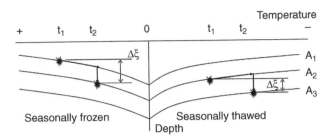

Figure 3.10 The effect of positive (warming) snow pack thermal offset on the depth of the active layer. $\Delta\xi$: t – temperature, A – temperature amplitude, t_1 – under snow, t_2 – without snow.

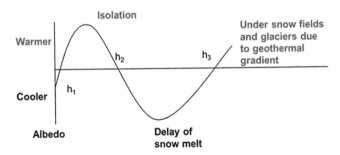

Figure 3.11 Diagram showing the effect of various depths of snow cover on the underlying ground temperatures.

The positive (warming) snow pack thermal offset of the MAAT affects seasonally frozen and seasonally thawed layers within the ground differently (Figure 3.10), the change being greater in seasonally frozen horizons.

In areas where the snow-pack is thin (less than h_1), the snow cools the soil (Figure 3.11) since its role as a reflector prevails, but reradiation continues. Snow is a cooling factor in this case. With an increase in snow depth from h_1 to h_2, it prevents excessive heat loss due to its insulating qualities.

Figure 3.12 shows the effect of the thickness of the snow pack on the mean winter ground surface temperature at Plateau Mountain, Alberta (Harris & Brown, 1978). Fifty centimetres is the critical lower thickness of snow cover to prevent significant freezing of the ground at an MAAT of about −2°C and a MAFI of around 2,000 degree days °C per year. The depth of snow required to have a similar effect in the Arctic Islands of Canada is estimated to be 110 cm (Alan S. Judge, personal communication, 1981). The critical depth varies during the year as the snow pack metamorphoses, as well as with the variations in MAAT and MAFI at a particular site. Thus Granberg (1973) found that 65–75 cm was the critical depth at Schefferville, Québec. It is the difference in the depth of snow cover that causes the considerable differences in permafrost distribution on the two sides of Hudson Bay, Canada. Mackay (1978) used snow fences to reduce the cracking of the ground during winter on Garry Island, and the same technique has subsequently been used elsewhere to reduce the winter cooling of the ground, or to prevent thermal contraction cracks breaking up the substrate.

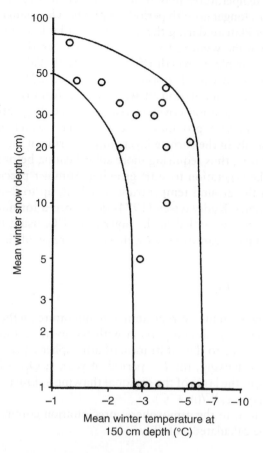

Figure 3.12 Effect of mean snow depth in winter on the ground temperature at 150 cm soil depth on Plateau Mountain (after Harris and Brown, 1978, Figure 9).

However, the formation of snow drifts or the piling up of snow around houses can destabilize the underlying permafrost (Chapter 13).

Duration of snow cover is also important since snow may persist for much of the period of the year when the insolation is at its maximum, thus reducing the amount of solar energy that can be used for thawing the active layer. Summer snow falls, such as often occur on the Qinghai-Tibet Plateau, greatly reduce the effects of summer heating (French, 1970). The first autumn snowfall uses up much of the heat in cooling the upper part of the active layer while adding a substantial quantity of water to the surface layers of soil. This becomes ice as the surface temperature drops below 0°C, but the ice increases the thermal conductivity of the now frozen surface layer, thus aiding the freezing process. During the winter, timing of arrival of the snow relative to the cold temperatures, drifting of snow and variations in snow density due to wind can cause tremendous variations in ground temperatures over short distances (Brown, 1972; Seppälä, 1982).

While warming temperatures may result in earlier melting of the snow, this does not necessarily mean a longer growth period for plants. The changes in air temperature on the Qinghai-Tibet Plateau during the last few decades have altered the phenology of the forbs. Although the winter snow pack melts earlier, there is negligible change in the time available for plant growth (Yu *et al.*, 2010). This has prompted various theories by biologists as to why this occurs (Chen *et al.*, 2011), but it is almost certainly due to increased accumulation of water in the form of ice in the soil during the winter. Lack of a vegetation cover increases soil cooling (Hu *et al.*, 2008) and the increased winter snowfall provides more water. This water moves to the coldest places in the soil, usually in the surface layer, and in spring, it has to be changed back from ice to water at 0°C, thus requiring more solar heating before the ground warms up sufficiently for the vegetation to start growing. Summer precipitation can have a significant effect on the ground temperature, reaching up to 2–5°C in areas with a warm and rainy summer. Kudryavtsev (1974) developed a formula based on the fact that the infiltrated water is cooled in the layer ξ to 0°C, using the entire reserve of heat to produce the phase transitions as well as an increase of the temperature of the ground:

$$\Delta t_{oc} = (Q_{pr} * \xi)(\lambda_g * T), \tag{3.8}$$

where: Δt_{oc} – an increase in the average annual temperature on the bottom of the layer of the seasonal thawing (t_ξ) in comparison with the average annual temperature on the ground surface (t_s) due to the infiltration of atmospheric precipitation, °C; ξ – the depth of the seasonal thawing, m; T – period of year, s; Q_{pr} – heat of atmospheric precipitation coming to the layer of the seasonal thawing ξ, J/m²; λ_g – effective thermal conductivity of the layer ξ, W/(m * °C).

The amount of heat in the atmospheric precipitation coming to the layer of the seasonal thaw can be calculated as follows:

$$Q_{pr} = C_w \sum V_{pr,i} * t_{pr,i} \tag{3.9}$$

where: $V_{pr,i}$ – the monthly (or decade) sum of summer atmospheric precipitation, kg/m²; $t_{pr,i}$ – the average monthly (or decade average) temperature of precipitation, approximately taken as the temperature of air (in the absence of data), °C; C_w – the heat capacity of water, J/(kg * °C).

Supra-permafrost water in the active layer inevitably imparts instability to sediments on slopes. The water perches on the frozen ground underneath and often causes slope instability on slopes. It also makes frost susceptible layers unstable resulting in cryoturbation, *etc.*

The winter precipitation tends to accumulate as the snow pack. There are losses into the cold ground as well as into the atmosphere by sublimation, but a substantial accumulation of snow is present in the spring. After deposition, the snow undergoes metamorphism, whereby the crystals lose their hexagonal shape and increase in size and density. This is particularly marked at the base where the grains take on a spherical shape, leaving large spaces under the main snow pack. It is this lack of contact with the underlying ground that causes the collapse of the snow pack when walked on in spring, as well as a tendency to produce snow avalanches.

3.2.6 Latitude and longitude

The position of the site on Earth is a major factor since the latitude determines the range of angles of incidence of the solar radiation at every location. While there is a similar total annual potential radiation throughout the tropics, the potential amount decreases towards the poles. The length of daylight compensates for this between about 54° and 60° latitude in summer, but reduces the incoming solar radiation in winter. The low angle of summer sun polewards of the Arctic and Antarctic circles is not adequately compensated for by the 24 hours of daylight, while at the winter solstice, 24 hours of darkness prevails. This intensifies the cooling in winter, and reduces the effectiveness of the low-angle sun in summer. Under conditions of equal MAAT, the permafrost thickness appears to be thinner on the Qinghai-Tibet Plateau than at 70°N, while ice wedge casts typically measure a maximum of about 2 m, compared with 4 m along the shores of northern Russia. The active layer also appears to be deeper. This may be due to the greater summer intensity of the solar radiation at lower latitudes due to the greater angle of incidence of the solar rays.

There are a number of factors influencing the lower altitudinal limit of permafrost, so the variation of the lower limit with latitude is complex, but follows certain patterns. Other conditions being equal, the relationship between the lower altitudinal limit and latitude can be expressed by a Gaussian curve (Figure 3.13; see Cheng, 1983; Cheng & Wang, 1992; Corte, 1988). This curve resembles curves of latitude versus

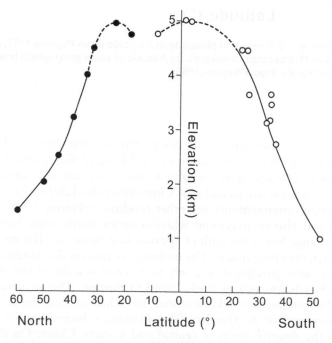

Figure 3.13 The lower limit of permafrost in the Andes of the Southern Hemisphere (open circles) (Corte, 1988) compared with the average in the Northern Hemisphere (closed circles) (Cheng, 1983a).

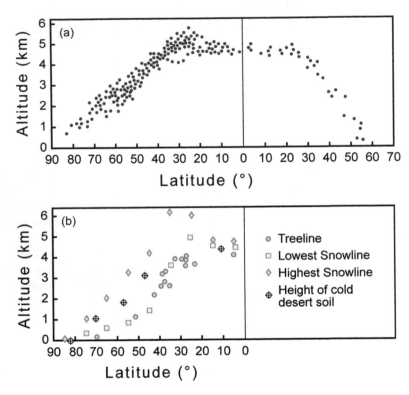

Figure 3.14 (a) Altitude of 0°C isotherm plotted against latitude (from Maejima, 1977, with permission of Tokyo Metropolitan University). (b) Altitude of some geographical boundaries plotted against latitude (from Wenyuan, 1980).

the 0° isotherm, snow line, alpine cold desert and tree line (Figure 3.14), indicating that these relationships have a close connection (Cheng & Wang, 1982). However the altitudinal limit of permafrost at the same latitude can vary considerably. Thus the lower limit of permafrost in east and west China differs by 1200 m (Zhou Guo, 1983). This is partly due to continentality and other conditions (Harris, 1988).

Longitude can also be important where a major north-south barrier such as a high mountain range lies in the path of approaching moist air. The air rises over the mountain barrier, dropping much of its moisture as rain on the windward side of the mountain (*orographic precipitation*), only to descend as a dry, warm air mass called a Chinook in North America or Foehn wind in Europe. The air mass can pick up more water when passing over lakes and seas, but when crossing the successive north-south mountains west of the Qinghai–Tibet Plateau, it becomes a dry air mass that contributes to the desertification of central and western China. On the east slopes of the Qinghai–Tibet Plateau, the shallow East Asian Monsoon only extends up to about 3600–4000 m elevation, so does not bring precipitation to the main plateau. This produces a north-south climatic divide.

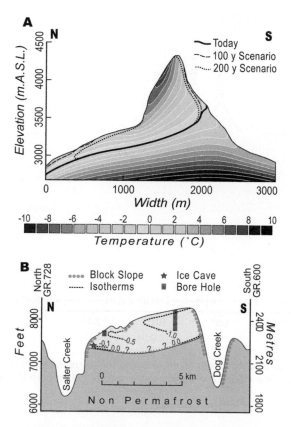

Figure 3.15 Comparison of A, the modelled temperature field for a north-south section of the Matterhorn showing present-day isotherms together with those projected for 2109 and 2209 based on present rates of climate warming (modified from Noetzli & Gruber, 2009), and B, Plateau Mountain with its relict permafrost remaining from when it was a nunatak surrounded by glaciers during the Last Major Glaciation (15–25 ka) (modified from Harris & Prick, 1997).

3.2.7 Topography and altitude

Topography is very important (Figure 3.4). In areas with relatively low maximum angles of incidence of solar radiation, there is normally a different vegetation on the north-east facing slopes to that occurring on slopes facing south-west. Other things being equal, permafrost will be found further towards the equator on the north-east facing slopes. This is one of the causes of discontinuous, sporadic and isolated permafrost. Again, the position on a slope matters, with isolated patches of permafrost first appearing at the foot of slopes. Steep-sided mountains cast large shadows for a considerable part of the year, this becoming more important with increasing latitude (Figure 3.15 and 3.16). There are also sun traps and cold spots, and there is usually an inversion producing warmer winter air temperatures about 100 m above a col or saddle in a Pass (Harris and Brown, 1982). Topography also causes local modifications

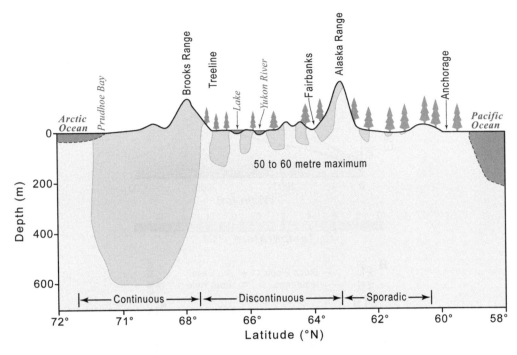

Figure 3.16 Section north-south across Alaska showing the position and thickness of permafrost to scale. Note the influence of aspect on the distribution of permafrost.

of the vegetation, winter snow cover and hydrology, which affect the distribution of taliks and the thickness of the active layer. Even in areas of continuous permafrost, the permafrost forms an outer cold layer in the upper part of the ground that tends to follow the contours (Figure 3.14a), though modified by aspect and by the hydrology of the area. In contrast, the base of relict permafrost in former nunataks tends to parallel the former limit of the ice sheet, being nearly horizontal on Plateau Mountain in spite of the topography (Figure 3.14b).

Air masses have a specific upper limit. Thus the Indian Monsoon is only about 5000–6000 m. high, so it cannot readily cross the Himalayas at the present time to bring appreciable moisture to the Qinghai-Tibet Plateau. Likewise, the East Asian Monsoon is only about 4000–4500 m high, so it does not bring large quantities of precipitation to the northeastern slopes of the Qinghai-Tibet Plateau (Harris & Jin, 2012). Some of the deeper permafrost on mountains such as along the Tanggula Range (5000 > 6000 m) is relict.

Altitude affects the distribution of permafrost because the air becomes colder with increasing altitude. The rate of change is called the *lapse rate*, and the average near-surface lapse rate is approximately minus 1.6°C/100 m rise in altitude. However, this is very variable and is particularly dependent on air mass distribution, time of day, season of the year, and ground cover (vegetation, cultivation, urban landscape, snow cover). The cold Mongolia-Siberian high pressure air mass is only about 4000 m thick

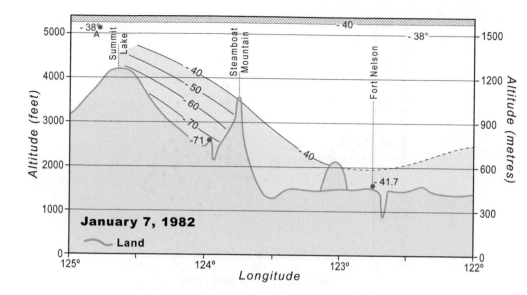

Figure 3.17 Anatomy of the cold air drainage west of Fort Nelson, British Columbia on January 7th, 1982, showing the uncorrected temperature distribution (modified from Harris, 1982, Figure 1). The coldest corrected temperature was −64°C.

on the northern flanks of the Plateau, so that the main portion of the Plateau above this altitude is affected by the local lapse rate in winter, and above 4500 m in summer.

Where there is warm air aloft, it will actually get warmer with altitude until the warmer air mass is reached, whereupon the lapse rate will become negative. Lower temperatures at higher altitudes are the cause of the southward extension of permafrost along the Eastern Cordillera of North America (Harris, 1986b), as well as the occurrence of permafrost on the Qinghai-Tibet Plateau. In extensive mountain ranges with enclosed deep valleys, cold air drainage may occur (see below).

3.2.8   Cold air drainage

In valleys and basins in large areas of mountainous terrain, the climate may be modified in winter by *cold air drainage*. When the sky is clear and there is negligible wind, there is very high reradiation of heat back into the atmosphere at night. Where there are differences in vegetation with altitude, differential heat loss by reradiation results in severe cooling of the tundra and the adjacent air, which becomes increasingly dense, compared with the forested slopes at lower elevations. When the difference in density of the air becomes high enough to overcome the resistance of the roughness of the terrain, the colder air moves downslope, displacing the less dense warmer air in the depressions (Figure 3.17, after Harris, 1983b). Once there, it is not readily removed. The cold air can remain in the lowland for up to a week, and warmer air simply rides over the top. The average number of events for the cold air drainage at Fox Lake, Yukon Territory, between 1987 and 1996 inclusive was 6.5 events per year. The exact threshold depends

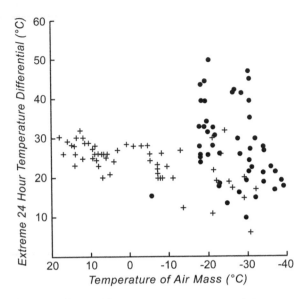

Figure 3.18 The temperature of the still cP air mass plotted against presence or absence of cold air drainage at Fox Lake, Yukon Territory for the winters of 1987–1996 (translated from Harris, 2007). The black circles correspond to data from cold air drainage events.

on the terrain and the temperature of the air mass, but in the Yukon River valley, it tends to occur when the ground surface temperature of the main air mass falls below −16°C (Figure 3.18, after Harris, 2006, 2007, 2010c). The cold air drainage causes a temperature inversion that shows in the MAAT at different elevations on a mountain side, *e.g.*, at Snag and Mayo, Yukon Territory, and again in the large mountain ranges of Siberia. The cold weather in the valleys results in the development of permafrost in areas with discontinuous or sporadic permafrost. MAGST in the valley is lower than on the mountainside above.

3.2.9 Buffering of temperatures against change in mountain ranges

One of the characteristics of these deep valleys in large areas of mountains in continental situations is that the MAAT in the valleys resists the climate changes being recorded in the surrounding lowlands (Harris, 2006, 2007, 2010c). It does not occur on the flanks of the mountains, nor in the lowlands subject to the influence of the ocean currents. In the case of the mountain valleys of central and southern Yukon and Alaska, this appears to be due to the combined effects of cold air drainage, inversions and steam fog (Harris, 2006, 2007). If the regional air warms up, and the warm air over-rides the cold air trapped in the mountain valleys, a thicker cloud layer develops at the inversion, thus producing decreased insolation at the ground surface. This counteracts the effect of the regional warming. The cold air drainage requires a certain minimum temperature to be exceeded at a given site for it to occur (Figure 3.18). Since a 2°C warming does not appreciably alter the period in winter when it is colder than this threshold, cold air drainage remains a major factor maintaining permafrost stability. In addition, most mountain valleys have extensive lakes and fens. When colder air overlies the open water in the early Fall, steam fog develops, cutting down on the insolation reaching the ground surface until the water becomes completely frozen over in early December. By then the incoming radiation at these northern latitudes is minimal. Regional warming causes the air to contain more moisture, which has the feedback effect of increasing the thickness and duration of the steam fog, These three processes can only occur in depressions in the mountains, and this buffering may explain why areas such as the mountains of Beringia and eastern Siberia had a decrease in MAAT of only 9–10°C during the last major glaciations, while western Europe endured a decrease of at least 19°C (Washburn, 1979; Isarin, 1997) Thus these buffered areas were very important refugia during glaciations.

3.3 TERRAIN FACTORS

3.3.1 Vegetation

The vegetation cover has an enormous effect on the microclimate in the forested areas underlain by permafrost, but this effect decreases towards the poles as the vegetation becomes more stunted, and is replaced by tundra plants (Brown, 1963, 1970, 1972). In both the Arctic and Antarctic, there are also polar deserts where even isolated cushion plants do not live. However, there are some lichens that can survive on rock. Removal of vegetation has an enormous effect (Moskalenko, 1998).

The most obvious effect of a continuous vegetation cover is in reducing the depth of thaw of the active layer by shielding the ground from insolation (Brown and Johnson, 1964; Lindsay and Odynsky, 1965; Jeffrey, 1967; Price, 1971). If moss is present, it greatly reduces the thickness of the active layer in a manner similar to peat (Greene, 1983; Harris, 1979b), *i.e.*, it acts as a thermal diode, as can a thick litter layer. The effect occurs throughout the year, and can result in a reduction of MASG of 2–4°C compared with soils without a moss cover (Kudryavtsev *et al.*, 1974).

Trees and dense thickets of shrubs shade the ground from insolation in summer and act as radiators of heat in winter. The reduction in solar radiation reaching the ground may be as high as 70% in a closed canopy forest in summer. Combined with the heat-moisture transfer in plant cover, the evaporation from the soil may be decreased several times compared to open sites. Turbulent heat exchange is always less in forests since the wind speed is reduced to 8–10% of that in open areas. In the Russian taiga, maximum cooling is seen in fir and pine forests, with a reduction of 0.5–2°C compared with open sites.

In the forest-tundra transition, the shrubs in the northern taiga of Western Siberia, the reduction in solar radiation is offset by the decrease in turbulent heat exchange. However, the much higher snow accumulation due to intercepting blowing snow causes better insulation of the underlying ground from the winter cold than in open areas which have a snow deficit. This results in a higher MAGT. The snow density is also 5–10% less in the bushes (Pavlov & Dubrovin, 1979).

Conifers intercept some of the snow, producing a low area with thin or negligible snow pack around their base, so altering the local heat exchange significantly (Viereck, 1975). In the dense, dark coniferous forests of Southern Siberia, there is a great reduction in snow cover, resulting in greater cooling of the ground. Deciduous trees only provide shade in summer, whereas conifers produce significantly different microclimates in winter, *e.g.*, the air temperatures beneath conifers in winter in the Rocky Mountains of southwestern Canada average 1°C higher than in neighboring ecosystems (Harris, 1976).

Beneath the canopy of the old growth Western Red Cedar forest east of Revelstoke, British Columbia, there is an air temperature of around 15–20°C when the air temperature in the adjacent more open second growth forest is over 30°C in summer. The air in the old forest is very humid so that ferns and mosses are abundant in the shade, whereas in the second growth forest, the air has a relative humidity of less than 60%. All trees modify wind velocities at the ground surface, thus reducing heat transfer. The ground under the trees tends to hold more moisture in summer than the soils under the more open grassland vegetation. All trees contribute litter to the ground surface, and since the rate of decay of vegetation is slow in the short summer season in areas with permafrost, there is usually a significant organic layer present. This tends to cause a weak thermal diode effect, so patches of permafrost may be present in the moist forested areas at the foot of slopes, *e.g.*, the northern slopes of the Tian Shan of China and in Switzerland. Williams & van Everdingen (1973) reported air temperatures several degrees colder in peatlands with stunted Black Spruce trees than in the surrounding vegetation types in northern Canada.

Removal or disturbance of the surface vegetation cover usually causes degradation of the underlying permafrost (Mackay, 1970), unless moss and/or peat are left protecting the ground (Linell & Johnston, 1973). Types of disturbance include grazing, fire, logging, cultivation of hay, or engineering constructions of all kinds. Seral evolution of the vegetation results in changes in the thermal regime resulting in growth or decay of permafrost. Thus the colonization by vegetation of sandy bars along rivers is usually accompanied by increasing development of permafrost, *e.g.*, on the Mackenzie Delta (Smith, 1975), but in Interior Alaska, permafrost is more widespread in the early successional stages of willow and Balsam Poplar (Viereck, 1970; Péwé, 1970; Van Cleve & Viereck, 1983). On lithalsa mounds at Fox Lake, Yukon Territory, the active layer was thinnest under willows and mature spruce forest (Harris, 1993), while Blok *et al.* (2010) have postulated that the expansion of shrubs in some Arctic regions (Sturm *et al.*, 2001; Tape *et al.*, 2006) may at least partly counteract the warming that causes the vegetation change (Lawrence & Swenson, 2011) by becoming a carbon sink.

On the microscale, the species of vascular and nonvascular plants present on the surface of the ground each produce different thermal effects on the organic soil underneath them at latitudes 61–64°N in the southern Yukon Territory (Harris, 1998b). Net radiation on bright sunny days in August represents 25% of the potential radiation. However, in the shade, net radiation is negative, even at 1400 hours. Heat flux into the substrate beneath different vegetation covers showed marked differences, depending on the plant species and vegetation structure. Single vegetation canopies or mats showed the least negative values and the highest diurnal amplitudes. Values were more negative beneath multiple canopies and exhibited the lowest diurnal ranges of heat flux. Sites in the shade of trees exhibited a negative net heat flux, even though the mean daily temperature (MDT) was 11.3°C based on the average of 20 minute observations. June measurements made immediately after melting of the snowpack at an MDT of 1.2°C showed higher heat fluxes and diurnal ranges than in August under all vegetation species except for *Cladina stellaris*, beneath which the heat flux scarcely changed. This may be due to the relatively small moisture content and open growth form of this species. Shrub-covered substrates always exhibited greater heat loss (Figure 3.3), probably due to shading of the ground and surface vegetation by their leaves. Multi-layer covers allow for even greater heat loss, *e.g.*, at Fox Lake, the mean annual ground temperature dropped 3–4°C once a two-canopy vegetation became established, while the active layer decreased from c.200 cm to 60–80 cm (Harris, 1993). Thus vegetation can insulate the underlying permafrost for short-term periods with an MAAT above 0°C.

Different surface vegetation covers have different albedos (Petzold & Rencz, 1975; Stoy *et al.*, 2012). At Schefferville, Québec, dry *Cladonia stellaris* had an albedo of 0.233 to 0.264 whereas the albedo of Black Spruce was 0.158, *Dicranum fuscescens* was 0.147 and sedges 0.071 (wet) or 0.109 (dry). Thus the light-coloured lichen (*Cladonia stellaris*) reflects more incoming solar radiation than other plants. Reindeer grazing can increase the surface temperature of the ground by reducing the effect of *Cladonia* species (Stoy *et al.*, 2012).

Vegetation also modifies the hydrology of a site. The effect of an O horizon in holding moisture and reducing run-off is well known, while the higher humidity beneath a closed tree canopy has already been described. *Sphagnum fuscum* can hold up to 2000% water by weight, while *Aulocomnium palustre* can hold up to 1600% (Harris, 1998). However, some mosses and lichens do not hold nearly as much moisture, and Kershaw & Rouse (1971, Figures 7 and 8) showed that different strains of the same species have different water retention properties. Thus it is necessary to test the water-holding capacity of lichen or moss mats before concluding what effect they may be having on the moisture availability in the underlying soil (c.f. Brown, 1969; Brown & Péwé, 1973). Blok *et al.* (2011) experimented with removal of the moss layer and found that evapotranspiration from the underlying soil increased, and the strong insulating effect of the moss layer was lost.

3.3.2 Hydrology

Woo (1986, 2012) describes this in detail. Water in permafrost soils occurs in three positions, *viz.*, supra-permafrost, intra-permafrost, and sub-permafrost. The supra-permafrost water is the most likely to be contaminated since the underlying icy permafrost has very low water permeability. Intra-permafrost waters are normally relatively saline cryopegs that are not suitable for consumption by most animals. Only the most salt-tolerant plants can use them as a water source. The underground drainage lines usually represent the last taliks in an area, and permafrost severely limits the locations at which water comes to the surface. Sub-permafrost water often contains salts washed out from the adjacent sediments. The main effect of sub-permafrost water is in the artesian springs that produce open system pingos and icings. A fourth category consists of the surface water bodies that are common in permafrost areas.

In arid areas such as in Mongolia and northeast Tibet, lines of surface and sub-surface water flowing over or through the permafrost table into inland drainage areas can create pingos (see Figure 1.19) and icings along their path (Kowalkowski, 1978; Froehlich & Slupik, 1978). In wetter areas, they tend to produce seasonal frost blisters. These lines have been called *water tracks* in Antarctica (Levy *et al.*, 2011), and consist of zones of high soil moisture which represent a route that ground water takes to move downslope. Water content of the soil and its form have a considerable effect on the thickness of the active layer. The thermal conductivity of ice is greater than that of soil, and changing ice back to water takes considerable input of energy before the ground can warm up. A similar removal of energy is required during freezing in the fall. Increased soil moisture results in a reduction in the depth of seasonal freezing and thawing. This is connected to the thermal offset, which is determined by the ratio of the coefficients of the thermal conductivity of frozen and thawed ground, and also by the value of annual heat exchange in the ground. The value of it ranges between 0.5°C and 1.5°C (Kudryavtsev *et al.*, 1974). If the thermal conductivity of the ground in the thawed state of λ_{uf} differs from the thermal conductivity of the ground in the frozen state of λ_f, then the average annual temperature of the ground changes with depth in the active layer, reaching a maximum at the bottom.

Where thermokarst mounds have developed, water flows through the active layer, down slope over the surface of the permafrost table. This water is generated from the snow pack, melting pore ice at the ice table beneath the water tracks, and thawing buried segregation ice formed during winter freeze-up. The water is enriched in solutes derived from chemical weathering of sediments as well as the dissolution of salts contained in the soil. Salt content is about four times that in normal runoff, and the quantity of water transported in this way is about twice as fast as in the adjacent wet soil. These intra-soil flows of water represent salt superhighways, redistributing water, energy and nutrients in the cold deserts of the Antarctic. The water ends up in lakes that are frozen over. Similar natural flow paths for water in the active layer have been identified along the Alaska Highway near Beaver Creek (de Grandpré *et al.*, 2010), where they create subsidence problems by degrading the permafrost beneath the Alaska Highway.

The effects of liquid surface water are seen in the more humid areas. Liquid water is a source of heat, and is an effective thawing and erosive agent, *e.g.*, Ling *et al.* (2012), especially when moving past ice-rich ground (Mackay & Black, 1973). This is particularly well demonstrated after fire or any other interference with the natural vegetation. Thus alteration of the water table by placing a culvert too low can produce spectacular results in areas with ice- wedge polygons (Figure 3.19). Conversely, the very dry air and strong winds on the Qinghai-Tibet Plateau result in extreme evaporation and cooling of the soil on valley floors with a high water table, which can result in development of permafrost. The thermal properties of water and ice mean that moist or wet ground has a much shallower active layer than dry ground under otherwise similar environmental conditions. Ice is the major constituent causing the formation of most of the characteristic, unique landforms found in permafrost regions, *e.g.*, rock glaciers and peat mounds. As a result, different landforms are found in dry permafrost areas to those found in poorly drained areas.

When the temperatures rise in the spring, the temperatures in the snow pack become isothermal at 0°C. As the snow slowly changes to water, the latter moves to the bottom of the snow pack, where it commences to flow down any available slope. The water-logged zone may be up to 40 cm deep and can have considerable erosive

Figure 3.19 Results of placing a road culvert too low in an area of ice-wedges along the Dempster Highway, Yukon Territory, after one year. The hydrological disturbance decreases with distance from the culvert. © S. A. Harris.

power, even on gentle slopes. Harris (1998c) described low mounds from the road in the North Fork Pass along the Dempster Highway, which had the normal mature tundra vegetation on the top of the mound but a modified vegetation on the sandy soils in the shallow depressions. The sand consisted of the remains of the sediment eroded from the underlying sediment by the sub-snowpack flow of water in the spring that had eroded the channels. Where the melt-water moves downslope on the surface of the ground in mountains, it warms in the sun and then sinks into the ground, so transporting extra heat into the active layer. In sediments, this tends to deepen the underlying active layer at that location, and can actually thaw a temporary vertical channel through the permafrost (Harris, 1999). The warmer water then drains into the unfrozen zone under the permafrost and the channel quickly refreezes (Figure 3.20).

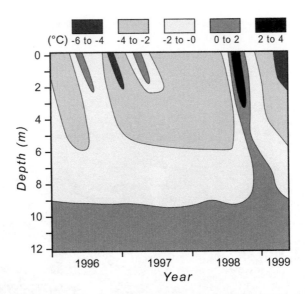

Figure 3.20 Ground temperatures (°C) in the upper 12 m between 1996 and 1999 inclusive at borehole Marmot Basin #2, Marmot Ski Area, Jasper National Park. Note the short-lived warming as meltwater thawed a narrow pipe through the permafrost, through which the water drained (modified from Harris, 1999, Figure 6).

Since the pattern of water flow changes from year to year, the moisture content and depth of the active layer vary widely from year to year at a given location on the slope.

In the higher mountains, there is often a temperature inversion, enabling melt water to penetrate into the underlying rocks and then move beneath the permafrost to provide a hydrostatic head that enables springs, icings, and open system pingos to form at the base of the slope (van Everdingen, 1990; Kane *et al.*, 2013). Fault lines in the rocks aid this process. Locally in limestone areas, this water causes seasonal flooding of depressions (van Everdingen, 1981).

Heat exchange can occur between the soil and the moving ground water, as well as percolating rain. The infiltration of summer warm precipitation can increase the MAGT significantly by at least 2°C. A method of evaluation of the influence of atmospheric precipitation on the average annual temperature on the bottom of the layer of seasonal thawing is proposed by V.A. Kudryavtsev (1974) who used a formula for estimating its effect, assuming that the infiltrated water is cooled in the layer ξ to 0°C, so using the entire reserve of heat to cause phase transitions together with some increase of the temperature of soil:

$$\Delta t_{oc} = (Q_{pr} * \xi)(\lambda_g * T), \tag{3.10}$$

where: Δt_{oc} – an increase in the average annual temperature on the bottom of the layer of the seasonal thawing (t_ξ) in comparison with the average annual temperature on the ground surface (t_s) due to the infiltration of atmospheric precipitation, °C; ξ – the depth of the seasonal thawing, m; T – period of year, hours; Q_{pr} – heat of atmospheric precipitation coming to the layer of the seasonal thawing ξ, J/m^2; λ_g – effective thermal conductivity of the layer ξ, W/(m $*$ °C). The amount of heat contributed by atmospheric precipitation affecting the layer of the seasonal thawing Q_{pr} can be calculated by:

$$Q_{pr} = C_w \sum V_{pr,i} * t_{pr,i} \tag{3.11}$$

where: $V_{pr,i}$ – the monthly (or decade) sum of summer atmospheric precipitation, kg/m^2; $t_{pr,i}$ – the average monthly (or decade average) temperature of precipitation, approximately taken as the temperature of air (in the absence of data), °C; C_w – the heat capacity of water, J/(kg $*$ °C).

3.3.3   Lakes and water bodies

Aqueous covers have a strong influence on the ground temperature beneath them. Where lakes and rivers do not freeze to the bottom in winter, a talik will be present. Large, shallow water bodies, thinner winter ice and deep winter snow packs result in a large talik (Feulner & Williams, 1967), as do dark-coloured bottom sediments (Brown & Péwé, 1973). Where the lake is sufficiently shallow to freeze to the bottom in winter, there will be permafrost beneath the lake. Even so, lakes and rivers can cause thawing of the surrounding sediments which can result in enlargement of the water body in ice-rich sediments, *e.g.*, during the formation of alas valleys in Siberia. Michel (1971) provides diagrams of the temperature regimes of deep (>10 m) and shallow freshwater lakes (Figure 3.21), based on studies of North American lakes. It assumes that there are no underground springs present, so the thermal stratification is closely related to depth. Note the effect of the change in density at 4°C. The energy budget and resultant temperature offset at the surface of lake ice in winter are shown in Figure 3.22 (from Liston & Hall, 1995). There are thermal offsets at both surfaces of the snow cover and at the base of the lake ice. Kudryavtsev (1954) used a different diagram and formula for calculating the temperatures below small lakes that freeze to the bottom in winter (see Figure 3.23):

$$t_H = 0.5 * \left[\left(1 - \frac{H}{H_i} \right) * t_{min} + t_{max} \right] \tag{3.12}$$

where: H – depth from the surface of the reservoir, m; H_i – the maximum for this region thickness of ice, m; t_H – annual mean temperature at the depth of H, °C; t_{min} & t_{max} – respectively minimum average monthly temperature on the surface of ice under the snow and the maximum average monthly temperature of the water, °C.

In deeper lakes and reservoirs, the effect of the water depends on its depth, its genesis, lifetime, size, and on the hydrodynamics of underground waters. Heat exchange at the surface, convective heat transfer due to local heating or wind-induced turbulence, and the presence of seasonal ice all affect the result. In deep reservoirs where the lower layers are not affected by convection (*i.e.*, generally deeper than 10–15m), the heat exchange is less important than the annual turnover of the dense 4°C water in the autumn. Input from underground streams can complicate the situation. In general, the lakes tend to thaw the surrounding permafrost (see Figure 2.12, after Ling *et al.*, 2012).

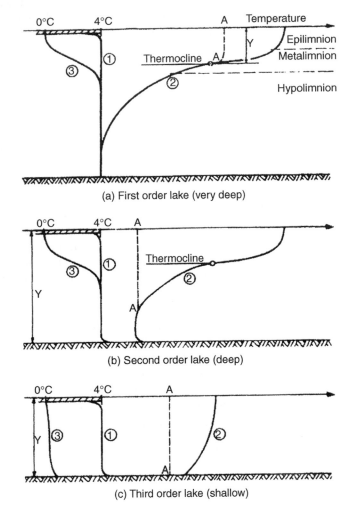

Figure 3.21 Temperature regimes in cold, freshwater lakes of different depths in North America (from Michel, 1971, Figure 30). 1 shows the isothermal condition in fall, 2 the water temperatures in summer, and 3 the winter conditions.

The depth where the average annual temperature is 0°C is called the *critical depth of lake* H_{cr}:

$$H_{kr} = H_i \left(1 + \frac{t_{max}}{t_{min}}\right) \tag{3.13}$$

If the depth of lake is less than H_{kr}, then permafrost exists under the lake. If H exceeds H_{kr} then average annual temperature will have positive value. In the interval of depths from H_{kr} to H bottom deposits will seasonally freeze (Kudryavtsev *et al.*, 1974).

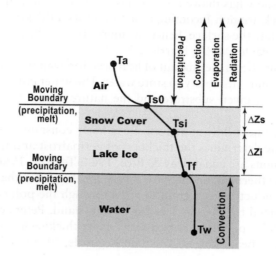

Figure 3.22 Energy budget and thermal offsets where lake ice occurs (modified from Liston & Hall, 1995).

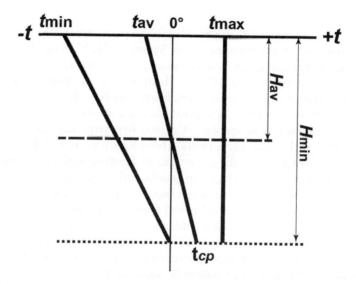

Figure 3.23 The influence of small lakes on the ground temperature (Dostovalov & Kudryatsev, 1967).

3.3.4 Nature of the soil and rock

There can be tremendous variation in the thermal properties of soil, rock or snow, resulting in considerable variation in permafrost distribution. Permafrost first develops beneath or within blocks of rock or peat due to the thermal diode effect. Dark colours absorb more incoming insolation. Reflectivity may be 90% for snow, 12–15% for rock and 15–40% for bare soil (Brown & Péwé, 1973; Brown, 1973a). This explains why there is a substantial change in rate of warming under a given amount of net insolation once the snow has thawed. In addition, water, ice and snow are translucent whereas soil and rock are not. Incoming radiation heats a thickness of these forms of water instead of merely the surface, hence the more effective heat absorbtion of water compared with soil (Pavlov, 1999; Harris, 2002a).

Moisture content affects the amount of heat used for evaporation, while the nature of the material determines its ability to store water. The water content greatly modifies the freezing rate for a given substrate and the nature of the freezing and thawing processes.

Since geothermal heat flow at a given site is fairly constant, the thermal conductivity of the substrate determines the thickness of permafrost and the overlying active layer if other conditions are equal (Day & Rice, 1964; Tedrow, 1966; Brown, 1973b). In massive rock, the thermal conductivity is very much greater than in sediments. In the latter, heat is conducted from grain to grain through the points of contact unless the grains are cemented together by ice. On Devon Island, Peters & White (1971, in Brown & Péwé, (1973, p. 790) reported active layer thicknesses of 70 cm on Stoney Arctic Brown Soils on beach ridges, 47 cm on Regosols, 34 cm in Arctic Brown soils, 24 cm in Gleysols, and 13 cm in peat.

3.3.5　Fire

Many natural fires are started each year within the permafrost areas of the Northern hemisphere, especially in the forested zones (Sykes, 1971; Viereck, 1973). Frequency of dry lightning strikes can be as frequent as once every 90–120 years along the east slopes of the Rocky Mountains in Alberta (Masters, 1990; Johnson & Larsen, 1991), and most of the zone of discontinuous permafrost has been burned at least once during the last 10,000 years. There is a great decrease in fire frequency north of the Yukon River in western Yukon Territory. Greatest fire frequency was probably during the Altithermal/Hypsithermal warm period about 6,000 years B.P. Fire frequency has almost certainly increased with the expansion of Human settlement into the region. The development of increasing MAAT in significantly large areas of continental permafrost is expected to increase the fire frequency, as well as lead to a change in both the weather and vegetation in directions that will favour increasing large-scale forest fires (Tchebakova et al., 2009). Fire can smolder for months in the spruce-lichen transition woodlands of the forest-tundra transition zone in Ontario and Québec (Rouse & Kershaw, 1971), but in the wet tundra, there is less woody fuel available and the effects are greatly reduced (Wein, 1976; Johnson & Viereck, 1983).

There are two kinds of forest fires, viz., crown fires and ground fires. The fast moving crown fires burn the trees over wet ground, but only the surface vegetation is removed, so that the peat plateaus and palsas survive. Thus in the Hudson Bay Lowland, only the top 2.5 cm of the vegetation mat is affected by fires and the underlying peat protects the icy permafrost (Brown, 1965; 1968).

In a study of 11 boreal forest fires in Interior Alaska, Yoshikawa et al. (2003) found that heat transfer to the permafrost by conduction during the fire was not significant. However, immediately following the fire, the ground thermal conductivity may increase by one order of magnitude and the surface albedo may decrease by 50% in cases of substantial burning of organic matter. The thickness of the remaining organic layer determines the effects of the fire. If most of the organic mat remains, or its remaining thickness is greater than 7–12 cm, there is little change in active layer thickness in spite of the change in albedo. Any significant change in the thickness of the organic layer increases downward heat flow resulting in an increase of the thickness of the active layer. After 3–5 years, the active layer will become so deep (up to 4.15 m) that the seasonal refreezing does not reach the base of the previous year's active layer and a talik develops above the permafrost. The return time of fire in the boreal forest in Interior Alaska is about 29–300 years (Yarie, 1981; Dyrness *et al.*, 1986; Kasischke *et al.*, 2000) and is probably decreasing.

Swanson (1996) found that the coldest and wettest parts of the landscape in interior Alaska (concave to plane, lower slope positions, and north-facing slopes) usually failed to thaw deeply after fires. In contrast, soils with permafrost on warmer and drier positions (convexities, crests and shoulders, and east-, west- and south-facing midslopes) thawed deeply if subjected to severe ground fires. Dry soils on gravel and sand do not redevelop permafrost under the present climate, and these provide less cover for voles or forage for moose.

Where conditions are dry at the time of the fire, *e.g.*, at Inuvik on 11th August, 1968, the vegetation and litter is completely burned, leaving blackened soil at the surface. Ice-rich frozen ground then thaws, resulting in surface meltwater and considerable thermal erosion before freeze-up in mid-September. Fire-breaks bulldozed to the permafrost table to control the spread of the fire suffer even greater thermal erosion. Subsequent studies showed that the greatest increase in depth of thaw occurred along the fire-breaks and continued deepening for years, in spite of the regeneration of the shrub vegetation (Heginbottom, 1971; 1973; Mackay, 1977). Similar effects were reported by Viereck (1982) from Alaska.

Smith *et al.* (2015) examined the effects of fire in 1985 further south in the central Mackenzie River valley. The intensity of the burn and removal of the vegetation was greater in the upper part of the sloping hillsides than in the valley bottoms, resulting in greater damage to the active layer. The removal of trees resulted in greater winter snow depths, as well as decreasing albedos in summer. The active layer doubled in thickness within three years, after which it stabilized. Active layer temperatures at the top of the slope started to cool before those at the bottom of the slope, all slopes having regrown a continuous vegetation cover by 2015. It is estimate that it will require 50 years before a full grown forest has redeveloped, and that this is unlikely to be a pure Spruce stand like the cover prior to the fire.

3.3.6 Glaciers

Glaciers develop where there is excessive winter snowfall that exceeds the ablation of the snow pack in summer, resulting in the persistence of snow and ice on the ground surface throughout the year. This snow metamorphoses into ice and as it thickens, it commences flowing downslope. Although it has been suggested that glaciers and permafrost should be treated as part of a single system (Dobiński, 2012), the dominant processes involved are very different. The relationship of glaciers to the mean seasonal freezing and thawing indices are shown in Figure 3.24 (after Harris, 1981b). The accumulation areas of glaciers normally only occur in areas with freezing and thawing indices similar to continuous permafrost. The only exceptions are where avalanches

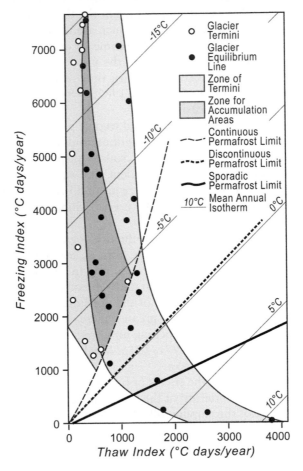

Figure 3.24 Relationship of the distribution of glaciers to the mean seasonal freezing and thawing indices and permafrost zones (redrawn from Harris, 1981b, Figure 4).

add extra snow to the base of a north-facing slope, *e.g.*, on the west precipitous side of Mount Chephren, Banff National Park, Canada.

Since the period of winter cold is longer than the summer, the glacier ice can sustain below-zero temperatures in both the glacier ice and in the underlying soil or rock if the climate is sufficiently cold and the glacier is relatively thin (*cold-based glaciers*), *e.g.*, the Kunlun Shan, Tien Shan and Tanggula Shan. Elsewhere, the base of the glacier is at the pressure-melting point (*warm-based* or *temperate glaciers*), and permafrost will be absent from the underlying substrate (Figure 3.26). The geothermal heat flow can vary considerably between tectonically active areas such as Kamchatka and the Qinghai-Tibet Plateau and stable, cores of continents, but it is this heat that warms the lower layers of glaciers.

The termini can occur inside or outside the permafrost zone depending on the total snowfall, size of the glacier and topography. The exact boundary conditions for the

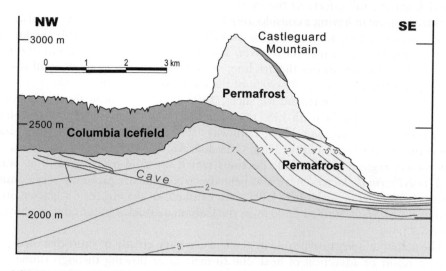

Figure 3.25 Permafrost distribution in relation to the Columbia Icefield in Banff National Park, Alberta. The Castleguard Ice Cave has developed in the unfrozen ground protected from extreme winter cold by the warm-based part of the glacier (see Muir & Ford, 1985). In contrast, the outer edge of the Columbia Icefield on the lower slopes of Mount Castleguard appears to be frozen to the underlying rock, making it a polythermal glacier.

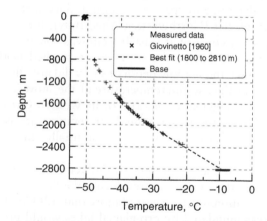

Figure 3.26 Ground temperatures in the borehole through the Antarctic ice cap beneath the Vostock station in Antarctica (Giovinetto, 1960, and Price *et al.*, 2002). Note the significant reduction in the pressure-melting point at the base, and the unusually low surface temperatures. Where mountain ranges are buried in Antarctica, permafrost is usually present beneath the base of the glacier.

formation of permafrost and glaciers depend on the local microenvironment, but the lower limit of permafrost rises in elevation in mountains where the equilibrium line on glaciers is at a low elevation (Harris & Brown, 1978). This is due to the insulating effect of a thick snow cover protecting the ground from the cold winter temperatures (Figure 3.11). The ice in glaciers often acts as a blanket protecting the underlying ground from the full effects of the local temperature regime, (Figure 3.25), but ice differs from snow in having a considerably higher thermal conductivity. It is also more translucent, so that incoming energy is absorbed through a considerable thickness of the ice after it loses its winter snow cover. This also depends on the amount of opaque sediment it is carrying. Above the firn line, the surface consists of snow at all seasons, resulting in reflection back of 80–90% of the incoming radiation.

Tests beneath the Greenland ice sheet, 150 km east of Thule showed a bottom temperature of $-13°C$ beneath 1387 m of ice indicating the presence of a cold-based glacier (Hanson & Langway, 1966; Langway, 1967; Robin, 1972), yet warm-based glaciers may also be found in the High Canadian Arctic (Müller, 1963). Glaciers with temperatures regimes that include portions which are cold-based as well as warm-based (*polythermal glaciers*) have also been described and are probably very common (Blatter, 1990; Blatter & Hutter, 1991). An example of the effect of a glacier on the distribution of permafrost at 2500 m in the Canadian Rocky Mountains is shown in Figure 3.25.

The actual temperature profile in glaciers can vary greatly in short distances, usually as a result of advection of heat due to meltwater flowing through tunnels and cracks, or through different snow and ice densities (Lüthi & Funk, 2001). Where the meltwater flows near the base of the snout of a glacier, it can thaw the permafrost close to its path (Moorman, 2003) in otherwise continuous permafrost.

Glaciers are very sensitive to changes in precipitation or temperature, either or both. Variations in weather can create considerable differences in the temperature profile of small, thin glaciers. Snowpack evolution and ice surface temperatures were studied on the Indren glacier (Northwestern Alps, Italy), for two winter seasons. It appeared that a deep snow cover of at least 100 cm was able to maintain the snow/ice temperature at around $-5°C$ until the snow cover reached isothermal conditions, whereas, during the winter, the shallow depth of snow did not allow basal temperature to reach an equilibrium value, so the temperature at the snow-ice interface oscillated between -2 and $-8°C$ (Maggioni *et al.,* 2009). In Western Canada, the glaciers are rapidly retreating although the mean annual air temperature is scarcely changing. However the winter snowfall has decreased substantially since the Last Neoglacial event.

Judge (1973) suggested that the present ground temperatures in Northern Canada are lower than the ice-bottom temperatures during the Pleistocene, based on an estimated thickness of the Laurentide Ice Sheet of more than 1200 m. During deglaciation, the permafrost in areas inundated by pro-glacial lakes would probably have thawed any remaining ground ice that was present. However, areas such as Plateau Mountain that were not protected by Pleistocene ice sheets developed deep, cold permafrost that could not be formed under the present climate.

3.3.7 The effects of Man

Modern living outside areas of permafrost involves considerable manipulation of the adjacent landscape. This is not compatible with preservation of permafrost. As a result, use of permafrost lands involves considerable modification of methods of construction of buildings, roads, railways and pipelines (Harris, 1986a). Extraction of minerals is also more difficult. Examples of the methods used are given in Part 3 of this book.

Chapter 4

Permafrost distribution

4.1 INTRODUCTION

As noted in the introduction to Part 1, frozen soils include temporarily frozen soils (lasting hours, days), seasonally frozen soils (lasting months) and permafrost (persisting for at least two consecutive years, often hundreds and even thousands of years). Between these categories of frozen soils can be intermediate forms. For instance, seasonally frozen soils which do not melt during a given summer and persist for several years are called *pereletok* in Russia. The area of distribution of frozen soils is called the *cryolithozone* (*temporal, seasonal* and *perennial*). The depth of soil freezing depends on thermal and ground conditions and reaches from a few centimetres to hundreds of metres. Permafrost thickness increases with a decrease in the mean annual air temperature (MAAT). Areas of temporarily and seasonally frozen ground are usually continuous and extend downwards a short distance from the ground surface (centimetres to metres). Permafrost occurs within the upper layers of the ground, below the zone of seasonal freezing and thawing (the active layer).

Permafrost is a major climatically induced condition of the ground that occurs in definite areas near the Poles and at higher elevations at lower latitudes. Figure 4.1 shows its approximate distribution in the Northern Hemisphere using zones representing the average abundance in the ground surface of areas. The southern extensions beyond the general latitudinal distribution are along north-south trending mountain ranges and plateaus (Figure 4.2). The zone of discontinuous permafrost that covers a wide latitudinal area on the relatively flat areas becomes narrow in the mountains due to the topography.

In Russia, if the top of the permafrost coincides with the bottom of the active layer, it is called *confluent permafrost*, but if its top is located deeper, it is referred to as *non-confluent permafrost*. The zone at the base of the active layer that periodically belongs either to the active layer or permafrost layers in a given year is called the *the transient layer* (Shur, 1988a; Shur *et al.*, 2005). Permafrost may consist of two or more layers, divided by thawed interlayers. In this case, it is called *layered* or *multilayered permafrost*. The deep layer of ancient or *relict permafrost*, formed in the late Pleistocene, was first discovered by Zemtsov (1957) in Western Siberia. The relict permafrost lies considerably south of the southern boundary of the modern permafrost (Berdnikov, 1970, 1986).

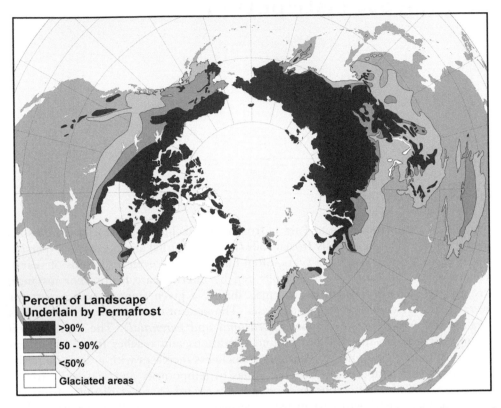

Figure 4.1 Approximate distribution of permafrost in the Northern Hemisphere, based on Brown *et al.*, 1997. See Brown *et al.* (1998) for a digital version.

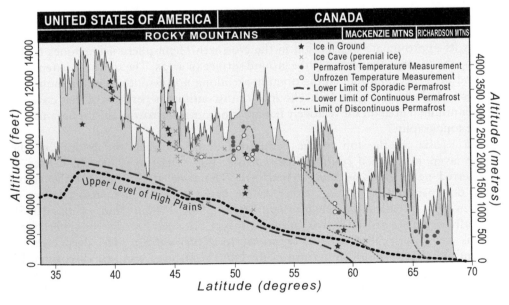

Figure 4.2 Distribution of permafrost zones along the Eastern Cordillera of western North America showing its relationship to the topography (after Harris, 1986b, Figure 1).

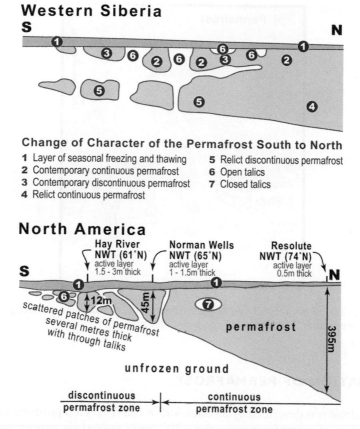

Figure 4.3 Diagrammatic north–south transect of permafrost in Western Siberia and in North America showing variations in thickness and the presence of relic permafrost (5) below that which is more or less in equilibrium with the present-day climate (2) in Siberia. Note the absence of relic permafrost in North America as a result of the permafrost developing after the last deglaciation.

Its existence, as well as the islands of permafrost in areas of non-frozen soils distribution, raise a question concerning the determination of the southern boundary of permafrost.

Figure 4.3 shows a diagrammatic north-south transect of permafrost in Western Siberia contrasted with that in North America. Note the double layer of permafrost towards the southern margin in Russia resulting from relic permafrost still remaining at depth and having been formed during earlier glaciations and the associated development of permafrost. In North America, this is rare because most of Canada was covered by the Wisconsin ice sheets (Brown, 1970).

Figure 4.4 shows diagrammatically the distribution of permafrost in relation to other soils along a north to south transect (after Dostovalov & Kudryavtsev, 1967). Obviously, some means of coping with the broad transition zone is needed.

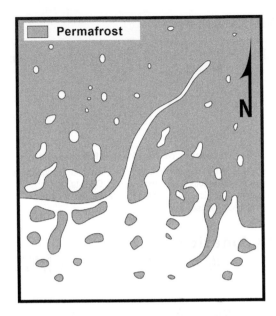

Figure 4.4 Distribution and relationship of permafrost with non-frozen ground along a north–south transect (after Dostovalov & Kudryatsev, 1967).

4.2 ZONATION OF PERMAFROST

After permafrost was described by Middendorf in 1867, G. Vil'd published a map of the outer boundary of permafrost using the −2°C mean annual air temperature isotherm that he presumed would approximate the permafrost limit (Nikiforoff, 1928). This did not work well, so Sumgin *et al.* (1940) developed the idea of *zonation*, separating out intensity of the development of permafrost using the ground temperature at 10–15 m depth as follows:

Continuous permafrost: $<-5°C$.

Discontinuous permafrost: $-5°$ to $-1.5°C$.

Sporadic permafrost: $>-1.5°C$.

Subsequently, these terms have been used internationally, but the definitions have been redefined to reflect the percentage of the area underlain by permafrost, since this is more useful to engineers. A fourth category, *isolated patches*, was introduced by Heginbottom *et al.* (1995).

The first map of permafrost in Canada (Brown, 1967) was also based on mean annual isotherms aided by only ten actual ground temperatures. As in Russia, this approach proved unsatisfactory, and it became clear that there was no simple relationship between MAAT and the actual ground temperature. As a result, mapping has depended on field measurements in spite of the vast areas occupied by permafrost,

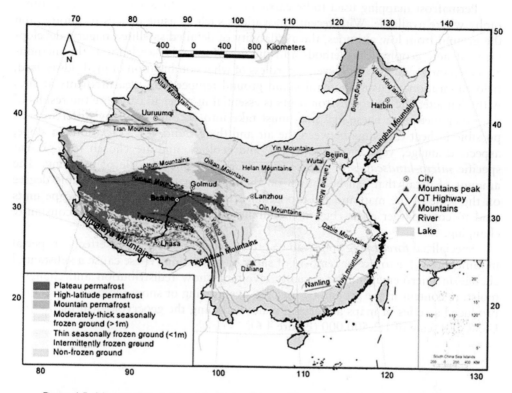

Figure 4.5 Nomenclature for permafrost zones in China proposed by Ran *et al.* (2013).

often in very remote locations. As noted in Chapter 2, much additional research has been carried out on the causes of the poor relationship between ground and air temperatures (Figure 3.2), and the main ones affecting the choice of map units will be discussed below after a brief summary of the variability of mapping units used in the literature. It should also be noted that the existence of permafrost at the ground surface does not provide information regarding its thickness, which depends very often on the depth to ground water, especially in the bottoms of valleys. The presence of icings, seasonal frost mounds and hydraulic pingos indicates ground water close to the ground surface under artesian pressure.

4.3 PERMAFROST MAPPING

The intent of *permafrost mapping* is to show on a map the regularities of the formation and development of seasonally and perennially frozen ground that are present in a given landscape. It depends on both the present and past history of the ground since the climate and associated vegetation and permafrost processes are dependent on these. In addition, the other factors discussed in Chapter 3 greatly affect the temperature distribution of the ground (Brown & Péwé, 1973; Kudryavtsev, 1981).

Permafrost mapping used to be carried out using field work and aerial photography where available. With recent innovations such as using drones to photograph the ground from low altitudes, the availability of detailed satellite imagery, development of new geophysical methods and the use of models (see below), the methods are constantly evolving. However, regardless of what combination of methods is used, field calibration by using boreholes and ground temperature measurements at sites with characteristic micro-environments is essential in order to calibrate the results of the other methods. These *key sites* must take into account as many differences as possible in heat exchange between the air and the ground, *e.g.*, variations in slope, aspect, drainage, vegetation, grain size, and geology, and are chosen to represent specific *micro-landscapes* or *landscapes*, depending on the scale of mapping. The amount of detail that can be shown on the map depends to a considerable degree on the scale of the map and the topography. In lowland areas, the landscape units tend to cover larger areas, but in mountains, the micro-environment is constantly changing.

Specialized *forecasting maps* can be produced for forecasting the effects of special modifications due to fire or removal of snow or vegetation. These cause a substantial change in micro-environment, usually involving snow redistribution and changing moisture content in the ground by logging, overgrazing or snow removal. Russia has produced a series of maps at varying scales including the geocryological map of the USSR at a scale of 1:2,500,000 (Figure 4.6).

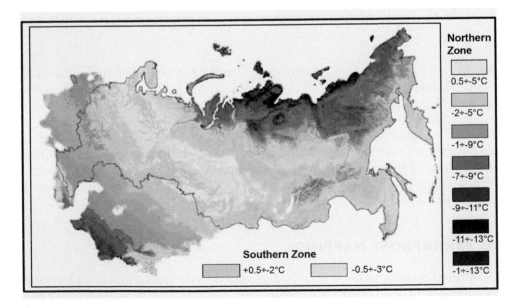

Figure 4.6 The current geocryological map of Russia and surrounded states (1:2,500.000) showing mean annual ground temperatures (after Kudryavtsev *et al.,* 1978).

4.4   EXAMPLES OF MAPPING UNITS USED

In actual practice, different authors in a given country use different definitions of the mapping classes (*e.g.*, Table 4.1 from Ran *et al.*, 2012). There are also great variations in their use between countries. This creates problems in creating regional maps crossing political boundaries. As a result, new classes of mapping units are appearing in recent publications. In the case of the zonation discussed above, Kudryavtsev *et al.* (1978, 1980) and Harris (1986) used 80% as the lower limit for continuous permafrost, whereas Brown *et al.* (1997) used 90%. The latter had the effect of causing areas such as the Mackenzie Delta to be classed as discontinuous permafrost due to the abundant water bodies there. However all the land there is underlain by thick, continuous permafrost. In the case of China, the boundaries of classes have been scrambled even more (Table 4.1). Each author picked boundaries that worked best in his study area. Ran *et al.* therefore suggested a different classification to overcome the problem (Figure 4.5, from Ran *et al.*, 2012). *Mountain permafrost* is permafrost occurring in complex mountain terrain affected by topography and cold air drainage. It has the normal zonation when studied in detail. *Plateau permafrost* refers to all permafrost occurring above about 4000 m on the Qinghai-Tibet Plateau. It includes both mountain permafrost and latitudinal permafrost. Land at this elevation is generally at a higher altitude than the thickness of the surface air masses affecting the climate in lower areas. The term has been applied inconsistently in the case of the permafrost in the

Table 4.1   Differences between the terminology used in China for permafrost zones in publications between 1988 and 2006 (from Ran *et al.*, 2013).

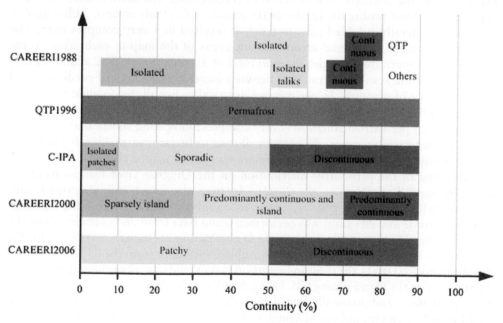

Rocky Mountains south of the 60° parallel, but not further north, though the zonation, processes, and mapping/cartographic problems are the same. *Latitudinal permafrost* includes all the other types of permafrost (sporadic to continuous) including mountain permafrost. This amounts to adapting the classification system to the local environment, which works well for individual regions but does not help in producing small scale maps of large areas crossing political boundaries.

4.5 MODELING PERMAFROST DISTRIBUTION

The idea of modeling permafrost distribution is not new. The models based on air temperature and snow depth by Wild (1882), Voeikov (1889), and Yachevskiy (1889) represent the earliest examples. With the development of powerful computers coupled with satellite remote-sensing technology, it has become possible to make models of permafrost distribution using layers corresponding to a number of the presumed factors controlling its distribution (Anisimov & Reneva, 2006). The number of layers used is usually limited, and the interpretation of the variation in a given factor is based on the available ground and satellite information. These layers are then superimposed on each other and produce a series of classes of probability of the occurrence of permafrost in a given area. In areas with limited variations in surface relief, this works better because mountains create many more local variations that are difficult to model accurately (Etzelmüller, 2013).

Following an earlier idea of Khrustalev & Pustovoit (1988) and Lewkowicz, Bonnaventure *et al.* (2012) used a "probability of permafrost" map, based on modeling some of the selected key factors in a study of the probability of occurrence of permafrost in the southern Yukon Territory (Figure 4.6). The scale of 0.0–1.0 is calculated by superimposing the results of the models (plus their errors). Although it is an area of generally subdued plateau relief, it resulted in a very complex map. The method introduces substantial errors in some areas of the map in each layer, so the resulting map inevitably includes the sum total of all such errors. The addition of an extra layer would result in a trade-off between increased accuracy of prediction and increased error of mapping. Hence it is not practical to use all the possible layers that could affect the permafrost distribution. Note that the boundary between the classes of the traditional permafrost zonation is very complex, indicating substantial local variability. In mountainous areas, the errors will tend to be far greater. By using complex data processing, the best apparent combination of layers can be chosen.

Mapping of the permafrost distribution on the Qinghai-Tibet Plateau has been aided by modelling (Nan *et al.*, 2002). To map the mean annual ground temperature (MAGT) on the Plateau, regression analysis of the relationship between MAGT, latitude and elevation was performed using data from 76 boreholes along the Qinghai-Tibet Highway. When the multiple correlation coefficient reached 0.90, the relationship was considered significant. Coupled with data from the TOPO30 digital elevation model (DEM), the resulting equation was extended to obtain a map of MAGT distribution on the Plateau using GIS. Using the MAGT isotherm of 0.5°C as the boundary of permafrost and seasonally frozen ground, the simulated map was compared with the Map of Frozen Ground on the Qinghai-Xizang Plateau (Li & Cheng, 1996), which had been digitized and organized with the same precision as the simulated map. It was found that the simulated map can effectively describe the distribution of frozen ground on the Plateau, although some differences occur in the southeast. These are probably due to other factors such as limited data and complex terrain. Based on the MAGT, the simulated map was used to divide the Plateau into five zones (stable, metastable, transition, unstable and extremely unstable), and the area of each zone was calculated. Permafrost thickness was determined using a simplified thickness-calculation equation and a simulated map of permafrost thickness was produced. The simulated MAGT map was also used for the prediction of the effects of future permafrost change on the Plateau during the next 50 years using a numeric-simulation prediction method (Li *et al.*, 1996). A 0.04°C/a increase in MAGT was assumed. The prediction indicated that there should be no large-scale loss of permafrost except around the edges of the Plateau. Gruber (2012) has produced a model to map the distribution of permafrost at a global and medium scale. He has to assume that >90% permafrost corresponds to an MAAT of −7.5 to −8.5°C while the <10% permafrost distribution corresponds to an MAAT of −1 to −2°C, following Bockheim (1995), King (1986), Péwé (1983a), Brown & Péwé (1973), and Nikiforoff (1928). The model includes adjustments for terrain ruggedness and MAAT, but he admits that it does not account for the warming effect of deep snow covers, aspect, shade, cold air drainage, etc.. He regards the results as an improvement on the International Permafrost Association map, but accepts that the results are difficult to validate.

4.6 ADVANCES IN GEOPHYSICAL METHODS

Since the 1980s, there have been tremendous advances in the use of geophysical methods of obtaining a picture of permafrost distribution and thickness, ice content and active layer thickness. Currently, most of these methods can only be applied to small areas but can produce very detailed information in 2D or 3D, provided that there is adequate information from boreholes (Fortier *et al.*, 2008; Hauck, 2013). Four methods (electric, electromagnetic, seismic and radar) have been improved to the point where they can produce very detailed information about the shape of the permafrost body, its ice and water content, and the distribution of ground temperatures, provided that good borehole data is available for the site. Figures 4.7, 4.8 and 4.9 provide examples of the results. It is obvious that ground temperatures, active layer thickness and even the presence and shape of the permafrost body can change in very short distances.

4.7 CAUSES OF VARIABILITY REDUCING THE RELIABILITY OF SMALL-SCALE MAPS

Obviously, producing reliable small-scale permafrost maps using a single classification for all areas of the world is proving problematic. It is necessary for planning purposes, but there is obviously too much variability to produce simple, reliable maps of permafrost zonation. The reasons are as follows.

Firstly, a fundamental assumption in mapping is that there is a smooth, continual change between the available data points in the study area. In actual practice, there are many different controls of permafrost temperatures that vary in short distances as

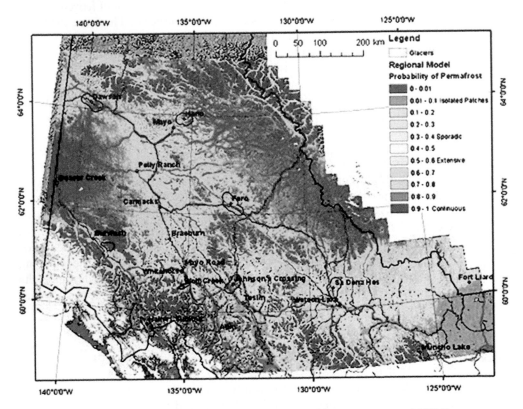

Figure 4.7 The "Probability of Permafrost" map of Bonnaventure *et al.* (2012).

indicated in the local variability of the mean annual offset. There is a natural variation in the thermal properties of the surficial materials in areas of sedimentary rocks. The moisture content, in particular, is always changing. Figures 4.7 to 4.9 illustrate the problem. No single borehole in one of these areas could produce a representative profile that would adequately represent the properties and distribution of permafrost and the active layer for the region. Note that this applies to all the attributes of the permafrost for all scales other than detailed maps. Areas of continuous permafrost will show lesser variability, while areas of sporadic permafrost will be the most variable, other things being equal.

Secondly, the weather and climate are constantly fluctuating, and usually there is a longer-term trend of change affecting a given place. These changes vary from place to place, depending on the proximity to the oceans or the trajectories of the major air masses.

Thirdly, there is the question of what can be shown on a map of a particular scale. Only on large-scale maps of small areas can the details of the variations be plotted. Even then, the map is two dimensional, whereas the landscape is three-dimensional. This can be overcome by using three-dimension representations produced by computers, but needs more data to be effective.

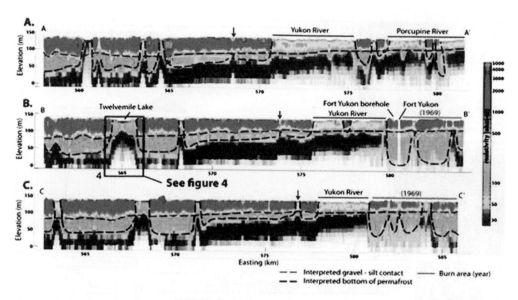

Figure 4.8 An airborne electrical resistivity traverse at the junction of the Yukon and Porcupine Rivers, Alaska (Abraham, 2011).

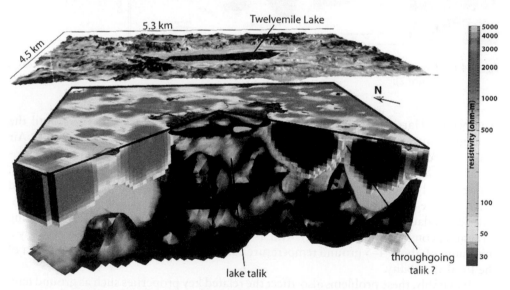

Figure 4.9 Detail of variability in ground temperature around Twelvemile Lake, Alaska, based on an airborne resistivity survey (Abraham, 2011).

Finally, there is the quality and quantity of the data. The longest monitored site is Shargin's well (116.4 m deep) in Yakutsk where regular temperature measurements have been made since 1848 (Kamensky, 2002). When it was found that permafrost in Canada was not readily predicted using mean annual air temperatures, a program of monitoring a series of boreholes along the Eastern Cordillera of western Canada

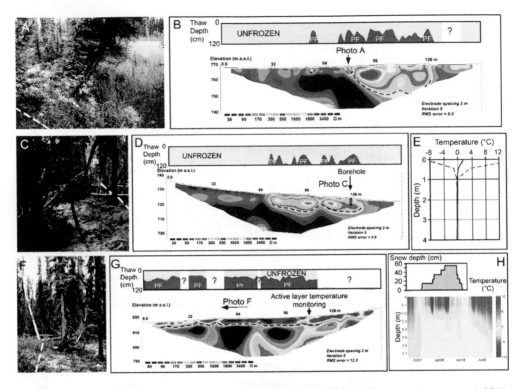

Figure 4.10 Detailed changes in permafrost temperatures in the Yukon (from Lewkowicz *et al.*, 2011, Figure 2).

was begun (Harris & Brown, 1978) and has continued for over 30 years until the ground temperature cables ceased to produce reliable data (Harris, 2008, 2009). Air temperature monitoring is continuing. Other networks followed, *e.g.*, along the proposed Foothills Pipeline route along the Alaska, Klondike and Dempster Highways. Unfortunately, there was only limited data collection from the commercially operated networks. Regular monitoring takes place along the Trans-Alaska Pipeline which provides good data for that transect across Alaska. It is essential to collect the data using a reliable, consistent methodology with as continuous a record as possible. Usually a minimum of about 4–5 ground temperature cables are needed to provide some idea of the local variability.

Inevitably, these problems also affect the related key properties such as ground temperature profiles, active layer thickness ice content, and thickness of permafrost. Thus small and medium-scale maps of permafrost properties must be used with caution. Hinkel and Nelson (2003) found that on the North Slope of Alaska, the end-of-season thaw depth was strongly correlated with local air temperature on an inter-annual basis. There was significant intrasite variation in thaw depth and near-surface moisture content within the grids, reflecting the local influence of vegetation, substrate, snow cover dynamics and terrain. Thaw depth was significantly greater in drained thaw-lake basins producing a bimodal distribution of properties. Foothill sites demonstrated much larger variability, making the mapping difficult, unlike the coastal plain. Mountain sites exhibit even greater variability, especially on benches below steeper slopes. Thus monitoring of ground temperatures at Marmot Basin #2, Jasper National Park produced a record of high environmental variability at the site rather than a clear picture of regional climate change (see Harris, 1999, 2001b).

4.8 MAPS OF PERMAFROST-RELATED PROPERTIES BASED ON FIELD OBSERVATIONS

This was the main way of producing these maps until computers made modeling of possible permafrost distributions. Several properties are usually measured to produce maps that are used in many practical applications. The distribution and aerial coverage of permafrost has been dealt with above.

4.8.1 Permafrost thickness

Hinkel & Nelson (2003) found that on the North Slope of Alaska, the end-of-season thaw depth was strongly correlated with local air temperature on an inter-annual basis. There was significant intra-site variation in thaw depth and near-surface moisture content within the grids, reflecting the local influence of vegetation, substrate, snow cover dynamics and terrain. Thaw depth was significantly greater in drained lake basins but thick beneath old, unglaciated land surfaces (Figure 4.11). The latest maps show a clear relationship between the Yakutian Shield and permafrost thickness. The metamorphic and granitic rocks of the Shield have a much higher thermal conductivity than the sedimentary rocks elsewhere, apparently producing the >1500 m thick relic permafrost during the various cold events during the last 3.5 Ma. Elsewhere in Russia, the permafrost is much thinner in the unglaciated, sedimentary permafrost areas. In Canada, most of the permafrost developed since the last glaciations so that permafrost thicknesses are related to the climate of the last few millennia. Maximum thickness is in the Arctic Islands and is only about 500 m thick except in unglaciated northern Alaska where it may reach 740 m. Thicker permafrost is also encountered on the isolated unglaciated mountain peaks along the margins of the last ice sheets such as Plateau Mountain (Harris & Brown, 1978; Harris & Prick, 1997). On the Qinghai-Tibet Plateau, permafrost appears to be less than 50 m thick, and relic permafrost is absent.

4.8.2 Maps of ice content

Maps of *ice content* are rare except for maps of areas of massive ice. The International Permafrost Association (IPA) permafrost map tried to show the probable distribution of areas of high ice content together with the permafrost zonation (Figure 4.12, after Brown *et al.*, 1997). It was based on limited field data, aided by separating out the zones of silty lacustrine deposits and peats of the wetter areas as being usually rich in ice (Mackay, 1971a, 1971b; Brown & Sellman, 1973). Currently, the National Snow and Ice Data Center (NSIDC) at Boulder, Colorado has a web site with the latest version of the IPA map, together with a link to the actual data on which it is based (https://nsidc.org/data/docs/fgdc/ggd318_map_circumarctic/). There are also maps of *permafrost temperature* on the same website.

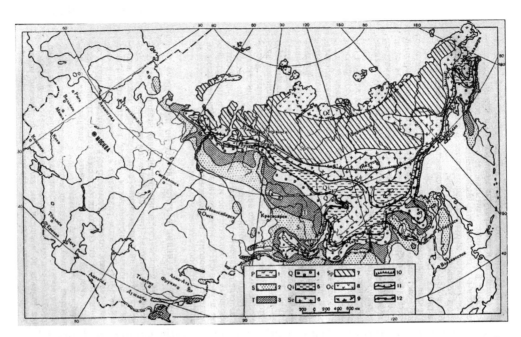

Figure 4.11 Permafrost in Russia (Tumel, 1946, modified): 1 – regions without permafrost among the territory with permafrost; 2 – regions of permafrost islands, thickness is less than 15 m; 3 – the permafrost regions, thickness is less than 35 m; 4 – the permafrost regions, thickness is less than 60 m; 5 – the permafrost regions, thickness is less than 120 m; 6 – the permafrost regions, thickness is less than 250 m; 7 – the permafrost regions, thickness is less than 500 m; 8 – the permafrost regions, thickness is more than 500 m; 9 – northern boundary of areas with temperatures at the depth of 10 m more than − 10°; 10 – northern boundary of areas with temperatures at the depth of 10 m more than − 5°; 11 – northern boundary of areas with temperatures at the depth of 10 m more than − 3°; 12 – northern boundary of areas with temperatures at the depth of 10 m more than − 1°C.

4.8.3 Water resources locked up in perennially frozen ground

In arid areas, water shortages are common, and the groundwater resources locked up in various *rock-ice features* are now being inventoried (Millar & Westfall, 2008). These include rock glaciers and ground ice. Corte (1976a, 1976b, 1987) and Corte and Beltramone (1984) have discussed the water storage in and under rock glaciers as a source of water for Mendoza, Argentina. Barsch (1996, section 10.4.1) discusses the estimated water storage in the rock glaciers in the Alps. For Swiss rock glaciers, he estimated a total water volume stored as ice amounting to 0.8 km^3 (Barsch, 1977). This represents about 1.5% of the water stored in the nearby glaciers, but excludes water held as perennial ice in other soils. In contrast, Buchenauer (1990) estimated the water content in the active rock glaciers in the Schober Mountains of the eastern Tirol (Austria) to be 34–45×10^6 m^3, *i.e.*, a quantity roughly equal to that held in the glaciers of the region. Water stored in mountain permafrost was estimated at 100–115×10^6 m^3 while the inactive rock glaciers were estimated to hold 20–23×10^6 m^3

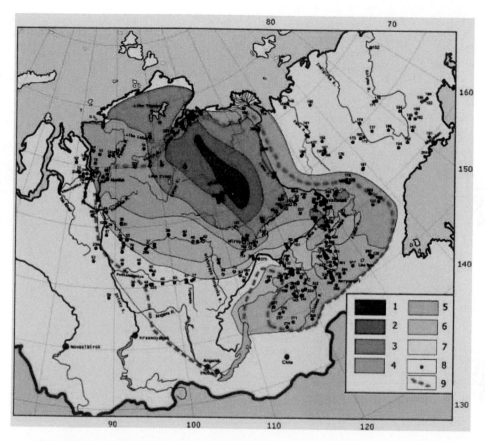

Figure 4.12 Permafrost thickness under the Russian Platform, underlain by the Yakutian Shield. 1 = Continuous permafrost >1000 m thick; 2 = continuous permafrost 800–1000 m thick; 3 = continuous permafrost 500–800 m thick; 4 = continuous permafrost 300–500 m thick; 5 = continuous permafrost 100–300 m thick; 6 = discontinuous permafrost <200 m thick; 7 = discontinuous and sporadic permafrost <100 m thick. Dashed red line (9) is the boundary of the Siberian Platform. © M. Zheleznyak.

of water. Haeberli (1985) regards these as only first-order approximations. Zhang *et al.* (1999) estimate the total ice content in permafrost to correspond to the sea-level equivalent of 3–10 cm of water, for a total of 11.37 to 36.55×10^3 km^3 of water.

Kotlyakov & Khromova (2002) estimated the total volume of underground ice in Russian permafrost as about 19,000 km^3. They recognized two types of water, *viz.*, cryptic, invisible water and ice in the form of bonded water around mineral grains, and secondly, visible ice occupying fissures and as stratified icy deposits. The first type is not available after thawing, but the second type joins the free gravitational water, mainly forming the water table.

There are estimates of the reserves of ground ice in the permafrost areas on the Qinghai-Tibetan Plateau in China. Thousands of boreholes have been drilled along the Qinghai-Tibetan Highway (QTH) to investigate the permafrost distribution during

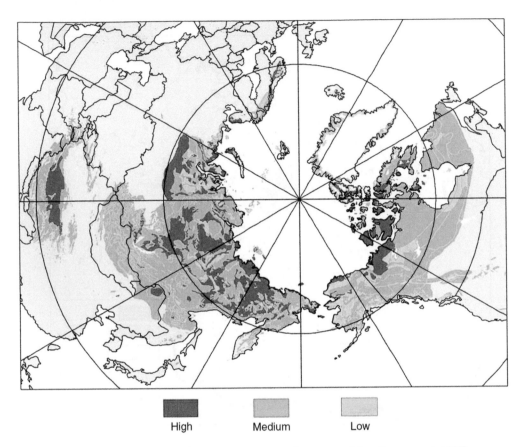

High   Medium   Low

Figure 4.13 Distribution of ice content in permafrost areas (after Brown *et al.*, 1997).

the past 50 years. Permafrost profiles from 697 boreholes, involving 9,261 actual measurements, were selected to analyze the characteristics of ground ice distribution. Based on the horizontal distribution of ground ice and water content revealed from the borehole profiles, the permafrost along the QTH can be divided into five types, *i.e.*, low-ice content permafrost, high-ice content permafrost, ice-rich permafrost, ice saturated permafrost and ice-layers with soil. The length of each type of permafrost along the QTH was calculated. The vertical distribution of ground ice was also calculated for three vertical sections, *i.e.*, within 1 m below the permafrost table, from 1 to 10 m below the permafrost table, and beneath 10 m below the permafrost table. The average thickness of permafrost along the QTH is about 38.79 m, and the average weighted water content is about 17.19%. Accordingly, the total volume of ground ice in the permafrost regions on the Qinghai-Tibetan Plateau is estimated to be 9,528 km^3. Examples of the borehole logs are shown in Figure 4.13, together with the details of the three typical profiles in Table 4.2.

Wu *et al.* (2015) have found gas hydrates in significant quantities in the Kunlun Pass Basin at a depth of about 250 m. This is a significant discovery since it indicates

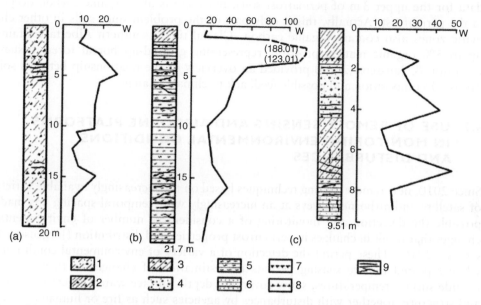

Figure 4.14 Stratigraphy of three typical boreholes on the Qinghai-Tibet Plateau (from Zhao *et al.*,
2010b). From left to right, Xidatan B-4, Fenghou Shan Pass B-1, and Tanggula Shan.
I = rubbly sandy loam; 2 = rubble loam; 3 = gravelly clay; 4 = pebble soil; 5 = mudstone;
6 = sandstone; 7 = permafrost table; 8 = permafrost base; 9 = ground ice.

Table 4.2 Details of the typical boreholes in permafrost regions along the Qinghai-
Tibet Highway. After Zhao *et al.* (2010b).

	Place	Depth of borehole m	Location	Elevation m (a.s.l.)
a)	Xidatan B-4	20.0	35°43′22″N 94°08′30″E	4427.5
b)	Fenghuo Shan Pass B-1	21.7	34°41′06″N 92°54′30″E	4950
c)	Tanggula Shan	9.51	33°30′40″N 91°58′12″E	5039

that the ice underground in these mountain ranges is thicker than previously realized
and contains potentially important gas hydrates.

4.8.4 Total carbon content

With the current concern over the effect of man-induced carbon emissions, a considerable amount of work has been carried out on *total carbon content* of permafrost
soils and the release of carbon dioxide and methane into the atmosphere. Tarnocai
et al. (2009) provide a map of the estimated organic carbon in the soils of the active
layer for the Northern Hemisphere. There is more carbon in these soils than in the
atmosphere. Kuhry *et al.* (2013) have improved the estimates using the available

data for the upper 3 m of permafrost soils, but there is also organic carbon deeper in the permafrost. Actually, this is only part of the problem since soils in other climatic zones also contain organic carbon. Glacial till in southern Alberta contains up to 5% organic matter, probably representing ground-up boreal forest tissues. Davidson & Janssens (2006) provided an overview of the relationship between soil carbon decomposition and possible feedback to climate change.

4.9   USE OF REMOTE SENSING AND AIRBORNE PLATFORMS IN MONITORING ENVIRONMENTAL CONDITIONS AND DISTURBANCES

Since 2010, new remote sensing techniques based on the increasingly available variety of satellite and airborne surveys at an increasingly wide temporal spacing has made possible the detection and monitoring of a considerable number of environmental changes that result in changes in permafrost properties and distribution (Jorgenson & Grosse, 2016). These permit the detection of a variety of environmental conditions, while repeated remote sensing permits the estimation of changes in them. These include surface temperatures, snow cover and depth, surface water, vegetation cover and structure, together with disturbances by agencies such as fire or human activity. Features indicative of growing or degrading permafrost can be monitored, *e.g.*, the growth of thermokarst lakes or pingos. This has the potential of allowing the assessment of annual variability which is currently poorly known, as well as the results of short and long-term climatic fluctuations. By choosing an appropriate scale and using frequent resurveys, it should be possible to determine whether the high degree of variability shown in detailed studies (Figures 4.8, 4.9, and 4.10) fluctuate or are stable. Some properties, such as soil moisture in the alpine active layer at the borehole Marmot Basin #2 in Jasper National Park, exhibit vast changes from year to year based on ground penetrating radar. This is due to variations in the distribution and thickness of the winter snow pack and the weather during the thawing of the snow.

4.10   SENSITIVITY TO CLIMATE CHANGE: HAZARD ZONATION

Sensitivity of permafrost to climate change has become an important topic. Harris (1986a: 1986b) produced permafrost sensitivity maps for the Eastern Cordillera of western Canada and the whole country based on permafrost temperature. As previously discussed in this Chapter, models have been used for over a century. With the realization that the climate is currently in a state of change at many locations in permafrost areas, there has been an explosion of attempts to model the effects of climate change on permafrost. It started with use of a 2 times CO_2 scenario and spatially referenced databases to identify areas of relative sensitivity to thaw in Canada (Smith & Burgess (1988, 1999). Subsequent work at large scales has included relatively simple assumptions about the nature and regional variability of both climatic change and geocryological conditions (Chizhov *et al.*, 1983; Vyalov *et al.*, 1993).

The zonation of risks had been used by the insurance industry for several decades. Carrara *et al.* (1995) introduced the term *hazard zonation* for landslides, and this was adopted by Nelson *et al.* (2002) using a series of models, to produce two hazard

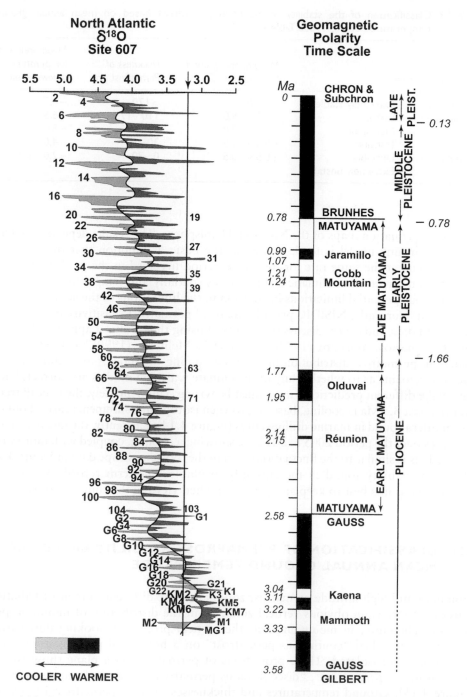

Figure 4.15 Oxygen isotope palaeotemperature record (Ruddiman *et al.,* 1986; Raymo, 1992) and geomagnetic polarity timescale (Cande & Kent, 1995). Black and white areas are normal and reversed polarity respectively. The arrow at the top indicates the mean Holocene oxygen isotope value. Numbers on the peaks and troughs are the isotope stages of Shackleton *et al.,* 1995.

Table 4.3 Classification of the stability of mountain permafrost based on mean annual ground temperature (Cheng, 1983a, Table 1).

Name of the zones		Mean annual ground temperature °C	Thickness of Permafrost (m)	Mean annual air temperature at lower limit °C
Upper zone	Extremely stable	<−5.0	>170	−8.5
Middle zone	Stable	−3.0~−5.0	110~170	−6.5
	Metastable	−1.5~−3.0	60~110	−5.0
	Transition	−0.5~−1.5	30~60	−4.0
Lower zone	Unstable	+0.5~−0.5	0~30	−2.0~3.0
	Extremely unstable			

permafrost zonation maps of the Northern Hemisphere. The results depend on which climate warming model is used, as well as the basic assumption that the warming will continue at the same rate for the entire period involved in the prediction. Since the rate and direction of climatic change varies spatially and temporally, these maps have obvious potential limitations in addition to the limitations of the actual database being used. Currently, NISIDIC provides maps of permafrost sensitivity based on the latest field data and others discuss the results of using the results of applying different climate scenarios. Anisimov (1989) and Street & Melnikov (1990) has predicted the change in permafrost distribution in Siberia assuming the models are valid. The big question remains as to whether any of the climate warming models are correct. Since they make different predictions, most must be wrong. Unfortunately, the overall trend over the past 3.5 Ma is cooling, based on oxygen isotope measurements carried out on foraminifera found in marine drilling cores (Figure 4.14). Currently, the mean air temperatures over the period 2005 to 2012 are showing a cooling, coupled with more variability. This is similar to the fluctuations seen in the oxygen isotope data in Figure 4.13 and the variability found in the currently available long-term records at weather stations. Thus it is best to keep an open mind when viewing the climate models.

4.11 CLASSIFICATION OF PERMAFROST STABILITY BASED ON MEAN ANNUAL GROUND TEMPERATURE

Mountain and high-latitude permafrost zones are quite different in their distribution patterns. There is an obvious vertical zonation in the distribution of mountain permafrost. Therefore, in mountain areas the vertical projection (looking down from the sky) of so-called "continuous permafrost" on a horizontal plane is usually not continuous, but isolated. Thus, projections of permafrost on a plane show islands of "continuous permafrost", "discontinuous permafrost", and "isolated permafrost" (Figure 4.15). Ground temperatures and thicknesses of these permafrost islands are quite different. Using a label of "continuous permafrost" in mountain permafrost research has caused confusion, and gives rise to difficulties in comparing permafrost conditions in different regions.

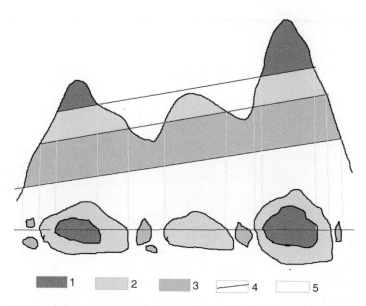

Figure 4.16 Sketch showing various permafrost islands. I – continuous permafrost; 2 – discontinuous permafrost; 3 – isolated permafrost; 4 – boundary of zones; 5 – seasonally frozen ground.

Based on the thermal stability of permafrost, Cheng (1983) devised a classification of mountain permafrost using mean annual ground temperature (MAGT) and related it to the mean annual air temperature (MAAT), rather than using the terms continuous, discontinuous, and sporadic (Table 4.3). Péwé (1983a) considered this method "workable" and Harris (1986b) compiled a map of the stability of North American permafrost based on MAGT (Figure 4.16). Vertical permafrost zones based on thermal stability are quite different in various climatic regions. In continental climate regions, there are three zones of permafrost. In subcontinental regions, there are two zones, and there is one zone in maritime regions (Figure 4.17). Various combinations of elevation, latitude, and distance from the ocean define these vertical zones, which allow a good description of the regional distribution of mountain permafrost. Ground temperature is the main parameter in the thermal regime of permafrost, and is important in not only estimating the distribution and thickness of permafrost, but also determining the hydrological system, the ecological environment, geomorphological processes, and carbon budgets. Mean annual ground temperature reflects the stability or sensitivity of permafrost to climate changes and human activities (Gravis & Melnikov, 2003). The ground thermal regime is an important component of a feedback mechanism into climate systems. Mountain permafrost has received increasing attention since the 1980s, due to concerns about the stability of linear infrastructure (such as railroads, highways, and transmission lines), and about the causal relationships between ground thermal regimes and geohazards in mountainous areas during a period of rising temperature and associated permafrost degradation (Etzelmuller, 2013; Gruber *et al.*, 2015). Recent new permafrost monitoring stations have increased our knowledge of

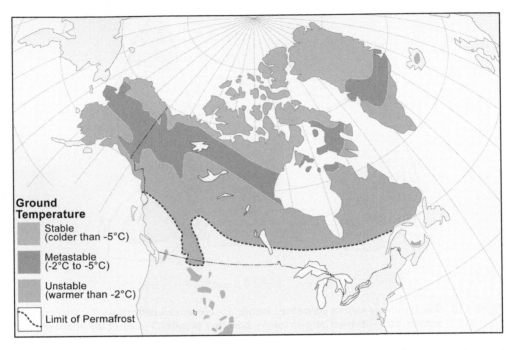

Figure 4.17 Stability of permafrost in North America based on the distribution of mean annual ground temperature (from Harris, 1986a).

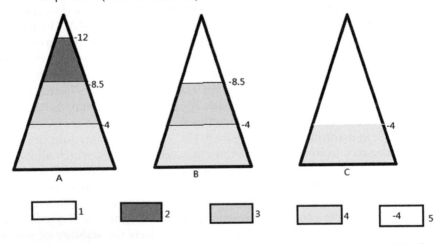

Figure 4.18 Vertical zones of permafrost in various climatic regions (Cheng, 1983, Figure 2). A—continental climatic region; B—sub-continental climatic region; C—maritime climatic region; 1—glacier; 2—upper zone; 3—middle zone; 4—lower zone. Temperatures are mean annual air temperature (°C).

the thermal state of mountain permafrost, and will provide important data on permafrost stability. The complex topography adds additional variability due to factors such as snow redistribution, aspect, slope angle and hydrology, making the permafrost more variable than on lowlands.

Part II

Permafrost landforms

II.1 INTRODUCTION

The Russian Geologist, Vasilii Vasielevich Dokuchaev was the first to realize that soils occurred in distinct zones across Russia that were closely related to climate (Dokuchaev, 1879a, 1879b, 1893). This was further developed by a close friend, Nikolai Mikhailovich Sibirtzev, who introduced the concept of soil zonation (Sibirtzev, 1895). He divided soils into zonal, intrazonal and azonal groups, and produced the first map of the soils of the world. The *zonal soils* are formed at well-drained sites and occur in zones closely following those of the vegetation and climatic belts. The *intrazonal soils* occur in lines or patches cutting across the zonal soil belts and are usually represented by flood plains of rivers and creeks or other places with poor drainage. The *azonal soils* do not show enough development of soil horizons to be classified in the other two groups. An example would be dune sand. The distribution of permafrost is similar to that of the soils, since permafrost processes are controlled by similar factors.

The permafrost zone exhibits soils including unique landforms not seen elsewhere in the different climatic zones. The landforms are mainly developed by processes related to the phase changes of water in areas affected by sufficiently cold climatic conditions, and the permafrost limits clearly cut across the boundaries of otherwise zonal soils. Permafrost may occur in maritime areas such as Norway, Iceland, Japan or New Zealand, or in extremely continental areas such as Central Siberia, Antarctica or east of the Western Cordillera in Canada.

In the 1950's, the systems of soil classification were modified by both the Food and Agriculture Organization (FAO) and the United States Department of Agriculture (USDA) to accommodate the wide variety of soils that was being discovered in many different environments including areas of permafrost. FAO introduced the term *cryosol* for soils with permafrost, while the USDA initially used the term *gelisol*. The former is now most widely used, and the latest system is discussed in Kimble (2004). These soils have permafrost in the upper 1 m (IUSS Working Group WRB, 2014), or within 1 m depth in soils lacking cryoturbation, or within 2 m of the surface in mineral soils with cryoturbation (Soil Classification Working Group – Canada, 1998; Soil Survey Staff, 2014). However, the WRB classifies all soils with continuous hard bedrock within 0.25 m of the surface or soils with <10% fine earth (<2 mm) as *leptosols*, regardless of the soil temperature regime. Obviously, the classification is still evolving.

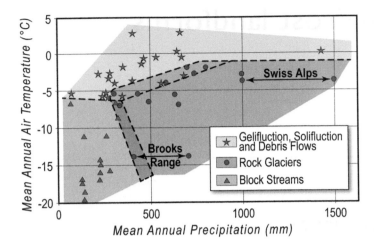

Figure 5.1   Changes in dominant slope processes on slopes underlain by permafrost in relation to mean annual air temperature and mean annual precipitation (after Harris, 1994a, Figure 13). These characteristic landforms are discussed in detail in Chapters 8 and 9. The overlap is due to variations between factors such as mean annual air and ground temperatures.

The last great German explorer and Physical Geographer, Karl Troll, thought that there should be a connection between the processes forming landforms in all parts of the world and the climate under which they developed. Troll (1944, 1958) discussed some of the processes occurring in frost-affected soils. After World War II, he led a series of expeditions to South America and Eurasia to try to determine the geographical distribution of geomorphological processes in relation to both climate and vegetation. The results of his expeditions were summarized shortly before his death (Troll, 1972), but, unfortunately, he never found the missing piece of the puzzle.

Subsequent studies in central Asia and in the Kunlun Mountains of the Qinghai–Tibet Plateau have demonstrated that there are different dominant processes occurring on steep mountain slopes in different climates (Harris, 1992). On slopes in wet, cold areas, rock glaciers are widespread, whereas in very cold, dry climates, they are replaced by block slopes (called kurums in Russia) and by block streams, wherever there is a suitable source of blocks (Figure 5.1). In warmer climates, gelifluction is the dominant process over permafrost ground. Where well-drained substrates such as limestone occurs in areas where there is negligible winter snow cover, block slopes may also develop, *e.g.*, Plateau Mountain, Alberta, Canada (Harris and Pedersen, 1998). More and more, the evidence is accumulating to indicate the importance of climate in determining which processes are taking place in the formation of cryogenic landforms.

Permafrost land forms exhibit *equifinality, i.e.*, where similar-looking looking landforms are produced by different processes in dissimilar climates. Precipitation tends to fall in largest amounts on mountains or hills, but gravity causes moisture to

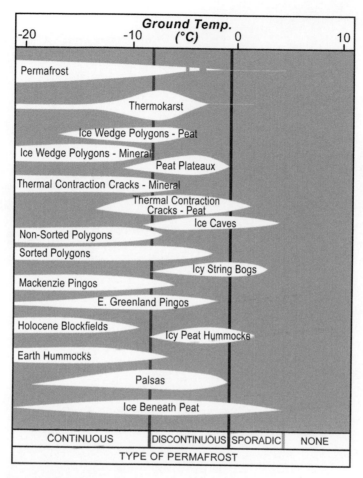

Figure 5.2 Distribution of individual cryogenic landforms with permafrost type (modified from Harris, 1986a, Figure 2.2).

migrate down-slope to the low-lying areas where it accumulates in the ground, often producing distinct landforms. In this part of the book, the various unique landform features found in permafrost areas are described, together with a discussion of their distribution and the processes that form them. Figure 5.2 shows the general distribution of the individual landforms within the main permafrost zones.

Chapter 5

Frost cracking, ice-wedges, sand, loess and rock tessellons

5.1 INTRODUCTION

Frost cracking is vertical or sub-vertical fracturing of the ground along a straight line as a result of thermal contraction (Figure 2.14), or as a curved line by stretching due to dilation cracking over an expanding mound (Figure 2.15). The latter are found on the slopes of growing mounds such as pingos or frost mounds, and are discussed in Chapter 7.

Thermal contraction cracking occurs in the case of the relatively soft, wet active layer subjected to very cold air temperatures (Leffingwell, 1915, 1919). When the ground temperature drops in the fall, the ground contracts, but after most of the moisture in it changes to ice, and the ground temperature falls below $-15°C$ (Dostovalov, 1952; Lachenbruch, 1962; Péwé, 1966), the material becomes very brittle. If the temperature continues to fall, the layer tends to contract, but this sets up horizontal stresses. These are partly offset by creep of the ice, but if there is a sudden decrease in temperature from $-20°C$ to below $-32°C$, cracking occurs (Allard & Kasper, 1998; Fortier & Allard, 2008). Mackay (1993b) suggested that the temperature drop producing the cracking should be about $1.8°C/day$ for at least 4 days. This is when these stresses induced by thermal contraction exceed the tensile strength of the ground, resulting in frost cracking (Figure 5.3). The thermal stresses are proportional to the rate of change in ground temperature, the gradient of temperature with depth, depth below the surface, modulus of elasticity, and coefficient of expansion of the soil components. Most minerals have a coefficient of thermal expansion of $2–12*10^{-6}1/°C$, whereas ice has values of $30–60*10^{-6}1/°C$, depending on the structure of the ice, the angle of the optical axis of the crystals, the temperature, and potential phase changes altering its volume. Air, which does not freeze, has a very high capability for expansion or contraction. With all these variables, it is not surprising that cracking is not readily predictable, is not entirely controlled by temperature, and does not happen every year along any particular crack (Mackay, 1974, 1989, 1992, 1993b; Harry *et al.*, 1985).

The first model of cracking was probably that of Dostovalov (1952), which allowed an estimation of the distance of the spacing of cracks and the depth of cracking. The latter is closely related to the average annual ground temperature, while the distance between polygons increases with lower average amplitude of ground temperature fluctuations. The spacing of the polygons varies from 0.5 to 10 m in areas in continental climates and 20–80 m in less continental situations. Width of cracks is 5–10 cm maximum, with a depth of 3–4 m. Grechishchev (1970) improved the model

Figure 5.3 An open frost crack on the Yamal Peninsula, Western Siberia. © A. Gubarkov.

by taking account of weakening of the stresses with time, but necessarily assumed a constant climate, and has shown that ice-wedges form mostly during temperature fluctuations varying from a few days to a one to two week period. On Mars, the size of contraction polygons reaches hundreds of metres, presumably due to the long periods of temperature fluctuations.

Western Siberia experiences cracking in peat at mean annual temperatures below −2°C, compared with −6°C in alluvial sandy loams. In Siberia, the southern boundary of significant frost cracking corresponds to the average annual air temperatures (MAAT) of −5 to −6°C (Romanovsky, 1974). Romanovsky (1977) recognized three basic zones of frost cracking. The southern zone has an MAAT > −3 to −4°C, and cracking is limited to the active layer or layer of seasonal frost. This cracking produces *soil wedges* (see below) consisting of mainly polygonal forms of 0.3 to 2 m diameter (Figure 5.3). The second, transitional zone has cracking in both the active layer and the upper part of the permafrost, while the northern zone has the cracking primarily in the permafrost. In this northern zone, the mean annual ground temperature is below −7°C. Present-day cracking produces polygons with diameters of 10–30 m and a depth of 5–8 m, but occasionally with cracking 3 to 4 m in diameter as far north as Arctic Yakutia. However, ice-wedges deeper than 4–5 m mainly occur in older sediments, *e.g.*, the Upper Pleistocene (Sartan time) that have undergone various types of thermal deformation (see yedoma, Chapter 6). Other regional studies of cracking of the ground

Figure 5.4 Low centre polygons with raised marginal ridges and a lower centre at Prudhoe Bay, Alaska. © S. A. Harris.

in Russia include Transbaical and Yakutia (Vtyurina, 1962) and also Central Siberia (Fotiev *et al.*, 1974).

In North America, the MAAT at the top of the crack was about −20°C in the cases studied in Northern Québec (Allard & Kasper, 1998; Fortier & Allard, 2008, 2013). This is consistent with the theoretical work of Lachenbruch (1962, p. 20) and Grechishchev (1973, p. 231). Not all the cracks will open in any one year, and if no ice or sand fills the crack during the winter or spring, it merely closes up as in Figure 2.14. In that case, iron staining may show the trace of the former crack. Once the ground has cracked along a given trace, any subsequent cracking in succeeding years tends to occur along this vertical/sub-vertical plane of weakness. However, the open cracks can be filled by any available material, *e.g.*, ice, loess, sand, rock, or any mixture of these. Mackay (1974, 1978, 1984, 1992, 1993a) has documented hundreds of cracking events where the cracks were developed in ice-wedges.

Cracking can also develop in peat (Vasil'chuk *et al.*, 1999), but in this case, the peat can sometimes accumulate at a very fast rate (4.5–5.0 m in 700–800 years) during the climatic Optimum, so that these Holocene ice-wedges are *syngenetic*. This occurred in the Central Yamal Peninsula, where Brouchkov observed cracking and the growth of an ice-wedge on the asphalt air strip with a sand base at Amderma airport on the Kara Sea coast. In Antarctica, the cracking on old land surfaces in sediments at Beacon Hill, Southern Victoria Land, produced fractures of 10–20 m diameter (Sletton *et al.*, 2003).

The cracking on flat areas occurs in a series of relatively straight lines forming a polygonal pattern or *tessellon* (Figure 5.4). The surface of the ground exhibits rapid

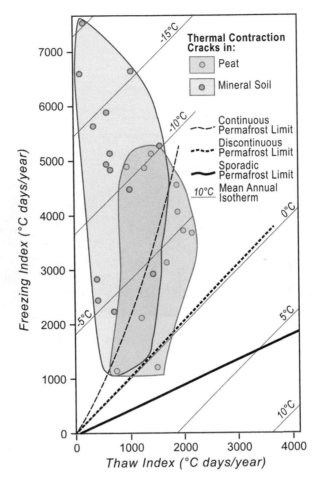

Figure 5.5 Distribution of thermal contraction cracking in peat and mineral sediments with freezing and thawing indices (after Harris, 1981c, Figure 4B; 1982a).

thermal contraction and expansion that correspond to changes in the soil temperature during the winter, resulting in corresponding variations in the width of the crack (Allard & Kasper, 1998). Mackay showed that the cracking usually starts at about 3 m depth near the top of the permafrost table on the Mackenzie Delta, and propagates up to the surface and deeper into the ground (Mackay, 1984b). The result is a polygonal pattern of cracking with diameters of 5 to 500 m, depending on the local conditions. Where the mean ground surface temperature is below −20°C, the cracking may produce an audible bang, similar to cannon shot (Mackay, 1993a) due to higher sound–propagation rates. The development of cracking is not a zonal permafrost feature since it can occur elsewhere in very cold winters (Figure 5.5). However, infilling of the cracks by ice or sediment is zonal because it requires repeated cold conditions for a significant number of years in order to develop an obvious landform. The exact number of sides to these tessellons depends on the topography and the degree of homogeneity of the materials present. Normally the polygons are 5–6 sided in homogenous sediments, and 4-sided along pre-existing structural lines, *e.g.*, rivers.

There are two systems of classification of infilled cracks that are used. In Russia, Romanovsky (1977) divided the polygonal structures into four types, *viz.*, original ground wedges, repeated ice-wedges, primary sandy wedges and pseudomorphs of repeated wedges. However elsewhere, it is usual to subdivide them into ice-wedges, ice-wedge casts, and tessellons. The tessellons can be filled by sand, loess, or rock fragments, with ice-wedges being a case where the infilling consists primarily of ice. Transitional forms between ice-wedges and tessellons with varying amounts of clastic sediment are also known. This second classification is used here with cross-referencing to the Russian system, since this emphasizes the differences in the environmental history that each type has endured. The massive, metamorphosed, mainly Pleistocene ice-wedges that occur as yedoma deposits are discussed in Chapter 6, together with other massive icy beds included in the *ice complex*.

5.2   PRIMARY AND SECONDARY WEDGES

Obviously there needs to be a distinction between wedges that are formed by the infilling of contraction cracks in cold ground in winter, and those formed by modification and infilling of ice-wedges during melting of the ice during warmer climatic conditions. Romanovsky (1977) used the name *primary wedges* for the wedges formed by contraction cracking, as opposed to the *secondary wedges* in which the filling formed during warmer times when the ice in the ice-wedges melts, resulting in infilling of the resultant space by slumping of the sides and the accumulation of both windblown and/or mass wasted material. In the following account, the term "*tessellon*" will be used for all types of infilled primary wedges to imply the polygonal form (AGI, 2005), with the exception of ice-wedges. The term, *ice-wedge,* is so widely used that it seems inappropriate to confuse readers by using the term *ice tesselon*, although that would be consistent with the terms used for the other primary infillings.

5.2.1   Primary wedges

These include *epigenic ice-wedges* that represent the best known primary wedges. However, in the absence of moisture to fill the polygonal cracks, any other loose sediment on the surface of the ground may form the infilling, ranging from clay to sand or rock. The different kinds are discussed below.

5.2.1.1   Ice-wedges

In areas with a winter snow pack, the melt-water in the lower part of the isothermal snow in the spring enters any contraction cracks formed during the previous winter and refreezes there, resulting in the formation of *ice-wedges* (Figure 5.6). These are two layer structures. During the summer, the upper part of the infilling within the active layer thaws, but the lower part persists within the main body of the permafrost. If the weather is sufficiently cold, any cracking tends to occur along the same wedge structure during the following winters, although the actual trace of the crack will vary

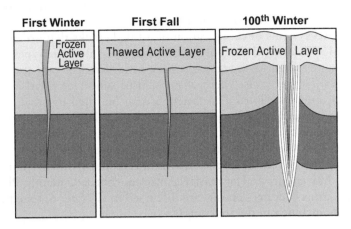

Figure 5.6   Diagram showing the development of ice-wedges.

from year to year (Dostovalov & Popov, 1966). This produces intersecting vertical foliations in the ice forming the ice-wedge, which are usually quite visible (Figure 5.7).

These individual vertical layers represent one year of cracking, but only some of the wedges crack in any one year. By counting the individual vertical layers in the wedge, it is usually found that the wedges may have been active for a very long time, often apparently for over 1000 years in Siberia. The hydrogen and oxygen isotopes in the ice infilling of individual cracks provide the opportunity for dating the cracking and providing an estimate of the temperatures during their formation using oxygen and hydrogen isotopes (Konyakhin, 1988). Radiocarbon dating has been used on peaty sediments and on organic remains such as peat or wood (Figure 5.8), although some layers of ice can have a different isotopic content due to subsequent additions to the wedges as indicated by the enclosed methane and carbon dioxide gases in the ice. In this case, the ice is polymorphic.

The rate of growth of ice-wedges increases southwards. The growing ice-wedges limit the thermal expansion in the lateral direction during summer as discussed above, producing upturning of the host sediment along the sides of the wedge ("shoulders" in the Russian literature), as well as "belts" along the upper margins of the ice-wedges (Figures 5.9). Ridges form along the sides of the surface of the wedge (Figure 5.4), often with water trapped between them and also in the centre of the polygon. This is the *"polygonal bead relief"* of the Russian literature.

Thermally induced seasonal movements of the active layer and subjacent permafrost have been measured in numerous ice-wedge polygons that have varied in age, type, crack frequency, and topographic location (Mackay, 2000). The field observations showed that, in winter, thermal contraction, which is inward, is constrained or vanishes at the polygon centres, but in summer, thermal expansion which is outward, is unconstrained at the ice-wedge troughs. This tends to result in a small net summer transport of the active layer, to varying depths, into the ice-wedge troughs, reducing the tendency to produce marginal ridges along both sides of the ice wedge. The movement was observed in all polygons studied. The slow net transport of material into

Figure 5.7 Ice-wedges of different ages cutting through part of the massive ice on Hershel Island. This is
an example of repeated ice-wedge formation. Note the vertical foliation inside the wedges,
as well as the upturning of the host sediments.

the ice-wedge troughs has implications for permafrost aggradation and the growth
of syngenetic wedges in some troughs, the palaeoclimatic reconstruction of some ice-
wedge casts, and the interpretation of polygon stratigraphy based upon the assumption
that the polygon material has accumulated *in situ*.

Repeated cracking of the ground at one location can result in two types of wedges.
Epigenetic wedges are developed by an increase in the severity of thermal conditions
in sediments deposited at an earlier time and when deposition of sediment has ceased.

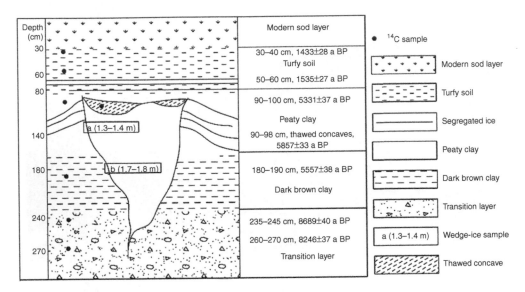

Figure 5.8 The southernmost inactive ice-wedge found at Yitulihe, N.E. China (50°37′N, 121°32′E) at 730 m elevation. The MAAT was −5.2°C in 1954–1955, with a snow cover of 30–35 cm. From Jia *et al.* (1987), Peng & Cheng (1990), and Yang & Jin (2011).

Figure 5.9 "Shoulders" (upfolding of host) and "belts" (banding in ice) of the left upper side of an ice-wedge. Central Yakutia. © A. Brouchkov.

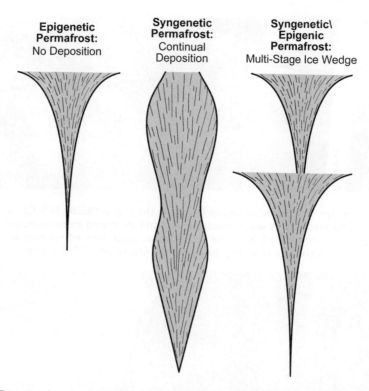

Epigenetic Permafrost:
No Deposition

Syngenetic Permafrost:
Continual Deposition

Syngenetic\ Epigenic Permafrost:
Multi-Stage Ice Wedge

Figure 5.10 Diagram showing the differences in shape between epigenetic, syngenetic and multi-stage (repeated) ice-wedges. The active layer has been omitted for simplicity.

Cross sections exhibit a triangular ice-wedge extending down to the layer of the annual fluctuations of temperatures (about 4 m). *Syngenetic wedges* are formed simultaneously with the accumulation of deposits and are normally up to 40–60 m high and 6–8 m wide (Figure 1.26). The type of ice-wedge is readily identified by the vertical differences in shape which is dependent on the stability of the land surface and climate (Figure 5.10). Any interruption due to climate amelioration may result in a period of deposition of sediment on the ground surface, and this may be followed by another cooling, resulting in recommencement of cracking of the ground. The resulting *repeated ice-wedges* (Shumski, 1955), *multistage epigenetic/syngenetic ice-wedges, or polygenetic ice-wedges* (the term used in Russia) show a larger wedge that is truncated, with a smaller, younger wedge growing on top of it (Figure 5.11). Repeated ice wedges can also be formed in Siberia where the ice-wedges are periodically covered by water carrying sandy loams and loams on aggrading flood plains at an MAAT $<-3°$C. This implies that the active layer has become thinner than its long-term (century – millennial) position. It normally contains less ice than the main permafrost table, though it represents a developing new permafrost table (c. decadal–century in age), unless something causes the active layer to deepen.

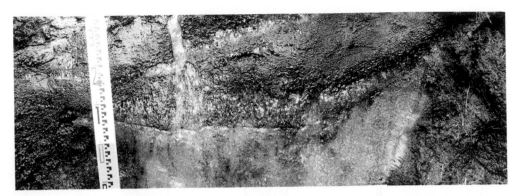

Figure 5.11 Multistage or repeated ice-wedge forming in the peat at Yitulihe, N.E. China. The cracking that resulted in the lower ice-wedge ceased and segregated ice developed in the overlying peat as it accumulated. Since then, cracking has recommenced, resulting in the growth of a second ice-wedge extending down into the first wedge. © Sizhong Yang.

Figure 5.12 Details of the margin of an active ice-wedge in Siberia. Note the up-turning of the sediments of the host rock and the vertical foliation in the wedge itself. © A. Brouchkov.

If ice-wedges occur on a slope undergoing erosion, the wedge is referred to as an ***anti-syngenetic wedge*** (Mackay, 1990). Over time, the length of the wedge becomes shorter unless the cracking is active, in which case, the new cracks extend further down into the host material. All ice-wedges show an upturning of the host sediment adjacent to the ice (Figure 5.12).

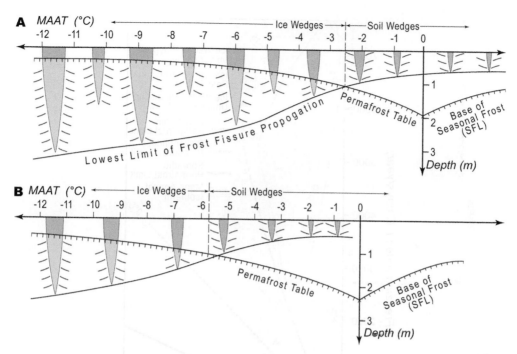

Figure 5.13 The relationship between soil and ice-wedges and mean annual ground temperature for A, clay substrates and B, sand and gravel (after Romanovsky, 1973). Note the upturning of the host sediment along the ice-wedges and down-turning around the soil wedges. Lithology of the sediments and amplitude of the temperature changes also affect the depth of frost action.

Since the depth of thaw is shallower at higher latitudes, there is a definite increase in size of the ice-wedge structures in the Arctic. Winter cooling also increases towards the Poles, resulting in greater depth of cracking. Those formed during the last glaciations along the Arctic coast of Siberia are up to 4 m deep. Ice-wedges are shorter at lower latitudes such as on the Tibetan Plateau, probably due to the high angle of incidence of the sun's rays in summer producing a thicker active layer truncating the wedge. Width of the wedge at any given depth perpendicular to the long axis depends on frequency of cracking and infilling, and length of time for growth.

Romanovsky (1973, Figure 7) showed that in Siberia, there is a strong relationship between mean annual ground temperature (MAGT), the formation of ice-wedges, and the grain size of the host ground. In clays, ice wedges can form at an MAGT of $-2°C$, whereas in sand and gravel, MAGTs as low as $-6°C$ are necessary (Figure 5.13). Note that both the active layer thickness and depth of cracking also change with temperature and texture of the ground. Thus the interpretation of fossil ice-wedge casts is more complicated than often interpreted in the literature (Washburn, 1985; Harry & Gozdzik, 1988; Murton & Kolstrup, 2003; *c.f.* Péwé, 1966). In general, the coarser the texture of the host ground and the greater the moisture content, the colder the temperature at which ice wedges begin to form in Siberia. (Romanovsky, 1977). Removal of a forest cover may cause ice-wedge growth in subarctic peatlands (Payette

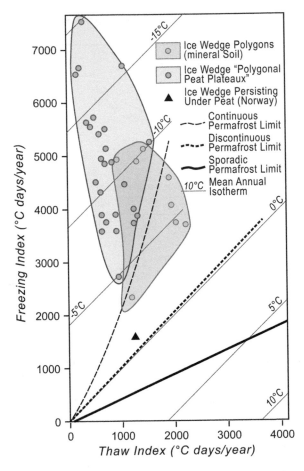

Figure 5.14 Relationship between the distribution of active ice-wedge polygons and freezing and thawing indices in peat and mineral substrates (after Harris, 1981c, Figure 4A, 1982a).

et al., 1986), while growth of a young forest may cause cessation of ice-wedge cracking due to deeper winter snow cover (Kokelj *et al.*, 2007) without there being a change in the MAAT.

Figure 5.14 shows the relation of active ice wedges to freezing and thawing indices. Repeated cracking and wedge formation depend on the substrate as noted by Romanovsky (1973) and shown in Figures 5.5 and 5.13. Wedges in peat can develop at warmer mean annual air temperatures (MAAT) than in mineral soils because of the thermal diode effect. However, peat can only form where temperatures are warm enough in summer for the production of sufficient biomass to exceed the rate of loss by decomposition. As a result, the lower limit of MAAT for active ice wedges in peat is about −11°C, whereas ice wedges can be active at much lower values of MAAT.

Active ice-wedges formed in peat can be found in the surface of all the main permafrost mounds including both peat plateaus and pingos, given cold enough conditions (see Chapter 7 and Figure 5.5). They are the most common permafrost feature of humid lowland areas, *e.g.*, the Mackenzie Delta, but become rare in dry, cold climates such as the Tibetan Plateau and in mountainous areas. On the Delta, the troughs may be filled by sediments and obscured by growth of vegetation. At four sites in the eastern Mackenzie Delta, over 85% of the trees within 1 m of ice-wedge troughs leaned towards these troughs (Kokelj & Burn, 2004). The mean angle of lean was 12° from the vertical, with some trees leaning by more than 25°. The angle of tree-tilt varied inversely with distance from the ice-wedge trough and most of the trees over 1 m from an adjacent trough leaned away from the ice-wedge. Trees near the troughs are susceptible to toppling because their root systems trail away from the troughs. Reaction-wood rings in cross-sectional disks obtained from trees leaning towards troughs indicated that progressive tilting has been sustained for decades to centuries. Long-term rates of tree tilting are estimated to be between 0.1 and $0.4°a^{-1}$. Progressive unidirectional tilting may eventually destroy the spruce trees. Thus in the Mackenzie Delta, where forest fire is infrequent, earth movements associated with ice-wedge polygons may be one mechanism driving forest change in old-growth stands.

Pleistocene ice-wedges along the Arctic Coast of Siberia are the largest (>10 m), but are syngenetic (Vasil'chuk *et al.*, 2000). They are the result of polymorphism, and the disappearance of the mineral sediments enclosing them. They are called y*edoma* or *ice complex* and represent extremely ice-rich, perennially frozen sediments with thick, polygonal ice wedges formed in accumulation basins during the Pliocene and Pleistocene glaciations (see Chapter 6), especially in the coastal lowlands (Romanovsky, 1993). They often contain greater than 90% ice by volume (Figure 1.25). The exact causes are unknown, but theories include fluvial deposition (Popov, 1953; Katasonov, 1954), aeolian deposition (Tomirdiaro *et al.*, 1984), or polygenetic formation (Sher *et al.*, 1987), modified by possible ice flowage and pushing up of the host sediment. However these large ice-wedges do not seem to occur everywhere, *e.g.*, the Yana River in Eastern Siberia (Popov, 1965) and on Chukotka Island (Kurdyakov, 1965). Equivalent complexes occur along the Arctic coast of Western North America (Shur *et al.*, 2009), but in far less saline host sediments. These are also discussed in Chapter 6.

Along the contact between the ice-wedge and the soil, high pressures are developed which can cause upward flowage of both the ice and the adjacent soil (Figure 5.12). The stresses decrease away from the wedge, reducing the tendency for upward flow. This may be connected to the development of the ridges along the margins of the wedges.

The stresses might be too small to cause deformations far away from the ice-wedge, but they are big enough to move soil particles and pore ice near it. Another reason for the deformed area being small is perhaps because of gradual movement of soil attached to the ice towards the surface, together with ice (Figure 5.15) caused by pressure and buoyancy.

Figure 5.16 shows the high variability in shape of some ice-wedges (Wilkerson, 1932; Péwé, 1962). Ice-wedge casts show much better shape, mimicking the original crack. The cause of the variability is unknown, though Popov (1965) suggested that they can grow due to freezing of water on the horizontal surface of the ice if it is cold enough to freeze on contact. This could explain complex shapes in syngenetic wedges. Other explanations include irreversible deformation of the soil by viscoplastic flow, particularly in saline soils (Tsytovich, 1973) as well as buoyancy effects causing diapirism and soil circulation (Hallet & Washington, 1991).

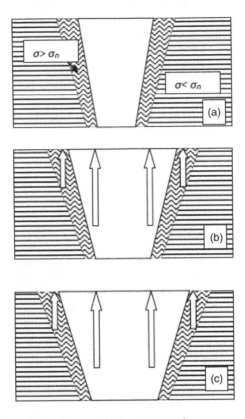

Figure 5.15 Stresses and resultant changes that appear to have occurred in ice-wedges on the Aldan River terraces shown diagrammatically. The vertical arrows show the direction of movement.

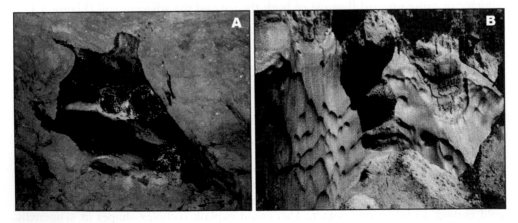

Figure 5.16 Irregular shapes of ice wedges A, in the wall of the tunnel in the old gold mine at Fairbanks, and B, Sirdah Lake, Yakutia. The pictures show an area of about 2 × 2 m.

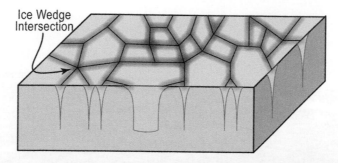

Figure 5.17 Block diagram of ice-wedges forming a tessellated pattern. The apparent width of the ice-wedge depends on the angle of the section relative to the length of the wedge.

Figure 5.18 An area of ice-wedges showing their distribution over the surface of a pingo in the Northwest Territories. Note the curving of the wedges over the mound resulting from tensional cracking on the domed surface, and the low centre polygons on the lowland.

Ice-wedges form a polygonal pattern on the landscape with diameters ranging from 5 m to over 50 m. It is important to note that the actual width of a given ice wedge must be measured at right angles to its orientation in the polygonal pattern (Figure 5.17). Diagonal sections across the wedge will make it appear considerably wider than it actually is. The number of sides varies from 3 to 8, with 4 or 5 being the most abundant in North America (Figures 5.18 and 5.20). Ground-penetrating radar provides a clear picture of the width and distribution of ice-wedges in the field.

Figure 5.19   Low centre polygons, Lena delta, Siberia. © M. Grigoriev.

It has been claimed (Dostovalov & Popov, 1966) that changes in the climate or deposition of sediment with different thermal properties can result in the development of secondary and larger orders of wedges within the primary pattern (Figure 5.21). Plug & Werner (2002) modelled this, assuming that the stress field over polygons does not change with time. Thus the late Holocene syngenetic ice-wedge polygons of Bylot Island (Canadian Arctic Archipelago) formed about 6000 years B.P. in outwash deposits from glaciers. They have been modified by the accumulation of wind-blown and organic sediments since about 3670+/−110 years B.P. The overlying sediments have a greater thermal contraction coefficient and have produced third and fourth order contraction cracks (Fortier & Allard, 2013). However, Burn & O'Neill (2015) noted that Mackay (1992) found that both primary and secondary wedges at Garry Island cracked about the same number of times during a 20-year period. They therefore concluded that as the depth of the primary wedges increases, changes in the stress field on the polygon due to increased snow accumulation in the hollows was a more likely cause.

There is an upturning of the sediments at the margins of the cracks due to the summer thermal expansion, which was measured at 2.5–$2.7 \times 10^{-5}/°C$ per thousand years in this case. Double ridges of upturned sediment are normal along present-day, active, syngenetic ice-wedges, being up to 0.12 m high and 4 m wide, bisected by troughs only 0.05 m wide and 0.09 m deep on the outer Mackenzie Delta (Morse & Burn, 2013). The centres are lower (Figure 5.19), and they often contain ponds, hence the name *low-centre polygons.*

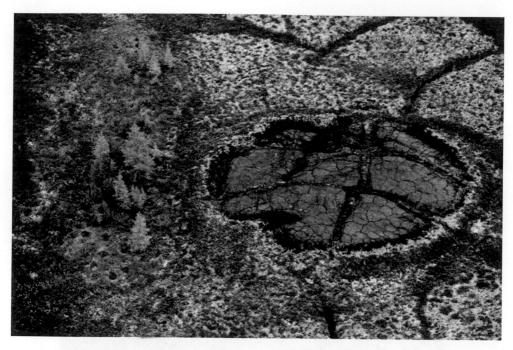

Figure 5.20 Polygons around and over a peat bog in Taymur peninsula, Siberia. © S. Fomin.

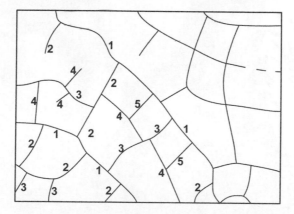

Figure 5.21 Typical arrangement of first, second and higher order ice-wedges.

Ice-wedges can survive climatic changes including interglacials if suitably insulated by peat or other suitable overburden (Figure 5.22). However they do not
undergo seasonal cracking unless the climate becomes suitable for this process to
resume. The oldest ice-wedge found in Canada so far (Figure 5.23) lies below a volcanic ash dated at 746 ka B.P. (Froese *et al.*, 2008). Older ice is probably present
in the unglaciated parts of Siberia, Katasonov (1977) reporting existing permafrost

Figure 5.22 The oldest ice-wedge dated so far in North America (c 750 ka) at Dominion Creek mining area in the Yukon Territory (Froese *et al.*, 2008, Figure 1). Its age was determined because it is overlain by a volcanic ash dated at 740,000 years B.P.

in Siberia dating back to the Lower Pleistocene, though he did not elaborate. Konishchev & Kartashova (1972) reported syngenetic wedges demonstrating the continuous presence of permafrost in the Yana-Indigirka lowland during deposition of sediments since the Middle Pleistocene.

Figure 5.23 Pleistocene sand tessellons west of Laramie, Wyoming, penetrating weathered metamorphic bedrock. © S. A. Harris.

5.2.1.2 *Sand tessellons*

These are one of the three types of primary wedges of Romanovsky (1977). They were first described from McMurdo Sound, Antarctica by Péwé (1959). In very cold, dry climates, ground cracking still occurs, but there is no moisture source to fill the cracks. Instead, wind-blown sand fills in the semi-vertical cracks in winter leaving a narrow depression at the surface of the ground (Berg & Black, 1966). Since the crack that develops is narrow and extends downwards for up to 6 m into the ground, the sandy infilling takes the form of intersecting vertical sheets (Figure 5.23). The latest sheet often cuts across many of the pre-existing sheets, so producing the vertical stratification (Figure 5.24). The rate of growth appears to be spasmodic. They were interpreted as being formed under extremely low temperatures ($<-12°C$) and with mean annual precipitation below 100 mm, resulting in the development of polygonal contraction so producing the vertical stratification (Figure 5.24). The rate of growth appears to be spasmodic. They were interpreted as being formed under extremely low temperatures ($<-12°C$) and with mean annual precipitation below 100 mm, resulting in the development of polygonal contraction cracks. Windblown sand infilled the narrow cracks. The host sediment at McMurdo Sound in Antarctica is sandstone with a variable cover of loose sand derived from it.

Confusion can occur between ice-wedge casts and these sand tessellons. Washburn (1979, Figures 4.34 and 4.35) provided an excellent pair of diagrams showing the

Figure 5.24 Details of a Pleistocene sand tessellon west of Laramie, showing the intersecting sand sheets filling the former contraction cracks. © S. A. Harris.

Table 5.1 The diagnostic characteristics of primary and secondary wedge structures, infilled thawed massive ice cavities and load-casting involutions in China.

Type of wedge-like structure	Name	Pattern viewed from above	Shape	Height: width ratio	Infilling	Internal structure
Primary	Epigenetic ice-wedge	Polygon	V-shaped	1.3–2.5	Ice	Vertically laminated
	Rock tessellon	Polygon	V-shaped	c.4->10	Platy rock fragments	Vertically laminated
	Sand tessellons	Polygon	Broader V-shaped	1.3->4	Sand	Vertically laminated
	Loess tessellons	Polygon	Broader V-shape	3->4	Windblown loess	Vertically laminated
Secondary	Ice-wedge cast	Polygon	Broad V-shape	1.3–3	Unbedded sediment with blocks of the host rock	Massive unbedded
Other	Former massive ice block	Somewhat circular	Vertically sided, shallow pit	<0.3–1.3	Sandy sediment	Massive unbedded
	Load casting	Various	Undulating tongueing	1.5–2	Alternating beds usually of sand and gravel	Cryoturbated involutions

vertical sheets of sandy infilling in sand tesellons. The sides of the latter are smooth and show no signs of slump structures, unlike ice-wedge casts. Hallet *et al.* (2011, Figure 1) report raised shoulders along the margins of the tessellons in Victoria Valley, Antarctica. There is always upturning of the enclosing beds. Romanovsky (1973, 1976) discussed apparent transitional forms in which both sand and ice occur in the infilling. These are apparently rare.

Unfortunately, some of the published literature has generally used the term "sand wedge" indiscriminately to describe both sand tessellons and other wedge-like landforms infilled with sand. This makes the interpretation of past climatic environments more difficult.

Table 5.1 shows the characteristics of some of these landforms for which the term, "*sand wedge*", is currently used in some parts of the world. The characteristics of the landforms representing primary and secondary wedges are described in this Chapter, whereas the formation and characteristics of former massive ice blocks and load casting are described in Chapters 6 and 3 respectively. The use of the height to width ratio was pioneered in China and is a very useful criterion for determining the origin of the feature, provided that the top of the wedge-like structure is still present. Ratios determined on eroded wedge-like features should not be used, although examples of their use are found in the past literature.

The formation of sand tessellons in permafrost regions requires a much colder and drier climate than that under which ice wedges form, so that differentiating them from ice-wedge casts has considerable palaeoclimatic importance. They imply the absence of blowing winter snow but the presence of blowing sand grains during their formation in winter and spring before the crack closes up.

Sand tessellons have subsequently been reported from other cold, arid parts of Antarctica (Berg & Black, 1966, 70–73; Tedrow & Ugolini, 1966), the Alaskan Arctic Coastal Plain (Carter, 1981, 1983), and from Prince Patrick Island, Northwest Territories (Pissart, 1968). Fossil sand tessellons have also been reported from the Pleistocene deposits of Wyoming (Mears, 1981), the Pleistocene Mackenzie Delta where they are reported to be 4–6 m deep (Murton & French, 1993), the Hexi Corridor, China (Wang *et al.*, 2003), the eastern slopes of the Qinghai-Tibet Plateau (Jin *et al.*, 2007; Harris & Jin, 2011) and Inner Mongolia (Vandenburghe *et al.*, 2004).

5.2.1.3 Loess tessellons

In the headwaters of the Yellow River at the village of that name in China, *loess tessellons* occur. Since they penetrate the bedrock, they are easy to recognize (Figures 5.25 and 5.26), although they are overlain by a younger loess. They have also been found along the Hexi Corridor, north of the Quilin Mountains at 38°6'N, 101°20'E at 2891 m elevation, penetrating Tertiary Beds. In that case, the loess tessellons are narrow and occur on the well-drained upper slopes of the rolling landscape, while ice-wedge casts are present on the lower slopes. In loess tessellons, the same sub-vertical, cross-cutting bedding occurs as in sand tessellons, as well as upturning of the host rock. Loess tessellons probably occur elsewhere but are difficult to recognize since one loess looks very similar to another. Careful checking for colour changes and differences in bedding would be necessary to find them in the loess deposits that are so common in many areas that probably had permafrost during cold events (glaciations/widespread permafrost). In the case of those occurring in rock in China, the infilling is over-compacted, and show the characteristics of a fragipan, in contrast to the overlying younger loess. Loess tessellons have the same requirement of cold, dry conditions as sand or rock tessellons except that fine sand and silt was being blown across the area in winter, *i.e.*, the nearby silty loess deposits were not totally frozen at the time the wedges were forming and were being moved about by the prevailing winds.

5.2.1.4 Rock tessellons

These are similar to sand tessellons, but the infilling consists of small plate-like fragments of fissile sandstone (Figure 5.27). The plate-like fragments are the result of very intense frost weathering of fissile sandstone exposed in an outcrop. If there is negligible snow and no blowing sand and silt, the loosened rock fragments can slide over an icy surface into any available frost cracks. For this reason, they may represent even colder conditions than sand or silt tessellons. They were first described from sections in a frost-fractured and cryoturbated fissile sandstone on the east flank of the Qinghai-Tibet Plateau (Harris & Jin, 2012). Subsequently, at least one more occurrence has been found about 1000 m lower in elevation.

5.2.2 Secondary wedges

These develop in two different climatic situations although both are formed subsequent to the surface of the ground developing polygonal cracks. Ice-wedge casts form when

Figure 5.25   Loess tessellon in bedrock at 4226 m in the village of Yellow River, Sichuan Province, China. There is a small soil wedge filled with younger loess in the top. © S. A. Harris.

the primary ice-wedge melts due to climate change, whereas soil wedges form the infilling of seasonal cracks which contain ice only for the winter season.

5.2.2.1   Ice-wedge casts

When the climate warms, the active layer becomes deeper and the ice in the tops of the ice-wedges thaws. Initially, this produces *high-centre polygons* at poorly drained sites

Figure 5.26 Detail of the vertical structure in a loess tessellon at Sandar, Gansu Province. © S. A. Harris.

(see Figures 11.6, 5.21 and 5.28). If the water in the trenches drains away, *thermokarst mounds* or *baydjarakhs* are produced by the deepening of the troughs as the ice disappears (Figures 3.14 and 5.29; see also Chapter 11).

The thawing commences at the top and the trace of the thawing ice wedge is often filled with water in lowland areas with poor drainage. Windblown dust, loess and sand accumulate in the hollow and gradually slump downwards. Chunks of wet sediment from the surrounding host deposits often slump into the infilling as thawing proceeds, producing an apparent widening of the former ice-wedge. At other times, enough windblown sediment will be deposited in the depression in the form of a cast (Figure 5.30). If there is negligible windblown sediment available, then slumping of the walls results in shallow, rather rounded depressions marking the outline of the former ice wedge (Figure 5.29). In between the trenches will be a mineral deposit in the form of a mound.

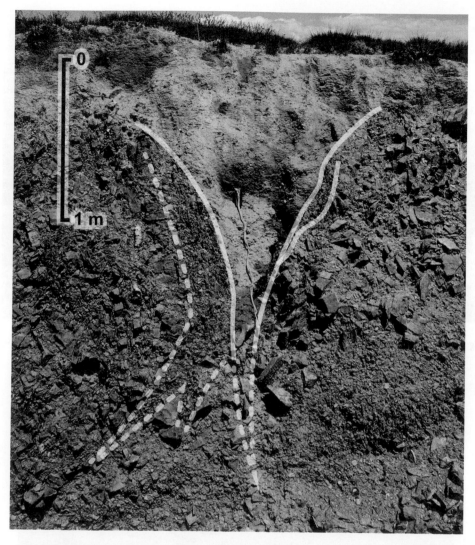

Figure 5.27 Rock tessellon in frost-shattered fissile sandstone, with subsequent development of an ice-wedge cast with an infilling of fine sand, eastern margin of Tibet (from Harris & Jin, 2012).

5.2.2.2 Soil wedges

Frost cracking can occur in soils in areas of seasonal frost during particularly cold winters (Svensson, 1977, 1988a, 1988b; Washburn *et al.*, 1963). These can result in downward slumping of surface sediment into the crack, producing small, shallow wedges of sediment, variously called *soil wedges* (Danilova, 1966; Jahn, 1983), *ground veins* (Katasonov, 1973) or *seasonal frost cracks*. They sometimes include some ice when investigated in cold weather in humid maritime environments such as Spitzbergen

Figure 5.28 High-centre polygons south of Prudhoe Bay, Alaska, showing water along the wedge troughs.

Figure 5.29 Thermokarst mounds (Baydjarakhs) in cleared larch forest in Siberia. © A. Brouchkov.

Figure 5.30   Ice-wedge cast of Younger Dryas age, Holland. Note the irregular margin of the infilling of loess as well as the inclusion of blocks of sediment. There is weak up-turning of the host cover sands. © S. A. Harris.

or Iceland (Figure 5.31), but otherwise, they consist of surface soil material penetrating into the lower soil horizons. Permafrost is absent in present-day examples, which form a tessellon or net that is usually readily apparent at the surface.

The Russians have recognized and described a number of different forms of cracking and infilling (Romanovsky, 1973, 1985; Melnikov & Spesiivtev, 2000), depending on the materials present in the host sediment, but they have rarely been noted elsewhere. An exception is French *et al.* (2003) in Southern New Jersey, where their young age,

Figure 5.31 Cracks filled with ice in Yamal peninsula, Western Siberia. © V. Samsonova.

shape, size and spacing differ from ice-wedge casts. In New Jersey, they are usually less than 2 m deep, between 20 and 60 cm wide and closely spaced (2–3 m). In the Yenisey river valley, shallow small cracks forming a 30–40 cm network that fill with ice are formed in winter. During the spring and early summer, the ice melts and soil flows into the void.

Figure 5.32 Cross-section of a soil wedge in Iceland. Note the down-turning of the host sediments. © S. A. Harris.

Soil wedges are particularly common in the Pleistocene deposits of southern Siberia (Katasonov, 1973; Katasonov & Ivanov, 1973), as well as in continental Eastern Europe (Dylik, 1966). The wedges vary in depth from about 15 cm in maritime climates up to 2 m in the fossil cracks in sands and gravels in northern Siberia (Danilov, 1973). In cross-sections, shallow soil wedges show an infilling consisting of a stubby wedge of sediment extending down into the upper soil horizons (Figure 5.32). The bedding in the host sediment usually exhibits down-turning adjacent to the wedge. Since similar structures can be produced by other means such as desiccation, they are not used as palaeoclimatic indicators unless there is much better proxy evidence for the climate under which they formed.

Soil wedges up to 5 m deep were reported by Marchenko & Gorbunov (1997) forming polygons 2–10 m diameter and 5 m deep in Miocene clays at the watershed of the Asy and Turen Rivers in the Tien Shan in Kazakhstan. Samples of humus from different parts of the wedge gave different C^{14} dates ranging from 7793 to 2904 (radiocarbon?) years B.P. While the height:depth ratio was that of a primary tessellon, the fact that they were developed in clays and lacked vertical bedding suggests that they may be infillings of desiccation cracks. Even so, they may indicate periods of cold climate since these were also periods of low precipitation under cold climatic regimes, but additional evidence for the palaeoclimate is needed to determine this.

Chapter 6

Massive ground ice in lowlands

6.1 INTRODUCTION

Much of the northern coast of the Northwest Territories in Canada, the European part of Russia and the north part of Western Siberia consist of lowlands. They are also a part of the permafrost realm that was partly glaciated about 20,000 to 30,000 years ago. However, large areas of the lowlands were the sites for fairly continuous deposition of cold-climate loess (Murton *et al.*, 2015). Along the Arctic coast of Russia, the eroding cliffs expose extensive outcrops of thick, horizontal beds of massive ice that is overlain by glacial till or other younger sediments (Figure 6.1). Often, the ice shows folded structures as well as signs of thrusting. Recently, massive icy beds have

Figure 6.1 Active thawing icy cliff of a retrogressive thaw flow slide in the river cliffs in Yakutia, Russia. © A. Brouchkov.

been found deep in bedrock. This raises the problem of their origin. Are they formed by accumulation of ice by cryosuction from the abundant groundwater, are they the result of burial of stagnant glacier ice that has not yet completely melted, or are they formed by some other process or processes? The problem is not entirely academic since any thawing of the ice would be disastrous for any man-made structures. The resulting slope instability prevents the construction of port facilities (see Chapters 11 and 12).

Another type of massive ice complex called *yedoma* is found along the eastern Arctic coast of Russia, *e.g.*, around the Lena Delta area (Figure 1.25), and in the unglaciated parts of northern Alaska. Around the Arctic Circle, it consists of the ice wedges formed in non-glaciated areas during earlier glacial events that have grown and expanded until they make up more than 60% of the sediment in the face of the cliff sections. This greatly enhances the rate of coastal erosion. It is claimed by some Russian authors that ice wedges of this type make up much of the material in the adjacent lowlands, as well as along the Lena River downstream of Yakutsk. They also occur in Alaska and the Yukon Territory (Murton *et al.*, 2015).

Additional thick, icy beds called *icings, aufeis, naleds, taryns in Russia* develop each winter along the bottoms of the smaller river valleys in areas of discontinuous permafrost. In western North America, they extend downstream for tens of kilometres towards the Arctic Ocean, whereas in Siberia, they tend to be found in the southern area around Lake Baikal. These aufeis bodies are often very thick (Figure 6.2), persist late into the summer and make the river useless for navigation. In winter, the build-up of ice can inundate roads, pipelines and houses, while their great length and uneven surface represent a significant barrier for transportation (Wrangel, 1841).

Massive ice can also occur in ice caves, some of which can be very large. Their origins are varied, but there is a definite cycle of evolution dependent on the medium to

Figure 6.2 Aufeis, river Artuk, in summer. ©V. Samsonova.

large scale fluctuations in mean annual air temperature. In this Chapter, the origins and characteristics of these massive icy beds will be examined, together with their suggested possible origins. In all probability, there have been multiple processes occurring, each under specific environmental conditions.

6.2 DISTRIBUTION OF MASSIVE ICY BEDS IN SURFACE SEDIMENTS

The main areas of massive icy beds in surface sediments that have been studied so far are indicated in Figures 6.3 and 6.5. The chief areas where they occur are located along the margins of the last ice sheets, while Figure 6.3 shows the areas of yedoma deposits in Northern Siberia and northwest North America. The two areas in which they occur were connected by the Beringian land bridge. In practice, the yedoma deposits can be divided into areas of continuous deposition of wind-blown *cryopedolith* (sediment that has experienced incipient pedogenesis with syngenetic freezing and ice-wedge formation) and other areas showing considerable modification of the disposition of the ice and sediment (Figure 6.4). For this reason, the origin of the massive icy beds will be discussed separately from the yedoma deposits.

The main areas of massive icy beds in surface sediments that have been studied so far are indicated in Figure 6.3. The chief areas where they occur are located along the

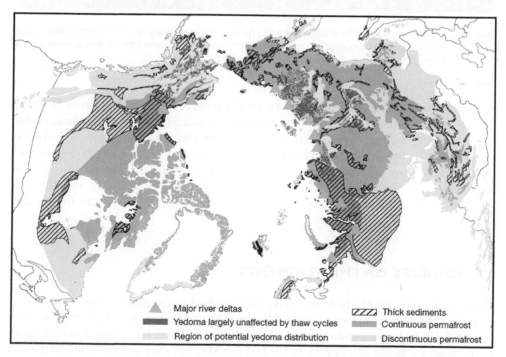

▲	Major river deltas	▨	Thick sediments
▬	Yedoma largely unaffected by thaw cycles	▬	Continuous permafrost
░	Region of potential yedoma distribution	░	Discontinuous permafrost

Figure 6.3 Distribution of yedoma, thick sedimentary deposits and permafrost in the Northern Hemisphere (USGS). See: https://prd-wret.s3-us-west-2.amazonaws.com/assets/palladium/production. Accessed 15/06/2017.

Figure 6.4 Massive ice in the cliffs of Big Lyakhovsky Island, Siberia, showing the expansion of the massive ice-wedge ice at the expense of the sediment. Obviously, some kind of metamorphosis has taken place, altering the original ice wedges. © M. Grigoriev.

margins of the last ice sheets. Also shown are the areas of yedoma deposits in Northern Siberia. The two areas in which they occur were connected by the Beringian land bridge. In practice, the yedoma deposits can be divided into areas of continuous deposition of wind-blown *cryopedolith* (sediment that has experienced incipient pedogenesis with syngenetic freezing and ice-wedge formation) and other areas showing considerable modification of the disposition of the ice and sediment (Figure 6.4). For this reason, the origin of the massive icy beds will be discussed separately from the yedoma deposits.

6.3 SOURCES OF THE SEDIMENTS

Many of the massive icy beds that have been studied in detail are found close to the margin of the maximum extent of the Wisconsinan ice sheets which is believed to have occurred 30–15 ka, depending on the local glacial history. In North America, there were about 16 major glaciations in the last 3.5 Ma, but they affected different areas (Harris, 2001; Barendregt & Duk-Rodkin, 2012), and the previous Illinoian glaciation was also very extensive. When a warm-based ice sheet moves across an area, it tends to entrain the upper layers of the underlying sediments and pluck rock fragments from the surface of any underlying massive rocks. This material is then transported towards the periphery of the ice sheet, becoming deposited there during ablation of the ice. Since the glaciations affected different areas each time, this sediment may have been moved several times, resulting in denudation of the main part of the Laurentian Shield and surrounding rocks. Along the way, the sediment has undergone considerable reduction in grain size producing considerable quantities of silt. The last few glaciations have been particularly widespread, and the resulting glacial sediments are now found in the south of Canada, the northern part of the United States, and in the lowlands of the Arctic coast. It is the sediments in the latter area that have the massive icy beds. Downwarping of the Arctic Islands complicates the situation, so that only the higher part of the former flood plains and glacial outwash areas are currently above sea level (Duk-Rodkin, 1994; Lemmon *et al.*, 1994).

In Russia, the area that was glaciated is now thought to have been relatively limited. Instead, the coastal plain in central and eastern Siberia has been regarded as consisting of alluvial, lacustrine and marine deposits. However, Murton *et al.* (2015) provide evidence for the occurrence of surface deposits over vast lowland areas consisting of windblown loess that has been deposited more or less continuously in a cold environment since at least 55k calendar years B.P. (the *Karginsky Interstadial*, and possibly the beginning of the *Zyryan Glaciation* (70–55k calendar years B.P.). These tend to have a high silt content, again providing suitable sediments for movement of water to freezing fronts. It is in these that the classic yedoma deposits have formed. They extended across the former Beringian land bridge into Alaska and the Yukon Territory. In western Europe, fluctuating air temperatures resulted in discontinuous deposition after the end of the *Sartan Glaciation* (25–15k calendar years B.P.), resulting in substantial formation of thermokarst landforms.

6.4 DEGLACIATION OF THE LAURENTIDE ICE SHEET

Figure 6.5 shows the ice fronts of the Laurentide and Cordilleran ice sheets in north America at their maximum about 18ka B.P., 10 ka B.P. and 7 ka B.P., together with the main ice stagnation deposits in southern Canada (after Prest *et al.*, 1968). About 15,000 years B.P., the dome of the Cordilleran Glacial rose above the tops of the Rocky Mountains and chinook winds caused the ice on the eastern side of the continental divide to retreat into the mountains. The weather remained cold until about 11 ka B.P. in the Foothills (Churcher, 1968) but this was quickly followed by substantial warming. This warming had been gradually taking place since about 12 ka B.P.

Precipitation had also decreased on the main Laurentide ice sheet, so that by 11.5 ka B.P., the outer zone of ice was becoming stagnant. By 10 ka B.P., the active part of the ice sheet lay in a distinct circle around Hudson Bay with a series of major lakes along its margins including Glacial Lake Agassiz (Figure 6.5). Although the area to the west had been largely deglaciated, large areas of stagnant ice remained stranded across the southern Prairie Provinces and south into North Dakota (Prest *et al.*, 1968; Stalker, 1973). Similarly in southern British Columbia, the clockwise circulation of air around the remaining Laurentide Ice Cap produced chinook conditions in that area, resulting in rapid downwasting of the Cordilleran Ice Sheet, producing ice stagnation in the interior valleys (Harris, 1985).

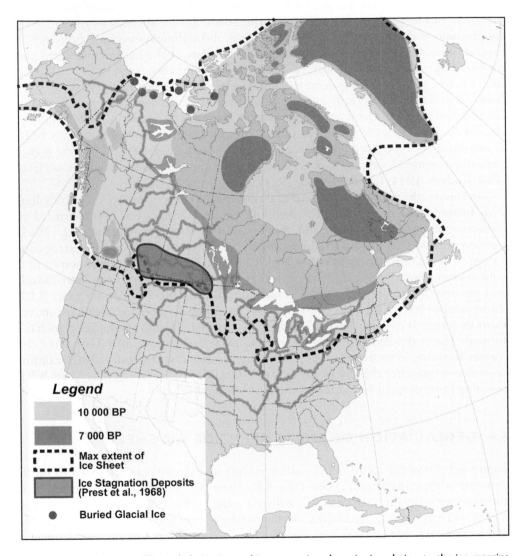

Figure 6.5 Distribution of buried glacier ice and ice stagnation deposits in relation to the ice margins and two retreat stages of the former Laurentide Ice Sheet in North America.

When ice carrying substantial quantities of debris stagnates, the melting of the surface ice produces a layer of supraglacial till that insulates the underlying ice from solar heating. Forests can grow on the surface (Figure 6.6), together with shallow lakes in depressions in which fish and waterbirds thrive. There are radiocarbon dates on organic remains of this surficial organic material, indicating that the ice persisted until at least 9,000 years B.P. in the large areas of stagnant ice in the southern Prairies (Clayton, 1967; Harris, 1985). The centre of the Laurentide ice sheet gradually thawed about 9,000 years B.P. (Dyke, 2004), resulting in the draining of the glacial lakes into Hudson Bay at about 8,400 years B.P. (Clarke *et al.*, 2004). However, the remnants

Figure 6.6 Photograph of the terminus of the Martin River Glacier in southern Alaska. The ice is stagnant (dead) and down-wasting slowly. A forest grows on top of the supra-glacial moraine that protects the underlying ice mass from rapid melting. © L. Clayton.

of active ice on the uplands persisted until about 7,000 years B.P. west of Hudson Bay and until 5–6,000 years B.P. in northern Québec. On Baffin Island, the retreat of the ice cap continued until about 4,000 years B.P. Since the Arctic coastal lowlands where the massive icy beds are found lie further north and closer to the cold Siberian air masses, the idea that some of the ice masses may represent stagnant glacial ice is not unreasonable.

6.5 METHODS USED TO DETERMINE THE ORIGIN OF THE MASSIVE ICY BEDS

To be glacial ice, the massive icy bed must occur in an area that underwent glaciation in the past. This rules out the eastern Siberian coastal plain, but it leaves the western coastal plain in the vicinity of the Yamal Peninsula and the Lower Yenisey River (Astakhov, 1986), together with the area immediately to the west as possible regions where buried glacier ice may be found. Similarly, the northern part of the Yukon and southern Banks Island could also have buried glacier ice. All these regions have a cold enough climate for the ice to have persisted under a cover of surface sediment, and they lie in the area of continuous permafrost today.

Mackay (1989a) summarizes a number of field criteria which can help in differentiating between the two origins, *viz.*, glacial versus ice segregation. A particularly successful technique has been to measure the oxygen isotope ratios of the ice in the unknown bed and compare it to the values from known segregated ice nearby. The value for ice formed during a glaciation should be noticeably lighter (more negative) than for the segregation ice (Souchez & Jouzel, 1984; Vasil'chuk, 1992). Some general characteristics of the massive icy bodies that have been interpreted as being glacial in origin are that they tend to overlie outwash sands, they have some banding or stratification, they usually contain glacial erratics of varying size and angularity, and are overlain unconformably by a diamicton with similar erratics in it. Gradational upper and lower contacts rarely occur in the case of buried glacial ice. Thus it should exhibit sharp upper and lower boundaries on ground penetrating radar outputs. Where intrusive ice is present, there are likely to be suspended fragments of the host sediment in it. Ice coatings beneath clasts indicate ice segregation. Glacier ice may have been subjected to melting and refreezing (*regelation*), in which case lumps of clear ice suspended in bigger (up to tens of cm), bubbly isomorphic ice will be found. The air in the bubbles is not of the "pipe-like" type, unlike those resulting from ice segregation. Vasil'chuk & Vasil'chuk (2012) pointed out that tundra pollen rarely occurs in polar glaciers or their snow cover, but cloudberry pollen and spores of horsetails and mosses and redeposited pollen, together with spores from older (Mesozoic and Palaeozoic) strata will often be present in non-glacial massive icy beds. Ice dykes formed by water intruded upwards under pressure along fissures are normal in segregated or intrusive ice.

Other types of ground ice such as ice-wedges are quite obvious, although multistage and syngenetic wedges are common in the older windblown sediments, *e.g.*, on the Alaskan coastal plain (Lawson, 1983). Likewise, injection ice is often present in the form of dykes and sills. Combined with deformation involving both folding and thrusting, these make the elucidation of the history and origin more difficult.

6.6 MASSIVE ICY BEDS INTERPRETED AS BEING FORMED BY CRYOSUCTION

Since the pioneer work of Mackay (1966, 1971b, 1972, 1973a) and Mackay & Stager (1966b), most authors have interpreted the massive icy beds in North America as being due to ice segregation (Mackay *et al.*, 1972; Mackay & Black, 1973; French, 1976), and they are sometimes referred to as *intra-sedimental ice* in North America.

Subsequently, both Kaplyanskaya & Tarnogradskiy (1977) and Mackay *et al.* (1978) pointed out that development of vein ice and burial of surface ice (glacier, lake or river ice) could also have occurred. These could be formed in either a cold glacial period or some could be the product of warmer intervals. This meant a re-examination of the earlier work using isotope geochemistry and ice petrology, as well as detailed stratigraphic studies.

Cryosuction is a dominant cause of the massive icy bodies on the North Slope of Alaska (Lawson, 1983) and is also important around the Mackenzie Delta, where buried glacier ice occurs (Murton *et al.*, 2005). Robinson & Pollard (1998) reported segregation ice forming massive ground ice within bedrock at Eureka Sound, Ellesmere Island. There, it is associated with large ice-wedges and discrete ice sills consisting of injection ice.

6.7 MASSIVE ICY BEDS THAT MAY REPRESENT STAGNANT GLACIAL ICE

Although it was obviously a possibility, the first interpretations of massive icy beds as representing buried glacial ice were only produced in the last 14 years of the 20th century by Kaplyanskaya & Tarnogradskiy (1986), followed by Astakhov & Isayeva (1988). Subsequently, others have reported evidence suggesting that buried glacial ice occurs at scattered locations in both North America and Russia. The actual locations are indicated in Figure 6.5.

In the eastern part of the Pechora Lowlands just west of the northern Ural mountains, Henriksen *et al.* (2003) found evidence that the glacial ice from the last glaciation (Early Weichselian, c. 90,000 years B.P.) survived beneath Lake Kormovoye until about 13,000 years B.P. The area lies north of the present-day boundary of continuous permafrost, but that boundary would have lain 20° further south during the last glacial event. Thus the climate in that area was suitable for the preservation of old glacial ice for very long periods of time.

Astakhov (1986) and Astakhov & Isayeva (1988) reported on a remnant of a massive ice body exposed in a retrogressive thaw-flow slide along the eastern banks of the Yenisei River between the towns of Turukhansk and Igarka. There is a sharp and unconformable upper thaw boundary with the overlying till. The ice consisted of an inclined sheet-like mass tilted downwards towards the northeast, and possibly reaching 70 m thick. The landscape includes numerous small hills, sometimes misinterpreted as being pingos. The ice is banded with layers of clear blue ice, milky-white ice and dirty dark ice. It has inclusions of glaciated angular boulders and lies between diamictons. The oxygen isotope ratio of the ice is −21.2‰ similar to the deeper layers of ice in the glaciers on Servenaya Zemlya, unlike the −14.1‰ for the nearby segregated ground ice. Subsequently, Arkhangelov & Novgorodova (1991) showed that the gases in the ice contained very variable contents of O_2, CH_4, and He, which would not occur if the ice was produced by ice segregation.

A second example of buried glacial ice has been described from Cape Shpindler, Yugorski Peninsula, just south of Novaya Zemlya along the coast of the Kara Sea (Lokrantz *et al.*, 2003). It occurs as a >10 m unit extending continuously for about 2 km along the cliffs. It is stratified and foliated, and is overlain unconformably by a

Figure 6.7 Shoreline at Tuktoyaktuk, Northwest Territories. The massive ice has ice-wedges up to 4 m deep in its surface and is believed to be of glacial origin. © S.A. Harris.

silty-clay till containing glacial erratics. In this case, the sediment protecting the ice included two tills and two interglacial deposits, making the ice appear to be from a pre-Eemian glaciation (c. 250,000 years B.P.) The bubbly ice has an oxygen isotope ratio of −18 to −25‰ compared to the vitreous and laminated ice with ratios of −17 to −19‰ (Ingólfsson & Lokrantz (2003). Lokrantz *et al.* (2003) interpreted it as being buried in a push-moraine at the front of the glacier, following the ideas of Astahkov *et al.* (1996). The latter noted that similar blocks of massive icy beds were present across the Yamal Peninsula and postulated deformation by younger glaciers moving over and deforming the pre-existing deposits. However another possibility is that it is sea ice, and similar ice occurs in the cliffs along the Yamal peninsula.

In Canada, evidence has been found for some glacially deformed massive ice occurring in the cliffs of the Eskimo Lakes Area and Summer Island on the east side of the Mackenzie Delta (Murton *et al.*, 2005). The massive ice interpreted as glacier ice is very thick, and yields oxygen isotope ratios of −25.6 to −35.7‰ with a mean value of −29.6‰. It includes segregated ice with lower oxygen isotope ratios. It has the ice petrography of glacial ice and is overlain unconformably by glacigenic sediments. French & Harry (1990) had interpreted a nearby massive icy bed as representing glacier ice, based on similar information. They concluded that glacial ice was present at the Sandhills moraine, southern Banks Island, and the massive buried ice in southern Eskimo Lakes region on the Pleistocene Mackenzie Delta also represented basal glacier (Figure 6.7). Lorrain & Demeur (1985) found massive ice with glacial isotopic signatures on Victoria Island that they regarded as being at least partly relict Early Wisconsinan ice.

The existing dead ice occurring beneath pro-glacial sediments that are associated with Holocene glacial advances have not been mapped, but are widely known. Buried ice masses occur in front of the Gulkana glacier in central Alaska, and similar features can be observed as far south as the Columbia Icefields in Jasper National Park. It is usually covered in supraglacial morainic debris, although deltaic deposition is recorded as burying glacial ice on Bylot Island (Moorman & Michel, 2003). Similar occurrences are found elsewhere throughout the glacial realm, *e.g.*, the Martin River glacier in Alaska (Figure 6.6), in Spitzbergen (Hambrey, 1984) and the Tasman glacier in New Zealand.

6.8 OTHER ORIGINS OF MASSIVE ICY BEDS

Obvious possibilities include ice formed by cryosuction from the abundant ground water or snow buried by mass wasting. Other origins include injection ice, lake ice, river ice and ice-wedges. *Injection ice* occurs as sills and dykes, as well as infillings of cracks in the other sediments, *e.g.*, on the Mackenzie Delta on the Alaskan coastal plain and in Western Siberia. It can be a particularly important component in massive ice bodies in bedrock, *e.g.*, at Eureka, Ellesmere Island (Robinson & Pollard, 1998). *Lake ice* is important along the Arctic coast of Alaska in situations where there are numerous lakes. The latter are believed to go through cycles of growth and demise, resulting in thawing of ground ice followed by refreezing of the lake bed and substantial segregation of ice in the lacustrine silts (Lawson, 1983). The ice content in the secondary icy beds is usually less than in the main mass of sediments along the Foothills of the Brooks Range.

Although *river ice* is possibly present in some cases, it has not been identified in North America. However, *ice-wedges* are a significant component on the Arctic Slope of the Brooks Range, Alaska, where they can be 10–15 m deep and 5–10 m wide on the uplands of the Foothills. They are regarded as being post Wisconsin in age and include both an inactive group and one that is presently active (Lawson, 1983; Shur *et al.*, 2009). They are presumably syngenetic, and probably result from progressive burial by alluvium and colluvium from the nearby mountains. Shur *et al.* (2009) also report similar large ice-wedges from the unglaciated parts of Interior Alaska and the Seward Peninsula. It is believed that they have developed from the processes involved in the formation of the yedoma deposits discussed below.

6.9 ICE COMPLEXES INCLUDING YEDOMA DEPOSITS

As noted in Chapter 5, ice-wedges tend to deform and grow after about 10,000 years. In some cases they keep the form of syngenetic wedges (Figure 5.8). but with the passage of time, they appear to be able to expand, largely squeezing out or replacing much of the host sediment, while growing in depth to as much as 50 m (Baulin, 1967; Baulin *et al.*, 1978; Trofimov & Vasil'chuk, 1983; Astrakhov, 1995). Examples are found along the unglaciated Arctic coast of Siberia (Figure 6.4), inland in the Arctic Lowlands around the middle Lena River, as well as along the north slope of the Foothills of the Brooks Range and in the Seward Peninsula of Alaska (Shur *et al.*,

2009; Kanevskiy *et al.*, 2014). They do not develop everywhere, but where they occur, the resulting thick sediments can consist of over 60% ice by volume. These sediments are referred to as the classical *ice complex*, or *yedoma*, a term used by the Russians for the geomorphological landform on the inland plains and in unglaciated area.

Vasil'chuk (2012) dates the main period of formation of these deposits as being prior to the Pleistocene-Holocene transition (10 ka B.P.), and extends back to beyond c. 20 ka B.P. Streletskaya & Vasiliev (2009) demonstrated that the thick polygonal ice-wedges making up the yedoma in the coastal sections of the West Taymyr had a very light isotopic composition (-24.3 to $-26.8‰$ of $\delta^{18}O$ and -185.0 to $-205.0‰$ of δD) compared with the composition of the Holocene ice-wedges (-19.0 to $-23.3‰$ of $\delta^{18}O$ and -140.0 to $-150.0‰$ of δD). This implies that the ice currently in them accumulated under much colder conditions than today. The width of the wedges, their vertical extent, and the total ice content of the yedoma increase with age of the sediments on the terraces along the coast, suggesting that there has been a gradual change or deformation taking place during periods of considerably colder weather. However, some areas lack these transformations.

Our understanding of the processes involved in the changes is still incomplete. The changes include evolution of the shape of the polygonal ice-wedges, polymorphism, and the disappearance of much of the host sediment surrounding the wedges. The processes believed to be involved during their evolution range from fluvial (Popov, 1953; Katasonov, 1954), aeolian (Tomirdiaro, 1983; Tomirdiaro *et al.*, 1984; Murton *et al.*, 2015), to polygenetic (Sher *et al.*, 1987).

In the following account, the mechanics of soil and ice flowage are considered as a possible cause of these transformations. The decrease in temperature in winter causes the soil to shrink and crack, permitting water to enter the new crevices during the spring thaw. When chilled by the cold permafrost, it freezes and expands. This is repeated year after year until the host soil is squeezed upward and is largely eliminated from around the ice-wedges. Ice is known to be able to flow under pressure, and it can probably more easily be pressed upwards than the soil. This raises a number of questions, such as why do we find evidence of the soil being pushed up and not the ice? What is the mechanical condition of the soil when it moves? Why are the soil layers not deformed to a greater extent (Figure 5.12)? Why is the shape of the ice-wedges often so irregular (Figure 5.16), and the basic vertical structure of the wedges tends to be destroyed.

6.10 CONDITIONS FOR GROWTH OF THICK ICE-WEDGES

Water freezing and expanding when it is chilled by permafrost can be expressed by the Clapeyron equation. Stresses are approximately 13.4 MPa per 1°C decrease in negative temperature. However, these heaving-out stresses occur under conditions of complete suppression of the strains. As the natural soils may not be considered as a closed, hard-to-deform system, the heaving-out component of the stresses does not predominate over the other components of heaving stresses and strains in the majority of cases. In the case of mechanical equilibrium, if horizontal stresses σ_z are equal in soil and ice, the heaving strain of about 9% of volume of freezing water ε_f is connected to mechanical compression of frozen soil ($d\sigma_z * l_{fr}/E_{fr}$) and ice

$(d\sigma_z * l_{ice}/E_{ice})$, being of l_{fr} and l_{ice} in size and having the strain modulus E_{fr} and E_{ice} respectively:

$$d\sigma_z = \frac{d\varepsilon_f}{\dfrac{l_{fr}}{E_{fr}} + \dfrac{l_{ice}}{E_{ice}}} \qquad (6.1)$$

If, for example, ε_f is 0.0045 m as a result of freezing of 0.05 m of water, $l_{fr} = 5$ m and $l_{ice} = 0.5$ m, and the long-temp compression modulus $E_{fr} = 20$ MPa and $E_{ice} = 50$ MPa, then stress $\sigma_z = 0.017$ MPa. In many cases, the value of ε_f is even less than 0.05 m, for example only 0.001–0.003 m in Barrow (Black, 1951b), and 0.002–0.01 m in Kolyma plain (Berman, 1965). Due to higher modulus values, the compression of the soil will reach 4.33 and ice 0.17 mm. The size of the deformed soil area varies, for example, near Fairbanks in the range of 0.3–3 m (Péwé, 1962). The lateral strains depending on Poisson's ratio will be less than the compression, but they will probably be more in the case of the soil than in the ice.

The stresses are small and perhaps unable to make considerable structural changes to the soil mass. Repeating this process a thousand times will result in an ice-wedge thickness of about 4 m. This is generally in agreement with the field evidence that the surrounding soil is pushed up during the formation of Holocene ice-wedges. However, it has been found that soil layers at a certain distance from ice-wedges are hardly affected at all (Figure 5.10). Sometimes there is no deformation of the soil near the ice-wedge, e.g., on the Yana River in Eastern Siberia (Popov, 1965), and on Chukotka (Kurdyakov, 1965).

An area of high density of deformed soil should be created at the contact with the ice-wedge to give the necessary space for the ice, and that area should gradually increase in size in accordance with increase of wedge thickness. One reason for the deformed area to remain small (Figure 5.13) is the distribution of stress in the soil mass. The stresses are not equal but become smaller with distance from the ice-wedge. Using q/unit length on the surface of a semi-infinite soil mass, or if the excess stress is according to the Boussinesq equation (Ahlvin & Smoots, 1988), the stress can be found approximately:

$$\Delta\sigma_z \sim \frac{\sigma_z}{z^n} I \qquad (6.2)$$

where: n changes from about 1 to 2, I – influence factor for the load, and z – distance from ice-wedge.

Formula (1) should be adjusted according to equation (2). If the horizontal stress is $\sigma_z = 0.017$ MPa near an ice-wedge, it is only about $\sigma_z = 0.004$ MPa at a distance of 2 m. Thus, the stresses might be too small to cause deformations very far from the ice-wedge, but they are big enough to move soil particles and pore ice in the vicinity of the ice-wedge. Another reason for the deformed area being so small is perhaps because of the gradual movement of the soil attached to the ice as it moves towards the ground surface (Figure 5.13) in response to pressure and buoyancy.

It might still be difficult to explain the deformation of the small soil layers in the case of thick ice-wedges with a size approaching 5–10 m (French, 1996), where the soil mass is generally replaced by ice. However, blocks of undeformed sediment also remain. Popov (1965) suggested a different, so-called "frontal" mechanism of ice-wedge growth. According to him, they can grow due to freezing of water on the horizontal surface of ice chilled in winter enough to cause freezing on contact. It is a possible condition in very cold climates, and synchronal accumulation of mineral deposits or peat, if the ice-wedges are syngenetic. The distribution of substantial ice-wedges shows that they form in areas where the mean annual air temperature is −6°C or colder (Péwé, 1966). Despite a list of other characteristic features (*e.g.*, the occasional absence of elementary vertical layers) and important consequences, the interesting hypothesis of Popov will merely be noted here.

6.11   THE MECHANICAL CONDITION OF THE GROWTH OF ICE-WEDGES AND ITS CONNECTION TO THE PROPERTIES OF THE SURROUNDING SEDIMENTS

An irreversible deformation of soil is caused by viscoplastic flow. The dependence of the steady flow rate $d\varepsilon/dt$ (or $\dot\varepsilon$) of icy frozen soil on the effective stress is expressed by (Tsytovich, 1973):

$$\dot\varepsilon = \frac{1}{\eta}(\sigma - \sigma_0) \tag{6.3}$$

where: η – is a time-variable viscosity coefficient that depends on the temperature of the frozen soil; σ – stress, MPa; σ_0 is the initial stress below which viscoplastic steady flow has not yet started, which is sometimes known as the creep threshold.

Experiments indicate that the creep threshold σ_0 is somewhat higher than the ultimate long-term strength, and equation (3) assumes the form of the Bingham – Schwedoff equation for the flow of plastoviscous bodies. The soil behavior is not always in agreement with the Bingham-Schwedoff equation, or viscoplastic flow cannot exist if stresses are less than the ultimate long-term strength. However, we still may assume that this applies for ice since $\sigma_0 = 0$, and for soil $\sigma_0 > 0$. It means that ice flows at any stress, but soils require some value of stress, above zero, which, taking into account the value of the stresses generated during ice-wedge formation, is unlikely to be reached. Sand is especially resistant, and thick ice-wedges are unknown in sand deposits. Therefore, thick ice-wedges can be formed more easily under conditions of low values of soil creep threshold σ_0. The creep threshold σ_0 value determines the area of soil deformation, so that the soil layers are folded if stresses exceed the creep threshold $\sigma > \sigma_0$. Moreover, we can suggest a hypothesis that the low values of soil creep threshold are a condition of growth of thick ice-wedges.

Ice wedges of greater thicknesses than a few metres are mainly found along the Arctic coast of Russia and in Yakutia. In both cases, the frozen soils are both saline and fine-grained. Salinization of permafrost soils in Central Yakutia varies from 0.1 to 0.5%. Where the ice complex is exposed on the right bank of the Lena River close to Moru Lake (Katasonov & Ivanov, 1973c), the salinization of the soils containing the massive ice-wedges varies from 0.06 to 0.12%, and is never below 0.05%. The coastal yedoma sediments contain 0.05 to 2.0% and belong to the marine type of salinization with a prevalence of the chloride ion (Brouchkov, 1998). These frozen saline soils have a very low bearing capacity (Velli, 1980). They are found between frozen and unfrozen soils because they freeze at lower temperatures and contain more unfrozen water than the frozen soils. Their *creep threshold* σ_0 values are low and close to zero, for example, frozen saline marine silt has the ultimate long-term strength of about 0.7 MPa at a salinization of 0.05%, and only 0.1 MPa at a salinization of 0.2% at $-2°C$. Under stresses of about 0.1 MPa (which is higher than the estimation made above ($\sigma_z = 0.017$ MPa) for ice-wedges), only saline soils may flow.

That provides an important reason for the wide distribution of thick ice-wedges in regions with saline permafrost. Previously, only paleogeographical conditions and particularly water supply were considered to explain the Ice Complex formation (Popov, 1953). Massive ice-rich permafrost also occurs in the coastal lowlands of the Mackenzie and Yukon River and saline soils are also common, but the ice-wedges are smaller there (French, 1996). However, the deposits are not as saline there (0.05–0.1% in Fox Tunnel, Fairbanks), and that could be the reason for their smaller thickness. If ice is pushed up during the freezing of water, the changes of shape and structure of ice are probably less obvious because of the gradual growth of the ice-wedges (about 50 mm increase against 0.17 mm of compression a year in the example above), as well as the melting of ice in the active layer. The situation on the North Slope of the Brooks Range and on the Seward Peninsula has yet to be described.

The flow of ice under horizontal stresses should change the shape and structure of ice-wedges. Since the first studies of ice-wedges (Popov, 1953; Katasonov, 1954), relatively little attention has been paid to the question of why many ice-wedges have irregular shapes. Wilkerson (1932) and Péwé (1962) have pointed out the irregular shape of underground ice and its unexplained foliated structure in the Fairbanks area. The foliated structure was commonly found in Siberian and North European ice-wedges (Popov, 1965), but vertical foliation of the ice is not always obvious. This is probably due to the change in ice structure during flowage. Tectonic, slope and glacial processes have also been claimed to be responsible for the dislocations and irregular shape of ice-wedges. Despite these obvious possibilities, there is another process which is caused by differences in the mechanical properties of soil and ice and is bound to change the shape of wedge drastically, *viz.*, buoyancy.

6.12 BUOYANCY OF ICE-WEDGES

Signs of diapirism and soil circulation are widespread in periglacial areas (Hallet & Washington, 1991). Buoyancy can be an effective driving force in the case of ice-wedges due to the different densities of frozen soil (1.5–1.7 and more cm^3/g) and ice (0.9–1 cm^3/g). If the size of polygons is either greater or equal to the height of ice-wedge h, and viscosity of ice η_i is much less than viscosity of the surrounding soil η_s the rate of vertical movement of ice-wedges v will be (Artyushkov, 1969):

$$v \sim \frac{\Delta \rho g h^2}{\eta_s} \tag{6.4}$$

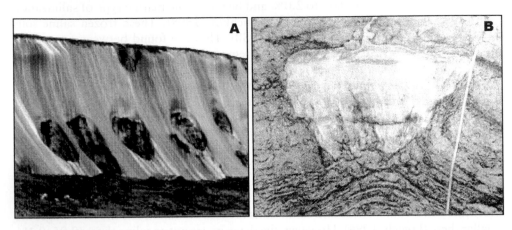

Figure 6.8 Ice-wedge structures featuring vertical flow. A. The ice complex exposed in a 20 m cliff on Big Lyakhovsky Island (© M. Grigoriev). B. An ice-wedge (0.8 m high) on the Yana River terrace showing deformation of the underlying beds (Kostyaev, 1965).

where: $\Delta\rho$ – difference of densities of soil and ice; g – gravity acceleration, 9.81 m/s^2; η_s – viscosity of surrounding soil.

Artyushkov assumed that the viscosity of ice is less than viscosity of the surrounding soil when the relationship is actually the opposite. The ice viscosity can be assumed as 10^{12}–10^{13} Pa $*$ s, and frozen soil viscosity as 10^{10}–10^{11} Pa $*$ s. However, simplifying these equations, the buoyancy of the ice-wedge and its vertical movement z can still be deduced from the similar Navier-Stoks equation:

$$z \sim vt \sim \frac{\Delta\rho g l^2}{18\eta_s}t \qquad (6.5)$$

where: t – time; l – width of ice-wedge.

If the width of the ice-wedge $l = 1$ m, and time t is 1000 years, then the upward movement of the ice can reach about 1.5 m. That value of vertical movement may change the shape of the ice-wedges drastically, especially in saline or high-temperature permafrost. Vertical orientation of rod-shaped air bubbles (Kurdyakov, 1965), "echelon breaks" (Péwé, 1962) and the foliated structure of ice can serve as indirect evidence for it. Sometimes soil layers near ice-wedge look as if they are "drawing in ice" (Péwé, 1962). Flexures of soil layers near ice-wedges are mostly directed upwards (Popov, 1965) in all probability due to the vertical ice flow. Eventual results of such a deformation can be observed (Figure 6.8). The view of the Ice Complex exposures on Big Lyakhovsky Island shows the flow of ice around the remains of soils which are surrounded by ice (Figure 6.8A). The unusual deformations of soil layers under ice wedges described by Kostyaev (1965) in the Yana River terrace may also be a result of the buoyancy (Figure 6.8B). Ice-wedges that flowed upwards by 1–3 m like salt-diapers, were described by Black (1983).

6.13 SUMMARY OF THE IDEAS EXPLAINING YEDOMA EVOLUTION

Obvious mechanical conditions resulting from the freezing of water in an ice-wedge and differences in buoyancy as a result of difference of density cause upwards movements of both ice and the host sediment. Simplified equations (1)–(5) used to estimate the possibility and range of deformations are far from accurate and comprehensive, but they show that vertical movement of soils and ice should take place over time. Saline soils and other deposits of lower strength and higher deformation ability will be particularly favorable for movement. Rozenbaum (1987) concluded that obvious evidences of the vertical ice flow is absent, but Konischev & Maslov (1969) suggested that the vertical movement of ice occurs as masses, or blocks of deposits, instead as a viscoplastic movement. Solomatin (1986) suggested ice recrystallization as the mechanism of ice movement.

Clearly, the mechanism of ice-wedge deformation is not fully understood. The studies of size and form of thick ice-wedges underline the necessity for long-term summer and winter observations of growth at sites where deformation can be registered through time (Mackay, 1992). The deformation of ice-wedges is connected to the behavior of the surrounded deposits, but the origin of the deposits is also not clear. Unfortunately there is no present-day analog of their behaviour (Danilov, 1996).

The thick ice-wedges can be formed more easily under conditions of low values of soil creep threshold σ_0 and their higher deformability since it is a condition of their growth by the cracking-filling-freezing mechanism. The stresses induced by freezing are small and perhaps unable to make any considerable structural change to the soil mass. However, creep threshold σ_0 values of frozen saline soil are low, and that gives an important reason for the wide distribution of thick ice-wedges in regions of saline permafrost. Ice is able to flow at any stress, and should be flowing upwards during the formation of ice-wedges. A number of features appear to be created during the evolution of the shape of ice-wedges due to flow, and irregular shapes of underground ice are common. Buoyancy can be another effective driving force in the case of ice-wedges due to difference of densities between frozen soil and ice. An estimation of the forces involved shows that the buoyancy of ice can reach substantial values and can result in substantial upward movement.

6.14 AUFEIS

Aufeis refers to the development of a sheet-like mass of ice consisting of successive layers of water freezing on the ground in winter, usually on floodplains or valley floors (Figure 6.2). The name was first used by A. F. Middendorf in 1859 (Leffingwell, 1919). Other names include *naleds* or *taryns* (Russia) and *icings* (North America), although the latter is best used as a descriptive term for a process (Harden *et al.*, 1977).

The water is usually supplied by springs discharging water on to the cold valley floor, or by rivers and streams freezing solid and then overflowing their banks. Ground water discharge is common in the alpine valleys with streams flowing down a fairly steep gradient. The discharge freezes, pressure builds up and then water breaks through and spreads out, only to freeze and form the next layer of ice (Hopkins *et al.*, 1955;

Figure 6.9 Aufeis along a river floodplain in August in the North Fork Pass, Dempster Highway, Yukon Territory. © S. A. Harris.

Carey, 1973; Anisimova *et al.*, 1973). This is repeated so that by the end of winter, the aufeis can be six metres thick (Williams, 1970; Péwé, 1973b). Frost boils (**bugry**) are common where the water breaks through the cover of ice. Harden *et al.* subdivide them into ground, river or spring aufeis, based on the whether the water comes from ground seeps, a river, or a spring. Ground aufeis is the result of a break in slope on a floodplain or river thalweg. Spring aufeis refers to icings on slopes where groundwater comes to the surface as a spring. In many cases, all three sources are involved.

In spring and early summer, the water from the melting snow must flow round them, so widening the river valley. During summer, the ice usually thaws, but sometimes, the largest aufeis may leave a small remnant which will act as the nucleus for the new aufeis next fall (Figure 6.9). They usually form in the same place each year (Hu & Pollard, 1997), and in summer, the stream course where they form is very wide and marked by braiding and willows. They can also form where man-made disturbances cause melting of permafrost or disrupt the frozen layer (Müller, 1947). Aufeis are found in Alaska (Leffingwell, 1919; Porter, 1966; Harden *et al.*, 1977; Kane, 1981), Arctic Canada (Veillette & Thomas, 1979; Reedyk *et al.*, 1995; Clark & Lauriol, 1997; Presnitz & Schunke, 2002), Spitzbergen (Åckerman, 1980); Russia (Alekseev & Savko, 1975; Sokolov, 1978) and Mongolia (Froehlich & Slupik, 1982). Figure 6.10 shows their relationship to freezing and thawing indices.

In North America, they are best developed on the arctic slope on the north side of the Brooks Range. Harden *et al.* (1977) provide a good summary of where and how they form. They start to form after the rivers have frozen over in October and growth often continues until spring. However in some cases, the aufeis stabilize part way through the winter, indicating that their water source has been cut off. The greatest

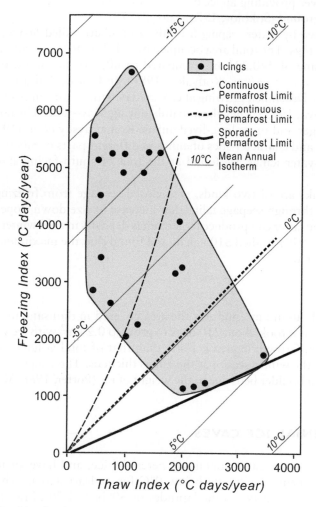

Figure 6.10 Relationship of aufeis occurrences to freezing and thawing indices (from Harris, 1981c, Figure 9B; 1982a).

length of aufeis is about 140 km on the Colville River. Small examples of aufeis also occur in the interior valleys of northern Yukon Territory (Figure 6.8, see Hu & Pollard, 1997), and apparently to a lesser extent in Keewatin, Northwest Territories (Veilette & Thomas, 1979).

In northeastern Siberia, aufeis is far more common, and can store 25–30% of the total volume of river discharge and up to 60–80% of the subsurface drainage (Alekseyev *et al.*, 1973). Their location and size depend on the available water sources, hydrostatic head and geological setting (Anisimova *et al.*, 1973). There has to be a source of flowing water beneath a layer of frozen soil, as well as something that causes the water to move towards the surface, *viz.*, impermeable strata, total freezing of a section of the river providing an ice dam, or reduced permeability downstream due to freezing of the ground (Sokolov, 1973; Carey, 1973).

Aufeis formed by water seeping from the ground are called *taryns*, as opposed to naleds or river icings. The total area occupied by naleds in Russia is about 101,165 km^2 for a total volume of 240.23 km^3, compared to 60,450 km^2 and a total volume of 252 km^3 in the case of naryns (Kotlyalov, 1997). In Central Yakutia, the aufeis first develops in early December, growing at c. 5 cm/day with an air temperature of −25° to −36°C (Gavrilova, 1973). They thaw mainly during May–June. The northern parts of Siberia have larger and more long-lived aufeis. Romanovsky *et al.* (1978) summarize the distribution and characteristics of aufeis in different parts of Siberia. The majority of the ground water icings are found at the foot of south-facing slopes (Kachurin, 1959).

Seepage taliks are of two kinds, *viz.*, closed seepage from freezing of the active layer, and open through seepage taliks that increase in size down slope (Romanovsky *et al.* (1978). The latter can produce larger aufeis deposits in cold winters with minimal snow. Sokolov (1973) studied 310 naleds and found that the maximum volume (*V*) at the end of winter was:

$$V = 0.96A^{1.09} \qquad\qquad (6.6)$$

where: *V* is in thousand m^3, and *A* is the area covered in thousand m^2.

He found that it took about 210 days to grow to 0.2 km^2 with ice volumes exceeding 4 million m^3. Growth in area is faster than that of total volume initially, but this reverses later in the winter season as the aufeis thickens. The icings can produce boulder pavements in boulder beds due to the weight of ice (Porter, 1966; Washburn, 1979; Åckerman, 1980).

6.15 PERENNIAL ICE CAVES

Perennial ice caves are caves containing perennial ice, and have air temperatures in at least part of them consistently below 0°C for at least a year. They can exist in a variety of situations between the latitudes of 30°N and 70°N (Mabliudov, 1985). Those found in or under glaciers, such as the Mutnovsky Ice Cave in Kamchatka, are not discussed here since they are not related directly to permafrost. However there are numerous caves located primarily in limestone karst or in areas of lava flows from volcanoes throughout the cooler parts of the world that have a climate conducive to the formation of ground ice. Figure 6.11 shows the relationship between mean annual freezing and thawing indices and some of the known localities with perennial ice caves. These caves can exist where MAAT exceeds 0°C with a maximum MAAT as high as 8.4°C in the Fuji Ice Cave (Ohata *et al.*, 1994a, 1994b). In the colder northern climates with continuous permafrost, the caverns in the rock are generally completely filled with ice. Ford & Williams (1989) referred to them as *non-conventional permafrost* and they represent a significant form of sporadic permafrost.

Thompson (1976) summarized the known karst caves in Canada, providing maps and descriptions of many of them. Many perennial ice caves are located in the western Cordillera as far south as New Mexico, U.S.A. (Halliday, 1954; Dickfoss *et al.*, 1997) in old lava tubes, while other caves are found in limestone karst along the Eastern

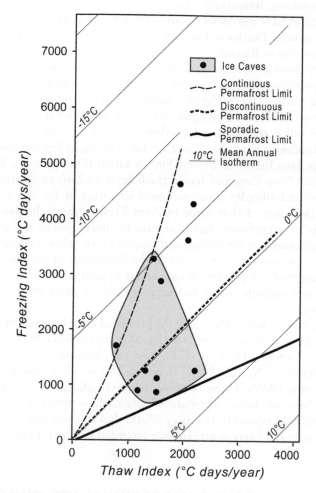

Figure 6.11 Relationship between accessible ice caves and freezing and thawing indices in North America (after Harris, 1981c, Figure 9A, 1982a).

Cordillera north to the Yukon Territory, *e.g.*, Old Crow. The majority of these western ice caves are found in the caves developed in Palaeozoic limestones. Barry and Gan (2011) report that perennial ice caves occur throughout the United Sates. Perennial ice caves are widespread in the limestone mountains of the Alps, Jura and Carpathians. The longest ice cave in the world is the Eisriesenwelt in Werfen, Austria (47°30'10"N, 13°11'23"E at 1656 m elevation). It is 42 km long and represents a major tourist attraction in summer in the Tennengebirge Mountains, 40 km south of Salzburg.

The mountains were uplifted during the last (*Wurm*) glaciation to form a karst plateau, and the cave is located on its north rim. Only the outer 0.7 km of the cave has ice. The passageways were eroded by the nearby Salzach River, and the ice consists of meltwater from the winter snow that has descended down cracks and then moves to the lowest part of the cave where it becomes frozen. The Schellenberger ice cave (at 1570 m elevation in Untersburg, Bavaria) has the greatest ice surface area, estimated at about $60,000\,m^3$. Ice caves have also been reported from Japan (Ohata *et al.*, 1994a, 1994b) and the Chili-Argentine border in the high Andes of South America. They also occur in the Ural Mountains in Russia.

Perennial ice caves in volcanic lava flows are usually located in empty lava tubes whose roof has partially collapsed. Examples occur in Idaho and New Mexico (Bandera volcano–Dickfoss *et al.*, 1997). The Surtshellir cave in Iceland is about 1.5 km long and is developed in vitrified layers of magma and basalt. The floor is covered in a perennial sheet of ice and fallen fragments of solidified lava. Large *speleothems* (*stalactites*, *stalagmites* and *columns* made of ice) are common along its length. About 9 perennial ice caves exist in lava tubes on Mount Fuji in Japan (Ohata *et al.*, 1994a, 1994b). The main one is 150 m long and had a thick layer (>2 m) of ice on its floor until 1990. There are undoubtedly some perennial ice caves in Russia, *e.g.*, the Kunger Ice Cave, near the town of that name in Perm Krai on the right bank of the Sylvia River. Inside the cave entrance, stalactites are common where the water enters the cave roof though cracks in the ceiling and freezes there, while stalagmites may form beneath them. Continued growth results in the formation of icy columns. However, the occurrence there of suitable limestone karst is limited, and the main volcanic areas in Kamchatka are extremely active, so that perennial ice caves primarily occur within or under glaciers.

There are few cases where the age of the perennial ice is known. Exceptions are the Silica cave near Prague, where there is archaeological evidence that the ice developed about 2 ka B.P. (Kunský, 1954, p. 176), and the Aiyansh caves in a lava flow in British Columbia (55°7′N, 128°54′W) dated at 250 +/−130 radiocarbon years B.P. (Sutherland Brown, 1969). Those in areas that were glaciated during the last major glaciation must have developed in postglacial times. Many probably formed in response to the Neoglacial cold period in the more southerly locations where they form a type of sporadic permafrost. There is no evidence recorded to suggest glacial ice is involved (Balch, 1900).

6.16 TYPES OF ICE FOUND IN PERENNIAL ICE CAVES

Figure 6.12 shows some of the basic forms of ice found in perennial ice caves. Icy stalactites often hang down from the ceiling near the cave entrance the water freezes as it moves through the cold air in the cave. Concentric banding occurs in the ice forming the columns with thin layers of cryogenic carbonate in it. Coupled with radiocarbon dates on organic remains trapped in the ice, these can enable speleologists to study the age and cave temperatures during ice deposition by analyzing the isotopic composition of the associated cryogenic carbonates deposited during the freezing of the water (Žák *et al.*, 2008; Laursen, 2010; Persoiu *et al.*, 2011; Žák *et al.*, 2013). Any excess water continues on to the floor of the cave, where it runs downslope, usually freezing as it moves. Banded ice usually lines the walls and is several metres thick (Figure 6.13) in the Werfen Ice Cave (Laursen, 2010). There is a definite seasonal fluctuation in ice thickness in response to the changes in the air temperatures and relative humidity during the year (Ohata *et al.*, 1994a). In summer, the upper part of the icy floor may thaw, but it refreezes downwards from the surface in the autumn, as indicated by the

Figure 6.12 Speleothems and icy slopes where water has entered the roof of the cold Werfen Ice Cave, south of Salzburg, Austria. The walls and floor are in permafrost, so the water freezes quickly after entering from the unfrozen rock above.

Figure 6.13 Banded ice coating the walls of Werfen Cave, Austria.

Figure 6.14 Plate-like ice crystal from the coating of the wall of Plateau Mountain Ice Cave in 1978. The crystal is about 12 cm in diameter. © S.A. Harris.

vertical crystal structure (Marshall, 1981). These icy floors occur in many ice caves, *e.g.*, the Werfen Cave (Figure 9.12 and 9.13). Ice falls (ice-coated steep slopes) are common, and again, the ice is generally banded. In caves in permafrost where the rock is isothermal at a temperature just below 0°C and there is negligible air movement, plate-like crystals of ice may form as hoar frost on the walls of the cave, *e.g.*, at Plateau Mountain, Alberta (Figure 6.14).

Currently, many ice caves are exhibiting a net loss of ice volume (Dickfoss *et al.*, 1997; Wisshak *et al.*, 2005). Most ice caves respond rapidly to any change in MAAT or precipitation. Any increase in either of these environmental factors has an immediate effect on the cave ice. Increases in MAAT (either from the ending of the Neoglacial period or from local warming of the climate) produces a reduction in ice volume. Increased snowfall results in increased ice in deep cavities in the ground, *e.g.*, lava tubes.

6.17 PROCESSES INVOLVED IN THE FORMATION OF PERENNIAL ICE CAVES

Ice caves are very dependent on the form of cavities in the rock, protection from direct sunlight, wind moving through cracks and crevices in the rock, and a suitable supply of water. The accumulation of snow blown into the caves and hollows in rocks in winter is also a very common factor. They can also be formed within permafrost, *e.g.*, at Plateau Mountain (Harris & Brown, 1978), but the Castleguard Cave in Banff National Park only has ice in the outer 1 km of its >13 km length, in spite of being situated in the continuous permafrost zone (see Figure 3.25). At one place, it ends in the glacier bed but the glacial ice does not enter the passage.

The key climatic factor is that there are normally seven months of winter temperatures and only five months of summer temperatures in the areas where these caves occur. The cave or hollow is protected from direct sunlight except at its entrance. Cracks and fissures in the limestone or volcanic rock may permit cold, dense air to enter the cave in winter (Balch, 1900), often also driven by wind. In summer, the less dense, warm air only produces a weak inflow into the cave (Ohata *et al.*, 1994a), and has difficulty moving the dense, cold air out of these caves. In the case of depressions and hollows, the Balch effect tends to keep the dense, colder air in place, resulting in significantly lower temperatures in some cases than the diode effect in adjacent talus (Edenborn *et al.*, 2012). Wind-blown snow tends to accumulate in the entrance and takes a considerable time to thaw in summer. The outer portion of the cave exhibits the greatest annual temperature fluctuations. Where there is a single source of wind, the temperatures in both the air and ground are coldest near the source and rise with distance into the cave. Ohata *et al.* (1994b) found that the wind speed of the air entering the Mount Fuji Ice Cave was proportional to the square root of the temperature difference between the air inside and outside the cave, with a maximum of more than $0.8 \, \text{m/s}^{-1}$.

Where the cave is located in permafrost, some of the water entering via cracks in the overlying rocks freezes forming a coating of ice on the rock, as well as tending to form speleothems where it enters the cave If the temperature of the cave is below 0°C (Wimmer, 2007). Excess water spreads out on the floor of the cave and freezes there. Once a cave or hollow has built up a mass of snow or ice, it takes extra heat to change the ice to water before the air temperature can rise much above freezing point. As a result, the ice in ice caves has a buffering effect, reducing the tendency of the cave to warm up during the warmer summers. Another effect of the water dripping into the cave is to cause the air to become supersaturated. When this happens, hexagonal plate-like ice crystals grow on the walls of the cave (Figure 6.13 – see LaChapelle, 1969). These can grow up to 15 cm in diameter. However, when the MAAT increases, the warmer precipitation in summer descending through the cracks in the overlying bedrock thaws some of the ice which had been largely blocking the cracks and joints, and there is a sharp increase in the amount of water entering the cave. This water can bring with it considerable quantities of heat. In the case of caves with air entering along the fissures and cracks in the rock, there is increased penetration of warm air in summer. Both of these effects result in thawing of part or all of the ice, depending on the degree of change and amount of ice buffering the cave from temperature changes. In the case of the Canyon Creek ice cave in Alberta (Harris, 1979), the increase in air movement in 1978 destroyed the ice in the cave in about three years. However, the ice returned immediately afterwards. On the other hand, the increase in water dripping into the Plateau Mountain ice cave has taken over 40 years to reduce the ice content, and some ice still remains in the deepest parts of the cave.

Shape of the cave or depression is important. Deep open cavities often carry perennial ice in the higher parts of the Rocky Mountains, while deep, vertical entrances to cave systems allow snow to accumulate in large quantities that cannot thaw during the following summer. Horizontal tubular caves require either cold permafrost, or else they need fissures allowing wind to enter to develop adequate cooling. Sloping caves such as the Werfen Cave have the warmer summer air escaping through chimney-type cracks at the upper end, while cold air enters at the lower end. Yang & Shi (2015a, 2015b) provide a feasible quantitative method of modelling the heat flow in the Ningwu Ice Cave in China using air and ground temperature data. Unfortunately, they had to use interpolated data from a weather station several kilometres away.

6.18 CYCLES OF PERENNIAL ICE CAVE EVOLUTION

The climate has fluctuated between colder glacial and warmer periods very frequently in the past (see Table 3.2 and Figure 4.14). During a cold period, ice caves would be developed along the southern portion of the permafrost areas, and these would degrade when the climate became warmer, while new ones would form further north. There are two basic types of caves, *viz.*, those consisting of a deep opening in the ground surface permitting snow to enter, and those consisting of a long tube-like cavern. The latter can be subdivided into those that are open at both ends, those that only have one main entrance, and those that have only one entrance but air entering via cracks and joints in the bedrock. Each has a distinct reaction to changes in climate.

6.18.1 Perennial ice caves in deep hollows

These caves have a deep opening in the ground that traps snow in the winter months (Dickfoss *et al.*, 1997). When the climate becomes colder, new openings in what have been warmer areas will start to accumulate snow, which then metamorphoses into ice. Figure 6.15 is a schematic diagram showing the movement of air and the accumulation of metamorphosed snow in the form of an ice mass near the entrance of the Candelaria Ice Cave. In winter, cold air settles in the cave and freezes any melted ice. Any warmer air exits the cavern to make room for the cold, dense air due to the Balch effect. In summer, the warm air cannot penetrate very far into the opening and sunlight cannot shine in for very long, due to the steep sides of the opening. However it carries considerable moisture, some of which condenses and freezes to the ice surfaces. In winter, the ice is continually melting, evaporating and/or subliming due to the dry dense air, and these losses have to be balanced by the winter snow accumulation plus any additions from water dripping from the roof and freezing. Ice cores in the upper 3 m of the icy mound show that the Candelaria ice cave was cold from about 1,800–3,000 years B.P. and again from 1,650–1850 A.D., *i.e.*, the last two Neoglacial periods. Since 1850, the ice has been fluctuating with increases in ice occurring from 1924–1936 and again from 1947–1991. In between, it was ablating (Dickfoss *et al.*, 1997).

Cooling of the MAAT and/or increased snowfall will cause an expansion of the cave ice, whereas reduction in winter precipitation, an increase in summer precipitation, or an increase in MAAT result in loss of ice. The latter may open up fissures permitting additional ice to form due to freezing of percolating meteoric water into the cave, as demonstrated by Wimmer (2007) in the Schönberg system, Totes Gebirge, Austria. If the water percolation is slow, the ice mass may increase, but if there are heavy summer rains, the warm water aids in melting the ice. In this way, the belt of ice caves along the Rocky Mountains change in ice content over time as controlled by the regional changes in climate.

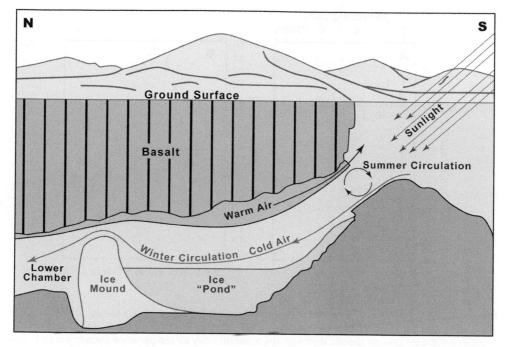

Figure 6.15 Diagram showing the pattern of air movement and ice accumulation in Candelaria Ice Cave, New Mexico (modified from Dickfoss *et al.*, 1995, Figure 5).

Speleothems are normally absent in these caves, as are coatings of ice on the walls. The main ice mass forms a steep-sided mound on the floor of the cave that may almost reach the ceiling. There may be limited ponding of melt water between the entrance and this mound in summer, but it freezes to form an icy floor in winter. There are no plate-like ice crystals on the walls.

6.18.2 Sloping caves with two entrances

The Eisriesenwelt Ice Cave in Austria is the best-known example of this type of cave, though others occur in the Rocky Mountains and in lava fields. The cave slopes gently upwards towards the Karst Plateau. The warm air can escape through the upper end, allowing cold air to move into the lower end. It is in this lower part that the ice accumulates due to the colder air temperatures there. Figure 6.12 shows the typical ice forms including thick, banded coatings on the wall, formed by ice accumulating as moist air entering the cave as it cools. Speleothems and ice flows have formed on the floor, resulting from water entering the cave along fissures and cracks in the limestone above the cave. Plate-like crystals are absent from the walls. Thaler (2008) reports that the temperature at the entrance follows the outside air temperature in winter, but in summer, the inside temperature is around 0°C from July to October as measured during 1996–2007.

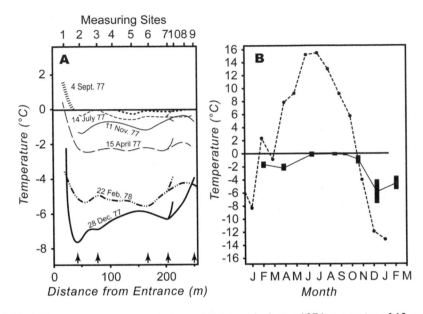

Figure 6.16 A. Temperatures measured at monthly intervals during 1974 at a series of 10 stations in the Canyon Creek Ice Cave, Alberta, and B., comparison of the average air temperatures in the ice cave on specific days with the mean monthly air temperature outside the cave. The arrows in A. show where air entered the cave through fissures in the bedrock. Redrawn from Harris (1979, Figures 7 and 6 respectively).

6.18.3 Perennial ice caves with only one main entrance but air entering through cracks and joints in the bedrock walls

In this case, the air entering the cave through cracks brings in cold air in winter, aiding in cooling the cave. An example is the Canyon Creek ice cave in the foothills of Alberta, west of Calgary. Figure 6.16 shows the temperature variation measured at 10 stations in the ice cave during 1974, and the effect of the incoming air is obvious. Since the winter freezing lasted for seven months of the year, the ice survived. This type of ice cave is ephemeral, the survival of the ice depending on the relative temperatures regimes and durations of successive warm summers. The ice completely disappeared by 1978, but was back again by 1982 with a complete range of banded ice, icy stalactites and stalagmites, columns, and ice-covered floors. Hexagonal plates of ice are absent. The ice has remained in the cave since then (*e.g.*, Yonge & Macdonald, 2014) since the climate change in the Calgary area has involved a roughly equal reduction in summer heating and winter cooling.

6.18.4 Perennial ice caves with only one main entrance and no other sources of cooling

In this case, there is a definite cycle related to climatic changes. Stage 1 is when the climate cools, and permafrost begins to develop in the ground. It gradually extends down to the cave. Meteoric water descending through cracks and fissures from the ground surface starts to deposit a coating of ice, gradually closing off the fissures and preventing further percolating water from reaching the cave. Until then, the water will tend to form speleothems, and any excess will form an icy coating to the floor and walls of the cave. If the fissures are very wide, air will enter the cave and aid in the cooling, as exemplified in the Canyon Ice Cave in Bragg Creek, Alberta in 1974 (Figure 6.15). However, continued cooling will result in ice closing off most of the fissures in the rock, which probably prevents the cave from becoming filled with ice. The entrance to the cave remains dry.

In stage 2, the influx of air and water has ceased and the cave steadily cools. As in stage 1, the air in the cave cools in winter, but not as much as the outside air. Under these conditions, animals use these caves as a shelter from the extreme cold outside. Prehistoric man also appears to have coveted these caves during past glaciations, since the remains of his activities and the skeletons of the animals he used for food have been unearthed in the floors of many caves that would have been ice caves during these climatic events in both Europe and at Zakoudian near Beijing, China.

Stage 3 begins when the climate starts to warm. The temperatures in the cave start to increase until the cave temperatures become isothermal all year round. An example of this stage is the Plateau Mountain Ice Cave prior to about 1970. The cave had icy coatings on the walls and flat icy floors. There may be speleothems present if they formed during stage 1. This stage is marked by the development of plate-like crystals of ice on the walls representing a type of hoar frost. No water is entering the cave from above because the fissures in the rock are still filled with ice.

Stage 4 begins when the ice in the fissures below the active layer thaws sufficiently for meteoric water to start to drip into the cave near the entrance. This adds extra moisture to the air and there is enhanced growth of plate-like ice crystals further in the cave (Figure 6.17). At first, the drips may form small speleothems (Figure 6.18), but these soon disappear. Gradually the amount of water entering the cave increases, thawing the ice in the outer cave. Eventually, enough water enters to flow on to the floor and thaw the ponded ice. The excess water drains through fissures in the floor of the cave. Meanwhile, the banded ice coating the walls of the cave starts to peel off and collapse on to the cave floor where it gradually melts. This degradation that started at the outer part of the cave, gradually works its way through the cave system as the rock steadily ceases to be cryotic. The Plateau Mountain Ice Cave is currently (2015) in the latter stages of this process. Eventually, air, meltwater and summer rain start entering the cave through cracks and speed up the demise of the ice.

6.19 ICE CAVES IN SUBTROPICAL CLIMATES

About 10 ice caves have been reported as occurring in the area bordering the southern boundary of continuous permafrost in north east China. The main one that has been partially studied is the Ningwu ice cave in Shanxi Province on the slopes of the mountains on the south side of the Hexi Corridor (Meng *et al.*, 2004; Gao *et al.*, 2005; Meng *et al.*, 2006; Shi & Yang, 2014; Yang & Shi, 2015a). It differs from caves elsewhere in the world in that it lies in the path of occasional summer monsoon rains, while the area also endures the very cold air from the Mongolia/Siberian high pressure air mass in

Figure 6.17 Plate-like crystals coating the walls of Plateau Mountain Ice Cave. © S.A. Harris.

winter. This cave is also sometimes referred to as the Luyashan ice cave (State Council, 2015), and has been known to the local population for a very long time. Villagers used to come there to take ice when there was a shortage of water (Sukha, 2014), as well as to cool their body temperature when they had a fever. Subsequently, the cave has been developed commercially, like many of the numerous karst caves elsewhere in China, *e.g.,* near Guilin.

The Ningwu Cave is the largest of the known Chinese ice caves and is situated along the north-facing mountain slopes within the larch forest above the steppe grasslands to the north. It also contains the greatest amount of ice among the small number of known ice caves in China. The small number is largely due to the low latitude where the caves are found and the effects of the East Asia Monsoon in summer. Meng *et al.* (2004) and subsequent authors argued for it to have an origin as glacial ice and assumed it dated from the beginning of the Ice Ages. However, the area has not been glaciated and it could not have survived the Hypsithermal/Altithermal warm climate that ended in this area about 6 ka B.P. (Yang & Jin, 2011). The Neoglacial events in this area began at that time and mean annual air temperatures were 4°C cooler than today, prompting

Figure 6.18 Stalagmite developed on the floor of Plateau Mountain Ice Cave by water dripping on to the cold floor of the cave in an area where the permafrost had not thawed and air temperatures were just below 0°C. © S.A. Harris.

Harris (2015) to conclude that it must have formed during this period. Yang & Shi (2015b) agree with this conclusion.

Unfortunately, there is no ground or air temperature data collected from this site. The shape of the cave is like a bottle with an opening at the top in a depression on the steep, north-facing, forested slope. Inside, the walls are encrusted with ice, with sheets of icicles dangling over its surface. This may be due to downward percolation of the summer monsoon rains. Yang & Shi (2015b) argue that there is limited summer warming on this site, balanced by considerable loss of heat by replacement of the warmer air inside the cave by cold dense air during winter, based on modeling. Any excess water is presumed to drain away into the fissures in the underlying rock. It is possible that the cave represents the remains of a former large underground ice mass that has been partially melted as a result of the warming of the climate since the end of the last Neoglacial cold event. Massive ground ice has been reported from two deep boreholes in similar strata in the Huole Basin, not far to the north (Wang, 1990). Taking the data on MAAT from a location at a lower elevation on the steppe to the north and applying an assumed lapse rate suggests that there are seven months of warm weather and only five months of cold conditions at the Ningwu cave, though the latter is in the larch forest which would tend to reduce air temperatures. However, Yang & Shi (2015b) seem to assume that the cave is in a state of thermal equilibrium based on the commercial development there.

Figure 6.19 Thaw pits left after the melting of large blocks of apparently segregated ice in the unglaciated area at Fairbanks, Alaska: A, in gravels; B, in poorly drained loess. © S.A. Harris.

6.20   MASSIVE BLOCKS OF ICE IN BEDROCK OR SOIL

It is quite common for segregation ice to develop small blocks or masses of ice lenses in a small patch of ground during winter in areas of seasonal frost. In the spring or early summer, the ice melts, leaving a pocket of water together with a thin air space above. If heavy vehicles drive over it, the overlying layers give way, producing a pothole. For this reason, it is usual to have load bans in place in North America, limiting the allowable weight of the vehicles. During the summer, the ground closes up the space by flowage of the surrounding soil, and the load ban is lifted.

In permafrost regions, these ice masses appear to grow to quite large sizes, *e.g.*, masses 2.5 m deep by 3–4 m in diameter have been thawing out, producing vertically-sided pits in both gravels and silts near Fairbanks (Figure 6.19). The holes that develop by thawing blocks of ice are usually dry in well-drained gravels (Figure 6.19A), but are filled with ice-cold water in silts (Figure 6.19B). As such, they can be a death trap for unwary pedestrians since they have vertical sides. Figure 6.20 shows massive ice reportedly in situ in peaty soils in an area of discontinuous permafrost in Alaska. When the ice thaws, they will produce similar pits to those shown in Figure 6.19.

Similar loess-filled cavities that are broader than they are deep have been reported as "sand wedges" on the northeast part of the Qinghai-Tibet Plateau in China (Cheng *et al.*, 2005a, 2005b, 2006). The infilling is younger, and has no obvious bedding, and they are quite different from the usual wedge-shaped ice-wedge casts and the loess and sand tessellons (see Chapter 5). The infilling is similar to the surface loess deposits. When viewed from above, they appear to consist of an approximately circular shallow hole in the underlying sediments that has subsequently been filled in with loess. They appear to have formed at the same time as the ice wedges, and it seems that they represent former large ice blocks similar to those thawing in Alaska.

In 1985, two deep boreholes in bedrock in the Huola River basin, just north of Mangui in Northeast China penetrated two layers of massive ice in Cretaceous shale and coal measures. The holes were located on the north side of Lunar Lake near its outlet (Wang, 1990). The drilling site was 1 m above the lake level (540 m a.m.s.l.)

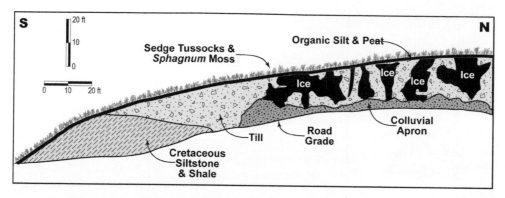

Figure 6.20 Massive ice blocks found in the SW side of a road cut along the Haul Road in Happy Valley, Sagavairktok Quadrangle, Alaska (modified from King & Reger, 1982). The area exhibits discontinuous permafrost.

and the ice mass was 20.7 m thick. In a second borehole, two massive icy layers were found between 40 m and 65.3 m depth, separated by a 7.35 m layer of carbonaceous mudstone. The ice consisted of colourless and transparent, relatively large crystals with clear contact boundaries between the grains. There were dark gray trace impurities in it, together with air bubbles elongated at right angles to the bedding planes of the host sediment. When the drilling reached the base of the massive ice, water confined in it rose to within 2.11 m of the top of the borehole, indicating a 64.74 m confined water head. The ice was interpreted as being the result of freezing of ground water intruded under pressure.

Chapter 7

Permafrost mounds

7.1 INTRODUCTION

A very important feature of the more humid permafrost areas is the existence of permafrost mounds, *e.g.*, the well-known **pingos** (Figure 7.1), although similar mounds have been reported from the drier locations and locations where springs may develop in otherwise arid regions, *e.g.*, Mongolia (Skyles & Vanchig, 2007), and the Qinghai-Tibet Plateau (see Figure 1.19). They range in size from over 250 m in diameter to small hummocks less than a metre across. Some have steep slopes while others have low slope angles. Most are permanent features of the landscape under stable climatic

Figure 7.1 Split pingo, Tuktoyaktuk Peninsula. It occupies a former thermokarst lake bed. As permafrost redeveloped in the talik, the remaining water body became compressed as the water changed to ice. When the hydrostatic pressure exceeded the overburden pressure, the water pushed upwards, only to freeze in the overlying cold soil and form the core of the pingo. Note the dilation cracks and crater at the top of the mound due to stretching of the covering of soil. © S.A. Harris.

conditions, whereas others are either unstable or seasonal. Some appear in a single winter while others develop over a period of one or more centuries. The substrate may be peat, mineral soil, or both.

The processes involved in the development of icy mounds are very varied, some developing in areas of essentially continuous permafrost, *e.g.*, closed-system pingos, while others are found in areas of seasonal frost, *e.g.*, *thufurs*. In this chapter, the main types are examined, largely omitting those decaying mounds resulting from thermokarst that are discussed in Chapter 11.

The mounds are sometimes grouped by size (Lunqvist, 1969) into large landforms, usually more than 2–10 m in diameter when fully developed and the smaller earth hummocks. In Russia, the larger mounds are often treated as a single group called *bulgannyakh* (Vasil'chuk *et al.*, 2008) or *hydrolaccolite*. However, *bulgannyakh* is the Yakut people language expression for *pingos* (mounds of injection type) inside of the common Yakutia thermokarst depression (*alas* in Yakut language) over the Ice Complex. The thermokarst depressions are called *hasyrei* in Western Siberia.

In more recent descriptions, field observations and laboratory studies indicate that many of the apparently similar forms can differ a great deal and actually exhibit quite different evolution and stability over time. There are also differences in the form of a given type of mound depending on the climate and the local geology and hydrology. Thus although the traditional classification can still be used, we must now subdivide a number of the landforms according to their evolution in response to the local microenvironment.

In this chapter, the main groups of pingos, palsas, lithalsas, peat plateaus, and earth hummocks will be treated, together with their apparent subdivisions and look-a-likes. Unfortunately, many published descriptions do not provide enough information to precisely determine which of these landforms was studied. Studies of a given landform over time have been the most revealing, and we can now classify the mounds by the dominant process as well as by size. This will be used in the following account, also describing the changes with fluctuations in the climatic regime.

7.2 MOUNDS OVER 2.5 m DIAMETER

These can be divided into at least three groups, based on the dominant process involved in their formation, *viz.*, mounds formed by injection of water, those formed by cryosuction, and those formed by accumulating a thick core of porous ice representing the thawing fringe.

Table 7.1 shows the main kinds of perennial icy mounds over 2 m high found in permafrost regions. Jahn (1975, p. 99) noted that in Russia, they were commonly treated as either pingos (*bulgannyakh*) or the remaining mounds (*bugor pucheniya*) resulting from cryosuction. This terminology is still common in the Russian literature. As will be demonstrated below, studies elsewhere in Europe and in North America indicate that there are fundamentally different processes operating in the mounds listed above, resulting in different rates of growth, internal structure and behavior when affected by climate change. Accordingly, they are treated individually in this book. In addition, there are a number of cases for which there is dominant cryosuction together with ice from the thawing fringe. Some injection ice may also occur.

Table 7.1 Comparison of the main types of mineral mounds over 2 m in diameter that occur in permafrost areas. It should be noted that some mounds also exhibit a complex history, and in Siberia, large frost heaves including aufeis may be present.

Landform	Topography	Height (m)	Area (sq. km)	Ice content (%)	Core	Dominant process
Pingo/ bulgannyakh	Mound	Up to 25 m, depending on water pressure	<0.3	80–95	Solid ice	Injection of water from below into cold sediment
Pingo plateau	Low, flat plateau	0.5–1.5 Low water pressure	<0.3	80–95	Solid ice	Injection of water from below into cold sediment
Palsa/Lithalsa*	Mound	Up to 10 m (palsa) or up to 18 m where water injection occurs	<0.5	40–65	Ice lenses	Cryosuction
Palsa/Lithalsa plateau*	Flat plateau	Up to 10 m	<2.5	40–65	Ice lenses	Cryosuction
Peat plateau	Low plateau, floating or anchored to sediment	<2 m high	Up to 14	<water in the adjacent mire/bog	Ice with air spaces	Accumulation of ice from the thawing fringe

*Palsas have an upper layer of peat or are developed entirely in peat, whereas the lithalsas only have a thin O layer (usually <10 cm thick).

7.2.1 Mounds formed predominantly of injection ice

The best-known examples are the pingos, that grade into seasonal frost mounds and frost blisters in areas with marginal or no permafrost. Note that lithalsas are composed predominantly of segregation ice, *i.e.*, there is no continuum with pingos (*c.f.* Gurney, 1998). Pingos come in two main topographic forms, *viz.*, pingo mounds of varying height and size, and those consisting of low mounds having a horizontal upper surface. The latter are found in the dry, cold steppes in Siberia. They will be dealt with separately, since it appears that different processes modify these mounds, though both are the result of injection of ice into the surface sediments.

7.2.1.1 Pingo mounds

Pingos or *bulgannyakhs* are usually large dome-shaped mounds having a core of ice primarily developed by injection of water due to hydrostatic or hydraulic pressure from below (Figure 7.1). In Russia, they are sometimes referred to as *hydrolaccolites*. The water under pressure freezes before it reaches the surface, resulting in the growth of the pingo (Mackay, 1973c). The growth results in stretching of the overlying soil so that dilation cracks and a crater usually develop near the top of the ice-cored hill. If the growth is too fast, the mound may explode, throwing blocks of ice across the surrounding area. If and when the crater at the top exposes the icy core, thawing of

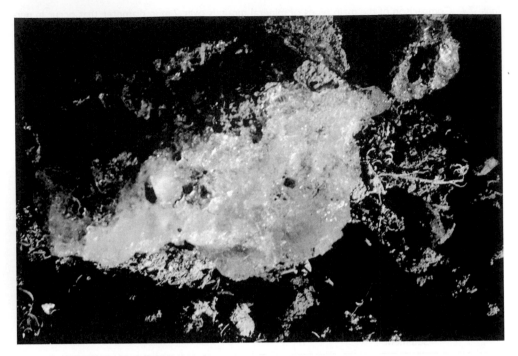

Figure 7.2 Solid ice with occasional soil inclusions obtained in drilling a pingo on the eastern slopes of the Qinghai-Tibet Plateau at 35°01′N, 97°34′E, at 4310 m elevation. © S.A. Harris.

the latter results in degradation of the mound, forming a pingo scar (Figure 7.3). The sediment overlying the core is usually 1–10 m thick. Some pingos have steep sides, *e.g.*, Ibyuk (58°), while others have gentle side-slopes. The overlying permafrost will have ice lenses in it, so that segregation ice in the surface layers may be found (Mackay & Stager, 1966). The pingo surface is also subjected to contraction cracking which usually results in the development of ice wedges (Figure 5.9), but the main mass of the interior of a pingo consists of massive injection ice with air bubbles with their long axes aligned perpendicular to the freezing plane, together with minor inclusions of the host sediment (Figure 7.2).

7.2.1.2 *Hydrostatic or closed system pingos*

Hydrostatic (Mackay, 1979a) or *closed system pingos* are also often referred to as *Mackenzie type pingos* since they have been thoroughly studied on the Mackenzie Delta by J. R. Mackay. First observed by Dr. John Richardson in 1851, they are the result of the freezing of a closed sub-surface water body, building up hydrostatic pressure as the water expands as it freezes. Eventually, this pressure exceeds the overburden pressure, resulting in heaving upwards of the overlying soil in the zone of lowest overburden pressure (Figure 7.1). This determines the width of the resultant pingo. Upward growth will normally continue until either the water body is entirely frozen, or until

Figure 7.3 Developing pingo scar, Mackenzie Delta. Note the ramparts resulting from the slumping of displaced soil downslope, leaving a depression in the centre. © J. R. Mackay.

the main ice body forming its core becomes exposed to the atmosphere by the star-shaped dilation cracks that develop at the crest of the mound. This results in either partial thawing of the core producing a one-sided dimple at the top of the mound or complete thawing producing a pingo scar (Figure 7.3). If the mound grows too quickly, the pressures developed within the mound may cause it to explode (Mackay, 1973c). During growth, the pingo grows higher, resulting in steepening of the sides. The highest closed system pingo on the Mackenzie delta (Ibyuk) has slopes as steep as 54°. Heights may reach over 60 m, with diameters sometimes exceeding 250 m. When the available water in the underlying talik has frozen, growth ceases, but the pingo will remain until erosion or climatic change causes the icy core to commence thawing. Mackay (1990) has shown that seasonal growth bands may be present in the solid ice core, consisting of alternating clear and bubble-rich bands, the latter being formed in summer. The hydrostatic pingos can develop in rock (St. Onge & Pissart, 1990), mineral soil, peat, or some combination of these substrates. They are found particularly in coastal deltas around the Arctic Ocean as well as on some Arctic Islands (Figure 7.1). On the deltas, they usually grow as a result of the freezing of the talik left when a thaw lake drains and freezing of the underlying ground occurs. On the Arctic Islands, they sometimes form where an oxbow lake is formed by a change in river channel (Craig, 1959; French & Dutkiewicz, 1976; Pissart & French, 1976). They also occur

Figure 7.4 Closed system pingo growing on the floor of an alas near Yakutsk, Siberia, formed as a result of refreezing of the talik formed under alases in Yakutia (see Chapter 11, and Figure 7.5). © S.A. Harris.

on the floors of alas valleys in Yakutia (Soloviev, 1973a, 1973b) where the talik developed under the former thaw lake is refreezing (Figure 7.4). The hydrostatic pingos are mainly found in areas of continuous permafrost (Figure 7.5). Mackay (1998) summarizes the stages of development, growth and decay of these pingos. Early growth rates are probably fastest, but the rate decreases with age. Ibyuk is over 1000 years old and is now growing at a rate of 2–3 cm/a (Mackay, 1986). When the water supply ceases, so does pingo growth, so it is taking more than a millennium for the talik underlying Ibyuk to freeze. It is estimated that only about 15 pingos will commence growth in a century on the Mackenzie Delta, and only about 50 are still growing. Mackay (1998) describes the cycle of growth and collapse.

7.2.1.3 Hydraulic or open system pingos

Hydraulic (Mackay, 1979a) or ***open system pingos*** are often referred to as ***East Greenland type pingos*** after the location from which they were first described (Müller, 1959, 1963). They represent the form of pingo found in areas of discontinuous permafrost where rain or melting snow feeds water through fissures in the permafrost on the nearby uplands, and feed the intrapermafrost or subpermafrost water table beneath the discontinuous permafrost This causes increased hydraulic pressures on the underground water at lower elevations, resulting in upward artesian movement

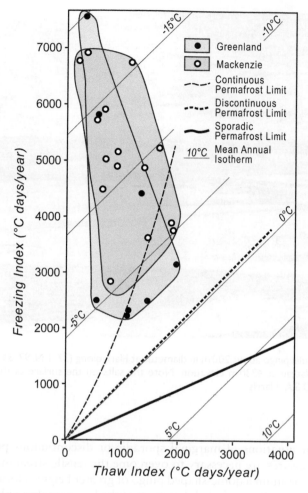

Figure 7.5 Relationship of hydraulic and hydrostatic pingos to freezing and thawing indices (after Harris, 1981c, 1982a).

through any cracks, faults or other planes of weakness in the permafrost on the valley floors. Instead of resulting in springs, the water freezes in the upper frozen layers of the ground (Figure 7.6), resulting in up-doming of the overlying layers to form a pingo. Normally, they occur singly on the floors of depressions or valleys floors (see Figure 1.19), and it is quite common to find a line of them in suitable situations. Unfortunately, they have not been examined as carefully as the hydrostatic pingos (Washburn, 1985), though their main characteristics are known.

In East Greenland, the pattern of groundwater flow and discharge suggest the influence of structural weaknesses in the underlying rocks (Worsley & Gurney, 1996). On the Qinghai-Tibet Plateau, their occurrence is sometimes related to faults in the underlying rocks. This may be due to relatively low hydraulic pressures which cannot always completely overcome the tensile strength of the frozen ground, weak as the

Figure 7.6 A hydraulic pingo about 200 m in diameter at Harigqiong (37° l'N, 97° 34'E) on the Qinghai-Tibet Plateau at 4330 m elevation. Note the salts on the surface of the soils around the pingo. © S.A. Harris.

latter may be in situations of marginal sporadic or discontinuous permafrost. Very high artesian pressures are necessary to overcome the tensile strength of a thick, cold active layer and maintain a dome-shaped pingo of greater height (Holmes *et al.*, 1968).

Holmes *et al.* (1968) concluded that most of the large number of hydraulic pingos in interior Alaska did not have enough artesian pressure to produce a large, high pingo, hence the lower form with gentler slopes and an oblique depression in the top. Similar lower mounds are seen in the valley floors and base of the mountains on the northeastern slopes of the Qinghai-Tibet Plateau (Figure 7.7). One or two large hydraulic pingos do occur in Mongolia (Sharkhuu and Luvsandagva, 1975; Yoshikawa *et al.*, 2013). Their growth can be intermittent, four major stages having been recorded at Mongot pingo since 8790 years B.P. (Yoshikawa *et al.*, 2013). On the Qinghai-Tibet Plateau, they are confined to areas with numerous springs occurring at the base of the mountain slopes but are rare (Figure 7.6). Hughes (1969) provides data on the distribution of hydraulic pingos in the Yukon Territory, Canada.

The artesian pressure is viewed by French (2013, p. 136) as needing to be steady but slow and continuous in order to sustain the pingo growth. Mackay (1973c) regarded the necessary conditions for continuous injection of groundwater as being the water pressure (determined by local environmental conditions), overburden strength (which varies with the seasons) and rate of freezing (dependent on the ground temperature).

Figure 7.7 Hydraulic pingo with a depression in the top at 34°26'N, 97°42'E on the Qinghai-Tibet Plateau at 4457 m elevation. © S.A. Harris.

Since any of these conditions can vary over time, it is possible to have periods of reduced or enhanced growth which can produce bands in the ice (Mackay, 1990a), as well as periods of formation of segregation ice. Thus Yoshikawa *et al.* (2013) reported evidence from the Mongot Pingo in Mongolia that had apparently experienced periods of open and semi-closed freezing conditions, based on chemical and isotope analyses of a core. It started with rapid growth as a hydraulic system pingo at least 8790 radiocarbon years ago, but experienced periods of reduced growth though the accumulation of ice continued to be formed by freezing of pressurized liquid. A solid ice core remains the dominant type in the mound.

On Svalbad, many of the larger pingos are fed by geological faults with groundwater recharge from melting ice from the base of warm based glaciers (Liestol, 1977; Yoshikawa, 1993, 1998; Yoshikawa & Harada, 1995; Jaworski & Chutkowski, 2016). These pingos are basically conical in shape, but often with a depression in the top, caused by the stretching and minor thermokarst. The pingo in the Kunlun Pass (Figure 1.19) is a good example. It has grown on the lobe of an alluvial fan overlying a major fault (Figure 7.8). Its oblong shape (40 m by 20 m) is elongated along the fault trace. It was 20 m high until it was blown up to provide water for the 62nd Highway Maintenance Squad. Subsequently it has started growing again. Each summer, springs flow from it. Drilling in the new pingo revealed 1.2 m of grey sandy loam overlying a 2 m layer of pure ice, typical in water injection, and underlain by an empty

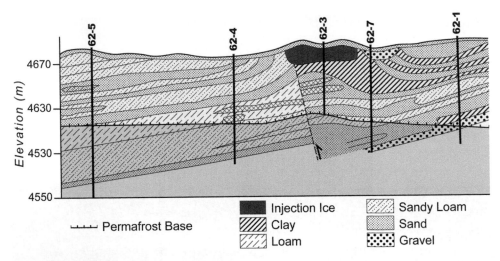

Figure 7.8 Section through the Kunlun Pass Pingo showing the artesian reservoir below the permafrost base in light blue.

cavity. Drilling deeper through the 55 m-thick permafrost, fountains of water 22–32 m high, erupted from below.

Figure 7.5 shows their relationship to freezing and thawing indices. They mainly occur in the areas of discontinuous permafrost. As in the case of hydrostatic pingos, the pre-existing permafrost may have segregation ice in the surface layers, and ice-wedge formation is possible if there is cracking of the surface in winter. However, clear, solid injection ice (Figure 7.1) should be dominant in the centre of the mound. It is usual to see air bubbles in injection ice due to release of air from the solute in water during freezing. If the air content is high, as occurs in the case of methanogenesis in water-saturated deposits, the mound can explode (Lyov, 1916; Solonenko, 1960; Vtyurin, 1975). A possible modern example of such an explosion is a crater in permafrost on the Yamal Peninsula (Figure 7.9) although gas hydrates are a more likely cause.

7.2.1.4 *Pingo plateaus*

A rare variety of the pingo is found on the eastern surface of the Qinghai-Tibetan plateau and in alases in Siberia that consists of an irregularly-shaped, low plateau-like mound up to 300 m in length with cusp-like thermokarst scars on some of its flanks (Figure 7.10) producing a very subdued form with a large surface area but low relief. Similar pingos have been reported from Yakutia with elevations between 10 and 20 m in alas depressions where the talik has commenced refreezing (Soloviev, 1972, 1973a, 1973b). They can be either hydraulic or hydrostatic pingos.

A drill hole through the centre of the mound penetrated 2 m of an icy core, with increasingly thick ice layers with depth. A pit dug in the mound showed a saline solonchak soil 124 cm thick, with a strong crumb structure and high salinity, overlying alternating silt and ice layers. The icy layers became progressively thicker with depth (Figure 7.11), so that they were up to 20 cm thick by the bottom of the pit (Figure 7.12),

Figure 7.9 The Yamal crater (photograph by Vasily Bogoyavlensky, The Siberian Times).

Figure 7.10 A flat-surfaced, open system pingo surrounded by water on the eastern slopes of the Qinghai-Tibet Plateau at 35°01′N, 97°34′E, at 4310 m elevation. It is only 1.5 m high but nearly 300 m long and overlies a fault which supplies water from below throughout the year. © S.A. Harris.

Figure 7.11 The upper 76 cm of the icy core of the pingo mound shown in Figure 7.10. © S. A. Harris. Note the increasing ice content with depth.

separated by thin silt layers. This suggests that each winter, a new ice layer is added to the base of the mound in the form of a sill, but ablation of the upper icy layers in summer due to intense insolation and low relative humidity results in an equilibrium height, and the accumulation of salts in the soil. No plants grow on the soil surface, and the

Figure 7.12 Part of the massive injection ice at 155–170 cm in the pit (Figure 7.10). © S. A. Harris.

processes forming it are still under study. It is in this environment that dry permafrost could occur. The pingo has developed along a fault line in lacustrine silts. The silts originated in a playa lake since this used to be a depression with inland drainage. Subsequently, a tributary of the headwaters of the Yellow River started draining the lake, reducing its level by c. 10 m, resulting in draining of this valley. The Chinese have since dammed the outlet, raising the water level several metres, resulting in poor drainage on the valley floor. In summer, a spring above the pingo produces a stream that flows past the pingo, resulting in erosion of its margins (Figure 7.10). In spite of that, the pingo survives the warmth of the summer quite successfully.

7.2.1.5 Seasonal frost mounds

Where there is a spring in an area of discontinuous permafrost at the foot of a mountain slope that moves its location from year to year, *seasonal frost mounds*, are formed (Figure 7.13) during the winter after the freezing of the active layer (Pollard & French, 1983, 1984, 1985). They are also called *perennial frost blisters* (Morse & Burn, 2014), and are characterized by a low seasonal mound with a core of vertically-oriented ice crystals beneath the surface soil layer (Figure 7.14). During the following summer or summers, they collapse as the ice melts leaving jumbled blocks of turf. They will form again in a different location during the following winter. Seasonal frost mounds are often closely spaced, with several developing within individual ice-wedge polygons (Morse & Burn, 2014, p. 208). Studies using remote sensing suggest that densities there

Figure 7.13 The people are standing on a seasonal frost mound at North Fork Pass, Dempster Highway, along the Blackstone River valley. © S.A. Harris.

may reach 1750 km^{-1}. While some disappear after 1–2 years, 49% can be identified after 4 years, with a few persisting for over 10 years. Some are circular but others may form elongate ridges. They have been described in detail from the North Fork Pass, Yukon Territory and the Mackenzie Delta, Northwest Territories, and have also been reported from Alaska, Scandinavia (Åkermann, 1980), Russia, Mongolia (Froehlich & Słupik, 1978), and from the Qinghai-Tibet Plateau of China.

7.2.1.6 Icing blisters

Icing blisters (Figure 7.15) are similar to the seasonal frost mounds but are the result of freezing of surface water above the active layer (French, 2013, pp. 140–141). However they occur repeatedly, year after year in the same place, and are therefore transitional to the hydraulic pingos, except that the icy core melts completely each year (Evseev, 1978; Van Everdingen, 1978, 1982). They can be up to 4 m high and 30 m in diameter, and usually exhibit good examples of dilation cracking (see Figure 2.15 and 7.15). When they thaw during the summer, there is usually minimal disruption of the vegetation, which may include trees, since they are surface features. If water escapes from the top, it produces icings over a large area. They are indicators of underground water transmission lines in the active layer above permafrost in winter (Froehlich & Słupik, 1978) and have been reported from the Yukon, Alaska, Northern Scandinavia, Tibet and Siberia. They are usually associated with icings (see Chapter 6). Åkermann (1980)

Figure 7.14 Cross section of a seasonal frost mound in the North Fork Pass, Dempster Highway. © S.A. Harris.

Figure 7.15 Icing blister and icing in Siberia along a river valley floor.

describes in detail the varieties found in Spitzbergen, including some with multiple phases of injection of water, resulting in a very complex internal structure (Åkermann, 1980, p. 179).

A modified form has been reported from Spitzbergen (Åkermann, 1980, pp. 180–184), consisting of a mound containing a cavity in the spring beneath an arch of elongate 25–40 cm long ice crystals. The dome is believed to have been developed over a pool of water in a depression on an icing. The upper surface developed vertical acicular ice crystals but the water was flowing in the interior of the low (1–2 m high) mound. In the spring, the water drained away into the underlying river gravels, leaving the empty low ice mound. They are therefore a variety of icing blister whose core never froze. They usually thaw in summer.

7.2.1.7   Perennial mounds of uncertain origin

Without drilling a hole in a mound or using geophysical methods to examine its structure, it is difficult to determine the cause of the feature. A number of such mounds that resemble pingos have been reported from the Canadian Arctic Islands. They can range from 1 to 13 m in height and may be up to 60 m in diameter (Pissart, 1967a, 1967b; Balkwill *et al.*, 1974). They are often associated with bedrock discontinuities between geological formations of various ages and are often labeled pingos, although their exact cause is uncertain. French (2013, p. 140) treats them as "other perennial mounds". Also in this category are elongate and partially collapsed mounds along river valleys that may be remnants of Pleistocene age (French, 1975, 1976; French & Dutkiewicz, 1976).

7.2.1.8   Similar mounds that can be confused with injection phenomena

The most common confusion is with the mounds developed dominantly by ice segregation (see section 7.1.2 below). Thus both active palsas and lithalsas are sometimes confused in the literature with pingos (*e.g.*, Gurney, 1998). As regards the past landforms, the remains are now interpreted as former lithalsas although they were previously called pingo remnants throughout many parts of Europe (Gurney, 1995, 2001; Pissart, 1963, 2000b, 2002, 2003, 2010). More studies by Ross (2006) and other authors now suggests that some may be kettle holes in areas that have been glaciated. Gurney (2001, p. 234) summarizes some of the requirements in terms of study of relict features in order to arrive at a reasonable conclusion as to their probable origin. Pissart (2003) provides an example of the kind of detailed observations that are necessary. Since they are important indicators or proxies for determining past climatic conditions, careful study is essential.

In the case of present-day mounds, it is necessary to carefully examine the microenvironment, drill boreholes in the feature, and describe the type of ice present in order to determine their origin. The mounds should also be monitored to determine how they evolve, in case they prove to be ephemeral. It is also advisable to install a suitable weather station, and to check on the air and ground temperatures along a transect from the valley floor to the highest point on the adjacent landscape to learn more about the microenvironment. Simple description of the surface features generally provides inadequate information to precisely determine their origin.

Figure 7.16   Stephan Vogel (right) and Dr. Fraser Smith (left) from Dundee University drilling into permafrost. To the right part of the ice-rich top of permafrost is seen. Scale below (cm) indicate depth below ground surface. Photos © H.H. Christiansen and O. Humlum.

Figure 7.17   Segregated ice in the form of crystals and interstitial ice in sandy and pebbly sediments in a Lithalsa at 35°01'N, 97°34'E at 4310 m elevation on the Eastern slopes of the Qinghai-Tibet Plateau. © S.A. Harris.

7.2.2   Mounds formed dominantly by cryosuction

These are the *palsas* and *lithalsas* whose permafrost core consists of either thin layers of segregated ice alternating with peat or mineral soil (Figure 7.16), reticulate ice veins, or interstitial ice that has accumulated in pre-existing sediment (Figure 7.17). Svensson (1964) and Pissart *et al.* (1998) give additional photographs of these. These types of ice are formed predominantly by cryosuction, *i.e.*, by water moving from the water table and the overlying unfrozen, moist soil to the freezing plane where it freezes, forming an ice lens. The change in volume of the water becoming ice results in uplift of the overlying layers of sediment, thus producing the mound.

Initial growth is quite fast (Seppälä, 1982), but gradually slows with time. Often, an equilibrium situation is reached, so that the mound does not grow any higher, *e.g.*, the palsas photographed in Varanger Peninsular, Northern Norway, had not changed appearance in 100 years until the 1970's. Wind erosion of the top of the mound and blocks sliding down the sides also limit the growth. However, there is probably significant spreading occurring in at least some types of continental palsas that may replace the losses in mass by slumping blocks of peat on its flanks. Harris (1988a, p. 367), Harris & Schmidt (1992), and Allard *et al.* (1996) found evidence that percolating meteoric water in humid regions enriches the ice content in the upper layers of palsas and lithalsas, but the main mass of ice in the mounds is believed to be the product of cryosuction (Pissart, 2002, p. 609). In Québec, the upper layer of permafrost can contain up to 80% of ice by volume, but overlies a layer with lower ice content (20–30% by volume). At the permafrost base, the ice content and thickness of ice lenses increase substantially (Allard *et al.*, 1996). This only occurs in cold, rather humid climates like Québec, since the lithalsas in the Yukon Territory in a cold but variable, drier climate with cold air drainage also have a thick, well-developed upper icy layer, but with an ice content of around 40–50% by weight in the middle of the mound. Those on the cold, arid *Kobresia* meadows on the Qinghai-Tibet Plateau completely lack the upper layer.

7.2.2.1   *Palsas*

The term *palsa* is a Fennoscandinavian term for peaty mounds commonly found in mires in Finland, northern Sweden and Norway (Seppälä, 1972). Originally, it was applied liberally to most of the larger permafrost mounds (*e.g.*, Åhman, 1977), but as more studies were carried out, it became obvious that there were significant differences between many of them.

First to be spilt away were the pingos, characterized by injection of water under pressure into the sediment where it freezes, as discussed above. Then Brown (1970) argued that the consistently different, extensive, low, flat-topped form called a peat plateau should be considered as representing a distinct landform, separate from the dome-shaped palsas, although both usually occur in peatlands. This terminology was followed by Zoltai (1972) and Zoltai and Tarnocai (1975) working in the peatlands of the northern part of Manitoba, although no significant difference in origin was described. However, there was a vastly different distribution in moisture in the permafrost in the mounds that was measured by Zoltai. The ACGR (1988) followed this lead, and subsequent studies showed that there was, in fact, a significant difference in the origin of the ice forming these mounds (Harris, 1998), warranting the *peat plateaus* being treated as a third type of permafrost mound based on the processes involved in their formation (see below). Harris (1993) described a suite of palsa-like mounds at Fox Lake, Yukon Territory, developed on lacustrine silts lacking a surface cover of peat, and suggested that they should be called *lithalsas*. The case for separating them from palsas is based on the fact that they require colder and drier environmental conditions for their formation (Pissart *et al.*, 1998). This restricted the use of the term "palsa" to mounds developed with a surficial layer of peat, where the diode-effect of peat caused the formation of permafrost mounds in areas otherwise lacking permafrost, as in Scandinavia. Earlier, some workers had distinguished between minerogenic palsas containing mineral sediment beneath the surface layer of peat and true palsas in which the permafrost was contained entirely in the peat (*e.g.*, Åhman, 1976; Pissart & Gangloff, 1984), but the significance of the absence of any peat was not really considered.

The classic Scandinavian palsas that were the first to be described are those developed entirely in peat, usually above tree line (Seppälä, 1972). Seppälä (1988, 2006) has argued that these palsas grow until the icy core enters the mineral substrate. He then hypothesizes that palsas decay by the melting of the icy core. This theory is accepted for Scandinavian cases by several other workers in the region (*e.g.*, Zudhoff & Kolstrup, 2005, in Sweden), suggesting that it may be a regional form related to the maritime climate of northwest Europe and Iceland. In the continental climate of North America and Siberia, similar landforms with a surface cover of peat are also called palsas, but usually are found in the boreal forest and have the permafrost extending deep down into the underlying mineral substrate without the tendency to decay, unless some external force, *e.g.*, increased level of the water table or climatic change intervenes. These palsas are found in colder, more continental climates. Accordingly, these two types are dealt with separately in the following account.

7.2.2.1.1 Palsas in maritime climates

Palsas are one of the most obvious forms of permafrost found in the glaciated regions of Scandinavia and Iceland. Maritime palsas are usually found above or close to tree line. The topography is undulating but rather subdued, and peat accumulates in the kettle holes and other depressions, though at a slow rate. The formation of *Sphagnum, Carex,* and *Eriophorum* peat often started in Finland about 9700–8000 radiocarbon years B.P. (Seppälä, 1971, Figure 11), but the formation of the present-day palsas mainly occurred less than 1000 years ago (Seppälä, 2005). Vorren (1979) gave ages of 450–1000 years B.P. in nearby Norway, while Ruuhijärvi (1960) concluded that palsa formation commenced in the Subatlantic period (after 5000 years ago) based on pollen in peat cores. Dating must be based on the age of the change in the peat in the mound from the mire vegetation of the surrounding fen to that of the ericaceous vegetation found on the surface of the mounds themselves.

Seppälä (1982, 1986, 1988, 2006) has described a cycle of growth and decay for these maritime palsas (Figure 7.18, after Seppälä, 2006). The initial formation of a palsa may be due to uneven snow cover due to wind eddies (Fries & Bergström, 1910), or a low winter snowfall over hummocks in the mire (Popov, 1953; Spolanskaya & Evseyev, 1973). The latter has been demonstrated in a field experiment (Seppälä, 1982). Some frozen ground will remain at depth at the end of the next summer, and will cause the peat above the ice to be a little higher than that in the surrounding fen. However a warm winter or thick snowpack at this early stage will prevent deep penetration of frost into the ground, resulting in the demise of the embryonic mound. If there are no sudden climatic fluctuations, these mounds can grow into palsa surrounded by a large embryonic palsa. This alters the drainage, so that ericaceous shrubs and *Cladonia* species of lichen appear on the surface of the mound. The latter have a grey colour that reflects back more of the incoming summer radiation, thus keeping the

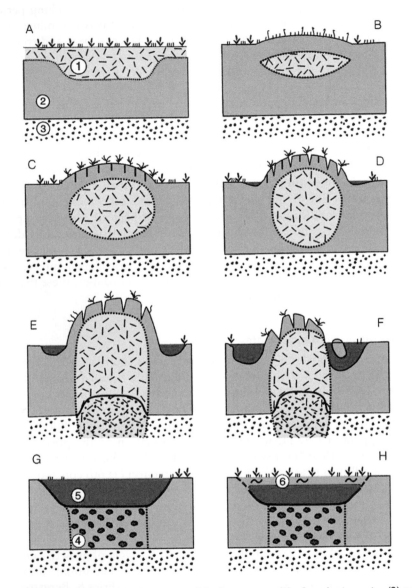

Figure 7.18 A general model of the formation of the frozen core (1) of a palsa in a mire (2) with a silty till substratum (3). A, the beginning of the thaw season; B, the end of the first thaw season; C, embryonic palsa; D, young palsa; E, mature palsa; F, old collapsing palsas surrounded by a large water body; G, fully thawed palsa giving a circular pond on the mire (5). The thawed peat is decomposed (4). H, new peat (6) formation starts in the pond (Seppälä, 1982, 1986, 1988). Reproduced from Seppälä (2006).

mound cooler in summer. This brings about an albedo change from about 13.8% to 21.3% (Railton & Sparling, 1973), who thought that this could be the cause of palsa formation. This cannot always be the case since large areas with these lichens exist in the boreal forest of Canada, but lack palsas. The conclusions of Seppälä have been supported by Worsley *et al.* (1995), Matthews *et al.* (1997) and Zudhoff & Kolstrup (2005).

The mound gradually grows, though the surface of the mound may be eroded by wind-blown ice crystals by as much as 40 cm in one winter (Seppälä, 1993). Blown snow tends to accumulate along the sides, thus limiting lateral growth (Seppälä, 1994). Since the thickness of peat is not great in most places, the maritime palsa mounds are only generally 1–2 m high (7 m maximum). The ice in the palsa is less dense than the water in the adjacent mire, so the growing palsa is floating and therefore not affected by changes in the level of the water around it. Once the permafrost enters the underlying mineral soil, thawing sets in from underneath the mound and the mound starts to sink. The exact cause of this is uncertain, but Gurney (2001) suggests that collapse of blocks of peat into the mire down the sides of the mounds, together with the effect of water thawing the lower parts of the permafrost may be involved. Sliding of the peat down the sides of the higher mounds may produce a low ring of peat (Seppälä, 2005a, Figure 5), though this is unusual. The lower the mound becomes, the greater the probability of the winter snow cover enveloping it, thus restricting winter cooling. In any case, the collapse of the mound results in mixing of peat of different ages in the resulting mire, resulting in problems for studying ages and pollen sequences (Vorren & Vorren, 1975). A further problem in dating the peat is the wind erosion of the usually bare crest of the mound during winter (Seppälä, 2003), which may also help in ensuring the height of maritime palsas is generally low. The highest palsas in Finland are 7–8 m high in Munnikurkkio, Enontekiö.

Another likely cause of the decay of the palsas after the permafrost enters the underlying mineral sediment is the fact that the palsa would become anchored instead of floating. Svensson (1962) noted that palsas were liable to degrade by melting of the icy core if the water level in the surrounding mire rose relative to the mound. The biomass added to the surface of the anchored palsa by the patchy vegetation cover on the mound is far less in unit time than that accumulated in the surrounding mire, so that there would be a gradual rise in the water level in the mire over time. A combination of wind erosion of the crest and rise in level of the peat and water table in the mire relative to the mound would result in a decrease in its height and eventual drowning of the mound, as in the case of humid, continental anchored peat plateaus (Harris & Schmidt, 1994). So far, there is negligible evidence for the lateral spreading of maritime palsas that has been reported from studies of cold, humid continental palsas.

Maritime palsas have been reported from Finland, Norway (*e.g.*, Åhman, 1967, 1977; Sollid & Sorbell, 1974; Matthews *et al.*, 1997), Sweden (Svensson, 1962; Zuidhoff & Kolstrup, 2000, 2005), and Iceland (Schunke, 1973; Priesnitz & Schunke, 1983). Typical dome-shaped palsas range from 0.5–5 m in height, 10–30 m in width and 15–150 m in length. Esker-like forms and palsa complexes are also present in Scandinavia (Åhman, 1977; Seppälä, 1988). Schunke (1973) reported a variety of shapes of palsas ranging from extensive *shield palsas* (the *plateau palsas* of Åhman, 1977 and Pissart, 2002) to small degradation mounds.

Åhman (1967) indicates that in northern Norway, there is a continuum between the maritime palsas formed entirely in peat, and those with a substantial frozen silty layer at depth, herein called *continental palsas*, though this is not mentioned by Seppälä. The areas in which maritime palsas generally occur have sporadic or discontinuous permafrost (Figure 7.19). The climatic requirements for the maritime

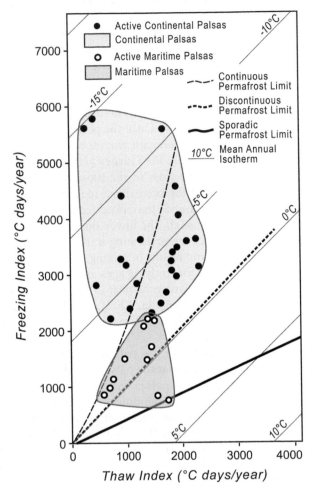

Figure 7.19 Distribution of freezing and thawing indices for some occurrences of active maritime and continental palsas. © S.A. Harris.

palsas are mean annual temperatures ranging from −0.6°C to −3.8°C, but with relatively low freezing and thawing indices (see Figure 3.5). It is the latter that probably explains the differences between maritime palsas and those of cold, continental climates.

7.2.2.1.2 Palsas in cold, continental climates

Palsas in cold, continental climates (***continental palsas***) have also been studied in considerable detail. They consist of a core of segregation ice developed in peat (15 to >150 cm) overlying silt-rich sediments. They usually occur in wet areas of discontinuous and continuous permafrost (Figure 7.18). Zoltai (1971, 1972) used them to map the southern limit of permafrost in Manitoba and Saskatchewan. They are present in the kettle holes in the northern glaciated parts of Canada (Dionne, 1984),

West Spitzbergen (Åkermann, 1980), and west Siberia (Spolanskaya & Evseyev, 1973; Åkerman, 1982; Vasil'chuk et al., 2012), but are rather different in internal structure from those in warmer, humid, maritime situations. Firstly, the dome-shaped palsas are generally higher and have much steeper side slopes. Secondly, the surface peat cover is usually quite thin (20–50 cm) and much of the underlying permafrost is developed in mineral sediments, usually lacustrine silts. They are found in the wetter areas of the boreal forest, so the tree rings can be used to determine the minimum age of the mound. Continental palsas can be found in any severe permafrost environments where slow decomposition of organic matter results in the formation of a surface layer of peat. Most of those found under less extreme climates developed during the last Neoglacial event, since some have trees on their surface dating back to 1700 A.D. They are very similar to lithalsas (see below) but the latter lack peat on the surface of the mound, and replace continental palsas entirely in the dry, cold climates in central Siberia and the Qinghai-Tibet Plateau. The similarities and differences in temperature regime between the areas with cold continental palsas and lithalsas in the southern Yukon Territory and southwest Alaska are shown in Figure 7.20.

Although typical cold, continental palsas are usually distinctly dome-shaped, Allard & Rousseau (1999) encountered low, flat mounds (>2 m high) with a somewhat undulating surface that had the ice content of the neighboring dome-shaped palsas. They referred to them as "peat plateaus" but they lack the low density core and the typical low elevation above the surrounding mire of true peat plateaus (see section 1.3 below). They represent the *palsa plateaus* of Schunke (1973), and have also been noted in Norway by Åhman (1977), in Finland, west of the Ural mountains, and Northwest Russia (Oksanen & Väliranta, 2006), and were shown to Alfred Jahn near Linné Kapp, Spitzbergen by J. Åkerman (Jahn, 1975, p. 99).

Under these climatic conditions with large freezing and thawing indices, the long winters and intense cold result in a very strong cold temperature wave moving down into the ground. The summer warming wave is relatively weak, and soil moisture moving to the freezing plane does not produce enough heat to cause degradation of the ice in the core of these palsas. As a result, the cold, continental palsas can even grow up to the maximum height theoretically possible (10 m) before the weight of the overburden prevents further upward heaving (see Pissart, 1987), e.g., at Tungsten, NWT. This results in relatively steep-sided, peat-covered palsa mounds forming clusters in suitable environments such as along flood plains or the sides of lakes (Figure 7.21), and implies that a similar initial growth mechanism to that of maritime palsas is involved. Floating palsas are rare in cold continental environments and probably quickly evolve into the usual anchored palsas, if they survive (Harris & Nyrose, 1992). The only large floating palsas described so far have had a short life due to being located in depressions in bedrock where a small pond has developed (Porsild, 1945, 1951; Kershaw & Gill, 1979). Either warming air temperatures or a decrease in precipitation can result in warming of the surrounding water, resulting in the gradual disappearance of these mounds, first described in MacMillan Pass, Yukon Territory by Porsild (1945). In the case of the embryonic mounds described by Harris & Nyrose (1992), the pond was held in place by a dam of mosses. A seagull built a nest on the palsa, but a grizzly bear discovered the nest. The bear ate the eggs and broke the dam, releasing most of the water in the pond. The increase in water temperature in the resulting smaller, shallower pool thawed the permafrost in the palsa. Beaver dams can produce similar

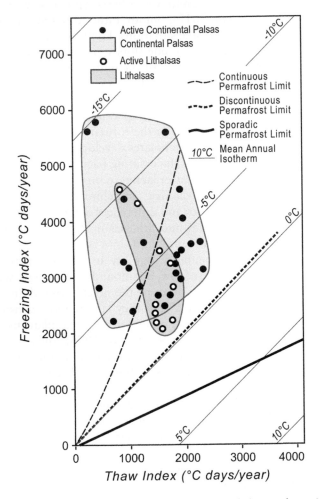

Figure 7.20 Comparison of the known distribution of active lithalsas and continental palsas with freezing and thawing indices. © S. A. Harris.

results in forested environments by raising the water level, so altering the hydrology around the palsas (Lewkowicz & Coultish, 2004).

The steep sides and height of the cold continental palsas produce increased cracking of the surface peat that can result in exposure of the permafrost inside the mound, resulting in decay of the mound. A second consequence is that the ice in the mound tends to flow laterally, increasing the width of the mound (Pissart *et al.*, 2011). As a result, sequential mapping or study of aerial photographs shows frequent changes in shape and a relatively short life of these mounds (Laberge & Payette, 1995). Brown (1968) regarded continental palsas as passing into an over-mature stage when the permafrost core melts and the palsa collapses. The collapsed palsas tend to form closed ridges of peat surrounding circular or oval ponds when viewed from above

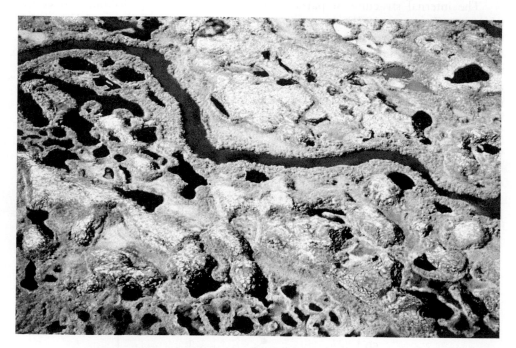

Figure 7.21 Palsas on a flood plain along the Sheldrake River, Québec. Both the dome-shaped and undulating flat palsa plateaus are present, as are the circular rings left when they degrade. © M. K. Seguin.

(Figure 7.21), but the ponds gradually become filled with new peat. These ridges may intersect one another or form a complex pattern of ridges and ponds. If the climate and microenvironment are stable, these changes are balanced by the appearance of new mounds which maintain the overall appearance of the area. Payette *et al.* (2004) found plateau-like palsas forming transverse ridges across a valley in Northern Québec that degraded into string fens. Presumably they represent the remains of palsa plateaus that originally developed from similar string fens under the cooler climate of the Little Ice Age.

Oksanen (2005) summarized the radiocarbon dates for permafrost aggradation in mires in northwest Europe, finding evidence for multiple phases including 0–600 years, 1900–2200 years, and 3000–3400 calendar years B.P. Okasanen & Väliranta (2006, Figure 4) show that organic carbon has been accumulating in palsa mires in Russian Pechora and Northern Finland from the time of deglaciation of the region, averaging about $14\,cm^2/^{a-1}$ until about 600 years ago, since when the rate has doubled. Laberge & Payette (1995) noted that while the palsas have tended to degrade since about 1945 both across North America (Samson, 1974; Thie, 1974; Kershaw & Gill, 1979; Laprise & Payette, 1988; Kershaw, 2003) and in Eurasia (Bobov, 1977; Badu & Trofimov, 1981; Anisomov & Nelson, 1996). New palsas are already forming in Northern Québec, together with new thermokarst ponds in which the new palsas can grow and which act as carbon sinks.

The internal structure of palsas can be studied by using drilling and coring (Zoltai & Tarnocai, 1971), by using geophysical methods (Tallman, 1973, 1975; Calmels *et al.*, 2008b), or by a combination of both. Calmels *et al.* (2008b, p. 39) found evidence of numerous small faults cutting across the ice lenses in the mounds near Umiujaq, northern Québec. Suggested causes were the upward growth of the mounds during the growth phase, thermal contraction cracking, and contraction cracking during the ice segregation process because of cryodessication of the soil induced by cryosuction.

In Manitoba, the moisture content beneath the active layer in a mature palsa usually reaches 90% by volume, but gradually decreases with depth after about 1.5 m below the permafrost table (Figure 7.22). This contrasts with 30–40% moisture in the active layer outside the mounds. Near the southern limit of permafrost in Manitoba, palsas standing 1.7 m, 1.45 m and 0.8 m above the surrounding water table had about 4 m, 3.3 m and 1.7 m of permafrost in them respectively (Zoltai, 1972).

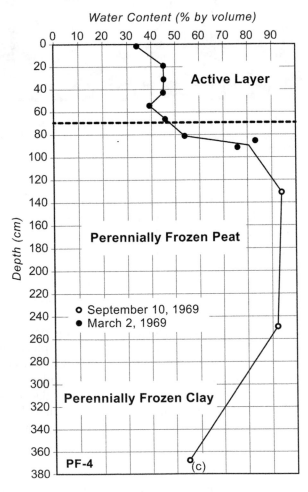

Figure 7.22 Water content (% by volume) at palsa PF-4 beneath woodland in Northern Manitoba. Redrawn from Zoltai & Tarnocai (1971).

Deeper permafrost profiles are found further north in colder climates beneath thinner active layers. Maximum height recorded for a continental palsa is 10 m at Tungsten, Northwest Territories.

Harris & Nyrose (1992) demonstrated that up to 50% of the ice in the core of the palsas in the MacMillan Pass was derived from meteoric water originating as snow or summer precipitation, since the groundwater contained twice the sulphate content of the ice in the mounds. Thus the continued upward growth of the mounds after initial formation involves addition of substantial quantities of moisture from precipitation.

Payette *et al.* (1976) and Allard & Rousseau (1999) emphasize the variations in the permafrost mounds according to the local ecological and lithological conditions. They demonstrated that the nature of the underlying sediment and the geological history of an area affect the type of mound that develops under a given climatic regime. Thus under forests in Québec, wooded palsas were dominant, but in the clay belt that has recently been uplifted by isostatic rebound around Hudson Bay, the dominant mounds are lithalsas (see below).

7.2.2.1.3 Lithalsas

Lithalsas are perennial permafrost mounds consisting of a core of segregation ice occurring in a profile showing a thin organic layer (<15 cm) over mineral sediment, usually high in silt (Harris, 1993). These represent the end member of the sequence with decreasing organic matter in the soil profile caused by a relatively arid, cold climate (Figure 7.23). They were originally described from the southern Yukon Territory at Fox Lake, where a whole developmental sequence of mounds representing the growth of the landform and the changes of vegetation were found. In fact these landforms are common in the dry valley floors in the rain shadow of the Wrangell-St Elias coastal mountains between Tok, Alaska, and Carmacks and Tagish in the southern Yukon Territory. All these valleys are subjected to cold air drainage in winter (Harris, 2007, 2009, 2010a), which provides the necessary cooling to develop the conditions suitable for cryosuction, producing a similar distribution of ice in the core of the mounds to that found in palsas. Table 7.2 shows the growth stages demonstrated at Fox Lake, together with sequence of disappearance after beavers had raised the water

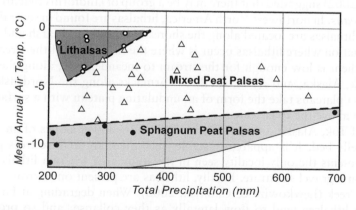

Figure 7.23 Distribution of the peaty mounds and lithalsas in relation to mean annual air temperature and precipitation in the Yukon Territory, Canada (after Harris, 1998c).

Table 7.2 Stages of development and demise of lithalsas at Fox Lake (modified from Harris, 1993).

Characteristic	Stage I	II		III		IV		V
Dominant vegetation	Grasses	Shrubs		Shrubs *Picea glauca*		*Picea glauca* regenerating		*Picea glauca* old
Height (m)	1.2	2.4		2.5		2.7		2.25
Active layer thickness (cm)	290–410	Shrub 60	Grass 80	N. facing 60	S. facing 80	N. facing 60	S. facing 120	130
Permafrost thickness (m)	1.9–3.1	5.4	2.7	5.5	5.0	5.0	4.8	4.2
LFH/A (cm)	0–2	20	20	60		40		15
Rapidity of collapse (years)*	1	8	5	>15	12	>15		6

*After the beavers had raised the dam 30 cm.

level in the surrounding lake by 30–50 cm. The high water table around the mound supplies the water for cryosuction, while some meteoric water is almost certainly involved.

The mature lithalsas have upright trees on them with rings dating back to 1700 A.D., so the lithalsas probably grew during the last Neoglacial event (c. 1450–1900 A.D.). Stages 3 and 4 have a thick (60–100 cm) transient layer, providing protection from warming climates. Ice-wedges have not been observed on them. Figure 7.24 shows a typical mound along the shores of a lake in the Southern Yukon Territory. These mounds are up to 3 m high with steep sides (>20°) and an undulating upper surface when mature. The active layer thickness depends on the vegetation, while the icy core is 5.5–7 m thick beneath the mature, tree-covered lithalsas. The average ice content of their core is 60–70% by volume, mainly in the form of thin lenses, as in Figure 7.16.

Pissart (2002) suggested that there may be a group of landforms that are best called *lithalsa plateaus*. In northwest North America, lithalsas are found in several situations. Most of the lithalsas are located along the shores of lakes, *e.g.*, Marsh and Fox Lakes. Another situation where lithalsas occur is where the floodplain of the stream is broad and the gradient is low enough for the stream to meander, *e.g.*, along the Tok River south of Tok, Alaska, and again, south of Carmacks along the Nordenskiold River. These lithalsa plateaus take the form of an undulating plateau with a surface typically about 3–4 m above the river water.

South of Tok, Alaska, one palsa is located on the lower slopes of an alluvial fan with a gravelly bed. It is apparently stable since there are mature trees growing on it. This represents the only locality seen in the study area where a lithalsa may have developed on a gravel substrate. Finally, lithalsas are present on narrow valley floors as at Wolf Creek (Lewkowizc & Coultish, 2003). When degrading at Fox Lake, the over-mature lithalsas tend to flow laterally as they collapse, and so produce small curved lunate ridges where mineral sediment has slid or flowed down the sides of the

Figure 7.24 A lithalsa beside Marsh Lake, Yukon Territory, showing the thin organic soil horizon overlying the core of icy lacustrine silt. © S.A. Harris.

degrading mound. This has not happened in the cases of degradation of the other stages at Fox Lake. This is probably due to the roots of the vegetation holding the peat together, and also the thickness of the active layer. The lithalsas with shrub or tree vegetation did not degrade during the first 15 years after the rise in water level of the lake.

Subsequently, lithalsas have been described from northern Québec (Delisle *et al.*, 2003; Calmels & Allard, 2004, 2008; Vallee & Payette, 2007; Camels *et al.*, 2008a, 2008b), the swampy forest ecotypes of the Great Slave Region, Northwest Territories (Wolfe *et al.*, 2014; Morse *et al.*, 2016), and the Hudson Bay Lowlands (Stevens *et al.*, 2013). The sediments there are similar to those at Fox Lake, and in one example, the lithalsas are developed on gravelly material (Pissart, 2000a). However, these sites were inundated by the sea during deglaciation and have sodium-saturated clays present that are notorious for their properties of lower freezing temperatures and poor bearing strength.

Calmels *et al.* (2008b) using tritium, demonstrated that the upper layers of permafrost in the core of a lithalsa at Umiujaq are not completely impervious to infiltration of meteoric water, especially near the surface. The oxygen and hydrogen isotopes in the precipitation and in the icy cores of the palsas in northern Québec suggest that the core ice probably formed during the colder conditions of the last Neoglacial event (1650–1900 A.D.) under temperatures c. 3°C colder than that today at Sheldrake,

6 km away. There was evidence of warmer temperatures during formation of the ice with depth, which could be interpreted as indicating that the oldest ice was at the top, though there was high variability between individual samples. Annual reversals of thermal gradients can also create a suction gradient that slowly drives water to move downwards from the top (Cheng, 1983; Allard *et al.*, 1996). At Umiujaq, the ice content in the permafrost was 50–80%, with a peak at 3 m. Gas contents of the ice varied from 0–3%, being highest in the upper 4 m.

Pissart *et al.* (2011) also suggested that these lithalsas may actually grow laterally. They differ from the lithalsas at Fox Lake in having lateral flowage of the sides of the mound. At Fox Lake, lateral flowage only occurs during degradation. This produces closed rings of sediment on the site of the former lithalsa that resemble small pingo scars. After seeing these, Pissart (2000a, 2000b, 2002, 2003, 2010) reinterpreted the periglacial structures on the Hautes Fagnes Plateau in Belgium together with similar Pleistocene structures elsewhere as indicating the former presence of lithalsas. Subsequent studies are supporting this interpretation, *e.g.*, Ross *et al.* (2006) at Llanpumsaint, Wales.

In the Himalayan Mountains of India, Wunnemann *et al.* (2008) have described the interaction between lake formation, permafrost activity and lithalsa development during the last 20,000 years. The studies in northern Québec provide examples of the demise of present-day lithalsas, whereas the Ladakh study discusses the conditions under which they have been forming in mountains at low latitudes over a considerable period of time.

Lithalsas have also been found on the northeastern slopes of the Qinghai-Tibet Plateau (Figure 7.25), where they form broad, low mounds in the alpine meadows on the valleys floors. The smaller mounds are oval with a long dimension of about 180 m and a height of 12.8 m. The south-facing side is steeper (up to 8°) than the others. They are developed in alluvial and colluvial sediments, and the soils around the mounds have a white efflorescence of salts on their surface, indicating a high water table and very intense evapotranspiration. Permafrost is absent under the nearby dry soils and on the surrounding hills, but is at least 25 m thick beneath a 1–2 m thick active layer in the centre of the mound. The ice in its core shows a reticulate structure (Figure 7.16), with interstitial ice in the pebbly bands (Figure 7.17). It appears that the evapotranspiration of groundwater has cooled the ground sufficiently in this dry cold, high-altitude (4313 m) climate to produce a local, thick pocket of permafrost in the form of a lithalsa.

Pissart (2002, pp. 613–614) speculated on the likely world distribution of lithalsas. He regarded them as occurring where the temperature of the warmest month is less than or equal to 10°C (the tree line) and where the permafrost is discontinuous (Pissart, 2000b). In North America, this suggests a distribution from Northern Québec, to the north of Slave Lake (Wolfe *et al.*, 2014), and in western Alaska. It also suggests their presence in Lapland and on both sides of the Ural mountains, east to the Yenissei River. They could also occur in Kamchatka and around the Sea of Okhotsk. So far, northern Québec, Lapland and on the lowlands east of the Ural mountains (Spolanskaya & Evseyev, 1973) seem to have the most lithalsas under the present climatic conditions, although significant numbers of lithalsas appear to be present in areas of suitable microenvironments in southern Yukon Territory and southeastern Alaska, as well as

Figure 7.25   Lithalsa (low mound, middle distance) with drilling crew on top, Kangqiang, 37° 1′N, 97° 34′E, at 4313 m elevation, source region for the Yellow River, China. It is similar in shape to some pingos on the Qinghai-Tibet Plateau and in Mongolia and these therefore require drilling or geophysical surveys to ensure proper identification. © S.A. Harris.

in the cold, dry upland arid regions of central Asia. Figure 7.20 shows the relationship of the known lithalsa occurrences to freezing and thawing indices.

7.2.2.1.4   Palsa/lithalsa look-alikes

The steppes of Central Asia are vast, and the higher areas bordering the many mountain chains have conditions suitable for the development of both pingos and lithalsas. Unfortunately, little field work has been carried out on the mounds that can be identified on aerial photographs and on satellite imagery. To determine the origin of these mounds, it is necessary to carry out studies of the internal structure of the icy cores by drilling and/or geophysical methods. Many "pingos" have been so-named without a careful study. It is now clear that lithalsas and hydrostatic pingos can occur together in a given region, but differentiating between them is difficult. If there is a depression in the top of the mound, it is likely to be a pingo, but otherwise, the main difference is whether the ground water is under hydrostatic pressure or merely represents an area with a high water table resulting in accumulation of an ice-rich core by cryosuction. The heat removed from the ground by evaporation of soil water can produce low mounds more than 200 m in diameter.

However, not all mounds in the region have a simple origin. In the Russian Altai, Iwahana *et al.* (2012) encountered mounds similar to lithalsas but with evidence of an unusually complex internal structure. They seem to have been formed in a complex glacial, lacustrine steppe environment with low winter precipitation but higher summer precipitation. There are four different kinds of internal structures, indicating a very complex history. The oxygen isotopes suggest limited climatic change, but clearly, the processes forming the mound have been complex and varied. Although Iwahana *et al.* treated them as lithalsas, they appear to represent a complex that is not well understood and is certainly different to the history of true lithalsas.

The discovery of low, plateau-like palsas around the Rivière Boniface region in Québec indicates that care must be taken in separating palsas from peat plateaus. On aerial or satellite imagery, they look similar, but in the field, the palsas are higher and have a more undulating surface topography. Floating palsas are much lower relative to the water table in the surrounding mire (see Figures 7.23 and 7.24). Drilling in the surface of the mound will provide evidence for a much higher ice content in the permafrost in the palsa, and the one studied by Allard & Rousseau (1999) showed a relatively thin layer of permafrost overlying a bed of sand. The latter was regarded as the cause of the stunted growth of the mound since it limited the thickness of silty layers from which water could move by cryosuction.

In Iceland, Saemundsson *et al.* (2012) have described "palsas" in an aeolian sedimentation environment north of the Hofsjokull glacier in a valley-like depression in the dry cold desert. The area is one of fairly continuous, spasmodic deposition of volcanic ash from the nearby volcanoes. The mounds are 70–200 cm high and up to 3000 m² in area. They have a flat top but a negligible organic layer at the surface of the andosol profiles. The active layer appears to be 43–81 cm thick, overlying up to 5 m of permafrost. Without a dome and peat, they do not seem to be the domed type of palsa and the general description fits that of small lithalsas, except for the flat surface, suggesting they are *lithalsa plateaus*. However, without details of the nature of the ice within them nor the actual ice content, they cannot be firmly identified. Palsas, lithalsa plateaus and peat plateaus are readily confused unless the moisture content of the mounds and the surrounding fen is measured.

7.2.3   Mounds formed by the accumulation of ice in the thawing fringe: peat plateaus

Peat plateaus are low plateaus of well-drained peat usually found floating in the peat bogs of the northern regions of the boreal forest in heavily glaciated lowlands (Figure 7.26). They usually have trees on them, in contrast to the surrounding fen, and can be up to 12 km² in area. Zoltai (1972) studied 26 palsas and peat plateaus in northern Manitoba and Saskatchewan, and found that the moisture content in peat plateaus was lower than in the surrounding fen. In the case of the palsas, all had a high volume of water in them by weight (Figure 7.21). This ice content explained the height of the floating palsas on the difference in density between the ice and water (Figure 7.27), but he could not explain the mechanism causing the peat plateaus to float. Subsequent work by Harris & Schmidt (1994) showed that the peat plateaus floated even higher than ice (Figure 7.28), and had a lower moisture content in the form of rotten ice with

Figure 7.26 Margin of a peat plateau along the Robert Campbell highway in the Yukon Territory. Note the sedge mire in the foreground and the abrupt edge to the peat plateau, with its light-coloured *Cladonia* lichen ground cover, ericaceous shrubs and Black Spruce trees. © S. A. Harris.

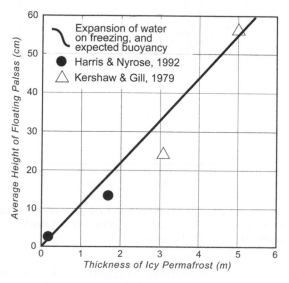

Figure 7.27 Comparison of the average height of a floating palsa with the thickness of icy permafrost in its core (from Harris, 1998c, Figure 2).

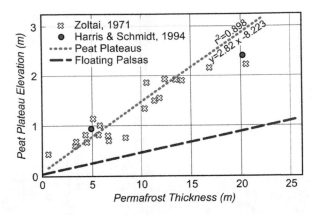

Figure 7.28 Discrepancy between the effect of expansion of water on freezing and actual elevation of peat plateaus above the surrounding bog (from Harris, 1998c, Figure 5).

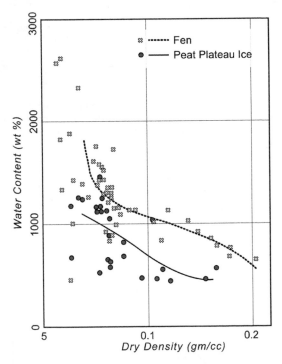

Figure 7.29 Comparison of moisture contents in relation to dry density for samples from the fen and the permafrost in the peat plateau, 5 m apart at km 167.0, Robert Campbell Highway, Yukon Territory (from Harris, 1998c, Figure 7).

through pores relative to the surrounding fen (Figure 7.29). The explanation for the higher level at which they floated had to be the presence of a substantial quantity of air in the ice. A pit dug in the surface of a peat plateau showed the presence of the ice with through pore spaces matching that of the frozen fringe. A further check was made

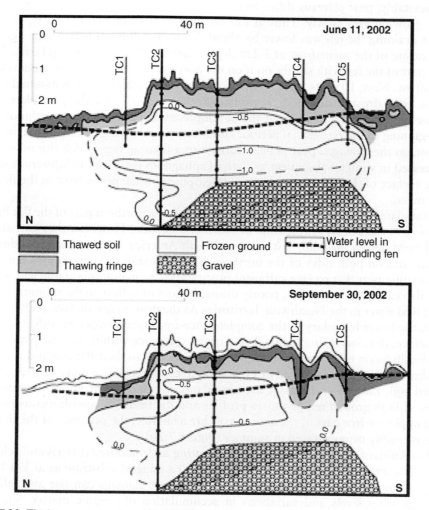

Figure 7.30 Thickness of the **thawing fringe** in the peat plateau in June and September, 2002 at km 167.0, Robert Campbell Highway, Yukon Territory (translated from Harris, 2005, Figure 2).

by comparing the depth to the ice with the depth of 0°C as indicated by thermistors at two different times of the year. The results showed that there was a thick zone of *thawing fringe* (rotten ice through which water moves in summer) with water draining through it in summer (Figure 7.30, from Harris, 2005). As more biomass is added to the surface, the lower boundary of the completely ice-free layer moves upwards parallel to the surface, thus increasing the thickness of the rotten ice within the peat plateau. They are subject to thermal contraction like any other surface, and in areas of continuous permafrost near Inuvik Airport, well developed ice wedge networks occur on their surface adjacent to the Dempster Highway.

Inevitably, peat plateaus differ from the other landforms discussed in this Chapter by growing very slowly. Thus at km 167.0 Robert Campbell Highway, the road culvert draining the fen was lower by about 20 cm, resulting in better drainage. The temperature of the permafrost at 3.2 m depth started to become lower (Harris, 2008) and a part of the fen with shrubs on it on the north side of the peat plateau developed permafrost. Now, 10 years later, this section of the peat plateau shows minimal uplift, but the permafrost is still there. It probably takes centuries for the peat plateaus to become mature, so they represent the development of permafrost in peatlands after the beginning of the Neoglacial period. Fire may cause thawing of the underlying permafrost, so the damaged peat plateau turns into a floating bog which can only safely be accessed in winter. Subsequent growth of sphagnum moss and *Eriophorum* sedges on the surface of the plateau results in redevelopment of the permafrost in the floating peaty mass.

Peat plateaus are particularly widespread across the northern part of the flat Prairie Provinces in the Boreal Forest zone of Northern Canada. However, they can also be found right across the cold, flat parts of North America, in Northern Scandinavia, Iceland, and on both sides of the northern Ural Mountains. They require flat mires with floating peat that receive sufficient precipitation to maintain a stable water level. They also can be found on flat, poorly drained floors of valleys in the mountains near Arctic Red River in the Northwest Territories As the peat on the surface of the mound grows, the lower boundary of the completely ice-free layer moves upwards parallel to the surface, thus increasing the thickness of the rotten ice within the peat plateau.

Identification of peat plateaus is relatively easy due to their flat top and relatively well-drained flat surface covered by plants and trees that require a moist but adequately drained soil. Lichens and mosses are abundant, and *Sphagnum* moss is common in swales. If both ground temperatures profiles and the results of simultaneous probing of the upper ice-free part of the active layer are available, the presence of the thawing fringe is readily demonstrated in summer (Figure 7.20).

Two kinds of peat plateaus occur, *viz.*, *floating* and *anchored* (Harris and Schmidt, 1994). The anchored peat plateaus are frozen to a mineral substrate as at km 167.0, Robert Campbell Highway whereas the floating peat plateaus can rise and fall with changing water levels and variations in accumulation of organic matter. Since the biomass production is usually lower on the plateau, the anchored peat plateaus slowly drown unless there is a reduction in water level in the adjacent fen due to natural or anthropogenic causes. Floating peat plateaus look attractive for construction, but as soon as the permafrost is destroyed, any heavy objects on its surface will tend to sink through the floating organic mat and end up at the bottom of the water body. This is quite different to the mature continental palsas and lithalsas that are relatively robust, since their permafrost cores often survive the passage of fire.

Figure 7.31 shows the distribution of peat plateaus relative to the mean annual freezing and thawing indices. Peat plateaus are generally developed in accumulating peat in shallow water bodies in northern Scandinavia (Kessler, 2013), the West Siberian and Hudson Bay Lowlands.

The main period of growth of existing peat plateaus seems to have been during the last Neoglacial event (1650–1900 A.D.). Subsequent warming has generally led to subsequent slow degradation. Late-season thaw depths in northern Sweden are primarily dependent on mean summer air temperatures and thawing degree-days (Sannel

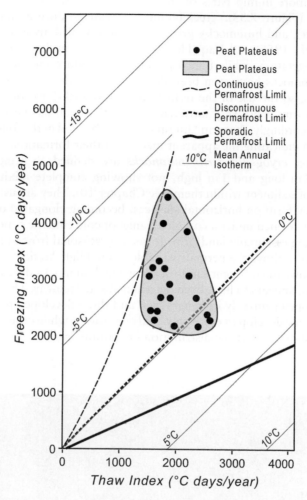

Figure 7.31   Relationship of stable peat plateaus to freezing and thawing indices (after Harris, 1981c, 1982a).

et al., 2016), whereas winter snow depths are directly related to minimum ground temperatures at or below 1 m depth. Variations in these climatic parameters control rates of degradation, other things being equal.

7.3   CRYOGENIC MOUNDS LESS THAN 2.5 m IN DIAMETER

These include the earth hummocks (Figure 7.32) and thufurs of many authors. Unfortunately, the term "earth hummock" has been used to describe a large variety of features that originate in quite different ways. There are the hummocks produced by ungulates as they move over wet ground, and by burrowing pikas that produce low circular mounds on the more humid parts of the east slopes of the Tibetan Plateau, often over permafrost (Figure 7.33). Trees falling over can produce hummocks (Beke & McKeague, 1984), and hummocky ground is found in arid areas such as Australia (Hallsworth *et al.*, 1955) and Iraq (Harris, 1958). Blowing loess or sand can form hummocks in vegetation, while bunch grasses can produce mounds of soil around them. Earth hummocks have even been described as products of volcanic eruptions and debris flows, so the use of the term "earth hummocks" for distinctive types of permafrost landforms is not very satisfactory, even though it has been perpetuated in the literature. Accordingly, the term "hummock" will be used in the following account, using a new term indicating the apparent method of their formation.

As used here, *cryogenic earth hummocks* are defined as being dome-shaped mounds up to 2 m long and 1 m high, not showing complete mixing and circulatory movement of sediment within them (see Chapter 10). They are usually circular in outline when they form on horizontal surfaces, becoming elongated on gentle slopes <6°. They can be formed under a complete range of cold climatic conditions, ranging from those that are essentially landforms formed by seasonal frost to those formed in the thin active layer above the permafrost table in the High Arctic.

Table 7.3 shows the different kinds of cryogenic hummocks developed in mineral soils whose origin appears to have been determined satisfactorily. They can be divided into those developed primarily in mineral soil and those developed primarily in peat. In each group, some develop in extremely cold conditions, while others are found only near the warmest margins of the distribution of sporadic permafrost. It is quite likely

Figure 7.32 Dr. Jerry Brown examining cryogenic earth hummocks at Hosvgul, south of Lake Baikal, Mongolia. © S.A. Harris.

that yet others may be found since relatively little research has been carried out on them in many parts of the world.

A major debate in the literature involves the initiation of mounds. This is almost certainly the result of a combination of initial micro-topography, lithology, and the

Figure 7.33 Pika mounds on dried lake bed over permafrost at 4636 m on the northeast slopes of the Tibetan Plateau at 34°23′N, 97°47′E. © S.A. Harris.

Table 7.3 Summary of the different kinds of currently known cryogenic earth hummocks developed in mineral soils.

Type of hummock	Relation to permafrost	Vegetation	Process
Oscillatory	Continuous and Discontinuous	Taiga	Build-up of ice lenses alternating with collapse after forest fires.
Thufurs	Sporadic/seasonal	Cool, perhumid Meadow	Movement of silt to the to the freezing plane soil moisture. Formed in stratified silt and sand.
Silt-cycling	Humid Temperate	Wet moors	Movement of silt with moisture to frozen sides and top of hummock. Formed in homogeneous silt loams.
Niveo-aeolian	High Arctic Tundra	Tundra	Winter movement of sand from top of hummock downwind in dirty snow.

distribution of both snow cover and vegetation. Often, there are small dimples or ridges in the freshly deposited sediment, and these can cause differences in winter snow cover and initiation of ground freezing. The lithology of the sediment is critical, and particularly its frost susceptibility and permeability. Even without dimples, the snow cover can vary considerably in very short distances due to eddies in the wind, the deeper snow cover providing insulation from the cold air temperatures.

The vegetation that grows on the soil surface may often be the most important factor, though it tends to be left out by geocryologists and geomorphologists. Each species of plant brings a different albedo and thermal offset to the underlying soil due to varying thermal and moisture retaining properties (Harris, 1998b). The vegetation cover is made up of a mosaic of species, some of which decay when dead, *e.g.*, the lichens. Others produce varying quantities of biomass in unit time which will result in the formation of an uneven surface. The mosses are particularly important. Tews (2004) reported that at Churchill, Manitoba, the species of moss on the mounds correlated with mound height, *Tomemthypnum nitens* occurred on low hummocks, *Hylocomnium splendens* on medium mounds, and *Pleurozium schreiberi* was found on mounds 60–70 cm in height. There were also other correlations with spacing of mounds, *etc.* Of course, it is impossible to determine the exact cause of each mound, but obviously, there are plenty of potential causes that can act singly or together to start the formation of hummocks.

Both cryogenic earth hummocks and thufurs consist of numerous small, closely-spaced mounds randomly distributed across a cold, rather humid landscape (Figures 7.32 and 7.38). Cryogenic earth hummocks indicate processes usually occurring in the active layer over frozen ground, while thufurs are found in cold, humid, maritime climates in areas lacking permafrost. Unfortunately, they have sometimes been regarded as being the same (Schunke & Tarnocai, 1988; Gerrard, 1992), but the treatment in this book follows Harris (1988b) in distinguishing between them. Four kinds of cryogenic earth hummocks are now recognized, although there are many fields of hummocks over permafrost whose origin is non-proven.

Pounas and string bogs developed in peat instead of predominantly mineral sediment also fall in this group. These are found in wet areas or fens. Only a proportion of these landforms have permafrost, and they represent the outer margin of the present-day permafrost in areas of seasonal frost.

7.3.1 Oscillating hummocks

Oscillating hummocks, commonly included in *earth hummocks* in the previous literature, and *jordtuva* in Sweden, consist of low vegetated mounds ranging from 20 cm up to 1 m high with a basal diameter of 50 to 150 cm (Figure 7.32), but tend to be elongated down slopes (Zoltai & Tarnocai, 1974). The name is based on the fact that they alternately grow higher and narrower until fire or some other disturbance occurs to destroy the vegetation, whereupon they collapse, becoming wider and lower due to thawing of the ice inside that caused growth of the mound. When the vegetation cover is rejuvenated, the MAGT in the mound decreases and ice lenses accumulate in it, causing the mound to grow again. They occur in regions with a clay or clay loam texture (Figure 7.34) in arctic and subarctic regions wherever the moisture conditions and soil texture permit their formation, and are found throughout the more humid

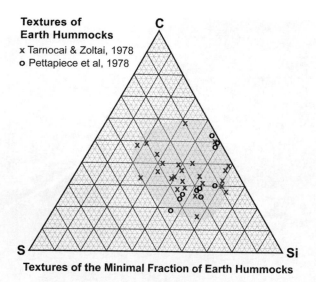

Figure 7.34 Textures of the mineral sediment in Canadian oscillating hummocks.

permafrost areas in the forest margins in central Asia (Cui & Song, 1992), and on the tops of mountains, *e.g.*, Mount Mahan, east of Lanzhou, central China, as well as on the Sunshine meadows in the Rocky Mountains of Alberta (Tarnocai & Zoltai, 1978), and can be found north to the Arctic coast throughout some parts of the permafrost areas of the Northern Hemisphere (Hopkins & Sigafoos, 1951; Mackay, 1958; Tedrow & Douglas, 1958; Tedrow, 1963).

Oscillating hummocks are often regarded as a type of non-sorted circle or net (Washburn, 1985; French, 1976, 1996), but cross-sections show contortions of the soil layers in the mounds with portions of the organic surface horizons being buried close to the permafrost table (Figure 7.35). Disruptions include faulting, smearing and thrusting of parts of the soil horizons, which are not the products of the churning motion seen in patterned ground. The mounds are always covered in vegetation. Accordingly, they are included in the permafrost mounds section in this book, rather than with patterned ground. In the more southern locations, the movement is not occurring today, based on negligible tilting of the larch and/or spruce trees growing on them, together with the absence of permafrost beneath them (Zoltai *et al.*, 1978). The hummocks have well-developed soil horizons that are undisturbed (Tarnocai, 1973).

Further north, tilted trees indicate that the movements are still taking place. Zoltai (1975) studying tree rings, found that these movements were not continuous, but occurred spasmodically at an average rate of once every 25 years. The cross-sections of the hummocks seem to indicate a form of disturbance of the soil horizons in which blocks of soil are moved about in the active layer rather than a thorough mixing (Zoltai *et al.*, 1978; *c.f.* van Vleit-Lanoë, 1988, 1991). This precludes a circulatory system in the soils (*c.f.* Mackay, 1979b, 1980), and cryoturbation (Hopkins & Sigafoos, 1951). Since the internal structure is the same in hummocks regardless of slope, downhill movement cannot be the sole cause of burying the organic matte (*c.f.*, Mackay, 1958).

Figure 7.35 Cross section of one of the oscillating hummocks at Lake Hosvgul (Figure 7.31). There has been some differential movement downslope to the right. © S. A. Harris.

The presence of permafrost beneath all the mounds with present-day movement precluded the relict mounds further south being formed during a former warm period (Tedrow, 1963).

The presence of blocks of intact organic matter provided Zoltai, Tarnocai and Pettapiece (1978) with the opportunity to date the organic matter and so get an idea of the antiquity of these mounds. The results (Figure 7.36) show definite periods of churning movements starting as early as 11,000–11,500 radiocarbon years B.P in the subarctic. The organic matter takes time to accumulate, so the dates are mean residence times for the organic matter before it was buried, *i.e.*, the soils are actually older than suggested by the dates. This mean residence time is likely to be about 500 radiocarbon years, since no younger dates were found, although the periodic disturbance is still occurring.

The fact that the first date (Figure 7.36) came from further south would be expected since much of northern Canada was beneath the Wisconsin ice sheets at that time. A second group of dates is clustered around 8,000–9,500 radiocarbon years B.P. and comes from the low arctic and high arctic, which had become deglaciated in many areas. The gap between 8,000 and 5,000 radiocarbon years in the low and mid-Arctic regions may mean the active layer was too deep for this type of cryoturbation to take place there. However, the main mass of dates spans the period from 6,500 radiocarbon

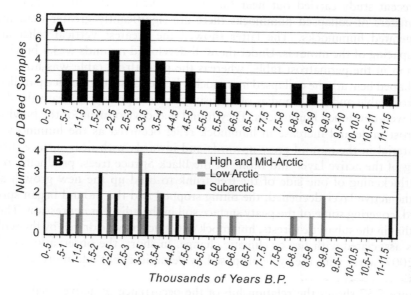

Figure 7.36 Radiocarbon ages of organic matter buried in oscillating earth hummocks from northern Canada (after Zoltai *et al.*, 1978). Burial has been active from before 11 ka during colder periods and especially during the Neoglacial events (<5 ka).

years B.P. until the present, with minor gaps in the last 500 years, 1,500–2,000 years, and 5,000–5,500 radiocarbon years B.P. The fact that these blocks of organic matter have not been totally mixed in with the other soil material appears to indicate that the movements are relatively rare, and they only affect parts of the active layer. The bulk of the movement has occurred during the Neoglacial cold events between about 5,000–3,000 radiocarbon years B.P. The authors concluded that there was periodic cryoturbation of some sort. This also suggested the idea that after an initial burst of activity, the earth hummocks become relatively stable and in equilibrium with the microclimate (Mackay & Mackay, 1976; Pettapiece, 1975).

After studying two soil profiles south of Inuvik in detail, Pettapiece (1974, 351–353) suggested an oscillating process brought about by episodic fires. The fire would destroy the vegetation and bring about deepening of the active layer to about 80–120 cm depth. The decrease in height of the mound due to melting ice as the upper part of the permafrost thawed would be accompanied by some collapse of the sides of the mound, thus burying some of the remaining organic matter in the depressions. As the vegetation (shrubs, lichens and mosses) become re-established, the surface would become drier, while the roots would provide additional drying and insulation. The mound would then grow higher, the sides would steepen, and the permafrost table would rise, providing a volume increase due to the phase change of water to ice that would result in the increasing mound height. This would continue for 100–200 or more years until a new equilibrium was reached. The active layer at maturity would be 60–90 cm.

A recent study carried out near Inuvik by Kokelj *et al.* (2007) indicates that the diameters of collapsed hummocks there are significantly greater than those of well-vegetated hummocks. The latter showed the greatest vertical relief and the widest spacing of the mounds. Under the well-developed mounds was a bowl-shaped depression in the permafrost table, whereas the permafrost table was horizontal or dome-shaped beneath the collapsed hummocks. Lenses of segregated ice parallel to the permafrost table and small bodies of intrusive ice were present beneath the developing and well-developed hummocks. They found that development of the bowl-shaped permafrost table and increase in hummock relief occurred as the hummocks grew. The material in the mound grew upwards and inwards with near-surface thrusting, thinning of the active layer and tilting of the Black Spruce trees, producing *reaction wood* (thickening of one side of the tree trunk to hold up the new growth above). When the active layer deepened, the tilting stopped and the mound began spreading outward, covering some of the peaty surface soil in the adjacent depression. They suggested that in the subarctic forests, hummock dynamics may be driven by the ecological changes associated with fire or climatic change. In short, the observations of Kokelj *et al.* (2007) agree with the cycle proposed by Pettapiece, but add more details of the process.

Figure 7.37 shows the relationship of the occurrence of active oscillatory hummocks to freezing and thawing indices. They are known from northern Canada, Mongolia, Russia (Kachurin, 1959; Karavayeva & Targulyan, 1960; Karavayeva *et al.*, 1965) and the alpine tundra of Europe (J. Lundqvist, 1962). Unfortunately, there are a number of localities where mounds resembling oscillatory hummocks have been reported but not studied in enough detail to classify them (see section 7.3.4 below).

7.3.2 Thufurs

Thufurs are also small mounds that are found in farmer's fields in cold, humid maritime climates mostly on volcanic tephra in places such as Iceland (Figure 7.38), East Greenland, Hokkaido (Nogami, 1980), Lesotho (Marker & Whittington, 1971; Boelhouwers, 1991; Grab, 1994, 1997; Backéus & Grab, 1995), France (Carbiener, 1970) and New Zealand (Billings & Mark, 1961). They were first described from Iceland by Gruner (1912) and Thoroddsen (1913) and occur in areas of alpine meadow that are subjected to seasonal frost but lack permafrost. They are of similar size to earth hummocks and render the field useless for agriculture. If they are flattened out, they reform in a matter of decades. In cross-section, they show the individual layers of volcanic ash that make up the soil, with thickening of the silty layers beneath the mound (Figures 7.39). The grain sizes of the sediment are coarse sands mixed with coarse silt, but little clay (Figure 7.41). Note that thufurs rarely form on nearby medium and fine sandy sediments, but mainly on frost sensitive volcanic tephra. Exceptions include Mahai Mountain, China, and Svalbard (Jaworski & Chutkowski, 2015).

Arturo Corte (1966) demonstrated in laboratory tests that silt particles can move with the liquid water to the freezing plane through the spaces between coarser mineral grains when a freezing front is descending into a sediment. Schunke (1977a, 1977b, 1977c, 1981) invoked this mechanism to explain the thickening of the silty layers in the mounds in East Greenland and Iceland (Figure 7.39 and 7.40). The climate

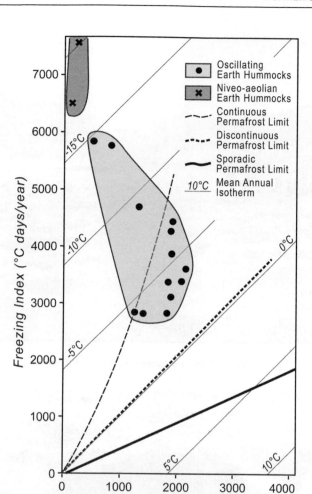

Figure 7.37 Distribution of oscillating and niveo-aeolian hummocks with freezing and thawing indices. The latter are only developed on the High Arctic tundra, whereas the oscillatory hummocks are mainly found in the Boreal Forest. © S.A. Harris.

is perhumid with a limited range of daily and annual air temperatures. The surface of the ground would have had slight undulations which initiate the lateral-horizontal movement of water to the freezing fronts beneath the highest undulations. Snow tends to fill the low spots, insulating them from the cold air (Schunke & Tarnocai, 1988, Figure 10.6). Where the snow is deeper, thufurs are absent (Thoraninsson, 1951). The silt tends to move with the abundant soil moisture moving towards the freezing front. The freezing front takes a very long time to penetrate to 30 cm depth (maximum of 40 cm) because of the substantial quantity of water moving to the freezing front and the relatively weak soil temperature gradient resulting from the relatively warm, oceanic

Figure 7.38 Thufurs in southern Iceland. © S.A. Harris.

Figure 7.39 Cross section of one of the thufurs shown in Figure 7.38. Note the thickening of the dark, silty layers under the mound. The postulated cause of the thickening is water moving to an uneven, downward-moving freezing front carrying finer grains of sediment with it (bottom right). These grains accumulate where the water turns to ice, and remain there when the ice melts (see Figure 7.40). © S.A. Harris.

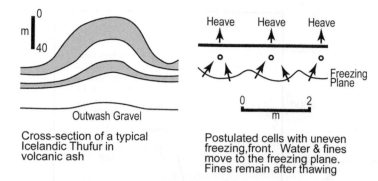

Cross-section of a typical
Icelandic Thufur in
volcanic ash

Postulated cells with uneven
freezing front. Water & fines
move to the freezing plane.
Fines remain after thawing

Figure 7.40 Postulated mechanism causing the movement of silt particles towards the centre of the mound, with the soil moisture moving to the freezing front in the mounds from the unfrozen soil below.

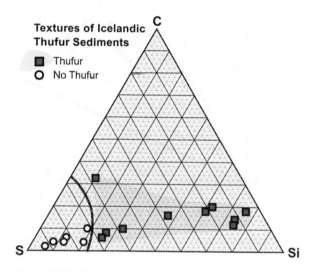

Figure 7.41 Grain sizes of sediments in Icelandic thufurs.

sub-zero air temperatures in winter. The silt moves with the original groundwater and is deposited with the ice in the freezing layer, thus raising the surface of lower areas. When the ice melts in the spring, the water drains away, leaving the silt, which remains in the elevated embryonic mound. Repeated annual cycles result in mound growth until an equilibrium is reached with the mass wasting processes.

Figure 7.42 shows the distribution of the main known locations of thufurs with both MAAT and freezing and thawing indices. They occur in areas of sporadic to discontinuous permafrost, but an oceanic or perhumid climate is critical to their formation.

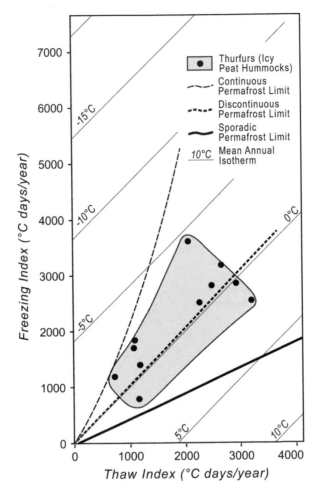

Figure 7.42 Distribution of thufurs in relation to freezing and thawing indices (after Harris, 1981c, 1982a).

7.3.3   Silt-cycling hummocks

Silt-cycling hummocks consist of mounds of different sizes with a fairly constant ratio of width to height, and with a range of different sizes present. Maximum width can reach 2.4 m, with a height of up to 35 cm, and they can be found on slopes up to 15° in Cornwall. They form in reasonably homogenous soils with a high content of silt and fine sand on moors (Killingbeck & Ballantyne, 2012), hilly grasslands Mark (1994), or on grassy mountain slopes (Grab, 1994). The soil organic matter decreases with depth as in normal soils, and there is no evidence of churning.

There is a continuum in climatic conditions between the extreme continental cold of the Arctic shores and the perhumid maritime conditions found in places like New Zealand and Southwest England. In the former, the change to cold weather is abrupt and rapid, so that the whole surface of the active layer freezes at once, and pressures from the phase change of soil water to ice during freezing of the ground cause circulatory movement, destroying part of the overlying vegetation and producing non-sorted circles (Chapter 10). As discussed above, cryogenic hummocks may also form at the better drained sites, usually in the taiga.

In maritime temperate climates, only the top and sides of the mound freeze at first, preventing the build-up of such cryostatic pressures, e.g., Iceland. In these perhumid maritime climates, the seasonal frost only penetrates to a depth of 30–40 cm in winter. The depressions may remain unfrozen through several freezing cycles in the fall (Mark, 1994; Grab, 1997, 2005b; Scott et al., 2008). During each cold event, moisture in the soil will tend to move towards the coldest part of the ground, so developing temporary ice lenses parallel to the freezing front as well as interstitial ice. During the shallow freezing of the ground, the hummock crests and sides freeze first, and particles of silt can be moved upwards with each freezing event (Corte, 1966; Schunke & Tarnocai, 1988). It is postulated that the mounds will become higher due to the multiple freezing and thawing events. Eventually, the increase in silt in the mound is balanced by gravity-induced processes moving soil down the sides of the hummock, resulting in the attainment of an equilibrium size (Mark, 1994; Grab, 1997; Killingbeck & Ballantyne, 2012). Killingbeck and Ballantyne describe an example from Dartmoor, where hummocks have developed in the last 3 ka on soils over sedimentary rocks. Where there is a lack of silt, e.g., in the sand-rich *grus* over weathered granite in the same area, hummocks are absent. Similar hummocks have been described from Cox Tor in Devon and East Cornwall (Bennett et al., 1996), and by Billings & Mark (1961) from the Old Man Range on the South Island of New Zealand.

Accordingly, these hummocks are called *silt-cycling earth hummocks* in this book. It remains to be determined how widespread these are, so the exact climatic range for their formation is unknown at this time. Both climate and grain size of the soils are critical for their formation. When the process of formation ceases, the hummocks should slowly be destroyed by mass wasting processes. At present, there are too few careful descriptions of these features to be able to know the actual distribution of these hummocks.

7.3.4 Niveo-aeolian hummocks

Lewkowicz & Gudjonsson (1992) and Lewkowicz (2011) have described *slope hummocks* from the Fosheim Peninsula, Ellesmere Island. They are defined as "decimetre-scale mounds present on moderate to steep slopes (8–27°) in permafrost areas, that are composed of stratified silty sand, have a relatively complete vegetation cover (>75%) and show limited amounts of cryoturbation" (Gudjonsson, 1992; Lewkowicz & Gudjonsson, 1992; Broll & Tarnocai, 2002). It is presumed that the sediment has limited frost susceptibility (Lewkowicz, 2011). The sediment exhibits stratification more or less parallel to the surface of the mound. The mounds appear to slide downslope over the underlying ground. On convex slopes, they move faster and exhibit a wider spacing and more gentle side-slopes than those on concave ground.

They are found on north-facing slopes where there is a combination of a thin active layer, two-sided freezing, an incomplete tundra vegetation cover, a winter snow pack that leaves the top of the ridges and mounds available for wind erosion, and

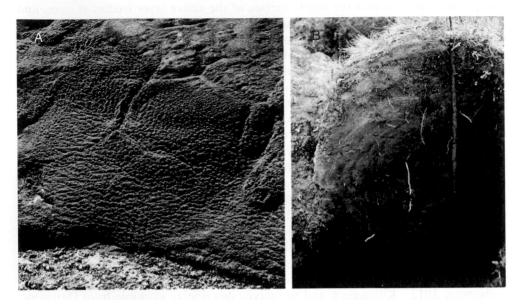

Figure 7.43 A, niveo-aeolian hummocks on the Fosheim Peninsula, and B, a cross-section of one of the hummocks moving slowly down slope to the left in the picture. © A. G. Lewkowicz.

localized deposition of the eroded sediment with the snow on the lower mounds on the slope (Figure 7.43). The slopes may be concave or convexo-concave, and the mounds are found on the down-wind side of the hills. Those on convexo-concave slopes are buried by niveo-aeolian deposition downslope of the snowbank remnant. The sand is deposited as part of dirty snow, having been eroded from exposed ground upslope. When the snow melts in summer, a thin coating of sediment is added to the ground surface.

Modelling suggests that the hummocks move 20–40 m downslope in about 2–4 ka, based on published C^{14} dates, time of site emergence from the sea, and measured rates of solifluction. So far, this type of mound has only been described from the cold, dry high Arctic tundra climate of the Canadian Arctic Archipelago (Figure 7.37). However, some of the "thufurs" reported from the upper slopes of mountains under tundra vegetation at lower latitudes may prove to be more examples of these when their origin is studied in more detail. Given the fact that they are formed by niveo-aeolian deposition and all types of hummocks can be found on slopes, the term *niveo-aeolian hummock* is used for them in this book.

The actual stratigraphy of vegetated mounds over permafrost is often not studied in detail. However, Kojima (1994) also described the relationships between vegetation, cryogenic earth hummocks and topography in the High Arctic on Ellesmere Island. The mounds were located on a leeward slope near its base, and while the mounds in the lowest topographic position had a complete vegetation cover, the latter decreased up-slope until only the bases of the mounds were vegetated. There was an abrupt end to the distribution of the sands in which hummocks developed, and they were absent further up the slope. The mounds exhibited an interesting temperature gradient

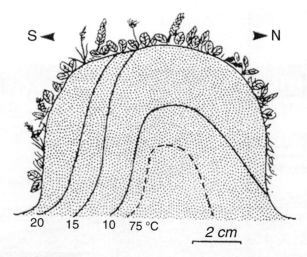

Figure 7.44 Temperatures in a hummock at 13.40 hours on 27th July, 1991 near Eureka (c. 79°N, 83°W) from Kojima (1994).

at 13.40 hours, local time on July 27th, 1991 (Figure 7.44), indicating the surprisingly marked influence of the low angle of the summer insolation near Eureka (c. 79°N, 83°W). The air temperature was 9°C at the time, and the warmer zone moves around the hummock with the direction of the sun in summer. A cross section of one of the lower mounds showed only minor disruption of the layers of fine aeolian sand and coarse silt. This sediment may not have been frost susceptible, although this was not established. The complete vegetation cover on the lower mounds indicates that negligible cryoturbation or churning of the type found in patterned ground had taken place. It is quite likely that the sediment was an aeolian deposit left when dirty snow melted in the early summer. The deposit ended abruptly up-slope. Thin salt crusts were found on the hummocks at the base of the slope, indicating that salts may also have been blown on to the slope with the snow from elsewhere, though the author did not check their composition. There was a special flora consisting of *Poa glauca*, *Potentilla vahliana* and *Poa hartzii* in this lowest, saline area, along with the ubiquitous *Salix arctica*. Kojima produced little hard evidence regarding the processes forming the mounds, and concluded that they were "a synergistic product of such factors as fine soil material, topography, snow, permafrost and vegetation." The description of the sediment as being fine aeolian sand and coarse silt suggests and both the location and microenvironment indicates that he was describing niveo-aeolian hummocks.

7.3.5 Similar-looking mounds of uncertain origin

Although there have been numerous descriptions of cryogenic earth hummocks in the literature, most lack enough information to determine the precise mode of formation. Examples include those that Grab (1994, 1997, 2005b) has reported from the upper

Figure 7.45 A string bog in Norway, with shrubs growing on the ridges and fen in between. © S.A. Harris.

slopes of mountains in East Africa, for which he uses several different names. Many of those, as well as the mounds described by Scotter & Zoltai (1982) need re-examination in the light of the recent work discussed above.

7.3.6   String bogs

String bogs are peaty mires with a pattern of elongated ridges, normally aligned parallel with the contours (Figure 7.44), though examples can be found where they follow the gentle slope downhill on the mire. They are developed entirely in peat and the ridges often provide better drainage for shrubs as well as the occasional tree (Figure 7.45). The ridges consisting of peat tend to freeze before the rest of the landscape due to the diode effect coupled with the being the highest parts of the ridges. Some of the higher parts may remain frozen in all but the warmest years, *i.e.*, these represent the outer margins of sporadic permafrost. The main requirement for string bogs is a very gentle, broad flat area occupied by a peaty fen or mire in a cold climate such as Norway, Sweden or Finland. They also occur in the northern parts of the Prairie Provinces of western Canada (Vitt *et al.*, 1994). Since peat covers the surface of the ridges, they tend to be the locus of thin permafrost at the outer margin of permafrost distribution,

Figure 7.46 Anastamosing string fen in northern Sweden showing how the ridges can sometimes form a network. Note the trees scattered along the ridges. © S.A. Harris.

although they can also be found in areas of sporadic or discontinuous permafrost in the otherwise unfrozen mires (Figure 7.46).

7.3.7 Pounus

Pouna is a local Finnish word for a small earth hummock composed of peat and/or mineral soils (Seppälä, 2005). It has also been used by Grab (2005a) to describe similar peaty mounds in southern Africa. The mounds are up to 1.2 m in height but less than 2 m long. They have vertical sides and are vegetated with species that can stand dry conditions. They occur in mires in the birch forest of northern Finland which have species of plants that need wet conditions to survive. Luoto & Seppälä (2002) found that about 60% of the pounus they examined had permafrost in them. The probability of permafrost being present decreased with increasing height of the mound. Unfortunately, they did not separate those in mineral soil from those developed entirely in peat, but presumably the peaty surface explains why they appear at the outer limit of the permafrost area in Scandinavia. Seppälä (2005) found only an insignificant difference in the seasonal frost heaving between the peat mounds with and without permafrost.

No indications of cryoturbation or mixing of the layers in the sediments were reported, so they would appear to represent peat-covered hummocks where the diode

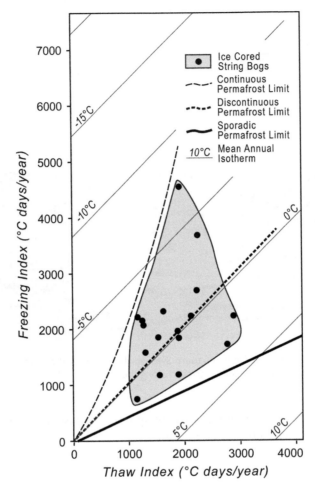

Figure 7.47 Distribution of string bogs in relation to freezing and thawing indices (after Harris, 1981c, 1982a).

effect of the peat produces increased cooling in some of the mounds. Mounds tend to freeze before the surrounding swales, so that water tends to migrate to the freezing plane, resulting in raising of the ground surface as ice lenses are formed.

Chapter 8

Mass wasting of fine-grained materials in cold climates

8.1 INTRODUCTION

Except in coastal marshes and other flat areas, gravity causes most deposits to move downslope. The movement usually involves a combination of processes that may operate together or at different times of the year in response to seasonal changes in the weather. There is an abundant body of literature on the subject from all climatic regions, and this chapter will examine some of the major processes and resultant landforms that are the product of the influence of seasonal frost or permafrost affecting predominantly finer grained material. The effects on coarse grained, blocky material will be dealt with in Chapter 9.

Seasonal frost may merely affect the upper surface layer of the soil, but where ice cements the sediment together, this causes much of the water from the melting snow pack to flow over the surface or through the soil above the frozen layers in spring or early summer. This causes erosion by sheet wash, soil flowage, as well as gullying. Warm rain on an isothermal snow pack can result in disastrous floods in the valley floors as has occurred in 2013 in both Calgary, Alberta, and Boulder, Colorado. These events, though random, occur about every 15 years in some part of the foothills of Southern Alberta. Diurnal frosts can also cause uplifting of objects, and rotation of stones in the frost-affected soils (see Chapter 1).

In areas underlain by permafrost, the freezing and thawing of the active layer can cause pressures to develop because of the phase change between water and ice, resulting in volume changes and churning movements (see Chapter 10). The permafrost table restricts water drainage, thus making the finer-grained soil more susceptible to mass wasting, and can result in detachment of the layers above the thawing front or permafrost table in spring and summer. Any large rocks anchored in the underlying permafrost tend to impede these downslope movements.

8.2 CLASSIFICATION OF MASS WASTING

Mass wasting is the collective name given to the processes involved in causing the movement of sediments and soils downslope under the influence of gravity. Often, multiple processes are involved. There is abundant but conflicting literature on the classification of mass wasting processes (*e.g.*, Sharpe, 1938; Gerasimov, 1941; King, 1953; Zolotarev, 1990; Spiridonov, 1956; Varnes, 1958; Kaplina, 1965; Sukhodrovsky,

Figure 8.1 Turf-banked terraces produced primarily by gelifluction on fairly steep slopes north of Nome, Alaska. The lobes are about 1 m high. Earth-banked lobes occur at the base of the unvegetated upper slopes. © S. A. Harris.

1979). In actual practice, these processes depend on the shape of the slope, its steepness, the structure and mechanical properties of the deposits on the slope, the hydrology (snow cover, precipitation regime, surface runoff, and underground water), the seasonal thawing, and the presence or absence of an active layer. Together, these decide the direction of movement of the mineral soil and associated water, the distance they move, their speed of movement, nature of the movement (mass movement or granular movement), the volumes involved, and the sequence of events. Some idea of the complexity involved will be obtained from French (2013).

Slopes can be divided into three kinds, *viz.*, convex-concave forms, *e.g.*, rectilinear debris-mantled slopes as in Figure 8.1, those with a near-vertical cliff, and those with distinct steps. The first type is found in rolling hilly country, typically with slopes less than 27°. Rectilinear debris-mantled slopes consist of an upper cliff consisting of sediments, *e.g.*, glacial till or marine deposits. Near-vertical cliffs usually are the source of blocky deposits (see Chapter 9), while the lower slopes may exhibit distinct steps including pediments and marine or river terraces. Formerly glaciated mountains often exhibit oversteepened slopes and are particularly liable to exhibit mass wasting processes. Rockfalls, snow avalanches and other forms of rapid mass wasting are common.

The surface materials can vary from clays to massive rock, or any mixture thereof. Water in saline sediments has a depressed freezing point, and in the frozen state, it has considerably less strength than nonsaline frozen sediments at a given temperature, other things being equal. Frozen clays have a very high unfrozen water content (sometimes >22% by weight) and tend to be plastic at low temperatures. During thawing of the active layer from above, the high liquid water content perched on the underlying frozen layer can act as a lubricant for a plane allowing slippage of the unfrozen surface material, or it can provide enough lack of cohesion in the surface material that the latter flows downslope.

In Russia, the term *deserption* is used for dry, temperature creep, as opposed to *defluction* for movement of wet soil. However the term deserption is also used in Physics for adsorption of gases, so should probably be avoided by geocryologists.

In this chapter, we will examine three main types of mass wasting and how they affect movement of fine-grained material down slopes in cold climates, *viz.*, slow flows, rapid flows, and slides. Slides are cases where the ground moves along a definite plane over a stationary surface. In flows, the surface soil moves fastest, with decreasing movement with depth until at greater depth, no movement occurs. Intermediate forms also occur and are dealt with under a separate heading.

8.3 SLOW FLOWS

These are either the result of gradual long-term flowage of ice-rich sediments downslope or consist of intermittent small-scale, ratchet-like movements of the surface layers of the soil in response to specific climatic situations (Washburn, 1979). The latter involve the formation and melting of ice in or on the surface of the soil on slopes, resulting in *cryogenic creep*. These can be in response to diurnal and/or seasonal formation of ice, followed by thawing. The formation of ice masses causes part of the soil to be raised normal to the slope, while melting results in the raised soil tending to move vertically down-slope under the influence of gravity. If there is a high concentration of ice melting out, the wet soil may contain more water than the liquid limit and can actually slowly flow a short distance until the excess water has drained away.

Matsuoka (2001b) provides a summary of the literature up to 2001 under the name of solifluction. He demonstrated that there were several distinct processes, although they often act together in producing the total annual downslope movement. He also included slow flows outside the realm of seasonal frost action. The following discussion will focus on the individual processes found under cryogenic conditions. Bondarenko (1993) discusses the prediction of stability on these slopes. Eichel *et al.* (2013, 2015a, 2015b, 2016) and Draebing and Eichel (2016) emphasize the importance of pioneer species, *e.g.*, *Dryas octapetala*, in altering the slope processes from erosional (sheet wash) to bound solifluction/gelifluction processes. Toposequences determine the actual processes operating which vary from ridge to valley floor.

8.3.1 Cryogenic creep

Creep is the slow deformation that results from long-term application of a stress that is too small to produce failure in the material (ACGR, 1988). Thus any granular material on a horizontal surface will tend to spread out laterally. This is well known in the case of gravel roads in Alaska (Esch, 1983), where the lateral movement is greatest on the north-facing slopes. His results suggest the additional influence of other factors such as ice content and ground temperatures in that case. On slopes, any disturbance of particles in the sediment tends to cause movement downslope under the stresses caused by gravity. Examples of such disturbances in warm climates include rain splash, heating and cooling, tree fall and the movement of animals.

In cold regions, *cryogenic creep* can occur in additional ways. In permafrost or seasonally frozen ground, *creep deformation* is mainly the result of the creep of pore ice and the migration of unfrozen pore water. In ice-saturated soils, most creep deformation is distortional with negligible volume change. In frozen soils with large unfrozen water contents as well as in unsaturated frozen soils, slow deformation due to consolidation as well as creep due to volume change may also occur. Usually, the bulk of the creep deformation is permanent (Vyalov, 1959, 1978; Ladanyi, 1972, 1981). Any objects on the surface or embedded in the soil are slowly moved downslope, although Matsuoka (1998, p. 132) noted that painted stones on the surface of the ground moved several times faster than the top of a strain gauge. Harris (1973) found that shallow-rooted trees moved at the same rate as the soil and could be used to obtain a good average of the movement taking place. Also at Kananaskis, he found evidence of *thermal creep* of soil peds in the fall when the soil was dry and subjected to diurnal heating and cooling. This process may be much more common than previously realized.

In addition to deformation of the ice in the ground, shorter-term freezing can result in two main types of episodic cryogenic creep. *Needle ice creep* is particularly effective on the ground surface in humid maritime climates, due to seasonal frost and numerous diurnal freeze-thaw cycles affecting the upper 10 cm of soil (Chapters 2 and 10). *Frost heave* is the most frequent cryogenic cause of creep in areas with colder climates, being due to ice lenses forming at depths down to >1 m in areas of seasonal frost as well as in the active layer in regions with permafrost. Cryogenic creep includes both frost creep and needle ice creep.

8.3.1.1 Needle ice creep

Also called *pipkrake* in Sweden (Krumme, 1935), this only affects the upper 20 cm of soil (Mackay & Mathews, 1974a, 1974b; Noguchi *et al.*, 1967; Matsuoka *et al.*, 2003). On Haleakala, Maui, an accumulation of silt at a depth of 2–3 cm beneath the surface aids in providing the flow of soil water to the freezing plane beneath the stones to produce the ice needles, in spite of being in a relatively dry climate (Noguchi *et al.*, 1987). The actual temperature of the ground surface tends to be colder than the air temperature due to reradiation, so that needle ice can develop at air temperatures just above 0°C. Critical conditions include the presence of silt and clay immediately below the surface and the presence of enough soil moisture in this layer to allow rapid movement of water to the freezing plane. The needles grow (Figure 2.13), thus raising stones on the ground surface or fragments of soil crust so as to cap the needles in cool, humid weather (Figure 2.6).

The height of the needles depends on moisture availability, the texture of the layer containing silt and clay, and the length of time available for the needles to form (Soons and Greenland, 1970; Lawler, 1993, Figure 10). Downslope movement is the result of melting of the pillar of ice by sunlight, resulting in stones and the overlying soil or stones moving downslope by rolling (Higashi & Corte, 1971; Mackay & Matthews,

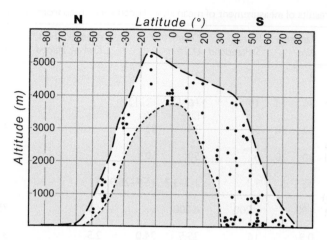

Figure 8.2 Distribution of reported cases of needle-ice formation, plotted against latitude.

1974b), by toppling, sliding, or in sediment-laden rivulets (Lawler, 1993, Figure 14). The actual downslope movement of soil caps in a given freeze-thaw event is often somewhere between the vertical and perpendicular to the slope (Schmid, 1955; Young, 1972, p. 50), probably due to cohesion effects. Repeated freeze-thaw cycles resulting in downslope movement can result in substantial annual movements of up to 0.52 m per year based on six years of measurements (Mackay & Mathews, 1984b). Large stones imbedded in the surface layers of soil in areas of seasonal frost reduce the amount of movement but do not stop it. Fahey (1979) showed that the proximity to the water table had an enormous effect on the annual amount of movement at a given site, though needle ice creep is limited to areas lacking a snow cover as well as a closed cover of vegetation. Maximum depth of displacement was 20 cm in the study of diurnal frost heave in the Japanese Alps.

Needle-ice creep is one of the most rapid and significant processes of creep of the surface of the soil (Dylik, 1969a, 1969b), though it only affects the top 5–20 cm of soil. It is found throughout the land areas of the world (Figure 8.2) where suitable weather occurs (Zotov, 1940; Troll, 1958; Jahn, 1961; Washburn, 1969), both in summer on the surface of the active layer over permafrost in polar climates, as well as at other times of the year in areas of shallow diurnal freezing (Hagerdorn, 1974; Pèrez, 1987; Boelhouwers, 1995; Matsuoka *et al.*, 2003). However, the total volume of sediment displaced is relatively small, even in temperate west coast climates such as the South Island of New Zealand (Gradwell, 1957; Soons, 1968), Japan (Ellenburg, 1974, 1976; Satake, 1977), and coastal British Columbia (Mackay & Matthews, 1974a, 1974b). Table 8.1 shows the results of measurements of needle ice creep from a variety of climates and microenvironments.

Needle ice is very disruptive to seedlings (see Pèrez, 1987), and can actually prevent the growth of vegetation on some tropical mountains (Troll, 1958; Coe, 1967; A. P. Smith, 1974). This results in increased erosion by wind and water, and it also aids in bank erosion along streams in temperate climates (Lawler, 1993).

Table 8.1 Some results of measurement of mean annual needle ice creep from different environments.

	Japanese Alps		Valleta, Switzerland	Colorado U.S.A.		Wieliczka Hills, Krakow, Poland	Hokkaido Mombetsu	Kitami
Length of study (years)	3		2	2	5	1		
Angle of slope (°)	30	14	10	16–18 Dry	12–13 Wet			
Diurnal Freeze-thaw days	52	50	55	34	33		3	4
Depth of Diurnal cycle (cm)	20	20	5	50	50			
Average rate of movement (cm/a)	45.5	5	1.1	1.2–3.7	16.9–25.9	5		
Seasonal freezing depth (cm)	100		c.200	c.200	>	c.50	75	85
Mean annual heave (cm)	0.9		16	25.5	34.0	2.5	3.3	5.7
Potential annual heave (cm)	0.9		2.8					
Source	Matsuoka, 1998		Matsuoka et al., 1997	Benedict, 1976		Olecki & Widecki 1970 (Figs. 4 & 5)	Kinisota 1969 Fig. 5c	Fig. 4c.

8.3.1.2 Frost heave and frost creep

Frost heave takes place when ice lenses are formed in the soil and result in an upward and outward movement of the surface on all but <2° slopes (ACGR, 1988). During freezing, the soil moisture moves through the unfrozen soil to the freezing plane where it gives out heat as it crystalizes to form ice. Provided the movement is sufficient to cause the heat of crystallization to neutralize the loss of heat to the ground surface, the freezing front will remain in place, resulting in the growth of an ice lens. The expansion on freezing results in uplift of the overlying layers at right angles to the slope (Figure 8.2). When the cooling exceeds the generation of heat, the freezing front descends until a new equilibrium is reached. There, a new ice lens will start to form. Changes in the rate of heat flow to the surface of the ground, together with moisture flow to the freezing plane appear to control the size and spatial distribution of the resultant ice lenses. Actual prediction of the amount of heave has proven to be difficult (Konrad & Morgenstern, 1984; Williams, 1986), but the heaving pressures may reach up to 300 kPa quite quickly (Williams & Smith, 1989, p. 197). This explains the role of the growth of ground ice in heaving stones out of the ground as well as rotating them (Chapter 1). Maximum thickness of the ice lenses is usually about 10 cm.

During subsequent thawing of the ground, the overlying surface materials will tend to move vertically downwards (Jahn, 1975; Washburn, 1979), thus migrating downslope in maximum proportion to the thickness of the ice lenses beneath (Figure 8.3A). This is called *frost creep* and produces a bending of flexible tubes or wooden dowels emplaced in the ground, normal to the slope, with movement at the surface and decreasing movement downwards until the lowest layer that grew ice lenses is reached.

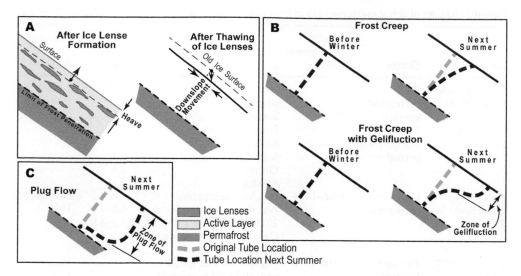

Figure 8.3 A, the mechanism of frost creep, and examples of its effect on flexible tubing or a column of dowels during a single winter, depending on whether B, gelifluction or C, plug flow occur as well as frost creep.

The actual movement is often less if cohesion interferes with the downslope movement. However, if the melting ice causes flowage of the upper layers of soil (solifluction or gelifluction), the annual downslope movement will be increased in and above the zone of flow. The latter is fairly common in soils rich in clay and silt, since the quantity of ice is usually larger in this case.

Frost creep and needle ice creep often both occur in the surface soils in cold climates. Table 8.2 provides examples of mean annual cryotic creep rates from a range of climatic situations. Frost creep dominates in areas of deeper seasonal frost, *e.g.*, Colorado (Benedict, 1970), as well as in the surface of the active layer at higher latitudes during colder weather in summer and fall. The amount of mean annual creep is small, but it affects a greater thickness of soil. This contrasts with needle ice creep which produces greater downslope movement, but only in the upper 20 cm. Actual volumes of soil moved are therefore rather similar in the two cases.

Three major factors decide the amount of downslope movement by cryotic creep in a given year. The most important is probably the texture of the soil. Sandy materials will have little movement of water to the freezing plane, resulting in negligible movement, whereas silty soils and those containing organic matter will tend to develop larger and more abundant ice lenses, other conditions being equal. Soils on slopes in the Japanese High Mountains consist of an upper frost-susceptible layer underlain by horizons which are not frost susceptible. Matsuoka (1996) found that only the upper part of the soil profiles underwent significant diurnal frost heaving, but the amount of heave could reach 3 cm in a given event. Rainstorms did not cause heaving in the lower soil layers.

Secondly, the moisture content of the soil decides how much moisture is available to move to the freezing plane. The frost heave strain may reach 20–60% in highly

Table 8.2 Results of actual measurements of frost heave in different environments in North. America.

Location	Climate	Period of measurement years	Site characteristics	Total heave cm/a	Reference
Cape Thompson Alaska, 70°N	Polar permafrost	n.d.	Frost boil, very frost-susceptible	32.5	Everett, 1966
Signy Is., South Orkney Is., 62°S	Lowland Temperate maritime with permafrost	1968	Sorted circle, highly Frost susceptible Surface Buried On stones At edge	4 0.4 3.6 2	Chambers, 1967
Mesters Vig, Greenland, 72°N	High Arctic permafrost	1958–1964	Very frost susceptible Wet slopes – 10 cm depth 20 cm depth Dery slopes – 10 cm depth 20 cm depth	0.0–1.0 1.5–5.8 0.5–0.7 0.8–1.4	Washburn, 1969
Inuvik, N. W. T. 68°N	Forested Lowland Permafrost	1976–1978	Highly frost susceptible with clay mud hummocks Undisturbed site (1977–1978) (Mean of 38 observations) Bulldozed site (1977–1978) (Mean of 9 observations)	10.39 14.09	Mackay et al., 1979
Kananaskis, Alberta, 51°N	Dry, alpine, seasonal frost	1970–1971	Frost susceptible: Wet slope, Dec.–March Dry slope, Dec.–March	0.0–1.8 0.0–0.4	Harris, 1971
Colorado Front Range, 39°N	Moist, alpine seasonal frost	1965–1966	Very frost-susceptible frost boil	25.0–29.5	Fahey, 1974
Colorado Front Range, 39°N	Moist, alpine seasonal frost	1976–1978	Very frost-susceptible Stone banked and turf banked terraces and lobes	0.5–38.0	Benedict, 1970

frost susceptible lake sediments in the Mackenzie Delta (Burn, 1990) whereas on well drained slopes on the Tibetan Plateau, it may only be 5–6% (Wang & French, 1994). In the case of diurnal frost heaving, rain storms add moisture to the surface layers of soil, which results in an increase in the frost heaving (Matsuoka, 1996). However, a winter snow cover reduces the depth of frost penetration into the ground (Fahey, 1973; Benedict, 1976). The third factor is the number of freeze-thaw cycles during the winter. Matsuoka and Moriwaki (1992) found that in Antarctica, the depth of thawing had to exceed 7 cm in order to allow the next nocturnal heave event to occur. Partial thawing to less than 10 cm in the Japanese Alps prevents heaving during refreezing (Matsuoka, 1996). The actual temperature reached during freezing does not correlate well with needle ice growth.

8.3.1.3 Gelifluction

Andersson (1906, p. 95) introduced the term *solifluction* for saturated soil flowing slowly downslope in the Falkland Islands. As such, it refers to all slow flows of surface

Figure 8.4 Gelifluction (light coloured) on the frozen ground at 4700 m on the upper north facing slopes of Kunlun Shan, China. © S. A. Harris.

soils over a stable base, though French (2013) uses it to include the sum of all the movements due to cryogenic creep in cold climates. However, Washburn (1979) introduced the term *gelifluction* for the actual flow component over a frozen substrate, so as to separate it from saturated soil flowing over an unfrozen but relatively impermeable soil layer. Gelifluction is therefore a special kind of solifluction where an icy layer underlies the slowly moving flow, and it is therefore used in place of solifluction in the following discussion (Figure 8.3B). In Russia, a distinction is sometimes made between slow solifluction (<1 m/a^{-1}) and fast solifluction (>1 m/a^{-1}) although this has negligible genetic significance. Gelifluction and solifluction can commence on slopes as little as 2°C. C. Harris *et al.* (2003) carried out two scaled centrifuge modeling experiments designed to simulate thaw-related gelifluction. The scales used were different, yet they produced virtually the same results. They therefore concluded that the movement was elasto-plastic in nature rather than a time-dependent, viscosity-controlled flow phenomenon. There was some minor (elastic) strain recovery detected in laboratory studies (Rudram, 1994) but these were an order of magnitude smaller than the movement by gelifluction.

Gelifluction is caused by the thawing of the surface of the ground with the addition of water from melting snow on the mountain sides at about 4800 m elevation in the Kunlun Shan (Figure 8.4). The area has very little snowfall, but it mainly falls in summer. The thin snow cover thaws briefly during the summer days, and the melt water thaws an icy layer about 2–5 cm deep. The combined quantity of water is sufficient to cause patchy slow flow of the surface sediment for up to 2.5 m before movement ceases forming small, low lobes elongated downslope. This is the result of one-sided thawing of the thin surface layer of soil after each minor overnight snowfall during summer. The mean annual downslope movement of the gelifuction material is estimated at 30 cm/a^{-1} at 4750 m, compared with about 3 cm/a^{-1} at 4650 m. In the Pamir-Alai and Tien Shan of Kazahkstan, the lower limit of gelifluction is between 2500 m in the north and 3800 m in the south (Gorbunov & Seversky, 1999). It occurs through a range of 900–1400 m elevation, producing block streams, gelifluction sheets and altiplanation terraces. Where the active layer is deeper, there may be either one-sided or two-sided thawing (Matsuoka & Hirahawa, 2000). One-sided thawing results in excess water occurring in the surface layers of the soil due to both melting snow and thawing ice lenses. This results in a deeper surface flow than in Figure 8.4, producing a modification of the results of heaving as in Figure 8.2. The actual amount of flow can be estimated from the modification of the bending of the dowels or flexible tubes, though this will only give a rough estimate. The plastic flow will tend to flow around the dowels and flexible tubes so the estimate will probably be low. In the case of two-sided thawing, the concentration of ice just above the permafrost table at the base of the active layer produces excess water in this zone, resulting in *plug-like flow* as in Figure 8.3C. This was first noted in permafrost by Mackay (1981), though, once again, the flow indicated by the dowels or tubing probably underestimates the amount of movement. There may or may not be additional near-surface plastic flow distorting the pattern of the dowels or shape of the flexible tubes. Dowels move more readily with plastic flow than the flexible tubing and give a better estimate of the amount and type of flow that has taken place.

On undulating tundra landscapes such as the Yamal Peninsula, deep thawing of the ice-rich active layer can result in fast gelifluction flows cutting a small network of shallow gullies, splitting the vegetation into separate blocks (Figure 8.5). The furrows, locally called *delli*, average 1–2 m in width, but expand to 2–3 m wide on the lower slopes. The gullies become vegetated with hygrophilous vegetation.

8.3.1.4   *Other creep-type contributions to downslope movement of soil*

The upper soil layer that becomes thawed in summer will be subjected to additional processes causing downslope movement when not frozen. When measurements of movement are only measured annually, the results of these other processes will be included in the overall results. Only when continuous recordings or regular measurements of movement are made at frequent intervals can the effects of the individual processes be differentiated.

Examples of movements not related to permafrost include changes in response to trampling by animals, heavy precipitation events, and changes in relative humidity of the air in the soil. Figures 8.6 and 8.7 show the evidence for the effects of the last two as measured by continuous recordings on a south-facing slope in Kananaskis, Alberta (Harris, 1973). They are essentially small diurnal rotational movements of the soil which complicate the overall downslope movement which is taking place. These processes are responsible for the downslope movement observed in hot arid regions such as Israel, but they can also contribute movement to the surface of the unfrozen active

Figure 8.5 Fast gelifluction in ice-rich silty sediments dissecting the tundra vegetation on the Yamal Peninsula. This usually develops on 5–20° slopes with a 0.4–1.0 m deep active layer. © A. Gubarkov.

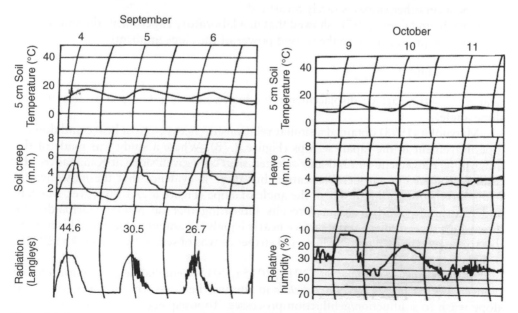

Figure 8.6 Evidence of thermal creep caused by diurnal fluctuations in ground temperature and humidity at Kananaskis, Alberta (from Harris, 1973, Figure 3).

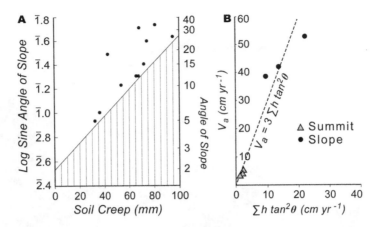

Figure 8.7 A: The movement recorded at random measuring stations on the south slope of Heart Mountain, Kananaskis. B: Downslope movements (V_s) recorded in the Japanese Alps. From Harris (1973, Figure 1B) and Matsuoka (1998, Figure 9B) respectively.

layer during summer in colder climates. In addition, rainfall splash and surface flow of rain water can add to the downslope movement in humid environments. Movement of wild animals and livestock also compounds the problem. As a result, the numerous studies of mean annual total downslope movement from a wide variety of environments cannot readily help to provide valid information on the magnitude of the contributions by each individual cryotic creep process in most cases. Furthermore, most are measurements carried out to determine the amount of movement in situations where downslope movement is obvious, *e.g.*, in the case of turf-banked terraces. The movement on other slopes is rarely measured.

Higashi and Corte (1971) showed that in a laboratory experiment, the soil surface movement is proportional to the second power of the slope gradient:

$$V_s = Ch \tan^2 \theta \tag{8.1}$$

where: V_s = the surface displacement; h = heave amount; C = a constant; θ = the slope angle.

Matsuoka (1998) obtained a similar relationship between frost heave and downslope movement in the Japanese Alps (Figure 8.7B). Where a study was made of the downslope movement on randomly selected sites on a given slope at Kananaskis, the results produced a considerable scatter, but there is evidence that the movement is related to the logarithm of sine of the angle of slope (Figure 8.7A). However, Mackay and Matthews (1974b) obtained results indicating that on Mount Garibaldi, the downslope displacement at the surface nearly equals the amount of heave. This may be due to movement by plug flow in sorted stripes as was observed by Matsuoka (2001b, p. 111, Figure 2d).

As noted earlier, Eichel *et al.* (2013, 2015a, 2015b) emphasizes the importance of pioneer species such as *Dryas octopetala* in altering the slope processes from erosional slope wash to solifluction/gelifluction processes. Toposequences determine the active processes occurring between ridge crest and valley floor.

Figure 8.8 Turf-banked terrace in the Rat Pass, Dempster Highway at the Yukon, Northwest Territories border, Canada. Note the cryopediment surface in the middle distance. © S. A. Harris.

8.3.2 Landforms produced by cryogenic slow flows in humid areas

The slow flows rearrange the sediments on slopes in humid areas to produce a variety of landforms (Matsuoka, 2001b, Figure 14). Since needle ice creep results in only shallow movements of soil, it can only produce small scale, thin lobes and micro patterned ground. Likewise, frost heaving moves a greater depth of soil downslope at a slower rate. Once again, this can produce small lobes as well as large scale patterns.

The process that alters this is gelifluction, when it occurs on a large scale each year. The end result is the development of lobes in alpine areas which may be tongue-shaped (Figure 8.8) or terraces that are more or less parallel to the contours as in Figure 8.1. The fronts of the lobes range in height from 0.2 to >2.0 m. If the lobes have a closed vegetation cover, they are called *turf-banked terraces* or *lobes*, as opposed to the *stone-banked terraces* or *lobes* that consist primarily of soil and rocks. Turf-banked terraces are found on meadow tundra slopes whereas stone-banked terraces are primarily formed in polar deserts that are devoid of vegetation or where there are large numbers of boulders in the soil that tend to become stranded along the front of the landform. In both cases, the surface layers move faster than the soil in front, resulting in overturning and overriding of the underlying soil. The soils on the rest of the slope move at a much slower rate (Benedict, 1970, 1976). As a result, the organic layer in front of the lobe is overrun by the movement, so that the rate of movement can be determined by digging a trench into the front of the lobe and radiocarbon dating samples of buried organic matter collected from various distances from the front of the lobe. Costin *et al.* (1967) first reported radiocarbon dates on buried wood in the Snowy Mountains of Australia, showing evidence of periglacial activity between 3000 and 1500 years B.P. Denton and Karlén (1973) have used this technique to determine possible world-wide cold events back to at least 3000 years B.P. Long-term average rates of movement from the colder parts of the Northern Hemisphere range from 0.6 to 3.5 mm per year except where autumn saturation and deep freezing occur. Under such conditions, Benedict (1976) has measured rates as fast as 24.4 mm per year for short periods of time. Radiocarbon dates as old as 6850 B.P. have been found, apparently representing earlier cold events on Mount Chavagl in the Swiss National Park (Furrer & Bachmann, 1972).

Turf-banked terraces are also widespread on tundra in the Vorkuta area of European Russia on slopes of about 30°. Further south, gelifluction features are rare in the tundra, due to a much more luxuriant and dense vegetation cover and a deeper active layer. The ice content of the active layer is also lower, while the Atterburg limits for the sediment are higher.

The processes involved depend on the microenvironment, Benedict concluding that the average rate of 3.7 mm per year of movement on active stone-banked terraces in Colorado were due almost entirely to frost creep. The adjacent soil being over-run by the lobe was moving at only 1.2 mm per year. However, the radiocarbon dates from buried organic matter indicate that these are relatively ancient landforms, and it is uncertain whether new landforms of this type are developing today. These features have been reported from Colorado (Benedict, 1970, 1976), Australia (Costin *et al.*, 1967); New Zealand, Spitzbergen (Åkerman, 1973, 1980; Rapp, 1960; Rapp & Åkermann, 1993), Alaska (Everett, 1966; Brown, 1969), Scandinavia (C. Harris, 1972), the Yukon (Price, 1973, 1991), Alberta (Smith, 1987, 1988, 1992); Switzerland (Matsuoka, 2001; Dreibing & Eichel, 2016), and Russia. In lowlands, the equivalent landforms are turf- or stone-banked terraces, but there is a continuum between lobes and terraces. C. Harris (1989) and Matsuoka (2001) discuss the mechanisms involved. Dreibing and Eichel provide a detailed field study showing the complexity of the processes involved (soil structure, thermal conditions and vegetation control) in the movement of turf-banked terraces on a Neoglacial moraine and suggest that there may be a cycle of development. One of the largest gelifluction deposit in alpine situations appears to be on the north-facing slope of Kunlun Shan, China at an elevation between 4800 and 4650 m (Harris *et al.*, 1998b). It extends laterally 1.6 km along the mountain front and has 16 tongues flowing down fluvial graded valleys towards the main strike valley below (Figure 8.9). Unlike rock glaciers (see Chapter 9), the lower front is flat (Figure 8.10) and consists of isolated boulders moving downslope in the form of a rock stream. However the main body of sediment consists of poorly sorted sandy loam with 10% boulders, probably formed under during warmer climatic conditions during the last million years. The average slope of the landform is about 19°C. The mean annual air temperature is −7 to −5° C and the mean annual precipitation is <300 m per year. Plants growing on the sediment extend their stems downslope to keep pace with the movement of the soil (see Figure 8.12).

Where a large rock is frozen into the underlying permafrost, it acts as a barrier temporarily impeding gelifluction (Figure 8.13). These are therefore called *braking blocks.* Where the blocks rest within the active layer undergoing gelifluction, they tend

Figure 8.9 The gelifluction deposit (white area) mantling the northern slopes of the Kunlun Shan, west of the Golmud Lhasa Highway, China. From Harris *et al.*, 1998.

Figure 8.10 The lower front of one of the gelifluction lobes on Kunlun Mountain, China. © S. A. Harris.

to move further and faster downslope, pushing a pile of sediment in front (Tufnell, 1996; see the references in Reid & Nesje, 1988). These **ploughing blocks** leave a groove or depression on the tundra surface behind them (Figure 8.12) and have a pile of soil in front. These are fairly common on mountain slopes where the surface sediments are underlain by permafrost (*e.g.*, Reid & Nesje, 1988).

Most of the terraces and lobes discussed above are in areas that have been glaciated, so that they have had limited time to develop. However, in areas which have not been

Figure 8.11 The large rock is frozen in the underlying permafrost and impedes the flow of the surface active layer. © S. A. Harris.

Figure 8.12 Elongation of the buried stem of a plant (*Saussura* sp.) on the gelifluction slope at 4760 m on Kunlun Mountain. © S. A. Harris.

Figure 8.13 Ploughing block in melting snow, Marmot Ski Area, Jasper. © S.A. Harris.

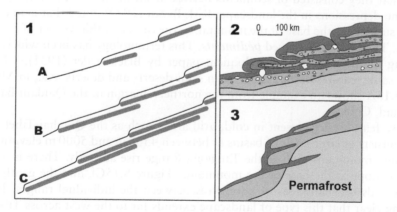

Figure 8.14 1, Diagram of the development of multiple buried organic layers in gelifluction lobes. 2, An actual cross-section of turf-banked terraces in northeast Siberia (after Savelev, 1962, Figure, 1962b). 3, Permafrost moving upwards into the lower stacked gelifluction layers.

subject to glaciation, *e.g.*, Siberia, they are far better developed. There we find multiple buried organic soils resulting from over-riding of successive terraces. Figure 8.14.1 shows diagrammatically the development of these layers over time. The pemafrost table has been omitted to simplify the drawing. Figure 8.14.2 shows a profile of superposed lobes with buried organic turf layers from the Belskiy Mountains, near Ust, Norheaster Siberia (Savelev, 1962a, 1962b). In this case, the active layer is over 160 cm thick. In the colder climates, the permafrost rises into the buried organic layers as the stacking process proceeds (Figure 8.14.3).

In Russia, gelifluction landforms are widespread in the low- to mid-mountain regions of the Taimyr Peninsula, in spite of the thin (0.4–0.8 m) active layer (Popov *et al.*, 1985). They consist of saturated friable weathering products moving on slopes of 3–4°. On the left bank of the Lena River near Chekanovsky, the river banks are 450 m high and gelifluction affects a zone several thousand square metres wide alongside streams and on terraces (Gravis, 1969). In the northern part of Novaya Zemlya, gelifluction develops in front of glaciers on ice-cored moraines (Kolomyts, 1976). Gelifluction also occurs on the low Medvezhyi Islands (Savichev, 1962), on Schmidt's Cape (Vityurin, 1975), and on the mountains surrounding the Anadyr Lowlands and on the Yamal Peninsula.

8.3.3 Landforms developed by cryogenic flows in more arid regions

Arid regions exhibit different landforms to humid landscapes. Rivers valleys and terraces are rare, and the mountains rise abruptly from the plains. The ideas about these landforms date back to the early 20th century when geomorphologists and geologists were exploring the deserts of the arid south-west of the United States. The gently-sloping plains (0.0–6.0° slope) extending down from the mountains to ephemeral playa lakes caused Tolman (1909) to introduce the Spanish term *bahada* or *bajada* for them, noting that they consisted of sediments derived from the mountains and spread out across the depressions in the landscape. Kirk Bryan (1922) noted that actually, some of these surfaces at the base of the mountains consisted of a thin veneer of gravel over bedrock. These he aptly named *pediments*. This terminology has been widely applied following the publishing of the eloquent paper by Blackwelder (1931), who noted that the same features occur elsewhere in semi-deserts and deserts, *e.g.*, in Mongolia. Figure 8.15A shows an example from the northern margin of the Qaidam Basin, east of Golmud, China.

These features are present in cold, arid areas such as the Qinghai-Tibet Plateau, which consists of large areas of basins at between 4500 m and 5000 m elevation, from which high mountains such as the Tanggula Range rise abruptly. There is a marked change in slope at the base of the mountains (Figure 8.15C), and the gently sloping basin extends for many tens of kilometres between the individual ranges. It is now becoming clear that this type of landscape extends far to the west across Tibet to the Pamir Knot. The mountains are underlain by continuous permafrost, while the flat, plateau surface is underlain by discontinuous permafrost in places. Climatic warming is increasing the number of thermokarst lakes. Unfortunately, little is known about the slow cryogenic creep processes that may be occurring, but frost heave and gelifluction are almost certainly taking place. The balance between slope wash on the rare occasions when it rains and the nature of the creep processes occurring today is unknown.

Undoubtedly these have changed with the changing climates during the last 130,000 years, since the area was being uplifted at a rate of c. 10 mm/a^{-1} until it became high enough to develop permafrost. It is likely that the landscape originally developed under warmer conditions, but the processes operating today have changed due to the colder climate.

Precipitation increases with altitude, so that the highest mountains have cold-based glaciers on their slopes. Below that zone lies the zone where gelifluction is rampant

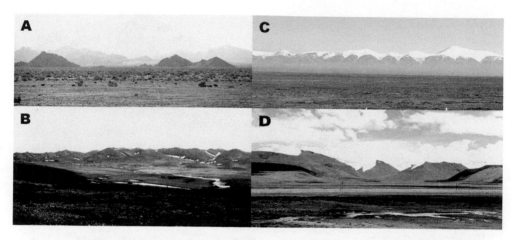

Figure 8.15 Comparison of pediments and bahadas: A, On the north side of the Qaidam Basin, east of Golmud, China (semi-desert without permafrost); B, In the Richardson Mountains, Yukon Territory (continuous permafrost); C, On the Qinghai-Tibet Plateau looking towards the Kunlun Mountains (continuous permafrost, 4700 m); D, The south exit from the Kunlun Pass (continuous permafrost, 4800 m). © S. A. Harris.

where ever there are finer grained sediments. The gelifluction deposit on the north slope of the Kunlun Range has already been described. Gelifluction is rampant and more effective than stream action in the upper valleys on part of the high altitude slopes of the Qilian Range, north of the Tarim Basin, wherever there are finer grained sediments and soils (Figure 8.15). Loss of the soils due to gelifluction results in the exhumation of unweathered rock masses and corestones as *tors* near the summits of the ridges (Figure 8.16). Tors are relatively unweathered blocks of bedrock, usually with a rounded form, indicating rounding by weathering agents. Similar features can be seen elsewhere in the semi-arid parts of Central Asia.

Cryoplanation refers to the presence of step-like benches or beveled surfaces in bedrock in cold climates (Eakin, 1916; Dylik, 1957; Demek, 1968, 1969a, 1969b; Péwé, 1970; French, 1976, pp. 155–166, 2016; Washburn, 1979, pp. 237–243). Originally called *altiplanation terraces* (Eakin, 1916), these *cryoplanation terraces* or *goletz terraces* were described from Siberia as cutting bedrock. Russian geologists proposed a cycle to explain these flat surfaces cut in bedrock (Boch & Krasnov, 1943), based on the terraces seen along the Ural Mountains. The concept involved gelifluction moving sediment downslope on an upper surface, a steep riser forming the drop to the next surface, with replication down to the lowland (Figure 8.17). The initial surface was presumed to be a structural bench, and this model was widely accepted in the central European literature for benches, the process being called cryoplanation (Waters, 1962; Richter, *et al.*, 1963; Demek, 1964, 1968, 1969a; Czudek & Demek, 1973; Czudek, 1990). Snow banks were believed to form on the risers, which on thawing caused erosion and retreat of the slope. Thus the lower risers gradually eliminate the higher ones. Demek (1969) regarded *cryoplanation terraces* as being best developed in continental semi-arid periglacial environments, while Reger and Péwé (1976) stated that they require permafrost for their formation.

Figure 8.16 Tors becoming exposed as gelifluction moves the soil and weathered bedrock downslope towards the pediment surface below the pediment angle, North Fork Pass, Yukon Territory. © S. A. Harris.

Figure 8.17 Altiplanation surfaces at Boundary, Yukon Territory. © S. A. Harris.

Surfaces corresponding to cryoplanation terraces occur in Alaska and north central Yukon (Cairnes, 1912; Eakin, 1916; Reger & Péwé, 1976; French *et al.*, 1983; Priesnitz & Schunke, 1983; Priesnitz, 1988; French, 2016). The main question is that of their age. Certainly they cut across the bedding in the underlying bedrock (Figure 8.15B), but French (1976, 2013) questions their age and conditions of formation, since they must have developed through multiple cycles of climatic change. Until adequate measurements of the rates of the processes currently involved in the changing landscape in these environments are available, their precise origin is nonproven. Textures of the surface sediments are usually sandy loams or loams and over-saturated clays, and the movement takes place throughout the summer.

8.4 CRYOGENIC FAST FLOWS

Large-scale fast flows such as debris flows are found in all environments where there is an accumulation of sediment on a slope that periodically becomes saturated with water, usually following heavy rains. Formerly glaciated mountains usually exhibit steep valley walls that are particularly vulnerable. Snow and slush avalanches and active layer detachment failures will be dealt with separately since they involve both flowage and sliding.

8.4.1 Cryogenic debris flows

A *debris flow* is where debris from high on a slope suddenly flows down the side of the slope of the hill or mountain, coming to rest as a fan at the base of the slope. It may also be called a *mudflow* if the material is fine-grained. Although most studies of debris flows have been carried out in areas lacking permafrost, they are a common landform in permafrost areas. Debris flow fans can be readily identified by their hummocky form, and the dense, unsorted debris of which they are composed. On aerial photographs, the deposits from successive flows can readily be seen. Numerous fans of this kind are found along the Slims River valley south of Kluane Lake, Yukon Territory (Harris & Gustafson, 1988, 1993). Curiously, these landforms have not been reported very frequently from other regions with permafrost, yet they are particularly well developed in the Coastal Ranges and Rocky Mountains of the northern part of the North American Cordillera. They are also known from the French Alps (Van Stein *et al.*, 1988) and from the Austrian Tirol (Aulitzky, 1970), and are also common in Russia on the Yamal Peninsula (Figure 8.18).

In the mountains of southwest Yukon Territory, there is a developmental sequence from retrogressive thaw slumps (see below), via debris flows, to the development of fluvial valleys in young glaciated mountain ranges (Harris & Gustafson, 1993). Early stages in the development of debris flow valleys are seen along the south side of the Summit Lake Pass, 100 km west of Fort Nelson (Figure 8.19). There, the sudden increase in mean annual air temperature by 5.8°C in ten years (Harris, 2007, 2009) has triggered thawing of the permafrost, resulting in the initiation of debris flow gullies along the steep slopes of the Pass in glacial tills.

Along the Slims River valley, melting permafrost of the headwalls of the steep side valleys results in the occurrence of numerous debris flows and associated fans

Figure 8.18 Mudflow in the Central Yakutia, Russia. There they occur on slopes ranging from 15–30°. © A. Brouchkov.

Figure 8.19 Embryonic cryogenic debris flows developing as a result of thawing of permafrost, Summit Lake Pass, Alaska Highway, British Columbia after an increase in MAAT of about 6°C. © S. A. Harris.

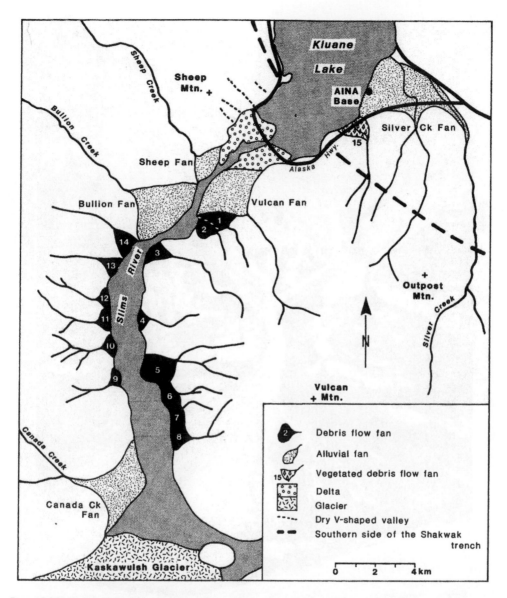

Figure 8.20   Debris flow fans along the Slims River valley, Kluane National Park, Yukon. From Harris and Gustafson (1993, Figure 2).

(Figure 8.20). A typical example is the Vulcan debris flow valley (Figure 8.21) where melting permafrost contributes both sediment and water to the accumulation area, and this is sometimes augmented by rainfall events. The timing of the flows on the nearby mature Sheep Mountain fan was approximately once in about 150 years (Harris & McDermid, 1998), based on eight debris flow events since the deposition of the White River ash at c. 1220 B.P. The thickness of debris deposited on the fan is about 5 m for

Figure 8.21 Oblique aerial photograph of the Vulcan debris flow fan in the Slims River valley.
© S. A. Harris.

each event, compared with c. 30 cm for the very frequent debris flows on the young Vulcan Debris flow fan.

The Vulcan debris flow fan is typical of a young, very active cryogenic debris flow fan (Harris & Gustafson, 1993). Figure 8.22 shows the variations along the valley.

Figure 8.22 Photographs of the Vulcan Debris Flow fan showing: A, the lower part of the accumulation area above the constriction; B, the levees along the sides of the constriction; C, the upper part of the accumulation zone; D, the debris flow sediment burying the bases of the dead spruce trees.

The upper part of the valley is V-shaped with 20° slopes cut into ice-rich sediments with negligible vegetation cover (Figure 8.22C). Thawing of the exposed ice allows large quantities of the sediment to accumulate above a narrow, steeper section (Figure 8.22A). When the sediment in the upper basin becomes sufficiently lubricated by water from rains, melting snow and permafrost, it loses cohesion and flows quickly through the constriction in the gorge below (Figure 8.22B), flowing out on to the valley floor below to form a debris flow fan. The water content of the flowing debris in the Vulcan debris flow ranged from 22 to 27%, *i.e.*, it exceeded the liquid limit for the sediment (plasticity index 3%). The events tend to occur during periods of hot weather and the rainfall events (Figure 8.23). In other cases along the same valley, the sediment comes from frost action breaking up the rocks in the headwall above the accumulation area, *e.g.*, in the case of the Sheep Creek debris flow.

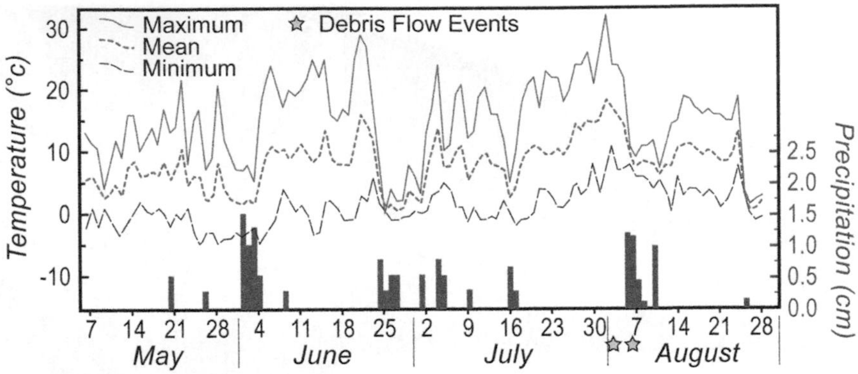

Figure 8.23 Timing of the movement of debris flows on the Vulcan debris flow fan in 1984. The solid bars indicate precipitation events. After Harris and Gustafson (1993, Figure 4).

Table 8.3 Characteristics of the five debris flows observed on the Vulcan Debris flow fan in 1994–5. From Harris and Gustafson (1993, Table 4).

Date of flow	*Volume (m^3)*	*Thickness on fan (cm)*	*Viscosity (poises)*	*Mean velocity (ms^{-1})*	*Peak discharge ($m^3 s^{-1}$)*	*Water content % (W_w/W_s)*	*Solid content % (W_s/W_t #)*
23 May, 1984	18	1–5	344	–	–	22.4–27.6	78.4–81.7
3 August, 1984	37.5	1–5	–	–	–	22.6	81.4
7 August, 1984	6000	15	8228	5.0–6.3	21–36	–	–
August, 1984, Fan 3	5250	1–30	–	–	–	–	–
6 May, 1985	0.375	–	1185	–	–	–	–

Where W_s = weight of solids – W_w = weight of water – W_t = total sample weight

Speed of movement has been measured to be at least 6.3 ms^{-1} along the Vulcan debris flow (Table 8.3), but higher speeds up to 14–17 ms^{-1} have been recorded in cases reported from the Austrian Tirol (Aulitzky, 1970). The debris flow gullies in the Sheep Creek come from an upper collecting basin filled with the debris derived from the headwall on Sheep Mountain. This sediment has a higher liquid limit averaging 28.8% water, which may partly explain why the frequency of debris flows there is less.

Table 8.3 shows the characteristics of the five flows observed on the Vulcan debris flow which exhibited frequent movements of debris (from Harris & Gustafson, 1993, Table 2). The flows varied considerably in volume and viscosity, but the water content of the moving flow was fairly consistent. Levees were left along the channel through the narrow gorge (Figure 8.22B), the levees showing the maximum height of flow during the last events. On emerging from the mountain front, the debris spread out in a thin layer extending 50–300 m down the fan. The larger stones tend to be concentrated at both the leading edge and along the levees of the flows. On the fan, the stones tended to be left behind as reported by Sharp (1942). There was no preferred orientation to the stones. The actual processes operating on the slopes around the Vulcan debris flow fan are shown in Figure 8.24.

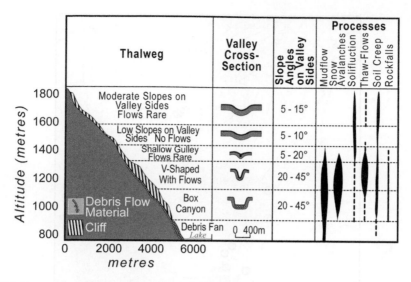

Figure 8.24 Processes operating along the thalweg of the Vulcan debris flow valley (from Harris & Gustafson, 1988, Figure 4).

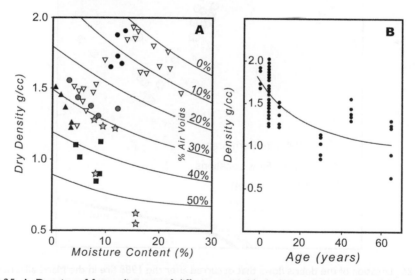

Figure 8.25 A: Density of fan sediments of different ages: black dots, one year old; open triangles, 5 years old; solid triangles, 10 years old; squares, 30 years old; open circles 45 years old; open stars, 65 years old. B: Dry density versus age for the same data from the Vulcan Debris Flow fan. From Harris and Gustafson (1993, Figures 5 and 6).

The deposits of debris on the Vulcan debris flow fan showed decreasing density with age (Figure 8.25). The trees that were in the path of the flow died within a year (Figure 8.22D), probably because the debris flow deposits set like cement and have no air voids (Figure 8.25B). Thus air and precipitation could not penetrate the ground. Gradually, the first seedlings start to appear in wind-blown loess on the surface of

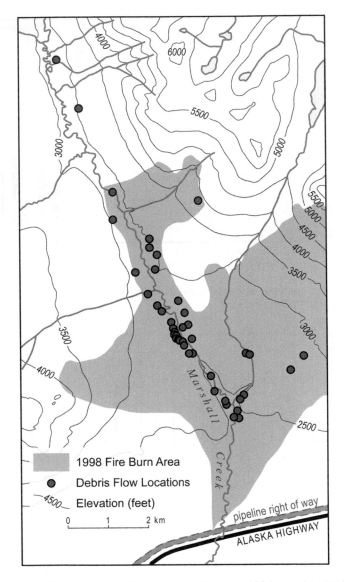

Figure 8.26 Location of the debris flows that occurred after the 1988 fire in the Marshall Creek basin, Yukon Territory (modified from Huscroft *et al.,* 2004, Figure 8).

the ground in the lee of logs after about five years. This is a common occurence, *e.g.*, Bardsley and Canon (1930); Finch (1937); Shroder (1978). The evergreens suffer more than willows and poplars. The vegetation had largely recovered on the Sheep Mountain fan, which had not had a major debris flow over its surface for over 100 years.

Permafrost degradation after the 1998 forest fire has caused numerous debris flows near the valley bottom in the Marshall Creek basin in the southern Yukon Territory (Figure 8.26, from Huscroft *et al.,* 2004, Figure 8). Similarly the increase in ground

Figure 8.27   Filter dams as defences against major damage by debris flows, Bormio, Italy. © S. A. Harris.

temperature following the fire caused by a lightning strike at Inuvik in 1968 caused a doubling of the thickness in the active layer during the next 20 years (Mackay, 1995).

Debris flows are particularly common where there is a source area with icy unconsolidated sediment. There is periodic activity after sufficient water is added. The movement takes place in the form of waves or pulses, usually 2–3 minutes apart, and they will continue until the supply of debris is exhausted. The levees are due to the bulldozing effect of the moving mass on the adjacent soft sediment, and to the loss of water from the margins of the moving mass. Large boulders can be transported and the resulting debris is unsorted. Movement can continue on slopes of <2° in the run-out zone.

Debris flows have often been reported from mountains lacking permafrost in New Zealand (Pierson, 1980), United States (Chalmer, 1935; Fryxell & Horberg, 1943; Sharp & Noble, 1953; Currey, 1966; Guy, 1971; Fairchild, 1987), Japan (Takahashi, 1981), Russia (Niyazov & Degovets, 1975), Scandinavia and Svarlbard (Rapp, 1985), European mountains, and western Canada (Owens, 1972; Naismith & Mercer, 1979; Hungr et al., 1984). They have similar characteristics to those found in areas with permafrost. The main difference is in the source of the water needed to trigger the movement. The sediment involved may consist of volcanic ejectamenta on volcanoes, e.g., Mount St. Helens, or an accumulation of loose, unconsolidated sediment near the top of mountains or hills. Debris flows in Russia, Italy and Chili are also sometimes generated when a rock or stone slide crosses a river. The water originates from precipitation or melting snow causing oversaturation of the sediment, resulting in it becoming a fluid that then flows down the valley.

Defences against debris flows include the use of a series of low dams across the valley floor near Bormio (Figures 8.27 and 8.28C). Other defences can include gently

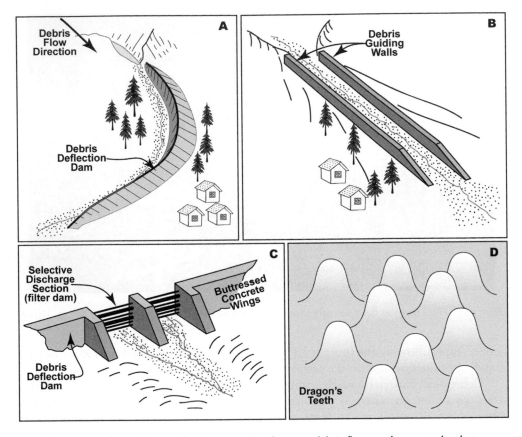

Figure 8.28 Several types of defences against frequent debris flows and snow avalanches.

curved levees leading the flowing debris away from a given area (Figure 8.28A), or concrete-lined channels to control where they move past a critical piece of infrastructure (Figure 8.28B). These two controls can also be used for redirecting snow avalanches, while Dragons Teeth (Figure 8.28D) can be used to make snow avalanches stop quickly.

8.4.2 Cryogenic slides and slumps

Slides are landforms where the upper part of the ground moves down-slope on a single, definite plane of movement, yet remains essentially intact. They are normally called *landslides*, though the plane of movement is often curved, in which case, they are also known as *rotational slides*. When the material involved consists of blocks of rock, it is referred to as a *rockfall*, due to the rock breaking up during the downslope movement over uneven surfaces. Cryogenic landforms involving ice or water movement over permafrost are common in cold climates, especially in glaciated areas. McColl and Davies (2013) described a case where the higher density rock along the side of a glacier in

New Zealand commenced squeezing a glacier along part of its course. It is well known that when glaciers retreat, rocks and landslides often develop along the over-steepened slopes of the valleys as a result of debuttressing. Likewise, cryogenic landslides on to the top of glaciers are common, especially after earthquakes. Fischer *et al.* (2013) have recorded cryogenic slope failures on the upper slopes of Monte Rosa in Switzerland, apparently induced by increased mean annual air temperatures. Nishu and Matsuoka (2012) described a rock slide in Japan that reactivated about six years after its initial movement. The result is episodic, large scale erosion.

Where there is more than one plane of movement in a slide, the feature is referred to as a *slump*. In the natural state, they can both be generated on any steep slope. The first stage is usually the development of cracks in the ground at the crown of the slide. This usually occurs on steeper slopes as a result of undercutting by rivers, *e.g.*, along the Takhini River just west of Whitehorse, Yukon Territory (Huscroft *et al.*, 2004), and in areas of human disturbance. Factors causing landslides include geological structure, seismicity, river migration, recent glacial retreat, extreme rainfall events and human disturbance. They occur throughout the hilly and mountainous areas with permafrost.

The geology is important since it determines the properties of the surficial materials. Silts and clays tend to have very high ice contents and are therefore prone to failure when partially thawed. Rocks with bedding parallel to the slope are also prone to slides along bedding planes, permitting the rocks above the bedding plane to slide if water lubricates a particular plane. Heavy rainfalls can trigger slides (Evans & Clague, 1989), as can destruction of vegetation by fire or removal of lumber. This increases the soil temperature, resulting in deeper thawing. Since ice tends to be concentrated at the surface of the permafrost table, deeper thawing releases large quantities of water that can cause slides, slumps or flows. Most mountains in cold climates have been under-going glacial retreat and the valley walls are over-steepened and therefore vulnerable to collapse. In addition, the mountains around the Arctic Basin and in parts of the North American Cordillera are still tectonically active.

Clague (1981) described a major cryogenic rockfall at the south end of Kluane Lake, Yukon Territory. Sheep Mountain contains permafrost in the ground and the slope is precipitous since it is located alongside the fault marking the south side of the Shakwak Trench. In addition, the slopes have been steepened by erosion by the late Wisconsin glaciers. Slumps are also particularly common along road embankments, as well as the shorelines of lakes, *e.g.*, the Beaver Creek embankment failure in September, 2006, whereas flows occur in loose sediments. Cryogenic rockfalls occur in solid rock.

8.4.3 Cryogenic composite slope failures

These include snow and slush avalanches, retrogressive thaw slumps and active layer detachment slides (Figure 8.29). In each case, the movement starts as a slide or slump, but evolves into a flow as the movement downslope takes place. Snow and slush avalanches occur throughout the colder mountain regions, but retrogressive thaw slumps and active layer detachment slides are unique features only found in permafrost areas. Together, these last two types of slope failure account for more erosion than any other geomorphological process acting in permafrost regions, especially during times of climate warming. They are the result of reduction in shear strength, and this is par-ticularly common in sediments with a high silt or clay content as well as large quantities

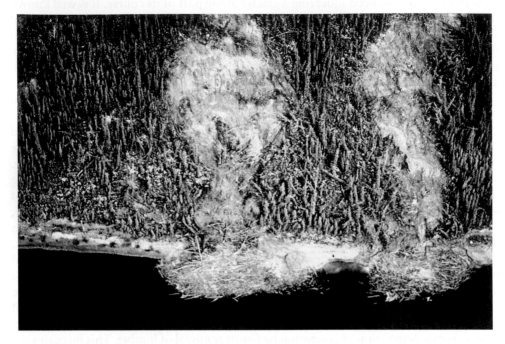

Figure 8.29 Elongate active layer detachment slides along Kluane Lake, Yukon Territory.

of segregated ice overlying ice-rich permafrost. They take place where the environmental conditions permit the moisture content of the thawed material to exceed its liquid limit, and in sediments with a high clay content; this can produce mudflows.

8.4.3.1   Active-layer detachment slides

An *active-layer detachment slide* consists of a mass of sediment on a slope that becomes detached from the surrounding soil and slides downslope on the surface of the still-frozen part of the active layer. As it moves, it tends to break up. The cause of the movement is the accumulation of water in the layer immediately above the still frozen ground becoming sufficient for the water content to exceed the liquid limit and thus starting to behave like a fluid (Lewkowicz & C. Harris, 2005a). It commences moving down-slope, taking the overlying drier but thawed soil with it. The movement ceases when it reaches flatter land, or when enough water drains from the soil to cause it to cease its motion. Along the way in elongate forms, the material breaks up so that lumps and blocks become stranded along its trajectory (Figure 8.30). The water comes from downward percolating surface water and from melting ground ice. When fire destroys the vegetation cover, it results in greater active layer thickness which aids in their formation (Harry & MacInnes, 1988), *e.g.*, at Haeckel Hill, 7.5 km northwest of Whitehorse (Huscroft *et al.*, 2004, p. 111). Climate warming can trigger their movement (*e.g.*, Carter & Galloway, 1981), as can major summer rainfall events (Cogley & McCann, 1976). The plane of movement is not usually the surface of

Figure 8.30 Remaining surface of the ground exposed by the passage of an elongate active layer
detachment slide.

the permafrost table except later during the summer. An alternative name for them is
a *skin flow.*

Two types of *active-layer detachment slides* occur, *viz.*, compact ones along
streams, and elongate ones that can develop anywhere on the landscape. Figure 8.29
shows examples of elongate active layer detachment slides. Length to width and length
to depth ratios are similar at all sites. The depth of the shear plane is closely controlled
by the local environmental conditions. Previous failure or weathering loosening the
ground tends to aid in their formation. Headwall recession can occur in some cases.
The actual surface left is often quite irregular and uneven. (Figure 8.30). C. Harris
and Lewkowicz (1993a, 1993b, 2000) provide more details of their form, internal
structure, and stability. Once formed, precipitation results in enhanced erosion until
the scar becomes revegetated (Kokelj & Lewkowicz, 1998).

Active-layer detachment slides can occur in areas of either continuous or discon-
tinuous permafrost, and there is an abundant body of literature describing examples
from northern North America (Anderson *et al.*, 1969; Hughes *et al.*, 1973; Mackay
& Matthews, 1973; McRoberts & Morgenstern, 1974a, 1974b; Hodgson, 1977;
Carter & Galloway, 1981; Stangl *et al.*, 1982; Mathewson & Mayer-Cole, 1984;
Harry & MacInnes, 1988, Aylesworth & Egginton, 1994; Aylesworth *et al.*, 2000;
Dyke, 2000; Lewkowicz & C. Harris, 2005a, 2005b). They have also been described

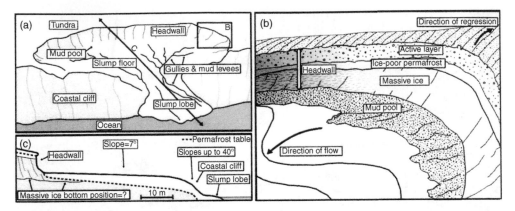

Figure 8.31 A polycyclic retrogressive thaw slump partly growing in the stabilized floor of a former retrogressive thaw slump. From Lantuit and Pollard (2008, Figure 2).

from the Yamal Peninsula in Western Siberia (Leibman, 1995; Leibman & Egorov, 1996; Leibman *et al.,* 2003, 2014) as a type of cryogenic landslide.

8.4.3.2 Retrogressive thaw failures

Retrogressive thaw flow slides (ACGR, 1988), now referred to as *retrogressive thaw slumps* (van Everdingen, 1998) are complex cryogenic features consisting of three main elements (Figure 8.31). First there is a vertical or subvertical headwall consisting of the active layer in ice-rich permafrost and ice-poor organic or nonorganic sediments. Below this is a headscarp at an angle of 20–50° which is retreating due to the ablation of ice-rich permafrost sediments due to solar radiation and sensible heat fluxes (Figure 8.32). In front of this is the relatively gently sloping floor consisting of the resulting sediments that are flowing downslope to the toe of the landform (Lewkowicz, 1987; de Krom, 1990). These sediments usually cover the *ice slumps* (Mackay, 1966), *thermocirques* (Czudek & Demek, 1970), *tundra mudflows* (Lamonthe and St.-Onge, 1961), *retrogressive flow slides* (Hughes, 1972), and *bi-modal flows* (McRoberts & Morgenstern, 1974).

They occur in two situations, *viz.,* where wave erosion along a coast exposes massive ice in coastal bluffs which undergo ablation (de Krom, 1990; Lantuit & Pollard, 2008), and along the banks of northern rivers and lakes (Brooker *et al.,* 2014). Figure 8.32 shows a typical example of a retrogressive thaw slump along the coast of the Yamal Peninsula in Russia, while Figure 8.33 provides an example of a polycyclic retrogressive thaw slump from along the Mackenzie River. They can also be initiated by forest fires (Johnson & Viereck, 1983) or by terrain disturbance associated with the use of seismic lines, oil fields, or by road construction (Bliss & Wein, 1971; Walker *et al.,* 1987).

Along the Mackenzie River system, they have often ceased to continue to erode the ground after 30–50 years during the 20th century, Studies of these landforms along the northern coast of the West Siberian Plain indicates durations of activity of

Figure 8.32   Headwall of retrogressive thaw slump along the Aldan river of the Central Yakutia, Russia. © A. Brouchkov.

at least 250–>300 years (Leibman *et al.*, 2014) and a more complicated cycle. The oldest dated retrogressive thaw slump there is c. 2250 radiocarbon years old. New studies by Kokelj *et al.* (2015b) on the Peel Plateau in northwest Canada demonstrate that the increasing summer precipitation in the area is resulting in greatly increased slump activity in ice-rich glaciogenic sediments and icy permafrost. There are increased numbers and sizes of slumps and debris tongue deposits. The slumps show diurnal pulses, while mega mass movement events occurring after rainstorms can displace streams and create new slumps.

Polycyclic retrogressive thaw slumps also occur along rivers (Mackay, 1966; Wolfe *et al.*, 2001). They consist of a new slump developed within the boundaries of a previous retrogressive thaw slump (Figure 8.33). As the original slump erodes the slope, supersaturated sediments and water form pools that move downslope in leveed channels, eventually flowing into the river or lake. Evaporation of water allows the sediments to dewater, leaving behind a thick layer of cohesive mud. Thus there is a definite zonation as it then develops into a dry mass. The plastic, dewatering zone extends about 30–40 m in very active slumps. The dry mass can then slowly be recolonized by plants. However, if erosion or other disturbance exposes the underlying massive ice, a new retrogressive thaw slump will develop.

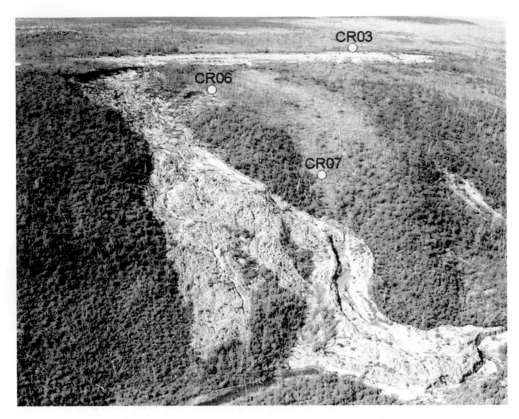

Figure 8.33 A polycyclic retrogressive thaw slump partly growing in the stabilized floor of a former retrogressive thaw slump. Reproduced from Singhroy *et al.* (2010, Figure 3).

In the case of retrogressive thaw slumps developed north and south of Inuvik along the Mackenzie valley, monitoring of regression rates of 18 slumps suggests that for the topography and ice content of the lacustrine sediments, the rate of retreat and the height of the head scarp are closely related (Figure 8.34). Wang (2011) found that the average maximum scarp wall recession rate of newly developed retrogressive thaw slumps was 12 m/a^{-1} but ranged from 8.5 to 15.5 m/a^{-1} for the 18 sites. The recession rate increased with the height of the scarp wall, though at a diminishing rate until about 15 m. Slope of the ground did not correlate with either the recession rate or the height of the scarp wall. The retrogression rate per year in m/a^{-1} (R) is related to the height of the scarp wall (H in m) as follows:

$$R = a + b * e^{(-H/c)} \tag{8.2}$$

where: a, b, and c are fitting constants with average values of a = 12.0, b = 28.5, and c = 2.6.

Figure 8.35 shows the two typical thaw slumps (G1 and G2) that were monitored. However, as the retrogressive thaw slumps extend landwards, the increasing distance results in reduction of the height of the headwall due to the need to maintain an

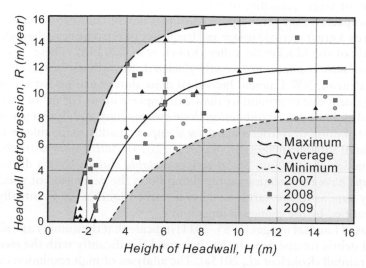

Figure 8.34   Headwall height relative to headwall regression during the early stages of the development of retrogressive thaw slumps along the lower Mackenzie River Valley (redrawn from Wang, 2011, Figure 3).

Figure 8.35   Retrogressive thaw slumps along the Mackenzie River valley (from Wang, 2011, Figure 1).

adequate slope (about 7°) for the saturated sediments to flow away. Regression ceases when the headwall is no longer in icy sediments, either due to their absence or after about 30–50 years. Their revegetation has been studied by Lambert (1972a) and Burn and Friele (1989).

In areas of large quantities of buried ground ice, increase in rainfall will have significant impacts on the geomorphology of permafrost landscapes. Since 1985 A.D., there has been a great increase in size and frequency of retrogressive thaw slumps along the west side of the Mackenzie Valley (Kokelj *et al.,* 2015b). These "mega slumps" (5–40 ha) are now common in formerly glaciated, fluvially incised, ice-cored terrain of the Peel Plateau, NW Canada. Individual thaw slumps can persist for decades and their enlargement due to ground ice thaw can displace up to 10^6 m^3 of materials from slopes to valley bottoms, reconfiguring slope morphology and drainage networks.

The distribution of retrogressive thaw slumps is readily mappable by using aerial photographs or satellite images (Brooker *et al.,* 2014). *Active slumps* are characterized by a wet, muddy slump floor. *Stabilizing* and *stable slumps* have a dry slump floor. *Relict slumps* have revegetation on the slump floor, the vegetation of which becomes increasingly similar to the surrounding undisturbed landscape. Eventually, they can only be identified by the change in topography.

Analysis of Landsat images (1985–2011) indicate that the number and size of active slumps and debris tongue deposits has increased significantly with the recent intensification of rainfall (Kokelj *et al.,* 2015a). The analyses of high resolution climatic and photographic time-series for summers 2010 and 2012 shows strong linkages amongst temperature, precipitation and the downslope sediment flux from active slumps. Thawing of ground supplies meltwater and sediments to the slump scar zone and drives diurnal pulses of surficial flow. This sediment is stored on the floor of the slump until heavy rainfall stimulates major mass flow events with the accumulated mud being carried down stream, sometimes resulting in a lateral shift in the stream course. This removal of sediments from the slump scar zone helps to maintain a headwall of exposed ground ice, perpetuating slump growth and leading to further retreat of the headwall. The development of debris tongue deposits divert streams and increases thermoerosion to initiate new slumps in adjacent valleys. These slumps also affect the chemistry of the salts in solution in the streams and rivers draining the area. The sulphate:chloride ratio in the streams changes with the quantity of water being supplied by the melting ground ice (Kokelj *et al.,* 2013).

On the Arctic coast of the USSR, retreat of the headwalls can reach 10 m/a^{-1} (Are, 1983), and rates along the shores of the Canadian Arctic Islands range from 6–8 m/a^{-1} recorded from E. Banks Island (French, 1974) up to 14 m/a^{-1} along the coast of southern Banks Island (Lewkowicz, 1987). In coastal situations, they occur where the rate of retreat of the headscarp exceeds that of the general coastal retreat. They are normally initiated along coasts by coastal erosion uncovering massive ice, resulting in increased ablation (Figure 8.36, after Lantuit *et al.,* 2012), but they can develop from active layer detachment slides along coastal bluffs (Lewkowicz, 1990). Inland, the rates can be as low as 1–3 m/a^{-1} on the Colville Delta, Alaska, ranging up to 16 m/a^{-1} near Mayo (Burn & Friele, 1989). Similar rates are recorded from the Yamal Peninsula (Leibman *et al.,* 2003).

8.4.3.3 *Snow avalanches and slushflows*

Although they are not directly related to permafrost, snow avalanches and slush flows are significant in that they move downslope soils, rocks, trees, and anything else that gets in their way. They erode the ground in the upper part of their path, depositing

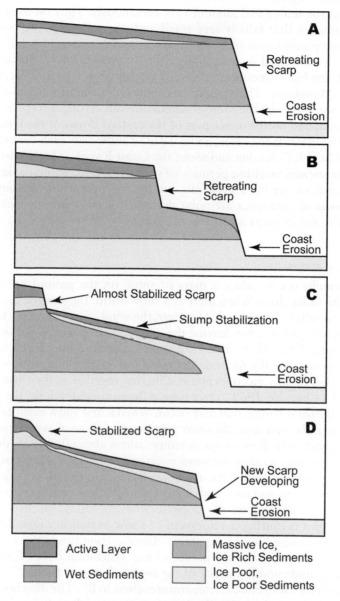

Figure 8.36 Cycle of slump initiation, stabilization and re-initiation of retrogressive thaw slumps along the Yukon coast. Modified from Lantuit *et al.* (2012, Figure 2).

the materials, mixed up with the snow in the runout zone. They are the reason that the most diverse vegetation in terms of numbers of species in a given landscape is found where they come to rest. They also cause great problems for the works of Humans in mountainous areas, *e.g.*, the Rocky Mountains, Alps and the Caucasus, as well as in areas of maritime permafrost such as Iceland (Decaulne & Saemundsson, 2006).

Snow falls throughout the cooler regions of the world that experience cryological conditions, but the timing and amount vary enormously. Thus on the Qinghai-Tibet Plateau, the amount that falls is very small and generally occurs in summer. Likewise, the winter precipitation in continental lowland climates is usually low, *e.g.*, on the Prairies in Canada and in central Siberia. The amount of precipitation increases enormously in the higher mountains, resulting in glaciers, *e.g.*, in Tibet, the Alps and in the Rocky Mountains. The highest precipitation is found particularly in coastal mountainous areas, though the Monsoon lands are not cryotic, except in the maritime Northeast China, and on the upper part of the eastern slopes of the Qinghai-Tibetan Plateau. Further towards the Poles, about 10 m of snow falls each year on the Selkirk Mountains of British Columbia and along the Coast Range of both British Columbia and Alaska. Areas with maritime permafrost such as Iceland, Scandinavia, Switzerland and the Caucasus receive large snowfalls, but the central part of the Canadian Arctic Islands and parts of Antarctica are polar deserts. It is the maritime climates that are particularly affected by snow and slush avalanches.

8.4.3.3.1 *Snow avalanches*

A *snow avalanche* occurs when a mass of snow on the ground breaks loose and descends to the valley floor. When there is a snowstorm, the snowflakes form a cover over the ground called a *snow pack*. However, the wind causes the snow to accumulate in depressions and gullies, often leaving the crest and steep windward slope of the hill or mountain bare. Where there are cliffs downwind of the mountain crest, a cornice of overhanging snow is usually formed. The wind compresses the snow, while the individual snowflakes rapidly metamorphose, adhering together as they lose their original crystal form (La Chapelle, 1969). Over time, a dense crust develops, the process being aided by any diurnal warming that may occur. When a new snow event takes place, the junction between the new and old snow represents a plane of weakness, over which the upper layer can slide. Rain or warm temperatures above 0° C, *e.g.*, during a warm chinook or foehn event weaken the snow pack and can trigger release of the snow. By the spring, the basal snow will have metamorphosed into round ice pellets, leaving an air space between them and the overlying snow pack. As a result, the overlying snow can break loose and come down the slope.

The conditions favouring development of snow avalanches include a steep slope with at least part exceeding 33°, suitable climatic conditions, the topography, the prevailing wind direction, and the nature of any vegetation cover. Climatic controls include the magnitude and frequency of the snowfall, *e.g.*, >10 m/a^{-1} at Rogers Pass in British Columbia with periodic temperatures close to 0° . The direction and speed of the wind controls the resulting distribution of snow, *e.g.*, filling depressions and gullies. Vegetation cover is important since trees tend to inhibit avalanching. Fire, destroying the tree cover, resulted in a 50% expansion in the area undergoing avalanching in the Vermilion Pass in the Banff-Kootenay National Parks, Canada (Winterbottom, 1974, see Figure 8.37).

Topography is also very important. Lee slopes with cliffs over 33°, mountain ridges at right angles to the prevailing wind (Figure 8.38), and gullies (Figure 8.39) are the places most prone to avalanches (Schaerer, 1972). In the Rogers Pass, 36 out of 67 sites are the result of steep cliffs releasing snow which then sets the snow pack

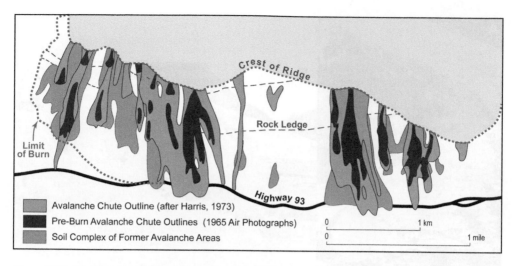

Figure 8.37 Distribution of avalanches before the 1966 burn, in 1973, and the distribution of avalanche soils from the last Neoglacial period (pre-1900) in the Vermilion Pass, Kootenay National Park, B.C. (modified from Winterbottom, 1974).

Figure 8.38 A. Lee cliff avalanches in the Vermilion Pass, Kootenay National Park, B.C. in winter, and B. in summer. Note in B, the three zones of acceleration over rock and soil, transit over shrubs, and the run-out zone. © S. A. Harris.

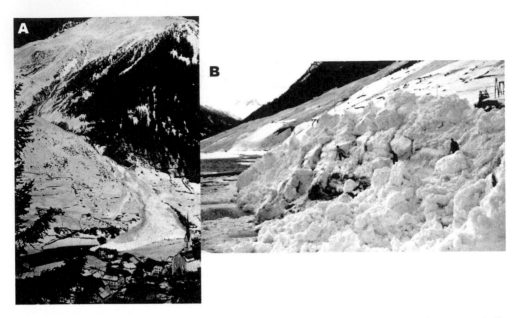

Figure 8.39 A. Wet gulley avalanche at Ischgl, Silvretta Alps, Austria, and B, the pile of giant snowballs in the run-out zone. © S. A. Harris.

below moving with it. Cornices on lee slopes are responsible for 9 avalanche paths, while release of deep snow in gullies cause 15 avalanche paths. The other 7 sites have multiple causes. The gully avalanches are particularly dangerous since they can have multiple releases during a period of avalanching, unlike the other types. The steeper the slope, the more frequent the avalanches (Figure 8.40). Avalanches commence slowing down and depositing their load on slopes lower than 28° (McClung, 2003). More detailed study of avalanches in several different Canadian avalanche areas by McClung indicates that avalanche frequency depends both on steepness of the slope as well as snow supply. Average avalanche magnitude appears to depend on terrain steepness, snow pack conditions in the starting zone, track confinement and the total vertical drop of the avalanche path, An increase in snowfall in the Rocky Mountains in 2013 and 2014 has resulted in the opening up of new avalanche paths, but an earlier decrease in snowfall did not alter the frequency of average avalanche events. Extreme snowfalls produce surprisingly large and frequent avalanches (Schaerer, 1988).

The loading of deep, fresh, powdery snow in the fall can cause avalanches, while fresh snowfalls over metamorphosed snow during winter often trigger avalanche release. In the spring, the snow overlying a pocket of air and ice pellets often detaches and comes down the mountainside. Typical avalanche slopes in the Rocky Mountains are shown in Figure 8.38 in winter and in summer. The part of the avalanche slope with shrubs (Figure 8.38B) indicates that the snow slides over the top of the vegetation which is packed with snow in winter. The actual path consists of three parts, *viz.*, the upper part (starting zone) where the snow accelerates and erodes the underlying ground, the middle part (track) where it keeps a fairly constant speed, and the run-out

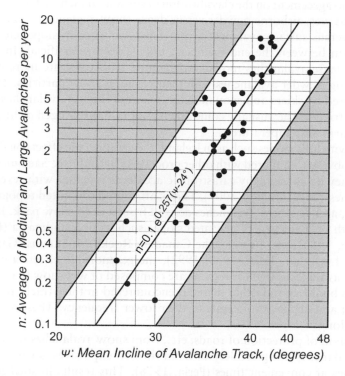

Figure 8.40 Relationship between mean slope angle and frequency of avalanches per year at Rogers Pass, based on 12 years of observations (modified from Schaerer, 1972, Figure 4).

zone where the snow and debris (rock, soil and vegetation) that it has brought down come to rest. The resulting deposits are characterized by a lack of soil formation and consist of the mixed-up material transported downslope by the avalanches. In wetter, more humid climates, the snow descends as giant snowballs rolling slowly downslope (Figure 8.39).

Speed of movement depends on the conditions. It can reach 90 m/sec. in the case of dry, powdery avalanches, but is least in the case of wet avalanches. In extremely wet, maritime climates, the snow may descend as giant snowballs, rolling slowly down slope to accumulate in a massive, jumbled pile in the runout zone (Figure 8.39B). In other cases, any slabs quickly break up with the movement and the snowy mass slides, tumbles, bounds and flows down the slope. Powdery avalanches tend to flow, engulfing anything in their path. They are shrouded in a cloud of blown snow crystals. Voellmy (1955), Meers (1976), Leaf and Martinelli (1977), Hopfinger (1983) and Platzer *et al.* (2007) discuss the avalanche dynamics in more detail. Wind is associated with the movement, moving at a similar speed to the snow due to the displacement of the air. This wind can blow over mature trees in front and around the avalanche path. Kotlyakov *et al.* (1977) reported an air-wave moving at greater speed than the main snow in the avalanche after the latter hit a dam. The acceleration effect on the air is discussed by Hopfinger and Tochon-Danguy (1977).

There is no agreement on the classification of snow avalanches. They can be divided into dry versus wet, airborne powder versus dense-snow avalanches, slab versus powder, ground versus surface avalanches, channeled versus open-slope, etc. Since there is a continuum between the end members, the classes are artificial. In general, the slower, wet, dense avalanches moving in channels tend to create more erosion than the fast-moving powder avalanches with their cloud of snow particles. The latter are readily triggered by skiers and snowmobilers whereas the wet avalanches are difficult for people to trigger. They usually come down after rain, prolonged melting by the sun or by very warm air temperatures.

In areas with hilly slopes such as in Iceland, the geomorphic effects on the landscape are not very obvious compared with other processes (Decaulne & Saemundsson, 2006, p. 90), but they do cause many catastrophic accidents together with severe economic costs (Jóhannesson & Arnalds, 2001). This is attributed to limited topographic range (300–700 m), and dry snow avalanches derived from a deep snow pack. In mountainous areas with over-steepened, glaciated slopes such as in Scandinavia (Rapp, 1960, 1995; Corner, 1980), the Alps and the Canadian Rocky Mountains (Gardner, 1970, 1983, 1989; Luckman, 1977, 1978, 1992), the geomorphic effects are much more obvious. The avalanches in these regions are composed of wet snow that. is capable of moving large boulders and other material encountered in their track, resulting in far more erosion and deposition in spite of their lower frequency. These areas also have significantly longer slopes

One method of protection of roads, etc., from snow avalanches involves bringing down the avalanches using artillery or hand-thrown explosives, so as to bring down the avalanches at convenient times (Perla, 1978). This results in smaller avalanches provided that they are brought down after each heavy snowfall. Avalanche sheds are placed where avalanches regularly cross roads or railway tracks (Figure 8.41). "Dragon's teeth", consisting of piles of earth up to 15 m in height can be built in a pattern in the run-out zone to cause the avalanche to stop more quickly. However, there is no real defence if a given winter produces a record snowfall.

8.4.3.3.2 Slush avalanches

Slush avalanches represent an extreme form of snow avalanche where the snow has become saturated with water, whereupon the mass flows down slope (Figure 8.42). This is usually the result of extreme precipitation events in the spring (Washburn & Goldthwait, 1958).

Because of their high density, they can entrain large boulders and other debris which end up in the pile when the mass comes to a stop (Figure 8.43). Moore *et al.* (2013) show that they can break material loose from below the snowpack. Quantities of material entrained annually varied from $70 \, m^3$ to $15 \, m^3$ in the Matter valley in Switzerland during a four-year period, corresponding to a denudation rate varying between 0.05 and 0.01 mm/a^{-1}. When the snow melts, this debris is left as an upstanding pile (Rapp, 1960, 1985, 1995). They are particularly common in gullies, often with low starting area inclinations where water tends to collect in the snow pack (Eckerstorfer & Christiansen, 2012). As they move downslope, they leave levees along the side of the track, and the mass comes to a halt as a distinct flow lobe as in the case of debris flows. Clearly, they are intermediate in characteristics

Figure 8.41 An avalanche shed in the Rogers Pass, Canada. © S. A. Harris.

Figure 8.42 Slush avalanches in the Austrian Tirol.

between wet avalanches and debris flows. Lichenometric studies at the Kärkerieppe slush-avalanche fan at Kärkevagge, Sweden indicate that almost all the blocks on the fan arrived there less than 50 years ago (Bull *et al.*, 1995). In the Rogers Pass and elsewhere along the Canadian Cordillera, slush avalanches are normally seen as the

Figure 8.43   Close-up of the run-out zone of a slush avalanche. Note the piling up of the snow at its terminus, and the snow levees along its track. The soil, etc., caught up in the avalanche will be left behind as a hummocky mound when the snow melts.

snow melts in the spring, usually aided by rain and the sudden warming of the air temperatures.

The nature of the underlying ground determines the amount and nature of the material that is transported downslope. Over solid rock, the load will be coarse and limited in amount. Over loose sediments such as volcanic ejectamenta, the quantity of material transported will tend to be much greater. They are relatively rare in the High Arctic, owing to the sparse frequency of extreme catastrophic events, *e.g.*, André (1990) working in Spitzbergen estimated that they mobilize 1300 to 7000 m^3 of rock debris only once in every 500 years, which matches the frequency estimated by Nyberg (1985) in Lappland.

At lower latitudes, they are relatively common in the spring, and similar amounts of debris are moved with each event. Bull *et al.* (1995) studied the lichens on the boulders left by slush avalanches on a slush avalanche fan at Kärkevagge, Sweden and found that at least 24 slush avalanches flowed down to the middle of the fan between 1790 and 1950 A.D. Since 1950, the frequency of slush avalanches amounts to c. 50 per century. Thus they are currently particularly effective causes of mass wasting at lower latitudes, both over permafrost, *e.g.*, Sweden and the Alps (Barsch *et al.*, 1993), as well as in non-permafrost areas, *e.g.*, the Japanese volcanoes (Morohashi *et al.*, 2007).

8.5   RELATIVE EFFECT IN MOVING DEBRIS DOWNSLOPE IN THE MOUNTAINS

The relative effectiveness of the mass wasting processes in mountains depends on the topography, climate and lithology. Slush avalanches tend to occur in small gullies during the spring thaw, whereas snow avalanches are dominant in larger valleys and operate in fall to spring. Snow avalanches occur in most years whereas slush avalanches occur less frequently, *e.g.*, 20–30 year intervals at Mont Albert in the Gaspésie Mountains of Québec (Laroque *et al.*, 2001), but every 500 years in Spitzbergen (André, 1990).

Likewise, the frequency of debris flows and rock slides varies enormously. Rapp (1985) showed that the rock type had an enormous effect on the rate of erosion caused by alpine debris slides and flows in northern Scandinavia and Spitzbergen. Granite was much more easily eroded than amphibolite, while mica-schist was the least eroded in terms of debris produced each year, based on measured rates of rock denudation extrapolated in mm/1000a, and by assuming a recurrence interval of 200 years. André (1993, 1995) concluded that in northwest Spitzbergen, average rates of erosion by snow avalanches on massive gneiss was $0.007 \, mm/a^{-1}$ compared with $0.08 \, mm/a^{-1}$ for fractured mica-schist.

Earlier measurements at Kärkevagge in Northern Lappland gave different results (Rapp, 1960), although the results were probably more spectacular since extreme events were involved. Recurrence interval could not be taken into account, but the results indicated that debris flows and slides were second only to transport of solutes by running water. Slush avalanches, transport of rock debris and rockfalls represented about one fifth of the denudation by debris flows and slides, while solifluction removed even less sediment. His rate of $0.005 \, mm/a^{-1}$ for snow and slush avalanches contrasts with the results of long-term studies on limestone by Luckman (1977, 1978) in the Canadian Rocky Mountains of rates of deposition by snow avalanches of $5 \, mm/a^{-1}$. Luckman also concluded that rockfalls from cliffs resulted in far less denudation in the relatively dry, cold eastern Cordillera of the Rocky Mountains. Likewise, Gray (1973) also found similar relatively low rates of erosion of quartz-monzenite in the Ogilive Mountains of the Yukon Territory.

It must therefore be concluded that the rates of denudation by the mass wasting processes discussed in this chapter are highly dependent on the local microenvironment. It is necessary to measure the actual rates occurring rather than trying to use the information obtained in the studies presently available. The nature of the underlying ground determines the amount and nature of the material that is transported downslope. Over solid rock, the load will be coarse and limited in amount. Over loose sediments such as volcanic ejectamenta, the quantity of material transported will tend to be much greater. They are relatively rare in the High Arctic, owing to the sparse frequency of extreme catastrophic events, *e.g.*, André (1990) working in Spitzbergen estimated that they mobilize 1300 to $7000 \, m^3$ of rock debris only once in every 500 years, which matches the frequency estimated by Nyberg (1985) in Lappland.

Chapter 9

Landforms consisting of blocky materials in cold climates

9.1 INTRODUCTION

In all areas subject to freezing and thawing, the rocks exposed at the surface will be subjected to additional stresses by the extreme temperature changes (diurnal and seasonal) that occur in these places. Moisture in the rock greatly increases the stresses by phase changes and by chemical reactions between the minerals and the dissolved gases in the water, *e.g.*, carbon dioxide. Dissolved salts in the soil water can crystallize out and exert additional pressures inside crystal grains due to the forces of crystallization. Ice lenses may develop in the sediment or rock, and can break the material apart. As a result, there is enhanced weathering and production of coarse detritus (sometimes referred to as *cryogenic weathering*), the products of which can then commence moving down the slope. They may also accumulate on mountain tops. In Russia, these are collectively referred to as *kurums*.

Potential processes involved in cryogenic weathering of exposed rock include freeze-thaw effects, salt weathering (Williams & Robinson, 1981, 1991), wetting and drying (Pissart and Lautidou, 1984; Prick *et al.*, 1993), hydration shattering (Hudek, 1974), thermal fatigue (Hall & Hall, 1991), chemical weathering (Dixon *et al.*, 1984), and biological weathering (Hall & Otte, 1990; André, 1995b). Several processes usually operate in any given microenvironment, and which ones are dominant in individual cases depends on the rock type, microclimate, topography and the presence or absence of vegetation, usually lichens.

9.2 SOURCE OF THE BLOCKS

There are six main sources of the blocks that are found in the landforms. Firstly, frost jacking can cause blocks in a finer grained matrix to be ejected on to the ground surface by the growth of ice lenses in seasonal frost or in the active layer during the winter (Chapter 1). This process is the reason that farmers working land on boulder-rich sediments have to remove the rocks periodically from the fields in order to continue farming. Glacial till containing substantial numbers of boulders or rocks is particularly prone to exhibit this process.

The second possible process is the deposition of blocks by glaciers as boulder fields, which appears to have happened occasionally. Flint (1955, pp. 85–87) describes an area like this in South Dakota, although the deposit is thin.

Figure 9.1 Talus slopes and the Murtèl I rock glacier coming out of a cirque in the Upper Engardine, Graubünden, Swiss Alps. Note the transverse compression-extension ridges on the tongue-shaped rock glacier as well as the active talus cones to the right of the rock glacier. The talus cones are fed from gullies in the rock wall above. © S. A. Harris.

The main source of blocks of rock is by detachment from bedrock. This can either be due to frost heaving on a relatively flat surface (Dyke, 1978, 1984), ultimately producing a boulder field or felsenmeer, or it may involve breaking away blocks from an exposed bedrock outcrop, usually a cliff (Matsuoka *et al.*, 1997; Koštak *et al.*, 1998; Matsuoka, 2001a). Prick (2003) reported that in Svalbard, the rock temperature dropped below 0°C and remained there throughout the winter, but moisture content was high, particularly in the fall. Thermal shocks were absent. Fluctuating rock temperatures above and below 0°C were only measured on about four occasions in a year, usually in the early fall. Coupled with the high moisture contents, these diurnal freeze-thaw cycles resulted in frost wedging in massive limestones, although porous sandstones were not affected.

At lower latitudes, more detailed studies by Matsuoka *et al.* (1997) in the Swiss Alps showed that southern exposures and north-facing ridges experienced multiple diurnal freeze-thaw cycles, while north-facing rock walls were affected by seasonal freezing and thawing. Hasler *et al.* (2011) give further details of the temperature variability on steep alpine rock and ice faces. Matsuoka *et al.* (1997) recorded appreciable joint widening at the onset of seasonal thawing, when meltwater percolated down through joints into the cold, frozen bedrock, where it froze. They therefore concluded that in that environment, the moisture availability determined the magnitude of the effects of frost action on massive, jointed rock. Diurnal frost heave and creep are dominant on ridge crests, while basal debris slopes experience large amounts of frost heave during seasonal freezing and thawing. This results in small sorted stripes predominating on crest slopes while solifluction/gelifluction occur on the basal slopes. Frost wedging produces blocks that detach from the cliffs to form slope deposits towards the base of the slope. Where there is a deep snow pack in winter, the early fall is the main time when freeze-thaw cycles affect the underlying rocks.

The concept of water penetrating cracks and developing enough force during freezing in an unconfined space to break rocks apart has been challenged by several authors (Davidson & Nye, 1985; Walder & Hallet, 1985; Hallet et al., 1991; Murton et al., 2006). Both theoretical considerations (Taber, 1929; Dash et al., 2006) and experimental work (Akagawa & Fukuda, 1991; Murton et al., 2006) have indicated that water can move along unfrozen films into interstices in the rock, where it freezes to enlarge pre-existing ice lenses. This puts the rock under stress, while thawing results in potential collapse of the expanded material. Dixon et al. (2002) found evidence of small interstices in rocks resulting from chemical weathering which could permit ice lenses to form and expand.

Fourthly, collapse of over-steepened rock walls when glaciers retreat can result in the development of piles of blocks (André, 1997). Once the ice is no longer pressing against the rock walls, stress relaxation causes rock bursts and landslides to occur. Similar ideas have been suggested for some block slopes in the Swiss Alps, Scotland (Ballantyne & C. Harris, 1994; Ballantyne & Kirkbride, 1987) and elsewhere. Rapp (1960a) and Francou (1988) noted that present-day rates of cliff retreat away from present-day glaciers were considerably lower than the overall retreat inferred for the last 10 ka.

The last two sources of blocks are snow and slush avalanches. These are important processes in all locations except in places which lack appreciable winter precipitation. The snow is moved about by wind to form cornices above cliffs, while snow accumulating on cliff faces readily breaks loose and causes the snow pack on more gentle slopes beneath to swell the moving mass of snow (see Chapter 8). Any loose soil or rock originally located above the plane of movement is transported to the runout zone. In spring, the early warm rainfall sinks into the isothermal snow pack, eventually turning it into a fluid which then flows downslope as a slush flow, taking any loose rock with it.

9.3 INFLUENCE OF ROCK TYPE

The nature of the rock has a great effect on the rates of production of blocks in a given permafrost climate (Table 9.1). Massive, monomineralic rocks such as quartzite are very resistant to both physical and chemical weathering. Granite is very susceptible to weathering in cold regions since it is composed of feldspars, quartz and mica crystals (Dixon et al., 2002). The feldspar crystals are decomposed by carbonic acid (carbon dioxide gas dissolved in water), producing stresses and tiny cavities in the rock, and then ice lenses can fill them and grow, so breaking up the rock.

Chemical weathering is very active in high latitudes and cold climates (Rapp, 1960a; Dixon et al., 2002). Rapp concluded that more material was removed in solution than by any other process. Pyrite in shales oxidizes to produce sulphuric acid, which then reacts with many of the surrounding minerals producing soluble salts and

Table 9.1 Typical results of measurement of average rate of rockwall retreat in northern environments since deglaciation. Note that present-day rates are likely to be different due to changes in climate.

Location	Lithology	Average retreat period (years)	Rockwall retreat rates (mm/1000 years)			Reference
			Min.	Mean	Max.	
Central Spitsbergen, Mt. Templet, Bjonahamma	Limestone	10,000	340		500	Rapp, 1960c
Swedish Lappland and North Norway	Amphibolite and Granite	9,000	0		900	Rapp and Rudberg, 1964
Ellesmere Island, Arctic Canada	Limestone	8,000	500		1,300	Souchez, 1971
Yukon Territory, Canada	Igneous (syentie, diabase)	10,000	7	30	18.5	Gray, 1972
	Metasedimentary (slate, dolomite, quartzite)	10,000	20	170	73	Gray, 1972
Northern Finland	Schist, dolomite, granite, quartzite	9,500	40		940	Söderman, 1980
West Greenland, Disco Island	Basaltic volcanics	7,000	500		1,500	Frich and Brandt, 1985
North Gaspé Peninsula, P. Q.	Shales, argillites, siltstones	10,000–13,000		32		Hétu, 1986
Alaska	Granite	1,000	5		20	Hall and Otte, 1990
Svalbard	Amphibolite		0		4	André, 1997
	Quartzite		10		158	André, 1997

a residue of red iron oxide. Some minerals can suffer hydration, resulting in expansion of the mineral. In the case of the clay mineral bentonite, it can expand to 20 times its dry volume, causing major contortions in the structure of any sediments in which it occurs. Gypsum and halite can flow under pressure, just like ice, and so cause disruption of the surrounding beds. Datsko & Rogov (1988) showed that the anions present in the principal clay minerals changed after being subjected to 100 freeze-thaw cycles.

9.4 WEATHERING PRODUCTS

André (1995b, 1997) pointed out that the products of weathering in cold climates include soluble salts, small granules and flakes from rocks such as amphibolite and granite, as well as the large angular rocks resulting from frost heaving (Dyke, 1978, 1984), rock bursts, or collapse of angular blocks from the cliff face. Initially, these rock fragments tend to be angular regardless of size, but weathering does not stop when the material comes to rest. Dunn & Hudek (1966) and Hudek (1974) have described the process of hydration shattering in argillaceous material, while Konischev *et al.* (1975, 1976), Konischev (1982), and Konischev & Rogov (1993) have shown the breaking up of particles of rock into predominantly silt-sized grains in both laboratory and field experiments. Silt-sized quartz particles will show a conchoidal fracture as well as stress cracks when photographed under an electron microscope (Minervin, 1982). These small particles can accumulate in the spaces between the coarser clasts over time, or may be removed elsewhere by wind or water.

9.5 BIOGENIC WEATHERING

This is a subject that has only recently been fully explored since it requires special equipment. Crustose lichens will tend to colonize the surfaces of exposed basic rocks, and McCarroll (1990) and McCarroll & Viles (1995) have demonstrated that the acids produced by their thalli (including oxalic acid) can cause spalling and flaking of the substrate. The thalli of some lichens penetrate any crack in the surface of the rock and grow there (Figure 9.2A), undermining the hard, relatively unweathered parts of the surface rock (Figure 9.2B), which ultimately flakes off the surface (Figure 9.2C). This process is less effective in the case of acidic rocks such as granite, limestone, and quartzite, which most lichens avoid. During this biochemical weathering of the substrate, the lichens absorb many elements into their thallus (Cooks & Otto, 1990), while the soluble chemicals that are produced are removed in the runoff of water when temperatures, *etc.*, are suitable. Some additional elements are believed to be absorbed from the atmosphere.

Lichens are very slow to colonize limestones and dolomitic rocks, so that when these rocks have a good lichen cover, they must have been in place for a very long time. Lichens die if the substrate is turned over, so a good lichen cover on blocks implies stability of the substrate. Over time, the blocks undergoing weathering become less angular since the weathering processes attack the blocks from three sides on corners, and two sides on edges. The result is a small decrease in size but a big increase in roundness.

McCarroll & Viles (1995) found that the minimum rate of surface lowering by *Lecidea auriculata* on gabbro bedrock on morainic ridges at Storbreen, Jotenheimen that had been dated independently by Matthews (1974, 1975, 1977) was 1.2 mm/ka^{-1}. This is 25–50 times faster than the weathering due to other processes in the same area. Lichens also thrive on bedrock outcrops and rock surfaces in the maritime, unglaciated areas of Antarctica (Ascaso *et al.*, 1990). Several saxicolous species are involved, some merely physically weathering the minerals in the substrate without producing any new minerals, *e.g.*, *Rhizocarpon geographicum,* in contrast to their action in more temperate areas. Other species produce new minerals including dihydrate calcium oxalate, calcite, various clay minerals and allophane, which were absent in the unweathered rock. By contrast, Moses & Smith (1993) found that when the lichen, *Collema auriforma* growing on Carboniferous limestone was subjected to wetting and drying cycles, the lichen thallus plucked rock fragments from the substrate during the drying phase. Thus the weathering action of lichens can be physical and/or chemical. Chen (2000) provides a review of the earlier literature on weathering by lichens.

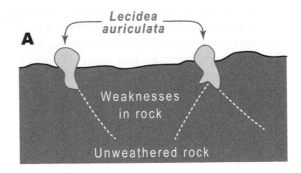

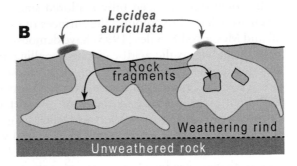

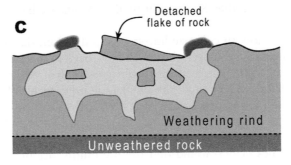

Figure 9.2 Weathering of rock by *Lecidea auriculata*: A, the beginning of growth, B, maturing lichen in the weathering rind enclosing rock fragments, and C, detached rock fragments (modified from McCarroll & Viles, 1995, Figure 2; see also McCarroll, 1990).

9.6 FATE OF THE SOLUBLE SALTS PRODUCED BY CHEMICAL AND BIOGENIC WEATHERING

The soluble salts produced by chemical weathering are mainly removed by the surface runoff or in the water percolating down to the water table in humid regions. However, some react with the minerals in the ground, *e.g.*, the presence of calcium carbonate causes montmorillonite to form among the clay minerals. Pyrite oxidizes to form limonite and sulphuric acid. The latter then reacts with a wide range of minerals, breaking them down. An area with sulphuric acid can be too acid for the growth of some plants, but caused the development of Spodosols near Kärkevagge in Swedish Lapland (Thorn et al., 2014). The reaction with other minerals can produce a white precipitate of relatively insoluble aluminium compounds along creeks there. At other times it produces crusts and efflorescences composed of sulphates. Sulphur dioxide and hydrogen sulphide make their way upwards along cracks and joints to the moist surface layers of the active layer, where they also oxidize to form sulphuric acid. This, in turn reacts with calcium carbonate to form euhedral gypsum crystals in areas with underground reserves of gas, e.g., the Foothills and Front Ranges of the Rock Mountains in Alberta.

In semi-arid areas, deposits of calcium carbonate form on the undersides of stones within the solum in the Foothills of the Rocky Mountains. In the arid steppes in China, a wide variety of salt deposits can be found forming saline lakes and dried-up playa lakes along the northern areas of the Qinghai-Tibet Plateau, e.g., at Golmud, Qinghai Province. The exact chemical composition depends on the rocks undergoing weathering in the surrounding mountains.

Under the extreme cold and arid climates in ice-free areas found in Antarctica, honeycomb weathering called *taffoni* is common. Salts also crystallize out in the interstices of granite resulting in granular exfoliation and disintegration. Selby (1971) describes the various forms of cavernous weathering that are found. The main salts present are dominated by sodium chloride, though this would not be expected to be produced by weathering of the granite. The granules of granite that are formed show no signs of weathering, so they cannot be the source of the salts. Since there are few freeze-thaw cycles and negligible soil moisture, it is postulated that the salts acted as nuclei for the snow crystals that get blown inland from the ocean and are deposited on the granite during the long, cold winter. The snow slowly sublimes leaving the salts deposited on the granite. During the Antarctic summer, the remaining snow briefly melts, permitting the salts to enter cavities in the granite in solution, where they crystallize out. The same process may occur along the coast in Northeast Greenland and in the cold arid coastal areas of the Canadian Arctic Islands. However, they do not occur in interior Yukon (c.f., French, 1996, p. 47).

9.7 RATE OF CLIFF RETREAT

In spite of all these processes, the rate of retreat of present-day cliffs is quite small, unless rock falls are involved (Table 1). The three principal processes operating on exposed rock are very slow, i.e., biogenic flaking due to lichens, moderate retreat resulting from frost shattering, and rapid retreat associated with post-glacial stress relaxation. André (1997) estimated these rates to be $2\,mm/ka^{-1}$, $100\,mm/ka^{-1}$, and $1000\,mm/ka^{-1}$ respectively, in northwest and central Spitsbergen. Overall rates varied from $0–1580\,mm/ka^{-1}$. Unfortunately, there are no other comparable studies of these three distinct processes at the present time. Matsuoka & Murton (2008) discuss the recent advances in frost weathering and suggest future directions for research, while Thorn et al. (2014) have argued that weathering and pedogenesis in cold climates are essentially an extension of these processes that are also encountered at lower latitudes with warmer climates.

9.8 LANDFORMS RESULTING FROM THE ACCUMULATION OF PREDOMINANTLY BLOCKY MATERIALS IN CRYOGENIC CLIMATES

These materials can accumulate on either flat or sloping surfaces. If there are substantial through passage-ways between the blocks, the diode effect of air moving in and out of these spaces can result in ground temperatures up to 6°C colder than in the surrounding soil. Harris & Pedersen (1998) provide details of the processes involved, and these processes are found to operate throughout Central Asia (Harris, 1996; Gorbunov *et al.*, 2004) as well as in North America. These spaces also permit the accumulation of large quantities of ice which can result in flowage of the mass of blocks on slopes in the form of rock glaciers. In areas of poor drainage, these materials facilitate the development of underground drainage ways that represent gently sloping streams of water flowing through more or less horizontal taliks that change position laterally, rather like rivers. In other cases, these spaces allow aeolian sand, silt and dust, as well as any finer products of rock weathering to accumulate at the bottom of the deposit or in discrete layers within the blocks. The ice in the permafrost in blocky deposits represents a very important water resource during periods of warming of the MAAT, *e.g.*, in the northern Tien Shan (Bolch & Marchenko, 2006).

On flat surfaces, gravity exerts little force, so the resulting sediments tend to remain in place, although they can produce patterned ground due to convection-type movements in the active layer (see Chapter 10). On slopes, there are a series of distinct processes involved in the downslope movement of coarse-grained, blocky materials which produce characteristic types of movement as well as landforms. These include various forms of creep, actual flowage of the ice holding the mineral grains apart as in rock glaciers (Figure 9.1), gelifluction, sliding, frost creep, actual rapid flowage of part of the surface sediment, and erosion by meltwater moving under the isothermal snowpack. Loess, mudflow deposits, or glacial till may be deposited on the surface, and snow or slush avalanches or landslides may add to and/or modify the deposits. Where the environmental conditions are suitable, ice can accumulate in the interstices, and when it exceeds about 50% of the deposit, the mass can start flowing slowly downslope. Usually several of these processes are involved, either together or at different times of the year. The result is a whole series of different landforms which can grade into one another as the balance of these processes changes across a landscape. In the following account, we will examine these landforms, one at a time.

9.8.1 Cryogenic block fields

Cryogenic block fields consist of a jumbled mass of broken-up rock distributed across a relatively flat surface, usually on a mountain top (Figure 1.16). If the slope is greater than 10°, the material is referred to as a *block slope, scree* or *talus slope* (see below). The term *felsenmeer* has also been used, though that name originated from Felsen, near Stuttgart, where inactive impressive block streams are found on mountain slopes in the nearby forest. The blocks appear to represent the results of humid tropical weathering. Thus this term has been appropriated and used for landforms with a significantly different origin, so it is best not used. *Talus* is the name used for the actual debris found on a talus slope.

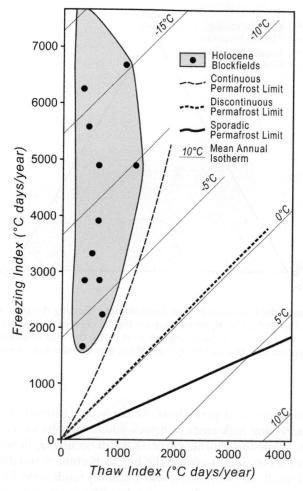

Figure 9.3 Distribution of active cryogenic block fields with mean annual freezing and thawing indices (after Harris, 1981c, Figure 2A, 1982a).

Block fields are relatively rare in young, heavily glaciated mountains such as the Rockies, but are more common in unglaciated areas and on nunataks, *e.g.*, Plateau Mountain in Alberta. Active *cryogenic block fields* are those in areas with permafrost (Figure 9.3), and relict block fields occurring in warmer climates are relict landforms that were formed during previous cold climatic events.

If the blocks have very little fine material in them, the fines get washed down to the permafrost table or to the base of the deposit, whichever is nearest the surface. In this case, the diode effect tends to accentuate the low ground temperatures in the upper layers above the fines. Frost action in the surface layer tends to form sorted macro-patterned ground (see Chapter 5) with the coarser blocks being found on the outside of the pattern. Once the pattern is formed, it will persist, even if the climate becomes

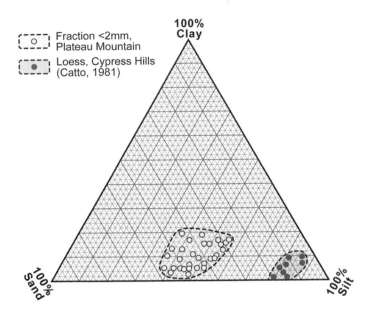

Figure 9.4 Grain size of the fraction finer than 2 mm diameter forming the sedimentary cover of the patterned ground under alpine meadow on the flat crest of Plateau Mountain, Alberta. The sediment contains a large percentage of sand, unlike the Late Wisconsin loess on the top of the Cypress Hills (Catto, 1981), or the flakes derived from frost weathering of bedrock.

unsuitable for the survival of permafrost. Under a cold climate, frost action in the surface of the underlying rock tends to detach additional blocks which can be thrust up to the surface. While this churning process is still occurring, lichens cannot readily grow on the blocks, but once it ceases, the blocks become a suitable substrate, *e.g.*, at Elliston, Newfoundland. Weathering also gradually renders the blocks less angular, and weathering rinds often form that can be dated by cosmogenic techniques. These alpine tundra deposits can persist for over a million years if the conditions on top of a mountain are suitable.

Where there is a considerable percentage of finer material in the blocky material, sorted or nonsorted patterned ground will develop (Figure 1.16). Often, a second, smaller pattern may form in the centre of the sorted polygon, though this may form at a later time when environmental conditions are suitable, as at Plateau Mountain. The fines may consist of windblown loess, flakes of weathered bedrock, or may consist of a former fluvial deposit left when the mountain top was the flood plain of rivers flowing from much higher mountains, as suggested by the grain size of the fraction of the sediment finer than 2 mm on the crest of Plateau Mountain (Figure 9.4). In this case, the surface is more hospitable for the growth of vegetation and the surface may be classified as alpine meadow.

In areas with limited winter snowfall, the snow is usually blown off to accumulate in the adjacent trees. As a result, the additional snow in the forest insulates the underlying ground from the cold winter air, and both the active layer thickness

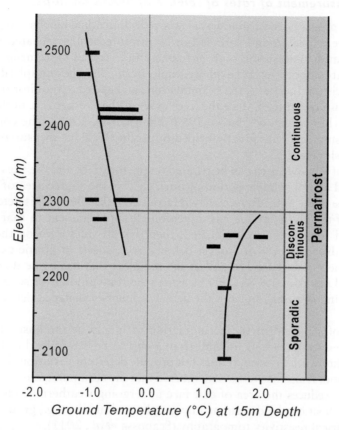

Figure 9.5 The difference between the ground temperatures at 15 m depth in forest (right-hand curve) and below In alpine meadow (left-hand curve) at similar elevations on Plateau Mountain, Alberta. The difference is even greater where alpine tundra is adjacent to forest due to the diode effect. Redrawn from Harris and Brown (1978, Figure 8).

and the ground temperatures are higher in the forest than the adjacent meadow tundra (Figure 9.5). This is even more marked in the case of an alpine tundra-forest interface.

Measurement of the thickness of the blocky deposit is usually achieved by drilling to bedrock. Attempts to use GPR to measure the thickness of the blockfield at Plateau Mountain were unsuccessful on alpine meadow deposits (B. Moorman, personal communication), but this technique apparently does work elsewhere in alpine tundra (*c.f.* Sass, 2006; Sass & Krautblatter, 2007). Siewert *et al.* (2012) found that electrical resistivity tomography was very successful in measuring the profile and frozen parts of talus cones at Longyeardalen, Svalbard, particularly when laboratory calibration is used. They were able to estimate the average rate of erosion of porous sandstone at 330 to 196 mm/1000 years, which matches the earlier estimate of 300 mm/1000 years by Jahn (1976).

9.8.1.1 Measurement of rates of release of blocks on slopes

This has been measured in various ways in the past. Individual blocks and their position on the surface of the ground surface can be measured using repeated triangulation from fixed stations measuring both horizontal and vertical movements. This is also possible on bare slopes using aerial photography or satellite photographs if they provide sufficient precision. In forests, the triangulation can be used to monitor the position of nails driven into tree trunks, since the trees move at the average rate of the underlying rooting zone (Blumstengl & Harris, 1988; Blumstengel, 1988). The coming of laser distance measurements has greatly helped to produce reliable measurements of both distance and slope angle.

Squares of a known area can be painted on the rock face and the sites revisited and photographed after a suitable period of time, e.g., on the source area of sediment for stratified scree in the Slims River valley (Harris & Prick, 2000). This determines what percentage of the given surface has undergone loss of sediment in that time period. If the painted fragments can be located, then this provides data on the distance of travel of the different sizes of blocks. Below screes, sheets of plastic can be laid out in the presumed runout zone to collect the sediment that has come down in a given time period (Luckman, 1978). This measures the total amount of sediment to reach that area during this time, but not the actual amount of material displaced from the cliff face.

Volume of sediment in the accumulation of blocks at the base of a rock wall can allow the average rate of rockfall from a cliff to be determined if the age of cliff formation is known. However, it does not provide details of variations in the rate over that time period. It also has to be remembered that the accumulated material below the cliff gradually reduces the area of cliff face undergoing weathering. The geometry of the blocky deposit can be determined by geophysical means, e.g., ground penetrating radar or electrical resistivity tomography (Scapossa et al., 2011).

Horizontal lines can be painted on loose debris to determine whether the debris is moving over time (Mackay & Matthews, 1974). It also shows whether the blocks are moving en masse or whether they move as individual blocks (Figure 9.18). Rotation of the larger fragments can also be observed (Harris et al., 1998a).

The development of new instrumentation is revolutionizing the study of slope movements, by using LIDAR and surveillance equipment that can record the movement of blocks. It can also record changes in shape of the slope that may occur before material begins to move (Abellán et al., 2009, 2014; Lato et al., 2009, 2012, 2014, 2015; Sturtzenegger & Stead, 2009; Kromer et al., 2015). This is allowing the remote monitoring of unstable slopes above railway tracks in the Fraser River Canyon in British Columbia, as well as allowing prediction of at least some rockfall events.

9.8.2 Cryogenic block slopes and fans

Block slopes, screes or talus slopes are accumulations of blocky deposits on slopes steeper than 10°, usually at the foot of rock walls. Normally, the upper part is at the maximum angle of rest for the blocks, i.e., at 33–37°. They take the form of either cone-shaped fans (Figures 9.1 and 9.6) or sheet-like slopes (Figure 9.6). The term talus is used for the material making up the landform. The blocks are usually detached by

Figure 9.6 Talus fans and a rock glacier beneath rockwalls along the side of a glacially eroded valley along the Haul Road on the North Slope of the Brooks Range, Alaska. © S. A. Harris.

weathering from rocky cliffs above, and tumble, roll, or slide downslope, eventually coming to rest when the slope angle is below 32°. The largest blocks usually travel farthest due to their momentum, and may even continue past the base of the landform. They can also come to rest at the bottom part of the landform, simulating a protalus rampart (see below). The overall thalweg (long profile) is usually concave.

Cryogenic talus fans consist of a cone of blocks derived from a gully in the rock wall forming the cliff (Figures 9.1 and 9.6), whereas *cryogenic talus slopes* are formed beneath linear cliffs (Figure 9.7). As the talus slope or talus fan grows, it will progressively reduce the rock wall being weathered. If the climate is stable, the rate of supply of weathered rock to the growing landform should slowly diminish. However, it is believed that colder periods such as during the Last Glacial Maximum, the early part of the Holocene, and the Neoglacial events resulted in an increased supply of debris, while the supply decreased during the Climatic Optimum (Andrews, 1961; Ball 1966; Ballantyne & Eckford, 1984; Wilson, 1990; Salt & Ballantyne, 1997; André, 1997; Wilson, 2007). In addition, the size of the blocks might be expected to increase during colder conditions and decrease during warming (Ballantyne, 1996; Sass, 2006). However Wilson (1990, 2007) notes that clast size is also controlled by the spacing of joints, an inherent property of each rock mass. On the whole, it has been found difficult to definitely prove whether a given layer in a talus fan or slope deposit formed under specific climatic conditions. Likewise, it is becoming clear that it is difficult to classify talus slopes as being proglacial, paraglacial, or post glacial (Wilson, 2014). Most are composite landforms.

Figure 9.7 Talus slopes on the south-facing slope of Mount Yamnuska, Alberta.

Further complications arise due to a variety of processes that can contribute material to the slopes (Van Steijn *et al.*, 1995, 2002). In addition to weathering of cliff faces, Harris & Prick (2000) and Sass & Krautblatter (2007) have shown that mudflows of varying sizes and frequency can occur (see the levees on the fan in Figure 9.8), loess can be deposited as distinct layers sandwiched between the weathered detritus (Harris, 1975), rockfalls may add substantial amounts of material instantly, rain-generated overland flow (Van Steijn & Hètu, 1997) and slush avalanches may occur (Caine, 1969), and snow avalanches will take place where there is enough winter precipitation. In areas with low winter snowfall, individual blocks may slide downslope when

Figure 9.8 Debris flow levees on two adjacent talus fans in Banff National Park, Alberta. © S. A. Harris.

coated with ice in winter (Hétu & Vandelac, 1989; Lafortune *et al.*, 1997; see block streams below). Inevitably, there will also be slow creep of the blocks on the slopes, especially on the upper part of the slope (Bertan & Texier, 1999). The end result is that the structure of both talus cones and talus sheets will depend on the processes involved in their growth.

Haeberli (1975), Lambiel & Pieracci (2008), Scapozza *et al.* (2011), and Siewert *et al.* (2012) found that permafrost usually develops in the lower two-thirds of the structures in cryogenic talus fans and slopes (Figure 9.9). If the ice content becomes great enough to hold the rocks apart, the icy mass can start moving as a rock glacier (Howe, 1909; Johnson, 1983, 1984a, 1984b; Hétu & Gray, 2000). Figure 9.10 shows a typical case of flowage in the St. Elias Range in the Yukon Territory.

9.8.3 Classification of cryogenic talus slopes

Besides consisting of either fans or sheets, the structure of talus slopes is best considered by dividing them into *coarse blocky talus* with limited amounts of fines and large through pore spaces, and *talus containing significant amounts of finer material* that prevents air from moving freely through the sediment. The first type is subject to the diode effect in areas with low winter snowfall, *e.g.*, Plateau Mountain (Harris & Petersen, 1998), while chimney effects occur in areas with a deeper winter snow pack (Von Wakonigg, 1996). All talus slopes consist of stratified layers, but although this is not obvious in coarse blocky talus deposits, it can be demonstrated using geophysical

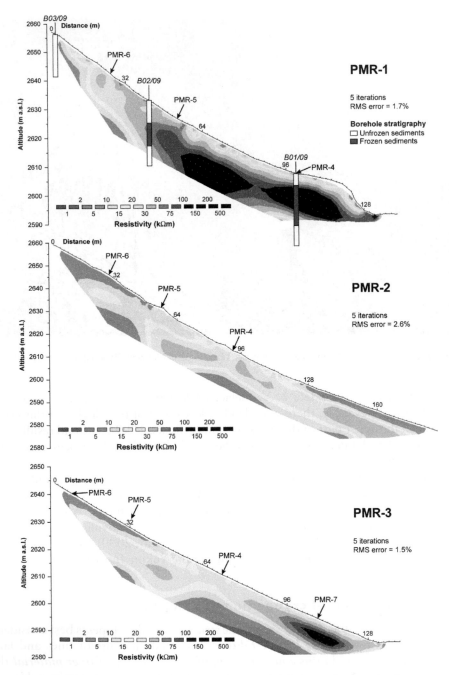

Figure 9.9 Longitudinal profile of the Petit Mont Rouge #1 talus cone showing the slope, permafrost distribution in 3 boreholes, and the ERT resistivity (from part of Scapposa *et al.*, 2011, Figure 9).

Figure 9.10 A major active, spatulate rock glacier formed at the terminus of a talus slope below cliffs of rock, Kluane National Park, Yukon Territory. © S. A. Harris.

methods, especially ground penetrating radar (Sass, 2006). However Sass found that the internal structure can be very complicated. The second type of sediment found in many talus deposits is usually stratified, and these deposits were named *grèzes litées* (Guillien, 1951) or *éboulis stratifiés* (Francou, 1988). The French geomorphologists have discussed the origin of the Pleistocene and Holocene fossil examples for the last 70 years. Since both types are stratified to a greater or lesser extent, the term "stratified scree" cannot satisfactorily be used for the sediments with significant quantities of fines.

9.8.3.1 Coarse blocky talus slopes

These are the classic talus slopes found beneath cliffs. In their simplest form, they consist of sheets of blocky sediment inclined at around 33–37° at the base of the exposed cliff, becoming 30–32° in the runout zone. The blocks tend to exhibit a rough sorting by size, with the smaller blocks tending to become stationary higher on the slope than the coarser blocks. However, snow avalanches in winter, slush avalanches in spring, debris flows and different kinds of creep can blur this pattern. The largest blocks are normally found on the ground in front of the fan or sheet. Over time, the deposit becomes thicker and covers a larger area, while the lower part of the rock wall becomes buried in the new layers. Thus the rate of growth of the fan would be expected to decrease with time under conditions of a stable climate.

Both geophysical studies and field observations of the structure of these blocky slope deposits show that the simple surface appearance often masks a much more

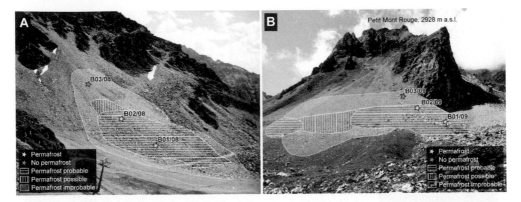

Figure 9.11 Distribution of permafrost on the lower slopes of a blocky talus fan (A), and a blocky talussheet (B). From Scapozza *et al.* (2011, Figure 11). The stars indicate the positions of boreholes confirming the interpretation of the geophysical layers in dry talus debris in the Eastern European Alps.

complex history. Thus Hètu & Gray (2000) found that blocky slope deposits overly marine and beach gravels along the south side of the St. Lawrence estuary because they started forming there as soon as the ice sheets retreated between 13.5 and 10 ka B.P., while isostatic rebound was still occurring. The establishment of a closed forest in the region about 7.25 ka B.P. did not stop their formation. Complete development of the forest has only been possible where the movement of talus on the talus slopes ceased. Sass & Krautblatter (2007) were able to identify several different types of bedding in several cases including the Kühtai Hinterkar talus (Sass & Krautblatter, 2007, Figure 9). However, this method only picks out major disruptions in the bedding.

As noted above, once the talus slope deposit is established, the diode effect results in significantly lower ground temperatures within the porous mass of blocks. Permafrost develops on the lower parts of the talus pile (Figure 9.11), whether it is a fan or in the form of a sheet across the lower slopes of a rock wall. This is a form of sporadic permafrost and can be found several hundred metres below the lower limit of continuous permafrost. Ice accumulates in the interstices of the blocks, and when it is abundant enough to hold the blocks apart, this icy mass starts moving down slope as a rock glacier (see below). Icy permafrost has been found by Gorbunov in blocky deposits under a forest vegetation near the Lower Glaciological Station in the Chinese Tien Shan, while Popescu *et al.* (2017) provide a description of permafrost in block slopes grading into active rock glaciers in the southern Carpathian Mountains of Romania.

9.8.4 Protection of infrastructure from falling rock

Cryogenic displacement of blocks can be disastrous. There are a number of ways in which the blocks can be intercepted, but ultimately, there must be a system of removal of the blocks before they overtop any barrier (Volkwein *et al.*, 2011). The most simple method is the use of *gabions* consisting of large stones inside a galvanized iron mesh cage. These hold back small rocks and permit the material behind it to drain to counteract flowage if finer material is present. A second method is the use of interlocking concrete blocks, often stacked in multiple levels. Sometimes a steel mesh is draped over the rock face that is breaking up, and this prevents the falling rock from spreading out too far. In these cases, the fallen rock must be removed regularly to protect the adjacent infrastructure.

Another method is the use of flexible metal barriers for protection against rockfalls. They consist of a metallic cable net, kept upright by structural steel posts, with the function of intercepting and stopping falling blocks (Gottardi et al., 2011; Gentilini et al., 2013). Loads are transferred through special connecting elements to the relevant foundations. A key feature is that these structures are relatively light, yet they are designed to deform, absorbing the kinetic energy of the rocks, rebounding back into their original position while dropping the rock behind the fence. They absorb the rather high energy levels through the development of large elasto-plastic deformations of the system. This is achieved by the displacement of the cable net and the activation of energy dissipating devices, mounted on the connecting cables.

9.9   TALUS CONTAINING SIGNIFICANT AMOUNTS OF FINER MATERIAL

The presence of a fine-grained matrix alters the processes occurring on talus slopes. They usually result in the development of distinct layers by a wide variety of processes (e.g., Hètu & Vandelac, 1989; Bertran, 1992; Bertran et al., 1992). A combination of snow, ice and wind produce at least 6 different processes, viz., several types of avalanches including a two-layer talus/snow avalanche triggered by rockfall, rapid snow creep in the spring producing grooves and push ridges, sliding of blocks on an ice crust, niveo-eolian deposits on the lower slopes, sliding of ice-coated rocks, and movement of gravel pushed by the wind onto the snow cover. Earlier, Guillien (1951) thought that slope-wash was dominant, but like the congelifluction-slope-wash theory of Journaux (1976), it does not fit the evidence from Charantes and other locations (Bertran et al., 1992).

Field observations by Sass and Krautblatter (2007) indicated that the actual processes occurring on the surface of the talus were much more complex. Resedimentation by large-scale debris flows was rare, and limited to slopes of around 14° (Fischer, 1965). The flows cut into the upper talus and resulted in considerable redistribution of material. Small-scale surficial flows were common, usually being initiated by overland flows originating in gullies on talus fans (see Figure 9.8), heavy rainfall events, or over-steepening of the talus slope in dry weather (dry grain flows of Lowe, 1976). Frost-coated grain flows (Hètu et al., 1994) also occurred. These processes produce local, more-or-less parallel, discordant layers which do not usually extend across the fan.

In areas with permafrost, similar deposits may be found. Harris and Prick (2000) documented the development of stratified talus in a cone below a weathering cliff at Kluane National Park. Frost action on the face of the cliff resulted in the detachment of a wide range of silt to gravel-sized rock fragments which then tumbled downslope. The finer grains came to rest on the upper part of the slope while the coarser grains accumulated towards the bottom in dry weather. Whenever it rained in spring and summer, the finer grained deposit became saturated with water and flowed down over

Figure 9.12 Stratified scree deposits near Kluane Lake, Kluane National Park. A, the 90 m high talus Fan, and B, detail of the thin layers with the late Professor Dr. Albert Pissart for scale. © S. A. Harris.

part of the coarser material, producing a more or less regular stratification (Figure 9.12) of a type found in fossil deposits in Belgium. The individual layers were not continuous across the entire landform and mainly occur on the lower slopes. Since these can also be formed in areas lacking permafrost (Coltorti *et al.*, 1983), independent evidence for the presence of permafrost in the area at the time of formation must be available before interpreting these deposits as being formed in a permafrost environment.

9.9.1 Rock glaciers

Rock glaciers have been defined as "a tongue-like or lobate body, usually of angular boulders that resemble a small glacier, generally occur in high mountainous terrain, and usually has ridges, furrows and sometimes lobes on its surface, and has a steep front at the angle of repose" (Potter, 1969, page 1, 1972, page 3037). They extend downslope largely through the deformation of interstitial ice. Examples are shown in Figures 9.1, 9.9, and 9.14. Barsch (1996) examines the earlier literature on them in detail. They were first discussed by Cross and Howe (1905) and named by Capps (1910, page 360). Alternative names include *blockgletscher* (Högbom, 1914), *glacier rocheux* (de Martonne, 1920) and *kamennye gletcery* in Russia.

It is usual to subdivide them into two main forms, *viz.*, *tongue-shaped rock glaciers* formed below cirques and *lobate rock glaciers* formed below talus slopes. Where a rock glacier terminates in a broader lower part (Figure 9.10), it is sometimes referred to as a *spatulate rock glacier* (Wahrhaftig & Cox, 1959) or a *piedmont rock glacier* (Humlum, 1982). Although these are generally considered to be tongue-shaped rock glaciers (Parson, 1987), they can also be found associated with lobate rock

Figure 9.13 Active tongue-shaped rock glaciers in the Zailijskiy Alatau, Khazahkstan, where the longest and largest rock glaciers occur (see Gorbunov, 1983). In the foreground is the Morenny rock glacier with the Gorodetsky rock glacier in the background. © A.P. Gorbunov.

glaciers beneath cliffs (Figure 9.8). This classification is based on the shape of the rock glaciers, the lobate forms having a length:width ratio <1 in contrast to the tongue-shaped rock glaciers having ratios >1 (Domaradzki, 1951; Wahrhaftig & Cox, 1959). Attempts to associate the lobate rock glaciers, *e.g.*, Figure 9.1, with talus beneath cliffs and rock walls are only successful in about 90–95% of the examples. They can also be found below small glaciers, having developed in morainic material. Tongue-shaped rock glaciers are normally found in valleys below the last stages of retreating glaciers (Domradszki, 1951; Foster & Holmes, 1965; Outcalt & Benedict, 1965), as in Figure 9.13. Some active Asian rock glaciers may extend down below the permafrost limit in the adjacent landscape (Bolch & Gorbunov, 2014), presumably due to the diode effect, although this has not been noted elsewhere.

The actual proportion of each type of rock glacier depends on the topography and whether the climate is suitable for the presence of rock glaciers. Thus in the cold, semi-arid Zailinjskiy Alatau in Kazakhstan, lobate rock glaciers make up 60% of the total, averaging 380 m long and 180 m wide (Gorbunov, 1983). They occupy about 70,000 m², whereas Luckman & Crockett (1978) estimated that lobate rock glaciers amounted to only 45% of the rock glaciers in the more humid area of Banff National Park, Canada. Headwall elevation and lower limit of active rock glaciers are similar for both types (Barsch, 1996), their lower limit approximating the position of

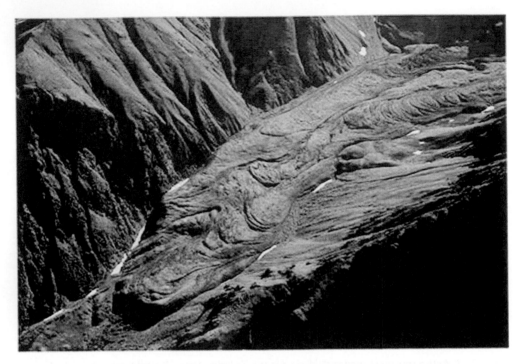

Figure 9.14 Typical complex rock glacier with overlapping and over-riding lobes.

the lower limit of continuous permafrost. They are normally found on north-facing slopes in the northern Hemisphere (White, 1979).

Barsch (1996) uses three classes, *viz.*, tongue-shaped, lobate and **complex rock glaciers** (Figure 9.14). The latter comprise those cases where one part is over-riding another, and the parts can be demonstrated to be of different ages (Yarnal, 1979; Johnson, 1980), or where the rock glacier either divides into more than one stream or is over-ridden by a glacier. Gorbunov (1983) and Gorbunov and Severskiy (2010) have described rock glaciers developed in piles of mining waste in various parts of Siberia, and Corte (1978) used the term **tecnogenic rock glaciers** for these. For more complicated classifications, see Barsch (1987) and Corte (1987a).

More importantly, all types of rock glacier may be active, inactive, or relict. The active ones are still moving forward at speeds up to $14 \, \text{m/a}^{-1}$, and the constituent blocks and finer sediment are held apart by the interstitial ice. It is the deformation of this ice by the pull of gravity exceeding the shear strength of the ice that causes the movement. During this process, individual blocks become tilted and rotated. Usually the fastest moving part of the rock glacier advances at under $2.5 \, \text{m/a}^{-1}$. In inactive rock glaciers, ice may still be present, but no longer holds the blocks apart. As the ice slowly melts, there is settling involving tilting and some limited downslope movement. The landform still looks fresh, and the boundary between these two kinds of rock glacier is difficult to determine without actual measurements of the creep downslope.

Relict rock glaciers show a softening of the features of the landform and the individual blocks usually have lichens growing on their stable upper surfaces. This is impossible in the case of the other two groups because of the churning movements during flow.

9.9.1.1 Sedimentary composition and structure of active rock glaciers

Rock glaciers develop in any suitable loose material. In tongue-shaped rock glaciers, the main mass often consists of glacial till, but snow and ice avalanches, debris flows and rockfalls from the valley walls provide additional sediment to the surface of the rock glaciers. In the case of lobate rock glaciers, the bulk of the material is usually from talus fans and slopes. In rare cases, loess can be incorporated into the mass, while they can also develop in piles of mine waste in suitable climates (Gorbunov, 1983; Gorbunov & Titkov, 1989).

Naturally-formed rock glaciers are generally two-layer phenomena (Barsch, 1996, p. 67). There is a boulder-rich upper *rock glacier mantle* up to 3 m thick in the case of Murtel I (Figure 9.1), which overlies an inner *rock glacier core*. The former includes the active layer, and when this layer consists primarily of blocks, it will be subject to the diode effect unless it lies in a high snowfall region. Often in tongue-shaped rock glaciers, the core lacks boulders but resembles glacial till. In Holocene lobate rock glaciers, the core will consist primarily of talus sediments. It generally has about 30–40% pore space, and ice volumes in excess of this indicate that it is *ice-supersaturated* and will be flowing slowly downslope. In the case of Murtel I (Haeberli, 1989), a drillhole on the rock glacier found that the rock glacier core consists of multiple layers, the upper one containing massive ice (3–15 m depth), overlying frozen, highly ice supersaturated silt, sand and gravel (15–28 m depth), then ice saturated frozen silt sand and gravel (28–32 m) resting on top of coarse boulders with ice-saturated layers of frozen sand (32–50 m depth) overlying bedrock. It formed during the Holocene in an area that was glaciated during the last major (Würm) glaciation.

Vonder Mühll (1992) reported seasonal variations in temperature between 52 and 56 m depth in Murtel I, and interpreted this as indicating deep groundwater flow near the base of the rock glacier. He estimated the total thickness of permafrost to be about 100 m. Using geophysical methods, it was found that the ice content within the rock glacier was highest in the centre of the second layer, which may explain why in unfrozen, relict rock glaciers, this part of the rock glacier deposit is lower than the outer ridges after the ice has melted.

Other means of determining the internal composition include deep seismic soundings (Barsch, 1971; Barsch & Hell, 1975; Vonder Mühll, 1993; Gerber, 1994) that can detect changes in rock density, which is least in highly icy material and highest in massive bedrock. Shallow hammer seismic soundings have also been used to measure the thickness of the active layer and the nature of the upper part of the rock glacier core (Potter, 1972; Barsch, 1973; King, 1976; Haeberli, 1979; Haeberli & Patzelt, 1982). Geoelectric soundings are also used (Fisch et al., 1978; Evin, 1983b, 1987b; Barsch & King, 1989), and can differentiate ice-rich pemafrost from the clear underground ice bodies that sometimes occur (King, 1982). Unfrozen debris has an electrical resistivity of around 1000 Ωm, compared to frozen debris at $10->100\,\Omega$m, and temperate glacier ice ($>10\,\Omega$m). Mixed results have been obtained using radio echo soundings, so this technique is not usually used (Haeberli, 1985a; King et al.,

1987; Vonder Mühll, 1993). The latter found that gravity soundings were more successful.

Measurement of the changes in inclination of the borehole walls with time has been used occasionally (Girsperger, 1973; Johnson & Nickling, 1979; Wagner, 1992). Johnson and Nickling found that there was a slow tilt of 2° towards the head of the rock glacier after seven years, but with varying, decreasing tilting with depth. The bouldery upper layer appeared to ride passively on the underlying core that was supersaturated with ice. Wagner found that the upper layers above about 30 m depth on Murtel I moved downslope at a rate of about 5–6 cm/a^{-1} over the underlying layers, which only moved a short distance. When the experiment was repeated more carefully in a second borehole (Haeberli et al., 1988; Haeberli, 1990a) using inclinometers with magnetic rings, it took over 9 months for the inclinometers to develop enough contact to demonstrate rates of movement. The upper layer was moving at a speed of an order of magnitude greater than the lower layer, while 75% of the deformation occurred between 28 and 32 m depth.

The thermal profile may exhibit changes in thermal gradient where some of the lithological discontinuities occur. Ground temperature measurements carried out over a period of six years in Murtel I (Vonder Mühll & Haeberli, 1990; Vonder Mühll, 1993) indicate that the depth of zero amplitude is at c. 20 m depth. The ground temperature gradient within the rock glacier was $69 \pm 11°C/km$ for the 20–30 m zone, compared with $47 \pm 5°C/km$ for the 30–35 m layer. The basal temperature above the bedrock was just below 0°C. More recent work demonstrates that rock glaciers in older landscapes whose rock glacier deposits have not been eroded by glaciers may contain the remnants of multiple periods of activity (Aubekerov & Gorbunov, 1999). Studies using ground penetrating radar in Svalbard (Isaksen et al., 2000) and southwest Colorado (Degenhardt, 2008) have produced evidence of multiple depositional lobes (Figure 9.15) which can be interpreted as representing multiple events. It is likely that glacial tills may have been buried by rock glaciers, and glaciers can over-ride rock glaciers without completely removing the sediments. Thus complex rock glaciers are widespread in all but heavily glaciated landscapes.

9.9.1.2 Origin of the ice in active rock glaciers

This is usually highly complex (Barsch, 1996; Bolch & Gorbunov, 2014). Sources will include winter snow and resulting infiltration of the thawed water in spring. Snow and slush avalanches add ice, water and sediment to the surface of the rock glaciers especially in the region near steep slopes. Underground springs may come out of the underlying bedrock under hydraulic pressure and either form a small stream within or under the rock glacier, e.g., under Murtel I (Vonder Mühll, 1992), or the water can add to the ice content by freezing, either as interstitial ice or as actual injection ice. Such layers represent discontinuities in the upper layers of a rock glacier that can result in localized faster movement downslope of the overlying layers or even movement of lobes riding over slower moving layers. Earthquakes may also be involved in the case of those on the Tien Shan (Bolch & Gorbunov, 2014).

Tenthorey (1992) carried out experiments with tracer dyes to determine the fate of water produced by the partial summer thawing of perennial névées in the Upper Val de Réchy, Valais, Switzerland. In two cases using active rock glaciers, the melt water

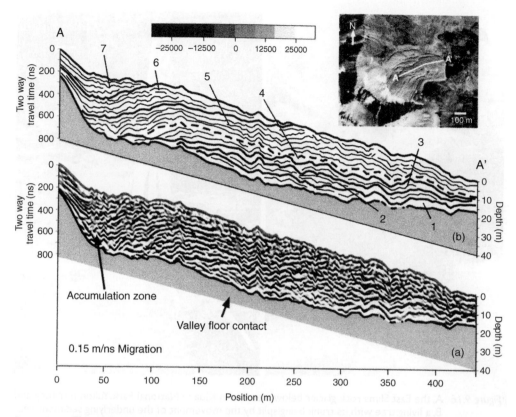

Figure 9.15 The results and interpretation of ground penetrating radar on the longitudinal profile of the Hiorthjellet rock glacier, Svalbard (see Isaksen *et al.*, 2000). A series of seven depositional lobes are indicated.

rapidly flowed downslope in the surface boulder layer of the rock glacier after disappearing from view, but remained in the upper boulder layer and travelled downslope quickly (3.5 m/minute). These results are similar to those obtained by Evin & Assier (1983) at the active Marinet II rock glacier in the Maritime Alps. In a third, inactive rock glacier, Tenthorey found that the water descended into the interstices of the main icy core and moved slowly through the substrate to the front of the landform (200 m in 15–30 days). This probably aids in the thawing of the icy core.

9.9.1.3   Relationship to vegetation

The usual vegetation cover of the surrounding area is tundra in both the polar regions and high mountains. There, lichens only grow on stable surfaces, though some arctic plants (mainly annuals) manage to go through their life cycle as isolated individuals in crevices and places with finer grained sediments. As such, they do not affect the movement of the rock glaciers. However in mountains, active rock glaciers can extend down into the boreal forest below treeline (Figure 9.16A), and the trees in western Canada have shallow roots being able to grow satisfactorily as long as the upper

Figure 9.16 A, the East Slims rock glacier below tree line in Kluane National Park, Yukon Territory, and B, a living tree with its trunk being split by the movement of the underlying sediment near the terminus of the rock glacier. © S. A. Harris.

blocky layer rides smoothly along on the underlying icy layer. Near the terminus, or where the icy core moves unevenly, split tree trunks (Figure 9.16B) and drunken forests (Figure 9.17B) occur.

9.9.2 Movement of active rock glaciers

Studies of the movement of active rock glaciers can be divided into two kinds, *viz.*, the horizontal component and the movements at the steep front. Usually, the horizontal movement is greatest near the terminus, decreasing towards the upper part of the rock glacier (Gorbunov *et al.*, 1992). At the front, the forward motion is considerably less than that on the rock glacier surface immediately above the front (Blumstengel, 1986; Blumstengel and Harris, 1988), so that the front gradually increases in height with age (Gorbunov *et al.*, 1992).

9.9.2.1 Horizontal movement

Horizontal movement of active rock glaciers can be made using aerial photographs taken in different years (Messerli & Zurbuchen, 1986; Barsch & Hell, 1975; Jackson &

Macdonald, 1980; Evin & Assier, 1982, 1983; Haeberli & Schmid, 1988, *etc.*). Measurements of movements on a forested rock glacier (Figure 9.16A) can be made by setting up a grid of measurements between nails hammered into tree trunks and carrying out repeated distance measurements in a horizontal direction, but there are usually significant vertical components, especially near the terminus. Although geophysicists may regard the upper layer of active rock glaciers as essentially moving as a sheet, the individual boulders can tilt and rotate during movement as changes occur in the volume of ice below, as well as in rates of movement of the layers at different depths. This results in split trees (Figure 9.16B) due to the differential movement in forested areas, and the development of obvious compressional and tensional ridges on rock glaciers without vegetation. The reduced speed of movement near the terminus is due to thawing of the interstitial ice due to increased mean annual temperatures at lower altitudes, so increasing the internal friction between the rock clasts. This, in turn, results in a piling up of the blocky sediment, thus increasing the height of the rock glacier front. Factors affecting the horizontal movement include the climate at that elevation, climatic change, flow of underground streams through the rock glacier, changes in gradient of the underlying surface, changes in width of the valley walls in tongue-shaped rock glaciers, and sudden addition of additional snow and rock by landslides, rockfalls, and snow and slush avalanches. Gorbunov *et al.* (1992) describe examples of some of these factors. Movement of groundwater is usually the cause of movements exceeding 2.5 m/a^{-1} (Gorbunov *et al.*, 1992). Degenhardt (2009) showed that movement in the fall was >2 times that in the spring, indicating the influence of climate, especially temperature.

9.9.2.2 Movement of the front

The steep fronts of rock glaciers can be up to 53° (White, 1976b, page 84) and may reach 122 m in height (Wahrhaftig & Cox, 1959, page 387). They are often light coloured (Figure 9.13), and the rocks lack lichens. The ridges and furrows may be arcuate and convex downslope or else longitudinal. These ridges represent the response to compressional and extensional movements within an active rock glacier brought about by variations in speed of movement and the effects of thawing of interstitial ice. The actual shape of the front depends on the mechanisms involved in its advance (Matsuoka *et al.*, 2005; Kääb & Reichmuth, 2005), the processes involved in the movement (Arenson *et al.*, 2002), the presence of water flowing in or close to the rock glacier, climate change, and the slope of the underlying surface (Gorbunov *et al.*, 1992).

The presence or absence of a vegetation cover (Figure 9.17) may also be related to the type of movement and is probably controlled by it to an unknown extent. Where material slides down the front, blocks of soil with juniper bushes on them may occur on the surface. Slope wash after rains may result in the development of outwash deposits at the foot of the slope, as on the north terminus of the Gorodetsky rock glacier. Isolated juniper bushes and spruce trees can grow on the surface of the rock glacier above tree line if the surface is fairly stable. Active rock glaciers may occasionally be found in the boreal forest, *e.g.*, along the east side of the Slims River valley, Kluane National Park, Yukon Territory (Figure 9.17A). In this case the forest extends down the front of the rock glacier (Figure 9.17B). Matsuoka *et al.* (2005) discuss the morphometry of

Figure 9.17 Comparison of A, the front of the Gorodetsky rock glacier under alpine meadow vegetation in Kazakhstan with that of B, the East Slims River rock glacier in the boreal forest of the Yukon Territory. © S. A. Harris.

the rock glaciers found in Switzerland, and provide details of the quantitative differences in morphology between the lobate and tongue-shaped rock glaciers in that part of the Alps.

Measurement of the movements at the front of the glacier is complicated by thawing of the ice in the landform as well as external processes such as fluvial erosion and

Figure 9.18 Front of a rock glacier advancing across a former road in the Tien Shan with Professor Dr. Zhu Cheng standing by his painted stones, each colour representing the remains of a painted line, painted in consecutive years. These prove the downslope movement of individual stones, probably sliding on ice. © S. A. Harris.

creep of the sediment down the steep front. These measurements should be made for several years (a minimum of three) and the results averaged to overcome variations in microenvironment from year to year. Actual forward motion can be determined in the same way as horizontal movement, but techniques are needed to check for other processes. Zhu Cheng *et al.* (1992) painted lines each year at a given height on the front

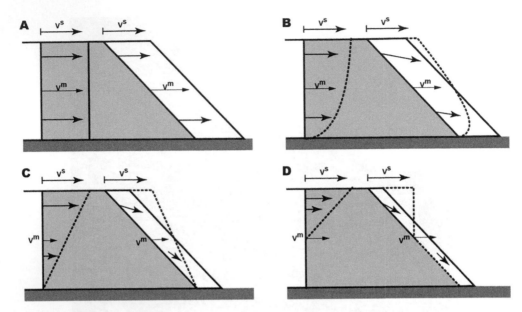

Figure 9.19 Possible mechanisms for the movement on the front of a rock glacier (modified from Kääb & Reichmuth, 2005, Figure 1). The black lines show what happens with an even forward movement, the red lines what happens if the movement varies, but because the front may become unstable, the green line shows the end product.

of the rock glacier advancing across a road in the Chinese Tien Shan (Figure 9.18). The individual rocks migrated varying distances downslope over time, indicating an obvious downslope component. Degenhardt (2009) used steel rods driven vertically into the crest of the front. These deformed over time, indicating a similar movement. Changes in inclination of inclinometers emplaced in drill holes behind the front allow the measurement of the movement of the surrounding sediment. These can then be compared with the actual horizontal movement of the front, and also provides a test of whether there are any abrupt changes in motion with depth.

In theory, there are at least four main possible forms for the movement of the front of a rock glacier according to Kääb & Reichmuth (2005) following Wahrhaftig & Cox (1959) in part (Figure 9.19). The initial vertical profile and movement relative to depth indicated by the inclinometers are shown in black on the left of each diagram. The dotted line in red shows the results of adding the movement with depth to the initial slope. The dashed line in green shows what the front should look like if the initial slope is maintained by rearrangement of the material by slope processes.

In actual practice, the amount of horizontal movement of the base of the front is less than that indicated in Figure 9.19 due to melting and ablation of ice, resulting in reduction in the mass of the material. The difference is regarded as being a measure of the minimum volume of ice that is lost. Individual rock glaciers show fronts corresponding to different theoretical cases. Thus the East Slims rock glacier (Figure 9.17B) has a front matching case B in Figure 9.19. The Gorodetsky rock glacier (Figure 9.17A) has a front on its north side corresponding to case C in Figure 9.19, though with an

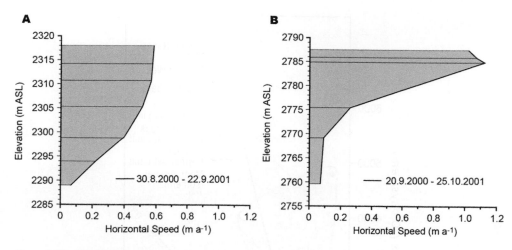

Figure 9.20 Actual measured movements of the fronts of A, the Suvretta rock glacier and B, the Gruben rock glacier in the Alps (modified from Kääb & Reichmuth, 2005).

outwash fan at its base. The Suvretta rock glacier (Figure 9.20A) has a similar pattern of vertical movement to the East Slims rock glacier but geomorphic processes resulted in a predominantly steep front, but with increased slope below 2310.5 m elevation. The Gruben rock glacier (Figure 9.20B) showed a frontal profile matching Figure 9.19D, because only the upper 12 m of its mass was moving relatively fast. Thus there is tremendous variety in both the processes and actual slopes of the fronts of individual rock glaciers.

White (1971a, 1971b, 1975, 1976, 1987) studied the Arapaho rock glacier intensively over at least two decades. He has documented debris falls along the steep front (White, 1971b), as well as the horizontal movement of the convex transverse ridges. Using triangulation between steel pins hammered into crevices in the ridges, he demonstrated that the compression and tensional ridges moved in a jerky pattern, widening and closing by up to $11–27 \, cm/a^{-1}$. The movement varied in direction, sometimes being transverse, diagonal or longitudinal. Other adjacent ridges only moved $0.5–4.0 \, cm/a^{-1}$ during the eight years of study. The cause is believed to be due to differential movement of the underlying layers.

9.9.3   Distribution of active rock glaciers

Most rock glaciers are found in continental mountainous areas with limited to moderate winter precipitation. They are rare in maritime climates, but can be found from Antarctica (Mayewski & Hassinger, 1980; Mjagkov, 1980; Barsch *et al.*, 1985; Linder & Marks, 1985), along the Andes of South America (Corte, 1976b; Marangunic, 1976; Corte & Espizua, 1981; Heine, 1976; Rangecroft *et al.*, 2014), the Cordillera of North America (Höllermann, 1983), north to Alaska, Greenland (Steenstrup, 1883; Humlum, 1982), Iceland (Martin & Whalley, 1987) and Svalbard (Kristiansen & Sollid, 1986; Berthling *et al.*, 1998), as well as the Alps (Barsch, 1996) and in the Tien Shan and Pamirs of Central Asia (Gorbunov, 1973). Active rock glaciers

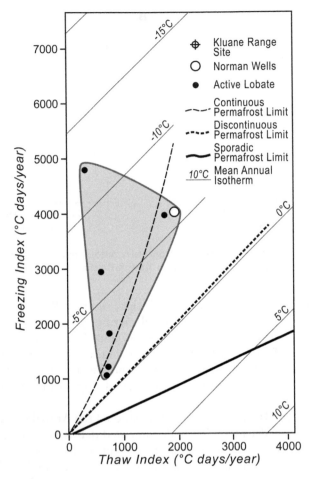

Figure 9.21 Relation of active lobate rock glaciers to mean annual freezing and thawing indices (after Harris, 1981b, Figure 6).

are also common in the Caucasus and in south central Siberia. In China, they are confined to the high mountains in southeast China near Lijiang, and they occur in a belt from there to Pakistan. They are only known as relict landforms in the mountains of Africa, *e.g.*, the Hoggar mountains (Rognon, 1967; Messerli, 1972), Mount Kenya (Mahaney, 1980), and the Drackensburg, South Africa (Lewis & Hanvey, 1993; Lewis, 1994). For detailed descriptions, see Barsch (1996, Chapter 5).

The wider the difference between snowline and timberline, the better the rock glaciers are developed if the climate is suitable (Barsch, 1996). Figure 9.21 shows the distribution of active rock glaciers below talus slopes to mean annual freezing and thawing indices, while Figure 5.1 shows the distribution relative to mean annual precipitation and mean annual air temperature. The greatest concentration of rock glaciers is found in the dry, high mountains of central Asia and the north-central Andes.

Modelling is now being used to try to predict where active rock glaciers may be found (Brenning & Trombotto, 2006; Frauenfelder *et al.*, 2008) using a number of external factors that depend on the local climate. These factors include rock-debris accumulation, hydrology, climate, and extent of glaciers. There is a transition from primarily glaciers and rock glaciers in the Andes of northern Peru, southwards to almost exclusively rock glaciers in southern Peru (Dornbusch, 2005). However a complication is the history of cold events in a given area. Some areas have suffered extensive glaciation in the past, *e.g.*, Argentina, but since then, the climate has favoured the development of rock glaciers. In the high mountains of Kazakhstan, there have been cold events since the Eopleistocene (3.5–2.8 Ma) involving the development of both glaciers and rock glaciers (Aubekerov & Gorbunov, 1999). Thus there are cases with several phases of reactivation of rock glaciers producing complex rock glaciers. It is only where the rock glaciers develop on a surface that was glaciated in the Late Wisconsin glaciation (30–15 ka) that the structure is relatively simple, *e.g.*, Murtél I. In some cases, glaciers over-ran rock glaciers but did not remove all the deposits, *e.g.*, on Marmot Mountain in Jasper National Park (Harris, 1999). On deglaciation, the rock glaciers sometimes redeveloped, burying both the old rock glacier deposits and the younger glacial till (Isaksen *et al.*, 2000; Degenhardt, 2009).

9.9.4   Inactive and fossil rock glaciers

These are rock glaciers with either less than the c.40% ice content to permit flowage of the interstitial ice (*inactive rock glaciers*), or no ice at all (*fossil rock glaciers*). The latter are usually the product of colder climatic conditions related to major glacial/permafrost events where the ice has subsequently thawed as a result of much warmer climates. The inactive rock glaciers are often the result of recent warming of mean annual air temperatures following Neoglacial or other Holocene cold events.

There is no movement of the surface of fossil rock glaciers other than minor creep due to temperature fluctuations or frost heave. There can be solifluction and slope wash affecting the finer grained sediment in between the blocks of rock. The overall feature has appreciably lower surface relief than when it was an active rock glacier, and the ridges and slopes are much more rounded after settling as the ice thawed and the resulting water drained away (Figure 9.22). Weathering tends to cause the surface blocks to become less angular, producing silt-sized fragments of rock. This silt, colluvium and loess tend to fill in the spaces between the blocks, permitting the revegetation of the material. Talus cones or sheets may be forming beneath cliffs, burying the adjacent parts of the former active rock glacier.

Inactive rock glaciers have intermediate characteristics between active and fossil rock glaciers. There is minor downslope movement of the gravitational component to the settling of the clastic sediments as the interstitial ice melts, but this is <0.2 m/a^{-1}. The surface rocks suffer tilting and become very unstable, but lichens tend to start to grow on the more stable surface rocks. Springs continue to flow from beneath the rock glacier as they do at the fronts of active rock glaciers. The settlement tends to result in lowering of the central part of the rock glacier more than elsewhere, presumably reflecting the higher ice content there when the rock glacier was active. The arcuate ridges remain near the terminus but become more rounded and exhibit lower

Figure 9.22 The Roc Noir rock glacier, France, viewed from the crest of the source of the rocks. Note the subdued, vegetation-covered lobes and the bare talus fan building out on to the source area of the fossil rock glacier. © S. A. Harris.

differences in relief. Movement stops when the ice has finally melted, and the streams carrying meltwater from the interstitial ice cease to flow.

9.9.5   Streams flowing from under rock glaciers

Often, there is a stream that flows out from under the terminus of a rock glacier. The flow in active rock glaciers can vary from 2–50 l/s in the case of the Gruben rock glacier (Haeberli, 1985), up to 50–200 l/s in the case of the East Slims rock glacier (Harris *et al.*, 1994). In inactive rock glaciers, the flow may be higher (90–270 l/s for the Hilda rock glacier (Gardner & Bajewski, 1987), although this may be due to a spring beneath the rock glacier. Figure 9.23 shows the measurement of the flow of water using a V-notched weir.

The quantity of flow varies according to the weather, time of year and precipitation events (Figure 9.24). The first peak occurs when the snow is melting in early summer, and is of the same order of magnitude as the response to summer storm events (Johnson, 1978; Jackson & Macdonald, 1980; Evin & Assier, 1983; Haeberli & Patzelt, 1982, Haeberli, 1985). Summer base flow is an order of magnitude less, and the water is always clear. Increase in discharge after storms is rapid and not much slower than surface flow in rock glaciers above tree line. Velocities range from 0.0–1.3 m/s^{-1} (Johnson & Lacasse, 1978), to 0.01–0.1 m/s^{-1} (Evin & Assier, 1993). Where there is a forest on its surface, Harris *et al.* (1994) found that there was

Figure 9.23 Measuring water flow at a spring at the front of the East Slims rock glacier, Yukon. © S. A. Harris.

a lag in peak flow of 10–14 days after a storm event on the East Slims rock glacier (Figure 9.24).

In rock glaciers, the water is in contact with rock for much longer than in glaciers, and so carries considerably higher mineralization (Souchez & Lorrain, 1991; Harris *et al.*, 1994). The degree of mineralization depends on the composition of the rock fragments, *e.g.*, around 204–209 mg l^{-1} around Mendoza (Corte, 1978) compared with 77–92 mg l^{-1} in the French Alps (Evin, 1993). Near the oceans, the conductivity of the water may reach 900 μS cm^{-1} during spring melt (Harris *et al.*, 1994). The water temperature is consistently below 2°C, and the pH values vary by less than 0.4 units. Tritium is stored in the rock glacier, so that some precipitation freezes in the upper layers of the landform and it appears in the stream flow over time.

There can be water flowing through the interior of rock glaciers (Evin & Assier, 1983; Haeberli, 1985; Tenthorey, 1992; Vonder Mühll, 1993). The best estimates of flow rate come from fluorescent die used at Becs de Boisson, Wallis by Tenthorey, estimated at 2.5×10^{-4} to 2.5×10^{-5} m/s^{-1}. It took between 1 and 12 months for the water to travel 70 m.

9.10 CRYOGENIC BLOCK STREAMS

Block streams consist of a stream of blocks on sloping hillsides resembling a stream deposit, usually on gully floors or slightly lower areas (Washburn, 1979; White,

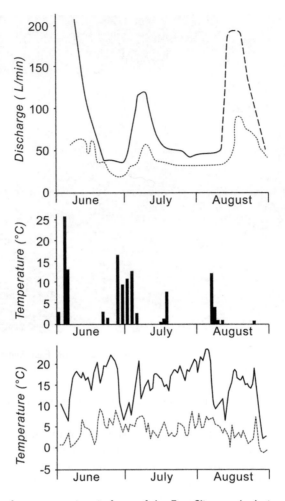

Figure 9.24 Discharge from two springs in front of the East Slims rock glacier compared with mean daily temperature and the precipitation times and amounts (Harris *et al.*, 1994, Figure 6).

1981; Harris, 2016a). The angle of slope is greater than 10°, but less than the maximum angle of repose of these sediments (Romanovsky & Tyurin, 1986), *c.f.* AGI (2005). Synonyms include *stone runs* (Darwin, 1839), *Blokströme* (Büdel, 1939), *rock streams* (Kesserli, 1941), and *rubble streams* (Richmond, 1962). In Russia, they are referred to as *kurums*. *Cryogenic block streams* refer to those formed in areas of permafrost, *e.g.*, Northern Siberia and Northeastern Tibet. Figure 9.25 shows a typical example from the eastern slopes of the Qinghai-Tibet Plateau, China. In both the Falkland Islands and Siberia, they usually originate in block fields higher up the mountainside (Romanovsky *et al.*, 1989). They represent a good example of landforms exhibiting equifinality, *i.e.*, where similar-looking landforms are produced by different processes in dissimilar climates. Although they do not necessarily occur in areas of

Figure 9.25 Block streams descending from a talus slope near Maqú (34°2′N, 102°30′E.), eastern slopes of the Qinghai-Tibetan Plateau at 4300 m. From Harris (2016a, Figure 1).

present-day permafrost, the blocky surface of all block streams can produce the diode effect on the ground temperature profile beneath them (Romanovsky & Tyurin, 1986; Harris, 1994).

Block streams were first noted in the Falkland Islands by Darwin (1839) and in Siberia by Middendorf (1867), and there have been numerous theories as to their origin. Recent detailed studies now provide significantly more data, and allow a better picture of their origin. Actively developing cryogenic block streams are found where there is a very cold climate, *e.g.*, on the mountain sides at high altitude in China (Harris, 1996; Harris *et al.*, 1998), and on the slopes of mountains at varying altitudes and latitudes in Central Siberia. The climatic conditions in areas where active block streams are among the dominant landforms are shown in Figures 5.1 and 9.26. They are characteristic of dry, cold regions with permafrost beneath or near them.

9.10.1 Characteristics

Active cryogenic block streams are found where the slope exceeds 10° but is less than the maximum angle of rest of the blocks, implying that some process or processes besides gravity is involved in the movement of the rocks. The source of the blocks varies from angular/subangular stones coming from frost shattered bedrock (Lozinski, 1912; Demek, 1960), lahars or debris flows (Matsumoto, 1970), or from frost-jacking of clasts from tills on the Qinghai-Tibet Plateau (Harris *et al.*, 1998). Corestones remaining from removal of the interstitial weathering products from past warmer climates have been described as forming block streams in Tasmania (Caine, 1968a, 1968b, 1972, 1983) and the Falkland Islands (André *et al.*, 2008), and are also present at Felsen, Germany (Büdel, 1937; Hövermann, 1949). The constituent blocks vary

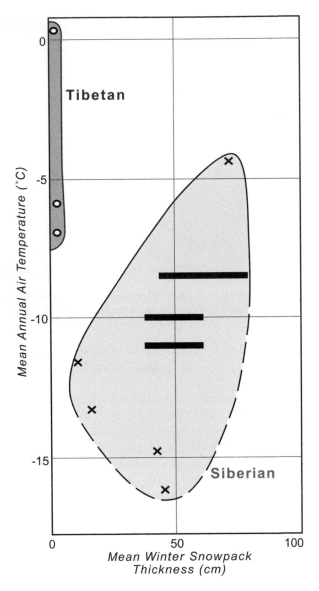

Figure 9.26 Relation of active Tibetan and Siberian dynamic block streams to both mean annual air temperature and mean depth of the winter snowpack From Harris (2016a, Figure 2).

from small angular stones ejected from till by frost-jacking or frost shattered bedrock (Harris *et al.*, 1998), up to large boulders over 2 m in length as at Felsen in Germany. These large boulders exhibit a subrounded to rounded shape, suggesting that they represent corestones produced by tropical weathering left by erosion of the interstitial finer sediment. At Felsen, a column, carved by the Romans from one of the blocks, shows little sign of frost action in spite of surviving two Neoglacial events.

Blocks produced by frost shattering of bedrock commonly have a roundness of 0.4 to 0.8 on the Krumbein scale (Krumbein, 1941), whereas blocks in lag block streams exhibit a roundness of 0.3 to 0.4 near the Chinese Upper Glaciological Station south of Urumqi, based on measurement of the shape of 100 clasts. Probable corestones such as those at Felsen, commonly exhibit even greater roundness (0.5–0.8 at Felsen).

In active block streams, either the blocks move while the underlying sediment remains relatively stationary (*dynamic block streams*), or the blocks may remain relatively stationary while the interstitial sediment is removed by running water (*lag block streams*). In the latter case, an alluvial fan is normally found at the terminus of the block stream, composed of material washed out from the matrix of the block stream (Harris, 2016a). Most active dynamic Siberian and Tibetan block streams are positive features, *i.e.*, they form a low deposit of blocks on the surface of the adjacent sediments, which they over-ride (Romanovsky & Tyurin, 1983, 1986; Harris *et al.*, 1998). Lag block streams are negative features on the landscape, generally occurring in gullies. Those blocks which move downslope normally show a fabric with the long axes of the blocks forming a primary mode indicating the direction of movement (Caine, 1968b; Harris *et al.*, 1998). The lag blocks either show the fabric of the parent material, or an essentially random fabric in the case of some corestones left from tropical or subtropical weathering.

On the Qinghai-Tibet Plateau, the ice content in winter in the actively moving blocks consists of a thin coating of ice formed by the melting and refreezing of a thin dusting of snow. In Siberia, burial by up to 45 cm of snow in winter results in the formation and diurnal thawing and refreezing of interstitial ice and ice lenses. Actual yearly movement of individual blocks has been measured as high as 120 cm in Tibet, although the average movement of the stones is 20 cm/a^{-1}. Thickness of the layer of blocks varies from a single layer up to 50 cm or more.

In Siberia, the thickness of the block streams varies from under 30 cm near the Arctic Ocean to 4 m in south central Siberia. Unfortunately, there is no data on rates of movement from there. Vegetation is normally lacking on all active dynamic block streams, in contrast to the surrounding tundra. In contrast, lag block streams are usually covered in lichens and occasionally partially by trees, indicating that they have remained stationary while the interstitial sediment is washed away. Block streams can also be formed under cold conditions lacking permafrost (Hack & Goodlett, 1960; Dahl, 1966; Ives, 1966; Rapp, 1967).

The occurrence of all types of actively forming block streams is controlled by the climate, lithological characteristics of the solid rock or the composition of the unconsolidated deposits, the ruggedness of the terrain, and the tectonic features of the area (Romanovsky & Tyurin, 1986), but have not been reported as forming today in warm temperate to tropical climatic conditions. Changing climate can result in cessation of the formative processes if the block stream becomes buried by other sediment or by an ice advance (Kleman & Borgström, 1990; Marquette *et al.*, 2004). Where buried relict block streams become exposed at the surface, renewal of the process of washing out of the interstitial sediment may take place, regardless of the local climatic conditions, provided there is water moving through them. Careful study of relict block streams often provides evidence for a complex history of the evolution of the climate and landscape at that particular location which may span millions of years, especially in the southern hemisphere.

Table 9.2 Characteristics of cryogenic active block streams over permafrost (from Harris, 2016a, Table 1).

Property	Dynamic cryogenic Block Streams		Lag cryogenic block streams
	Tibetan	Siberian	
Location	Tongues descending from cliffs over other slope deposits		In gullies and on narrow valley floors, usually ending in alluvial fans
Source of the blocks	Till or bedrock		Bouldery till, rock glaciers, rocky sediments
Roundness (Krumbein scale)	0.17 to 0.2	No Data	0.3 to 0.4
Lichens present	No		Well-developed except on limestones
Positive or negative landforms	Positive		Negative

9.10.2 Classification

Table 9.2 shows a genetic classification of actively forming cryogenic block streams using the currently available data. Cryogenic block streams can be active or relict. As noted above, the actively-developing block streams are divided into two groups, *viz.*, those dynamic block streams where the blocks actually move downslope, and the lag block streams in which relict blocks remain stationary while the interstitial material is slowly removed by flowing water, primarily in the spring.

Cryogenic dynamic block streams result from two distinct sets of processes that are dependent on the climate. The Tibetan dynamic block streams of the Qinghai Tibet Plateau are formed in areas with minimal winter snowfall (Harris *et al.*, 1998a), while the Siberian dynamic block streams occur in the Ural mountains and in northern Siberia, where the winter precipitation is higher than on the Tibetan Plateau (Romanovsky, 1985; Romanovsky & Tyurin, 1983, 1986; Romanovsky *et al.*, 1989) and the mean annual air temperature (MAAT) is lower (Figure 9.27).

Cryogenic lag block streams have been reported from the Chinese Tien Shan (Harris *et al.*, 1998a), and from the more southerly parts of Siberia where the permafrost is relict (Perov, 1969; Tyurin *et al.*, 1982; Romanovsky, 1985). They are found where there is a considerable source of water moving downslope during part of the year. It is important to note that lag block streams can also be active in cold climates where permafrost is absent.

Relict block streams are found scattered around the world in a variety of situations. So far, there are too few descriptions and detailed studies to produce a genetic classification. Instead, several examples will be described, showing the wide range of variation in past history of the landscape in which they are found. They occur at scattered locations across North America, Europe, Asia, Australia, and on the Antarctic Peninsula. They usually represent relict deposits formed on the remains of older landscapes, although dating them has only recently been achieved by cosmogenic dating

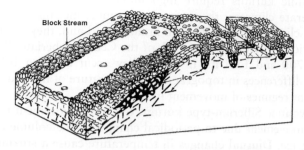

Figure 9.27 Drawing showing the typical location of active block streams in Yakutia (from Tyurin, 1979).

methods (Marquettte *et al.*, 2004; Gray *et al.*, 2005). Rea *et al.* (1996) and Whalley *et al.* (1997, 2001) have used clay content and thickness of related soils to argue for Pre-Quaternary ages of blockfields in Northern Norway. However, the most compelling arguments are based on the underlying sediments, pioneered by Caine (1968a) on the block slopes in Tasmania as discussed below.

9.10.2.1 Siberian active dynamic block streams – kurums

These were first noted by Middendorf (1867), but detailed studies were not carried out until 1911 (see Romanovsky Tyurin, 1983). Initially, Lozinski (1912) showed that both climate and lithology were closely connected to their area of occurrence in Siberia. They occur as "filmy" layers about 30–40 cm thick close to sea level in the humid Arctic areas such as the Polar Urals (Matveev, 1963) but become thicker southwards, reaching 1 m in northeast Russia and Northern Siberia (Romanovsky and Tyurin, 1986). In Central and Southern Siberia, they continue to become thicker (up to 4 m thick), although many of these landforms are relict, being apparently formed during previous cold events corresponding to major glaciations in more humid areas. They occur in a region where there is a shallow snow pack in winter which results in several processes causing the movement down slope. They tend to occur on slopes around block fields that cap the tops of rock outcrops on hills (Figure 9.27). A good summary of the past work on these will be found in Romanovsky & Tyurin (1996). Tyurin (1985) described their interference with road construction in Siberia.

Klatka (1962) and Glazovskij (1978) carried out critical work in determining how they form. Lozinski (1912) proved that both climate and lithology determined where they formed in the region. Romanovsky and Tyurin concluded that climate, lithological features of solid rocks, composition of unconsolidated deposits, ruggedness of the topography and tectonic features all affected their distribution in Northern Asia. They are found from the arctic region, south to South Yakutia and North Zabaikalye. In the south, they average 3–4 m thick and are both very old and inactive today, whereas on Wrangel Island, Novaya Zemlya and Severnaya Zemlya, they are active, but only 3 to 40 cm thick.

Active dynamic kurums require high humidity in the Polar Urals (Matveev, 1963). In temperate climates, they occur near timberline, but they occur in different belts in each climatic region. Thus in Franz Josef Land, they occur in a belt from 50–160 m elevation, whereas in the north of the Central Siberian Highland, they are found between 700 and 1500 m. In Tuva, they occur between 1600 and 3500 m. Combined with differences in topography and the nature of the bedrock, the climate produces different regimes of movement in each area.

All active dynamic Siberian-type kurums owe their movement to a slow settling and downslope movement due to periodical changes in the volume of the individual blocks including ice. Diurnal changes in temperature cause a stirring of the blocks, but contribute little to the downslope movement. However, frost heaving occurs when growth of ice causes the ground to rise normal to the slope and is followed by vertical settling as the ice thaws, resulting in ground subsidence. The end product is movement downslope as in frost creep. However, the process is more complicated than in finer grained sediments.

Ice starts to form in the active layer in late autumn by crystallization of meteoric water (rain or snow, or dew) on the cold surface of the blocks. This produces limited heaving, and continues until the whole soil profile is frozen, which occurs approximately at the same time as a stable snow pack is developed. Most of the active layer is air-dry, and the ice tends to form an icy cryostructure, 5–15 cm thick above the permafrost table, although this layer varies in thickness and even in its presence from year to year.

Little changes during the winter but in spring, the meltwater from the thawing snow on the adjacent slopes enters the ground and forms an impermeable, near-surface, icy crust. The debris mantle fills with ice, beginning in the upper part of the slope but gradually extends down slope. The thickness of ice depends on the heat capacity of the blocks. The ice grows diurnally due to the daytime melting, pushing the stones apart, while excess water flows down the section to freeze where it meets negative temperatures. At night, the icy-covered blocks cool to negative temperatures close to the ambient air temperature.

The amount of ice built up depends on the size of the blocks and the thickness of the active layer. Smaller blocks accumulate more ice than coarse blocks. Where ice only partly fills the interstices lower down the slope, part of the water flows to the bottom of the active layer where it completely fills the voids.

The accumulation of water as ice leads to heaving corresponding to the expansion of water changing state to ice. Vedernikov (1959) showed that the larger blocks cool faster at temperatures below 0°C. This is why icy coatings form preferentially on them, and then the remaining water becomes concentrated in small, closed spaces. Pressure generated by crystallisation of ice then forces the blocks further apart, resulting in further heaving. Total downslope movement can reach several centimetres a year, provided no obstacles interfere with the movement.

Laboratory experiment show that moisture retention by the blocks is good on slopes up to 15°, but results in limited speed of movement due to the low slope angle. Annual movement increases as the slope steepens to 20°, but after that, the water flows too rapidly through the interstices in the blocks and the amount of movement decreases. On slopes above 40°, gravitational forces take over and the annual movement increases with increasing slope angle (Ospennikov, 1979). Cryogenic deserption (creep) of the blocks results in movement of the blocks as separate parts with a downslope component instead of movement perpendicular to the slope. This is only possible at the surface. At depth, the expansion in all directions is impossible due to the lateral space problem. When surface thawing occurs, the icy upper layer thaws first due to the high thermal conductivity of ice, and this exposes the honeycomb structure in the underlying layers. The thawing front takes the form of cells, and the fragments of rock slip downslope in their individual cells (Cigir, 1977). In the Arctic, this collapse is observed every 1–2 years, whereas in the deeper blocky active layers further south, it may only occur once in every 5–6 years.

Night time cooling results in refreezing of the water films resulting in uplift of the blocks perpendicular to the slope. Thawing by day results in sinking of the material in the block stream in a vertical direction at their new, downslope location. The larger blocks have thicker water films, so move further in any given freezing event, but the smaller blocks freeze more often, which may result in greater total movement in a season. Any movement of the lower fragments as they thaw causes all the layers above to move down slope. Thus the overall movement decreases with depth in the blocky active layer as in frost creep.

Maximum downslope movement is seen in the Arctic filmy kurums where the entire active layer becomes saturated with ice. Southwards, the thickness of the blocky active layer increases and the zone saturated with ice decreases. The warmer climate to the south is also involved, though this is less important than the amount of water turning into ice. However there is a decrease in movement as the thickness of snow pack increases, resulting in open taliks. The usual diode effect occurs where there is a poor snow cover, but is ineffective where the snow is deeper.

9.10.2.2   The Tibetan type of active dynamic block streams

The dynamic Tibetan block streams are relatively thin, active block streams found on the eastern parts of the Qinghai-Tibet Plateau at an elevation of 4300–5100 m amsl, and cannot form if the snow pack is deep (Figure 9.26). The difference in the winter snow cover clearly causes different processes to produce the same block stream appearance of the landforms as in Siberia. So far, transitional environments with active block streams have not been reported.

Parent materials are partly supplied as blocks ejected by frost jacking from tills (Harris et al., 1998a), but they are also supplied by frost action breaking blocks off rock outcrops of jointed rocks forming cliffs or crags upslope under cold climatic conditions (Lozinski, 1912; Demek, 1960). The geomorphological position of the block stream tends to make each occurrence unique. River erosion or the development of icings can modify the form and processes involved in their formation.

The thin active dynamic block streams of the Qinghai-Tibet Plateau on the south-facing slopes of mountains at about 4800 m were studied by Harris et al. (1998a) on the Kunlun Mountains at 35°50'N and 94°05'E. The site consisted of a 15 cm thick layer of blocks moving down a slope of 31° over a sandy loam of lacustrine origin (Figure 9.28). The blocks were derived from both frost shattering of exposed bedrock at the crest of the slope and by ejection of blocks from a till capping the hill. No water flows in the block stream and its surface is only slightly raised above the adjacent soils. Mean annual air temperature is −6°C and mean annual precipitation is about

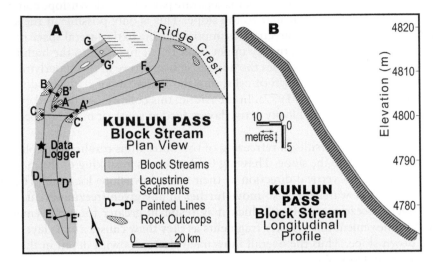

Figure 9.28 Plan (A) and cross-section (B) of the Tibetan-type block stream studied in the Kunlun Mountains (from Harris *et al.*, 1998a, Figure 3A and 3B).

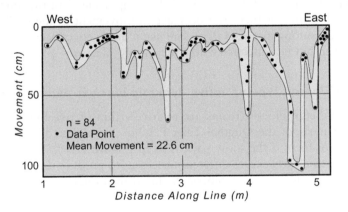

Figure 9.29 Movement of the painted line block stream along line A-A' between 24th July, 1990 and 19th August, 1991 (from Harris *et al.*, 1998a, Figure 3C).

320 mm/a^{-1}. The few plants growing between the blocks exhibited elongated stems and/or roots, indicating active downslope movement of the blocks. The blocks are gradually extending downslope.

Rates of movement of the blocks were measured using painted straight lines across the rock stream (Figure 9.28A). The movement was found to occur primarily in winter, in spite of minimal snowfall. The rocks tended to acquire a coating of ice, either from the formation of an icy coating when warmer air passed over very cold blocks, or when the limited snow cover thawed and then refroze on the surface of the blocks. Any disturbance caused the individual blocks to slide downslope varying distances (Figure 9.29), often rotating as they moved. The average annual movement was 22.6 cm/a^{-1}, with the greatest movement of individual stones being $105\,cm/a^{-1}$. Movement rates varied both laterally and longitudinally on the surface of the block stream. No liquid water was involved in the movement and the rock stream differs from talus by always being on a slope lower than the maximum angle of rest of the blocks. There is no sorting by clast size, unlike talus slopes. The rates of movement suggest that the block stream commenced its growth about 500 years ago at the lower limit of the zone of active block streams, probably in response to the cooling at the beginning of the last Neoglacial event. Those at higher elevations are much longer and probably date from earlier cold phases of the Neoglacial.

Mean annual ground temperature beneath the blocks averaged 7°C lower than in the adjacent soil, implying that cold air was entering and being trapped in the fabric of the rocks (Harris, 1996). Permafrost is present in the sandy loam at about 1 m depth, though it is absent in the adjacent loamy soils. The ground also has a higher moisture content beneath the blocks, which explains the occasional plant growing on the landform.

9.10.2.3 Active cryogenic lag block streams

These show a good covering of lichens on their upper surface, unlike the clean surface of the stones in active dynamic block streams. Good examples of these are found in the Chinese Tien Shan (Figures 9.30 and 9.31), near the Upper Glaciological Field Station of the former Lanzhou Institute of Glaciology and Geocryology (now called CAREERI). There, they form wherever there is a bouldery till with fast flowing water flowing over or in it that can erode the interstitial fine material in which the boulders are embedded. This is occurring today where meltwater and runoff descend a 30° slope from bedrock above (Figure 9.31), or where the entire drainage from the hanging valley known as the empty cirque, flows down the over-steepened slope of the main valley of the Urumqi River (Figure 9.32). An additional area with an inactive lag block stream is present in front of a moraine on the valley floor just to the east. The moraine is low, but formed in front of a cold-based glacier with minimal sediment in the ice. It must represent a long stand-still stage of the glacier during which, great quantities of meltwater eroded the interstitial sediment in the main drainage area below the moraine. The Glaciological Station is located at 43°04′North and 86°30′East, at 3550 m elevation. The mean annual temperature is −5.4°C, and the MAAT remains below 0°C from September to May. The average precipitation is 420 mm/a. The active lag block streams are found between 3700 and 3800 m elevation. The resulting exposed boulders are more rounded than those found in the other active Tibetan and Siberian type block streams, and are covered in lichens (Figure 9.32). That would be impossible if the boulders were moving. The boulders exposed at the surface have a shape of 0.3 to 0.4 on the Krumbein scale.

9.10.2.4 Inactive, relict block streams

These have been reported scattered from locations on all the continents except Africa. Unlike active block streams, the boulders are often subangular to subrounded. In the case of the Odenwald rock stream at Felsen, Germany, the boulders along the margins of the stream have trees growing over them (Figure 9.33). A similar inactive block stream with subrounded boulders occurs at Vitosha Mountain in Bulgaria.

Figure 9.30 An active lag block stream in a gully above the access road to the Upper Glaciological Field Station, Chinese Tien Shan. Note the alluvial fan on the more gentle slope below the main cliff. From Harris (2016a, Figure 3).

Relict block streams can occur in a variety of situations. The simplest case is where the climate has changed, becoming warmer and drier. In North America, such relict block streams are found in the bottoms of gullies, and are presumed to have been active during part of the earlier cold glacial events, *e.g.*, the Hickory Run boulder field in Carbon County, Pennsylvania (Smith, 1953). They are also known from Elliston, Newfoundland. Narrow boulder streams in gullies also occur in parts of the Western Cordillera, *e.g.*, west of Laramie, Wyoming, although they have not been studied in detail. The widespread glacial erosion and deposition during the Late Wisconsin event is presumably responsible for their rarity in Canada. Snow melt and rain may or may

Figure 9.31 An active lag block stream on the front of a hanging valley below the Empty Cirque above the Upper Glaciological Field Station in the Chinese Tien Shan. The drainage from the valley flows rapidly down the steep slope, eroding the interstitial sediment between the boulders in the glacial till. From Harris (2016a, Figure 4).

not continue the erosional process under the present climatic conditions, exposing more blocks, depending on the local climate and base level.

Kleman & Borgström (1990) have suggested that block fields and other blocky deposits have been preserved beneath cold-based protective ice covers in west-central

Figure 9.32 Lichen covered subrounded to subangular boulders in part of the active lag block stream on the front of the slope below the Upper Cirque in Figure 9.31 (from Harris, 2016a, Figure 5).

Sweden, while André (2004) has described evidence for this kind of preservation from Norway. At the Marmot Basin Ski Area near Jasper, Alberta, a block stream is reappearing from beneath a thin covering of glacial till in the valley above the Middle Chalet (Figures 9.34 and 9.35). The lower part of the block stream has emerged completely except where the Ski Hill operator has put gravel over the southern margin for the access road to the upper ski lifts. Higher up the valley, the till can be seen gradually being washed into the interstices of the blocks, exposing the ancient block stream.

A pit excavated in the blocks exhibited about 90 cm of blocks overlying similar blocks with a sandy matrix. Water flows on the surface of the sand and is collected in a small dam at the snout of the block stream built 40 years ago to supply the resort with water the year round. The absence of a delta in the pond behind the dam suggests that there is negligible erosion of the sandy matrix occurring at the present time.

In the Southern Hemisphere, relict block streams have been reported from both Tasmania and the Falkland Islands. Both areas suffered limited glaciation, so that the land surfaces are old and were subjected to cryological processes during the cold periods that affected Antarctica instead. However, the climate of Australia has also been influenced by northward movements away from Antarctica, due to plate tectonics. Caine (1968a) carried out a detailed study of the block streams at Mount Barrow, Tasmania by digging trenches across them. At the bottom of the trenches was a layer of residual clays with pseudomorphs of dolerite rock, typical of tropical soils. This

Figure 9.33 Part of the inactive block stream at Odenwald, near Felsen, Germany. Note the lichens and the location along the floor of a gully situated today on a mature, forested mountainside. Mean shape is between 0.6 and 0.7 on the Krumbein roundness scale, based on measurements of 50 blocks (Harris, 2016a, Figure 6).

was overlain by residual clays with an admixture of sand but no pseudomorphs of cobbles, which in turn are overlain by a layer of residual dolerite cobbles in yellow-brown clay. These deposits graded upwards into the yellow-brown silty sand with few cobbles, while above this was a layer of muck with blocks and cobbles present. Finally at the surface was the open-work mass of blocks ranging between 40 and 150 cm in thickness. Caine interpreted the middle layer as being the remains of a temperate soil, separated from the tropical soil and the overlying blocky deposit by unconformities. Thus the sedimentary sequence represents the results of a series of different climatic events, culminating in the formation of the block streams. Subsequently, Caine (1968b, 1972, 1983) was able to demonstrate the range of soils from valley bottom to mountain top over the entire landscape.

Numerous theories have been advanced about the origin of the block streams and block fields in the Falkland Islands. However, André *et al.* (2008) recently carried out a similar study to that of Caine, and found that the block streams overly a similar sequence of layers to that found at Mount Barrow. It started with chemical weathering under subtropical conditions, *i.e.*, the production of a Tertiary regolith. This was followed by a period of mass wasting, and the development of a temperate forest soil. Then regolith stripping occurred with downslope accumulation and matrix removal, possibly corresponding to the Quaternary cold stages. Subsequent periglacial reworking of the remaining blocks produced the block streams. Today, bioweathering of the

Figure 9.34 The relict block stream formed prior to the last glacial advance down the east-west trending valley above the Middle Chalet, Marmot Mountain Ski Area, Jasper National Park, Canada. Erosion has removed part of the thin covering of glacial till, and water is flowing within the block stream at a depth of about 1.5 m. Below this is clean sandy matrix to the boulders. The lower limit of permafrost is about 40–50 m upslope, and the water is collected behind a dam at the block stream terminus and supplies the ski resort with water throughout the year. From Harris (2016a, Figure 7).

Figure 9.35 Close-up of the margin of the reappearing rock glacier above the Middle Chalet Marmot Ski Area, Jasper National Park. © S. A. Harris.

Figure 9.36 The relatively flat surface of the blocks on the block fields of the Falkland Islands.

boulders is producing limonite that stains the upper layers of the underlying sediments. A Tertiary palaeosol was described from nearby by Halle (1912).

The age of the surface blocks in relict block streams has also been studied using cosmogenic dating of surface blocks on slopes in the Falkland Islands (Wilson *et al.*, 2008). The ages range back at least as far as 731 ka but are inconsistent within a given valley, varying with location of the deposit. Larger stone runs appear to have formed over a longer period of time, but this has yet to be checked. They suggest multiple cold events causing accumulation of the angular-subangular blocks. This does not necessarily mean that the area was not glaciated.

Alternatively, the spread in ages could be due to spalling off of flakes due to weathering by lichens. However, OSL dating of the fine sediments below the blocks provides age-estimates of 16–54 ka (Hansom *et al.*, 2008). Thus dating of the block streams by this means is still problematic.

9.11 SURFACE APPEARANCE OF BLOCKY LANDFORMS

Unlike rock glaciers, the surface appearance of block streams and block fields is variable. Most do have a rough surface appearance as at Odenwald, Germany (Figure 9.33) or Plateau Mountain, Southwest Alberta, but others such as the block streams of the Falkland Islands and Marmot Basin have a relatively flat surface (Figures 9.34 and 9.35). That is why the British Army could cross the block streams in the Falkland Islands during the war with Argentina. Pissart (1992) encountered a similar phenomenon with the occurrence of boulder pavements in streams in front of snowbanks in the French Alps (see Figure 9.37). He concluded that the weight of the snowpack and associated ice pushed the boulders down into the underlying water-saturated soft sediment to produce the flat surface that he named the boulder pavement (*"un dallage*

Figure 9.37 A boulder pavement along the headwaters of the river flowing southwest from the MacMillan Pass, Yukon Territory. © S. A. Harris.

des pierres"). This may also have taken place in the Falkland Islands at some time in the past. The same phenomenon is observed on the surface of the block streams in Newfoundland, and on some block fields and block slopes elsewhere. An alternative explanation is that at some time in the past, aufeis formed and was thick enough for its weight to press the stones down into the underlying silts (Porter, 1966). At Marmot Basin, the flattening of the surface of the block stream was probably accomplished by the weight of the subsequent glacier that has since disappeared. Thus any heavy weight that ends up on a former block stream or block field for a significant length of time may result in the flattening of the surface of the blocks.

Chapter 10

Cryogenic patterned ground

10.1 INTRODUCTION

As used here, *cryogenic patterned ground* refers to the arrangement of stones or disturbances in the vegetation cover producing distinctive patterns resulting from convection
cells in the active layer or seasonally frozen ground due to freezing and thawing.
A typical example is shown in Figure 10.1. The term, *patterned ground* is descriptive, and has also previously been applied to patterns generated by wetting and drying
(Hallsworth *et al.*, 1955; Harris, 1958; Hunt & Washburn, 1966), erosion during
rainstorms (Kelletat, 1985), as well as patchy distribution of vegetation. It is therefore
a name of convenience, acting as an umbrella for a variety of different features produced by several different processes. Accordingly, the term must have the adjective,
cryogenic, to adequately describe the subject matter of this chapter.

Figure 10.1 Sorted stone circles 2–3 m in diameter with gravel borders about 0.25 m high at
Broggerhalvoya, northwest Spitzbergen. © B. Hallet.

Figure 10.2 Change in shape of sorted polygons down a slope in the Falkland Islands. © M. Chambers.

Washburn (1956) summarized the 19 processes suggested to cause patterned ground up to that time, using his broad definition. The processes involved often act together in different combinations. Subsequent field studies have demonstrated that there are certain specific conditions under which the patterns develop. Most convective processes do not produce significantly large topographic forms, as will be demonstrated below. However, these processes have the potential to interfere with engineering projects due to the strong forces involved in the movements.

In this book, those processes producing hummocks without involving seasonal convection-type movement of soil have been dealt with in Chapter 7 as being a form of mound. Likewise, the patterns related to thermal cracking of the ground are dealt with in Chapter 5.

10.2 FORMS OF CRYOGENIC PATTERNED GROUND

Cryogenic patterned ground comes in a variety of sizes, the microforms consisting of circles or polygons <1 m diameter, occurring in areas with seasonal frost, while macroforms (2–50 m in width) are found in areas underlain by permafrost. In both groups, there can be isolated circles and polygons which are formed on essentially flat surfaces. These patterns grade into nets, and then into steps or stripes as the slope increases (Figure 10.2), the exact slope angle for the transition depending on the microenvironment. Matsuoka *et al.* (2003) found that there was a continuum in diameter from microforms to diameters of 1.5 m for the sorted stripes and circles on the upper mountain slope in the Upper Engardine, Switzerland. The forms larger than 20 cm diameter were called *intermediate forms,* and are regarded as being formed by the same processes as the macroforms. This interpretation agrees with that of Haugland (2004) in studies in the Jotenheimen in Norway.

The macro-stripes range up to 20 m across, while microforms are typically spaced at 10–20 cm intervals. In addition, the circular forms include low, hummock-like forms under extremely cold conditions (MAAT $<-5°C$) with a very shallow active layer.

Sometimes, the sediment shows negligible sorting and the feature appears as a break in the vegetation cover. In other cases, there is obvious sorting of the finer material from the coarser rocks and pebbles. Usually the coarser material is accumulated at the margin of the circles, polygons or nets, or it forms a narrow band running down slope. Occasionally, the coarser material is found as islands in the surrounding sediment.

10.3   FACTORS AFFECTING THE DEVELOPMENT OF CRYOGENIC PATTERNED GROUND

Obviously the first requirement is a period of freezing temperatures, diurnal or seasonal, resulting in freezing of at least part of the soil surface. The more intense and prolonged the freezing, the more likely the development of the landforms. Another key climatic factor is the presence of adequate moisture in the upper part of the soil, since the uneven growth of ice lenses in this layer and their partial or complete thawing is a fundamental part of most of the processes involved. Wang & French (1995) noted the limited heaving due to growth of ice lenses on the Qinghai-Tibet Plateau and suggested that the desiccation of the middle of the active layer was responsible for the absence of patterned ground there. However, nonsorted circles are active today on the summit of Plateau Mountain in spite of the existence of a dry zone from 1.0–1.7 m in depth within the active layer (Harris, 1998). The position of the movement relative to the dry zone when the latter is present will probably be a component of any genetic classification. The grain size distribution of the sediment is also fundamental. If there are no pebbles or stones in the sediment, the sorting process cannot take place (Goldthwaite, 1976).

Macro-scale sorted patterns only develop on mixed parent material composed of a wide range of clast sizes. A high silt content aids in the movement of moisture and finer particles to the freezing plane in the active layer (Corte, 1966b), and favours the development of substantial ice lenses. When these thaw, they can result in liquefaction of part of the surface layers, resulting in flowage if they are under compression. Sometimes, the wet, fluid soil flows out of the surface as a mudboil (Egginton & Shilts, 1978; Shilts, 1978; Dyke & Zoltai, 1980; Egginton & Dyke, 1982; Swanson *et al.*, 1999). This liquefaction aids in any circulatory movements. The low liquid and plastic limits of fine-grained soils aid in any movement caused by build-up of pressures within the active layer (Shilts, 1978). Even the finer grained forms of volcanic ejectamenta can develop small-scale patterns in perhumid climates due to the high relative humidity and moisture content of the ground (Figure 10.3). The hydrology of the microenvironment is also critical. There must be enough moisture in the zone of freezing and thawing to allow ice formation, even if it is only a few centimetres thick (Noguchi *et al.*, 1987).

Figure 10.3 Small scale (12 cm width) patterned ground in Katmai National Park, Alaska. © B. Burton.

Once segregation of the coarser clasts has taken place, the blocks forming the margins of the polygons act as drainage channels for the summer precipitation (Goldthwaite, 1976). It has long been recognized that variations in moisture gradients are much more effective in driving convection cells than differences in ground temperature (Mortenson, 1932; Gripp & Simon, 1934). Segregation of ice in silty soils develops concentrations of solid water that modify thawing rates and during the spring thaw, provide enough localized moisture with its lower density in the active layer to initiate the convection process (Hallet & Waddington, 1992; Hallet, 2013). The presence of salt in areas recently uplifted from the sea alters the thresholds of freezing and makes the soil more susceptible to movement, though this has not been mentioned in the bulk of the literature.

Sorted circles are particularly common in shallow ponds where they seem to form very quickly. This may be due to the abundant water reducing the strength of the sediments. Figure 10.4 shows a typical result in a dried-up seasonal pond in Alaska. Actual rates of formation have not been measured, but are probably much more rapid than in the case of sorted circles on mountain tops. So far, there have been no observations of rates of movement of the circulation under water, nor of the beginning of the sorting. Sorted circles that are either still in the process of undergoing sorting of the clasts or where the sorting process has either ceased or been reduced due to changes in

Figure 10.4 Typical mature sorted circles in a dried-up ephemeral lake bed in Alaska. The coarser stones have been moved to the junction between adjacent circulatory cells. The knife is 20 cm long. Note the even distribution of larger stones across the surface of the finer grained core. © S. A. Harris.

the microenvironment appear to show a range of abundance of coarser clasts increasing towards the margins of the central finer-grained core, as well as a poorly defined boundary between the coarse and fine sorted sediments (see Křížek & Uxa, 2013, Figure 5).

Vegetation cover is very important since the patterned ground normally occurs in windswept locations. The roots of plants act as anchors, normally resisting movement of the rooting zone. Since the pressure of the outward soil movement decreases with distance from the centre of the convection cell where the motion is occurring, the outer margin of the circulating soil in the cell can extend a considerable distance beyond the beginning of the vegetation cover in non-sorted circles (Harris, 1998a). However, some plants are specifically adapted to surviving movement by elongating their upper part of their roots to keep pace with the movement of the surrounding soil (Figure 8.13). They may grow as single plants on an otherwise bare, mobile soil. Once the surface movement of the bare soil ceases, gelatinous algae or cyanobacteria (Figure 10.3) harden the otherwise bare soil surface outside the High Arctic. This explains the hard, smooth surface of the finer grained surface sediment that resists wind and water erosion on many bare patterns.

As already noted, the shape of the cryogenic sorted pattern changes as the slope becomes steeper. Figure 10.5 provides the results of measuring the slope angle and shape of inactive sorted patterns on Plateau Mountain, Southwest Alberta

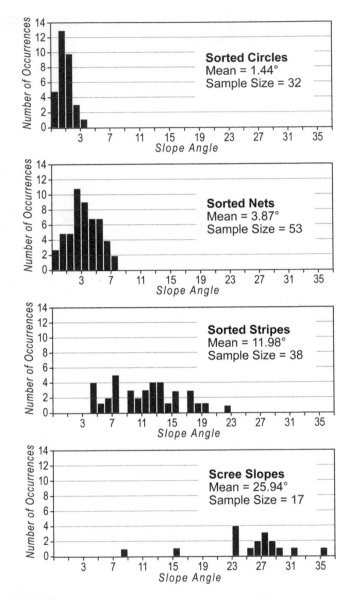

Figure 10.5 Statistical distribution of inactive sorted patterns of different shapes in relation to slope angle (in degrees) on the summit of Plateau Mountain (Woods, 1977, Figure 11).

(Woods, 1977). Also shown are the slopes occupied by talus or block slopes. There is some overlap between the slopes on which individual shapes occur, but the patterns are normally absent on slopes steeper than 23°. The slope angles at which the changes take place vary with the climate on the mountains in the Cordilleras of North America, the angle required for change increasing with decrease in mean annual precipitation (Table 10.1). Table 10.2 shows the rates of downslope movement of the coarse and

Table 10.1 Measured ranges of limiting slope angle for cryogenic sorted patterned ground and areas of block fields.

Location	Stone circles or polygones	Stone steps	Stone stripes	Block fields	Block slopes	Maximum stone diameter	Reference
Plateau Mountain Alberta, 50°N	0–8°	2–16°	4–23°	0–8°	8–36°	25 cm	Woods, 1977
Mesters Vig. 72°N Greenland	0–7°	10°	>3°				Washburn, 1956, 1969
Alaska	0–5°	5–15°				30 cm	Sharp, 1942
W. Snake River Plain, Idaho				2–5°			Malde, 1964
Potter County, Pennsylvania	0–3°		2–5°	0–10°		15–90 cm	Denny, 1956
Lake District, England			9–19°				Caine, 1963
Mount Washington, New Hampshire	0–5°		5–7°		>25°	60 cm	Antevs, 1932
Xeric Andes of Bolivia & Ecuador			20–40°				Graf, 1976

Table 10.2 Measured rates of downslope movement of the coarse and fine sections of sorted patterned ground as related to type of slope processes involved.

Location	Period (years)	Rates (cm/year) Coarse	Fine	Processes	Slope angle	Sorting	Reference
Chambeyron, Basse Alpes	10	0.2–3.0	2.0–6.0	Gelifluction & Frost creep	7°	Active	Pissart, 1972
		0.18–0.5	4.0–7.0	Gelifluction & Frost creep	6–9°	Active	Pissart, 1977
Colorado Front Range	3	1.3–3.9	2.6–7.8	Gelifluction & Frost Creep	12–13°	Active	Benedict, 1970b
Cinder Cone, Garibaldi Park, B.C.	10	15	35	Needle-ice Creep	10–15°	Active	Mackay & Matthews, 1974b

finer bands as related to the processes currently involved. The movements of the finer bands can reach nearly 8 cm/a^{-1} and affect much deeper layers than those produced by needle ice. In spite of this, a larger volume of sediment is moved when gelifluction and frost creep are involved. The sediment is described as moving in a corkscrew fashion down the mountain side.

10.4 MACROFORMS OF CRYOGENIC PATTERNED GROUND

These are formed under conditions of continuous permafrost, discontinuous permafrost, or even seasonal frost. However, when the climate warms up in permafrost areas, the patterns usually remain unless the ground suffers from erosion, thus providing evidence of a former cold climate. Since there were many cold events during the last 3.5 Ma (Figure 4.13), determining exactly when the pattern first formed can be difficult. The typical churning movement will tend to occur in the finer grained soil every time it is subjected to a suitably cold climate, *e.g.*, during the Neoglacial episodes (Becher *et al.*, 2013), though any segregated stone margins may or may not move. Where organic soil horizons become buried, dating of the climatic events causing the burial is easy. In the case of the sorted polygons on Plateau Mountain (Alberta), there are no organic soils, but there are depressions over 1.5 m deep in the rocky borders of the patterns suggesting that ice wedges may have been present in the past. The stones are covered in large lichen thalli indicating that they have not moved for a very long time. This may hinder accurate cosmogenic dating due to spalling of the surface of the stones by biogenic weathering.

There is considerable debate as to how to classify these forms (Washburn, 1956; 1979; Mackay & MacKay, 1976; Tarnocai & Zoltai, 1978; Shilts, 1980; Mackay, 1982; *etc.*). They range from a one-time churning of the soil layers at a single point in areas without permafrost, to extreme seasonal churning taking place in the thin active layer in the polar regions. In northern latitudes, the movement takes place involving the entire thickness of the active layer, but towards the Tropics, it occurs in an upper ice-rich zone in the surface part of the active layer. Since there is an ice-rich zone at both the top and bottom of the active layer which are separated by a dry zone (Mackay, 1980a), the movement may involve most of the active layer, or merely the upper ice-rich zone. Further south, it takes place where permafrost is absent. Obviously, these are best dealt with in turn, although Walker *et al.* (2004) have argued that these landforms grade into one another, based on the results of a study of soil morphology in nonsorted patterns along a north-south transect in northwestern North America. Chernov & Matveyeva (1997) provide a similar description from the Russian Arctic, while Peterson & Krantz (2003) and Peterson *et al.* (2003) discuss possible models of the processes involved. Unfortunately, actual studies of the processes involved in specific cases are rare, though these are described in turn in the following account. Studies providing some important information but lacking precise studies of actual movements are also described.

10.4.1 Cryogenic nonsorted circles

Cryogenic nonsorted circles are a type of patterned ground showing some form of seasonal convective type movement as a result of freezing and thawing in either the entire active layer or in seasonally or diurnally frozen ground as a result of hydrostatic or temperature gradients that produce differential movements in part of the active or seasonally frozen layer. They are usually divided into perennial *mudboils, frost boils, nonsorted circles* and *plug circles*. Mudboils consist of circular patches of bare ground in areas with a tundra vegetation where fluidized sediment has come to the surface and spread out as a patch or circle of bare ground, surrounded by vegetation. Frost boils are frost-churned patches of bare soil found primarily in the subarctic zone of alpine tundra, as well as in forested areas lacking permafrost. The subarctic forms are perennial, but those in areas of seasonal frost are one-time events, the surface of which becomes revegetated within a short period of time. There remains a need for

Table 10.3 Comparison of the characteristics of the four main types of cryogenic nonsorted circles excluding plug circles discussed in this paper. LL is the liquid limit and PL is the plastic limit. Two other kinds of nonsorted circles (maritime nonsorted circles and plug circles) have been partly described, but their characteristics need more study.

Property	Arctic mudboils	Subarctic mudboils	Zeric nonsorted circles	Frost boils
Topography	Marshy lowland	Plateaus and flat hilltops	Mountain tops	Mountains
Vegetation	Wet tundra	Tundra	Alpine tundra-Shrub tundra	Alpine-Boreal forest
Soil	Silt-clay	Silt-clay LL <20%; PL <10%	Loams	Silty sediment
Drainage	Wetland	Relatively good	Good-very good	Good
Cryological condition	Active layer <40 cms thick	Active layer c. 150 cm	Active layer 1.5–2.5 m	Seasonal frost
Process	Slow convection throughout the active layer	Diapir breaking through the central desiccated active layer	Slow convection above the desiccated active layer	One-time diapir at random location
Reference	Mackay, 1980	Shifts, 1978	Harris, 1998	

further study of nonsorted circles developed in deep active layers in humid climates since there appear to be no actual studies of the mechanics of their movements, despite partial studies of some of their attributes and descriptive studies of their distribution in northwest North America. Plug circles are where partial upward movement of a plug of subsoil causes an updoming resulting in low mounds in extreme Arctic climates, *e.g.*, at Resolute.

Climate and hydrology are the main controlling factors determining the processes involved, together with the presence or absence of a vegetation cover. Mackay (1971) introduced the concept of three layers in the active layer with different moisture regimes, and this is used in this paper to develop a classification of most of the cryogenic nonsorted circles studied in detail to date. The nomenclature used here attempts to highlight the critical differences between the processes involved in each environment. Table 10.3 shows the characteristics of the different kinds of mudboils and nonsorted cicles in the classification used in this monograph.

10.4.1.1 Cryogenic mudboils

Cryogenic mudboils form a random pattern in lowlands with silty and clayey sediments around the Arctic Basin, associated with tundra vegetation (Egginton & Shilts, 1978; Shilts, 1978; Dyke & Zoltai, 1980; Egginton & Dyke, 1982; Mackay, 1980a, 1980b). They can be divided into two groups, *viz.*, the *Arctic mudboils* occurring in wet lowlands in areas with active layers shallower than 40 cm and the *Subarctic mudboils* occurring on better drained undulating tundra with active layers of the order of 150 cm deep. These will be dealt with in turn below. The nature of the transition/boundary between them has not been adequately determined.

10.4.1.1.1   Arctic mudboils

Arctic mudboils are characterised by low height and small width, and occur in wet tundra, or more rarely, in the northern margin of the boreal forest around Arctic Red River, Northwest Territories. They are also widespread across the wet lowland tundra areas of Northern Canada. Synonyms include *frost boils* (Walker, 2008), *hummocks* and *mud hummocks* (Mackay, 1979b, 1980b). They are round to elongate in shape, up to 0.5 m in diameter, 10–15 cm high, are perennial, and often have bare soil patches on the surface that mark places where mud has flowed on to the ground surface. The sediment in them has a high silt and clay content with a liquid limit less than 20% moisture and a plastic limit lower than 10% water. This means that they liquefy readily. In the cases discussed by Mackay (1980b), the churning extends down to the permafrost table. Undoubtedly, similar forms occur in the upper parts of deeper active layers, but these remain to be examined in detail. On slopes, the mudboils become elongated.

Figure 10.6A shows the relationship of the distribution of Arctic mudboils to freezing and thawing indices. These mudboils disrupt the natural surface runoff in the Arctic, where they are ubiquitous in wet lowlands. Surface water has to wend its way along the maze of depressions between the hummocks (Quinton & Marsh, 1998), while the shallow permafrost table prevents appreciable water movement through the soil. However, the mudboils facilitate shrub expansion in Low Arctic tundra (Frost *et al.*, 2013).

Early theories regarding their origin involved cryostatic pressures in pockets forcing the soil to form a diapiric movement when the ice lenses thawed (Crampton, 1977), although he thought summer precipitation could add moisture to aid in the process. The frequent occurrence of extrusion of mud through cracks in the surface in summer indicates that at least sometimes, significant pressures are built up during the summer between the dry surface layer and the permafrost table.

After considerable study, Mackay developed the equilibrium theory that is currently accepted today (Figure 10.7). The opposite curvature of the top and bottom of the active layer aids in the postulated process. Ice lenses form at the top and bottom of the active layer in winter and gravity-controlled, cell-like movement occurs during thawing. This theory explains the grain size distribution, radiocarbon dates of the organic inclusions in the hummocks as well as the upturned tongues of saturated material in late summer (Zoltai & Tarnocai, 1974; Zoltai *et al.*, 1978). It also does not require substantial cryostatic pressures (Mackay & Mackay, 1976). The fact that these sediments around the Arctic shores were under the sea during deglaciation and have since been uplifted may indicate that salinity may also be a factor in some places. Ray *et al.* (1983) and Gleason *et al.* (1986) have suggested that the density difference between the water at about 0°C at the base of the thawed zone and the 3–4°C temperature of the water at the upper surface of the hummock may be sufficient to induce free convection of water in the mudboil. This could explain the concave (upwards) shape of the permafrost table, though the effect of summer warming of the larger surface area presented by a mound could also be involved. Shur & Ping (2003) have discussed the initiation of the mudboils, starting with cracking. However, their assumption that organic matter moves progressively down the marginal cracks is not in agreement with the old radiocarbon dates obtained from buried organic matter that sometimes date back several thousand years (8000–10,000 radiocarbon years – Douglas & Tedrow,

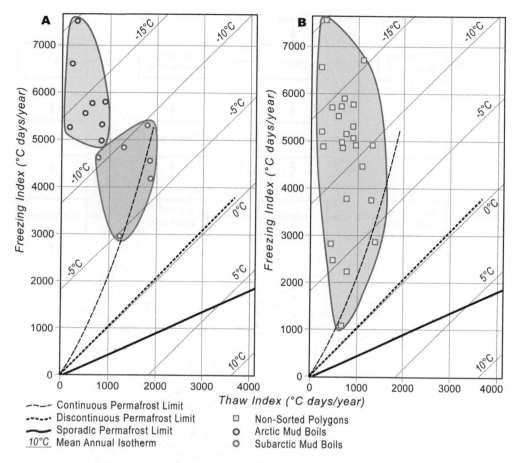

Figure 10.6 A, the relationship of arctic and subarctic mudboils to freezing and thawing indices, and B, the relationship of all reported nonsorted circles to freezing and thawing indices, omitting the subarctic mudboils.

1961; 4750 radiocarbon years – Trautman, 1963; 2140–3800 radiocarbon years – Dyke & Zoltai, 1980). Clearly the buried organic matter represents the products of past environments and climates, rather than those operating today. However, they provide a clue as to the antiquity of these landforms.

Active Arctic mudboils have been reported from Alaska (Hopkins & Zigafoos, 1951), the Arctic Islands (Washburn, 1947, 1950; Mackay, 1953), Greenland (Jahn, 1948), Sweden (Lundqvist, 1962), Antarctica (Vieira & Ramos, 2003), as well as from northern Siberia. A systematic study of these features was carried out from the northern Canadian Arctic Islands to Happy Valley along the Haul Road to Prudhoe Bay. Daanen *et al.* (2008) showed that the vegetation cover and thin snow pack provided heterogeneous insulation to the soil surface. During soil freezing, the non-insulated tops of the bare mounds froze first, producing preferential ice accumulation in these areas. This ice prevented expansion of the vegetation cover, stabilizing

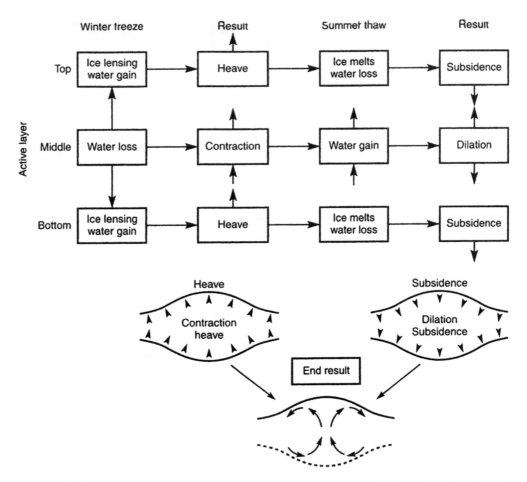

Figure 10.7 The formation of nonsorted circles (mudboils) in the Lower Mackenzie Valley, Canada, by the equilibrium model of hummock growth (from Mackay, 1980b, Figure 4).

the non-sorted circle pattern (Peterson *et al.*, 2003). The actual effect of these factors in the far north was far lower than that at the Happy Valley site, which is an example of the Subarctic mudboil type. The availability of water was critical for the sustenance and stability of the mudboils in all cases. Climatic variations did not seem to affect the freezing process. This is not surprising considering the extremely cold winters in the areas in which they occur. Michaelson *et al.* (2008) described the changes in soil properties along the same transect, together with the limited variations in carbon storage. The amount of vertical heave was 4 to 5 cm in the coldest locations in the northern Arctic Islands (Raynolds *et al.*, 2008). The time taken for a complete cycle of movement under the present climatic conditions was not measured, nor was the transition from the Arctic mudboils directly overlying permafrost that are discussed by Mackay (1980b) to the Subactic mudboils of Shilts (1978) described below.

Figure 10.8 Subarctic mudboils at the Arctic Circle along the Dempster Highway. A, general view, and B, a cross-section of the circle in the middle of the picture. © S. A. Harris.

10.4.1.1.2   Subarctic mudboils

Also called *stony earth circles* (Williams, 1959), these are found on the tundra near the tundra-forest boundary in North America. Figure 10.8 shows typical examples from the Dempster Highway at the Arctic Circle. The bare mud circles are larger (0.7–3 m) and higher (0.15–0.4 m) than the Arctic mudboils, and are surrounded by a higher vegetation cover of about 90% as at the Happy Valley site on the Arctic Transect. They were first described in detail from the better drained sites at Rankin Inlet (James, 1972).

Tyrell (1897, p. 50F) remarked about "little clay flats or discs surrounded by raised rings of grass" in the Canadian north, and they have since been examined and discussed by many workers. Shilts (1978) carried out the most comprehensive study made so far while studying the area in the Central Keewatin area, west of Rankin Inlet. He found mudboils with raised, bare surfaces developed in soils with high silt and clay contents and low liquid (<20%) and plastic limits (<10%). The sites were better drained than those of the Arctic mudboils, and the vegetation consisted of subarctic tundra, just north of the tree line. The active layer was generally about 0.8–1.5 m deep and consisted of an upper ice-rich zone up to 70 cm thick, a central desiccated dry zone about 30–40 cm thick, and a lower ice-rich zone above the permafrost table, similar to the threefold zonation of the active layer described by Mackay (1971). Raynolds *et al.* (2008) reported up to 20 cm of heave of the bare soil by the end of the winter. Figure 10.9 shows the arrangement of the layers and the ice lens content within the active layer in winter.

The excess ice comes from water moving over the surface of the mineral grains to the freezing fronts at the top and base of the active layer. There is usually a perched water table at the latter. The net result is desiccation of the middle layer of the active layer, which develops into a dry, hard carapace, imprisoning the lower part of the active layer and its excess ice. When the latter thaws, the result is a fluid that tries to escape to the surface along any planes of weakness such as cracks. Shilts also noted that the pressure in the active layer under the dry carapace can also be relieved by subsurface flow downslope over the permafrost table or bedrock, in which case, mudboils will not form.

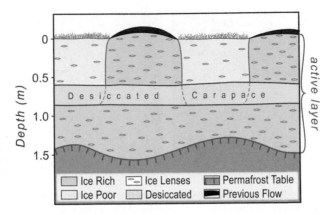

Figure 10.9 Diagrammatic cross-section showing the structure of subarctic mudboils in winter.

The upward movement of the plug may be a form of load casting (Kling, 1997). To cause the upward movement by load casting, Van Vleit-Lanoe (1991) suggested that the cryogenic structure in the layer that flows must be completely collapsed and the water content must be more the 10% above the liquid limit of the sediment. The weight of the overlying layers presumably provides the necessary pressure to cause the upward movement.

Figure 10.8B is a photograph of a cross-section of a mudboil along the Dempster Highway showing the lack of soil horizon development in the interior of the land-form. Small stones tend to become stranded on top of the mound as the finer material is washed/blown off the surface of the hummock. There is no sign of sorting of the clasts. Measurements made at Happy Valley, Alaska indicate about 26% greater heav-ing in the boil compared with the surrounding vegetated ground, indicating more ice segregation in winter, and this is accompanied by a drastic increase in soil instability. Similarly, the winter soil temperatures will be lower, causing preferential accumula-tion of ice lenses. This results in a greater tendency to continue movement once the surface of vegetation is disrupted. Mudboils are generally absent in the boreal forest, the ground of which has a continuous vegetation cover.

A fire destroyed the vegetation around these active mudboils near Ennardi Lake, Northwest Territories, and exposed a superficial ring of large stones around the bare patch that had been previously overlain by vegetation (Shilts, 1978). The stones did not extend downwards into the ground as in the case of sorted circles. Near Abisko, Northern Sweden, similar landforms have been described by Kling (1997), exhibiting definite evidence the upward movement of rocks and their accumulation on the ground surface in the form of a superficial layer along the margins of the bare areas. The structure of the stone accumulation is different to that of sorted circles, apparently supporting the conclusion that these are formed by processes distinct, though similar to those operating in active sorted circles. He also found evidence of the widening of the plug of bare, extruded sediment with time in the upper 50 cm of the ground. However, no circle of surficial stones was seen at the occurrences at Happy Valley,

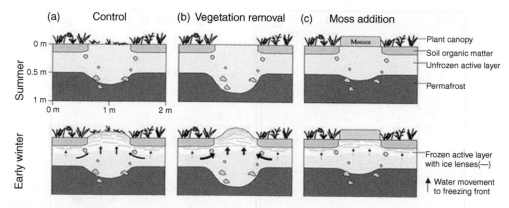

Figure 10.10 Effect of modification of the surface of sparsely vegetated frost boils developed in a thin active layer over continuous permafrost in northern Alaska (from Kade & Walker, 2008, Figure 6).

Alaska, nor along the Dempster Highway near the Arctic Circle, so the frequency of their occurrence is uncertain. They must correspond to an intermediate stage in intensity, trending towards the development of sorted patterned ground.

Kade & Walker (2008) carried out an experiment involving the addition of moss over the bare soil and also complete removal of the sparse vegetation over the top of the mudboil (Figure 10.10). The removal of the vegetation promoted deeper thawing in summer but more development of ice lenses with resultant surface heaving in winter. Addition of mosses caused a cessation of the differential ice lens development, terminating the movement. Theoretically, frost boils can develop into mud boils if the vegetation is not allowed to recolonize the surface of the disrupted soil, although no such cases have been described.

Kaiser *et al.* (2005) present a good account of carbon and nitrogen mineralization in the soils of mudboil systems in Siberia. Two-thirds of the carbon sequestering is in the troughs. Nicolsky *et al.* (2008) have modeled the biogeophysical interactions of mudboils on the north slope of the Brooks Range. Subarctic mudboils may also occur in the South Shetland Islands in Antarctica (Serrano & López-Martinez, 1998). The actual rate of churning is unknown. Drew & Tedrow (1962), working in Alaska, noted nonsorted stripes on moderate slopes. These are believed to form in similar ways to the circles, but with the additional effects of gravity on the movements on the slopes. Nonsorted circles are active in areas of discontinuous permafrost as well as continuous permafrost, though they are found in slightly colder regions than some sorted circles (Figure 10.8B).

10.4.1.2 Xeric nonsorted circles

These nonsorted circles are developed in the upper moist part of the active layer on the cold, well-drained tops and upper slopes of mountains at lower latitudes. Churning does not extend down into the middle drier zone of the active layer. The tundra vegetation is sparse in well-drained continental situations such as the Rocky Mountains of

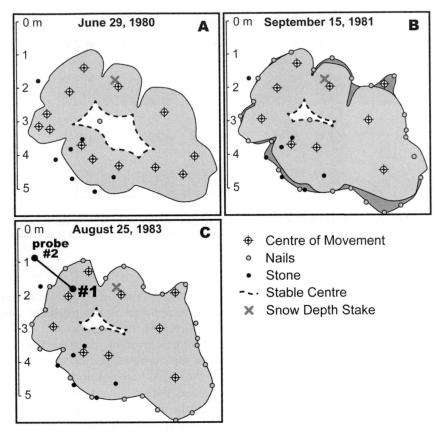

Figure 10.11   Change in shape of the mud flowing from the centres of circulation of embryonic non-sorted circles on Plateau Mountain, Alberta between 1980 and 1983 (after Harris, 1998a, Figure 4).

Alberta, but similar features can also occur in maritime areas such as northern Sweden if the weather is sufficiently severe (Becher *et al.*, 2013). At Plateau Mountain (latitude 50°55′N) in southern Alberta, the active layer is over 2.5 m deep, and the circulatory movements only operate to about 80 cm depth (Harris, 1998a). Thus they are occurring in the upper part of the active layer, separated by the central dry zone from the lower zone of high moisture content immediately above the permafrost table.

At 2519 m in alpine tundra on the flat summit of Plateau Mountain, Alberta, there were no obvious nonsorted circles present when climate measurements began in 1974. There was only limited soil formation in the soils (Bryant & Scheinberg, 1970), and the active layer is about 3 m thick. By 1979, the mean annual air temperature had dropped by 1.5°C, and areas of outflow of fluid, fine-grained soil appeared around centres cutting through the vegetation cover during early summer (Figure 10.11), killing the vegetation beneath them. These centres were often close together and the flows coalesced. The outflows continued until ceasing in 1984. Two telescoping, nested rods were then dug into the centre of one cell, with a second rod located well outside the flow

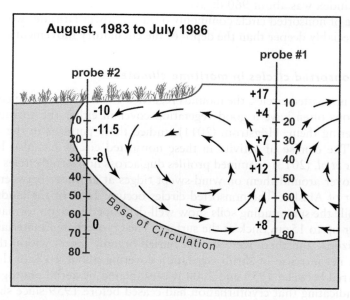

Figure 10.12 Apparent average rates of circulation (mm/a^{-1}) of the soil in a xeric nonsorted circle on Plateau Mountain, Alberta (after Harris, 1998a, Figure 7).

(Figure 10.11C). The rods were examined every month for three years and the vertical movements recorded. The movements started with the first snowfall in late August-early September, continuing until December, and then reappeared in the spring. The results of three years of measurements of the relative movements of the different segments of the rods indicated the pattern of movement shown in Figure 10.12. There was a clear circulation taking place in the upper 80 cm of the soil and it extended laterally considerably beyond the surface mudflow. The base of the centre of the circulation was convex, unlike the circulations in the active layer studied by Mackay.

Harris (1998) proposed the name *xeric nonsorted circle* to distinguish between these lower latitude forms from the mudboils of the Arctic. Average rate of movement appears to be about 12 cm/a, requiring about 480 years to complete a full cycle at the present speed. Meanwhile the surface of the mudflow had been stabilized by cyanobacteria and gelatinous algae. It is very dense and the surrounding higher plants have not colonized it, even after more than 30 years.

Beyond its rim, the vegetation has managed to survive, presumably because the zone of movement is below their shallow roots. The source of the moisture must be snow in the fall that lands on the warm surface of the active layer. There it melts, penetrating into the upper part of the active layer, where it freezes. Additional water is drawn upwards to the freezing plane from the underlying part of the active layer. Periodic chinooks may cause some additional meltwater to penetrate the soil during winter, but there is another big input of moisture from the isothermal snowpack in the late spring. Apparently, thawing of the ice lenses in the upper part of the active layer results in sufficient moisture in pockets in the soil to cause the buoyant, convective, circular movement. The freezing index was about 1700 degree days/a^{-1} while the thawing index was about 960 degree days/a^{-1} in 1983 (Harris, 2009). It is likely that this type of nonsorted circle commonly occurs at lower latitudes where the active layer is appreciably deeper than the depth of the circulatory movements.

10.4.1.3 Nonsorted circles in maritime climates

Unlike xeric nonsorted circles, the mounds regarded as being nonsorted circles in maritime situations often have a good vegetation cover in either the surrounding area, or even covering them. Högström (2011) studied these patterns in the Abisko area in Sweden. The demise of activity in these nonsorted circles has also been studied there. Becher *et al.* (2013) examined profiles dug across nonsorted circles with a good vegetation cover around them on wind-swept ridges at altitudes between 400 m and 1150 m, east of Abisko. The nonsorted circles occupy 16% of the landscape above tree line, while the surrounding soils show well-developed horizon formation with an organic layer up to 15 cm thick at the surface. The cryoturbated materials cut across the soil horizons, and there were several buried organic layers within the disturbed sediment. A net increase of shrub vegetation covering about 10% of the nonsorted circles occurred between 1959 and 2008 was estimated by aerial photographic interpretation, indicating that cryoturbation had ceased before 1959 since shrubs cannot grow on soils in areas that are affected by cryoturbation (Jonasson, 1986; Makoto & Klaminder, 2012).

^{14}C and ^{210}Pb dating of buried organic layers from ten profiles in the cryoturbated sediment correspond to three main time periods, *viz.*, 0–100 A.D., 900–1250 A.D., and 1650–1950 A.D. These correspond to periods that have been interpreted as corresponding to cooler climates by other proxy data (Karlén, 1988), while the period between 1250 and 1650 A.D. is believed to have been warmer, based on modelling of long term trends (Johansson, 2009). Becher (2015) interprets the surface organic horizon as having formed in warmer times after the climate had begun warming, and suggests burial of the organic horizons is the result of flowage of mineral sediment over the surface organic layer under warmer conditions. Formation of other distinct soil horizons is slow in this climate The unmixed B-horizon in the undisturbed sediment at Ridunjohka A is 1000–6000 years old, based on chronosequence studies for well-developed B-horizons to form in boreal environments (Ellis, 1980; Protz *et al.*, 1984; Barret & Schaetzl, 1992). The slow rate of soil formation may be partly due to low precipitation (Klaminder *et al.*, 2009).

This evidence clashes with the theories involving the evolution of earth hummocks from nonsorted circles that have been recently suggested by scientists in Alaska, *e.g.*, Shur *et al.* (2008). These were based on modeling and the use of photointerpretation, rather than actual field evidence. It seems that xeric nonsorted circles do not evolve into earth hummocks, at any rate in northern Sweden.

Where stones are present on top of relict nonsorted circles in northern Sweden, the rate of revegetation by lichens can be measured using the mean size of *Rhizocarpon geographicum* on noncalcareous stones at different distances from the centre of the circle (Figure 10.13). The age increases towards the margins, and in the case examined in Figure 10.13, the range of ages was from almost zero in the centre, ranging up to about 250 years on the margins of the bare part of the circle. In the past, this has been interpreted in two ways, either as an indication of repeated upfreezing in the centre

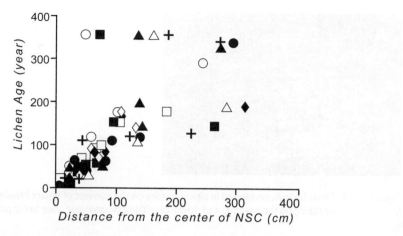

Figure 10.13 Increasing age of *Rhizocarpon geographicum* phalli towards the margin of relict nonsorted circles in northern Sweden (from Makoto & Klaminder, 2012, Figure 3).

with colonization occurring as the soil moves outwards, or as progressive reduction of the frequency of past fatal cryogenic disturbances towards the margin of the circle (Washburn, 1980; French, 2007). Makoto & Klaminder (2012) also found that herbs colonized the centre while shrubs were only found on the outer portions of the relict circle. The highest plant diversity occurred at about the 150-year location, which is consistent with other studies of plant colonization that showed that plant diversity is at a maximum in the middle of a primary successional gradient (Aubert *et al.*, 2003; Malacska *et al.*, 2004; Zhu *et al.*, 2009). Makoto & Klaminder (2012) therefore concluded that the first theory was most applicable.

An alternative interpretation, not discussed by any of these authors, is that colonization by vascular plants begins at the margin of the circle after the main rotational churning diminishes sufficiently, and then spreads progressively towards the centre. The obvious source area for recolonization would be the surrounding vegetation, and the appearance of shrubs first on the margins of the circle tends to support this interpretation. It is also consistent with the interpretation of Becher *et al.* (2013), discussed above. It implies that the churning motion decreases from the margins inwards during warming. Supporting this is the fact that the ^{14}C ages of the organic inclusions of Becher *et al.* (2013) suggest rotational movement as late as 1650–1950 years A.D., while the evidence from the lichen growth studies suggest that movement started to decrease 250 years ago, *i.e.*, about 1750 A.D. (Makato & Klaminder 2012).

10.4.1.4 Frost boils

In areas with sporadic permafrost, or where permafrost is absent, the main features are *frost boils*. These consist of a fresh area or slight mound less than 50 cm across, exhibiting subsoil material and plant roots forming a break in the natural vegetation cover which is usually boreal forest (Figure 10.14A). Unlike subarctic frost boils, they do not always recur in the same location in subsequent years. Material from all the

Figure 10.14   Frost boils: A, developed in alpine tundra on the summit of Heart Mountain, Alberta, and B, breaking up the pavement of downtown Yakutsk, one year after being paved. © S.A. Harris.

underlying horizons of soil are mixed together in the bare sediment extruded through the vegetation (Van Everdingen, 2005). The forces involved can be powerful enough to lift up paving and break it up (Figure 10.14B). The vegetation around frost boils soon colonizes the disturbed area, so the evidence for their former action becomes concealed from surface inspection. However, in cross section, the disruption of the soil horizons in a vertical column is obvious due to the mixing of the soil horizons with fragments of roots. They are a form of nonsorted circle, and similar eruptions can be found breaking through the vegetation cover in most moist, cold climates from areas with only seasonal frost, polewards, to areas of continuous permafrost.

10.4.1.5   Plug circles

Washburn (1952, 1956, 1979, 1997) described upward movement of sediment without any obvious sorting in the High Arctic that he classified as nonsorted circles. He spent many summers studying them at Resolute, Cornwallis Island, and in the Northwest Territories, and found that they can develop within 20 years in disturbed soil. Others can be 9 ka old on the highest marine terraces. He concluded that they were a result of upward movement of ground in a fluid, plastic, or frozen condition by freezing-induced causes. They are therefore a type of upward movement of sediment as discussed in Chapter 2, and shown in Figures 2.16 and 2.18. They usually have a low mound as surface expression, and the plug-like upward intrusion shows clearly in cross section. He included them as being nonsorted patterned ground, although the intrusive material may or may not be sorted.

Plug circles consist of a circular variety of patterned ground, usually at flat, poorly-drained sites, characterized by a fine clay/silt-rich soil that continues downward as a plug. It may be cylindrical or more irregular in form, and is composed of stony or stone-free sediment, either sorted or nonsorted. The sorted ones have a sharp, even contact with the surrounding sediment. In spite of the designation as being a type of nonsorted pattern, they are virtually the same as the upward-extruded sediment at lower latitudes, commonly regarded as being the result of upward-moving sediment, intruding into or through the overlying soil. As such, they represent a high latitude form of cryotubation

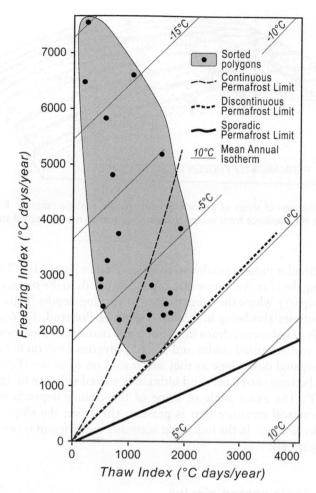

Figure 10.15 Relationship of active cryogenic sorted polygons to freezing and thawing indices (from Harris, 1981c, Figure 5B, 1982a).

in the general sense. Without knowing more about the conditions causing the intrusion, their relationship to the other forms of nonsorted circles is unclear. They are also known from Greenland, Spitsbergen, and the Eurasian Arctic.

10.5   CRYOGENIC SORTED PATTERNED GROUND

Cryogenic sorted patterned ground is the name given to patterns where there is a distinct separation of the larger stones surrounding a fine grained mass of sediment in the circles, with the boundary between them descending up to 3 m into the ground, or else these two materials occur as alternating broad, thick bands running down slope. Figure 10.15 shows the relationship of active cryogenic sorted and landforms to freezing and thawing indices. When active, the finer-grained material in the sorted circles is

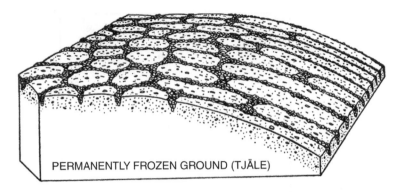

PERMANENTLY FROZEN GROUND (TJÄLE)

Figure 10.16 Modification of shape of patterning with change in slope (after C. F. S. Sharpe, 1938). Note the sequence from sorted polygons via sorted nets to sorted stone stripes.

in motion in a circular pattern similar to that depicted in Figure 10.12 during thawing or freezing during the year. Any stones are carried upwards to the surface and then move towards the periphery where they accumulate to varying depths (Figure 10.4, above), apparently sometimes also being in motion (Hallet & Prestrud, 1986). On slopes, the individual bands of sediments churn in a corkscrew manner as they move slowly down slope. As noted earlier, sorted circles and sorted polygons form on horizontal surfaces but become elongated downslope as they are traced on to slopes (Figure 10.16). The resulting forms become sorted nets and ultimately sorted stripes with increasing slopes up to about 20°. The exact angle of slope of the change depends on the moisture regime. The more soil moisture that is present, the lower the slope angle at which circles grade into nets, *etc.* In the following account, the different types of patterns will be discussed separately.

10.5.1 Cryogenic sorted circles

The most informative studies of sorted circles are those by Hallet & Prestrud (1986) and Hallet *et al.* (1988) studying the classic sorted stone circles in western Spitzbergen for over 20 years (Figure 10.1). Developed on raised beaches, they are regarded as the best examples in the world. The movement occurs in summer with little change during the rest of the year. Figure 10.17 shows the apparent movement of sorted circles on the raised beach in West Spitzbergen, A, showing the circulation of the sediments as originally interpreted by Hallet & Prestrud (1986, Figure 7) and while 10.17B shows what Pissart (1990) considered to be a more realistic pattern. The circles are consistently circular (3–4 m diameter), active, and largely unvegetated. The upper surface of the finer material in the centre is convex with a height 5–10 cm above the contact with the coarser ring. The border is abrupt, and the gravel outer ring tends to rise higher than the centre of the fines, ranging from 0.5 to 50 cm in height. Excavations suggest that the contact between the two sediments dips outwards at about 20° at first, becoming nearly vertical below 20 cm. The active layer is just over 1.5 m thick in the study area. The outer border ranges from 0.5–1.0 m in width. Markers

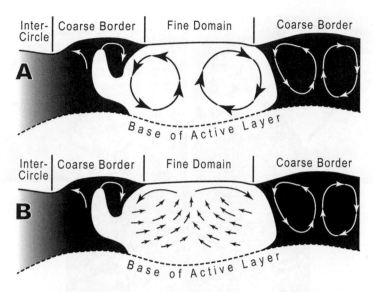

Figure 10.17 The apparent movement of sorted circles on the raised beach in West Spitzbergen as interpreted by Hallet and Prestrud (1986, Figure 7) and by Pissart (1990).

in the finer sediment indicate vertical displacements at 25 cm depth ranging up to 10 mm/a^{-1}. The soil is moving horizontally at rates of 4% a year. The circles seem to be in a state of equilibrium and not changing in size. The beaches were formed 3–4 ka B.P. Stiff nylon rods buried in the fine domain showed progressive outward tilting with distance from the centre after 20 years (Hallet, 2013), while the surface speed of movement was dependent in part on the slope of the land surface. Calculated recycling times based on the measured surface displacement rates is about 500 years for a 1 m thick active layer and a radius of 1.5 m for the fine domain. In Spitzbergen, it is assumed that the circulation extends down to the permafrost table, based on the results from continuously recording apparatus. Heaving takes place during September and October and amounts to between 5 and 12 cm, starting at 0.4 cm/day, and then slowing down. This is due to ice lenses forming in the top 20 cm of the active layer, and the expansion continues for up to one month after the ground freezes. Rotation of the gravel in the coarse rims indicates another possible rotational cell there. Details will be found in Hallet (2013). The latest data suggests that the movement in the sediment below 50 cm depth in the centre has less movement as might be expected in the drier central portion of the active layer. It is this that supports the interpretation in Figure 10.17B.

Figures 10.18A and 10.18B show the clear separation of the finer material from the stones on the margins of typical sorted circles in Alaska. The large clasts move upwards to the surface of the central finer material and will then be maintained there by the upfreezing process. Freezing occurs earlier in the coarse borders of the pattern than in the central finer sediment (Schmertmann & Taylor, 1965), resulting in a heat flow towards the borders. In an established sorted circle, the stones on the surface will either be carried by the underlying fines towards the borders or will slide down

Figure 10.18 The results of excavating A, the coarser material along the borders of a polygon and B, the finer material in the centre in Alaska. This is the polygon with the knife in Figure 10.19C. Note the clean separation between the two materials composing the circles. © S. A. Harris.

the slight slope of the mound to the edge, where they accumulate to form the stony border. In this case, the stony border is believed to remain stationary, but the smaller stones circulate with the finer fraction. This is the currently accepted explanation of the segregation, and is why the range of sizes in the parent material determines whether a nonsorted circle or a sorted circle evolves. Thus the sorted circles in West Spitzbergen are probably representative of the buoyancy-driven convection occurring throughout the deeper active layers (c. 0.5–1.5 m thick) discussed above. Similar segregation of the finer fraction from the coarse clasts occurs in all sorted patterns, *e.g.*, in ephemeral lakes (Figures 10.4 and 10.18C). In that case, instead of centuries to achieve the sorting as in Svalbard, the sorting is completed in a few seasons, though the exact rate of formation is not known.

The reason for the deep penetration of the segregated stones into the active layer seems to have been largely ignored. At Plateau Mountain, Alberta, the relict sorted circles have the segregated stones penetrating at least 2.5 m into the active layer. Ground cracking along the colder stony columns may be involved. The columns also exhibit linear pits along part of their course that can be up to 90 cm deep. This seems to indicate that ice wedges formed along fissures in the stony columns at sometime in the past, but have subsequently melted, leaving a depression in their surface.

Ray *et al.* (1983a, 1983b) developed a model for sorted pattern ground regularity. They concluded that the width:depth ratio for sorted polygons on land should be 3.81, which compared with 3.57 based on the average of 18 locations at which field work was carried out. However, the sorted circles at Plateau Mountain exhibit sorting to a greater depth than predicted in their model. As noted above, it is now thought that the patterns are self-organizing, but it implies a distinct limit of depth for the convective circulation. In the High Arctic, this is unimportant since the thickness of the active layer is the limiting factor. Southwards, the active layer deepens and soon exceeds the lower limit of the convective circulation.

10.5.2   Cryogenic sorted polygons, and nets

Sorted polygons are also found on flat surfaces. When the centres of convection interact, many of the circles become distorted to form *polygons* with 4–5 straight sides, but the same processes operate as in the adjacent circles. As they are traced on to slopes, the polygons become modified to form *nets* with the longer axis aligned downslope, or *steps* with their longer axis parallel to the contours. This is the same for micro-patterns, macro-patterns, and for the intermediate sizes of Matsuoka *et al.* (2003). The exact angle for the change varies with the local microenvironment, and there is an overlap between the distributions of each form (Figure 10.5). Telescoping tubes similar to those used on Plateau Mountain indicate that the soil is churning as it moves down slope. However more observations are needed to determine the patterns of movement involved.

10.5.2.1   Sorted stripes

In the case of nets, these change into *stripes* of sediment with different textures aligned down the slope (Figures 10.3, 10.16 and 10.20). There is usually minor vegetation growth on the finer grained sediment between the coarse stripes. Figure 10.19B shows the partly vegetated macro-forms on Devon Island in the Canadian Arctic, while the sharp boundary between the micro-stripes due to needle ice sorting is shown in Figure 10.19A.

Not only does the shape of the pattern change, but in the stripes, the individual stripes are thought to "corkscrew" down the slope with the finer material circulating and leaving any blocks in the coarser stripe. This is merely a modification of the circulatory pattern seen in the circles due to the addition of the effects of gravity. The coarse stones are elongated downslope and tend to be on their side. Although stripes have been reported from many places, *e.g.*, Antarctica (Hall, 1983), they have not been studied in such a way that the exact directions and rates of the movements have been measured.

There have been many studies of the actual rate of movement of the individual bands of the micro-stripes down slope (Table 10.2). The rate of downslope movement depends on the prevailing slope processes. The finer sediment always seems to move faster than the coarse material, contrary to the assumption by Benedict (1970) that it could be otherwise. The measured surface movement is far greater when needle ice is the cause, but this only affects the top 10–20 cm of soil and is always related to microforms. Both macro- and micro-stripes presumably result in the sediments

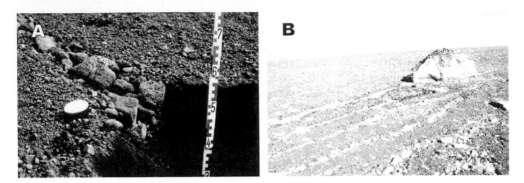

Figure 10.19 A: Cross-section of a sorted micro-stripe developed on a slope in volcanic ejectamenta in Iceland. B: Sorted stone stripes on Devon Island, Arctic Canada.

Figure 10.20 Intermediate sized stone stripes developed on volcanic ejectamenta in Iceland. Note the lichen growth on the stones at the surface of the coarser stripes. © S. A. Harris.

gradually extending further downhill with the passage of time, as long as the processes keep operating. When the processes causing the segregation stop, both coarse and fine material move downslope due to other mass wasting processes, usually both moving at about the same speed.

10.5.2.2 Stone pits

In areas with sediments that include occasional large blocks, *stone pits* may be found (Figure 10.21). Gallin & Erdenbat (2005) used the term *stone islands* for them. They

Figure 10.21 A stone pit in the Swiss Alps. Note the signs of sorting forming rudimentary stone steps. © S. A. Harris.

consist of an isolated pit of large stones separated by a sharp sorted boundary from the surrounding sediments that also show evidence of buoyancy convection cells. They are relatively rare, but Gallin & Erdenbat (2005) describe them from Mongolia. They occur where stones are in short supply in a finer-grained matrix. This agrees with the model of Kessler & Werner (2003) which suggested that the stones were first exhumed by a combination frost heave and frost jacking (Washburn, 1979). Once at the surface, the stones creep downslope towards the loci of areas of low relief. If enough stones accumulate, they can move laterally to form *stone polygons*. In the example in Figure 10.21, the pit seems to co-exist with rudimentary sorted stone steps on a 2° slope.

10.6 IDENTIFICATION OF ACTIVE VERSUS INACTIVE FORMS OF MACRO-SORTED PATTERNS

Care needs to be used in interpreting the climatic implications of the formation of macro-patterned ground. Active sorting is indicated if the stones lack lichen growth and appear clean. Lichens on the stones in the otherwise unvegetated area indicate probable cessation of the sorting process, although that may not necessarily imply the complete end of buoyancy movements at depth in the finer centres. Lichens take a long time to become established on calcareous substrates, whereas they readily colonize siliceous clasts. Ballantyne & Matthews (1982) found that sorted macroforms developed immediately after deglaciation in front of glaciers in the Jotenheim, Norway, presumably due to cold, katabatic winds, but as the glaciers continued their retreat, development ceased on all except wind-swept moraine crests.

10.7 MICROFORMS OF CRYOGENIC PATTERNED GROUND

These have similar shapes to the sorted macroforms, but are found in cool, humid climates subject to seasonal frost. They reflect separation of stony soils into coarse and fine domains by differential frost heave and/or needle ice growth. The sorting takes place in or on the upper 5–10 cm of soil (Matsuoka et al., 2003), producing sorted features with a diameter of 15 cm or less (Figures 10.3 and 10.14A). The width of the individual stripes of different textures is related to the grain size of the parent sediment.

Ballantyne (1996) carried out an experiment on the growth of sorted nets near sea level in Scotland. A pattern was evident after only eight freeze-thaw cycles, with a minimum surface temperature of $-5.3°C$ and freezing extending down to 17 cm depth. After two years, a miniature stony net with cell diameters of 56–172 mm and stony borders 4–9 cm deep had developed. Growth of needle ice under stones raised them above the ground surface. During the same freezing process, updoming occurred where there was a concentration of fines in the top 5 cm of the sediment. During thawing, the needles bend, resulting in a net of the stones towards the margins of the dome. This is repeated during each cycle. This is similar to the experience of Chambers (1967) on Signy Island in the South Orkney Islands.

Why there is a concentration of fines in a suitable pattern to produce the updoming in these microcircles has not been explained. Meentemeyer & Zippin (1981) have shown that a clay-silt content of 12–19% is optimal for needle-ice growth. A second problem is the development of the marginal space where the stones accumulate. Chambers (1967), Pissart (1973b, 1977) and Ballantyne & Matthews (1983) ascribe it to frost cracking, though Ballantyne argues that the irregularity of the net is inconsistent with frost cracking in this case. There was no evidence for inclined freezing planes nor a buoyancy circulation as has been suggested by some authors. Maximum size of the diameters of the coarse clasts is believed to be 15 cm.

Matsuoka et al. (2001) studied differential frost heave in the Upper Engardine, Swiss Alps. Where there is a granule mantle 5 cm thick, overlying loam that is subjected to periodic, shallow freezing, the granules heave to a smaller extent and for shorter time than the loam. This results in a lateral movement of the loam towards the granules, which can lead to a lateral sorting of the top 5 cm. There was no development of needle ice in this case. This is therefore a second potential cause of the sorting of the upper 5 cm of the soil during diurnal freezing cycles.

Microforms of cryological patterned ground are therefore not related to permafrost distribution. Instead, they occur in areas which undergo diurnal freeze-thaw cycles with surface ground temperatures colder than $-2°C$, that have a high soil moisture content together with a soil texture conducive to rapid migration of soil moisture to the freezing plane (Outcalt, 1971a). Troll (1944, 1958) reported that needle-ice formation is widespread in the high equatorial mountains, subtropical mountains, continental mid-latitude uplands and oceanic polar environments. The review by Lawler (1988a) indicated that needle ice only occurs near sea level between latitudes 34–41°N in Japan and the east of North America, and 52–65°N in northwestern Europe. He noted that the upper limit of the microforms appears to coincide with the lower limit of permafrost. They can potentially develop in the finer textured centres of inactive sorted circles.

Ballantyne (1996) also provides a list of locations where microforms have been reported. Figure 8.3 shows the differences in distribution of the development of seasonal needle ice with latitude in both hemispheres. These differences are due to the great height of the mountains in the northern Andes, the climatic distribution of suitable climatic conditions, together with the occurrence or otherwise of suitable land areas at a particular latitude.

Chapter 11

Thermokarst and thermal erosion

11.1 INTRODUCTION

Initially, the term *thermokarst* was introduced by M. M. Ermolaev (1932) to describe disturbed terrain formed by melting and destruction of ground ice along the coastal lowlands of northern Siberia (Figure 11.1). Ice-rich permafrost is widely distributed

Figure 11.1 Thermokarst produced by thawing of ice wedges in the Central Yakutia. The icy cliff retreats at about a few metres per year, with the resulting sediment flowing downslope to the river (retrogressive thaw slump).

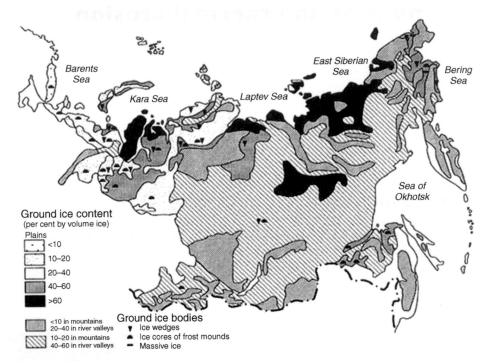

Ground ice content
(per cent by volume ice)
Plains
<10
10–20
20–40
40–60
>60

<10 in mountains
20–40 in river valleys
10–20 in mountains
40–60 in river valleys

Ground ice bodies
▼ Ice wedges
▲ Ice cores of frost mounds
– Massive ice

Figure 11.2 Ground ice content of the permafrost areas of Siberia. On materials of Cryolithology and Glaciology Department of Moscow State University, http://www.rusnature.info/geo/06_3.htm.

throughout the permafrost realm, and the use of the term thermokarst has now expanded to cover any modification of the ground surface by all the processes involving melting of all the kinds of ground ice (Soloviev, 1973b). As such, it can be produced by any alteration of the microenvironment that results in thawing of ice in the ground. These disturbances may be natural or man-made. Jones *et al.* (2013) provide an annotated bibliography emphasizing its effects on habitat and wildlife.

Thermal erosion or *thermo-erosion* refers to the erosion of ice-rich permafrost by the combined thermal and mechanical action of moving water (Walker and Arnborg, 1966; Mackay, 1970; Are, 1978; Newbury *et al.*, 1978; ACGR, 1988). It involves thermokarst, but with the added effect of the mechanical action of moving water.

Thermokarst features occupy up to 40% of the landscape in Siberia, most of which was unglaciated. The high ice content is partly due to the large amount of ground ice there that has developed through successive glaciations during the last 3.5 Ma (Figure 11.2), but it can potentially occur throughout the areas that undergo seasonal freezing of the upper layers of soil in winter accompanied by the development of ice in the ground. Thus the development of blocks of ice during winter under roads causes a pocket of water to develop during the spring thaw. If a vehicle puts too much weight on the asphalt surface above the water-filled pocket, a pot hole develops. This is the reason for the weight limits being imposed for vehicles during the spring thaw in northern lands. Again, the maritime palsas described in Chapter 7 go through a cycle of growth and decay, the latter also being a form of thermokarst.

In Western Europe, the El Niño current passed through the opening between North and South America and kept the north part of the North Atlantic warm until about 2.4 Ma (Harris, 2013a). Cold periods commenced about 1.8 Ma (Ehlers and Gibbard, 2008). Unfortunately, there is a rather incomplete record of the sequence of the earlier climatic events in Western Europe, though there are abundant remains of both thermokarst and cold events that are preserved in Eastern Europe (*e.g.*, Jahn, 1975). In Russia, it is widespread, especially in Western Siberia and along the southern margins of the permafrost region. There was a particularly active period of thermokarst development during the Altithermal/Hypsithermal/Megathermal warm period during which the extensive areas of alases were formed in eastern Siberia.

In North America, thermokarst is less widespread in the younger, post-glacial landscapes because the permafrost could only develop in most areas after deglaciation of the landscape. The permafrost is thinner except in the unglaciated areas, primarily in Alaska and the Yukon Territory. However, a full range of thermokarst landforms occur, with the possible exception of alas valleys.

In Antarctica, thermokarst features are essentially confined to the relatively temperate maritime climate of the South Shetlands. Elsewhere in Antarctica, the climate is too cold and dry, and this also limits the development of thermokarst features in South America.

In China, the permafrost in the far north and north-east is the southern extension of the Siberian permafrost in Russia. On the Qinghai-Tibet Plateau, permafrost only developed in the last 140 ka in response to the rapid uplift that is currently occurring. Continuous and discontinuous permafrost occur above 4500 m. The earliest remains of permafrost are ice-wedge pseudomorphs in the gravel layer of fluvial and alluvial sediments dated at 135.7 ka B.P. on the Da'Heba sandbar in Xinghai, Qinghai Province at 35°50′N, 99°40′E, at 3350 m elevation (Pan *et al.*, 1997). Other remnants of the Penultimate cold event have been reported from loess at the North Yangsigeguzui village, Junger Banner in Outer Mongolia (39°59′N, 111°18′E, at 1231 m), dated at 132 ka (Zhou *et al.*, 2008). The last glaciation produced widespread permafrost, as indicated by features interpreted as being buried pingo scars (Xu *et al.*, 1984), "sand wedge casts" and "sand-wedges", soil-wedge casts (Wang, 1989), ice-wedge pseudomorphs, frost-crack polygonal nets, and rock tessellons (Harris and Jin, 2012), all dating between 7.05 ka and 33.4 ka (Cui *et al.*, 2002; Chang *et al.*, 2014). The "sand-wedge casts" in at least one area were found to correspond to two different sets of dimensions, one having a depth to width ratio between 1.7 and 2.7 probably corresponding to ice-wedge casts, whereas the other exhibits ratios ranging from 0.6–1.3 (Cheng *et al.*, 2006), typical of infilled, large, shallow ice-block casts.

Three conditions are necessary for the development of thermokarst. Firstly, there must be ground ice below the permafrost table. The more ice per unit volume, the more spectacular will be the changes in the landscape. Secondly the depth of thaw must descend into the upper icy layers of the permafrost. Lastly, there must be release of the melt water which may either pond on the surface of the ground forming a shallow marsh or lake, *e.g.*, the northern Seward Peninsula (Parkesian *et al.*, 2011), or drain away underground or along river valleys.

11.2 CAUSES OF THERMOKARST

The main natural causes of thermokarst development include changes in mean annual air temperature, mean annual precipitation, local hydrology, changes in vegetation, erosion, and lightning strikes resulting in fire. Anthropogenic causes include wild-fires and disturbance of the landscape during development, *e.g.*, agriculture, forestry, construction of transportation routes, etc.

In the unglaciated areas of Siberia and Beringia, repeated cycles of cold glacial periods and warm interglacials beginning as early as 3.5 Ma (Figure 4.14) resulted in alternating periods involving the formation of icy permafrost and at least partial destruction of the ground ice by thermokarst processes. The thawing of ground ice would have been complete towards the extremities of the former distribution of permafrost, but were incomplete in the high latitudes and tops of the highest mountains. In the north, this has resulted in the development of the yedoma deposits discussed in Chapter 6. It also produced the various thermokarst landforms described in this Chapter.

The first wave of warming at the end of the last glaciation produced lakes of glacial origin as the glaciers retreated. In cold, mountainous regions such as the central Altai Mountains, pingos developed in these lakes (Blyakharchuk *et al.*, 2008). During the warm mid-Holocene event, more water entered the lakes from thawing of ground ice, drowning these pingos. Subsequently the climate has cooled during the Neoglacial and a new set of pingos has developed in the broad fen on the margin of the younger lake, where they are found today.

The warming of the climate from the Last Neoglacial event that ended in the first half of the 20th century has resulted in the development of a new wave of thermokarst activity that is aided by the anthropogenic changes to microenvironments over large parts of the permafrost realm. Each time there has been a warming of the local climate, thermokarst develops in the permafrost lands, its rate and extent depending on the amount of warming. Duration of the warmer weather is a major factor in determining the amount of permafrost degradation that takes place, as is latitude.

There is a fairly rapid response to any appreciable change in mean annual air temperature. Thus in North America, the Altithermal/Hypsithermal warm period may have only have involved a 2°C increase in MAAT in Alberta (Harris, 2002c), but it was enough to produce substantial thawing of permafrost along the Mackenzie Valley (Mackay, 1975b). The onset of warmer conditions varied from place to place, but now appears to have started by about 7000 ka, and perhaps about 8200 years ago further north (Mackay and Terasmae, 1963). The main phase of warming in Eurasia occurred in the mid-Holocene (late Atlantic) reaching about 1.5°C (Velichko *et al.*, 1997). Melt water collecting on the ground surface in depressions accelerated the process in most cases. In Siberia, Kachurin (1958, 1961) and Fukuda *et al.* (1995) showed that much of the extensive development of alas valleys occurred during these warmer periods.

About 6 ka in northeast China and 4500 B.P. elsewhere, the climate cooled, and the thermokarst development largely ceased except during the warmer interludes during the Neoglacial events. Mackay (1975b) was able to document the successive changes in permafrost distribution, growth and decay over the last 10 ka along the lower Mackenzie valley, Northwest Territories.

Soloviev (1973b) has shown that relatively little thermokarst had developed after the end of the Last Neoglacial event in Russia until 1973, while Lubomirov (1987)

Table 11.1   Characteristics of the transient layer on the Abalah plain on the right bank of the Lena river, Yakutia (Brouchkov *et al.*, 2004).

Landscape	Depth of ice wedges, cm	Active layer depth, cm	Thickness of the transient layer, cm	Water content of the transient layer (% by volume)
Larch forest 130–150 years old	200	140	60	19–39
Larch forest 80 years old	195	130	65	18–37
Larch forest 50 years old	185	155	25	18–35
Larch forest 15–20 years old	200	135	65	20–35
Grassland between alases	210	190	10	17–33
Thermokarst depression	225	220	5	66

demonstrated that increased surface water does not always lead to increased ground temperatures and greater active layer depths. Part of the reason may be that the surface water freezes in winter, and the frozen ice-saturated soils have a higher thermal conductivity than unfrozen soils. Another factor is the unstable water balance in the region which results in ponds of a certain size tending to dry up (Grigoriev and Baranovsky, 1990). In Yakutia, the salts in the water are left in the soil when the water evaporates. If there is negligible winter snow cover, the cold weather more than offsets the effects of the shorter period of summer heating. The considerable variation of the active layer thickness from year to year is what caused Shur (1988a, 1988b) to introduce the term *transient layer* for the frozen zone between the base of the active layer in years of shallow active layers and the top of the ice wedges which correspond to the maximum active layer thickness. Table 11.1 provides data on the active layer thickness and depth of the transient layer under different vegetation covers on the Abalah plain of the right bank of the Lena River. It is marked by a very ice-rich layer.

Shallow surface water acts as a particularly efficient material for absorbing solar radiation, since it is translucent. As a result, it absorbs approximately five times as much incoming radiation as the adjacent bare ground (Grave, 1944; Pavlov, 1999; Harris, 2002a), but only reradiates heat from the surface. Convection and wind-driven currents cause a mixing of the water, which is much more efficient than the movement of heat through soil by conduction. Thus once a puddle of water appears on the ground, the increased heat absorbed by the water results in thawing of any ice in the adjacent sediments in summer. This results in the expansion of the thermokarst water body until it is no longer in contact with icy sediments or it drains into a stream channel (Harris, 2002b). This is the basic mechanism involved in the development of thaw lakes and alas depressions (see below).

An increase in summer precipitation not only brings extra heat to the surface of the ground but it also tends to develop both surface water puddles in depressions as well as adding to the perched water table on the upper surface of the permafrost. The latter increases the thickness of the active layer. Where the permafrost has gone and there are no relatively impermeable layers in the soil to cause high moisture contents, the mean annual ground temperature of the soil will rise.

Streletskiy *et al.* (2015) describe the results of climate warming on the stable isotope composition of the water in streams draining an area of degrading permafrost in the Lower Yenisei river basin, Siberia. Lowering of the water table results in increased water storage and pathways for water to discharge into the river. Decreased seasonal frost over the last 40 years has resulted in increased storage capacity of permafrost-affected soils, and a higher contribution of groundwater to winter stream discharge. Timing and the quantity of late summer precipitation affects the contribution of groundwater to winter stream flow as well as its isotopic composition.

Changes in vegetation alter the thermal conditions on the ground. They can result from changes in climate or hydrology, or through natural seral changes in vegetation (Smith, 1975; Viereck, 1970, 1973). One of the most important causes of temporary or permanent vegetation change is fire. Dry lightning strikes result in wildfires which are a fairly common occurrence south of 65° latitude in North America, as well as in the vast Taiga of Russia.

In the boreal forest, there is a substantial difference between the effects of crown fires as opposed to ground fires. In crown fires, the fire moves rapidly through the tree tops, but there is usually limited damage to the ground vegetation and soil. The effect of the tree and shrub canopies on the thermal offset is lost, while the blackened soil has a different albedo. Layers of peat usually survive, so that the diode effect of peat still operates. As a result, there is relatively limited disruption of the active layer thickness and any underlying layers of ice. Thus at Fox Lake, Yukon Territory, a crown fire took out the trees and shrubs, but the lithalsas remained relatively unaffected except for warmer ground temperatures. The trees and shrubs are gradually growing back.

Ground fires and fires on the tundra are another matter. An example is the forest fire close to tree line that occurred near Inuvik in 1968 (Heginbottom, 1973; Mackay, 1977). The trees were spread out and not very tall. Within four years, the active layer had increased in thickness by an average of 40 cm. There were substantial differences across the site, the worst effects being along the Man-made fire breaks cut by bulldozers trying to limit the extent of the fire. The active layer was still becoming deeper in places even after a decade, but where the vegetation became re-established in relatively well-drained areas, the depth of the active layer started to decrease.

11.3   CAVITY DEVELOPMENT IN PERMAFROST

Cavities in permafrost have been reported from permafrost areas since the work of Halliday (1954) and Kachurin (1959), and play an important role in the subsurface hydrology. They have been reported from drained frost blisters (ACGR, 1988), gas-domed mounds (Mackay, 1965), cultivated fields in ice rich permafrost (Péwé, 1954), as well as in ice-cored moraines Healy (1975). When stable, they are capable of transmitting large quantities of water through otherwise relatively impermeable sediments (French and Harry, 1988). If they become unstable, thermokarst features are produced (Péwé *et al.*, 1990).

The stability of cavities depends on their size, shape, overburden thickness, and strength of the surrounding sediments (Huang and Speck, 1989). The latter is dependent on time, stress and temperature. The strength of sediments decreases rapidly at warmer ground temperatures, so any warming of the ground is liable to cause thermokarst development and dislocation of the underground drainage system (Walsh, 1991).

Hyatt (1992) studied the development and stability of cavities in icy permafrost at Pangnirtung, Baffin Island, following a large rainfall event. He divided cavities into *subcritical* (stable due to the arching support of the roof) and *supercritical* (where the roof support was inadequate to maintain stability). This system follows the use of these terms in coal mining (Whittaker and Reddish, 1989, p. 51). Supercritical cavities have ground-arch heights equal to or greater than the thickness of the overburden under constant ground temperature conditions below the depth of zero amplitude. In this case, the roof will collapse. Subcritical cavities have lower heights. Where the seasonal fluctuations in the ground temperature make the cavity unstable for part of the year by reducing the strength of the sediment, the cavity is referred to as *variable subcritical* or *variable supracritical*, as opposed to *constant subcritical* and *constant supracritical*. Hyatt goes on to discuss the measurement and use of the tunnel parameters, but in general, constant subcritical cavities are found at lower elevations where large quantities of water flow. Supercritical cavities are more common at upland sites.

11.4 EFFECT OF THERMOKARST ON SOIL

Mackay (1970) showed that when the upper layers of permafrost start melting, there is a subsidence of the surface of the soil which becomes stable after a new equilibrium permafrost table is attained and the excess melt water has drained away (Figure 11.3). This new permafrost table will remain essentially static until there is another change in the thermal microenvironment. The total loss in height during thawing of the upper part of the permafrost table correlates better with climate warming than the new depth of the active layer.

Mineral soils in the active layer usually undergo overcompaction due to repeated freezing and thawing (diurnal and seasonal) resulting in increased density, increased thermal conductivity, and even more subsidence of the soil surface (Brouchkov *et al.*, 2005). This reduces the permeability of the soil, resulting in increased surface runoff and erosion. In extreme cases, fragipans may be produced (see Chapter 1). Roots of plants often have difficulty penetrating the compact ground (Harris, 1960), so this produces changes in the vegetation cover, while the lack of roots increases the effects of compaction. The change in density is especially marked in the upper organic layer, where it is present.

There may also be changes in the moisture regime in the ground. Potentially, this can result in increased leaching of soluble substances such as plant nutrients if the increased moisture drains away. Organic matter that has been locked up in the permafrost for centuries becomes vulnerable to decomposition. Organic matter decomposes faster in terrestrial soils at higher temperatures, and the soluble organic matter becomes more labile (Schuur *et al.*, 2008).

Where flooding occurs, methane bubbles come to the surface of the margins of thaw lakes in increasing amounts (Walter *et al.*, 2006), but where there is an increase in surface wetness accompanied by an expansion of bogs, the growth of peat acts as a net carbon sink, *e.g.*, along the Hudson Bay Lowlands, Northern Ontario (Dyke and Sladden, 2010).

When large quantities of melt water and warm summer precipitation cross the relatively impermeable frozen ground below the snow pack, *thermo-erosion* occurs.

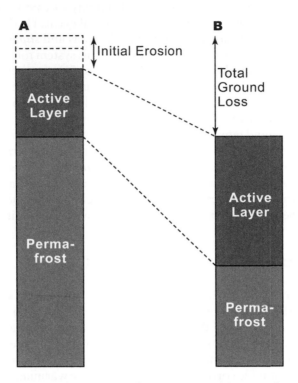

Figure 11.3 Subsidence resulting from thawing of the upper part of the permafrost table (modified from Mackay, 1970, Figure 3). This diagram assumes minor initial soil erosion but no change in density of the material in the active layer.

This term is used for warm water causing ice in the permafrost to thaw at the point of contact. Where ice-wedges are absent, wide (1–3 m), shallow (5–25 cm) erosion channels may develop forming an anastamosing pattern on gentle slopes as on the rounded slopes of the mountains in the Ogilvie Range, Yukon Territory. The channels are marked by an upper sandy layer and by pioneer species of plants. The intervening islands have mature soils with an undisturbed climax vegetation cover. Where ice-wedges are present, thermo-erosion of the tops of the ice-wedges produces gullies that follow the pattern of the ice-wedges along the lowest part of the slope (Fortier *et al.*, 2007; Godin and Fortier, 2012a, 2012b; Morgenstern *et al.*, 2013). The exact pattern depends on soil texture, peat content and the topography.

Where fire initiates the development of permafrost, the soil properties have already been changed. Fire affects the energy balance and water budget, as well as the soil properties. These include the surface albedo, soil density, and soil moisture, infiltration and evaporation rates as well as the soil thermal conductivity (Figure 11.5) and soil heat capacity (Viereck, 1982; Hinzman *et al.*, 2001). The ground is warmer in summer (Pavlov, 1984), resulting in active layers in mineral sediments averaging 1.6–1.7 m thickness in forest near Yakutsk, but 1.8–1.9 m thickness in the open, burned areas. Presence or absence of a moss layer also alters the summer soil temperature, though

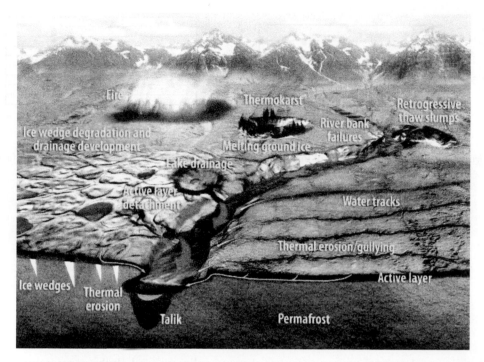

Figure 11.4 Diagram showing the response of an Arctic landscape to development of some thermokarst features (from Roland *et al.*, 2010). Sediment and solid constituents follow the pathways indicated by the tan arrows.

not in winter due to its high ice content. This results in greater risk of development of thermokarst.

A characteristic of the soils of Yakutia is the presence of frozen saline soils of continental type. It is caused by the preponderance of evaporation over precipitation, resulting in the accumulation of carbonate and sulphate ions. Frozen saline soils are defined as those containing >0.05% by weight of soluble salts in dried soil. The forest soils contain minimal salt content, but salinization increases in the alases. This results in changing vegetation resulting in a positive feedback. Saline soils freeze at lower temperatures and have different thermal properties to nonsaline soils (Table 11.2). They are often produced during the formation of thermokarst (Desyatkin, 1993).

11.5 THERMOKARST LANDFORMS

Thermokarst landforms can be subdivided into a series of different types, each produced by the thawing of the ice in different permafrost landforms (Figure 11.4). The fact that water absorbs five times as much incoming solar radiation as soil results in the thawing of the ice being essentially unstoppable once water appears at the surface. This has been called *self-developing thermokarst* by Aleshinskaya *et al.* (1972). Only

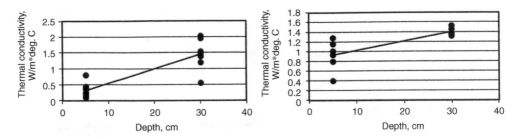

Figure 11.5 Thermal conductivity of the soil at 5 cm depth, A, before and B, after an experimental fire in the larch forest near Yakutsk. Note the different scales.

Table 11.2 Calculated mean annual temperatures of saline and nonsaline soils and the resulting thickness of the active layer.

Salinization, %	Thermal conductivity of frozen soil, W/m*°K	Active layer depth, m	Annual mean temperature, °C
0	1.39	1.6	−2.0
0.5	1.25	1.9	−1.6

draining of the surface water or a great change in climate can halt the process until all the ice has melted. Thermal erosion occurs where thawing of ground ice is combined with mechanical erosion, *e.g.*, the shoreline or along banks of large rivers.

11.5.1 Thermokarst pits

Examples of these from near Fairbanks, Alaska, are shown in Figure 6.18. Large blocks of ice are fairly common in permafrost in unglaciated areas though their origin has not been examined carefully. A good example of ice masses in permafrost is shown in Figure 6.19 from a section along the Alaska Pipeline haul road. Numerous examples have been described from Siberia. Increase in ground temperature is the general cause of the melting of the ice, although the example in Figure 6.18A was the result of clearing the trees in the yard of a house built on gravel. Although the grass was long, the loss of the trees resulted in increased ground temperatures. The water produced from the thawing ice drained away through the gravels, and the overlying soil collapsed. The resulting pits have a ratio of depth to width of <0.2–1.3 since they can be quite deep. In poorly drained silts, the water remains in the cavity, but when the soil collapses into the water, it produces a vertically sided pool that may be 2 m deep, filled with ice-cold water (Figure 6.18B).

On the Qinghai-Tibet Plateau, these depressions have been referred to as "sand wedges" since they have been infilled with loess or blown sand. They have the same depth to width ratio as those in Alaska (Cheng *et al.*, 2006), which is quite different to the nearby ice-wedge casts. The last main phase of their infilling in China has been dated at about 13.6 ka.

Figure 11.6 The result of removing the surface soil over ice-wedges along the upper Blackstone River valley, Dempster Highway, Yukon Territory. Note the rapid thawing of the exposed ice resulting in mounds of frozen soil in between the wedges. © O. L. Hughes.

Thermokarst pits can also be produced by collapse of parts of the roofs of drainage tunnels. The collapse is usually brought about by warming of the ground through the agencies of warm summer precipitation (Hyatt, 1992), changes in the vegetation cover, warming mean annual air temperatures, increased snowfalls, fire, or modification of the microenvironment by Human activity. These pits may be elongate, following the trace of the original drainage-way.

11.5.2 Thermokarst mounds

When ice-wedges degrade, the surrounding soil slumps into the empty space producing the rounded *thermokarst mounds* that indicate their former presence, as described in Chapter 5. In Russia, these are known as *baydjarakhs*, and examples are shown in Figures 3.16 and 5.20.

Figure 11.6 shows the initial pattern of ice-wedges that have been exposed unexpectedly two days earlier during road construction along the Dempster Highway, Yukon Territory. The ice has already started to thaw, and the engineers had to quickly cover them up to try to curb the development of thermokarst. Several kilometres further on, placing a drainage culvert too low resulted in the development of excellent thermokarst mounds in a single year (Figure 3.16). These are examples of interference in the microenvironment by Man. Péwé (1954) provided one of the first detailed accounts of the effects of attempts at farming on ice-rich soils near Fairbanks. The Agricultural Station there was closed after they started losing tractors into thermokarst pits resulting from the beginning of the development of thermokarst mounds.

Thermokarst mounds develop quite naturally where the microenvironment changes and the underlying ground containing ice-wedges undergoes warming. Inactive wedges remain stable at ground temperatures below °C until conditions are suitable for renewed active cracking of the ground, or until the sediment around the top of the ice-wedge warms above 0°C in summer. Once this occurs, thawing of the surface of the wedge results in the commencement of the formation of thermokarst mounds.

The developmental stages of thawing of ice wedges were discussed in Chapter 5. Whether active or inactive, stable ice wedges (Figure 5.17) have a low centre with two low ridges on either side of the ice-wedges (*low centre polygons*). Once thawing of the ice in the wedge commences, subsidence along the trace of the ice-wedges results in high centres with troughs marking the position of the thawing ice-wedges (Figure 5.21). These are called *high centre polygons*. If the thawing ceases, they can remain in this state for long periods of time until a change in microclimate occurs. Figure 11.7 shows a photograph of the Siberian larch forest showing high centre polygons with trees collapsing into the troughs as the thawing proceeds.

The original ice-wedge is a called a *primary wedge* whereas the ice wedge cast representing the infilling of the cavity vacated by the melting ice is referred to as a *secondary wedge.* In the process of infilling the cavity created by the melting of the ice in the upper part of the ice-wedge, the structure tends to widen as material falls off or flows from the walls of the primary wedge to be incorporated in the filling. This results in a depth to width ratio decreasing to 1.7 to about 2.7 on the Qinghai-Tibet

Figure 11.7 Early stages of thawing of ice-wedges in a Siberian larch forest. © V. Popov.

Plateau, where the infilling that occurred c. 13 ka B.P. consists mainly of loess or wind-blown sand (Cheng *et al.*, 2006). Primary wedges lacking ice such as rock, sand and loess tessellons in the same area have a ratio of 3.5 to >10. The measurements must be made only on those wedge structures that have not suffered from erosion of their surface. When the height to width ratio of the top of the feature is combined with the presence of vertical stratification of the sediment in the primary wedges, separation of primary, secondary (ice wedge casts) and thaw pits is quite easy, though it has rarely been done in the past. As a result, the term "sand wedge" includes several different landforms in the literature, both in China and elsewhere, and is best avoided (see Chapter 7).

11.5.3 Pingo, palsa and lithalsa scars

Pingo scars are circular ridges resulting from the thawing of the icy cores of pingos (Figure 7.3). Hydraulic (open system) pingos are located where artesian water moves towards the surface from higher ground. The flow of water that freezes usually fluctuates with the season, but will be continuous over long periods of time. The pingo usually grows bigger until the icy core becomes exposed by cracking of the soil cover at the apex of the mound. Once the ice is exposed, it starts to melt during the summer producing a small lake in the crater (Figure 11.8). This then absorbs more heat in summer, resulting in a gradual thawing of the icy core. Eventually, the side of the lake collapses and the water drains away. Meanwhile, the sediments and soils on the

Figure 11.8 Pond in the thawing summit of a hydraulic pingo immediately south-west of Dawson City airport, Yukon Territory. Note the collapsing sides of the mound, caused by water from the melting core descending the sides of the mound, thawing more of the ice inside the pingo. © R. O. Van Everdingen.

slope slump down the sides to produce the circular low mound called a *pingo scar*. However, if enough soil and sediment cover the remaining ice, the mound can continue to exist and even grow for a short time until further up-doming exposes the ice again. Ultimately, this leaves a low circular rampart with a shallow depression in the centre. Since more water will continue coming towards the surface, the mound will redevelop as long as the artesian flow continues. Thus these pingos go through a definite cycle of growth and collapse. In contrast, hydrostatic (closed system) pingos can only grow so long as water welling up from the reservoir below is freezing into ice. When the supply of water under pressure ceases, the mound stops growing. Like the hydraulic pingos, they can go through the cycle of thawing if the icy core becomes exposed but the upwelling continues. Both types of pingos can explode if the growth is fast enough to develop severe internal stresses in the mound, resulting in the icy core being broken up and scattered across the surrounding area.

Seppälä (1982, 1986, 1988, 2006) describes the cycle of growth and decay of maritime palsas. Since they develop in peat and decay once the permafrost bulb reaches a mineral substrate, they can only form relatively weak, peaty circular ramparts during their decay, though these sometimes can be seen (Seppälä, 2005b). Usually, sphagnum mosses and sedges will generate enough peat to fill the depression and soon make the rim indistinguishable from the surrounding mire, so they are short-lived. However, continental palsas can leave a rim when the ice-rich core thaws (Brown, 1968) and these *palsa scars* are fairly common in Northern Québec. The rim consists of both peat and mineral sediments and they tend to occur in mires, so that they quite obviously produce a different result from decay to maritime palsas.

Schunke (1973) reported that there was a decay sequence involving large, maritime palsa plateaus in Iceland involving the thawing of channels through the structure resulting in the production of several smaller palsas, ending up in palsa scars, and finally the complete disappearance of all visible traces of the original landform (Figure 11.9). There should be evidence of this past history in the stratigraphy, although he did not check this. In a study of the growth and decay of a maritime palsa in Sweden, Zuidhoff and Kolstrup (2000, 2005) described the splitting of one palsa into several small mounds as it decayed. Where there is no splitting up of the mound, the end product is often a low, short-lived ring of peat surrounding a shallow pool in the mire.

Payette *et al.* (2004) have documented a case of thawing of a continental palsa that they originally regarded as being a palsa plateau in northern Québec. As it degraded, the palsa plateau split into a series of sub-parallel elongated palsas that stretched across the valley. Over the last 50 years, these have developed into a weak string bog pattern within the mire occupying the valley floor.

Continental lithalsas in poorly drained situations such as northern Québec can also form *lithalsa scars* (Camels *et al.*, 2008a, 2008b). These consist of rims of mineral soil surrounding a shallow depression up to 2 m deep when the icy core thaws. In this case, outward spreading of the sides of the mound may also take place, as indicated by splitting of tree trunks (see Chapter 7). Due to the wet silty sediment in which they have developed, such outward creep/flow is readily understandable.

To differentiate between these three types of scars, it is necessary to try to determine the microenvironment in the area when they formed. It is this that has caused the confusion in the earlier literature between pingo, palsa and lithalsa scars. The

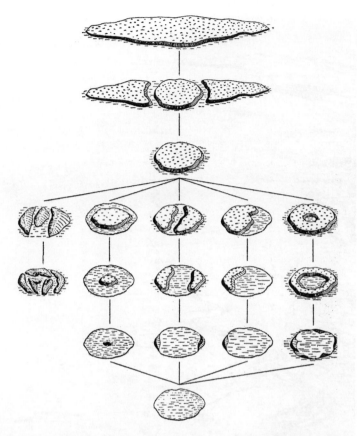

Figure 11.9 The sequence of decay of plateau palsas in Iceland (from Schunke, 1973).

latter are now believed to be rather widespread in Pleistocene deposits, both in lowlands, on the slopes of mountain sides, *e.g.*, Wales, or on plateaus such as the Hautes Fagnes in Belgium at locations where formation of pingo scars would be unlikely.

11.5.4 Beaded streams

Beaded streams show up on aerial photographs or satellite imagery as bead-like widenings of streams scattered at fairly close but somewhat irregular spacings along a small but permanent stream in a lowland area underlain by ice-wedges (Figure 11.10). The individual beads are the result of heat from the flowing water melting the ice in the ice-wedges in the banks of the stream. This is an example of the combined effects of thermokarst and thermal erosion due to thawing by transfer of heat from the flowing water to the ice in the ice-wedge combined with limited mechanical erosion.

Figure 11.10 A beaded stream in the Northwest Territories. © W. W. Shilts.

11.5.5 Thermokarst lakes

These are small lakes occupying a closed depression in permafrost areas resulting from the thawing of ground ice. They occur where localised thawing of ice in sediments containing limited amounts of ice takes place, resulting in the development of a small, shallow pond. Localised subsidence of the ground as ice changes to water can provide the necessary depression to allow ponding of the water, since the underlying permafrost is essentially impermeable at these sites (Soloviev, 1973a, 1973b). Once the water appears at the surface, it absorbs more solar energy than the surrounding soil, resulting

Figure 11.11 Thermokarst Lakes on the Mackenzie Delta northeast of Inuvik. © S. A. Harris.

in thawing of the frozen ground along its margins. Annual rates of retreat of its banks were estimated to be between 15–20 cm/a^{-1} by Wallace (1948) in Alaska, although this varies considerably (Hinkel *et al.*, 2012b). Niu *et al.* (2014) discuss recent advances in thermokarst lake research.

The underlying permafrost may be segregated ice, tops of ice-wedges, or large blocks of ice. A slight depression due to deeper thawing of the active layer may allow precipitation to accumulate on the ground surface in summer, and this can initiate the development of a thermokarst lake. The initial shape is irregular, but soon becomes rounded or elongate (see below), due to thermal erosion of the lake margins. The pond and subsequent lake are normally shallow, often being only 30–50 cm deep near the shore. The only exceptions are where it is the top of an ice-wedge or thawing of a block of ground ice that initiates the pond. Except for the cases where the pond may reach 9 m deep (Hinkel *et al.*, 2012a), the water becomes deeper by conduction of heat into the underlying permafrost in summer, resulting in deepening of the lake to 1–2 m. In addition, a talik develops under the lake, which may penetrate through the permafrost with time, resulting in draining of the lake underground. The lakes rarely exceed 2 km across before either drying up or draining into a lower part of the landscape. Permafrost then regenerates in the upper layers of the sediments in the former lake talik, and if the talik did not penetrate though the underlying permafrost, a pingo may develop (Figure 11.12). The same sequence occurs in the Northwest Territories where Mackay drained a lake at Illisarvik in 1978, and the recovery stages of the permafrost together with the sequence of development of the typical landforms are being documented (Burgess *et al.*, 1982; Mackay, 1980a, 1981b, 1982, 1986b, 1997, 1998; Mackay and Burn, 2002).

Figure 11.12 Typical pingo (hydrolaccolithe) on the floor of a former thaw lake in Western Siberia. © A. Gubarkov.

In Alaska and the semi-arid parts of Siberia and China, the evaporation from the surface of the lake may eventually be greater than the input of water, resulting in drying up of the lake. This then initiates the growth of terrestrial vegetation and the redevelopment of permafrost. The expansion due to freezing of the soil water causes the surface to heave, until it eventually regains its former elevation (Hopkins, 1949). During the recovery, the vegetation changes due to diminishing winter snow depths and the resulting colder ground temperatures until the original vegetation re-establishes itself.

Climatic gradients, meteorological conditions and basin characteristics affect the actual lake temperatures. Hinkel *et al.* (2012a) report that the lake bed starts to warm to between 1–4°C as the ice cover thaws in spring in Alaska. This is followed by rapid warming to c. 13°C at inland sites on the Arctic Coastal Plain in Alaska, but only c. 7°C near the coast. The water temperature changes with the air temperature, and the water is well mixed and generally isothermal. Weak stratification can occur during calm, sunny days. Lakes in peat have a more uneven bottom than those developed in mineral sediments. In lakes deeper than 6 m, the lower part of the water column cools as it loses heat by conduction into the underlying permafrost.

Thermokarst lake morphometry depends primarily on the underlying substrate and the quantity of sediment eroded from the margins of the lake (Hinkel *et al.*, 2012b). The low-relief coastal area of Alaska consists of marine silts which are ice-rich in the upper 1–6 m. Thermokarst lakes there are usually about 2 m deep at maturity. Inland, ice-poor aeolian sands result in c. 1 m deep ponds with 3–5 m deep central pools. This results from the considerable erosion of the lake margins, redepositing sediment near the shore, and is associated with deep penetration of the talik into the underlying permafrost. In the Arctic Foothills, yedoma produces relatively deep lakes over the underlying enlarged ice wedges. Depths of 6 m can be found inshore over former ice-wedges, but there is only gradual deepening offshore.

The thermokarst lake cycle in Alaska is believed to be relatively rapid, Black (1969a) suggesting that it may be completed in 2–3 ka in the case of the smaller lakes. Dating of organic matter suggests that most thaw lakes on the coastal plain in Alaska developed after 8 ka B.P. (Black, 1969a; Tedrow 1969). Bockheim and Hinkel (2012) found that excess ice excluding ice-wedges increased from 20% in young basins <50 years old, to 40% in ancient basins as a result of accumulation of meteoric water. This makes the older basins susceptible to differential thaw and subsidence (Pullman et al., 2007; Jørgenson and Shur, 2007). The numbers of ice-wedges in the drained lake basins also increases with age (Hinkel et al., 2003).

Where the thermokarst lakes develop in undulating areas of discontinuous or sporadic permafrost, the outcome of the growth of the lakes is usually different. In this case, they develop as before, but when the through talik develops, the lake drains, but the talik may remain. Ice-wedge polygons provide increased rates of development in the through talik. This is a major reason for the shrinking of thermokarst lakes and a reduction in their area during warming by changing mean annual air temperatures. Until the climate changes, the talik often acts as a relatively permeable conduit for surface waters to drain into the rocks below (Yoshikawa and Hinzman, 2003). If there is a lower regional water table than the base of the permafrost table, the overlying soils will dry out, aiding in the deepening of the active layer. The vegetation cover becomes more sparse and this permits additional warming of the ground due to changing surface energy balances (Carr, 2003). In the streams, there are increased winter stream flows (Yang et al., 2002), decreased summer peak flows (Bolton et al., 2000), changes in stream water chemistry (Petrone et al., 2000) and other fluvial processes (McNamara et al., 1999), e.g., river icings.

In peat areas in lowlands, the results of thawing of permafrost are different. Payette et al. (2004) examined the changes that have occurred along the lowlands of the east coast of Hudson Bay and demonstrated that as the permafrost decreased, areas of fen and thermokarst ponds increased in abundance. The permafrost is believed to date from the last Neoglacial event, and its degradation appears to be a fast, one-way, self-organized system that began during the last 50 years. Being close to sea level, the melting ice produces the ponds in which peat forms. Subsequent studies demonstrate a vegetational sequence involving the appearance of trees on the higher parts of the peat and the gradual replacement of the ponds by swamps (Bouchard et al., 2014). The biomass has increased dramatically. Thus the area has become an important carbon sink. Similar conditions are found across the Hudson Bay Lowlands of Northern Ontario (see the articles in Arctic, Antarctic and Alpine Research 46(1)).

11.5.6 Oriented lakes

When thermokarst lakes are elongated in a particular direction, they are called *oriented lakes* (Figure 11.13). These were first described from the arctic coast of Alaska by Black and Barksdale (1949), but have recently been shown to also develop in thermokarst lakes developed in peat in northeast European Russia (Sjöberg et al., 2013). In peatlands, the shores are generally steeper and have more

Figure 11.13 An oriented lake in the Northwest Territories. © J. R. Mackay.

cracks. Oriented lakes vary in both their length:width ratio and actual shape. Clam-shaped lakes with one straight edge have been reported from the Great Plain of Kaukdjuak in eastern Baffin Island (Bird, 1967), while luminescent, oval, triangular or eliptical lakes were described from the Tuktoyaktuk Peninsula by Mackay (1963). They are not confined to permafrost regions but occur in many other climatic zones (Price, 1968). The currently accepted origin for those on the Arctic Coastal Plain is that lateral expansion occurs because of wind-driven erosion, thaw slumping and thaw subsidence along the lake margins (Carson and Hussey, 1960; Osterkamp *et al.*, 2009). The sediment contains considerable quantities of sand that tends to form beaches protecting the lee sides of the lake. Wave action is only effective if the wind has a fetch greater than 30 m (Hopkins, 1949). The wind causes waves and currents that result in erosion at the ends of the lakes at right angles to the direction of air movement (Livingstone, 1954; Mackay, 1956, 1963).

On the Old Crow Flats, Yukon Territory, the thermokarst lakes surrounded by taiga vegetation are irregular in shape since the tree roots protect the underlying sediment from erosion (Roy-Leveillee and Burn, 2015). However, the numerous thermokarst lakes surrounded by tundra vegetation with ice-wedge polygons exhibit an elongation parallel to the prevailing wind direction. This is ascribed to the fine-grained texture of the glaciolacustrine sediment which is not coarse enough to accumulate as a sandy beach near the shore on the leeward side of the lake, leaving the bank vulnerable to thermo-mechanical erosion by waves.

Other theories have been suggested including the effects of sand dunes (Fürbringer and Haydn, 1974; Sellman *et al.*, 1975), snow cover distribution (Sturm and Liston, 2033; Seppälä, 2004), geology of the area, and the overall slope and aspect of the landscape (Pelletier, 2005).

11.5.7   Alases

Alases are the result of the development of large thermokarst lakes in very ice-rich permafrost, often along wide valleys (Bosikov, 1991), that have subsequently drained (Soloviev, 1973b). They are of critical importance to the Yakutian people because they represent the only areas with minimal permafrost that can be used for agriculture in Central Yakutia. They occur over wide areas of Central Siberia, south to Mongolia and have a characteristic cycle of development (Popov *et al.*, 1966; Shumskiy and Vityurin, 1966; Czudek and Demek, 1970; Katasonov and Ivanov, 1973).

Most of Central Siberia was not glaciated, so thick permafrost developed during the cold events of the last 3.5 Ma. Although thawing occurred along its southern margin during the intervening warm periods, large areas north of latitude 40° gradually built up substantial thicknesses of very icy sediments to a depth of 20–25 m in the lowland areas and undulating hills. Along the Lena River, the active layer is 0.8–3.0 m thick and overlies 400–500 m of continuous permafrost. The Quaternary deposits are 5–100 m thick and overly sandstones and argillites. The soils in the south are sands and sandy loams, but these grade northwards into ice-rich clay loams with numerous ice-wedge polygons. The original vegetation was larch forest, but this has undergone burning by forest fires and logging, so a secondary birch forest is now taking over.

Mean annual air temperature is c. −10.5°C, but there are three months of hot (30°C+) air temperatures in the summer. Although the summer precipitation is small (190–230 mm), the warm nights mean that crops can grow very well in the absence of a great thickness of permafrost. Winter snow cover is 0–40 cm, although it can reach 60 cm. The larch forest keeps the ground 4°C colder in summer than in burned areas due to the insulative properties of the organic surface layer under the trees.

In Central Yakutia, the dry climate and negative water balance prevent any widespread development of thermokarst today. Gavrilova (1969) reported a mean annual precipitation of 240–320 mm, with mean annual evaporation being several times larger. Today, infilling of any alas lake by erosion of sediments along its shores is faster than the increase in lake depth, so those alas lakes started by present-day forest fires tend to dry up before they become very large.

The main time of formation of the numerous alas valleys was 5–11 ka B.P. (Kachurin, 1961; Bosikov, 1991; Fukuda *et al.*, 1995; Velichko *et al.*, 1997). Warming at the end of the last major cold event culminated in the mid-Holocene with the mean annual air temperature reaching about 1.5°C warmer than today (Velichko *et al.*, 1997). This also appears to have been a time with a wetter climate than today (Bosikov, 1989), thus favouring alas development. Suhodrovsky (2002) has suggested that without a corresponding increase in precipitation, an increase in temperature leads to increased evaporation, drying up the lakes and halting alas formation. Initiation of

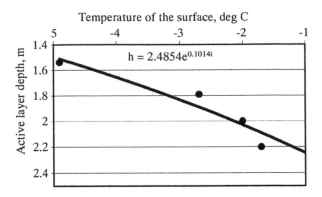

Figure 11.14 Calculated depth of the active layer as a function of the mean annual air temperature in Central Yakutia.

alas formation appears to be the result of surface disturbance and changes in the temperature and moisture regimes. Thus the causes would be climate change, incidence of forest fires, and human impact on the environment.

The fire return interval averages 25–70 years near Yakutsk, although there is a very large variation from 7–15 years up to 250–300 years (Global Forest Fire Assessment, 2001). Accepting the estimate that 40% of the land has undergone alas formation in the Holocene, and assuming an average fire frequency of 200 years, the probability of thermokarst appearance due to fire is about 1%. Sometimes thermokarst appears several years after fire, but often not. In other words, the role of fire is unlikely to be more important than the disturbance of the vegetation by logging or climate change.

Although Central Siberia is regarded as a classic area for finding the results of thermokarst, this is misleading. It is actually rather rare today in this largely forested area, due to substantial variations in the thickness of the active layer from year to year, resulting in a thick icy transient layer (Table 11.1). Fire causes an increase in the temperature of the permafrost of about 2°C while the active layer thickness increases by about 30–40 cm (Figure 11.14). Under forest, the transient layer is generally thicker than this, so preventing thermokarst from developing (Vasilyev, 2002). This is aided by the substantial depth of the tops of the ice-wedges. This contrasts with the situation with shallow active layers further north towards the Arctic in the subarctic woodland, and the 5°C increase in ground temperature there after a ground fire (Rouse, 1976). This probably explains the increased thermokarst further north where fires recur about once every seventy years on average.

11.5.8 Cycle of alas formation

In the forest, the development of an alas goes through a series of stages (Brouchkov *et al.*, 2004; see Figure 11.15). It commences with the development of a wet depression with hillocks about 10–15 m wide and 1.0–1.3 m deep (stage 1 or *bilar*, Figure 11.16A). The surface settles at a rate of 2–3 cm/a^{-1}. These then join together to form a small depression containing a shallow pond up to 0.5 m deep, but surface settlement remains about 10–30 cm/a^{-1}. This then expands at a rate of 0.8–4.0 m/a^{-1} (Bosikov, 1998)

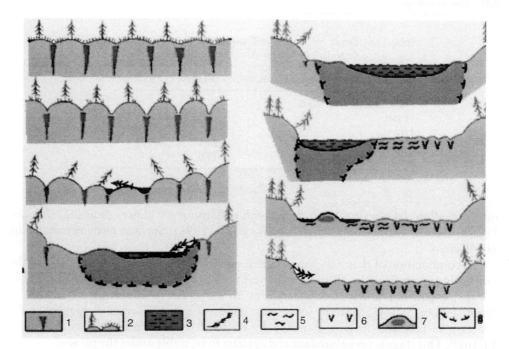

Figure 11.15 Sequence of development of an alas in Yakutia in the larch forest (modified from Desyatkin and Desyatkin, 2006). Symbols are: 1 – syngenetic ice-wedges; 2 – trees and grass; 3 – water; 4 – landslides; 5 – alas sediments; 6 – epigenetic ice-wedges; 7 – pingo; 8 – permafrost table.

Figure 11.16 Alas formation at the Yukechi site, right bank of the Lena River, near Yakutsk: A, the first **bilar** stage; B, the second **dujoda** stage; C, the third **timpi** stage; D, the last, stable **alas** phase. © A. Brouchkov.

Table 11.3 Depth of alases and thickness of the ice-complex deposits in the Lena-Amga region (Bosikov, 1991).

Site	Depth of alas, m	Thickness of ice complex, m
Hara-Soboloh	5.0	6.1
Hooro	5.0	12.8
Hannah	7.0	10.5
Haiagatta	17.0	24.5
Oner	30.0	45.0

into a shallow lake up to 2.5 m deep, with collapsing icy shores (stage 2, *dujoda*, Figure 11.16B) within 40 years (Gavrilova *et al.*, 1996). Surface settlement remains 10–30 cm/a^{-1}.

A combination of thermal erosion of the shoreline and conduction of heat into the underlying permafrost result in an expansion in area and the development of higher collapsing cliff-like shores with a height up to 2.5 m above the water level (stage 3, *timpi*, Figure 11.16C). Rate of subsidence increases to 50–70 cm/a^{-1}. Stage 4 is the decrease in lake size leaving a small lake on the former lake bed (stage 4, Figure 11.16D). This stage is very abundant and appears to be stable under the present climate. For comparison, interdepression sites where thermokarst is active due to warming of the climate show a surface subsidence rate of 2.6–5.4 cm/a^{-1}. On well-drained flat, inter-alas surfaces, the rate of subsidence averages 0.5–0.8 cm/a^{-1}. Often a low domed pingo (Figure 7.4) or pingo plateau may be formed due to partial refreezing of the water in the talik beneath the former lake. The pingos will be the hydraulic type if the talik penetrates through the permafrost, or of the hydrostatic type if the talik is surrounded by permafrost.

The drying up of the lake stops the process producing the "latent" phase until the water level drops again. This stage is marked by old shallow lakes and small dry depressions. If there are no changes in temperature, moisture, or any drastic vegetation changes, the alas is in the "stable" phase. Bosikov (1989) reports that there is a repetitive cycle of 150–180 years with increased wetness causing renewed alas formation. The entire cycle of alas formation takes 100–200 years, based on the rate of expansion of the shorelines and size of the stage 4 alas. The depth of the alas (D, m) is related to the depth of the ice complex (H, m) underground (Table 11.3):

$$D = 0.701H - 1.069 \tag{11.1}$$

Thus the time required for complete thaw of the ice complex does not appear to exceed 200–300 years. The "active" phase of alas formation is short in comparison to the geological history of the Holocene (Brouchkov *et al.*, 2004), and the alas formation appears to be a short-term catastrophic event, as suggested by Grave (1944).

Desyatkin and Desyatkin (2006) describe the sequence of changes in the soil profiles during one cycle of alas formation from larch forest to the stable marsh stage (Figure 11.17). Two extra layers are deposited on the ground surface, the surface one usually being peat. Table 11.4 shows the differences in the carbon, nitrogen and carbonate content of the soils under the different microenvironments.

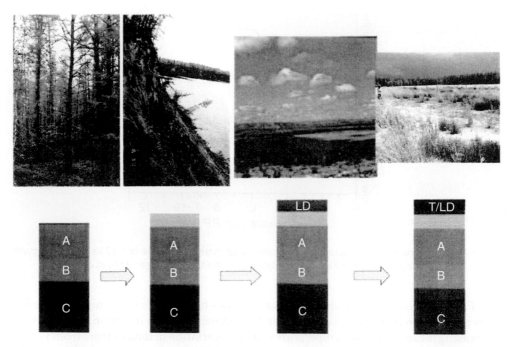

Figure 11.17 Changes in stratigraphy of the soil during one alas cycle (Desyatkin and Desyatkin, 2006).

Table 11.4 Accumulations of carbon and nitrogen in humus in various soils together with carbonate content in the alas, meadows and forests around Yakutsk, Siberia (from Desyatkin and Desyatkin, 2006).

Section	Vegetative cover	Site	Capacity of an active layer	Substratum	C, kg/m²	N, kg/m²
AS-1	Larch forest	High site	96 cm	Fall	1.79	0.04
				Humus	4.67	0.47
				Carbonates	2.01	–
AS-2	Larch forest	Slope	66 cm	Fall	1.16	0.03
				Humus	3.92	0.27
				Carbonates	0.0	–
AS-5	Motley grass meadow	Edge of forest	90 cm	Fall	3.26	0.28
				Humus	36.53	3.17
				Carbonates	2.25	–
AS-4	Wet meadow	The bottom belt of alas	80 cm	Fall	0.39	0.02
				Humus	65.19	5.39
				Carbonates	6.79	–
AS-6	Real meadow	The middle belt of alas	137 cm	Fall	0.32	0.02
				Humus	30.21	2.39
				Carbonates	4.0	–
AS-3	Steppe meadow	The upper belt of alas	260 cm	Fall	0.12	0.01
				Humus	16.26	1.79
				Carbonates	15.7	–

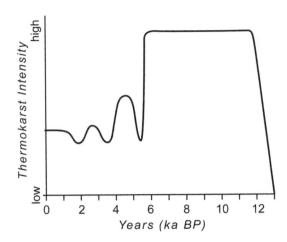

Figure 11.18 Fluctuations in thermokarst intensity over time during the last 13 ka on Kurungnakh Island, central Lena River delta, Siberia (modified from Morgenstern *et al.*, 2013).

Frequency of active alas development increases northwards to the higher latitudes. This is due to an increase in ice content in the sediments in this direction, coupled with decreasing mean annual air temperatures and less variation in temperature from year to year. As a result, the thickness of the active layer decreases, as does the thickness of the transient layer. Incoming solar radiation is less, but the difference in soil temperature between burned and unburned areas increases to about 5°C (Rouse, 1976).

Morgenstern *et al.* (2013) examined the Holocene history of Kurungnakh Island (72°19'N, 126°12'E) in the south-central Lena River Delta, Siberia, in the zone of continuous permafrost and subarctic tundra. Icy permafrost with yedoma/ice-complex deposits had been accumulating during the cold events during the Pleistocene when glaciations were occurring elsewhere. Warming at about 13 ka B. P. resulted in intense thermokarst activity producing a large alas lake. This lake persisted until 5.7 ka B.P. when it drained abruptly. Thereafter, thermokarst intensity was greatly reduced, although it showed three short, further decreases probably signalling Neoglacial events (Figure 11.18). The draining resulted in a >20 m deep alas with residual lakes. Subsequently, the minor climatic fluctuations have resulted in weak periods of alternating permafrost growth and decay, including the growth of polygonal ice-wedges and a small pingo. Since the massive ice can be up to 50 m thick, multiple alas events are possible. Kotov (1998) reported ice-complex deposits overlapping lake deposits.

Generally, it is suggested that the thermokarst action was very intense during the warming corresponding to the Boreal period (9–7.5 ka), but since then, the landscape has continued to be similar until today (Romanovsky *et al.*, 2004; Kaplina, 2009). Previously, various authors have suggested that the complete thermokarst lake cycle has been repeated several times during the Holocene (Hopkins, 1949; Tomirdiano, 1978; Billings and Peterson, 1980; Hinkel *et al.*, 2003), but evidence is lacking to support the idea of several complete cycles (French, 2007; Jorgenson and Shur, 2007).

Southwards, the Siberian larch forest extends into northeast China, but ends just south of the southern border of Russia, and also extends into northern Mongolia. Thick

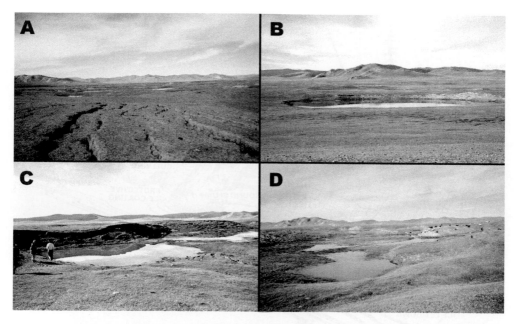

Figure 11.19 Sequence of the development of an alas on the grassy steppe of Mongolia. A, the initial cracking and subsidence of the ground, B, the appearance of a small lake, C, enlargement of the lake to link up with other lakes, and D, the elongate lake developing into an elongate valley that eventually dries up, often by drainage into a nearby river (Harris, 2002b).

ground ice occurs in northern Mongolia under a steppe grassland vegetation. Further south, the semi-desert areas and deserts of northern China are slowly encroaching on the grasslands. Where the Mongolians keep moving their herds on so as to avoid over-grazing, the permafrost landscape is stable, but when the herdsmen overgraze the land, alas formation commences (Harris, 2002b). The stages in the development of an alas in this environment are shown in Figure 11.19. Initially, deep, parallel, curved cracks develop in the ground as thawing takes place and the overlying soil caves in (Figure 11.19A). This is followed by the appearance of a small pond where the subsidence began (Figure 11.19B). The water level in the pond is often 5 m below the surrounding landscape, thus indicating the amount of ground ice that has thawed. Next, the individual ponds start to coalesce into shallow, elongate lakes while sinking lower into the landscape (Figure 11.19C). These, in turn, link up to form an elongate valley (Figure 11.19D). Eventually, the lake drains into a stream, or dries up due to an imbalance between the supply of meltwater from the permafrost and the evaporation from the large lake surface into the dry semidesert air. Alas are relatively rare but represent a significant source of destruction of the icy permafrost in the region, left over from the succession of major cold events during the late Pliocene and Pleistocene. They leave a meandering dry valley underlain by a talik, or else no permafrost in the otherwise gently rolling permafrost landscape. Permafrost can then reappear once the vegetation is flourishing if the climate is suitable. Even pingos may develop if a source

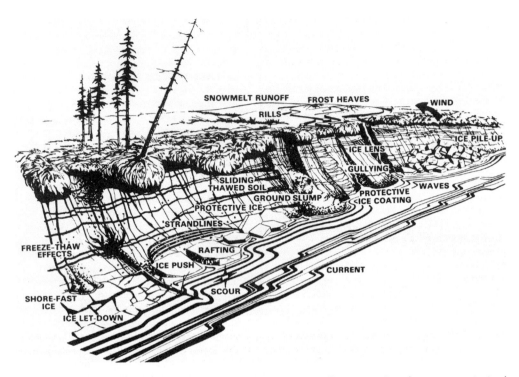

Figure 11.20 An idealised sketch of the processes and features affecting erosion along streams incised into icy permafrost (from Larson, 1983).

of water under pressure is formed. This scenario is regarded as having taken place at least twice during the Holocene in the formerly glaciated areas of the central Altai Mountains (Blyakharchuk *et al.*, 2008).

11.6 THERMOKARST AND THERMAL EROSION ALONG RIVER BANKS

Figure 11.20 shows the processes involved in the loss of ground ice along stream banks (from Lawson, 1983). The water in the stream represents a source of heat when it is flowing, and can undercut the bank by thermal erosion. However, landslides, slumping banks, piles of ice blocks, broken ice sheets that covered the stream in winter and outflow of sediment from retrogressive slumps help protect the banks from erosion during peak spring runoff when the water level is very high. As a result, there is limited erosion unless the climate warms or the vegetation cover is disturbed by fire. Local gullying will occur from snow melt and runoff from precipitation. Any thawing of the icy permafrost will add extra water and sediment to the runoff. Where ice wedges are intersected, slumping of the ground usually occurs forming beaded drainage or slumps with increased erosion.

Figure 11.21 Thermal erosional notch along the bank of a river. The undercutting eventually causes the bank to fall in or slide into the stream.

During the summer, the water flowing past icy sediments in the banks will cause thermal erosion and limited under-cutting producing a weak water-cut notch (Figure 11.21). Eventually, the sediment that is undercut falls into the stream, increasing its width at that point and contributing to meander development.

Several processes eventually result in either active layer detachment slides, retrogressive thaw slumps or rotational slumps (Figure 11.22). If a rotational slide occurs in areas of continuous permafrost, it is usually developed along ice-wedge segments more or less parallel to the bank. Thawing of the ice-wedge with attendant accumulation of water in the resulting crack provides the lubrication of the upper backwall of the slide. Excess interstitial water as a result of thawing ice and/or heavy summer rains can also produce the movement. Landslides on slopes can also develop during the melting of ground ice (Bolikhovskii and Kyunttsel, 1990).

11.6.1   Ice jams

Another important factor in erosion of the river and stream banks is the development of ice jams. There can be minor ice jam formation due to frazil ice accumulating at certain points along the waterway in the fall during the freezing process but the spring break-up of ice cover produces the most striking results.

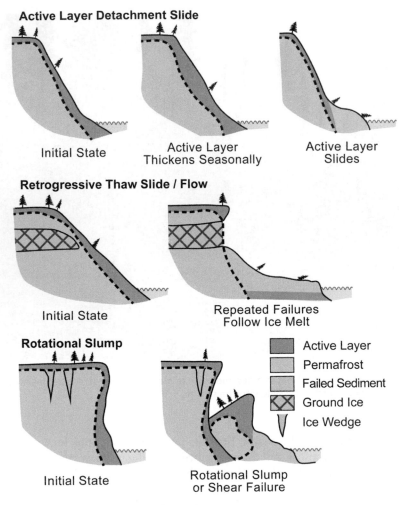

Figure 11.22 The three different kinds of slope failure found along river cliffs in areas with icy permafrost (after Church and Miles, 1982).

Most rivers in the northern regions flow north into the Arctic Ocean. Thawing starts in the south and moves north. As a result, the broken pieces of ice floating north, tend to flow limited distance before piling up on the still-frozen ice or where obstacles impede their movement. The four main locations where these accumulate are, sharp meanders, narrowing of the river channel, gravel bars, or islands. Channel restrictions are the most common cause of the ice jams. There is no apparent correlation between channel width and height of the flood. The actual spring flood consists of two peaks, one when the ice breaks up and one corresponding to the melting of the snow pack on the surrounding mountains (Gerrard *et al.*, 1992). There can also be summer floods resulting from extreme rainfall events. These also inundate the flood plain if there is sufficient precipitation. Thus the water level in the stream is not stable but fluctuates,

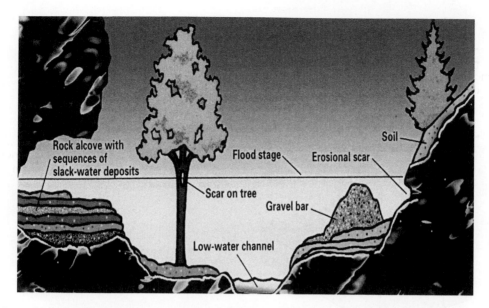

Figure 11.23 A diagrammatic cross section of a northern valley that has been affected by repeated ice jams (after Jarrett and England, 2002).

complicating the stability of the banks as well as the identification of past ice jam events.

A diagrammatic cross section of a northern valley is shown in Figure 11.23, after Jarrett and England (2002). Summer floods and spring ice jams raise the water level substantially, resulting in flooding of the valley floor and deposition of characteristic slack-water deposits, first recognised by Bretz (1929). Key evidence for the effect of ice jams includes ice-rafted pebbles in the silty slack-water sediments. Summer floods produce deposits lacking ice-rafted pebbles, although layers of rounded gravels may occur if the channel changes shape. Figure 11.24 shows a diagrammatic longitudinal section of an ice jam.

Livingston (2004) has reviewed the past literature on the ice jams along the Yukon River between Dawson City, Yukon, and Circle, Alaska, and has carried out a detailed study of the general ice jam frequency. The river flows in a relatively narrow valley on which the settlements are located between the mountains of the Ogilvie Range. Major ice jams occurred on average every 34 years over the last 2 ka, but there is evidence for short-term cycles of 12–66 years in shorter, more detailed records. There was also evidence of increased ice-jam frequency during the warmer period from 1.5–1.0 k calendar years B.P., the cause of this being unknown. The ice-caused tree scars range up to 9 m above river level, while Figure 11.25 shows the icy landscape left after an ice jam. The highest ice jams resulted in water levels exceeding those of summer floods (Figure 11.26).

Ice jams can cause severe damage to settlements, bridges, dams, or any other obstacles in their path. Conversely, release of water from hydroelectric plants causes

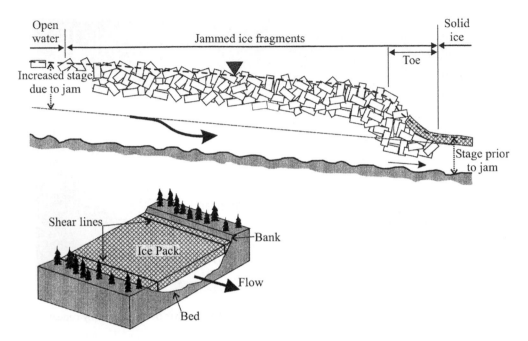

Figure 11.24 The longitudinal profile of an ice jam, together with a block diagram showing the shear lines near the shore (after Environment Canada, 1996).

Figure 11.25 An inundated settlement resulting from flooding by an ice jam on the Yukon River (USGS).

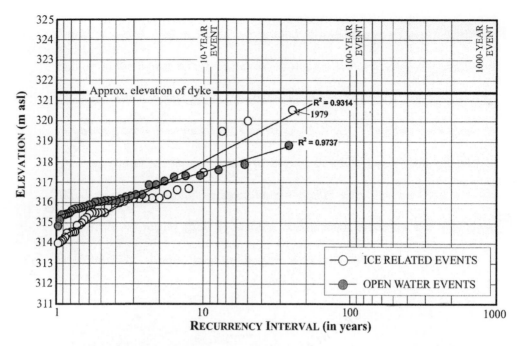

Figure 11.26   Recorded stage-frequency of ice jam events versus summer floods for the Yukon River at Dawson City between 1896 and 1999 (Environment Canada, 1996; Livingston, 2004). Note that the data for ice jams is incomplete.

fluctuations in water level that aids in breaking up the ice cover immediately down-stream even in winter. The fragmented ice is then carried generally northwards down the valley until it reaches the solid ice covering the river. It then piles up forming an ice jam during the winter. The height of such ice jams can reach 11 m. They are about five times as frequent as summer floods in Alberta. The ice jams can destroy any dams or other structures in their path unless special precautions are taken, *e.g.*, building suitably high dykes lined with large rocks. The construction of bridge piers changes the evolution of ice jams downstream (J. Wang *et al.*, 2016).

11.7 THERMAL EROSION AND THERMOKARST PROCESSES ALONG SEA COASTS

About 70% of the world's coastlines are affected by ice in some way. In Canada and Russia, the rates of coastal erosion are extreme, currently reaching up to c. 18 m/a^{-1} (Jones *et al.*, 2009). Figure 11.27 shows the distribution of erosion rates around the Arctic Ocean where data is available. It is obvious that the high rates of erosion are limited to certain regions, and this also shows up again in studies of small areas (Figure 11.28). In general, the shorelines were much further north when the sea levels were 100 m lower during the last glaciation.

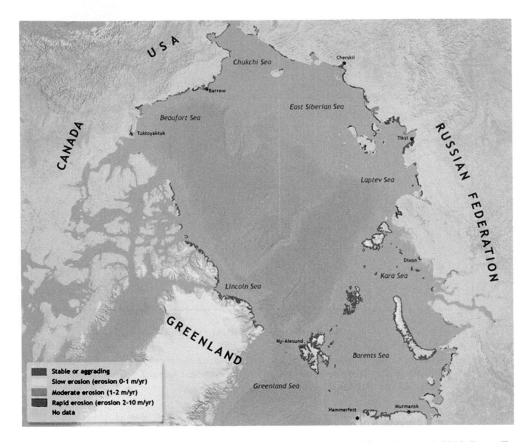

Figure 11.27 Circum-Arctic map of coastal erosion rates (from Lantuit, Overduin, *et al.*, 2012, Figure 7). Note the high variability, with some areas being essentially stable or aggrading.

11.7.1 Effects of seasonal sea ice

During freeze-up, blocks of ice broken up by wind can pile up along the shore (Figure 11.29) or as offshore bars. The ice blocks result from break-up along pressure ridges with an onshore wind moving the blocks inland 90 m or more (Alestalo and Häikiö, 1976; Taylor, 1978; Kovacs and Sodhi, 1980; Kovacs, 1983). The larger the blocks, the further they can travel. Pile-up heights of 10–20 m are common, but they can be much bigger. Sverdrup (1904) reported 36 m from the Canadian Arctic, while Zubov (1945) measured over 50 m in height along the Siberian coast. Offshore, piles of ice can move sea floor sediments down to a depth of 40 m. Reimnitz *et al.* (1988, 1990) have suggested that these ice blocks can carry sediment entrained in the ice from as deep as 5 m of water about 50 m off the beach front. These piles of ice blocks protect the shore from wave action for at least part of the summer.

During thawing in the spring/early summer, blocks of ice can float away carrying beach sediment with them. Figure 11.30 shows the maximum size of clasts that can be moved by ice floes of different sizes. Likewise, Dickins (1987) found that sediment

Erosion Rate (m a⁻¹) ● > 10 ● 5 to 10 ● 2 to 5 ○ 0 to 2 ○ Deposition

Figure 11.28 Mean annual erosion rates for a 60 km segment of coastline of the Alaskan Beaufort Sea: A, from 1955–1979; B, from 1979–2002; and C, from 2002–2007 (modified from Jones *et al.*, 2009, Figure 2). Note the increased erosion rate in the last block of measurements.

on the shallow sea bed can be entrained in the sea ice, which can then be thrust up with the icy rubble. These processes can create severe problems for any anthropogenic structures along shorelines, both on large lakes and the seashore.

The largest shore ridges are found around *polynias* (perennially open areas of water) and the western shores of the Canadian Arctic Islands, *e.g.*, Prince Patrick Island (Hudson *et al.*, 1981; Forbes *et al.*, 1986; Taylor and Hodgson, 1991). In many cases, repeated ice thrusting on the shores results in a steady growth and movement inland with time.

The Arctic coast of northern Canada is undergoing isostatic rebound at a considerable rate. This results in raised beaches being widespread along the coast around Hudson Bay and the Arctic Islands of northern Canada, as well as resulting in the

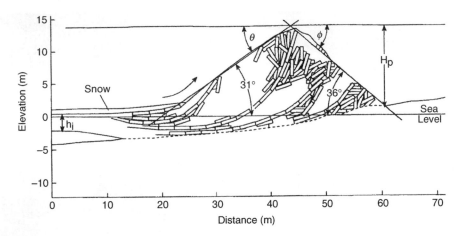

Figure 11.29 Cross-section of an ice pile-up at Cape Kellett, Southern Banks Island, Northwest Territories, modified from Kovacs and Sodhi (1980) and Forbes and Taylor (1994). The internal structure is conjectural.

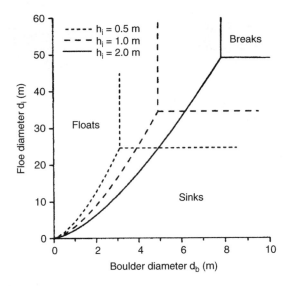

Figure 11.30 Conditions for rafting spherical boulders of diameter d_b by circular ice floes of diameter d_i, together with conditions of ice failure, assuming representative sediment and water densities of 2600 and 1030 km m^{-3} and a bearing strength of 1×10^5 kg (from Drake and McCann, 1982, as modified by Forbes and Taylor, 1994, Figure 7). h_i is the thickness of the ice floe.

development of terraces along the rivers. Ice-push ridges can be found along the former shorelines, extending inland along the outer margins of some of the river terraces (Figure 11.31). Many of the terraces are unpaired due to the rate of uplift exceeding the rate of terrace formation.

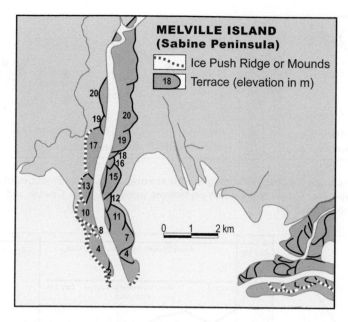

Figure 11.31   Ice-push ridges on the Sabine Peninsula of Melville Island, North West Territories (modified from Forbes *et al.*, 1986, Figure 2). Note the ice push ridges and mounds on the western margins of the terraces.

11.7.2   Effects of Geology

Where massive igneous and metamorphic rocks form the shore, erosion is minimal. In the case of massive icy beds, erosion is most rapid due to thermal erosion aiding mechanical retreat of the shoreline. Sedimentary rocks always contain some ice that bonds the mineral particles together. The presence or absence of ice-wedge polygons is also important. Rate of erosion under weak wave attack is proportional to the quantity of ice present, other things being equal. However under strong wave action, rates of erosion are similar, though different processes and cliff heights are involved.

11.7.3   Topographic effects

The higher the cliff, the more material needs to be removed. In the case of retrogressive thaw slumps as well as slides, the higher the cliff, the larger the slump or slide zone tends to be. Where the ice-rich land is virtually at sea level, the rate of retreat can be very fast, *e.g.*, on old alas beds. Jones *et al.* (2009) demonstrated a 25 m retreat between the 16th July, 2007 and the 20th July, 2008 in the absence of a westerly storm event along the coast in Alaska. Figure 11.32 shows the stages in retreat of a cliff undergoing erosion of permafrost with frozen, ice-poor sediment at tide level, overlain by both yedoma-type ice wedges and ice-rich permafrost.

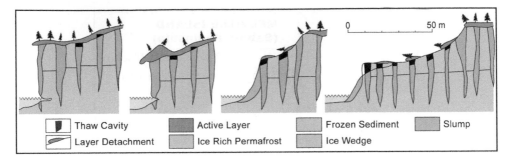

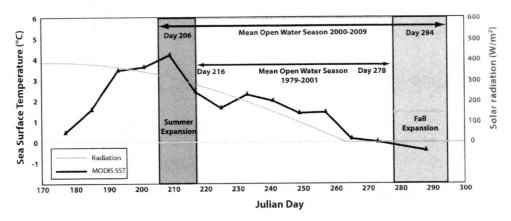

Figure 11.32 Evolution of a coast undergoing erosion in frozen, ice-poor sediment, overlain by yedoma-type ice wedges and ice-rich permafrost (modified from Czudek and Demek, 1970, Figure 3, p. 106).

Figure 11.33 Mean sea surface temperature (2000–2009) versus day of the year (black line), together with solar radiation and change in mean open water season (from Overeem *et al.*, 2011, Figure 4).

11.7.4 Sea conditions

Sea conditions are critical. Figure 11.33 shows an example of the mean sea surface temperature between 2000 and 2009 on the coast of the Beaufort Sea in Alaska (Overeem *et al.*, 2011). Since about 2000 AD, the area of open water each summer has been increasing, as has the mean open water season. This has resulted in greater fetch and stronger waves that have greatly increased the rates of erosion (Figure 11.34). All types of shorelines now have similar rates of erosion and the shorelines are straightening out (Jones *et al.*, 2009).

Height and direction of the waves, together with the slope of the bed of the sea all affect the amount of erosion. The sea level along the Arctic coasts is also changing with time due to isostatic and eustatic movements, as well as hanging sea levels due to melting of glaciers and changes in the volume of the ocean basins. There is also considerable variation in the sea levels due to localized tectonic action, *e.g.*, the faulting

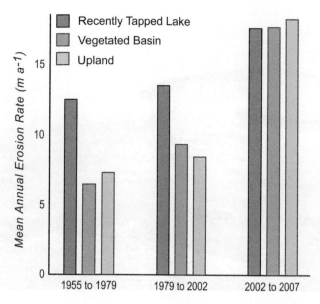

Figure 11.34 Annual erosion rates for three distinct shorelines along the Alaska Beaufort Sea shore (modified from Jones et al., 2009, Figure 3). Note the change to roughly equal rates of erosion after 2002.

occurring spasmodically along the Laptev Sea bed resulting in the development of horsts and graben (Figure 1.30), and the sinking of the Canadian Arctic Islands.

11.7.5 Deposition of sediments

Localized aggradation is occurring in places. The Lena and Mackenzie Deltas are obvious examples although the outer margins of these deltas are undergoing erosion. The eroded sediment is redeposited offshore (Aré et al., 1999), producing the shallow Beaufort and Laptev seas, while the coarser material is deposited on beaches and as offshore bars where the wave action and currents permit. Dallimore et al. (1996) found that the long-term sediment budget (1947–1985) for northern Richards Island in the Beaufort Sea shows a near balance between headland erosion and coastal deposition when ground ice and offshore erosion are accounted for.

11.8 PROCESSES INVOLVED IN THE EROSION OF ICE-RICH ARCTIC COASTAL SEDIMENTS

The same three processes of bank erosion that affect river banks (retrogressive thaw slumps, active layer detachment slides and landslides) occur on shorelines (e.g., Grigoryev, 1996). However, the changes in water level with tides and the steady retreat of the shoreline modify the results. Thermal erosion is especially pronounced where there is thick, massive ice exposed along the shore. The cliff along the northwest coast of Yukon exposes 10–15 m of ice-rich silts with up to 60–70% segregated ice by volume

Figure 11.35 Retrogressive thaw slumps along the northwest coast of the Yukon Territory. Note the ice-wedges in the sediments at the top overlying the massive ice forming the lower part of the cliffs, and the sloping surfaces of older retrogressive thaw slumps. © J.-S. Vincent.

beneath a 2 m deep thaw unconformity (Mackay, 1966; Harry *et al.*, 1985; Pollard, 1998). It contains obvious polygonal ice-wedges in relatively ice-poor sediments overlying massive ice, possibly over 40 m thick (Lantuit *et al.*, 2012b). The permafrost itself exceeds 600 m in thickness in places (Smith and Burgess, 2000). King Point is part of an ice-push moraine formed during a standstill or readvance of the Sabine Phase (18–22 ka B.P.) of the Laurentide Ice Sheet. The result is a shoreline marked by retrogressive thaw slumps (Figure 11.35).

Aré *et al.* (2005) described situations where rejuvenation of retrogressive thaw slumps was indicated by the presence of **thermo-terraces** (old slump floors) occurring above the new, rejuvenated slump. By projecting the sloping surface of the thermo-terrace towards the coast, they were able to show that the terraces formed when the coastline was further out to sea. The average headwall retreat rates on nearby Herschel Island are 9.6 m/a^{-1} compared to coastal retreat rates of 0.6 m/a^{-1} (Lantuit and Pollard, 2005). 15–20 m of ice or ice-rich sediments may be exposed in the headwall of the retrogressive thaw slump, so the headwall retreats rapidly. The outer edge of the sediment formed by the slumping material is undergoing erosion by the sea, and eventually, marine erosion exposes a new icy cliff and the process starts over again, leaving the earlier slump floor as a thermo-terrace (Figure 8.35, 8.36, and 11.35). Radiocarbon dates suggest that at King Point, the thermo-terrace was active until about 1940, but others on Herschel Island date back 223–305 years B.P. Cases of up to three thermo-terraces were observed in one place on the Island, the oldest dating back to 550–660 years B.P., though these dates are based on AMS dating. The result is block erosion, consisting of the sliding of the centres of truncated ice-wedge polygons into the sea

Figure 11.36 Wave-cut notch on the shores of the Laptev Sea. Note the ice wedge and the collapsed block of ice-rich sediment that has broken off from the cliff. © Benjamin M. Jones.

Figure 11.37 Block erosion due to failure along ice wedges in coast erosion (from Hoque and Pollard, 2007, Figure 1).

(Figure 11.37). Hogue and Pollard discuss the stresses leading to the various types of slides found along the Arctic shores of the Yukon Territory. A relatively infrequent spectacle is an active layer detachment slide on a coast (Figure 11.38). These occur where there is a concentration of soil moisture in the active layer during the

Figure 11.38 An active layer detachment slide descending on to the beach from a valley behind the cliff. Note the massive ice forming the cliff and the wave-cut notch at water level.

Figure 11.39 Results of thermal erosion along the northeast coast of Alaska (USGS). Note the ice along the beach consisting of the remnants of ice-push ridges. These protect the shore in the early part of the summer.

spring/summer thaw, together with a sufficient slope towards the cliff. These conditions are usually found where a narrow valley is truncated by coastal retreat and results in the upper layer of the valley floor sliding/flowing on to the beach.

Cliff erosion in sediments with considerably less ice is usually initiated by the development of a *wave-cut notch* or *wedge* undermining the overlying cliff (Figure 11.36). Eventually, the stresses exerted by the unsupported rock exceed its tensile strength and the unsupported mass collapses into the sea (Hogue and Pollard, 2007). Alternatively, if the sediment has joints parallel to the shore, failure may result from the effects of this plane of weakness. A modification of this occurs where large ice-wedge polygons divide the sediments into blocks.

11.9   IMPORTANCE OF COASTAL EROSION OF SEDIMENTS CONTAINING PERMAFROST

Coasts developed in permafrost sediments represent some of the most inhospitable coasts in the world. Rate of retreat is very high, and any man-made structures near the shore, *e.g.*, oil wells, buildings, docks, *etc.*, are generally fated to fall due to the encroaching cliff (Figure 11.39). There is also the problem of ice-pushed ridges and enhanced erosion along low-lying coasts, making development of these coasts very difficult. Recent research is examining the use of rocks and bags of sediment to reduce erosion (Artières *et al.*, 2010), and the results look promising for protecting quays and docks.

Part III

Use of permafrost areas

III.1 INTRODUCTION

It is obvious that the use of permafrost areas by Mankind is far more difficult if we want to have a similar standard of living to those living in warmer areas. We need natural resources, and increasingly, the remaining reserves of minerals and forests are located in regions with permafrost. Accordingly, it is becoming vitally important to determine the best techniques for developing these areas and bringing the resources to the regions where most people live. The realisation that climate changes are constantly occurring in addition to short-term fluctuations brings added complications to construction of facilities that have not been considered in most previous developments in permafrost areas (Bobov, 1977; Vyalov et al., 1993, 1997; Pavlov, 1994, 1996; Anisimov and Nelson, 1996; Anisimov et al., 1997; Balobaev and Pavlov, 1998; Trofimov et al., 2000).

The longest experience in utilizing these cold, permafrost areas is by the Russians (Harris, 1986; Afanasenko et al., 1989) who conquered the northern part of Siberia between 1490 and 1692 (Tsytovich, 1966; Fullard and Treharne, 1972). Initially, the Muscovites used the old ways of the indigenous peoples, but the limitations of permafrost were noted in the early military reports of Glebov and Golovin as early as 1642. The failed attempt by Shargin to find water in a well in Yakutsk triggered the study by Academician A. F. von Middendorf which resulted in the first measurement of ground temperatures (Middendorf, 1848, in Tsytovich, 1966). Subsequent studies were carried out by military companies, many of whom were from what is now Poland. The most successful of commander of these was Major Nikolai Mikhailovich Preswalski, after whom a medal was named by the Russian Geographical Society. The first known map of Siberian permafrost was made by G. Vil'd in 1882 (Nikiforoff, 1928; Baranov, 1959) using the $-2°$ isotherm. This put the southern limit too far north in Siberia and too far south in Europe. In 1895, I. V. Mushketov with V. A. Obruchev and others wrote the first "Instructions for investigation of frozen soils in Siberia". N. S. Bogdarov and A. N. Lvov described the experiences of the engineers building the trans-Siberian Railroad and searching for suitable water supplies for the steam engines, while M. Sumgin (1927) published the first textbook on "Permafrost within the U.S.S.R.". This established the study of *geocryology* as an independent branch of Science, and Cold Regions Engineering is now recognised as a speciality (Washburn, 1979; Johnston, 1981; Yershov, 1990; Smith and Sego, 1994; Kamensky,

Figure 12.1 Results of frost heaving on a concrete apartment building in Vorkuta, European part of Russia.

1998; Senneset, 2000). Andre'eva *et al.* (1995) have summarized a method used in Russia to reduce the complexity and uncertainty in making Arctic Resource decisions.

Initially, settlers in North America also used the methods of the indigenous people in the far north. After war broke out between Japan and the United States of America, North Americans quickly marshalled the available information from the USSR (Muller, 1943, 1946) and began carrying out extensive studies to support the building of the Alaska Highway, together with military bases in the northwest of Alaska and Northern and Arctic Canada. This was followed by the development of systematic research to facilitate settlement of the northern lands and the use of its valuable mineral resources. The discovery of the latter, and in particular, the areas of natural gas and petroleum deposits both on-shore and in the shallower parts of the Arctic Ocean has resulted in a rapid increase in the northern populations and in related resource development throughout the region. Other important high-priced resources include diamonds and uranium. These are valuable enough that they can economically be mined provided that the ore can be processed on site. This requires the provision of electricity. Large copper-zinc-lead ores are known in places like Faro, Yukon Territory, but getting them to smelters and the market is currently too costly. However the mines at Norilsk are close enough to the coast to be viable and supply Eastern Europe and Russia with a large proportion of its copper-zinc-lead needs.

A relatively new area of research involves the gas hydrates. These pose problems for drilling in cold regions, yet represent vast reserves of natural gas if only a method can be found to obtain them safely at a controlled rate.

In China, long distance communications are critical to its economic growth, together with the provision of adequate energy supplies. Accordingly, hydroelectric power and transmission lines are very important. Fibre-optic cable has been laid across vast distances, freeways, railways and high speed rail lines have been built, and pipelines bring oil and gas from Western Siberia. All of these must cross areas of permafrost, and have had to be built to keep pace with the economic growth of that country.

Part 3 of this book examines how the engineers have developed techniques to enable development in permafrost areas. Construction alters the local microenvironment which must be taken into account if construction is to be successful (Afanasenko *et al.*, 1991). Chapter 12 discusses the mechanical properties of soils that affect their response to the seasonal heating of the ground in permafrost regions. It also discusses the problem of determining the frost susceptibility of the soil. It is followed by Chapters dealing with foundations, linear transportation systems, the problems facing the oil and gas industry, mining, electricity grids, water supply and waste disposal, and finally agriculture and forestry. All these require special techniques to be successful. Unfortunately, when exploitation of a resource ceases, there has usually been inadequate remediation and restoration of the landscape to a satisfactory condition.

Chapter 12

The mechanics of frozen soils

12.1 INTRODUCTION

Mechanical processes in soils are the processes generated by internal stresses of various kinds resulting in *elastic, viscous, or plastic strains* occurring with or without breakage of the continuity of their material (Goldshtein, 1952; Vyalov, 1978). Internal stresses in frozen soils are due to a change in the main thermodynamic parameters. These are the external pressure (P), temperature (t), and volume (V). In this connection the stresses may be divided into two groups. The first group of stresses is associated with an application of an external pressure. The mechanical processes (compressive, tensile, shear strains) developing in this case may be considered *baromechanical processes*. The second group of stresses is generated inside a soil body as a result of *nonuniform changes in the elements of its volume* due to physical and chemical processes, such as desiccation, moistening, heating, cooling, moisture phase transitions, and migration with changes in volume (Taber, 1930; Fedosov, 1935; Edlefsen and Anderson, 1943; Yershov, 1986; Henry, 2000), *etc.* Non-uniform changes in the volume elements (V) related to each other occur due to dissimilar changes in the parameters of the soil body, such as *temperature, liquid moisture content* (W), *ice content* (W_i), *etc.*, resulting in temperature strains, swelling strains, heaving-out and some other processes. One can recognize two types of physico-mechanical processes in this group of strains and stresses. These are the *volume-gradient stresses and strains* in frozen soils induced by a change in their negative temperature (*thermo-mechanical processes*), and the *volume gradient strains and stresses* in freezing-thawing soils induced by moisture phase transitions, migration, and processes of texture formation (*aggregation, dispersion, coagulation*) of soil particles.

12.2 STRAINS AND STRESSES IN THE FREEZING AND THAWING OF SOILS RESULTING IN FROST HEAVING

Freezing of water-saturated soils results of soil heaving due to the 9% expansion of the volume of water during freezing (Beskow, 1935; Bozhenova and Bakulin, 1957) when the moisture has no way of migrating into the frozen ground or escaping into the underlying thawed soils. Particles, aggregates, and soil debris are forced apart, resulting in an increase in the total volume of the frozen soil body of the sample by a few percent (usually not more than 3–4%). Aggregates and soil debris can be separated by thin ice layers, or sometimes lenses of ice are found scattered throughout the frozen soil. In closed systems (with no possibility of lateral or vertical soil expansion) these can result in the generation of pronounced stresses at the contacts between soil particles due to the crystallization pressure of ice reaching 2,200 kg/cm during freezing, their conglomeration, squeezing of particle aggregates, compression, reorientation, and plastic movement (Fedosov, 1935). The strains and stresses are not found in the thawed portion of fine-grained soils if there is no moisture migration. They can occur only in the presence of the suppressed strains in a frozen block of soil.

During freezing of oversaturated permeable soils, *e.g.*, water-saturated sands, gravels, *etc.*, the excess water moves downward (the *piston effect*). In the underlying thawed soils, partial or complete breakage of the structural bonds occurs as a result of the hydrostatic pressure that is generated. In the presence of an aquiclude, artesian suprapermafrost or intra-permafrost waters are accumulated. If these waters are forced to the surface along the cracks or some other weakened zones of the frozen soil, they result in icings, injected frost mounds, pingos, and other phenomena. They are also observed in the seasonally thawing layer, when closed volumes of an oversaturated soil form, and then freeze.

In the thawed zone, the cryotexture depends upon the rate and degree of desiccation due to migration of moisture to the frozen zone, resulting in the accumulation and coagulation of the soil particles in the *desiccation zone* (Figure 12.2). Larger blocks and aggregates form in this case (*peds*), while the particles pack more closely together, the porosity decreases, and the soil density increases. As a result, stresses and strains due to shrinkage develop. The desiccation zone may be divided into two portions. The first portion (occurring closer to the frozen zone) is characterized by more pronounced changes in moisture contents. The thickness of this narrow zone is rather small (the first few centimetres). The second portion is much wider (a few tens of centimetres). It is characterized by smaller gradients in moisture contents. The width of these portions depends upon the freezing rate. With decreasing freezing rate, the second (wider) zone increases in thickness, while the first (narrow) zone decreases in thickness. Clearly, the soil shrinkage only increases to a particular density, and thereafter, an equilibrium is established between the capillary forces and the film forces tending to draw the particles together, which resist the forces generated by the texture. The soil moisture content corresponding to this density is the moisture content of the *shrinkage limit* ($Wshr$). The soil desiccation continues up to the shrinkage limit in the thawed zone close to the freezing front. Later on, the soil serves as a transit zone for moisture transfer. The maximum shrinkage strains occur in the vicinity of the freezing front in the portions of the greatest desiccation of the thawed zone and the frozen zone of the soil. Ice appears in the *frozen fringe*, and the heaving occurs due to the growing ice layer. The ice layer growth is provided by the frozen zone heaving upward as well as by downward shrinkage of the predominantly thawed zone undergoing desiccation. The total amount of the shrinkage strain in the vertical direction is an order of magnitude more than that in the horizontal direction. This may be explained by the fact that the shrinkage of the thawed zone of the soil in the horizontal direction is "suppressed" due to adherence between this zone and the frozen zone retarding the strain. As a result, during freezing of water-saturated fine-grained soils with active moisture migration and ice accumulation, a zone ("neck") forms in its frozen portion, which resembles

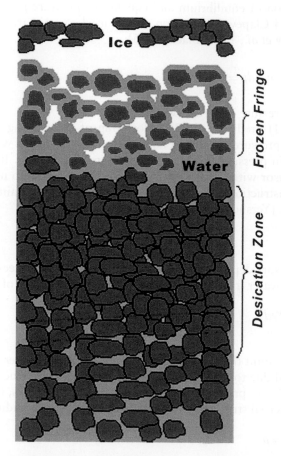

Figure 12.2 Results of freezing fine-grain soil showing the frozen (freezing) fringe and desiccation zone below.

the drying portion of the thawed zone and the zone in the freezing sample undergoing dehydration under decreasing negative temperature. When the shrinkage stresses reach the breaking values, cracks form.

Freezing of water in a closed volume (for example, in some soil pores) can induce pronounced stresses in conditions of strain suppression. However, these heaving-out stresses occur at values just below the complete suppression of the strains. As the natural soils may not be considered as a closed, hard-to-deform system, the heaving-out component of the stresses do not predominate over the other components of heaving stresses and strains in most cases. This component may be considered as some addition to the swelling stresses developing at the expense of the wedging-out effect of thin films of migratory water. It is possible that in nature, the crystallization pressure (which is many orders of magnitude higher than the soil strength, its actual value depending on the counterforce) relaxes quickly due to deformation of the plastic frozen soils in the freezing zone.

Under condition of equilibrium and equality of pressure $p_{ice} = p_w$ in the closed system a simplified Clapeyron equation can be used to estimate stresses in freezing soils (Grechishchev et al., 1980; Black, 1995):

$$Q\frac{dT}{T} + (v_{ice} - v_w)dp = 0 \qquad (12.1)$$

where: p – equal pressure of ice and water.

Equation (12.1) predicts a 0.074°C lowering of the melting temperature for a 1 MPa increase in pressure equally on both the ice and the liquid water phases. There is a large change in pressure for a small change in temperature. If this pressure p is measured by a sensor with a factor of frigidity k_g (or directed to massive ground, or to engineering construction), the measured value depends on strain x caused by frost heaving as follows (Yershov, 1985):

$$dp = k_g dx \qquad (12.2)$$

Thus, the pressure depends on the increased volume of the freezing soil, or on the flow of migrating water q and water volume changes at freezing of 9%:

$$dp = k_g q * 0.09 d\tau \qquad (12.3)$$

where: τ – time.

Taking into account the fact that ice lenses play a role in determining the bearing capacity of the soil due to their sizes and position in the soil structure, and unfrozen water layers are always present between the ice and soil particles, the pressures of ice and water and external stress should be equal in equilibrium conditions:

$$p_{ice} = p_w = \sigma_t = p. \qquad (12.4)$$

The flow of migrating water q should be proportional to a difference in chemical potentials $(\mu_w - \mu_{ice})$ and can be expressed in the simplified case $p_{ice} = p_w = p$ as follows (Grechishchev et al., 1980; Deryagin et al., 1985):

$$dq = k\left[-Q\frac{dT}{T} - (v_{ice} - v_w)dp\right] \qquad (12.5)$$

where: k – a factor.

Then the flow of migrating water in the constrained conditions, which are defined by the factor of rigidity k_g, could be expressed as follows:

$$dq = -k\left[Q\frac{dT}{T} + (v_{ice} - v_w)k_g q * 0.09 d\tau\right] \qquad (12.6)$$

If free water outflow occurs ($p_w = 0$) and the process is long enough, we are getting another formula for the equilibrium (Schofield, 1935), similar to (12.1):

$$Q\frac{dT}{T} + v_{ice}dp_{ice} = 0 \qquad (12.7)$$

According to (12.5) a lowering of the melting temperature of ice is greater by an order of magnitude than predicted by (12.4), because:

$$\frac{v_{ice} - v_w}{v_{ice}} \approx 0.1. \tag{12.8}$$

In this case, the ice pressure is free to vary with temperature, while the liquid water pressure remains zero or constant. Distribution of ice and water pressures and effective stresses is not as simple as that presented in the model of Miller (1972, 1978) nor in similar models due to the surface effects of soil particles, the complex geometry of soil pores and the different and changing mechanical properties of ice and soil particles (Drosf-Hausen, 1967; Chistotinov, 1973; Grechishchev et al., 1980; Deryagin et al., 1985; Golubev, 1988; Henry, 2000). A reasonable approach could be the simplified physics of frost heaving. If it is considered that the stresses in soil are a result of the suppressed strain $\varepsilon_s = \varepsilon_h - \varepsilon_a$, where ε_s and ε_a are the values of suppressed and allowed strains; ε_h is the value of potential frost heaving strain without the strain suppression. The pressures do not increase if there is a change in volume due to water flows or phase transfer during the freezing occurring without resistance. Conditions of the shrinkage suppression are of decisive importance for the generation of the heaving stresses and the shrinkage stresses. The more pronounced is the suppressed strain, the higher are the stresses noted in the freezing soil. The increase in the gauge rigidity results in a decrease in the allowable sample strain.

The mechanism of the heaving stress formation taking into consideration interaction between all of the parts of the freezing soil can be presented as follows (Brouchkov, 1998). For the freezing zone, defining the development of the heaving stresses where the suppression of its deformation occurs, the following condition must be true (Fig. 12.3):

$$dp_h = f(\varepsilon_s) \tag{12.9}$$

If we simplify and assume a linear approximation for the function $f(\varepsilon_s)$ as it is shown on Figure 1:

$$dp_h = k_{gr}d\varepsilon_s = k_{gr}d(\varepsilon_h - \varepsilon_a) \tag{12.10}$$

where ε_s and ε_a are the values of suppressed and allowed strains of the freezing zone; ε_h is the value of potential free frost heaving strain in the case of an absence of strain suppression. The coefficient k_{gr} shows the stresses per unit suppressed strain of the freezing part.

Further studies could estimate the function $f(\varepsilon_s)$ and probably give a better approximation than equation (12.6).

Together with the heaving strains, the stresses form in the frozen part of soil. Mechanical stresses forming at the surface of freezing soil, and effects on a gauge or an engineering structure are a result of the interaction between the unfrozen and frozen parts and the gauge or engineering structure retaining the strain. The nature of the forces generated in the freezing zone aside, we can see the dependence of stresses dp_h on the value of the "suppressed" strain in the linear form (substantiated above) (see Eq. 12.6). In the case of mechanical equilibrium, if stresses p_h are equal in all parts of the sample and equal to external stress $p_h = \sigma_t$, given the limited heaving strain on

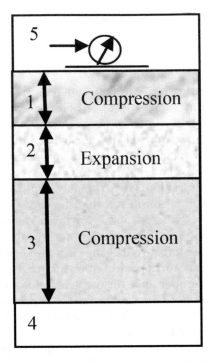

Figure 12.3 Mechanical interaction of frozen (1), freezing (2) and thawed (3) zones underlain by a rigid horizon (4) if frost heave forces are measured by a sensor (5).

the soil surface, the allowed strain ε_a of the freezing zone (frozen fridge) is defined by the gauge or engineering structure rigidity $((1/k_g) * dp_h)$, together with the mechanical compression of both the unfrozen and frozen parts of the soil sample ($dp_h * l_t/E_t$ and $dp_h * l_{fr}/E_{fr}$ respectively) due to the formation of the stresses dp_h being of l_t and l_{fr} in size and having the strain modulus E_t and E_{fr}, respectively, and given the value of the unfrozen part's shrinkage $d\varepsilon_{sh}$:

$$\varepsilon_a = \frac{1}{k_g}dp_h + \frac{l_t}{E_t}dp_h + \frac{l_{fr}}{E_{fr}}dp_h + d\varepsilon_{sh} \tag{12.11}$$

On rearrangement of (12.6) and (12.7), the relationship between strains, stresses and properties of soil and gauge can be written as (Yershov, 1985):

$$dp_h = \frac{d\varepsilon_h - d\varepsilon_{sh}}{\dfrac{1}{k_{gr}} + \dfrac{1}{k_g} + \dfrac{l_t}{E_t} + \dfrac{l_{fr}}{E_{fr}}} \tag{12.12}$$

where: ε_h – is the value of potential free frost heaving strain in the case of an absence of the strain suppression; $d\varepsilon_{sh}$ – shrinkage of the unfrozen part due to dehumidification of the soil; k_r – rigidity of sensor (or object to affect, for example, pipe); k_{gr} – factor

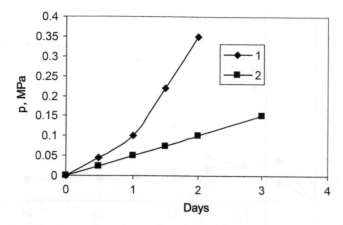

Figure 12.4 Stresses developed during the freezing of glacial silt at −2.0°C; I, in an open system with water inflow, and 2, in the closed system. The frigidity of the sensor was 1700 MPa/m. The samples were 5 cm long and 5 cm in diameter.

of decreasing of freezing pressure in soil, if the soil is allowed to expand; p_h – pressure (stress) acting in the system; l_{fr} and l_t – size of frozen and unfrozen part respectively; E_{fr} and E_t – compression modulus of frozen and unfrozen part of soil respectively.

From equation (12.8), it follows that the heaving stresses being measured increase in proportion to the potential heaving strain ε_h and decrease with increase in the allowed strain ε_a being inversely related to the total deformability of the soil system (denominator in Eq. 12.8). Rheology of unfrozen and frozen soil, stress relaxation, the complex character of the function $dp_h = f(\varepsilon_s)$, values of pressure p_h described by the Clapeyron equation, and feedback of stresses on frost heaving strains in the freezing zone are not taken into account. Thus, the wedging out effect of the thin, unfrozen film causing its migration plays a leading role in the formation of the heaving strains.

A study of freezing of five-centimetre soil samples with an outside moisture inflow and with no outside moisture inflow has shown that the values of free heaving strains were 5 and 2 mm, respectively, while heaving stresses were of 0.4 and 0.15 MPa (Figure 12.4). Heat and mechanical conditions for stress formation associated with water crystallization in a limited volume were nearly the same in both cases. Therefore, an increase in heaving strains in the presence of outside moisture with water inflow, and 2, in the closed system. The frigidity of the sensor was 1700 MPa/m. The samples were 5 cm long and 5 cm in diameter. Inflow and active segregated ice streak formation in the given experiment may be explained only by the wedging out effect of the migratory moisture films. If these stresses were associated with water crystallization, *i.e.*, if the crystallization pressure were higher than the wedging out effect of the thin water films, the stresses would be the same in both cases.

Decrease of water flow during all-dimensional freezing in comparison to one-dimensional freezing leads to decrease of stresses (Figure 12.5). The small size of the samples, the size of their frozen and unfrozen parts and of values l_{fr} and l_t respectively are among the reasons for larger values of registered stresses p_h.

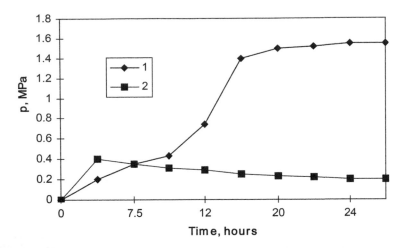

Figure 12.5 Stresses in the soil during freezing of water-saturated glacial silt at −2°C: 1, one-dimensional, and 2, all-dimensional freezing. The frigidity of the sensor was 1777 MPa/m. The samples were 5 cm long and 5 cm in diameter, and there was no outside water flow.

The more pronounced the suppressed strain, the higher the stresses noted in the freezing soil. The increase in the gauge rigidity index in this case is responsible for a decrease in the allowable sample strain.

The experimental results show that a decrease in size of the unfrozen part or its compression is of considerable importance in the formation of heaving stresses. In the early stage of freezing, the unfrozen part is truly desiccated and easily deformed under a load. However, later on, the mass of the unfrozen part is less able to compress and the total capacity of the soil system to resist loads increases, resulting in accumulation and increase in the unrelaxed stresses (Figures 12.4, 12.5). This provides an explanation for the fact that heaving stresses increase with time with a deepening freezing front, reaching the maximal values by the end of freezing. The values k_{gr} for clay soils calculated from the laboratory data are about 0.4 MPa on average for a frozen layer of 1 cm in thickness under a "suppressed" strain of 1 mm.

The heaving stresses increase as a whole with fine and silt particle contents of the completely water-saturated soils due to an increase in the initial water content, migratory moisture fluxes, and heaving strains ε_h. The heaving strains ε_h in the initial stage of freezing decrease in a descending sequence from kaolinite via polymineral to montmorillonite clay, while their final values increase. This may be explained by greater heaving strains due to active moisture migration and ice formation in the initial stage of freezing in the kaolinite clay, smaller heaving strains in the polymineral clay, and even smaller heaving strains in the montmorillonite clay. A great amount of remaining "nonredistributed" water freezing in the final stage of freezing in the montmorillonite clay is responsible for high strain increments, and hence, high heaving stress increments.

The stresses decrease sharply with an increase in the degree of water saturation from 0.8 to 1, due to a decrease in the formation of ice lenses and heaving strains, as well as due to an increase in the deformability of the soils E_{fr} and E_t. Surprisingly, the highest values of stresses were found at the beginning of freezing of the sample of lower dry density; since the stresses were developing more slowly than in samples of higher dry density. It could be due to greater moisture migration and accordingly ε_h at the beginning of freezing in the sample of low dry density. Biggest stresses at the end of freezing were registered in the fully water-saturated sample, possibly, due to large values of ε_h, E_{fr} and E_t. Both free strains ε_h and stresses p_h of frost heaving were higher in a sample where the original structure was maintained in comparison to samples which were prepared by mixing dry soil and water. The calculations of frost heaving stresses in the lower part of equation 8 are presumed to be a part of the analysis of the interaction between freezing soil and engineering construction. Values of ε_h, ε_{sh}, E_{fr}, E_t and especially k_{gr} can be determined by experiments, and then frost heaving forces p_h can be estimated.

Thermodynamics can only be used to estimate the maximum frost heaving forces in a condition of equilibrium (Edlefsen and Anderson, 1943; Deryagin et al., 1985). However, distribution of ice and water pressures and effective stresses might also not be as simple as is presented in models, e.g., those of Miller (1972, 1978), due to the surface effects of soil particles, complex geometry of soil pores and different and changing mechanical properties of ice and soil particles. Mechanics of interaction of soil phases has its own problems. The rheology of unfrozen and frozen soil, stress relaxation, feedback of stresses on frost heaving strains during the freezing of soil and other questions remain to be studied. Conditions of the suppression are of decisive importance for generation of the heaving stresses: the more pronounced the suppressed strain, the higher will be the stresses that are found in the freezing soil.

Frost heaving, i.e., the volumetric expansion of the soil and uplift of the ground surface is the most characteristic process during soil freezing (Taber, 1930; Fedosov, 1935). It is observed throughout the region of seasonally and perennially frozen soils, but is non-uniformly distributed over these areas due to local features, e.g., the soil composition, structure and conditions of freezing (the microenvironment). The contributions of the massive heaving-out, migratory ice accumulation, and shrinkage to the overall value of heaving depend on the particular conditions under which fine-grained soil freezing takes place, e.g., the formation of frost mounds, pingos and other mounds ranging from tens of centimetres to several tens of metres in height. Comparatively uniform heaving strains which are not clearly defined in topography are, however, of prime importance for the stability of engineering structures and are of the greatest interest for practical purposes. The spatial lack of uniformity of these strains is a serious hazard, rather than their absolute values. Percentage of the difference between the heaving value of one point and the heaving value of the second point together with the distance between them is taken as the **nonuniformity criterion**. In nature, the values of the coefficient of the heaving nonuniformity can range from 3–4 to 10–15%. The range of the change in the absolute values of the spatial heaving strains can vary depending upon the natural conditions from fractions of centimetres to a few tens of centimetres. The contributions of the massive heaving-out, migratory ice accumulation, and shrinkage to formation of the overall value of heaving depend on the particular conditions for fine-grained soil freezing.

Development of the heaving strains of water-saturated soils mainly at the expense of expansion of pore moisture by 9% of its volume during freezing (massive heaving) is typical of sandy soils with no pronounced moisture transfer in the frozen zone, as well as for the conditions of rapid freezing, when the segregated ice layers have no time to grow due to the high rate of movement of the freezing front. In this case, the frost heaving is:

$$h_f = 0.09(W_\xi - W_{uf})\xi \tag{12.13}$$

where: ξ is the freezing depth, W_ξ is the volumetric moisture content at the interface from the side of the thawed zone; W_{uf} is the amount of thawed water.

Massive heaving of freezing, water-unsaturated soils will occur depending upon the degree of filling of the pores with moisture. It has been found during numerous investigations that ice accumulation in the freezing soils due to moisture migration to the frozen zone is of great importance in formation of the total value of their heaving. Therefore, ice accumulation and heaving increase with an increase in density of the migratory moisture flux in freezing soils (due to high potential gradients ($\mathrm{grad}\,\mu_W$), or increasing hydraulic conductivity along films), and also with a decrease in the freezing rate. Nearly all the calculations and procedures of approximate quantitative estimates are based on this.

The density of the migratory moisture flux J_W is usually expressed as follows (Yershov, 1985):

$$J_W = -\lambda_W\,\mathrm{grad}\,\mu_W = -K_W\,\mathrm{grad}\,W_{uf} = -K_W\frac{\partial W_{uf}}{\partial t}\frac{\partial t}{\partial x} \tag{12.14}$$

where: λ_W is the hydraulic conductivity; K_W is the moisture diffusivity; W_{uf} is the volumetric moisture content due to unfrozen liquid layers; t is the temperature; x is the coordinate.

Thus the heaving strain due to migratory ice accumulation is:

$$h_W = 1.09 K_{il} K_W \frac{\partial W_{uf}}{\partial t}\frac{\partial t}{\partial x}\frac{\xi}{v_\xi} \tag{12.15}$$

where: K_{il} is the coefficient approximately equal to the cosine of an angle of migratory ice lenses, $\frac{\partial W_{uf}}{\partial t}$ is the thermogradient coefficient; ξ is the depth of the freezing; and v_ξ is the freezing rate.

Accordingly, the heaving value due to migratory ice accumulation depends on the moisture-conducting soil properties, the peculiar features of liquid films and their temperature dependence, temperature gradient, and the freezing rate. The shrinkage strains depend upon the soil composition and structure, while the conditions for its freezing can be of significant importance. Some special cases may occur when the freezing fine-grained soils will not increase but actually decrease in volume, due to the shrinkage strains being higher than the heaving value.

The shrinkage is defined from the expression:

$$h_{sh} = \beta(W_{in} - W_\xi)\xi_{sh} \tag{12.16}$$

where: β is the coefficient of the volumetric shrinkage; W_{in} is the initial volumetric moisture content; W_ξ is the volumetric moisture content in the desiccated zone; ξ_{sh} is the depth of the shrinkage zone being equal to the freezing depth plus the zone of desiccation under the freezing front.

The total frost heave will be (Yershov, 1985):

$$h = h_f + h_W - h_{sh} \tag{12.17}$$

The contribution of the migratory ice accumulation to heaving of the freezing kaolinite clay is about 80–95%, but does not exceed 50–60% in the freezing sample of silts, loams or sand. On the other hand, the contribution of the massive heaving is not more than 20% in clay soils compared with 70–80% in loamy sands due to much lower contents of bonded water in the latter. At the same time, an increase in the shrinkage strains is observed in the range from loamy sand to loam to clay, and is accompanied by more pronounced similarity in the heaving strains. Thus, the heaving strains of the soil surface decreased by 20% in the freezing kaolinite clay due to shrinkage strain during desiccation of the thawed zone and by 10% in loam. The shrinkage is hardly discernable in the loamy sand soils.

The freezing conditions are of great importance for the development of heaving strains. The maximal values of the migratory flux are observed in clay soils, although the heaving value in them is often smaller than that in loams. The increase in the freezing rate in all of the cases is accompanied by a decrease in the total ice accumulation and heaving. Soil freezing in conditions of an "open" system, *i.e.* with moisture inflow from a water-bearing horizon is often accompanied by a sharp increase in the heaving strains as compared to the "closed" system. Presence of an overload on the freezing soil, *i.e.*, freezing under a pressure, causes a decrease in the heaving due to a decrease in the migratory moisture flux density in the frozen portion of the soil. At high freezing rates, high temperature gradients form in the frozen zone and are responsible for the more pronounced migratory moisture flux. However, the value of the segregated and migratory ice accumulation is insignificant due to the short period of moisture migration to the frozen zone resulting from higher freezing rates. This results in uniform ice content distribution in the freezing layer depth.

In winter, a pronounced redistribution of moisture and ice content together with rearrangement of the cryogenic structure occurs after complete cracking of the freezing layer and the perennially frozen soils. The rate and value of heaving decrease from top to bottom. The heaving strain gauges occurring in the subsurface layer experience the most pronounced displacement.

A further complication is the presence or absence of soluble substances in the soil or water (see Chapter 1). Many soils in permafrost areas contain salts left after uplift from the sea or due to evapotranspiration in a relatively dry climate (see Fig. 1.21). These soils freeze at lower temperatures and have different, usually lower heaving properties (Chamberlain, 1983; Brouchkov, 1998). Artificial additives have similar effects, *e.g.*, Voroshilov (1978) experimenting with coagulators in Northeast Siberia. Although kaolinite gave 22.19 mm heaving, this diminished to 0.78 mm after a treatment of 1.5% by weight of sulphite alkalis which are a waste product of nearby cellulose factories. Urea formaldehyde reduced the heaving to 3.42 mm, but is a carcinogenic substance. Thus the addition of non-toxic chemicals can reduce the amount of heaving.

12.3 RHEOLOGICAL PROCESSES

These are processes in frozen soils induced by external loads. They manifest themselves as creep (strain development with time), stress relaxation (rheological processes decrease with time), and long-term strength (a decrease of the soil resistance to load with time). These processes are due to peculiar features of the internal bonds among the soil components, especially due to the presence of unfrozen water and ice, where cementing particles represent a perfect fluid (Tsytovich, 1973).

The rheological properties of the frozen soils result in their deformation continuing with time under an applied (even small) load that persists for a long period of time. The elastic, viscous, and plastic strains can develop in combination in this case (Tsytovich, 1973; Morgenstern and Roggensack, 1978). The character of the deformation depends upon the value of the applied load. Thus, the deformation proceeds at a decreasing rate (attenuating creep) under small constant loads (Figure 12.6). If the loads increase, or when the stresses in the frozen soil exceed a certain limit, the relative strains develop at an increasing rate (viscous creep).

The schematic curve for creep may be divided into several portions representing various deformation stages (Figure 12.7). Depending on the load value, this deformation can be elastic, or elastoplastic. It disappears after the load is completely or partially removed. The portion AB corresponds to the deformation at a decreasing rate, or the transitional creep (stage I), or primary creep. At this stage the frozen soil deformation disappears after the load is partially removed with time (plastic after-effect), since it includes reversible, irreversible and plastic deformation. The attenuating creep continues until some final value is reached, which depends upon the stress value, or develops continuously at a decreasing rate. The deformations are not stabilized in this case, but increase indefinitely (a so-called permanent creep). In both cases the deformation rate under attenuating creep approaches zero. If the deformation is continuous, the constant strain rate creep (viscoplastic flow) occurs (stage II, or secondary creep), which is completely irreversible. The tertiary creep, or the breaking stage (stage III, portion CE) occurs under high stresses. The deformation rate increases, resulting in brittle or viscous fractures of the frozen soil. This can be divided into two stages. In the first stage (portion CD), the plastic deformation continues and there is no fracturing, whereas in the second stage (DE), active minute cracks are formed, resulting in an extremely rapid increase in the deformation which cause failures.

The duration and function of the particular creep stages depend upon the load value. The bigger the load, the shorter the stage of secondary creep, resulting in the stage of tertiary creep beginning sooner. The function and importance of the various stages of deformation also depend upon the soil properties and temperature. All three creep stages develop in frozen soils. The more plastic the soil and the higher its ice content, the greater are the effects of stages II and III. The permanent creep in ice occurs under any stress, so that the constant rate creep and the creep developing at an increasing rate are of the main importance. The creep can be described as follows:

$$\sigma = A(t)\varepsilon^m = \xi t^{-\alpha}\varepsilon^m \tag{12.18}$$

where: σ – stress; $A(t)$, m, ξ, α – coefficients; ε – strain.

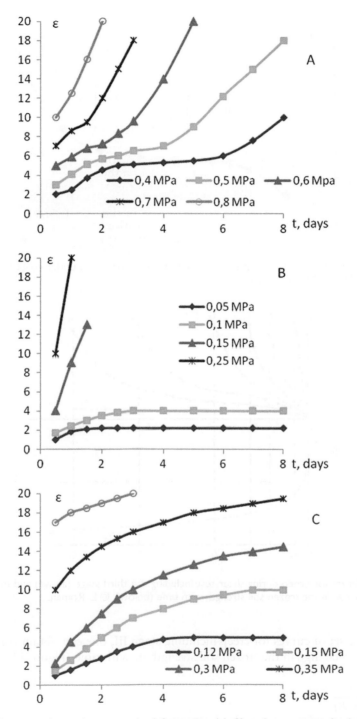

Figure 12.6 Frozen soil samples one-axis deformation (ε), % under constant loads: A – sand at Dsal $= 0.1\%$, t $= -2°$ C; B – sandy loam at Dsal $= 0.2\%$, t $= -2°$; and C – loam at Dsal $= 1\%$, t $= -3.8°$ C.

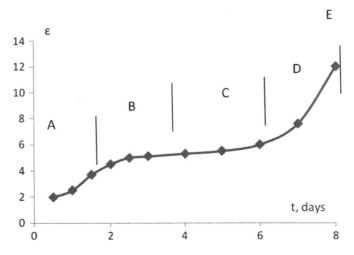

Figure 12.7 Creep stages of frozen sand at one-axis deformation (ε), % under constant load 0.4 MPa at Dsal $= 0.1\%$, t $= -2°C$.

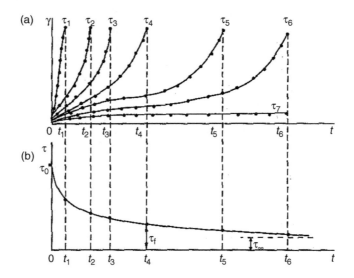

Figure 12.8 Curves for creep during shear test including the third stage (above) and the curve for a change in the frozen soil strength with time (below). © L. Roman.

Using the set of curves for creep including stage III, one can obtain (Vialov, 1978) the curve for a change in the frozen soil strength σ_t with time (Figure 12.8):

$$\sigma_t = \frac{\beta}{\ln(t/B)} \tag{12.19}$$

where: β, B – coefficients; t – time.

Criteria for bearing capacity (long-term strength) of frozen soil could be used depending on the type of failure. In Figure 12.8, failure is determined by actual breaking up of the sample. However, failure criteria could be a beginning of tertiary creep, or the breaking stage (point C at stage III, portion CE, Figure 12.7). If the tertiary creep is not observed (Figure 12.6, portion C), failure criteria could be reaching a value of 20% deformation ($\varepsilon = 20\%$).

When considering the behaviour of various bodies and materials under a load, it should be stressed that it is adequately described by the stress-strain relationship (Hook's law) for solid bodies. As for viscous bodies, the strain can increase with time under a constant load and the stress-strain relationship loses its significance. In this case the "stress-strain rate" relationship is usually used. S. S. Vyalov (1978) points out that it is more applicable, since parallels between these diagrams of deformation of solid and viscous bodies (Fig. 8.28) are observed. According to the data of Vyalov (1978), the change to viscous creep occurs under the stresses comprising 25 to 50% of the conventionally-instantaneous strength of the frozen soil in the case of shearing or compression of the frozen soils, and under the stresses of 8 to 10% in the case of tension.

N.A. Tsytovich (1973) recognised three main mechanisms of ice deformation. First is ice flow under slow shear parallel with the basal planes of the crystals, with no change in the ice structure. Disruption of the ice crystal lattice takes place in the second stage due to molecular disintegration, recrystallization, intercrystalline shifts, and chips (Voitkovsky, 1960). The last stage is when the ice melts under high stresses. The ice recrystallization and crystal basal plane reorientation parallel to the shearing force occurs under shearing stresses higher than 0.1 MPa. The crystal size of the ice decreases by several orders of magnitude in this case. The presence of specific ice-cement bonds and structures (of cryogenic type) in frozen soils are of great importance for the rheological processes proceeding in them, along with a presence of ice and unfrozen water. Thus, a few contacts of fundamentally different types occur simultaneously in frozen soils (unlike in unfrozen fine-grained soils). These are contacts between the soil particles, including contacts and cohesion of mineral particles with ice-cement and ice layers, ice binding with film water, *etc.* Non-uniform frozen soils (character of contacts, porosity, cryogenic structure and texture) and soils with more imperfections in its texture will change the soil strength, resulting in more pronounced rheological properties. Consequently, the frozen soil deformation is a result of a displacement of individual soil particles and micro-aggregates relative to each other along the bond water films and ice inclusions separating them. Taking into consideration the presence of various imperfections scattered throughout the frozen soil, the plastic deformation caused mainly by dislocation development can change to the brittle failure at the expense of the active minute cracks and through-the-thickness crack formation.

The nature and mechanism of the slow deformation (creep) vary with the different stages of the creep (Vyalov, 1978) (see Figure 12.7). At the first stage (primary creep) only slight changes in the frozen soil texture are usually observed. The inter-aggregate porosity decreases, air exclusion and some redistribution of ice and moisture contents occur, "healing" of old bonds, the formation of new inter-particle bonds, and soil strengthening are observed. As a result, strengthening of the frozen soil is observed, and is responsible for the deteriorating character of the creep. At the second stage of deformation (secondary creep or viscoplastic flow at a constant rate), there is more pronounced rearrangement of the texture of the frozen soil. Disintegration and breaking of the aggregates commences. These aggregates are reoriented by the basal planes along the vector of the shearing stress or at right angles to the compressing stress, decreasing the resistance of the soil to any applied load. The breaking of the bonds along the most weakened portions of the frozen soil increases, and the textural imperfections grow. The new texture formation leads to strengthening of the soil but is of only minor importance in this case. The unfrozen water migration to the zones of shearing or tensile stress concentration is of dominant influence at this stage of deformation. The period of deformation time is sufficient for the results of the moisture migration to be observed as a gradual and more or less uniform increase in the soil ice at these sites, together with a decrease in the strength of the soil.

12.4 FROST SUSCEPTIBILITY

Frost susceptibility is defined as ground (soil or rock) in which segregated ice will form under suitable conditions of moisture supply and temperature (van Everdingen, 1976, 1987; Chamberlain, 1981; Fukuda, 1982; Konrad and Morgenstern, 1983; Williams, 1984). This results in a change in volume of the soil causing frost heaving. Frost susceptible ground gradually becomes ice-rich regardless of the quantity of water initially available. It also implies that the ground becomes susceptible to *thaw weakening* effects when the ice thaws.

The criterion most commonly used to predict frost susceptibility is grain size (G.H. Johnston, 1981; T.C. Johnson *et al.*, 1984). Minimum grain size, amount of fines, degree of mixing (grading), mineralogy, and the presence and type of salts in both water and sediment are all important. Given an adequate water supply, Casagrande (1932) stated that considerable ice segregation should be expected in non-uniform soils containing more than 3% of grains finer than 0.02 mm diameter, and in very uniform (well graded) soils containing more than 10% finer than 0.02 mm. No ice segregation was observed by Casagrande in soils containing less than 1% of grains finer than 0.02 mm diameter, even when the water table and frost table were coincident.

This was regarded as being the best general system (Linell and Kaplar, 1963) and the *frost design classification* developed by the U.S. Corps of Engineers is based on it (Table 12.1). The soils are listed in their general order of increasing frost susceptibility. Thus the F1 sediments have the highest bearing capacity during thaw in spite of having similar volumes of ice prior to thaw. Groups F3 and F4 have the greatest frost susceptibility and greatest loss of strength during thaw.

In reality, some gravelly soils with 1% material finer than 0.02 mm show significant heave, while other sandy sediments with up to 20% of material finer than 0.02 mm are stable. This is due to other factors such as the mineralogy and permeability. As a result, it is standard practice to test the rate of heave of the soils under standard conditions to aid in choosing appropriate designs. A *classification of frost susceptibility* based on laboratory heaving tests (Table 12.2) has been developed by the U.S. Corps of Engineers (Linell and Kaplar, 1963). The tests are based on a laboratory frost penetration rate of 6–13 mm/day^{-1}. These tests take about two weeks but a reasonable indication of rate of frost heave can be obtained using a rate of frost penetration of 76–200 mm/day^{-1} in about 2–3 days (Kaplar, 1971). Currently, the American Standards for Testing and

Table 12.1 The U. S. Corps of Engineers frost design soil classification.

Frost group	Soil type	Percentage finer than 0.02 mm, by weight	Typical soil types under Unified Soil Classification System
F1	Gravelly soils	3 to 10	GW, GP, GW-GM, GP-GM
F2	(a) Gravelly soils	10 to 20	GM, GW-GM, GP-GM
	(b) Sands	3 to 15	SW, SP, SM, SW-SM, SP-SM
F3	(a) Gravelly soils	>20	GM, GC
	(b) Sands, except very fine silty sands	>15	SM, SC
	(c) Clays, PI > 12	–	CL, CH
F4	(a) All silts	–	ML, MH
	(b) Very fine silty sands	>15	SM
	(c) Clays, PI < 12	–	CL, CL-MI
	(d) Varied clays and other fine-grained, banded sediments	–	CL and ML; CL, ML, and SM; CL, CH, and ML; CL, CH, ML, and SM

Table 12.2 U.S. Corps of Engineers classification of frost susceptibility based on laboratory heaving tests.

Average rate of heave (mm/day)	Frost susceptibility classification
0.0–05	Negligible
0.5–1.0	Very low
1.0–2.0	Low
2.0–4.0	Medium
4.0–8.0	High
>8.0	Very high

Materials has an up-dated protocol for these tests (ASTM D5918-13), which replaces ASTM D5918-6. In Norway, the grain size criteria for frost susceptibility are based on 0.002 mm, 0.02 mm, and 0.2 mm sizes (Knutson, 1993). The soils are classified into four groups ranging from T1 (no frost susceptibility) up to T4 (high frost susceptibility). Figure 12.9 shows examples of grain sizes of representative examples of the four classes. However, Finnish studies indicate frost susceptibility is highly dependent on the amount of material passing the 0.0063 mm sieve (Nurmikolu and Kolisoja, 2008) and its quality (Nurmikolu, 2010), meaning mineralogy and weathering characteristics.

Several factors (Figure 12.10) actually affect the test results, (Penner, 1960; Fukuda, 1982) showing that freezing rate was probably the most critical, since rate of heaving and rate of frost penetration are inter-dependent. The standardized frost penetration rate provides good comparative data for different materials, but for proper design purposes, the rate of frost penetration should be matched to that expected on site for the best results. However this rate is usually unknown, is variable from

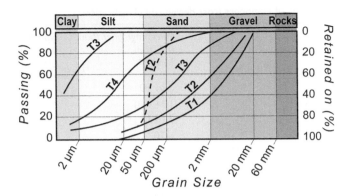

Figure 12.9 Examples of the grain size distribution of the four different classes of frost susceptibility according to the Norwegian classification.

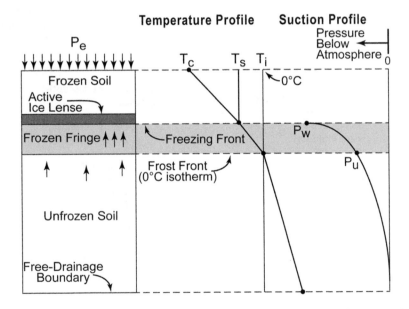

Figure 12.10 Diagram showing the conditions of temperature, active ice lens formation, frost front, freezing front and suction profile during ice lens formation (modified from Konrad, 1999, Figure 1). The arrows in the soil profile (left-hand diagram) show the direction of water migration.

year to year, as well as from place to place (Puzakov, 1960; G. H. Johnston, 1981, pp. 142–143) due to local variations in the microenvironments.

The key elements determining frost susceptibility are now thought to be the quantity of unfrozen water in the pore space which is closely related to the quantity of fines, their arrangement with respect to the coarser fraction which controls the quantity of through pores, the degree of packing of the soil particles (related to soil density), and the surface area of the grains. Also involved are the relative amounts of capillary and adsorbed water which is dependent on the mineralogy of the clay fraction. The latter also affects the actual fabric of the soil due to the varying inter-particle forces and their action during soil formation. Finally, the overburden pressure modifies the frost susceptibility of any given soil, fabric and moisture regime. Thus it is necessary to know the grain-size distribution and content of fines, the clay mineralogy, the soil fabric and overburden pressure to successfully predict frost susceptibility. Where soils or water are saline, this must also be measured (Brouchkov, 1998) since the physical properties of the soil change with increasing salinity. There is depression of the freezing point and a significant decrease in strength of the soil. The specific surface area may also be needed (Rieke *et al.*, 1983).

The Russians use a similar system in their Design Manual (Puzakov, 1960; Tsytovich, 1973), using grain size, natural and critical moisture contents, liquid limit, plastic limit, and a dimensionless coefficient which is numerically equal to the mean winter air temperature at the open soil surface during freezing of the soil. Sokolova and Gorkovenko (1997) have discussed the problems of assessing the frost susceptibility of coarse-fragment soils with a silty-clay filler as used in road beds in that country.

Chapter 13

Foundations in permafrost regions: building stability

13.1 INTRODUCTION

Present-day construction on permafrost ground utilizes one or several specialized techniques to minimize the problems encountered with thaw-sensitive and frost-susceptible soils. Various combinations of these basic techniques are used in almost all construction projects carried out in permafrost regions. Choice of the method or methods used for buildings depends on the type of structure, the weight of the objects to be stored in it, the properties of the substrate, the climate, and the drainage conditions. Along linear right-of-ways, the choice of method will usually vary according to the changing microenvironments that are being traversed.

This chapter will describe the problems to be overcome and the way in which they are usually dealt with. It includes case studies. This is followed by a systematic description of the basic techniques used to minimize the problems. However, it is important to understand that in all construction on permafrost, the methods chosen will only work satisfactorily for a given length of time, which is based on the current environmental conditions prior to construction, including the climate and hydrology (Table 13.1). Any changes in the environment including climatic change will either shorten or rarely lengthen the design life of the structure.

Table 13.1 Table showing the number of buildings showing structural damage in various Russian towns and cities located in areas of permafrost that will eventually result in the buildings being uninhabitable (from Velli, 1973, 1977; Brouchkov, 2003; Parmuzin, 2008).

Town or City	Deformation rate (%)
Chita	60
Pevek	50
Amderma	40
Dudinka	35
Dickson	35
Yakutsk	27
Tiksi	22
Anadyr	20
Salekhard	>10
Labytnangi	>10
Vorkuta	>8

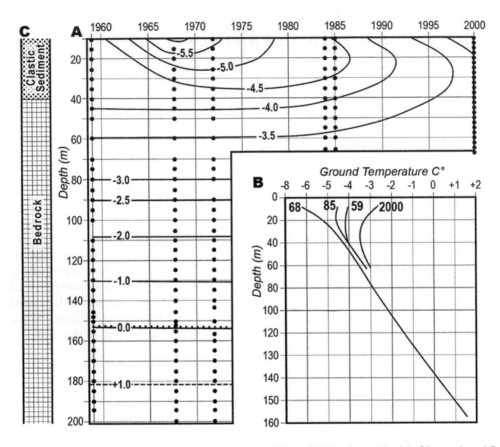

Figure 13.1 Changes in ground temperatures from 1960 to 2000 A.D. at Norilsk, Siberia: A and B, with depth, while C shows the lithology with depth. On materials of Norilsk Permafrost Service.

Vorkuta city in European Russia is a good example of the effects of climate warming. The area of permafrost has decreased by about 25% over the last 60 years and this has affected some buildings (Belotsercoskaya, 1986; Belotsercoskaya *et al.*, 1989). Outside the city, the thickness of the active layer has increased from 7–8 m to *c.* 10 m, while under buildings, it increased to 15–20 m. This has caused some buildings to fail, although the actual temperature of the permafrost has only decreased in 15% of the urban area during this time. However, the actual rates of failure are only 1.8 times that for cities built on ground lacking permafrost in that region, based on an analysis of 979 buildings over a 30 year period (1945–1977).

The rates are far higher in Siberia (Table 13.1). Thus the rate of failure has increased in Norilsk since 1980 due to an increase in mean annual air temperature in the city centre of about 0.6°C (Figure 13.1). Permafrost temperatures are about −2.4°C. The trend is blamed on surface disturbances associated with expansion of human activity and the city, the urban heat island effect, snow clearance with piling of snow in certain places, and the increase in mean annual air temperatures across the region.

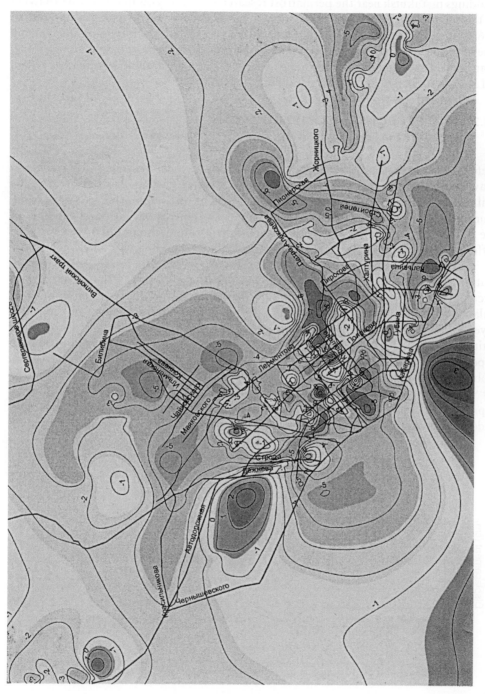

Figure 13.2 Variation in permafrost temperatures in and around Yakutsk in 2015 on the Lena. Note the effect of construction using strong cooling of the ground to prevent thawing of the underlying permafrost. The unfrozen areas (taliks) are where the river or lakes are found. © P. Semenov, Courtesy of Sakha (Yakutia) Republic Construction Ministry.

In contrast, Figure 13.2 shows the effects of using cooling of the foundations of new buildings in Yakutsk near the permafrost research station to counteract future thawing of the underlying permafrost. This cooling of the underlying permafrost amounts to 4–6°C and is accompanied by a thinner active layer beneath the buildings and heaving due to the accumulation of additional ice in the ground. It is considerably more cooling than needed to counteract the warming of the mean annual air temperatures (see Figure 2.13) during the design life of the buildings.

13.2 THE EFFECT OF CONSTRUCTION ON PERMAFROST STABILITY

Deformation or damage to structures is the result of seasonal freezing of soil water and thawing of the ice in the underlying soil or surface water. During seasonal freezing, the soil becomes bonded to the foundation, which will be lifted up during ice segregation. During thawing, it will sink. If the differential movements vary around the base of the building under the structure, the structure will suffer tilting and/or stress. If the differential movement exceeds the design strength, the building will fail.

In its natural state, the ground is covered to a greater or lesser extent by vegetation which modifies the heat flow in and out of the ground. When construction occurs, both the vegetation and surface of the soil are disturbed, resulting in the heat balance changing, producing a warming of the ground and deepening of the active layer (G. H. Johnston, 1981). When an unheated structure is built on the ground at latitude c. 64°N, the walls intercept differing amounts of insolation depending on their aspect, resulting in uneven thawing of the underlying permafrost (Figure 13.3 from Lobacz and Quinn, 1966). The foundation consisted of two concrete slabs separated by 90 cm high piers with a 1.71 m gravel pad beneath the lower slab. The movements represent both seasonal frost heaving and thawing of the active layer (<5 cm amplitude) and progressive long-term thawing of the permafrost (up to 10 cm in 10 years).

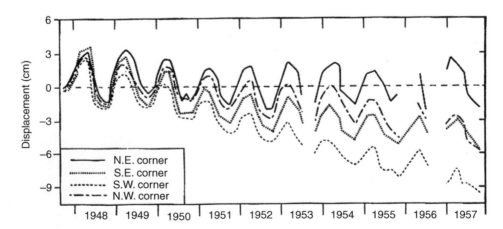

Figure 13.3 A, differential settlement and heave at the four corners of an unheated building with time (Lobacz and Quinn, 1966).

Note that the long-term differential settlement eventually dwarfs the seasonal changes. The amount of differential settlement will vary with latitude, type of foundation, ice content of the soil and the resulting susceptibility of the soil to annual heaving and settlement. Other conditions being equal, the amount of differential movement will be less at lower latitudes, *e.g.*, Tibet, but it is still important (He, 1983).

If the structure is heated, the heat from the building develops a ***thaw bulb*** beneath it (Figure 13.4). This will result in additional differential soil settlement if the permafrost has a low ice content. Where the ice content is high, thawing of the ice may result in actual flowage of the unfrozen sediment, increasing the amount of differential settlement (Figure 13.5). In any case, the strength of the soil usually decreases

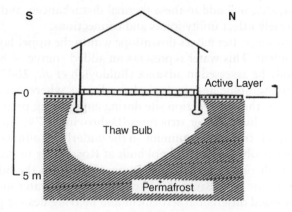

Figure 13.4 Cross-section of the foundation of a heated building after three years (after RIGCDR, 1975, Figure 5).

Figure 13.5 Result of the development of a thaw bulb beneath a wooden house in Amderma, Kara sea coast. There may also have been flowage of the underlying soil resulting in the bowed deformation. © A. Brouchkov.

after thawing. The depth of the thaw bulb will eventually reach an equilibrium condition, whereas the changes shown in Figure 13.1 will continue to increase the differential settlement. This is the cause of the eventual failure of most buildings (see Figure 5.1).

Jumikis (1978) and Zhao and Wang (1983) provide examples of graphs and formulae for calculating the probable thaw depth and temperature distribution for a given set of environmental conditions. Today, this is carried out by computer modelling (Khrustalov and Pustovoit, 1988). These assume two-dimensional heat flow with negligible advection of heat. Ponding of water alongside a structure will cause additional problems since the water absorbs approximately five times as much insolation as the adjacent soil. Thus regular, careful grading of the surrounding surface is essential. Utilities, pathways, *etc.*, will add to these thermal disturbances, and settlement of the building may adversely affect utility joints and connections.

The fact that groundwater moves downslope within the upper layers of the ground adds a further problem. This water represents an added source of heat, the effects of which cannot readily be assessed in advance (Buldovich *et al.*, 2002). It is also essential to watch for topographic and geological conditions where underground streams of water may occur at the construction site during engineering preparation, construction and occupancy of buildings or structures (Dubrovin, 1974; Grebenets, 1999). In Yakutsk, underground streams are common in the underlying alluvial sands and sometimes cause stability problems. The school built at Ross River in the Yukon Territory in the 1990s was built on a gravelly fan sloping down from the hills situated to the south, northwards to the Pelly River. It is likely that groundwater moving through the gravels may have caused differential settlement and heaving despite precautions based on the calculated change in two-dimensional heat flow at the surface of the ground. Addition of numerous thermosiphons failed to stabilize the foundations, so the building had to be replaced by one built on deep piles. The latter appears to have overcome the problem.

13.3 CHOICE OF METHOD OF CONSTRUCTION

There are two main kinds of methods used to construct structures on thaw-sensitive substrates. The *active method* (*principle 2* in Russia) involves deliberately causing the permafrost to thaw prior to construction. This is used in areas of sporadic or thin permafrost and involves removal of the vegetation and exposure of the underlying dark, organic topsoil. The woody vegetation is burned, so that insolation warms up the soil due to the heat balance changing because of the new low albedo. The land is then left for a sufficiently long time for the permafrost to thaw entirely. This pre-thawing is then followed by pre-consolidation of the existing substrate, and frost susceptible material such as peat can be removed and replaced with more stable material such as gravel. Then construction can proceed as in warmer climates. The only problems that may arise are if the local climate experiences either climatic cooling or changes in hydrology.

The *passive method* (*principle 1*) involves using techniques designed to minimise the thermal changes during the design life of the structure, while keeping the changes within acceptable limits. These methods are usually used in areas of discontinuous or continuous permafrost. They work best in areas of stable permafrost or on most coarse substrates where the permafrost temperatures are constantly below $-3°C$. However, the inevitable thermal changes will result in progressive settlement and loss of strength of the soil, thus limiting the potential design life. Changes in climate or other environmental factors will modify the length of time until failure. Thus the subsequent building of a pipeline and high-speed railway along the Qinghai-Tibet transportation corridor has changed the surface hydrology of the area, adversely affecting the original highway road bed and pipeline, decreasing their lives.

Structures are usually designed to be entirely above ground. Construction of trenches or basements entails excavating into the permafrost, resulting in inevitable changes in ground temperature alongside the excavation. This results in drainage and settlement problems. The only exception is on dry, well-drained, dense, thaw-stable substrates.

For engineering design, conservative values for ground temperatures are used based on mean ground temperatures measured at depth. These measurements should be based on a minimum of one or two years of data, and a safety margin should be allowed for. Even so, the climate can vary considerably from year to year, especially in areas of cold air drainage. Unfortunately, there are few places where the ground temperatures have been monitored long enough to obtain a good idea of the variations and long-term trends. Exceptions in Canada include Inuvik (since 1955), Thompson, Manitoba (G.H. Johnston, personal communication, 1984), the Mackenzie Delta (J. R. Mackay), and the Front Ranges of the Rocky Mountains (Harris, 2009).

The Global Terrestrial Network for Permafrost (GNT-P) was developed in the early 1990s to obtain data from as many stations in permafrost as possible throughout the Northern Hemisphere. The first component involved studying "The Thermal State of Permafrost" (TSP) using temperatures measured in deep boreholes (Romanovsky et al., 2002). These give a general idea of the thickness of permafrost and any temperature trends in the permafrost, though it must be remembered that both the thermistor strings and observers have a limited life, and funding for monitoring has become difficult in Russia and North America. China currently has the most active on-going borehole temperature measurement program.

13.4 BUILDING MATERIALS

In the Boreal Forest or Taiga, wood is readily available. It has reasonable insulating properties and is relatively cheap. Wood has the disadvantage of needing a lot of maintenance for the parts that are above ground, while the wood in the active layer slowly decays. Steel is stronger but generally expensive to import. It rusts over time unless suitably protected, but is the strongest building material. However, it also has high thermal conductivity, as does concrete. The latter has been used in the majority of modern building foundations, etc., since it can be mixed at the site. However, it gives out appreciable heat of hydration if cast in place (Zhang and Lai, 2003). It is stronger if reinforcing bars are used, but these must be brought in from elsewhere. Unfortunately, concrete has micropores that water can penetrate, where it freezes and opens up cracks with time (Figure 13.6). Problems identified include hydrostatic pressure, osmotic pressure, critical water saturation, micro ice crystal lens growth

Figure 13.6 Deterioration of a concrete retaining wall in the dry permafrost climate of the Qinghai-Tibet Plateau, 7 years after being built. © S. A. Harris.

and the effects of salt crystals (Yan *et al.*, 2014). Thus it has a limited lifespan even when carefully maintained. The most effective measures to improve durability are controlling the water-cement ratio, mixing in an air-entraining agent into the concrete, incorporating resistant mineral admixtures and steel, polypropylene, carbon (fly ash), ethylene glass and plant fibres into the concrete.

Soils can be reinforced by adding solid composites consisting of two or more different structures in which their identities are preserved. Kazemian *et al.* (2010) list their advantages as being higher specific strength, stiffness, and lower thermal conductivity. The composites may or may not be permeable to water. Orakoglu and Liu (2014) discuss the effects of these reinforcing agents on the thermal conductivity of soils.

13.5 TIMING OF CONSTRUCTION

This is usually done in winter when the ground is frozen so as to cause minimal surface disturbance. By planning ahead, the amount of time that the permafrost is exposed can be minimized. Construction in summer tends to cause severe thermokarst development (Figures 11.6 and 13.7), that renders the site unusable. On peaty or soft ground, a mat of logs forming a corduroy road may be used, and in the absence of suitable trees, artificial wooden mats about 2 × 4 m in size may be laid end to end to protect the ground surface. These mats are removed after construction of the foundation is

Figure 13.7 Result of the development of thermokarst on a parking in Yakutsk.

complete, and the footings are left for about two years to ensure freeze-back prior to application of a load. In northern China, water is sprayed on to the ground in winter to produce an ice road and work-pad. When this melts in the spring, the vegetation recovers satisfactorily. In Russia, it is now mandatory to install thermistor strings with the foundations so that the progress of freeze-back can be monitored. These strings are also used to monitor deepening of the active layer around the footings to provide early warning of any excessive thawing that could compromise the structure.

13.6 TYPES OF FOUNDATIONS

The following is a description of the main types of foundations that are used singly or in combinations. The basic form is constantly being improved for specific purposes, and new methods are being added from time to time.

13.6.1 Pads

Pads are the basic foundation used for roads, airfields and railways, and are exposed to cyclical seasonal temperature fluctuations. When used alone, they should be thick enough to exceed the calculated thickness of the new active layer after the structure is completed and the new thermal equilibrium has been attained. Other techniques are

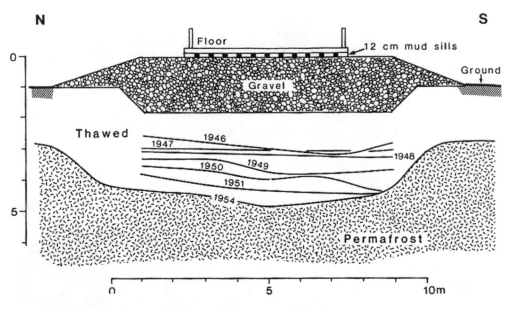

Figure 13.8 Progressive degradation of permafrost beneath a 1.83 m gravel pad near Fairbanks Alaska (after Lobacz and Quinn, 1966). The base of the active layer at the end of each annual thaw period is indicated.

usually used in combination with them to reduce the rather large thickness of gravel required to provide stability.

The problem of attaining a thick-enough gravel pad to insulate the underlying ground from thawing is greater under buildings (Figure 13.8). In that case, the original active layer was 0.9 m thick prior to construction in 1946, but increased to about 1.75 m after clearing the trees and shrubs. The building was unheated but the active layer was approaching 5 m after eight years, in spite of the gravel pad. A similar study at the Zhaohui Field Station in northeast China using a thinner gravel pad with a 75 cm air pocket beneath the floor of a building yielded even greater thawing (Figure 13.9). The original active layer was 1.6 m deep. The thaw bulb is steadily descending, but is more symmetrical than that at higher latitudes due to the higher angle of incidence of the sun's rays (see Figures 13.3 and 13.4).

13.6.2 Slabs and rafts

Slabs and rafts are rigid floors, usually made of reinforced concrete. They are commonly underlain by a gravel pad of suitable thickness and provide a smooth, uniform base on which to build. The weight of the building and its contents is spread over the whole floor area so the structure is capable of supporting greater loads, provided that the slab is strong enough. Differential movements result in tilting of the slab, but unless the concrete fails, the walls will not be subjected to extra stresses, even though

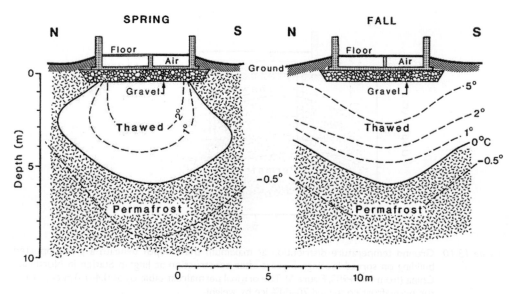

Figure 13.9 Isotherms below a 75 cm air space and sill foundation overlaid a 90 cm gravel pad beneath a building at the Zhaohui Field Station in Northeast China (from He, 1983, Figures 3 & 4).

the building eventually becomes difficult to use. Pads are often used in pairs, separated by piers as in Figure 13.9, which shows the ground temperatures in spring and fall below a heated building overlying two slabs, separated by a 75 cm air space with piers, that are placed on top of a 90 cm gravel pad (after He, 1983). The figure ignores the tilting of the corners demonstrated by Lobacz and Quinn (Figure 13.3), though this will have been less at the more southerly Chinese location. Without the air space and piers, the thawing would have been at least three times as great due to the relatively high thermal conductivity of concrete.

Slabs are commonly used in power houses, warehouses, garages, hangers and unheated buildings. Additional protection such as air ducts and insulation is desirable to reduce the thawing of the underlying permafrost.

13.6.3 Sills

These are the simplest form of foundation, consisting of beams laid on top of or just below the surface of a compacted pad or ground surface. They can tolerate some seasonal movement as well as settlement with time. Examples are the log houses seen in Siberia and Alaska. When the windows start to sink below the ground (Figure 3.3), the building is raised up on a gravel pad (Harris, 1986, plates V and VI).

Wood is the most usual building material in the Boreal Forest, and most houses today normally have a ventilated air space between the pad and the floor, as well as insulation if the building is to be heated. Regular caulking of the spaces between the logs is necessary in log houses. In timber-frame construction, the walls are bolted to the floor to allow for packing when settlement occurs. However, strong winds can

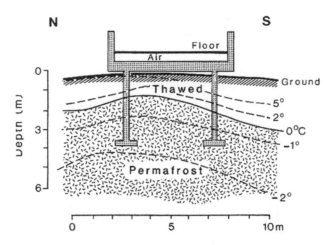

Figure 13.10 Ground temperature distribution at maximum thaw depth beneath a pier-supported building on spread footings, 5 years after construction at Jingtao Station in Northeast China (from He, 1983, Figure 5). The original permafrost table lay at 0.4–1.0 m depth and the permafrost contained 70–80% ice by weight.

blow the buildings off their foundations, so such buildings must also be anchored in the ground.

Despite their limitations, these buildings represent low cost storage structures using local materials in rural areas. There, they are usually unheated, but are generally being replaced by other types of foundations.

13.6.4 Spread footings

These consist of reinforced concrete, wood or steel footings shaped like an inverted "T". They are used mainly in areas of continuous permafrost where there is a thin active layer that allows rapid refreezing of the ground after disturbance. They are placed in the ground at a depth The footings may be one continuous wall, or a series of piers attached to a horizontal base. In Figure 13.8, the temperatures in the active layer had stabilized in 5 years but there was about 6 mm of differential seasonal frost heave affecting the foundation that will eventually make the building difficult to use due to tilting.

To reduce the depth of thaw, the footings may be placed in a suitable gravel pad laid on the original gravel surface, often with a layer of insulation placed above the footings so as to prevent the thawing of the permafrost in the fill above the flat base of the footings. In Figure 13.10, the building has an air space between the floor of the building and the ground. The raised floor should be at least 0.7 m above the ground with adequate insulation placed beneath the floor boards to minimize downward heat flow into the foundation. The long-term heaving of the ground and consequent floor settlement must also be allowed for in the design. Alternatively, a compact, frost-stable granular fill is placed in the excavation both under and over the spread footing, and/or a greased collar is put on the vertical part of the footing to minimize the frost jacking

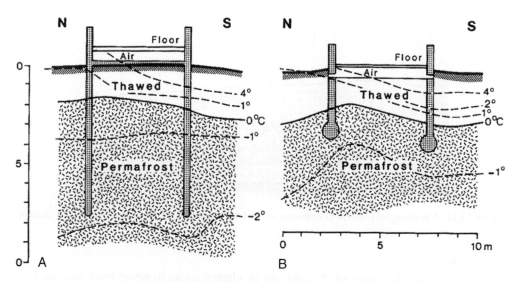

Figure 13.11 Maximum ground temperatures beneath buildings constructed on piles in permafrost in Northeastern China at Zhaohui Field Station (after He, 1983, Figures 7 & 1). Figure 13.11A shows the thermal regime beneath drilled pile foundations, while 13.11B shows the pattern around explosively enlarged pile foundations.

forces. Peat or other insulation may be placed on the ground surface to protect the footings.

Successful construction of buildings of various sizes using spread footings has been reported by Dickens and Gray (1960), Harding (1962), and He (1983) at inland locations. Crory (1978) reported problems when it was used for the foundations of the Hospital at Kotzbue due to complex conditions near the coast. These included saline soils, ground water problems and very variable grain size and stratigraphy. Piles are usually used instead in that situation.

13.6.5 Piles

These are the most widely used foundations for buildings and some linear structures (rail lines and pipelines) in permafrost areas (Vyalov, 1984; Ding, 1984; Ladanyi, 1984) since about 1980. They consist of relatively thin vertical supports with a substantial air space between the base of the structure and the ground surface. These vertical supports are embedded in the permafrost and can support very heavy loads if thoroughly frozen in to a strong frozen bed prior to loading. They must be spaced in a suitable pattern in order to adequately carry the load. For buildings 24 m wide in Russia, an air space of at least 1 m must be used beneath the floor of the building (>1.5 m for wider buildings). The floor of the buildings must be properly insulated to protect the piles, and the gravel pad should be 1–2 m wider than the building. He (1983) provides two examples of the thermal regime beneath pile-supported buildings in Northeast China (Figure 13.11).

The piles may be made of wood, steel H beams, steel pipe, or precast reinforced concrete as in Figure 13.11A and are placed in holes drilled into the round to the

Figure 13.12 A building on piles on continuous permafrost at Harasavei,Yamal Peninsula. © A. Osokin.

required depth. The shape of the pile can be altered so as to resist frost jacking better (Figure 13.11B), though this requires a greater diameter drill hole or explosive enlargement of the lower part of the hole. Drilling is usually carried out in winter so as to reduce the thermal disturbance to the ground, and the piles are left without a load to reach thermal equilibrium with the surrounding ground. Figure 13.12 shows a large apartment building in Russia built using reinforced concrete piles. In some recent construction projects, self-jacking piles are used (Figure 13.13).

Steel H beams can be driven into the ground if the latter consists of warm, plastically frozen, fine-grained soils such as are found along the Arctic coast. They have great vertical and lateral strength, but need a pile driver. Drilling a small pilot hole and filling it with water warmed to a temperature above 66°C for 30 minutes in sands and 60 minutes in clays prior to pile driving greatly speeds up production rate (Manikian, 1983). Pile driving minimizes thermal disturbances so that construction can start earlier.

Steel pipes fill drilled holes better than H beams. The pipes can be filled with concrete or sand to carry heavier loads, or they can readily be modified to allow thermal piles to be installed in them at a later date. This saves extra drilling for thermosiphons if the latter are subsequently needed for remedial cooling. Hydraulic jacks can be used to adjust the height of the top of the piles to match differential heaving and subsidence (Figure 13.11). The main problem with steel is its high thermal conductivity, which results in warmer and deeper active layers in summer.

In Russia, it is usual to use concrete piles except where steel can readily be brought in (Tsytovich, 1973). The piles are fabricated on site if good, clean gravel is available. They must have adequate reinforcing in the form of reinforcing bars to provide good strength to handle compression. Concrete piles are weak under tension, and are expensive and subject to long-term weathering as discussed above. If they are cast in place in a hole, the heat of hydration warms up the surrounding ground and greatly lengthens the time required for suitable freeze-back prior to construction. Cold temperatures in the ground also prevent proper curing as well as the development of adequate strength of the pile. The outside of the pile becomes scaly and will not freeze properly to the soil.

Figure 13.13   Piles with hydraulic jacking capability at Prudhoe Bay, drilled through a gravel pad. The jacks can be adjusted to match the differential vertical movements as necessary. © S. A. Harris.

Any salts present in the soil will adversely affect the strength of the concrete. There must be adequate drainage for the intended design life of the building.

Wooden piles are commonly used in forested areas, *e.g.*, along the Mackenzie Valley. The main zone of rot is in the zone above the permafrost table. Creosote is normally used as a preservative, in spite of being carcinogenic. This, and other preservatives reduce the adfreeze *strength* of the section of the pile to which they are applied by up to 29%, thus combatting frost jacking (Aamot, 1966; Parameswaran, 1978). Sometimes a greased collar is installed around the upper part of the pile in the zone of the active layer as a protection against excessive heave during freeze-back, or they can be enclosed in an outer casing with grease between the pile and the outer tube.

Piles are usually installed from a gravel pad or timber corduroy mat to prevent damage to the soil. Alternatively, a snow or ice pad is used. The site must be adequately drained to prevent ponding of water around the piles (Figures 13.10 and 13.11). The holes are made by steam-thawing, augering, boring, or driving into undisturbed frozen ground. Steam jetting used to be used in Canada (Pihlainen, 1959; Johnston, 1966) and is often used in Russia. Unfortunately it puts too much heat into the ground. The hole becomes too big in ice-rich or stony soils. If water is used as the lubricant in drilling, it also adds extra heat that must be dissipated before the piles can be loaded. The wide holes may be over 8 m deep and need to be backfilled with a slurry made from local soil, once the water has been removed. In contrast, pile driving to 15 m depth scarcely

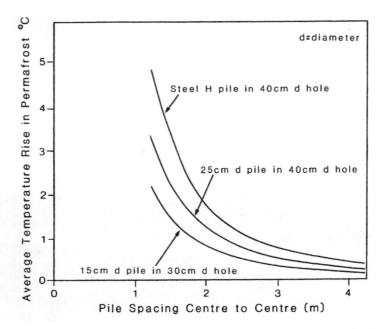

Figure 13.14 An example of the effects of slurry and pile spacing on temperature rise in the adjacent soil. Each situation will be different (from G. H. Johnson, 1981, p. 310). Reproduced with the permission of the National Research Council of Canada.

modifies the ground thermal regime. In China, boreholes may be enlarged by using explosives so that shaped piles can be emplaced (Wang *et al.*, 2014).

The holes should be 10–20 cm wider than the pile to allow for correcting the verticality of the pile during installation. A vibrator is used to ensure compaction of the slurry around the pile. The slurry consists of sands to silts in texture, and Figure 13.14 shows an example of some freeze-back times. The latter depend on ground temperature, spacing of the piles and the diameter of the hole. Ground temperatures need to be below −3°C to ensure reasonably rapid freeze-back, and the length of exposure of the permafrost in the excavation should be as short as possible.

The *adfreeze strength* refers to the strength of the freezing of the pile to the adjacent soil. In the active layer, it results in potential frost jacking as ice lenses grow during freeze-back. In permafrost, this strength provides a resistance to frost jacking. The forces acting on the pile are shown diagrammatically in Figure 13.15. The expected thickness of the active layer is the critical factor, after allowing for any compaction beneath the work pad. Its estimated thickness should be increased by 0.3–0.6 m, and the pile should extend twice as far into the permafrost as in the active layer at the end of the design lifetime plus this safety margin. Roughening or notching of the piles makes relatively little difference in the long-term (Muschell, 1970). However, Andersland and Alwahhab (1983) have demonstrated that lugs on piles placed in a sandy matrix do improve adfreeze strength. If warming of the climate is anticipated, this requires

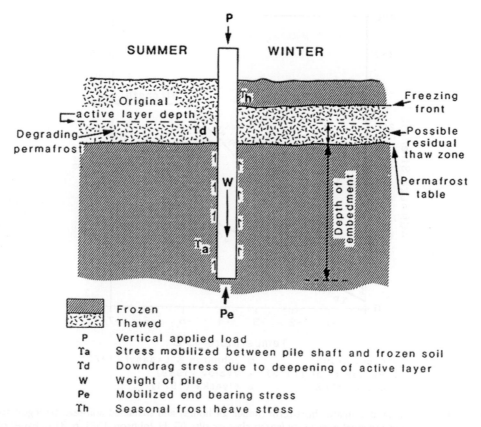

P   Vertical applied load
Ta Stress mobilized between pile shaft and frozen soil
Td Downdrag stress due to deepening of active layer
W   Weight of pile
Pe Mobilized end bearing stress
Th Seasonal frost heave stress

Figure 13.15   Diagram of the forces acting on a pile in permafrost (after G. H. Johnson, 1981, p. 313). Reproduced with permission of the National Research Council of Canada.

an appropriately greater depth of penetration of the pile into the permafrost (Vyalov *et al.*, 1993).

Figure 13.16 shows the adfreeze strengths expected for short-term and sustained loading of wooden or steel piles. Short-term refers to wind or seismic loads lasting less than 24 hours.

The spacing of the piles needs to be adjusted to make sure that the anticipated load on each pile is not too great for long-term stability. Failure to make the right calculation can result in catastrophic differential movements. If the load that is expected is very great, *e.g.*, for a dance hall, allowing extra air space between the insulated floor of the building and the ground will permit more cooling in winter and greater adfreeze strength. Removal of winter snow from under the building will also help. The original surface soil should be left intact unless a gravel pad is used. Figure 13.17 shows the consequences of differential movement of piles under an apartment block in Yakutsk. Dokuchayev and Gerasimov (1988) discuss the anchoring capacity of piles subjected to tangential forces produced by uneven frost heaving of soils.

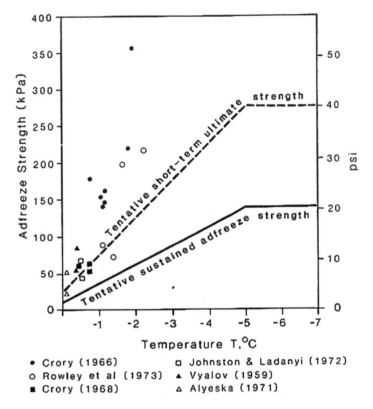

Figure 13.16 Estimated ultimate short-term (under 24 hours) and sustained adfreeze strengths for wood and steel piles in icy frozen clays or silts (G. H. Johnson, 1981, p. 314), based on actual tests by various authors. Reproduced with permission of the National Research Council of Canada.

13.6.6 Thermosiphons

Thermosiphons are natural convection systems utilizing devices driven by the difference in temperature between the ground and the air. They have no moving parts, require no external power, and only operate when the air is colder that the ground. For individual units such as those used on the Trans-Alaska Pipeline (Figure 13.18), the cost used to be *c.* $250 per unit. However, they must be monitored regularly using infrared sensors to determine whether they are still working. First developed independently during the 1950's in Alaska and Russia (Long, 1966; Gapeyev, 1969), they are sometimes divided into two basic kinds, *viz.*, *thermoprobes* and *thermosiphons* (Holubec, 2008). Thermoprobes are constructed above ground as in the case of the Trans-Alaska Pipeline (Figure 13.20). Millar (1971) and Luscher *et al.* (1975) showed that thermal piles could counteract the effects of increased snow accumulation around structures. They are also used to counteract the heat produced at pumping stations, and to keep the ground frozen around piles. Thermosiphons are used to support structures on piles

Figure 13.17 Failure of an apartment building in Amderma, Kara sea coast due to an increase of temperature and loss of bearing capacity of its piles in saline permafrost. ©A. Brouchkov.

within frozen ground. They are used to counteract the heat produced at pumping stations as well as the thawing of the margins of palsas along railway lines in Northern Manitoba (Hayley *et al.*, 1983). Figure 13.20 shows their use in pairs along the raised portion of the TransAlaska pipeline.

Goering and Saboundjian (2004) found that thermal piles with a bend in the below-ground section were more useful in cooling roadbeds. The bend enables better cooling under the road (Kondratiev, 2010). These *sloping thermosiphons* were also placed along the shoulders of the Golmud-Lhasa road on the Tibetan Plateau in 2013 (Figure 13.21), to try to stabilize the road. Unfortunately, this also entails putting in a crash barrier on each side of the road to protect the thermal piles from damage by passing vehicles (Figure 13.21). This has the effect of making the passing of trucks an adventure.

The ground adjacent to the pipe may be cooled by 4°C in winter, but this is decreased to about 1°C due to heat conduction along the metal pipe during the summer months. They are often emplaced inside steel piles to speed up the cooling process after drilling. Afterwards, they may be removed and reused. They can have a bend in them so that they cool a diagonal zone beneath a roadway or building, although installation and removal is more difficult.

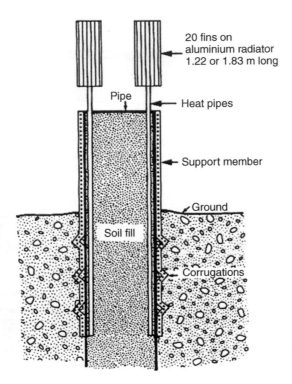

Figure 13.18 Diagram of a typical vertical support member (VSM) use along the Trans-Alaska Pipeline (not to scale). From Harris (1986a).

Thermosiphons can be either one or two phase systems (Figure 13.22). The single phase systems (Figure 13.22B) are filled with either a liquid or air, and work by convection within the fluid moving heat from the ground upwards into the colder radiator section of the thermosiphon in winter. Unfortunately, they have not proven to be as efficient as the two phase systems (Babb *et al.*, 1971; Jahns *et al.*, 1973; Reid *et al.*, 1975; E. R. Johnson, 1981). However, they have provided adequate heat removal from beneath buildings built over ice-rich continuous permafrost at Point Barrow, Alaska (Cronin, 1983). Two phase thermal piles (Figure 13.18 and 13.22a) consist of a sealed tube containing the working fluid in both liquid and gaseous form Long (1966). Propane, carbon dioxide and ammonia have commonly been used. At low air temperatures, the gas condenses on the inner surface of the radiator, giving out heat of vaporisation. The liquid then trickles down to the main body of the tube to accumulate in the reservoir at the bottom. The change in volume causes lower pressure in the tube so that the liquid boils and the vapour then rises up the tube to the radiators to repeat the process. The heat transfer begins when the radiator cools to as little as 0.006°C below the temperature of the evaporator.

Jahns *et al.* (1973) showed that the performance depends on a combination of depth of the pipe below the ground surface, the texture of the soil and on the length of time the heat pipe is operating. Cooling improves with length of time of operation,

Figure 13.19 Thermosiphons cooling a part of the foundation that is failing due to thawing of the underlying permafrost in Vorkuta, European part of Russia. © A. Brouchkov.

Figure 13.20 Thermosiphons in pairs cooling the piles supporting the TransAlaska Pipeline north of the Yukon River. © S. A. Harris.

Figure 13.21 Sloping thermosiphons (thermoprobes) used to stabilize the Golmud-Lhasa road in an area of warm permafrost. © S. A. Harris.

and is most effective in clays. The latest innovation (seasonally active cooling devices – SCD's) comes from the Yukon (Holubec, 2008), and is now used in Russia to cool the foundations of oil and gas facilities (Rilo *et al.*, 2013) as well as in Nunavut. They use liquid carbon dioxide in a two phase system with a network of coils of heat exchange pipes placed in or under the pad of buildings, with either horizontal, sloping or vertical evaporator thermosiphons (Figures 13.22 and 13.23, and Table 13.2). The alternative gases used are ammonia or propane.

The first Canadian use of a *sloped thermosiphon* was at the old Ross River School in the Yukon Territory. The limitation is that the slope under a building must be only about 5%. It makes slab construction more difficult and produces uneven cooling below the building. By using a larger diameter pipe, a leak can be repaired by placing a second narrower pipe within it. For cooling a roadbed (Figure 13.21), an angle of about 30° is used in sloping thermosiphons. Figure 13.24 shows their use beneath a building.

Flat loop thermosiphons were developed at Winnipeg during the winter of 1993–1994. It cools 1.4 times the volume of soil cooled by the sloped thermosiphon and so has come into frequent use in permafrost environments (Figure 13.25). The loop is installed over the ground on a granular base, often including insulation. It is similar to the experiment of Cronin (1973) who placed thermosiphon tubes in the concrete base of an experimental building. Holubec provides examples of the application of the new

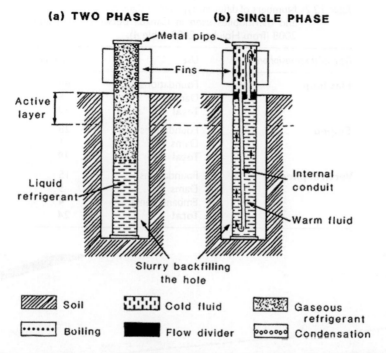

Figure 13.22 a, the two phase system and b, the single phase system of thermosiphons (after Johnson, 1971). Reproduced with permission of the National Research Council of Canada.

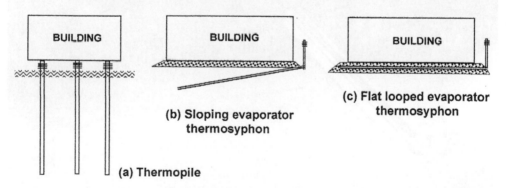

Figure 13.23 Three thermosiphon designs beneath buildings, comparing **thermosiphons** with **thermopiles** (after Holubec, 2008, Figure 10).

designs, although the best methods for their use have yet to be determined in areas with warm permafrost.

Thermopiles require larger diameter holes, so are not used much in Canada for building foundations. Their main use has been around transmission towers (Chapter 17, Figure 17.9) and around railway embankments in Manitoba. They are also commonly used for pipelines around gas plants in Western Russia. They can also be used below the crawl space where the building is raised above the underlying ground.

Table 13.2 Numbers of different types of designs of thermosiphons and thermopiles used in Canada between 1994 and 2008 (from Holubec, 2008, Table 4).

Type of thermosiphon	Use	Number
Flat loop	Foundations	69
	Dams	6
	Total	**77**
Sloped	Foundations	28
	Dams	1
	Total	**29**
Vertical piles	Foundations	15
	Dams	5
	Embankments	4
	Total	**24**

Figure 13.24 Sloping thermosiphons under a sports complex in Yakutsk causing frost heave due to insufficient precooling of the ground before construction. © A. Brouchkov.

Services can then be hung from the ceiling of the crawl space instead of having to be installed in special conduits or corridors installed in the gravel between the loops.

A fundamental fact that is often ignored is that it takes several years to produce the full cooling effect on the underlying soil by either piles or thermosiphons. Impatience

Figure 13.25 A flat loop seasonal cooling system cooling the base of oil storage tanks to stabilize the underlying permafrost in the Western Siberia. ©V. Samsonova.

commonly results in significant problems. It is quite common to see unattended piles sitting on a construction site waiting for the ground to cool adequately to allow safe loading.

13.6.7 Artificial refrigeration

This can be used wherever a source of cheap power is available, if this is economically feasible. Crory (1973) reported that it was first used as coils around newly installed piles to speed up initial freeze-back. They have also been used to provide adequate cooling of the foundations of the pumping stations for the Trans-Alaska pipeline and in Russia (Sadovsky and Dorman, 1981). They are more efficient than passive piles, but they are expensive to operate due to continuous operating costs and periodic replacement of the moving parts. Along pipelines, there is a continuous, plentiful supply of fuel for electricity generation. In North America, they are also used to stabilize the portals of mines, as well as keeping the air circulating through the mines at a suitable temperature to avoid problems of stability of the walls. They were also used to stabilize the foundations of important infrastructures in the Arctic, *e.g.*, radar stations (Fife, 1960). Combinations of thermosiphons with artificial refrigeration are called *hybrids*.

Figure 13.26 Ventilation ducts emplaced in the embankment of the Golmud-Lhasa railroad near Beihu'he, Qinghai Tibet Plateau. Note the gradual ejection of the short lengths of pipe at the ends of the ducts, ultimately resulting in closing off of the duct. © S. A. Harris.

13.6.8 Ventilation ducts

This is one of the cheapest methods of cooling foundations and involves the use of ducts or culverts emplaced within the pad or embankment so as to catch the prevailing cold wind in winter. The ends are opened whenever the ambient air temperature falls below −3°C (Reed, 1966; Jahns *et al.*, 1973). The ducts or culverts should be made of a single long duct, otherwise frost jacking can cause the ejection of the short sections from the embankment after about 7 years (Figure 13.26). The ends of the pipes should be closed in warmer weather to prevent accumulation of heat (Figure 13.27). The Balch effect does the work in winter, *i.e.*, cold, dense air displaces the warmer air in the duct by gravity flow.

Besides linear transportation embankments, this technique is used for aircraft hangers, oil storage tanks and maintenance garages at Inuvik, Northwest Territories. These buildings must be heated and placed at grade to permit ready access. The floors are usually slabs of reinforced concrete overlying a gravel pad so as to be able to withstand heavy loads that are concentrated at certain points (R. H. Williams, 1959; Sanger, 1969). Where necessary, the ducts can be connected to vertical stacks that are equipped with fans that are turned on in suitably cold weather. These are turned off in summer. Figure 13.27 shows the results of either closing the air vents on horizontal air ducts in summer or leaving them open (Sun *et al.*, 2014).

Beaulac (2006), Beaulac and Gore (2006a, 2006b) and Jorgensen (2012) describe the experimental use of tubular heat drains in cooling the shoulder of road embankments (Figure 13.28). They are effective in cooling the shoulders, but not as effective as using geocomposites. The latter are more practical since they can be placed as a sheet beneath the surface material, whereas a very large number of tubular heat drains would be needed per kilometre of road. Also, the risk of blocking the top of the drains with snow, as well as vehicles damaging the tops is considerable. In summer, runoff from precipitation could drain down the tubes, reversing some of the winter cooling. More

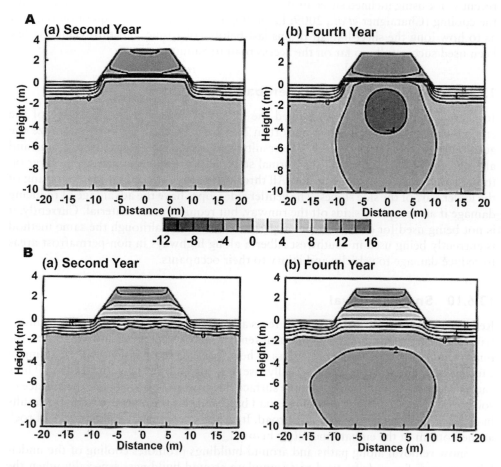

Figure 13.27 Results of A, closing the air doors on vents open in summer, and B, leaving them open throughout the year in a composite embankment (modified from Sun *et al.*, 2014, Figure 6).

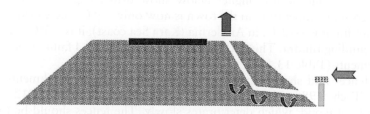

Figure 13.28 Diagram showing how the experimental tubular or sheet-like heat drains cool the shoulder of the road or airfield embankments (modified from Beaulac, 2006; Beaulac and Dore, 2006a, 2006b). Recent experiments using sheet-like air conduits are proving to be more effective (Chatagner *et al.*, 2009; Lamontagne *et al.*, 2015).

recent work using inclined sheet air drains in the sides of an embankment can improve the cooling (Chataigner *et al.*, 2009; Lamontagne *et al.*, 2015). The question remains as to how long the sheet-like air drains last before collapsing, but this technique has been used successfully so far on the access road to Salluit airport.

13.6.9 Angle of slope of the embankment sides

Jørgensen and Doré (2009) examined the effects of reducing the side slope of the embankment to 8:1 at Tasiujac airport in Northern Québec, so as to minimize the accumulation of snow in winter. The results were similar to tubular heat drains and also the use of coarse rock on the normal steep slopes with an impermeable layer on top together with ventilation drains. All three methods produced far better cooling of the lower part of the side slopes. The gentler side slopes have the advantage of reducing damage if an aeroplane skids off the runway, but requires more material. Currently, it is not being used for embankments in permafrost regions, although the same method is currently being used in southwest Alberta along highways in non-permafrost areas to reduce damage to vehicles and injury to their occupants.

13.6.10 Snow removal

Removal of snow from around structures is an important way of cooling the ground in winter. Placing roads on an embankment tends to reduce costs of snow clearance, the removal of snow resulting in better loss of heat from the road base in winter, aiding in causing the permafrost table to actually rise into the base of the embankment. There can be some additional heaving of the surface due to water moving over the surface of soil particles to the coldest part of the road base, but this is only a problem if it results in differential heaving of parts of the road. It can be overcome by using blocks of rock at the bottom of the embankment (see below).

Snow removal along paths and around buildings promotes cooling of the underlying ground. Snow drifts tend to accumulate around buildings, especially when the buildings form clusters close together. This was typical of the Russian Arctic coastal settlements constructed in the period 1930–1950. The snow blows in off the frozen surface of the Arctic Ocean and tends to accumulate around any obstacles. Ground temperatures are significantly higher below snow drifts, *e.g.*, in Tiksi (Laptev Sea coast), the ground temperature in the town is now only $-7°C$ whereas in the area surrounding tundra, it is $-12°C$. In Amderma (Kara Sea coast), it is $-3°C$ versus $-4.5°C$ in the surrounding tundra. This is the reason for the enhanced failures of buildings in these settlements (Table 13.1).

Snow fences and sheds can be used along sections of embankments subject to snowdrifts (Esch, 1988; Beaulac and Doré, 2004; Lautala *et al.*, 2012). These are usually made of wood, which deteriorates slowly. The fences should be located 20–50 m up-wind of the embankment. Snow sheds keep both the summer sun and winter snow off the ground, resulting in an overall reduction in ground temperatures and the thickness of the active layer. They have been used at Bonanza Creek in the Yukon Territory and were found to reduce temperatures on slopes from $3.9°C$ to $-2.3°C$ beneath the sheds. However they have been deemed impractical because of high cost and low durability (see also awnings shading the slopes, below).

13.6.11   The diode effect: use of rocks

The *diode effect* (*thermal semi-conductor effect* – Cheng, 2004) of both peat and rock offers a means of cooling foundations and so reducing the thickness of the underlying pad. In practice, it is often difficult to find an adequate supply of frost-stable material. Peat moss was used under roads and railways in 1903 in Scandinavia to control frost heaving in seasonal permafrost (Skaven-Haug, 1959; Solbraa, 1971). Experiments in Alaska showed that it underwent considerable compaction unless compacted blocks are used. Wood chips and bark can be used, but obtaining an adequate supply is difficult without damaging the landscape. They also decompose over time.

Rock can often be found in great quantities in cuttings, and most rock types withstand weathering reasonably well. The Skovorodino Permafrost Station of the former All-Union Railway Institute showed the effectiveness of blocks of rock in cooling the underlying ground in 1969 and 1970 (Mikhailov, 1971). In 1973, the former Chinese Academy of Sciences LIGG built a 2.7 m experimental embankment of coarse rocks (0.3 m diameter) over an ice-rich permafrost section at the Reshui Coal Mine in Qinghai Province (Cheng and Tong, 1978; Cheng *et al.*, 1981). An obvious cooling was observed. Subsequently, the Department of Mechanical Engineering of the University of Alaska at Fairbanks completed a series of computer simulations on heat convection in porous media. An experimental embankment demonstrated the cooling effect and the Alaska Department of Transportation named it the "air-cooled embankment" (Goering and Kumar, 1996; Goering, 2003). Harris (1996a) reported permafrost under block streams in the Kunlun Pass in China while the surrounding soils were only subjected to seasonal frost. The four main processes involved were finally determined in a year-long experiment at Plateau Mountain (Harris and Pedersen, 1998). The blocks in a block slope produced a decrease in mean annual ground temperature of about 6°C compared with that in colluvium, 10 m away. Lai *et al.* (2006b) showed that when the rocks were covered by other material, the diode effect operated in winter beneath a colder surface layer while inefficient warming occurred in summer. The net effect was cooling of the embankment. Without a cover, winds cause warm air to enter the blocks in summer, thus reducing the annual ground cooling.

Cheng (2005) summarized the previous work and discussed the history of road construction over permafrost and the resulting problems. Thus along the Qinghai-Tibet Highway, 85% of the problems were due to frost damage, and 15% due to frost heave (Wu *et al.*, 2002). He discussed the physics behind the processes known at that time, and carried out experiments with thermosiphons and thermopiles, embankment ventilation with control structures, sodding the slopes of the embankment, and the use of rocks piled along the side slopes. He also discussed the use of dry bridges and shading boards to shield the permafrost from insolation (see below). He then proposed to the Chinese Government that all these proactive measures could be tried to cool the embankment beneath the proposed road and rail beds over warm permafrost. Accordingly, a full scale embankment was constructed at Beihu'he on the Qinghai-Tibet Plateau, trying out all the known methods of cooling the ground. The results confirmed the processes observed by Harris and Pedersen, and were used in designing the Qinghai-Tibet Railway embankment.

Cheng *et al.* (2007) found that all the processes could be explained by theoretical physics. Cheng *et al.* (2008), Gu *et al.* (2008), and Wu *et al.* (2009) report the results

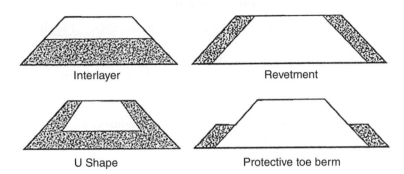

Figure 13.29 The four types of configuration of rock layers tested for use in building the Golmud-Lhasa railway (Cheng *et al.*, 2008, Figure 8).

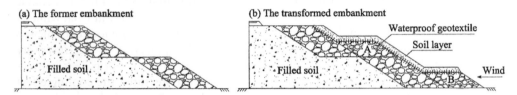

Figure 13.30 The new design of the embankment using waterproof geotextiles and two toe berms on the side-slopes with a surface covering of soil (modified from Liu *et al.*, 2014, Figure 2).

of a series of different configurations of rock layers as part of a full-scale test to find a suitable method of controlling permafrost degradation. The first major use was for the Golmud-Lhasa Railway built over the permafrost along the transportation corridor over the Qinghai-Tibet Plateau. Figure 13.29 shows the configurations of rock used in the experiment. Crushed rocks, ventilation ducts and thermosyphons were used. The interlayer method with an overlay of soil produced a considerable cooling providing the overlay was not more than 5.5 m (Wang *et al.*, 2005). Crushed revetments on the side-slopes improved the cooling, and Lai *et al.* (2006) demonstrated that doubling the thickness of the rock on the south-facing slope produced symmetrical cooling. Blocks of 40–50 cm diameter were the most effective, apparently due to wind-forced air exchange (Cheng *et al.*, 2008). The U-shaped form produced the best results, and the layer of blocks of rock decreased the probability of the slopes of the embankment sliding or slumping Figure 13.30a). Nui *et al.* (2015) examined the results from the first 10 years of operation of the Qinghai-Tibet Railroad and concluded that the U-shaped use of rocks (Figure 13.29) caused the least interference with the ground thermal regime beneath the tracks, whereas rocks covering the embankment produced greater interference and vertical movements. If the height of the embankment exceeds 4 m, the cooling effect of the blocks decreases (M. Zhang and Niu, 2009). Jørgenson and Doré (2009) showed that the addition of rocks on slopes with a waterproof textile under the surface rocks improves the cooling of the side slopes. Subsequently, Liu *et al.* (2014) have demonstrated that by using a waterproof textile immediately beneath a soil layer on the side slope and configuring the blocks as in Figure 13.27b, the warming effect of melting snow and summer rains can be largely eliminated (Figure 13.31). This

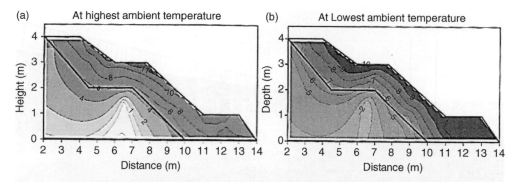

(a) At highest ambient temperature   (b) At Lowest ambient temperature

Figure 13.31 The temperature fields in the embankment with the design shown in Figure 13.23 measured at the times of maximum and minimum ambient temperatures (Liu *et al.*, 2014, Figure 8).

Figure 13.32 Sod being used to cover the impermeable membrane on the side slopes of a freeway under construction north of Madoi, China. © S. A. Harris.

does eliminate lateral air exchange with the blocks, though this is no worse than the snow plugging up the voids between the blocks in winter in areas of appreciable winter snowfall. On the eastern slopes of the Qinghai-Tibetan Plateau, the soil covering used consists of blocks of sod (Figure 13.32) to reduce wind erosion. Further experimentation indicates that the addition of a perforated air duct in the middle of the layer of rock in the interlayer configuration further improves the cooling, provided air doors are used (Sun *et al.*, 2014).

When blocks are used in dam construction, the convection of air within the spaces between blocks creates warming of any rock layers in frozen dams built on permafrost.

Figure 13.33 Shading of the piles along the Golmud-Lhasa railroad using the raised track as a shield from the sun. It also prevents trespassing by herdsmen and animals. The gravel pad raises the top of the piles above the surrounding ground, giving good drainage. © S. A. Harris.

The heat comes from the water, and the convection carries the heat into the surrounding materials in the dam and can cause serious problems.

13.6.12 Shading

At lower latitudes such as on the Qinghai-Tibet Plateau, the sun is high in the sky for much of the day. Under these conditions, a structure such as a wide, raised railway structure on narrower piles shades the base of the piles from the sun for most of the summer, thus decreasing the effect of insolation (Figure 13.33). The use of piles emplaced into the permafrost from a gravel pad has proved quite successful on the Golmud-Lhasa railway in flat terrain with good drainage. Figure 13.34 shows the detail of the support with the space to allow for expansion and contraction at each pile.

At higher latitudes along the Baikal-Amur-Manchurian (BAM) railway, the Russians experimented with placing white sun-precipitation awnings over the side slopes of the embankments (Kondratiev, 1996, 2010, 2013). It produced a significant improvement in the cooling of the side of the Tommot-Kerdern Railroad in Russia, and 1,841 km of the BAM right-of-way through larch forest uses these to protect the embankment. Winter snow helps to hold the covers in place (Figure 13.35).

The experiment was repeated by Feng *et al.* (2012) using covers on the embankment slopes of the Qinghai-Tibet Railway, and the net radiation was significantly reduced between March and October relative to the unshaded slope. Monthly net radiation was less than 20 W/m^2 in the shade compared to 60–130 W/m^2 on the normal slope. The main problem with it was keeping the covers in place in spite of the

Figure 13.34 Detail of the support underneath the rail bed with shaded piles. © S. A. Harris.

Figure 13.35 Experimental shading awnings along the BAM railway. ©V. Kondratiev.

wind and air currents generated by passing trains. While it helped in reducing the inso-lation in summer, the winds blew the covers away, so this method has been abandoned in this area of negligible precipitation.

13.6.13 Insulation

In order to reduce the quantities of frost-stable gravel needed for pads, insulation is often used. Peat moss has been used in Alaska under the centres of road beds (Berg and Aitken, 1973; Clark and Simoni, 1976; McHattie and Esch, 1983), as well as under roads and railways in Norway to control frost heaving in non-permafrost areas since 1903 (Skaven-Haug, 1959; Solbraa, 1971). However, it compacts unless used in the form of compacted blocks. Frozen peat has twice the thermal conductivity of unfrozen peat, and peat is only available in suitable quantities in very humid, cold locations. Wood chips and bark have also been used in areas of discontinuous permafrost.

Accordingly, artificial insulation materials have been developed, although they should have high compressive strength and not absorb moisture. The latter requirement eliminates rigid board or foamed-in-place polyurethane. The two main materials now in use are expanded polystyrene boards (XPS) and cellular concrete. They perform best if covered above and below by a waterproof geotextile sheet to reduce moisture movement and to keep the boards dry. If the insulation is not kept dry, there is a slow loss of performance as moisture accumulates in the boards (Lautala et al., 2012). Shifting of boards due to creep can be a problem, but is largely overcome by using at least two layers with the joins staggered (Gandahl, 1979). Where XPS is used, it must be placed below 50 cm of the surface to avoid hoar frost (Horvath, 1999).

Insulation has been used to suppress the "edge-effects" that are found along embankments (Ashpiz et al., 2010), and Nurmikolu and Silvast (2013) have described the use of XPS boards to upgrade older railway embankments so as to bring them up to standard for high speed trains (see below). However aprons of frost-resistant blocks of rock are the latest successful method (see Figure 13.25). Their main use is under roads, runways and taxiways, as well as under heated structures such as oil storage tanks (Davidson and Lo, 1982). They can be used for stabilizing landslides and steep slopes (Pufahl and Morgenstern, 1979; Ashpiz and Khrustalev, 2013). Nixon (1973) provides a formula for calculating the thickness of insulation needed beneath a building in order to reduce heating of the permafrost to acceptable levels.

13.6.14 Use of geotextiles and waterproof plastics

Waterproof plastics are widely used to reduce or eliminate water movement in a foun-dation (Stark et al., 2004). Han and Jiang (2013) provide a summary of some of the available geosynthetics, their properties and behaviour when used in earth structures in cold regions, while Horvath (1999) provides case studies of their use. If moisture cannot move into the ground and cannot move to the freezing plane in large quanti-ties, the seasonal heaving and settlement can be greatly reduced, thus increasing the potential life of the foundation (Figure 13.26 and Figure 13.27). In actual practice, light-weight plastics such as polyethylene are too easily punctured although there have been some experiments testing it. Instead, nonwoven textiles can be used. They should

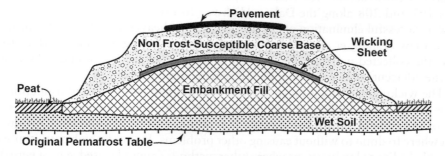

Figure 13.36 Cross-section of the road embankment showing the location of the moisture-wicking material beneath the pavement.

Figure 13.37 Wicking fabric being laid down along the Dalton Highway Alaska to remove moisture from the embankment beneath the pavement in an area with a high water table (from Rosen, 2014).

be placed on a sloping surface so that any water entering the embankment from above can drain laterally over it (Figure 13.33).

Geogrids have become very important in strengthening railway beds (Petryaev and Morozova, 2013). They possess rigidity, high tensile strength, high module of deformation (*i.e.*, low elongation when broken), and increased resistance to temperature fluctuations and chemical and biological attacks. Their use is discussed further in Chapter 14.

The most recent innovation involves *moisture wicking* by placing a synthetic, moisture-wicking, porous sheet as a continuous layer in the embankment below the gravel base beneath the paving (Lai *et al.*, 2012). Figure 13.36 shows the position of the layer in the cross-section of the Highway, while Figure 13.37 illustrates the method of construction. The Alaska Department of Highways has tested it between miles 197 and 208 along the Dalton Highway to Prudhoe Bay, and the first results show a successful elimination of the development of heaves and frost boils during the spring thaw (Zhang *et al.*, 2014; Rosen, 2014). By having a sloping surface to the wick, the collected moisture drains away to the side of the embankment by gravity, and the adjacent surface layers of the fill remain comparatively dry.

The wicking material is more expensive than other synthetics and insulation, but seems to do a good job in reducing the moisture to acceptable levels in the upper part of the embankment fill. However, it is only useful if the collected moisture has somewhere to drain to without causing other problems. Due to its high cost, its use will probably be limited to locations where other methods cannot be used, *e.g.*, where there is a limited supply of non-frost susceptible coarse gravel or rock to prevent moisture moving upwards from soils with a high water table.

Chapter 14

Roads, railways and airfields

14.1 INTRODUCTION

Canada, China and Russia are very large countries so linear transportation systems are vital to their economy and future development. However, building them on permafrost presents serious problems. The main kinds of systems must be based on preventing heaving and subsidence in the right-of-way, and this is particularly important in the case of construction of high speed railways. The total length of those in China exceeds the total length of such lines in all of Europe, and a substantial part of their route on the Qinghai-Tibet Plateau is built over warm permafrost.

Harris (1986, Figure 5.1, 5.2 and 5.3) provides maps of some of the main road and rail networks and the locations of airfields in Canada, Alaska and Russia. Year-round roads are only just being extended into the areas of continuous permafrost, and many start as gravel roads which are slowly upgraded until they are stable enough to pave. The rest of the roads are usually winter roads. Similarly, most railways only cross the zone of continuous permafrost where there is a strategic need, e.g., the Baikal-Amur (B.A.M.) railway, and the Golmud-Lhasa railway. They are also justified where there are important bulky goods that need transporting to smelters and/or markets, e.g., the Labtnangi line servicing mines in the Urals and the port of Salekhard. Finally, the railway at Norilsk was built to carry ore to the coast for shipment to Murmansk, while the Winnipeg-Churchill line is required for exporting grain from the Canadian Prairies. High value products such as diamonds are transported by aircraft, as are the workers. Lautala et al. (2012) provide an invaluable summary and evaluation of past work on railway embankments.

14.2 THE PROBLEMS

These include the effects of modifying the heat and moisture regime along the right-of-way, together with compaction of the soil, especially peat. Whereas areas of continuous permafrost tend to have cold ground temperatures and so are less prone to damage, construction over warm permafrost (usually defined as ground with temperatures warmer than $-2°C$ at the surface of the permafrost table) requires special care. Since most of the population of the world lives in areas without permafrost, this warmer permafrost must be traversed before reaching the more stable, cold permafrost lands. The more ice that is present in the permafrost and the more frost susceptible the underlying ground, the worse are the construction problems. Hilltops and ridges are less vulnerable to problems with underground streams that are carrying appreciable quantities of heat in taliks (Afanasenko and Volkova, 1989; Boitsov, 1998). These underground drainage-ways appear to move laterally over time, creating considerable problems, e.g., near Beaver Creek along the Alaska Highway. Where dealing with bedrock, the weathering characteristics of the rock can be very important, e.g., the siltstone at the Rat Pass on the NWT-Yukon Border along the Dempster Highway breaks down into silt as a result of frost weathering (Legget et al., 1966). Likewise, the rocks used as masses of blocks to modify the temperatures of an embankment must be resistant to frost action and weathering. Saline soils have less strength and are often plastic (Aksenov and Brouchkov, 1993), but by choosing optimal sand-clay mixtures, these problems can be minimized (Aksenov and Petrukhin, 1991).

14.3 TYPES OF ROADS

There are several types of roads found in permafrost regions. The simplest type is the winter road that is a temporary road only used when the ground is adequately protected from damage in winter. Gravel roads are year-round roads that use embankments and other methods to protect the underlying permafrost. They are often subject to frost heaving and need regular grading. Paved roads can be used once the embankment on which a gravel road is built becomes sufficiently stable, but paving alters the albedo, causing the temperatures in the road bed to rise. A national freeway is currently being constructed across the Qinghai-Tibet Plateau.

Where the roads are constructed along a designated transportation corridor, the interactions of the various components must also be considered. Thus the gas pipeline, oil pipeline, railway, standard road and freeway run parallel to one another across the Qinghai-Tibet Plateau. The embankments alter the surface drainage and snow distribution, and this is causing landslides and mudflows in places that have disrupted the old buried pipeline. Subsidence occurs where the snow accumulates in winter and the rainwater from summer precipitation collects, resulting in thermokarst depressions parallel to the roadway (Figure 14.2). This, in turn, causes rotational-type sliding of the margins of the embankment, coupled with longitudinal cracking of the pavement (Figure 14.1). This cracking is closely related to the height of the embankment and its aspect (Figure 14.3).

At least one major problem remains, i.e., coping with the heating effects of underground taliks carrying water beneath the right-of-way. Since water supplies considerable quantities of heat and moisture to the underlying frozen ground and the position of this talik can move over time, these represent a significant problem where roads or railways must traverse lowlands across which underground drainage travels from higher land nearby, e.g., the Alaska Highway near Beaver Creek, Yukon Territory (Borisov, 1984; Buldovich et al., 1991, 2002). Accordingly roads and railways should be sited along drainage divides where possible, e.g., the section of the Dempster Highway in the Yukon crossing the Ogilvie Mountains.

Figure 14.1   Longitudinal cracking of the road embankment of the Bovanenkovo to Harasavei road on the Yamal Peninsula. © Geosystem – A. Osokin.

Figure 14.2   Subsidence alongside the Golmud-Lhasa Highway resulting from thawing of the permafrost due to spring runoff from melting snow and insulation by snow drifts in winter. © S. A. Harris.

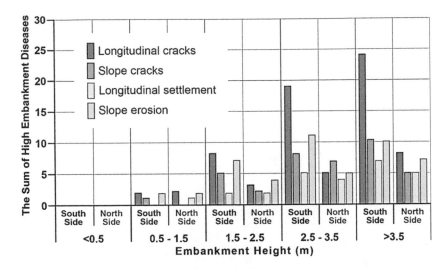

Figure 14.3 Relationship of permafrost problems on the original Golmud-Lhasa road to embankment height during the first 30 years of study (modified from Wang *et al.*, 2009, Figure 2).

14.4 EXPERIMENTAL EMBANKMENTS

The study of permafrost on the Qinghai-Tibet Plateau began with the construction of the old Golmud-Lhasa road. The embankment was heavily instrumented and the results of thirty years' experience were summarized by Wang *et al.* (2009). They also highlighted the main problems and successes. In order to try to design the expressway embankment for the Golmud-Lhasa railway and motorway, the Chinese built full-scale test embankments to try out different combinations of the construction techniques discussed in Chapter 13 (Figure 14.4). The difficulty is that at that latitude, insolation is high throughout the year. And the temperatures at the permafrost table are generally warm. The region is cold and arid, and at a high elevation. Thus the choice of the methods to be used in construction was critical to avoid disasters. The same basic construction methods are used for roads, railways and airfields.

The test embankment was fully instrumented with thermistor strings that provided data on the ground temperature along cross-sections of the different treatments, together with measurement of the moisture profiles (Gu *et al.*, 2010). The experiment has continued to be monitored and is also being used in the design of the new Qinghai-Tibet Expressway. The installation includes a meteorological station. Additional experiments have been carried out to minimize the problem of thermokarst developing along the sides of the embankment and to stabilize the slopes (see Sun *et al.*, 2014; Liu *et al.*, 2014).

By using this full scale test area, the Chinese have greatly reduced problems with embankment failures, though the presence of other structures such as roads and pipelines in the same transportation corridor creates other problems such as interfering with the natural surface drainage causing landslides.

Figure 14.4 The experimental embankment constructed just north of the Beihu'he Research Station adjacent to the Golmud-Lhasa road, China. Different methods of cooling the embankment are being tested with thermistor strings measuring the performance of each kind of treatment (Gu *et al.*, 2010). © S. A. Harris.

14.5 WINTER ROADS

The simplest form of transportation route across permafrost terrain is the winter road. It ranges from a simple trail across snow-covered frozen ground or the surface of a frozen river to a specially surfaced compacted snow or ice-water mixture along a suitable route. In both Canada and Russia, the outlying northern settlements rely on these to bring in bulky, low-value goods and essential supplies. However, if the ground and rivers do not freeze soon enough, such settlements are in trouble.

The main advantage is that winter roads do not require a special bed to be used, but the less sophisticated the road construction, the more specialised the vehicle must be that are using them. Tyre pressure is the main limitation, so the weight of the load must be spread over large tyres or tracks to minimize damage to the underlying ground or to prevent them from going through the icy cover of a body of water.

Ninety percent of winter roads in Canada in 1978 were *winter trails* which follow a cleared right-of-way through the boreal forest/taiga. The ground surface must be completely frozen to at least 20 cm and covered with more than 10 cm of snow prior to their use. This is only achieved in January near Yellowknife, compared to October in the Canadian Arctic Islands. Bulldozers with wedges fitted to the base of the blade (mushroom shoes) are used to remove the upstanding vegetation. Only tracked vehicles or those with balloon tires have a low enough bearing pressure to be used, though the Russian troika does the same job. All trailers and heavier vehicles must be fitted with skis to reduce the bearing pressure.

In Russia, when a suitably thick layer of ice has formed on the large rivers flowing north without rapids or waterfalls, these are used as winter roads. Similarly, when large lakes become frozen, these can provide adequate roadways, provided that the surface is smooth and the ice is thick enough.

Snow roads are winter roads where the snow cover has been increased and modified to allow the road to be able to carry greater loads. They require regular maintenance, so are only used where there is too little snow or heavier loads need to be transported. Compacted snow roads are built by piling snow on the right-of-way and then compacting it by rolling and dragging. Minimum acceptable snow density is 0.4 g/cc although values approaching 0.6 g/cc are preferred. The resulting surface is smooth and compact and will support vehicles with tire pressures up to 550–620 kPa. Drifting of snow is greatly reduced.

Processed snow roads are constructed by blowing snow on to the road and then compacting it. The snow crystals break up and recrystallize in a denser form. Even so, they cannot withstand the use of heavy trucks for any length of time. This can be overcome by spraying the surface with 2.5 cm of water, thus producing a hard icy cap which supports medium weight vehicles. However, these *ice-capped* roads eventually develop ruts. The water is preferably obtained from lakes, not rivers.

Where there is insufficient snowfall, *manufactured snow roads* can be used by making snow using water sprayed into the air in cold weather. However, these are the most expensive to build and are only used to connect snow roads in arid areas.

Ice roads are made entirely of ice. Water can be sprinkled on to the ground surface in a series of sheets to provide a smooth surface covering all the depressions and vegetation It is strong but very expensive. Another method involves using chunks of ice chipped from frozen surfaces, but needs 10 cm of water to bind the aggregates together to form a smooth surface. Ice roads have been the main type of winter road used to support exploratory drilling in the National Petroleum Reserve on the North Slope of Alaska.

Whole river ice roads are mainly found in Russia where the snow is cleared along a 60 m wide right-of-way along a frozen river (Figure 14.5). Icings, wind-driven pressure ridges, springs and ice jams can be a problem. This method is also used on large lakes, *e.g.*, Lake Athabasca. The advantage of ice roads is that they can support heavier loads and allow higher speeds than snow roads. A modification is where the snow is pushed into windrows and bowsed with water. The mixture is then quickly spread by a grader to form a smooth surface like asphalt. These can be used for several months without damaging the underlying ground.

Ice bridges can be built over rivers by spraying layers of water along the desired route to develop a thick enough layer of ice to stand the weight of the traffic. They are emplaced where ferries are used in summer, *e.g.*, Dawson City, Fort MacPherson, Fort Providence in northern Canada. This leaves only about six weeks in the spring and fall when there is no means of crossing the rivers.

Figure 14.5 Trucks crossing the Albany River, Manitoba, on an ice road.

14.6 ENVIRONMENTAL EFFECTS OF WINTER ROADS

Harris (1996a) discusses this in detail while Steever (2001) discusses revegetation after the vegetation on the right-of-way on a winter road has been damaged. The disturbance is of three kinds, *viz.*, change in terrain morphology, change in vegetation and compaction of the peat cover. The tops of the hummocks are chopped off and the intervening depressions tend to widen. If there is a sparse vegetation cover, ruts may develop as in the eastern Canadian Arctic Islands, but the low precipitation and mean annual temperatures inhibit runoff (Radforth, 1972). The roads should run straight up and down slopes and avoid ice-rich or poorly drained terrain. Some increase in erosion and gullying is unavoidable. Guyer and Keating (2005) have shown that one or two years of use of ice roads and ice pads have not produced obvious damage to tundra ecosystems on the Petroleum Natural reserve in Alaska.

The change in snow thickness and density alter the thermal properties of the surface. The snow cover on compacted snow roads will thaw later, thus delaying warming of the ground (Mackay, 1970; Kerfoot, 1972). Minimal change occurs with winter trails and maximum change with ice roads.

Vegetation disturbance is the most pronounced effect. Removal of trees concentrates the thermal heat exchange at the surface of the ground (Haag and Bliss, 1974a). Trees act as heat radiators in winter, while increased snow depth gives better insulation to the ground. Tree removal also reduces evapotranspiration causing higher moisture

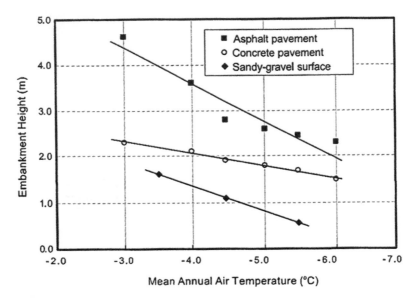

Figure 14.6 Relationship between minimum embankment height and mean annual air temperatures for different roadway surfaces (after Zhang and Wu, 1999).

contents (Haag and Bliss, 1974b). Although the disturbance tends to loosen or uproot plants, the majority of the species are rooted in the sides of the hummocks where snow packs in and protects them (Lambert, 1972b). The plant succession on mudflows is described by Lambert (1972a).

The key to preventing damage is to protect the layer of peat found in these areas. It has a high water content and specific heat capacity, but lower thermal conductivity than mineral soils. The latter warm up more quickly in the spring, while the drying peat has lower thermal conductivity in summer. Removal of peat inevitably results in a deeper active layer and thawing of ice in the underlying permafrost, as well as the effects of the increased albedo of mineral soil. Clearing by hand does less damage than using a bulldozer. Compaction of the peat will occur if the tyre pressures are too high.

If the peat only contains pore ice, melting of the ice only results in minor subsidence (Mackay, 1970), but ice-rich peat will shrink in volume producing double the thickness of the original active layer. Draining the resultant water is critical if thermokarst is to be avoided. The subsidence is permanent and irreversible in the short-term, but the permafrost will slowly aggrade, accumulating segregated ice lenses.

14.7 EMBANKMENT HEIGHTS

One of the outcomes of the old Golmud-Lhasa road embankment is the discovery of the fact that there are maximum and minimum heights of an embankment to protect the underlying permafrost from thawing (G. D. Cheng *et al.*, 2004). The *minimum embankment height* is closely related to the surface material on the roadway and the mean annual air temperature (MAAT, see Figure 14.6). When planning the

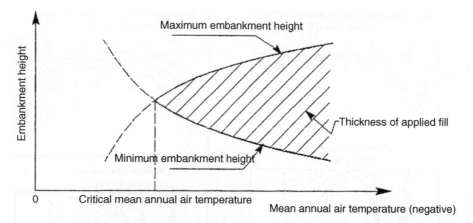

Figure 14.7 Diagram showing the application method for determining the critical MAAT and range of embankment heights that will protect the underlying permafrost in a given situation. On-site experiments are needed to establish the actual values (G. D. Cheng *et al.*, 2004).

embankment height, it is essential to decide on the road surfacing to be used during the design life of the road, and to use the height appropriate for the surfacing requiring the highest minimum embankment height.

Ding and He (2000) concluded that there is a *critical MAAT* ($-3.8°$C in the case of the Golmud Lhasa transportation corridor) above which it is impossible to prevent some thawing of the underlying permafrost, regardless of the embankment height (Figure 14.7). The actual critical value decreases from sandy-gravel surfaces, through concrete pavements to asphalt pavement.

The *maximum embankment height* is the height above which there is negligible increase in the protection of the temperature of the permafrost. Winter construction improves the attainment of equilibrium conditions in the embankment and underlying ground. Where the MAAT is changing, the amount of change expected in the design life of the embankment must be allowed for in calculating the minimum and maximum embankment thicknesses. The nature of the embankment fill may also affect the calculations, hence the need for experimental data to determine the critical MAAT and the maximum effective embankment height.

14.8 UNPAVED EMBANKMENTS

In areas with seasonal frost and those with continuous permafrost, gravel-surfaced roads are adequate and cost effective. They are also the basic form of embankment used for railways and emergency runways, and can be upgraded to paved roads or runways, once they have stabilized. The method of construction of the embankment depends on the availability of gravel, rock, sediment with less than 50% silt and clay, the presence or absence of peat or organic soil, and the climate. Using a gravel embankment, 5 m thickness is needed to achieve stability on permafrost colder than $-2°$C or 2 m in Alaska if insulation is used. Esch (1983) found that even thicker pads

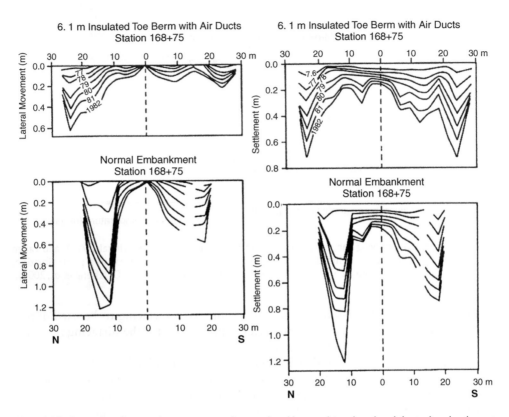

Figure 14.8 Lateral and vertical movements of normal and bermed, insulated and ducted embankments measured from reference points left and right of the centreline of a gravel road in Alaska (from Esch, 1983, Figure 9).

were needed to achieve stability over warm permafrost ($<-2°C$). Partial protection is usually used to cut costs. Test sections should be instrumented and monitored to avoid unpleasant surprises. Construction on permafrost should be carried out in winter.

Once the route has been chosen, current North American practice is to bulldoze any peat or organic soils to the side and put in a layer of rock or gravel to stop capillary movement of moisture into the overlying material. Usually, a waterproof geotextile is emplaced above it. Then the main embankment is built up, either using alternating layers of gravel and clay, mixed together and compacted by grading and vehicle traffic or by construction machinery. The resulting surface must be graded regularly to avoid the formation of deep ruts by the traffic.

Where the embankment does not run parallel to the drainage ways, culverts must be installed at a distance and location suitable for ensuring that there is no ponding of water along the sides of the embankment. Toe berms are often used on the higher embankments to minimize erosion and creep. Calcium chloride solution is often used in place of water prior to grading so that the gravel-clay mixture retains some moisture for a much longer time to minimize deterioration of the smooth surface. Even so, appreciable creep of the gravel occurs, together with considerable subsidence (Figure 14.8).

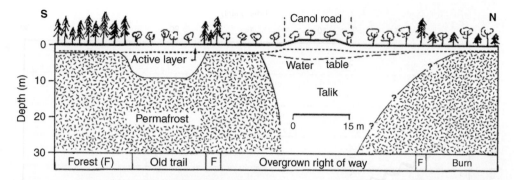

Figure 14.9 Diagrammatic cross-section of the Canol Road in 1972/1973 at Heart Lake, Northwest Territories. It shows the effects of minimal pad construction on the permafrost table, 20 years after the road was abandoned (from Harris, 1986a, after Kurfurst and Van Dine, 1973).

The subsidence is proportional to the height of the embankment and additional material must be added to the surface of the road each year in order to maintain its height. Initially it is emplaced in 15–30 cm thick layers and then compacted before more material is added. Without an adequate embankment, severe thawing of the permafrost takes place resulting in subsidence, the effects of which will remain for decades (Figure 14.9 and 14.10). The cooling of the lower part of the embankment does not increase as much when the embankment is higher than 4 m (G. D. Cheng *et al.*, 2003). Load weight must be limited during thawing of the active layer, but has little effect during the other seasons. However, ice build-up in the active later is closely correlated with lateral creep and heaving during winter. Chapin *et al.* (2009) discuss how to apply road restrictions most effectively, while X. Yu *et al.* (2009) describe the results of monitoring the ice and water present in the thawing road bed using a new method.

The older methods of construction of embankments used in the 20th century in Russia are similar. However, the layer of rock or gravel to prevent capillary movement of water is mainly used in wet areas. It represents a layer of draining material whose thickness equals the height of capillary rise of water plus 20–30 cm. On top is placed soil containing less than 50% silt and clay, with a layer of nonwoven geotextiles at its base and every 0.5–1.0 m above it. The organic surface layer of soil is left in place. Side slopes on gravel have a gradient of 1:1.5, but 1:3 is used in other cases. Where the height of the embankment exceeds 6 m, toe berms are installed for railway embankments, whereas the angle of the side slope is decreased in the case of roads. These methods were used until recently when a study of the embankment problems encountered in Siberia prompted new methods of construction using rocks and the modifications of the new designs that are being more successful in China on the Qinghai-Tibet Railway (Isakov, 2013; Kondratiev, 2013). These will be discussed below following a description of the new Chinese construction methods.

An added complication is the fact that in places along the B.A.M., the permafrost is aggrading in areas of discontinuous permafrost. The permafrost is about 50 m thick, but as it becomes thicker, it heaves the overlying ground, deforming the rail bed. Areas with long-term thawing of the permafrost suffer from gradual subsidence.

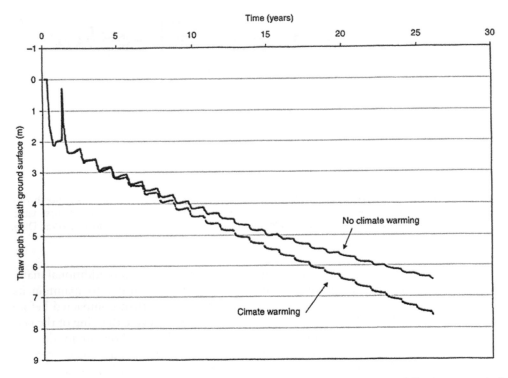

Figure 14.10 Results of modelling the long-term changes in ground temperature following removal of vegetation along the route of the Norman Wells pipeline, both under the present-day climate, and with a climate warming rate of $0.08°C/a^{-1}$ (from Oswell, 2011, Figure 6).

Over ice-rich permafrost in northern Russia, at least 2 m thickness of embankment is required to minimize thawing of the ice lenses. The fill should be premixed and end-dumped to avoid unnecessary disturbance to the underlying soil. Layers of insulation may be emplaced in the layers above the waterproof geotextile, and another layer of the geotextile is laid on top before more fill is added. The objective is to cause the permafrost table to rise up into the free-draining base of the embankment to minimize heaving and subsidence.

Unfortunately, the rising of the permafrost into the base of the embankment prevents water moving from one side of the embankment to the other. Any surface water accumulating along the side of the embankment will cause thermokarst subsidence (Figure 14.2). To avoid this, large culverts must be emplaced at the level of the ground. In winter, these tend to fill with snow and ice, so they must either be reopened by steam being played on the ice from a truck, or else a heating wire must be placed in the culvert and attached to a power source. Without this, water from melting snow in the spring thaw ponds, and then washes out the road (Figure 14.11).

Another effect of altering the surface drainage is that water becomes concentrated near the culverts and this can cause failure of pre-existing structures nearby that were not designed to cope with the new drainage pattern. This is proving a problem along the Qinghai-Tibet Transportation Corridor. Even if there is no erosion of the embankment,

Figure 14.11 Erosion at a culvert on a local road in the Bovanenkovo gas field, Yamal Peninsula. © Geosystem – A. Osokin.

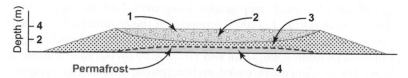

Figure 14.12 Consolidation of the embankment on a local road in Yamal. The change of volume was less than 10% in area 1, less than 5% in area 2, small in area 3, and unchanged in area 4 (permafrost). Redrawn after Geosystem – A. Osokin.

the water adds heat to the embankment adjacent to the culvert, so creating a deeper active layer in that zone (Périer *et al.*, 2014). This then allows more seasonal heaving and subsidence in that zone, while the lack of permafrost and the presence of a substantial amount of water aids in producing washout at these locations.

Use of large diameter corrugated culverts helps, but the water must be led away from the embankment so that it does not impact embankment stability. In areas with acid waters, it may be necessary to use plastic or wooden culverts because of rapid destruction of steel or aluminium culverts (MacFarlane, 1969). Livingston and Johnson (1978) showed that insulation of the culverts plus use of boulders in the openings to reduce cold air circulation helps to reduce the problem. Placing covers over the entrances in winter is even better, though they must be removed before the spring thaw (Sun *et al.*, 2014). Lining of the drainage way with rocks may also be necessary.

The amount of vertical compaction as a result of seasonal temperature changes and traffic varies in the different layers of the embankment (Figure 14.12). It is greatest in the upper layer of gravel, decreasing with depth. There will be no compaction in the underlying permafrost.

Producing stable embankments on the Qinghai-Tibet Plateau is more challenging. Figure 14.13 shows the route across the area with warm permafrost on the Plateau, together with the mean annual ground temperatures. The lower latitude means higher net radiation per unit area, which is enhanced because of the altitude, which means there is greater transparency to the incoming radiation (Cheng, 2003). The dry climate ensures bright sunshine for most of the year so the daily ground temperature range is higher than in other parts of China (Gong *et al.*, 1997; Chou, 2008). Since the transportation corridor runs northeast-southwest, the solar radiation received by the embankments is not symmetrical. 85% of the longitudinal cracks develop on the sunny slope, and gradually grow, resulting in large-scale damage to the road (Cheng and Ma, 2006; Chou, 2008). The difference reaches a maximum when the embankment trends at 135°, and is least on the shady side (Zhang *et al.*, 2003). Maximum net radiation is absorbed by the embankment when it is aligned north-south (Hu *et al.*, 2002).

The sediments are generally sandy so there is little problem in obtaining suitable fill. In flat, low-lying areas, the soil is saline, and saline lakes are common. These areas have to be crossed, but the railway is located on the sides of hills where possible (Figure 14.14). Rock is brought in from quarries in bedrock, and its liberal use under and along the sides of the embankment successfully counteracts the problems of building on warm permafrost (Figure 14.11). In the foothills of the mountain ranges, the methods of construction using rocks to cool the embankment are used as discussed in Chapter 12 (Figure 14.15). A tunnel was cut through the Fenghuoshan Mountains to reduce the need for additional steep grades going over 5000 m. Figure 14.16 shows the details of the north portal. Li (2009) reports that the ground had some ice lenses in the sediment. The first 2.5–3.0 m was cut with Φ25 mm threaded anchorages at 1.0 m intervals, followed by adding Φ6.5 mm and 10 cm × 10 cm bar mat reinforcement one by one, followed by welding the threaded anchorages together and spraying with 20 cm thick concrete. The ground around the entrance was grouted. Excavation of the tunnel where it was below the water table was by micro-vibration smooth blasting after installing a steel network of the outline of the tunnel in holes drilled into the sediment, followed by grouting. Further details will be found in Li (2005), Zhan and Kuang (2006), and Jiang and Wang (2006). Zhang and Sun (2003) analyzed the potential effect on the permafrost temperature for design purposes while Tang and Wang (2007) describe the temperature controls on the tunnel construction.

Vegetation cover is sparse, and certain areas of the railway are affected by blowing sand (Zhang *et al.*, 2011; Yang *et al.*, 2012; Xie *et al.*, 2015). Figure 14.17 shows a longitudinal dune west of the railway in the Tuotuohe section, together with fences erected to minimize sand movement. The strongest winds blow from the west-north-west at Cuonahu Lake (Yang *et al.*, 2012), moving sand across the plateau, and the current trend towards more widespread desertification is likely to increase the problem which affects both the old road, the new expressway and the railway. Yang *et al.* (2012) found that upright sand resistance fence barriers including nylon net barriers, upright concrete sand barriers, movable-board sand barriers and sand interception ditches placed at a suitable distance up-wind of the railroad were needed to help to reduce the effects of blowing sand (Figure 14.17). Gravel and rocks laid out in a 1 × 1 m grid

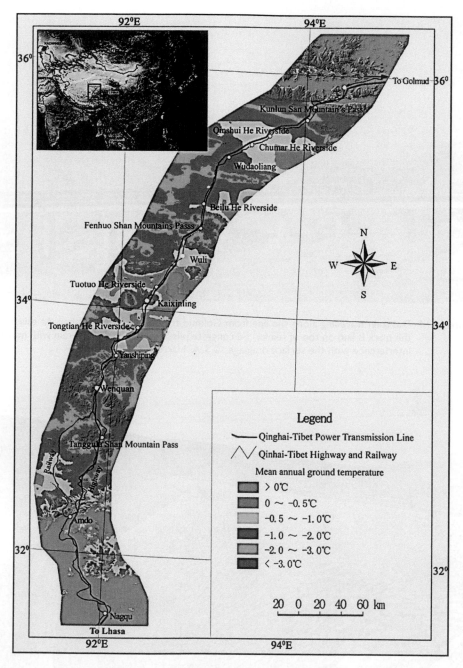

Figure 14.13 The route of the Qinghai-Tibet Railway across the permafrost areas of the Qinghai-Tibet Plateau, together with the mean annual ground temperatures.

Figure 14.14 The train travelling along the line from Golmud to Lhasa at Fenghoushan. At this point, the track is laid on top of reinforced concrete piles to cross gullied terrain with minimal interference with the surface drainage. © S. A. Harris.

Figure 14.15 Typical layout of the embankment of the Qinghai-Tibet Railway in the north slopes of the Fenghoushan Mountains. In the background is the original road. © S. A. Harris.

Figure 14.16 The north portal of the Fenghoushan Tunnel on the Qinghai-Tibet Railway. © S.A. Harris.

Figure 14.17 Sand dunes and fences to reduce problems with blowing sand along the Qinghai-Tibet Railway at the Honglianhe River section, near Tuotuohe. © S. A. Harris.

pattern upwind and downwind within 200 m of the railway embankment can stabilize the sand surface. This grid pattern works well in trapping sand and aiding the growth of vegetation, preventing sand from blowing across the railway line and forming barchan dunes 10–30 m long and 5–20 m wide (Duan, 2002; Xie *et al.*, 2015). The aeolian sand accumulates on the rail ballast, often covering the rails and the side slopes of the embankment. It can halt the operation of the trains, as well as causing accidents on the adjacent old Golmud-Lhasa highway. Wind erosion also erodes the embankments of both the highway and the railway. The sand grains sand-blast the rolling stock, rails and signals and shorten the life of the railroad. At Tuotuohe, the strong winds blow for three months during the winter, blowing sand across the rails. In summer, a north-easterly wind blows the sand back against the embankment. Aeolian sand problems affect about 270 km of the railway in the relatively flat areas of the Plateau. M. Zhang and Niu (2009) have modelled the effect of a snow and/or sand cover over the blocks, altering their cooling effect on the embankment between the mountain ranges.

14.9 MAIN PROBLEMS WITH EMBANKMENT STABILITY

The most obvious one is compaction and differential heaving and subsidence of the material in the embankment. Most compaction occurs in the surface layers with neg-ligible compaction in the underlying permafrost. This is why it is desirable to have the permafrost table rise up into the base of the embankment so that the underlying icy soils remain perennially frozen (Fang *et al.*, 2013). The new methods employed on the Qinghai-Tibet Plateau try to achieve this by using rocks and geosynthetics to prevent moisture moving up into the sediment making up the embankment by capillarity as well as descending down from the surface.

For roads, relatively small differences in heaving and subsidence can be tolerated on gravel roads by limiting load weights during the spring thaw. Regular grading will smooth out the surface. However, once the road is paved, the thermal regime of the road changes (see below) and these differences become a problem. On railway tracks, the problem is even worse, especially with the use of high speed trains. In the case of a train travelling at 220 km/h, the limit for vertical variation in track level is 7 mm in Finland (Nurmikolu and Silvast, 2013, p. 364), and both rails must remain at the same level within these limits.

Where the Russian railway embankments were built prior to about 2006, both these requirements and the 21st Century methods were unknown, and considerable differential heaving and subsidence occurred. In the case of the Baikal-Amur Rail-way (B.A.M.), approximately 80% of the line overlying permafrost has had to be rebuilt. About 1000 km of the track suffers from settlement and lateral spreading of the embankment. The amount of yearly settlement is rather consistent at between 4 and 12 cm/a^{-1} (Dydyshko *et al.*, 1993). Figures 14.18 and 14.19 provide examples of these problems, and more examples are illustrated in Kondratiev (2013). The same problems have been encountered in embankments in areas of seasonal frost such as in Finland (Nurmikolu and Silvast, 2013).

Identification of the location of the problem in the embankment is achieved by systematic in-depth track computer analysis of the track geometry and high quality GPR data (Silvast *et al.*, 2013). Silvast *et al.* (2010, 2012) describe the GPR methods used in identifying frost susceptible areas on railways.

Several problems may be involved. These include increased absorption of solar radiation into the subgrade compared with the natural, undisturbed surface, infil-tration of precipitation through the embankment, increased snow cover along the margins of the embankment and in areas of drifting snow, and migration of water

Figure 14.18 Problems with rail lines in Siberia. ©V. Kondratiev.

Figure 14.19 Differential subsidence and lateral creep of the embankment along part of the railway in Siberia. ©V. Kondratiev.

Table 14.1 Proposed countermeasures for reducing plastic flow on Russian railway embankments (from Ashpiz and Khrustalev, 2013, Table 3).

No.	Measure	Target	Note
1	Construction of berms of drainage soils dumped onto geotextile.	Reduce the unstable area, prevent pumping of soft soils to the daylight surface.	–
2	Construction of stone rip-rap on the slope.	Freeze soft soils of the foundation.	Used if thickness of snow on the slopes does not exceed 0.4 m.
3	Construction of rock-fill trenches by the embankment foot.	Block horizontal displacement of the soil.	Used if depth of the unstable area does not exceed 2 m.
4	Construction of stone forcipate yokes on subgrade slopes.	Freeze soft soils of the foundation, block horizontal displacement of the soil.	Used if thickness of snow on the slopes does not exceed 0.4 m and depth of the unstable area does not exceed 2 m.
5	Construction of ventilated shields on the slopes.	Freeze soft soils of the foundation.	–
6	Installation of thermosiphons on slopes or berms of the subgrade.	Freeze soft soils of the foundation, block horizontal displacement of the soil.	–

into the base of the embankment sediments when the railway line is built on a side slope (Kondratiev, 2013). To these, the Finns add insufficient thickness of non-frost-susceptible layers (ballast and sub-ballast) to prevent the freezing of frost-susceptible material, use of frost susceptible material in the embankment, and fouling of ballast making it frost susceptible. The ballast material must not become broken up during frost weathering. Inconsistent frost protection at transitions from cuttings and embankments, culverts, level crossings and switches can cause problems. Many of the materials used in embankments which were judged visually to be non-frost-susceptible are turning out to be frost susceptible (Pylkkänen and Nurmikolu, 2011; Silvast *et al.*, 2010, 2012).

Seasonal plastic flow or creep of the soils in the active layer in the embankment also occurs (Ashpiz and Khrustalev, 2013). Table 14.1 lists their proposed counter measures for reducing plastic flow of thawed soils in embankment foundations, while Figures 14.20 and 14.21 show some of their suggestions to avoid these problems using stone rip-rap, stone berms and vertical thermosiphons. Thermosiphons have been used successfully in Canada on the railway to Churchill, Manitoba (Hayley *et al.*, 1983). The success of these methods in Russia is still being determined, but it is probable that it will partly depend on whether the permafrost table rises into the base of the embankment. Any variations in the frost susceptibility of the soil in the active layer will cause differential heave and subsidence. Whether the material in the embankment is truly nonfrost-susceptible will also be crucial. In the case of slow speed trains, *e.g.*, on the Anchorage-Fairbanks line, the track was connected to the railway ties by shoes that can be adjusted in height up to 20 cm without the use of packing. This overcame the smaller fluctuations in level of the track, but involves considerable labour. It is usually in depressions where the greatest settlement problems occur (Figure 14.20).

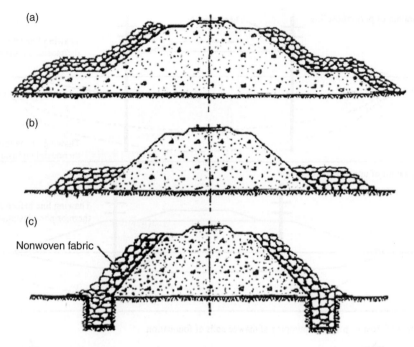

(a)

(b)

(c)

Nonwoven fabric

Figure 14.20 Suggested use of A, rip-rap, B, stone berms and C, forcipate yoke to try to stabilize the railway embankments in Siberia (after Ashpiz and Khrusyalev, 2013, Figure 4).

Use of fine round gravel is widely used in the subgrade of high speed railways in the cold regions of China. When fines are few, less frost heave occurs, but good compaction is more difficult (Yue *et al.*, 2013). Even with the care used in constructing the Qinghai-Tibet railway, some heaving and settlement occur. A special locomotive-car combination has been instrumented to detect differential movements (Figure 14.23), and it is run along the track at regular intervals to minimize the risk of accidents. Vibration from the passage of trains aids the movements of the material making up the embankments. The greatest problems occur in the transition from the frozen to the thawed state in spring (Petryaev and Morzova, 2013). The greatest influences are hydration, freeze-thaw cycles and the vibrodynamic impact of the trains. The latter is dependent on the load, frequency, and duration of the vibration and is cumulative (Su and Huang, 2013). Figure 14.24 shows the bearing capacity of the subballast consisting of non-cohesive material in frozen, thawing, and thawed state, compared with that for the geogrid installed immediately below the ballast. The geogrid spreads the load from the sleepers to the intervening embankment surface and dampens the effects of the vibrations (see Chapter 13). Z. Y. Wang *et al.* (2013) monitored the vibration generated by trains at different seasons of the year on seasonally frozen ground in the Daqing area of northeast China. The vertical vibration was greater than in the longitudinal and lateral directions, and the vibration decreased with distance from the rails. The attenuation was most marked in winter when measured on a frozen embankment surface, and the vibration from freight trains was greater than for passenger trains

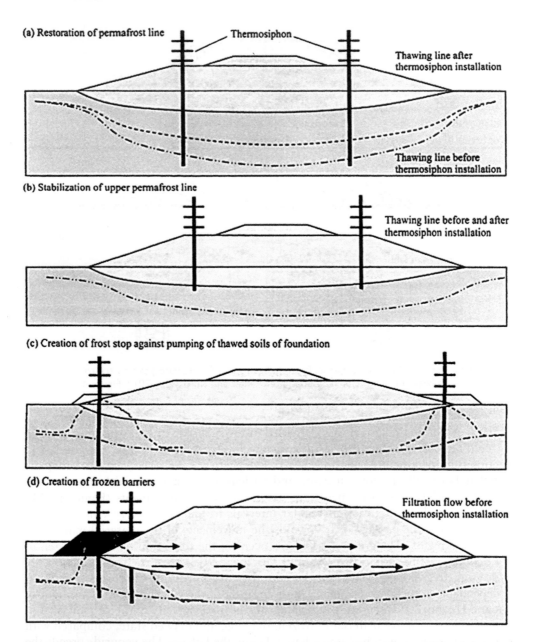

Figure 14.21 Suggested use of thermosiphons to counteract embankment flowage (after Ashpiz and Khrustalev, 2013, Figure 6).

with a similar number of cars. This is believed to be typical for the 5.14 million km² area with seasonal permafrost in China.

The Chinese "Code for Design of High Speed Railways" (MRPRC, 2009) demands strict controls of post-construction settlement of embankments on soft ground.

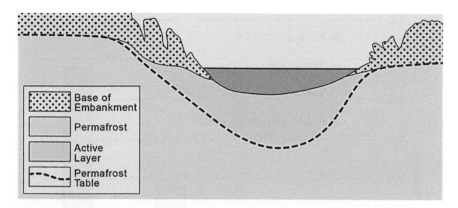

Figure 14.22 Development of thermokarst and thermoerosion on the Yamal railroad (after Geosystem – A. Osokin).

Figure 14.23 The locomotive equipped to measure differential heave or settlement along the Qinghai-Tibet Railway. © S. A. Harris.

To achieve this on the soft, saline finer-grained soils in the flatter areas of the Qinghai-Tibet Plateau, a composite foundation consisting of a pattern of plain piles with a strength grade of C10 to C20 is installed beneath the embankment to increase support for the weight of the loads and to overcome stresses. Su and Huang (2013) tested the effectiveness of these, and showed that the loading and stresses are sufficiently large to displace these piles and to bend them. The amount of damage depends on the location of the piles, the compressive modulus of the soft soil around the pile and

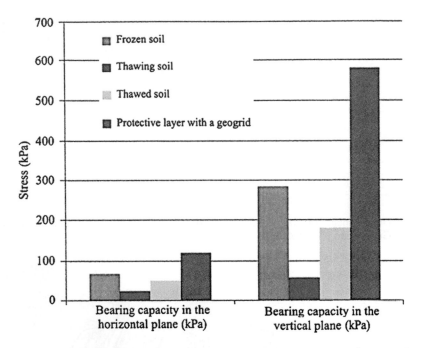

Figure 14.24 Change in bearing capacity in vertical and horizontal directions of a non-cohesive soil from the frozen state, through thawing, to the thawed state, compared with the use of a layer of geogrid under frozen conditions that has been installed directly below the ballast (from Petryaev and Morzova, 2013, Figure 5).

loading conditions. Greatest displacement occurs between 30 and 80 mm underground. The piles fail first at the toe of the embankment and failure progressively moves to the centre line. Testing is advisable to estimate the correct spacing to be effective against deformation.

The radius of curvature of the high speed railway tracks must be less than those limiting traffic speeds on freeways. The length of each rail is also important, and rail length has been increased to improve performance. Rybkin *et al.* (2013) have studied the effectiveness of these longer rails and finds that a new design of the pre-stressed concrete sleepers improves resistance to lateral shift of the continuous welded rails on curves with a radius of 300 m or less.

14.10 CONCRETE VERSUS BALLAST RAILWAY TRACKS

A logical solution to the problems associated with the traditional ballast track is the use of concrete tracks (Figure 14.25). These are used in Korea (Lee *et al.*, 2013) because of their operational stability and a significant reduction in maintenance costs. The load is spread over the entire concrete surface, the tracks being anchored in shoes set in the concrete without the use of sleepers. A disadvantage is that the initial cost is greater than for a ballast track.

Figure 14.25 Concrete tracks for a high speed railway in South Korea. © I. W. Lee.

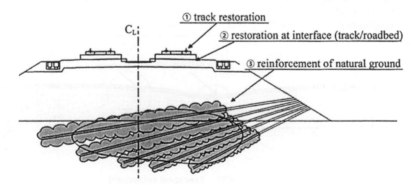

Figure 14.26 The methods of correcting the level of the rails 1. By adjusting the shoe, and 2. By modifying the interface between track and railbed, and 3, injection of the grouting cement diagonally beneath the track (from Lee *et al.*, 2013, Figure 2).

The main problem encountered is differential settlement of the embankment, which is reflected in the level of the concrete pad. The allowable residual settlement is 10.0 cm on the Kyungbu to Honam high speed railway. For small settlements, adjustments to the shoes will correct the movement, but for larger movements, the concrete pad needs under-pinning. This is done without interfering with the operation of the line by using special types of very quick hardening and even faster middle-hardening cement-grouting injected into the underlying ground beneath the cement pad under the track after hydraulic jacking (Figure 14.26). The middle-hardening cement is sufficiently strong after 1 hour to allow rail service to resume. The grouting can also be

Table 14.2 Methods of adjusting the rails for settlement on concrete tracks (from Lee *et al.*, 2013, Table 2).

Location	Methods	Details
① Track	Adjusting fastener	The level can be adjusted using a thin fastener pad. In general, 20–40 mm should be restored.
② Track/trackbed interface	Hydraulic jacking	Voids should be filled after lifting the track with hydraulic jacking.
	Pressurized rapid-hardening cement grouting (PRCG)	Pressurized injection and rapid-hardening cement mortar should be used.
	Urethane reinforcement	The expansion pressure of urethane foam is considered.
③ Natural ground	Natural ground grouting	General grouting methods such as DGI (double grouting injection) are used.

Table 14.3 Advantages and disadvantages of the methods of dealing with track subsidence (from Lee *et al.*, 2013, Table 3). ○ Good. △ Normal. × Poor.

Methods	Cost	Safety	Construction	Durability	Reliability
Hydraulic jacking	○	△	△	○	△
Urethane reinforcement	○	△	○	△	△
Pressurized rapid-hardening cement grouting (PRCG)	○	○	○	○	○
Natural ground grouting	×	△	△	○	△

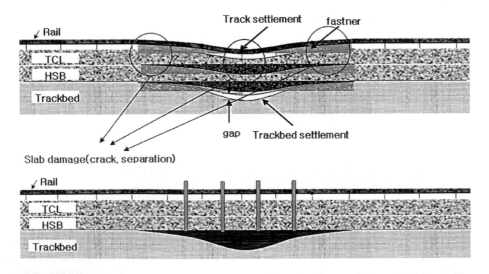

Figure 14.27 Concrete track rails before and after grouting (from Lee *et al.*, 2013, Figure 3).

used in the rail bed (Table 14.2). Table 14.3 shows the advantages and disadvantages of each of these methods, while Figure 14.27 shows a diagram of the track, before and after restoration. The concrete pad overcomes problems of welded rail deformation on curves tighter than 300 m radius.

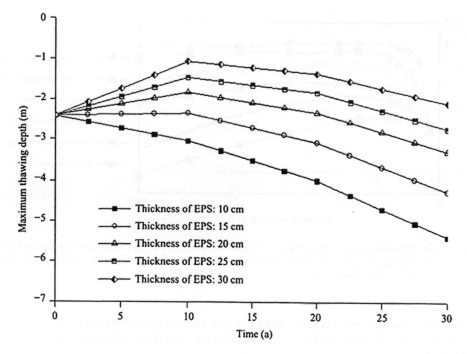

Figure 14.28 The predicted maximum thaw below the centre line of embankments using 10 cm thick EPS sheets on Highway G214, Qinghai Province (from Tian *et al.*, 2013, Figure 5).

14.11 PAVING OF ROAD AND AIRFIELD RUNWAYS

Gravel-surfaced embankments should not be paved until it is established that the underlying ground is stable. Addition of asphalt increases the mean annual temperature of the underlying embankment for a standard width road with two lanes by about 3.5°C in China (Zhu *et al.*, 1989). The buildup of heat on the pavement increases greatly with increasing width of the pavement, reaching about 6°C at the centre line on freeways, 24 m in width in Qinghai Province (Tian *et al.*, 2013). This must be taken into account before paving is begun. In the case of the G214 Highway between Yushu City and Gough County, these authors modelled the consequences of using different thicknesses of extruded polystyrene insulation boards (EPS) in the embankment to counter the extra heating. Figure 14.28 shows the results. Since the Qinghai-Tibet Highway lies at even lower latitudes, it may need 30 cm of EPS in the embankment to be effective. Regular monitoring of the ground temperatures in the embankment across the road right-of-way at regular distances is necessary to ensure that there are no embankment problems after paving. If climate change is occurring, the presumed ground temperature at the end of the design life of the road must also be allowed for.

Recent studies have shown that warm mix asphalt saves energy, takes longer to harden, and produces an equally dense pavement (Goh and You, 2009). The fact that it takes 27 minutes longer to harden extends the available haul time by 27 minutes.

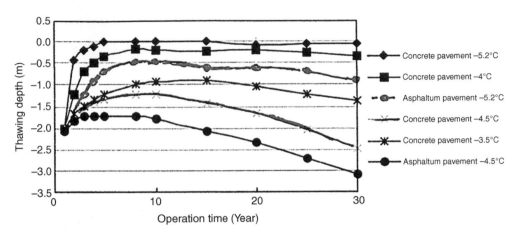

Figure 14.29 Maximum thickness of the active layer in the embankment of the Qinghai-Golmud road during 30 years with either asphaltic concrete or a flexible asphaltic surface (from Wang et al., 2009, Figure 3).

While the tensile strength is lower than hot mix asphalt pavements, they are less likely to develop ruts. Robjent and Dosh (2009) describe the three different types and their performance.

Porous asphalt is a newer type that has been reported to be successful in the slow release of captured storm water, even over a poor-draining subsoil (Tiggelaar and Lisi, 2009). Another study by Roseen et al. (2009) compared pervious concrete with porous asphalt pavements in an attempt to determine which was best in managing storm water. It was found that even after frost had penetrated 45 cm into the ground, surface infiltration capacities in porous concrete surfaces exceeded 500 cm/hour. 75% less salt was needed on the pavement, black ice rarely formed, and skid-resistance was 12% greater. This produces less salt contamination of the drainage ways and is considerably more environmentally friendly.

A study of the temperatures in the subgrade below pervious Portland Concrete Cement surfaces and hot mix asphalt showed that the subgrade was up to 4°C warmer than beneath impervious Portland Concrete Cement paving (Rohne and Lebens, 2009). Freezing cycles were reduced by 60% over a three-year period, probably due to trapped air in the pore spaces.

The other variable is the strength and durability of the asphalt (Wang et al., 2013). Methods of determining this vary between countries, and the best technique for determining the type of asphalt in hot summer climates such as China is still being studied. At higher latitudes, the mean annual temperature range on asphalt surfaces is less, and the necessary research on the method of choice of the asphalt is better known. There are two kinds of asphaltic surface, viz., *asphaltic concrete* forming a rigid surface, and the *flexible asphaltic surface* which bends with differential movements of the embankment. The asphaltic surface absorbs greater heat than the asphaltic concrete (Figure 14.29), so that the active layer is approximately 0.5 m deeper after 10 years (Wang et al., 2009).

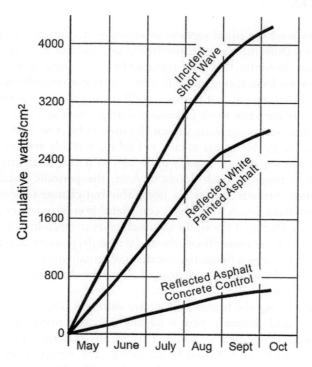

Figure 14.30 Cumulative incident and reflected solar radiation along highway test sections comparing white-painted asphaltic concrete with a control section (from Berg and Aitken, 1973, Figure 4).

14.12 USE OF WHITE PAINT

The effect of changing the colour of the asphaltic concrete surface was first investigated by Berg and Aitken (1963) and Fulwider and Aitken (1963) in Thule, Greenland and on the Farmers Loop, northeast of Fairbanks Alaska. The permafrost degradation was least under the white-painted road surface, and the difference was shown to be due to the amount of shortwave radiation reflected by the painted surface (Figure 14.30). At Thule, repainting of the white-painted runway only needed to be carried out once every five years, but under highways, it would have to be done at least once a year due to the heavier traffic. Jørgensen and Andreasen (2007) and Jørgensen (2009) showed that at Kangerlussuaq Airport in western Greenland, the white-painted asphalt caused a c. 1 m rise in the late summer depth to the permafrost table (3.5 versus 4.5 m). The black asphalt surface lost its snow cover earlier in the spring.

Unfortunately, subsequent studies indicate that the paint at Thule runway reduces the braking ability of aircraft and dramatically increases the maintenance and labour costs (Bjella, 2013). The paint also loses its pristine white colour, becoming off-white. Subgrade insulation and excavation of ice rich soils are more cost-effective. As a result, the white painting is never used on roads and only infrequently on Arctic runways.

14.13 BRIDGES

Originally, ferries were the usual method of crossing streams and rivers. The ferries had to be removed from the water before freeze-up, and put back in after the spring floods and ice jams. In winter, they were replaced by ice bridges, so that there were only gaps of about 4–6 weeks in spring and fall when it was not possible to cross the water. This is obviously inconvenient, but ferries are still commonly used in both Canada and Siberia. It obviously does not work in semi-arid areas such as Tibet, where there are broad small streams occupying wide stream channels subject to flash flooding.

The obvious big problem this entails is finding a stable footing for the bridge supports. Rivers and streams are usually underlain by a talik with saturated unfrozen sediments. Even in zones of flash floods in Asia, the periodic flooding affects the ground temperature. Stream courses are not stable but change over time, and this is particularly true of flash floods which have considerable erosive power and flow across relatively flat areas. During a flood, the channel tends to straighten so as to carry the increased volume of water more efficiently, and it also deepens its channel temporarily. During the waning phase of a flood, the stream may remain in one of the new channels, depending on where the sediment is deposited along the channels as the speed of flow of the water decreases.

The problems facing the bridge builder are different, depending on the climatic regime. In humid, cold climates, there is usually only one major flood in the spring, but it may be accompanied by ice jams. This means that the bridge deck must be high enough to avoid being removed by ice during the spring thaw. On continuous permafrost, the construction is easiest and involves piles driven into the permafrost. Figure 14.31 shows a cross-section of the Eagle River at the location of the Bridge constructed along the Dempster Highway (Johnston, 1980). The permafrost on the north side was 90 m thick and had a mean annual ground temperature (MAGT) of −3°C. Beneath the river is a deep thawed talik extending down through the permafrost at an angle towards the south under the river flood plain. On the south side, the river sediments have been aggrading, and the permafrost is only 8–9 m thick with an MAGT of −0.4°C. Construction was carried out in winter from cribs built on the frozen river. Each abutment consisted of 15 steel piles. On the north side, the piles were driven to a depth of 12 m, but on the south side, they extend down to 30 m to provide comparable bearing capacity. Subsequent monitoring shows that the single-span bridge has been successful. A similar problem is described by Titkov and Grebenets (2009) for a railway bridge in the South Yamal Peninsula.

Crory (1985) provides an example of the problems encountered with footings for bridges in areas of discontinuous or sporadic permafrost. The talik underneath the water course is usually large, and the disturbance of putting in the embankment and footings for the bridge increases the size of the talik. Any permafrost present is usually very ice-rich. Fluctuations in the flow of the stream together with erosion during spring runoff as well as damage by ice jams tend to cause the streams to change their course. The bridge supports have to be deep enough to support the weight of the loads in the unfrozen ground, while the bridge decks must be sufficiently high to allow for the passage of boats and ice during ice jams. The upstream side of the piers usually has steel armour and is shaped so as to present low resistance to the flow of water and ice. Figure 14.32 shows the old bridge at Goldstream Creek discussed by Crory while

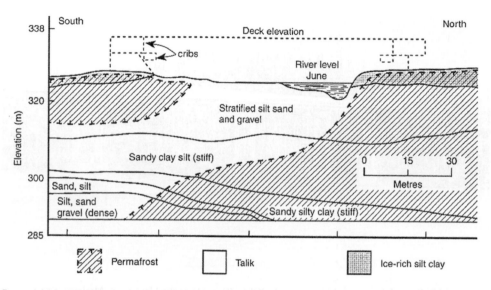

Figure 14.31 Cross-section of the Eagle River at the Dempster Highway bridge (redrawn after Johnson, 1980). Note the armoured upstream side of the supports on the Yukon River where the average annual level of ice jams in spring thaw is 3 m.

Figure 14.32 The old Bridge at Goldstream Creek, Fairbanks, Alaska. © S. A. Harris.

Figure 14.33 shows the new bridge over the Yukon River on the Dalton Highway, Alaska. The old bridge requires annual shimming of the tops of the supports to correct for their uneven subsidence, whereas the new bridge is stable.

In the higher latitudes, problems arise due to ice jams in the spring thaw since most of the rivers flow north to the Arctic Ocean. The piers must be strong enough to

Figure 14.33 Part of the new bridge across the Yukon River along the Haul road, Alaska. © S. A. Harris.

withstand the pressures by the ice, and T. L. Yu *et al.* (2009) discuss the compressive strength of floating ice at various temperatures, and provide a method of calculation of the ice force on bridge piers during these events. Care must be taken to build the bridge deck well above the expected extreme level of any ice jam.

In low latitudes, cold, high elevation permafrost sites such as occurs on the Qinghai-Tibet Plateau, and bridges crossing the wide, braided stream flood channels are a problem. The ground has a high water table underneath a thin layer of permafrost. The soil making up the active layer is usually saline, so that the bearing capacity of the ground is less than in areas lacking salts. Where there is a perennial stream, there is usually a through talik, and the area is in a tectonic zone where earthquakes are frequent. The original construction on the old highway followed the accepted guidelines for normal, non-saline soils, but did not allow for the thermal effects of flash floods or earthquakes, nor the lower bearing strength of the saline soils. As a result, causeways show zones of significant subsidence (Figure 14.34), while one section of a key bridge collapsed due to too great tilting of a support (Figure 14.35). The new bridges over the streams in the area use much deeper piles to overcome the poor strength in the upper layers of the ground.

14.14 ICINGS

Where there are springs, pingos, seasonal frost mounds and icings occurring along river valleys, there is usually trouble with icings affecting any embankments in the area. These icings can grow to 2.4 m or more in height unless remedial measures are taken (Van Everdingen and Allen, 1983). On roads and airfield runways, graders with

Figure 14.34 Buckling of a causeway across a dry stream bed along the old Golmud-Lhasa road. © S. A. Harris.

Figure 14.35 Tilting of a pile support resulting in the collapse of a section of bridge, Qinghai-Tibet Highway. Note the new, successful railway bridge in the distance. © S. A. Harris.

grooved blades must cut away the ice formed every 24 hours, or else serious accidents can occur (Van Everdingen, 1982). The natural surface of the icing slopes away from the source, and vehicles or aircraft will slide off the surface even with chains on their wheels. As the ice accumulates, it grows from the base producing significant dilation cracks if left alone. Obviously trains will be derailed if the ice covers part of the track, so icings must be avoided if possible. Figure 14.36 shows the usual situation on sloping land alongside road embankments in regions with seasonal frost. Since water also moves downslope underground either beneath the permafrost table, or in taliks parallel to the slope, these water bodies can also produce significant seepages and springs that result in icings developing in winter.

Although they are necessary to deal with overland drainage, culverts are a problem in several ways. In summer, the warm water flowing through the culvert is an additional source of heat that causes a deeper active layer around it (Périer *et al.*, 2014). This makes the adjacent section of the embankment vulnerable to erosion. Secondly, surface water trying to flow through it in the fall freezes in the culvert slowly blocking it (Figure 14.37). The surface water then builds up an icing on the upstream side of the culvert which eventually overtops the embankment causing the icing. When the spring thaw begins, the culverts block the flow of melt water which then flows over the embankment, eroding it and producing a washout.

High embankments do not necessarily solve the problem. Even though several large diameter culverts are installed at different levels at one place along the Dempster Highway (Figure 14.38), there is a build-up of ice upstream of the embankment, resulting in the icing extending across the road in winter. This is caused by ice forming in the culverts, filling them up, and producing a small glacier-like mass behind the causeway. This eventually fills the valley and ice forms on the road as water moves across it. Then the grader with a toothed blade must remove the ice each day for the remainder of the winter weather.

Methods of combatting icings can be divided into passive methods and active methods (Thomson, 1963). Passive methods are those minimizing the effects of icings. Three methods have been used to de-ice a passageway through the culvert. The most common is a daily visit by a truck that produces *steam* from a boiler. A steel pipe is installed in the culvert in summer with a vertical extension on one side of the road that is high enough not to get snowed in. The cap is removed for sending steam through the pipe to thaw the adjacent ice. A second method, probably unique to North America, is the *"moose warmer"* (Figure 14.39). It consists of a 200 litre drum of diesel fuel that slowly feeds fuel to a burner in a second drum containing a coil through which the overflow water passes. The heated water then keeps a passageway open through the culvert. The fuel consumption is about 40–100 litres/d. The cheapest device consists of *electrical heating tape* in the culvert attached to solar panels (Sweet, 1982). The panels may be installed in the spring but removed during the other seasons to prevent theft. Alternatively trucks carrying generators can supply the electricity to thaw open the culvert.

These methods are expensive, and obviously the active methods, avoiding the development of icings, are best. These involve carefully studying all sources of information about the area of the proposed embankment including aerial and satellite photographs (Harden *et al.*, 1977; Åkerman, 1980, 1982), and locating all icings that may have developed the previous winter. Unfortunately, they do not always occur in the same place each year (Van Everdingen, 1982; Slaughter, 1982). The construction of the embankment will act as a dam, holding back surface drainage, thus intensifying the icing problem. The shadows of bridges also act as loci for icings in winter. Use of *oversize culverts* helps reduce the problem, but putting in a diversion dyke upslope of the road is another method of control. *Freezing belts* consisting of a zone upslope of

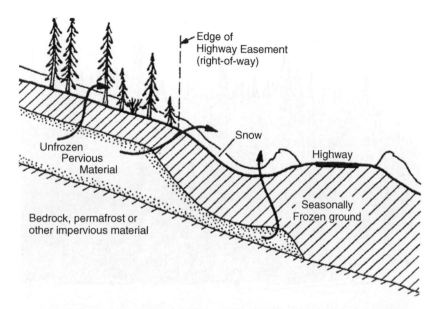

Figure 14.36 Diagram of a typical icing affecting a road embankment on a slope with springs under conditions of seasonal permafrost (from Carey, 1973).

Figure 14.37 Ice blocking the outlet of a culvert along the Alaska Highway, Yukon Territory, after partial thawing with a moose warmer. © S. A. Harris.

Figure 14.38 Embankment on the Dempster Highway crossing a small stream. Note the multiple large culverts intended to slow the onset of icings crossing the road in winter. Once the icing has filled the space behind the embankment, the icing grows across the right-of-way. © S. A. Harris.

the embankment that are cleared of vegetation and kept clear of snow can also work well on slopes. The cleared surface allows the ground to freeze and collect any surface water that would otherwise flow across to pond against the embankment. In spring, the vegetation mat must be replaced to keep the ground cold and to prevent erosion.

Figure 14.39 "Moose warmer" keeping a space open in a culvert in spring along the Alaska Highway, Yukon Territory, in 1993. It kept the culvert partly open so that water from melting snow and ice could drain away, thus avoiding icings crossing the right-of-way. © S. A. Harris.

In Russia, trenches or berms may be constructed along the freezing belt to enhance its efficiency. *Hessian, canvass* or *plastic net fences* placed up-slope of the embankment can collect the water as ice before it reaches the road. Stones can be placed in the entrance to the culverts to reduce snow blowing in. On the Qinghai-Tibet Plateau, thermosiphons have been used around springs to stop the outflow of water.

Insulated subdrains were suggested by Livingston and Johnson (1978) but did not stop icings forming on roads in Denali Park, Alaska (Vinson and Lofgren, 2003). Where XPS boards are used in the embankment, they must be placed 50 cm below the surface. If not, *differential hoar frost* will appear on the embankment surface over the insulation when it does not form elsewhere (Refsdal, 1987; Horvath, 1999; Stark *et al.*, 2004).

Climate is also a factor in the development of icings, *e.g.*, in Denali Park (Vinson and Lofgren, 2003). When the maximum snow accumulation exceeds 50 cm, icings are rare, whereas they are common with a thinner maximum snowpack. As stated by Carey (1973), if heavy snowfalls occur during the first two or three months of winter weather (October to December), icings are rare or nonexistent. This is probably due to the insulating effect of the snow cover preventing the ground from becoming very cold. If all else fails, the best defence is to modify any stream channel so that the water travels down a steep, narrow channel with rapid flow.

14.15   CUT SLOPES

One of the major problems in permafrost regions is how to deal with *cut sections* in ice-rich permafrost (Smith and Berg, 1973; Pufahl *et al.*, 1974; Shu and Huang, 1983). In competent bedrock, the only problems are normally with perched water tables above the permafrost table, and with blocks of rock falling from the 4:1 face. These can normally be coped with since there should be a 2.5 m wide drainage ditch at the base of the slope. Pufahl (1976) described four possible methods of stabilizing cut slopes in ice-rich permafrost (Figure 14.40). In North America, the normal practice in ice-rich cohesive sediments (Berg and Smith, 1976) has been to cut slopes nearly vertically and then to allow the slope to thaw and slump. The overlying vegetation mat drops down and forms an overhang and prevents the underlying ice-rich material from flowing out during the thaw. Generally, the slope stabilizes when it has changed to about 1:1.5, although on occasion even this slope is too unstable. Good examples can at present be seen along parts of the roads in central Alaska along the Trans-Alaska Pipeline. Although problems with gullying can occur, the slope will revegetate naturally in as little as three thaw seasons. Insulation or an impermeable membrane can be added to reduce thawing (Liu *et al.*, 2014). Other methods include the use of gabions or concrete retaining walls, *etc.*, see chapter 13. In China, the slope is covered with sod cut from the ground nearby almost as soon as it becomes exposed.

14.16   AIRFIELD CONSTRUCTION

Airfields can be divided into three groups, *viz.*, temporary, limited use, and permanent. They need a similar protective base or embankment, but it is essential that they can stand the pounding of the landing gear consisting of one or more pairs of wheels that take the full weight of the aircraft during landing. Thus all except temporary airstrips are normally paved with asphaltic cement of sufficient thickness according to the tire pressures of the aircraft. Table 14.4 provides some examples of the required minimum thickness of asphaltic concrete and runway base and sub-base courses for some tire pressures. The runway embankment should be built on non-frost susceptible materials such as beach gravel or rock, if possible. Where the underlying ground contains ice wedges, these may melt out as a result of the disturbance of the thermal regime (French, 1975). Moving sediment from the active layer and placing it in depressions results in subsidence problems from ice rich ground thawing and subsiding in the area from which the soil was taken. There must be good drainage along the sides of the embankment.

Johnston (1981) describes the construction of the Inuvik airport runway, carried out in winter. Rock was brought from a quarry 0.8 km to the west and placed on the ground after the shrubs and trees had been removed by hand. The base was added during the next summer and insulation was used to reduce the quantity of fill needed (Brown, 1970a). Above this was placed a stony sub-base, overlain by the clean gravel base course. This served as the runway surface until paving occurred.

The snow cover and thaw depth have been monitored throughout the year since 1958 (Figure 14.41). Initially, the active layer in the embankment was 2 m deep, but this stabilized after about 5 years. In 1969, the runway was paved with asphaltic

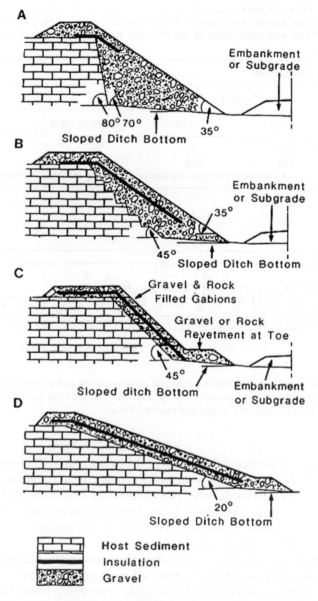

A

Embankment
or Subgrade

80° 70°   35°
Sloped Ditch Bottom

B

Embankment
or Subgrade

35°

45°
Sloped Ditch Bottom

C

Gravel & Rock
Filled Gabions

Gravel or Rock
Revetment at Toe

45°
Sloped ditch Bottom

Embankment
or Subgrade

D

20°
Sloped Ditch Bottom

Host Sediment
Insulation
Gravel

Figure 14.40 Some possible methods of stabilizing cut slopes in ice-rich cohesive sediments (after Pufahl, 1976; Pufahl and Morgenstern, 1979).

concrete, and since then, the active layer has increased to just over 2 m in summer. Sufficient height of the embankment has prevented problems, although warming of the mean annual air temperature by about 2°C caused the permafrost table to descend after the 1970's (Hoeve and Hayley, 2015). Seasonal thaw is now extending below the embankment materials in some places. Since part of the runway was constructed over a wet area with peat, additional drainage along the embankment has been required after

Table 14.4 Typical minimum thicknesses of asphaltic concrete, base and sub-base courses used in Canada for aircraft with various tire pressures (after G. H. Johnston, 1981, p. 391, reproduced with permission of the National Research Council of Canada, and Harris, 1986a).

Minimum Layer Thicknesses (mm)

	Design Aircraft Tyre Pressure (kPa)			
Layer	<415	<690	690 to 1030	1035 to 1370
Asphaltic concrete	50	65	90	100
Base	150	230	230	305
Sub-base	– As necessary to provide adequate bearing strength in flexible pavement.			

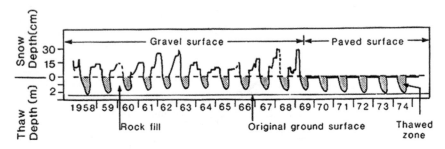

Figure 14.41 Snow cover and thaw depth below the runway centre line at the west end of the Inuvik airstrip (after G. H. Johnston, 1981, p. 391, reproduced with permission of the National Research Council of Canada, and Harris, 1986a).

a slab of the concrete surface subsided up to 0.5 m for about 10 m rather abruptly in late summer 2013. It was due to thawing of icy permafrost in the peaty ground beneath the fill producing subsidence. In other places, airstrips have been built in winter over frozen ponds and shallow lakes up to 1.2 m deep without problems, provided that a suitably thick embankment is used to ensure that the permafrost table is higher than the former water surface (Harwood, 1966).

If saline sediments occur in pockets below the runway, then subsidence will occur at these locations. Jørgensen (2012) identified such pockets beneath the runway at Tasiujaq in the southwestern part of Ungava Bay (Nunavik), using GPR. The bulk of the runway is built on non-frost susceptible river gravels, and the saline marine silts were not identified when the runway was built. Now the silts will have to be dug out and replaced with gravel, or else these sediments will have to have special cooling devices installed to provide better strength in the embankment.

The airfields built over saline marine sediments are particularly problematic. At Amderma on the Arctic coast of Siberia at about 5 m above sea level, the permafrost table in sands has a temperature of about −3°C, but the permafrost is only 2–10 m thick. The underlying sands are saturated with cryopegs containing 40–120 g/l of salts. Inevitably, the runway is affected by freezing and thawing, together with frost cracking.

As a result, it has had to be reconstructed. Similar problems with substantial heaving over frost susceptible material have been found on the Kola Peninsula and in the Kurill Islands in Russia, while Fortier describes a similar problem at the airport in Umiujak in Nunavik (Québec). The volcanic ashes along the Pacific Rim are particularly problematic since they are exceptionally frost susceptible. In Russia, the use of heating machines to heat the surface of the runway in winter subjects the underlying material to extreme thermal gradients, shortening the life of the concrete.

Warming climates have also taken their toll. Thus at Tasiujaq, the mean annual air temperature has increased by 3.2°C between 1992 and 2002 (Fortier and Savard, 2010). The active layer thickness increased from 1.3 m in 1995 to 2.0 m in 2005. An increase of this magnitude had not been allowed for during construction, so heat drains and geocomposites placed on the shoulders of the embankments are now being used to cool the embankment (Jørgensen and Doré, 2009; Jørgensen 2012). As noted in chapter 13, the geocomposites are proving more effective and are certainly more aesthetically pleasing. Jørgensen and Doré demonstrated that reducing the side slope of the embankment to no more than 8:1 achieved the same results. This could also help save lives and damage to aircraft if they skid off the runway, though this is not being considered at this time.

Similar changes in air temperature have taken place at Kangerlussuaq Airport in Western Greenland. There, the use of white paint on the asphalt has raised the permafrost table about 1 m (Jørgensen and Andreasen, 2007; Jørgensen and Ingerman-Nielsen, 2008). However, it is more costly in terms of maintenance since it has to be repainted annually, it can cause localized icing on the runway surface, and can dazzle the pilots during landing (Beaulac and Doré, 2006). The latter examined the status of all the airfields built on the warm and thin permafrost in Nunavik, finding that 5 were stable, 4 were acceptable, and 3 were vulnerable.

Chapter 15

Oil and gas industry

15.1 INTRODUCTION

Modern society and industry use large amounts of oil and gasoline. Extensive reserves of these fuels are found in permafrost areas in the Northern Hemisphere including Alaska, Canada, and Russia (Figure 15.1). Furthermore, offshore continental shelves around the Arctic Basin and along the northwestern coasts of Norway have important gas and petroleum fields. The oil can be retrieved relatively easily whereas the gas, primarily in the form of gas hydrates, has proven difficult to obtain safely.

It is expensive to work in these northern regions, and when the oil or gas is brought to the surface, it must then be transported to the main locations where people live outside the permafrost areas (Figure 15.1). This can be done by using tankers filled with oil or liquefied gas, or by using pipelines. Tankers have the advantage that as soon as one is built, it can go into service, earning revenue. However, offshore terminals along the Arctic coast are difficult to construct (see chapter 9) and the tankers can only operate in the short season of open water. While the warming of the Arctic provides a longer open season for transport, coastal erosion is also increasing, making the successful operation of tanker terminals more problematic. Thus there is a limit to the quantity of gas and crude oil that can be moved in this way. The use of offshore islands for drilling platforms, production facilities and tanker terminals was tried in the 1970–1980 period (Harris, 1986a), but has largely been abandoned. Floating drilling rigs are now used and can either be sitting on the bottom of the sea floor or can be tethered in deep water until the drilling is completed.

Pipelines cannot go into use until the last compressor station is completed, so they require a much bigger initial capital outlay. However, they can operate year round as long as they are properly built and maintained. They are the main workhorse of the oil and gas industry. However, in most countries, environmentalists are very concerned about spills when pipelines leak, so there is pressure to take as few risks as possible in most places.

15.2 OIL AND GAS EXPLORATION

The first observations of oil in the northern parts of North America were by Sir Alexander Mackenzie who described oil seeps along the Mackenzie River near Norman Wells in 1789. Subsequently, Leffingwell (1919) reported similar oil seepages near Cape

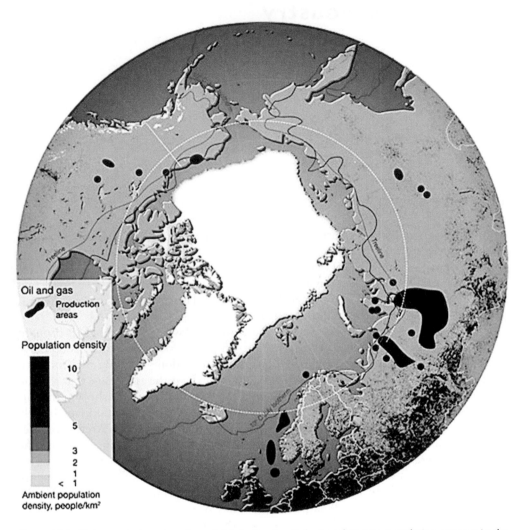

Figure 15.1   Main producing oil and gas fields in the Arctic basin relative to population centres in the Northern Hemisphere (Hugo Ahlenius, UNEP/GRID-Arendal).

Simpson, Alaska. Early exploration was carried out by dog-team and boat, and the first well was drilled near Norman Wells in 1924. Float planes helped further exploration, and the oil reserves were estimated using modifications of the methods used in non-permafrost areas to the south. The oil reserves of the North Slope of Alaska were not seriously examined until after World War II, when access was easier due to the building of the Alaska Highway.

Most modern exploration begins with seismic work to identify the underlying rock structure. It is mainly carried out on land in winter using winter roads along lines cut through the forest or across swamps and tundra. The ground must be snow or ice-covered and the underlying soil frozen to a sufficient depth to avoid serious damage to

Figure 15.2 Pipelines carrying gas to Western Europe from the gas fields north of Nadym, Western Siberia. The pipelines are in the ground but frequently appear at the surface through flotation and frost jacking in spring. Note the environmental damage in the foreground. © S. A. Harris.

the environment. Even so, scalping the tops of earth hummocks to place the geophones on a firm surface can do long-term damage (Adam and Hernandez, 1977; Adam, 1978), and cause the development of thermokarst. In North America, there have been precise regulations regarding the use of seismic trails since the early 1970s, but their abundance has seriously impacted wildlife habitat by cutting the undisturbed lands into small blocks. Less care appears to have been taken in Western Siberia, probably due to the much larger area being exploited. Lato *et al.* (2012) discusses the use of oblique photography from small drones to obtain data on slopes and geology in planning the optimum location of pipelines in mountainous terrain. This type of technology can also be used in monitoring.

These same winter roads can be used to take in small drilling rigs to confirm geological structures, obtain samples and carry out near-surface geotechnical site investigations. The latter involve determining ice content, stratigraphy of the surface layers, and the frost susceptibility of the materials. The best sites with frost-stable materials are chosen for deeper drilling, and this involves added expenses such as building suitable access roads and buildings, finding an adequate water supply, and choosing a suitable location for a large sump for the waste drilling fluid. Ground penetrating radar is used to detect icy layers.

15.3 DRILLING RIGS

The drilling rigs used in oil or gas exploration on land in North America are very large. It takes 50 large trucks to move a single rig after it has been disassembled. Gravel roads require regrading after the passage of every four loads in summer. Thus winter roads are preferred since they are cheaper and more environmentally friendly. The other alternative is to fly the drill rig and workers in by aeroplane, but this involves constructing a suitable airfield to handle the traffic. Air transport is therefore mainly used in exploratory work in new areas such as the Arctic Islands.

A major development in the last decade is the ability to drill at an angle and even bend the drill hole so as to drill horizontally (see Meyer, 1999). This angled drilling was first developed to enable several drill holes to be emplaced from a single offshore island, but the same technique can reduce the number of drilling pads needed to develop a given oil or gas field. Likewise, the number of access roads and pipelines is reduced, leaving the landscape in a more natural state. The use of fracking is usually limited to non-permafrost areas.

The basic requirements for drilling include construction of a stable pad which will not move during drilling (Metz et al., 1982; Metz, 1984). The drills use a circulating fluid to lubricate and cool the bit, so there must be a suitable water supply in a large reservoir. The size of the reservoir increases with increasing drilling depth, so that the final depth of drilling, together with any lateral or horizontal drilling to be done from the pad must be determined in advance. It is extremely expensive to increase the depth to be drilled. The reservoir must be excavated or blasted in the frozen ground as close to the drilling rig as possible, but thawing of its walls can cause the drilling pad to be eroded if it is in ice-rich sediment or excavated too close to the rig. It must include enough fluid to allow for any seepages that may occur. The soil cover and other excavated material must be stored to be used in filling and revegetating the infilled reservoir at the end of the work. The drilling fluid is normally water with various additives such as heavy barium sulphate (barytes) to provide enough weight in the drill hole to prevent blowouts and to reduce freezing. The sides of the hole must be sealed to avoid losses of fluids. Calcium chloride, potassium chloride or sodium chloride may be added to prevent freezing of the fluid in the upper layers of the ground, although this facilitates thawing of the sides of the drill hole if it is not protected from escaping.

On completion of the drilling, the drill and associated buildings must be disassembled and removed, and the sump filled in to minimize the environmental effects and restore the area to as close as possible to its original state. The fluid must be monitored even after it has become part of the permafrost. Likewise, the drilling pad and other ground disturbances must be revegetated as far as possible, as discussed by Steever (2001). A decision must be made as to whether the borehole is to be developed as a producing well, or be plugged and abandoned, when they are referred to as "keepers". The plugs are emplaced in such a way that they can be drilled out and the well put into production at a later date. All production wells will undergo far more long-term thaw subsidence than abandoned wells due to the heat escaping from the oil or gas. The oil leaves the ground at 40–80°C at Prudhoe Bay.

Offshore, drilling is accomplished from a suitable drilling platform. This may be a movable drill ship, an artificially thickened ice platform, or a man-made island.

The latter is expensive to build, easily damaged by ice pushing or large ice islands, and can only be economically constructed in shallow water (Harris, 1886a). As a result, ice islands are not currently being built. Water is readily available but must be stored on the drilling platform. Used drilling fluids are generally discharged into the sea.

15.4 PRODUCTION AND KEEPER WELLS

On land, drilling is preferably carried out in winter to protect the environment in North America. Anon. (1982, pp. 4.50–4.65) discusses the methods in drilling through the permafrost, together with the main problems encountered. Figure 15.3 shows schematically the special precautions taken for casing producing wells on land where permafrost is present in the upper borehole that is surrounded by permafrost to minimize thawing of the surrounding ground ice.

The first problem to be overcome is the presence of thaw-sensitive ground. Adequate pads must be placed under any roads and airstrips as well as under the actual drilling pad on which the rig stands. Often in North America, a coating of ice is placed over the existing vegetation and ground to protect it (Anon., 1982; Figure 4.4 to 4.6). Insulation is essential, and wooden pallets are laid around the buildings and other structures to protect the ground. The sump must be constructed and is often lined with both insulation and an impermeable membrane. The surface soil should be saved for the restoration process and the excavated material stock-piled for infilling the sump during restoration of the area. The temperature of the drilling mud is kept near freezing by cooling it as it returns to the surface. Often it is recycled. Special cements are used for cementing the upper casings to the wall of the borehole that set at low temperatures (Goodman, 1978b). The casing must remain intact, even if thawing and subsidence of the permafrost occurs (Goodman, 1978a). In offshore drilling, the salty sea water aids in thawing the permafrost around the well head and the platform above the well sometimes sinks as the sediment supporting it thaws (Goodman, 1977).

Since the oil or gas is warm, there will be thawing of the permafrost around the casings as the soil deforms (Goodman and Mitchell, 1978). This thaw subsidence produces stresses in the casings along with ground settlement. The designer must determine the properties of the permafrost before and after thaw, the radius of the thaw zone for the design life of the well, and the stress/strain properties of the casing. The design will only work if the casings can withstand the stresses from the thawing around them. If not, the casing and tube may need to be insulated, and there may have to be refrigeration of the well head (Goodman, 1978c). Where there are multiple wells drilled from a given pad, they will all tend to cause warming of the permafrost beneath it. Goodman et al. (1982) discusses the necessary calculations required to estimate the total warming to be expected in the case of offshore islands.

In the upper 1500 m of the borehole, gas hydrates may be encountered. The gas in the drilling fluid must be monitored for gasification. If there is an increase in gases in the fluid, there are probably gas hydrates in the wall of the borehole and the weight of the drilling fluid must be increased to stop the decomposition of the hydrates (Goodman and Franklin, 1982). Chilling of the drilling fluid and slower drilling must also be used to change the hydrate equilibrium conditions (see Figure 1.26) on the walls of the hole so that further decomposition is prevented. This is particularly important during

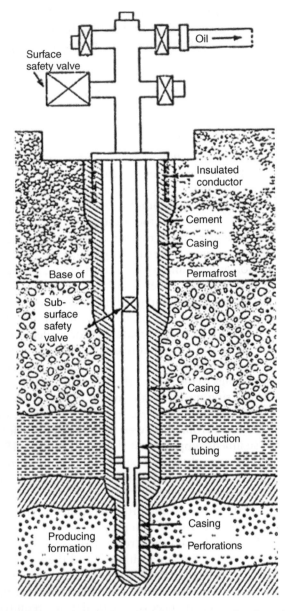

Figure 15.3 A schematic diagram showing the special precautions taken in casing production wells in permafrost (From Anon., 1982, Figure 4.4–6).

tripping and setting the well casing. The slower drilling produces less heat. Where the gasification occurs, the zone is cased in a cement collar at least 34 cm thick.

Problems may also occur with water under high pressure trapped beneath the permafrost. It may flow out of the borehole under artesian pressure and must be

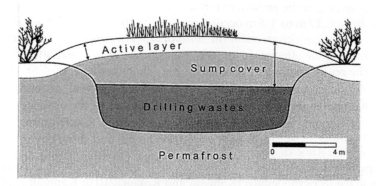

Figure 15.4 Diagram of the parts of a successful sump (from Jenkins *et al.*, 2008, with permission).

allowed to do so until flowage ceases. If necessary, a second relief well may need to be drilled to keep the pressure low enough that drilling can proceed to greater depths.

15.5 SUMP PROBLEMS

Figure 15.4 shows the ideal cross-section of a successful sump (Dyke, 2001; Jenkins *et al.*, 2008; Kokelj *et al.*, 2010). The top of the drilling mud should be significantly lower than the permafrost table in the surrounding ground, and there should be a gently sloping dome of fill capping the mud and extending laterally on to the surrounding ground. Initially, it was assumed that permafrost was impermeable.

Early research by French (1978, 1980) indicated that there were problems showing up in 20–25% of the wells drilled in northern Canada by the mid-1970s. These included underestimating the volume of drilling fluid, decisions to increase well depth during drilling resulting in the necessity to build additional sumps, and thawing of the ground around the sump that can engulf the drilling rig. There were also problems with leakage of drilling mud into streams, as well as slope failures, *e.g.*, Caribou N-25, Peel Plateau, Yukon Territory. These problems were largely overcome by lining and insulating the sumps prior to use.

The performance of these old sumps on the Mackenzie Delta has been re-evaluated since 2001, when Dyke reported contaminants from the sumps appearing in the surrounding ground and streams (Dyke, 2001; Kokelj and GoNorth Ltd., 2002; Kokelj and Kanigan, 2010; Kokelj *et al.*, 2010). Permafrost can provide containment for drilling wastes provided that the entire body of drilling mud in the sump remains frozen. Unfortunately in areas of permafrost near tree line, the heaving of the overlying fill and drilling mud during freezing steepens the slope of the dome, causing increased snow accumulation down-wind, which results in increasing ground temperatures and active layer thickness. In modelling simulations under conditions of warm (−3°C) and cold (−6°C) permafrost, the snow normals for Inuvik (south of tree line) and Tuktoyaktuk (Coastal Tundra) maintained the sump wastes at ground temperatures between −1.5°C in warm permafrost and −3°C in cold permafrost. However the climate is warming around the oil facilities at a rate of about $0.09°C/a^{-1}$, and there is a natural sere ending in the growth of tall shrubs over the sumps. A gradual increase in snow depth from 0.17 m to 1.5 m occurs, resulting in the change in vegetation producing thawing of the sump contents in the third decade. In contrast, the effect of climate warming, omitting the effects of vegetation changes, was to commence thawing after 35 years in the region of warm permafrost, and to warm the ground to –2°C after 40 years in the area currently consisting of cold permafrost. Vegetation changes, however caused, result in more rapid and far-reaching changes in ground temperature than the current rates of climate change.

There are over 150 legacy sumps on the Mackenzie Delta (AMEC Earth and Environmental, 2005). Time of year in filling the sump had no effect on the performance of the sumps, but thawing results in the accumulation of salts in the thaw depressions around the margins of the sumps (Kanigan and Kokelj, 2010; Kanigan *et al.*, 2010).

The implication of these results is that the storage of drilling muds is not environmentally safe in areas of tundra vegetation underlain by permafrost in areas with ground temperatures currently above –5°C. Alternative disposal/storage methods need to be devised to replace the use of sumps in these areas.

15.6 PIPELINES

Having brought oil or gas to the surface, it is necessary to move it to a central processing plant and on to the consumer. This is normally done by using pipelines. Figure 15.5 shows some of the proposed pipelines in Canada, prior to 2011, and more have been proposed since then. Oswell (2011) provides a good discussion of the methods and

Figure 15.5 Pipeline routes proposed for construction in Canada by 2011 (J.D. Mollard and Associates, Ltd., in Oswell, 2011, Figure 2).

importance of carefully choosing a pipeline route. Both buried and elevated pipelines can be used, as explained below.

15.6.1 Buried mode

This is the most common method of transporting oil and gas across the land, particularly in Russia (Figure 15.2). It is also used on the smaller projects in areas of North American permafrost, e.g., the Norman Wells pipeline. There are one hundred and fifty thousand kilometres of pipelines in Russia, where buried gas and oil pipelines have been constructed and operated in permafrost areas since the 1960s. Underground pipelines there are considered economically reasonable, protected, and safe by GASPROM.

Trenching is carried out in three ways, viz., backhoes, bucket wheel trenchers and chain trenchers (Hayley et al., 1984; Blanchet et al., 2002). Table 15.1 summarizes their advantages and disadvantages, as determined along the Arctic Pipeline Project in northern Canada (Oswell, 2011, Table 5). The bucket wheel ditching rate of excavation in the vicinity of Norman Wells is strongly influenced by the ice content (Hayley et al., 1984) as well as both the soil type and terrain type (Saunders, 1989).

Failures of the underground pipelines are more frequent than those pipelines constructed above ground on piles, for example in the Norilsk area. It was found that the thawing of soil under the influence of a thermal flow from the gas pipeline would result in loss of the bearing capacity of the soil. Surrounding areas are also disturbed, thus the interaction between pipeline and the underlying permafrost is the reason for the widespread damage to pipes in both Russia (Čigir, 1977; Čigir et al., 1997; Remizov

Table 15.1 Summary of the advantages and disadvantages of the main Arctic trenching methods (after Oswell, 2011, Table 5).

Trenching method	Advantages	Disadvantages
Backhoes	Relatively independent of terrain and soil conditions. Can trench around corners and bends.	May require pre-treatment of ground (ripping, blasting). May not trench bedrock. Rough and irregular trench geometry. Spoil needs processing prior to backfilling. Moderate depth range (to approximately 5 m).
Bucket wheel trenchers	High productivity in uniform fine grained soils. Can trench soft bedrock. Very uniform trench geometry. Spoil can be used for backfill without processing. Productivity increases with ice-content.	Low productivity or restricted use in soils with cobbles and boulders greater than 200 mm diameter. Trench only in straight lines. Sallow to moderate depth range (approximately 4 m).
Chain trenchers	High productivity in some soils and bedrock. Can trench most bedrock. Very uniform trench geometry. Spoil can be used for backfill without processing. Trench depths to 7.5 m. Higher forward velocity than bucket wheel machines.	Trench only in straight lines. May have lower productivity in soft or ductile soils.

Figure 15.6 Deformation of the uninsulated, high-pressure gas pipeline in the Medveje gas field. © Gennady Griva.

et al., 2000; Seligman, 2000) and North America (Rowley *et al.*, 1975; Nixon *et al.*, 1983; Carlson & Butterwick. 1983; Burgess *et al.*, 1993; Razaqpur and Wang, 1996). New projects like the Bovanenkovo-Ukhta gas pipeline include chilled gas pipelines; however, this may also cause problems with river crossings and taliks. Statistically, about 40% of all failures occur during freezing (Kamensky, 1988) in the autumn and early winter (Figure 15.3) in Eastern Siberia. However, in Western Siberia, 65% of all failures occur between July and September, with another 10% occurring in June.

An example would be the 120 km long high pressure gas pipeline with a diameter of 1420 mm on the Medveje gas field that was constructed on discontinuous permafrost in the south, and on continuous permafrost in the northern area. The temperature of the frozen soil varies from 0°C in the south down to −3.0° to −5.0°C in the north, and ice content of frozen soil changes from 0 up to 30–50% by weight and more. The pipeline was not insulated, and areas of thawing developed with depths of 5–6 to 13 m. Actual vertical displacements of the pipe reached 5–10 m (Figure 15.6).

Oil or gas can be transported at warm (>0.0°C) or cold (<0.0°C) temperatures. Either way, there will be problems in buried pipelines. In the case of warm oil or gas passing through a pipeline crossing blocks of icy permafrost in otherwise unfrozen

Profile of Warm Pipeline Crossing Frozen Span

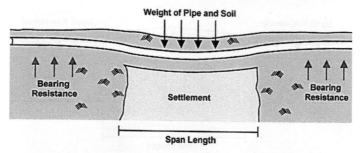

Cross-Section of Warm Pipe in Frozen Soil

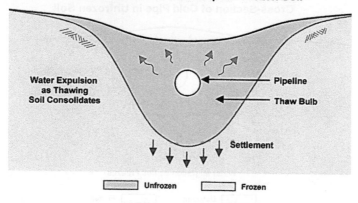

Figure 15.7 Forces acting on a warm pipe buried in discontinuous permafrost or in continuous permafrost.

ground, the heat from the pipe causes the development of a thaw bulb in the permafrost (Figure 15.7). As the ice thaws, the ground sinks due to the volume change resulting from ice melting to form water. The water decreases the bearing strength of the ground, and some of it may drain away causing further subsidence, and sagging of the pipeline. Where the warm pipe passes through permafrost, a thaw bulb develops around the pipe, also inducing subsidence. The resultant water, often combined with surface runoff, tends to form a wet depression (Figure 15.6). In both cases, the weight of the pipe, the internal fluid and the soil overburden push the pipe down. The stiffness of the pipe provides some resistance.

Where a cold pipeline is involved (Figure 15.8), any patches of unfrozen ground are subjected to additional cooling, resulting in freezing. Moisture moves in thin films over the surfaces of the grains to the coldest area adjacent to the pipe, where it freezes. Ice lenses form just behind the freezing front and the resulting expansion on freezing causes upward heaving of the overburden and the pipe. This movement is resisted by the frozen soil around the pipe in the adjacent span (Nixon and Hazen, 1993), but it can cause

Profile of Cold Pipeline Crossing Unfrozen Span

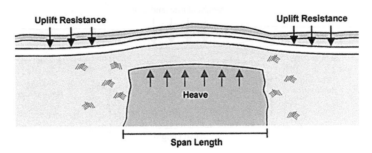

Cross-Section of Cold Pipe in Unfrozen Soil

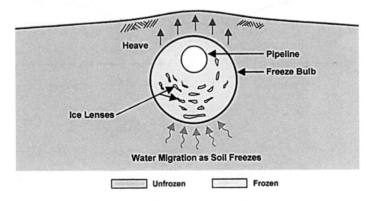

Figure 15.8 Forces acting on a cold pipeline crossing patches of unfrozen ground or buried in unfrozen sediments.

floating pipelines (Figure 15.9). The length of floating section of the Urengoy-Nadym-Punga gas pipeline is about 150 km out of a total of 250 km. The extent of emerged sites of gas pipelines belonging to Gasprom has reached many hundreds of kilometres, and the annual increase of the length of floating gas pipelines is the effects of tensile strength, creep and modulus of deformation on the development of floating pipelines. An additional complication to both interpreting field observations and geothermal modelling is that of thaw consolidation and generation of excess pore-water pressures in the thawing soil. The thaw consolidation theory was first described by Morgenstern and Nixon (1971) and Nixon and Morgenstern (1973). When thawing ice-rich soils at a rapid rate, excess pore-water pressures are generated that can weaken the soil and induce instability on slopes, cause additional buoyancy forces around pipelines, and weaken subgrades under foundations or backfill over pipelines. Figure 15.10 provides an example of excess pore-water pressure generated as a result of thawing of ice-rich permafrost. It is based on a surface insulating layer of 0.76 m of woodchips, a mean annual temperature of c. $-1.0°C$, and climate warming at a rate of $0.08°C/a^{-1}$. This is typical of conditions along the Norman Wells pipeline.

Figure 15.9 Gas pipeline that once was buried and now is floating in a marsh, resulting partly from thawing of the permafrost. © Gennady Griva.

Where the calculated factor of safety for a particular situation is less than desirable, some form of mitigation is necessary. This may mean altering the slope angle, installing thermosiphons, or using insulation to slow down the thawing, so reducing the pore-water pressure to more acceptable levels. Hanna and McRoberts (1988) discuss this in more detail. As usual, moisture tends to move downslope, making the problems greater on the lower lands.

Several kinds of insulating materials were tried for the Norman Wells pipeline (Table 15.2). Straw bales were by far the most successful, though they rot over time. Wood chips were the chosen insulation for the pipeline because of their ready availability, but they provided rather limited protection. The thermal resistance (R_T) is the ratio of insulation thickness to its thermal conductivity. Similar results were obtained from a test site north of Wabasca, Alberta (Oswell, 2011). There was relatively small variation in degree of thawing that was related to slope aspect (Burgess *et al.*, 1995). Figure 15.11 shows the results of modelling long-term ground temperatures along the Norman Wells pipeline with the use of different insulation R_T values and no surface insulation where the initial ground temperature was $-2°C$ with a climate warming rate of $0.08°C/a^{-1}$.

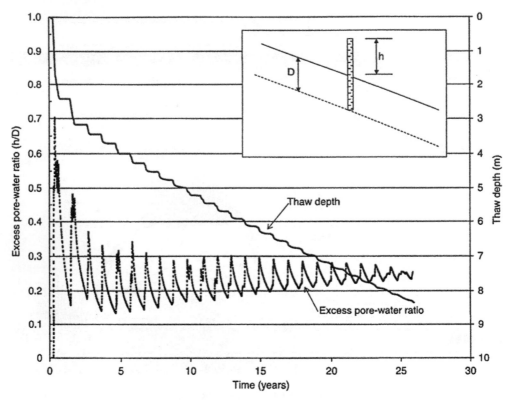

Figure 15.10 Modelled excess pore-water pressure generated as a result of thawing of ice-rich permafrost at a location along the Norman Wells pipeline (Oswell, 2011, Figure 8). D is the thaw depth while h is the pore pressure head above ground surface.

Table 15.2 Comparison of different insulation materials tested for thermal resistance for the Norman Wells pipeline.

Material	Thermal conductivity (W/m°C)	Thickness (m)
$R_T = 1.1\ m^2 \cdot °C/W$		
Extruded polystyrene	0.03	0.03
Wood chips	0.265	0.3
Straw bales	0.1	0.11
$R_T = 1.7\ m^2 \cdot °C/W$		
Extruded polystyrene	0.03	0.05
Wood chips	0.265	0.45
Straw bales	0.1	0.17
$R_T = 2.9\ m^2 \cdot °C/W$		
Extruded polystyrene	0.03	0.09
Wood chips	0.265	0.76
Straw bales	0.1	0.29

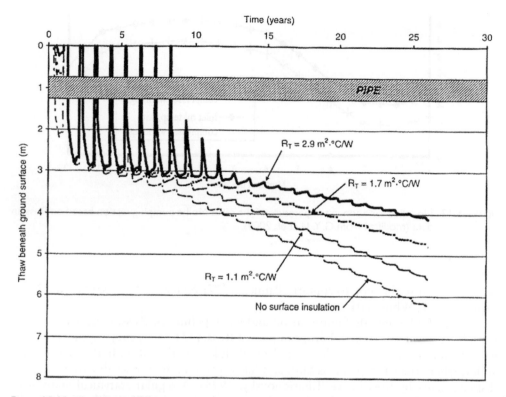

Figure 15.11 Modelling of Geothermal changes along the Norman Wells pipeline right-of-way showing the effects of different insulation values, an initial ground temperature of −2°C and a climate warming rate of 0.08°C/a⁻¹ (from Oswell, 2011, Figure 7).

Where there is a choice between dealing with frost heave or thaw subsidence, designers usually choose to confront thaw settlement (Oswell, 2011). Designers also try to route the pipeline around geological hazards rather than trying to mitigate problems. Nixon *et al.* (1991) found that the number of frozen/unfrozen interfaces found by geophysical methods greatly exceeded the number identified by visual methods along the Norman Wells pipeline project.

Particular problems are encountered in all areas of continuous permafrost where the pipeline crosses large rivers that have a talik beneath them (Meyer, 1999). This is a case of the pipe going from permafrost to talik and back again in a short distance. A buried chilled pipeline must be placed low enough to avoid damage when the river deepens its channel during spring runoff. Meanwhile, the pipeline tends to rise due to formation of ice lenses (Figure 15.8). In these cases, the pipeline is often brought out of the ground and suspended on suitable support structures above the river.

In both Figures 15.7 and 15.8, variables determining the amount of expansion include depth of freezing, moisture content, grain size of the sediment, temperature gradient and soil pressure. Very considerable pressures may be developed on the pipe

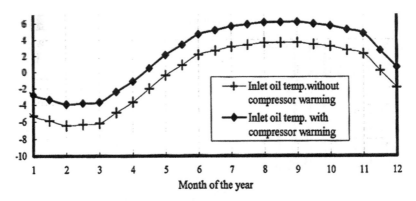

Figure 15.12 Modelled variation in temperature of the oil at Mo'he pumping station based on Russian data (modified from G. Y. Li *et al.*, 2009).

which can cause the pipe to rupture (Palmer, 1977; Nyman, 1983; Kim *et al.*, 2008; Oswell, 2011; White, 2013).

About 50 large-scale failures at oil and gas pipelines in Russia occur every year (Seduh, 1993). In the North European part of Russia, 21 emergency ruptures of main gas pipelines have been recorded since 1993 (Salukov *et al.*, 2000). In the Yenisei river valley, where the gas pipeline is buried, it has ruptured a few times due to 50–60 mm displacements by frost heave (Kharionovsky, 1994). A regular statistical analysis of failures of main pipelines in Russia was begun at the end of 1970 (Sedih, 1993). In the north part of Western Siberia, about major 65 failures were recorded during the last 20 years. During the same period, about sixty-five failures occurred on the permafrost pipelines Taas-Tumus-Yakutsk and Mastakh-Yakutsk (Konstantinov and Gurianov, 2001). The gas pipelines with a diameter of 1020 mm have the worst incidence of failure. This agrees with the experimental data of Oswell (2011, Figure 15) which demonstrates that pipeline buoyancy on permafrost consisting of thawing icy clay increases in frequency with increasing diameter of the pipe. Uplift or buckling of the pipe can be the result of temperature changes. When the pipe is installed at −35°C, and then operated at +5°C, the steel pipe will undergo thermal expansion due to the 40 degree change. For steel with a thermal expansion coefficient of 11×10^{-6} m/m/°C, a 500 m segment of the pipe would lengthen by 0.22 m. If there is insufficient restraint supplied by the backfill, the pipeline will expand upwards through the backfill (Palmer and Williams, 2003). Examples have been reported from small diameter oil pipelines on the Qinghai-Tibet Plateau (He and Jin, 2010) and in northern Alberta (Oswell *et al.*, 2005) during the summer months. The TransAlaska fuel pipeline has developed similar upheaval buckling problems (Nixon and Vebo, 2005). Elsewhere, the development of a frost bulb around a chilled pipeline would provide additional rigidity that would tend to prevent this happening.

Part of the complexity of planning and designing an oil or gas buried pipeline is illustrated by the summary of the thermal characteristics of the oil and the surrounding ground along the China-Russia Crude Oil Pipeline (CRCOP) that is designed to transport Siberian crude oil from Skovorodino in Russia via Mo'he to Daqing in China (G. Y. Li *et al.*, 2009). The 953 km route within China crosses a wide variety of terrain including extensive areas of permafrost differential frost heave and thaw subsidence, frost mounds and icings. Even during the last Neoglacial event, the amount of cooling of the mean annual temperature in the region is believed to be 4°C compared with 2°C in southwest Alberta. This is probably due to the core of the Siberian winter high pressure centre moving southwards during this event, producing colder ground temperatures and the survival of ice wedges at Yitulihe (see chapter 5, Figure 5.9). The proposed range of oil temperatures within the pipeline was originally planned to be rather variable between –6 and 10°C, due partly to lack of control of the heating during compression at the pumping stations, as well as friction as the oil moved along the pipe.

After changing its plan, the PetroChina Daqing Oilfield Engineering Company (2009) proposed using an 813 mm diameter pipe carrying 300,000 barrels of oil annually. This allows for an estimated seasonal range in oil temperature of –6.41 to 3.65°C at Mo'he, as estimated by Russians (Figure 5.12). This range is part of the cause of frequent failure of the oil pipelines in Western Siberia. Isolated patches of talik occur in the north along about 441 km of the total length of the Chinese section of the China-Russia Pipeline. The seasonally frozen southern section is 552 km long. The pipeline will face formidable challenges from differential frost heave and thaw subsidence, frost mounds and icings.

G. Y. Li *et al.* (2009) examined the estimated thaw depths using the new design in Figure 5.13, assuming the temperatures of the oil shown in Figure 5.13 were correct. In Figure 5.13A, the authors modelled the change in oil temperature after both 1 year and 50 years along the pipeline assuming an annual flow of 1.5×10^7 tons (0.3 mbpd) using the combined heating effect of the compressor stations and friction along the pipe. At the Jiageaqi compressor station at Km 403, the pipeline leaves the permafrost and the temperature of the oil in the pipe increases by 2°C, so that the oil temperature is always above 0°C, and frost heave of the pipe becomes unlikely until its arrival at its final destination. The effect of friction on the temperature of the oil in the pipeline is shown in Figure 5.13B. The heat generated by friction significantly alters the temperature of the oil.

In Figure 5.14, the results of modelling various other attributes of the changes in ground temperature are depicted for a 50 year time span. Note that the maximum freezing depths beneath the pipeline in Figure 5.14C and 5.14D decrease with time and increasing water content of the soil. Comparable estimates of depth of thaw with time for the TransAlaska Pipeline are shown in Figure 5.15.

15.6.2 Pipelines on piles

In parts of the Norilsk region of Russia and along the TransAlaska Pipeline, the pipeline is carried on piles as far as possible when crossing permafrost. The most widely documented project is the TransAlaska Pipeline, details of which are discussed in Allen (1977), Alyeska Pipeline Service Company (1977), Roscoe (1977), P. J. Williams (1979) and Metz *et al.* (1982). This pipeline is important since it is the first major pipeline to successfully avoid many spills, and so stands as a model for this type of construction. The reason this elevated type of construction is not widely used is that its initial cost

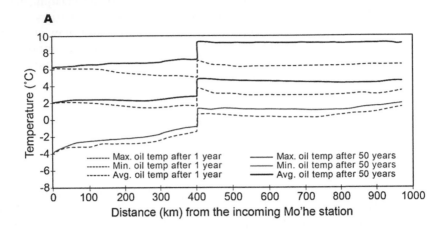

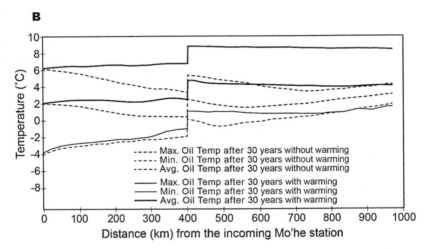

Figure 15.13 Modelled variation in temperature of the oil along the pipeline with an annual flow of 1.5 × 10⁷ tons, A, including the effects of the compressor stations and the frictional heat; B, with and without the effects of the frictional heat in the 30th year based on Russian data (modified from G. Y. Li et al., 2009).

is greater than the buried mode. However, it is far more environmentally friendly, and when properly monitored and maintained, has proved to perform well even after its designed life span is passed.

The pipeline was built between 1969 and 1977 and extends for 1,300 km. There was an eight-fold increase in costs between the start of construction and its completion, the final cost totaling over $7 billion. The pipe has a diameter of 1.22 m with a wall thickness of 1.3 cm. Four thousand drill holes were made in 1968 so as to delineate the areas of icy permafrost, and a further 2,000 were drilled in critical areas during construction. The pipeline was to be built on piles over permafrost wherever possible (70% of the route), while it was buried in ice-free areas. During trenching

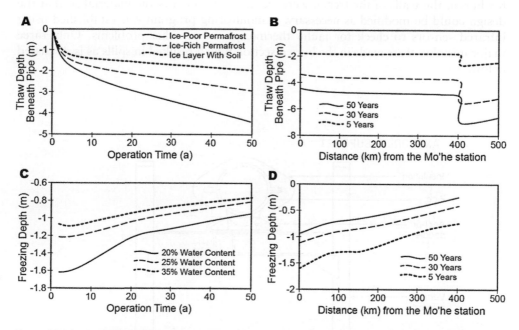

Figure 15.14 Modelled results of A, variation of the maximum thaw depth beneath the pipeline in warm permafrost with different ice contents over time near Mo'he station; B, the variation of the maximum thawed depth along the pipeline with time; C, change in maximum freezing depth at different moisture contents beneath the pipeline in taliks near Mo'he Station; and D, the variation in maximum freezing depth along the pipeline with time (modified from G. Y. Li et al., 2009).

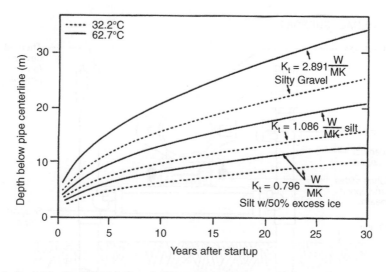

Figure 15.15 Probable thaw depth below the TransAlaska Pipeline in buried mode as a function of time, temperature of the oil, and thermal conductivity of the ground (K_t) (after E. R. Johnson, 1981; see Harris, 1986a, Figure 7.16).

for burial, the walls of the trench were examined to identify the material so that the design could be modified as necessary. A monitoring program was established using infrared sensors to check for faulty thermosiphons or other problems. Only three major mistakes were made in the buried section which resulted in spills as is discussed below.

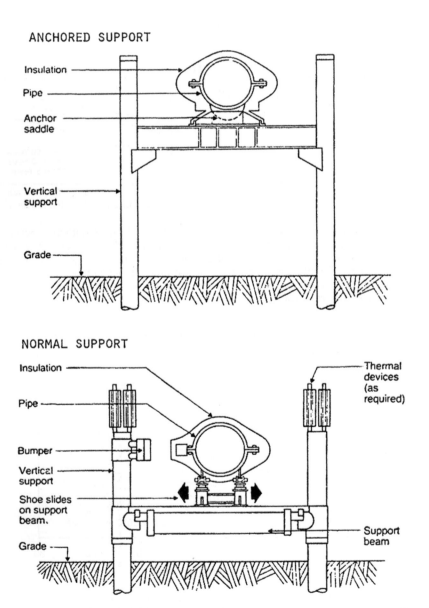

Figure 15.16 Types of vertical support members used for the elevated sections of the TransAlaska Pipeline (after Alyeska Pipeline Service Company, 1977; see Harris, 1986a, Figure 17.15).

15.6.2.1 Design parameters

Over permafrost and thaw sensitive ground, the pipeline is carried on vertical support members (VSM's, see Figure 5.16). The pipe is covered with 10 cm of resin-impregnated fiberglass insulation inside a galvanized steel jacket to keep the oil warm during any shutdowns, as well as minimizing heat loss. Typical thaw-stable materials along the route were dense, frozen, clean sand and gravels and frozen competent rock. However, most surficial materials were ice-rich silts and clays containing up to 70% ice. The thaw stable sands and gravels were defined as having less than 6% water by weight, passing the ASTM 200 sieve (0.074 mm diameter), and having a dry density greater than 1.92 gm/cc.

The VSM's are spaced at either 22 or 25 m apart. On the anchored support which is placed at every 245–550 m, the pipe is clamped on the cross beam which is usually supported on four piles. Thermal piles are used in pairs wherever there is a risk of thawing of the underlying permafrost. The supporting pipes are placed in the drilled holes which were then backfilled with a sand-cement slurry. Artificial refrigeration was used to refreeze the thawing ground and slurry, and then the thermosiphons were installed to replace the mechanical refrigeration system. Only 0.6% of the 122,000 thermosiphons settled more than 9 cm. Blocked thermosiphons were replaced. Corrosion sometimes produces hydrogen inside the thermal piles, stopping the heat radiation in winter. By using the piles in pairs, if one failed, the other still worked. Infrared scanning in winter from a helicopter shows the defective piles as dark instead of red. Design life of the piles was 30 years compared with 25 years for the pipeline. However, the pipeline is still functioning satisfactorily after 38 years with careful maintenance. To allow for expansion and contraction, the pipeline zig-zags with flexible bends (Figure 15.17).

Figure 15.17 The bends in the TransAlaska Pipeline to allow for expansion and contraction.

Figure 15.18 A typical example of the pipeline being carried across a river on piles in Alaska. © S. A. Harris.

The normal supports use a shoe that can slide back and forth as necessary to adjust the pipeline length. Over the Denali fault line, the cross beam is much longer to deal with movements caused by earthquakes (6 m horizontal movement and 1.6 m vertical movement).

15.6.2.2 Construction methods

The construction was carried out in winter so as to minimize environmental change and to provide maximum strength as quickly as possible after construction. A 25 cm natural gas pipeline was added along the northern 200 km to provide fuel to pump stations #2, 3, and 4 to push the oil over the Atigun Pass across the Brooks Range. A haul road was constructed north of Goldstream to provide access for pipe and equipment (Metz, 1984). It consisted of 2.5 m of thaw-stable gravel laid over 5–10 cm of board insulation in winter. Revegetation of the disturbed area was begun immediately (Johnson, 1981). Two roads were constructed, one for hauling freight and one alongside the actual pipeline.

15.6.2.3 Failures in the buried section

Two failures occurred in the buried section of the pipeline. At Mile 166 in the Atigun Pass, a narrow valley on the north-facing slope at the head of the Pass made it difficult to use the raised mode of construction due to snow avalanches. A buried mode was therefore employed with an extra 50 cm of high density polystyrene insulation, but the pipeline was partly located in ice-rich bedrock (Figure 15.19). As a result, the pipeline

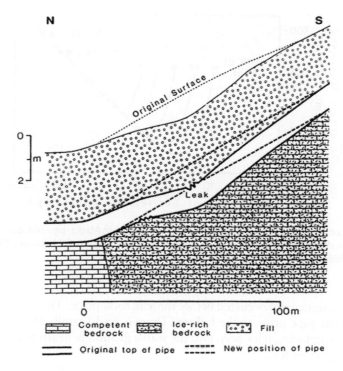

Figure 15.19 Conditions at the time of failure of the TransAlaska Pipline at mile post 166 on the North Slope of Atigun Pass, Brooks Range, Alaska (after E. R. Johnson, 1981; see Harris, 1986a, Figure 17.17).

sagged as the ice thawed, resulting in failure in June, 1979 after twenty-four months of operation. Sixty boreholes were drilled around the leak and the surface of both the ground and the pipe were relevelled during the excavation. Thaw settlement of up to 1.22 m had occurred and resulted in subsidence of both the ground surface and the pipe along a 122 m length. It was also discovered that the pipe was laid on bedrock with ice masses up to .3 m thick in it, showing a visible ice content of 50–90% by volume. This occurred at 3–7 m below the ground surface. Below 31.8 m, the rock was ice-free. Maximum subsidence designed for was 0.91 m, so the pipe had performed as expected. The leakage took place on the upper buckle with a bending of 3°.

Subsequent drilling in 1979 in a similar area on the south side of the Pass revealed excessive thawing, thaw settlement and ground water flow. Surface water from the crest of the Pass was disappearing underground, much of it draining through the trench containing the pipeline. This was solved by better surface drainage and grouting, followed by mechanical refreezing of the ground (Stanley and Cronin, 1983).

At Mile post 734, a leak developed a week later at the southern limit of sporadic permafrost. Although the ditch digger reported "thawed materials" in the trench bottom in May, 1975, drill-holes after the leak showed that frozen, highly unstable permafrost existed along a 91 m length of the pipe. There was up to 75% visible ice present,

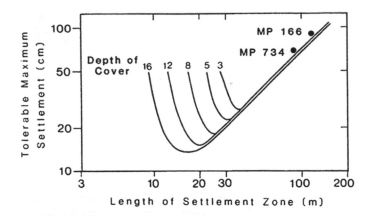

Figure 15.20 Maximum subsidence designed for along the TransAlaska pipeline compared with the conditions of failure at mile posts 166 and 734 (after E. R. Johnson, 1981; see Harris, 1986a, Figure 17.18).

and over 2 m of settlement occurred before the leak took place. Thermokarst had developed around the pipe prior to the failure (Thomas and Ferrell, 1983). The problem was solved by emplacing piles deep into the stable ground beneath the permafrost and mounting the pipe on top.

15.7 MONITORING

After the failures in 1979, a regular program of monitoring of the pipeline was begun, since this is cheaper than making repairs. Settlement rods were attached to the buried pipe at 1,100 locations and extra boreholes were made to check out the soil conditions. Regular levelling of the pipeline and rods is used to detect any embryonic problems, while regular infrared photography of the thermosiphon vanes is used to detect any that fail. These can be changed as necessary, but otherwise, when major problems are detected by monitoring, the pipeline is shut down and a new loop constructed, *e.g.*, at the Dietrich River crossing.

15.8 COMPRESSOR STATIONS

In the oil and gas fields, compressor stations collect the fluid from the wellhead via local pipelines and process it. Since gas contains water, this must be removed before the gas is compressed so that it can readily flow along the main pipeline to the consumer. Failure to remove the water may result in blockages in the pipe due to the formation of gas hydrates (Hammerschmidt, 1934). Compressor stations are located at intervals along the main pipeline in order to adjust its temperature and to keep the gas or oil moving. The main collecting station may also separate the fluid into its main components

Figure 15.21 The successfully repaired section of the TransAlaska Pipeline in Atigun Pass using closely spaced thermosiphons to keep the buried pipe from thawing the surrounding permafrost. © S. A. Harris.

(sweet gas together with sour gas consisting of hydrogen sulphide and sulphur dioxide) and one or more oil fractions. These are then sent in separate pipelines to the consumer.

In North America, compressor stations are usually built on reinforced concrete slabs that are cooled by artificial refrigeration. At all the compressor stations, the temperature of the fluid is adjusted after compression to a carefully selected value

Figure 15.22 A major compressor station and gas processing plant, Yamburg, Western Siberia. © S. A. Harris.

for which the pipeline is designed. This prevents too rapid thawing of the adjacent ground by heat loss from the fluid in the pipe in transit. Gas is used as a fuel for the compressors.

In Siberia, the main period of construction of the compressor stations was in the 1970–90 period. Fifteen percent of them have been working for more than 25 years. More than 30% of the compressor stations need to be refurnished in Western Siberia.

Figure 15.22 shows a typical compressor station in the middle of a gas field. The underlying ground is cooled using modified thermosiphons of the type shown in Figure 13.20. These thermosiphons are used liberally under the buildings that are raised off the ground on piles. Unfortunately, there is still some significant frost jacking of these piles since the construction took place before the ground had fully cooled. The compressor stations collect the gas from the pipelines coming from the well heads, remove the water in the gas and compress the gas so that it flows along the main pipeline to the next compressor station. Seligman (1998, 1990) noted that there was no temperature control and regarded this as being one of the main causes of the flotation problems of the pipeline in the area. Now, the pipelines have monitoring systems to warn of any temperature changes in the ground.

Figure 15.23 shows the variations in temperature along the Mackenzie pipeline in the Northwest Territories with only five compressor stations when it was brought into use, contrasted with 15 compressor stations when the project was complete. Compression of the gas heats up the fluid which cools as it moves along the pipeline. Cooling needs to be applied at the compressor station to keep the heating within acceptable limits, and the closer the spacing of the compressor stations, the better the control of the range of temperatures of the gas as it moves along the pipeline.

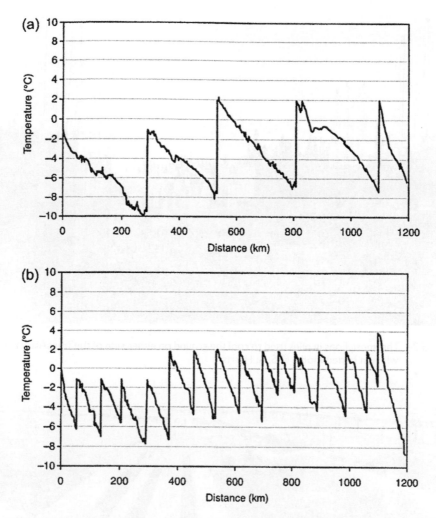

Figure 15.23 Pipeline temperature profile proposed for the Mackenzie gas pipeline (modified from Colt-KBR, 2003; see Oswell, 2011, Figure 5). (a) represents the temperature profile along the pipeline when it started with only five compressor stations. (b) represents the temperature profile at full capacity using 15 compressor stations. Note the drop in temperature between stations due to Joule-Thomson decompression cooling.

15.9   PIPELINE CROSSINGS

Where roads have to cross the Alaska pipeline, two different methods are used. Along the main pipeline, the pipeline may be buried with an array of thermosiphons around it to prevent excessive thawing of the underlying permafrost (Figure 15.24). Where collector pipelines are involved in the oilfield, the pipeline is placed within a culvert that is then covered in gravel (Figure 15.25).

Figure 15.24 Thermosiphons in piles used to cool the ground on either side of the TransAlaska Pipeline when in buried mode. © S. A. Harris.

Figure 15.25 Production pipelines going through larger culverts so that air surrounds them to avoid thawing the permafrost, Prudhoe Bay, Alaska. The gravel cover permits vehicles to cross the pipe. © S. A. Harris.

15.10   EFFECTS OF HEAT ADVECTION FROM PRODUCING WELLS

When oil or gas is brought to the surface from a producing well, it has the temperature of the producing horizon in the ground. In the case of the wells on the North Slope of Alaska, this averages over 40°C, but the mean annual temperature is only −12 to −15°C. At Prudhoe Bay, Alaska, over 17 trillion barrels of oil have been brought to the surface and pumped through the Trans Alaska Pipeline in the first 38 years of operation (Trans-Alaska Pipeline System, 2013). This represents an enormous amount of heat energy that is dissipated into the surrounding environment. A similar situation occurs in the oil-producing areas of Western Siberia.

Considerable precautions are taken to minimize the effects of this energy on the ground temperatures, but no attention was paid to the air temperatures around the oil facilities, nor to the possible effects on the adjacent sea. As soon as the Trans-Alaska Pipeline came into use in 1977, the mean annual air temperature along the pipelines and at facilities at Prudhoe Bay started to rise, while the ice cover on the adjacent Arctic Ocean started to retreat (Harris, 2016b). Romanovsky *et al.* (2014) have monitored the changes alongside the production facilities, which are small along the vicinity of the actual pipeline, but close to the pumping station at Prudhoe Bay, the mean annual air temperature rose 3–4°C until stabilizing after 2003 when the output had declined considerably. The rate of retreat of the Arctic ice cover increased exponentially before reaching a quasi-stable state after 2008. The mean annual air temperatures on land away from the oil operations show no evidence for similar warming. Provision of a method of utilizing this advected heat energy could have decreased or avoided this damage to the adjacent environment. A similar advection of heat must be occurring in Western Siberia that will be causing warming of the region adjacent to both the producing wells and the oil field facilities.

15.11   GAS HYDRATES IN PERMAFROST ICE

Decomposition of organic matter in cold climates produces both oil and gas. The composition of the gas varies considerably and can be divided into sweet and sour gas. The sour gas consists predominantly of both sulphur dioxide and hydrogen sulphide. These can be collected and used to produce sulphur, though this is mainly carried out as an economic proposition using the gas near transportation centres. Since the molecules of sulphur dioxide are relatively large, they are not usually found as major components of gas hydrates, but tend to occur as pockets of gas in the ground.

The gases in sweet gas that are used as a fuel or for making plastics, consist of much smaller molecules, *i.e.*, predominantly methane, together with propane, butane, carbon dioxide and hydrogen sulphide. These can also occur as pockets of gas, but they often form inclusion compounds with ice in which the small molecules are enclosed in the nearly spherical cavities in the crystal lattice of the ice. They are commonly referred to as *gas hydrates* (Davidson, 1973), although technically, they are called *clathrate hydrates*. Actually they are solid solutions of gas in solids (Englezos, 1993), and are only stable under specific conditions of low temperature and high pressure, *e.g.*, Figures 1.26 and 15.26 show the stability of the methane-water-gas hydrate system. The first realisation that they could occur in the ground was by Strizhez in 1946 (Makogon,

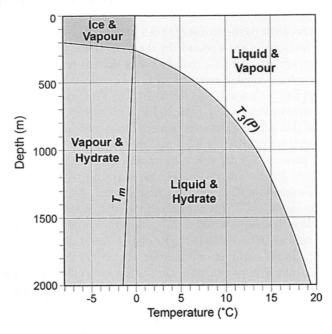

Figure 15.26 Stability of the methane-water-gas hydrate system with pressure plotted as an equivalent depth assuming a hydrostatic gradient of 10^4 Pa m^{-1} (after Buffett, 2000).

1982), and this proved correct in the case of the blowout of the Markhinskaya well in Northwest Yakutia in 1963. Since then, gas hydrates have been found at the shallower depths in permafrost throughout the Arctic and on the Qinghai-Tibet Plateau (Wu *et al.*, 2015), see Figure 15.27, and are also known from similar high pressure – low temperature environments of the ocean depths along continental margins by the Lamont-Columbia group (Figure 15.28). They also occur in Martian ice caps (Miller and Smythe, 1970).

Clathrate hydrates can store large volumes of gas. Both ice and water can store gases such as methane (the most abundant gas), but it is the ice that can store enormous quantities of these gases. The actual amount that can be stored depends on the ice structure, with types I and II having two cavity sizes and type H having three sizes (Ripmeester *et al.*, 1987). Structure I is the most common type, but structure II and H can store larger molecules as well (Sassen and MacDonald, 1994; Brooks *et al.*, 1986). Figure 15.29 compares the three structures. In addition, molecules of gas can be trapped between individual ice crystals.

The actual composition of the gas as well as the presence of salts in the host sediment, alter the pressure-temperature phase boundaries of the hydrates (Figure 15.30). This makes prediction of the potential zone in the permafrost profile where they may be expected more difficult. The first sign of the presence of clathrate hydrates during drilling is the appearance of bubbles of gas in the drilling fluid, together with gas "kicks" (Davidson *et al.*, 1978). Cooling the fluid and adding heavier mud including barytes normally controls the decomposition of the hydrates. The hydrate zone is

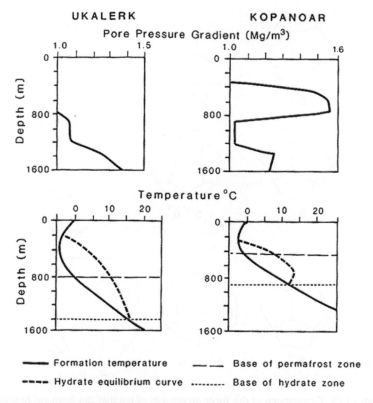

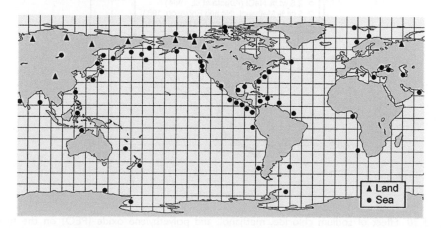

Figure 15.27   Prediction of the *in situ* hydrate zone using formation temperatures and pore pressure gradients (after Weaver and Stewart, 1982). Reproduced with the permission of the National Research Council of Canada. Note that clathrate hydrates may occur below the permafrost base.

Figure 15.28   Locations of some known and inferred gas hydrates. The circles denote marine occurrences while triangles show permafrost occurrences (modified from Kvenvolden *et al.*, 1993). They can also occur in cold land areas lacking permafrost.

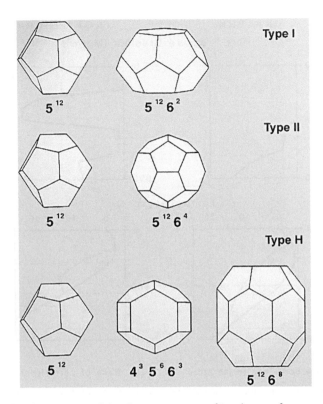

Figure 15.29 Comparison of the three structures of ice that can form gas hydrates.

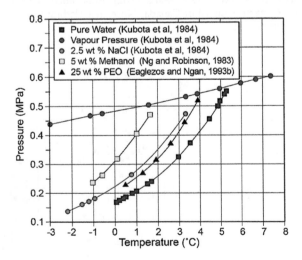

Figure 15.30 Effect of sodium chloride, methanol, and polyethylene oxide (PEO) on the hydrate-aqueous liquid-vapour phase equilibria (modified from Kubota *et al.* (1984), Ng and Robinson, 1983, Englezos and Ngan, 1993, and Englezos, 1993).

normally less than 1,600 m thick in permafrost, but can cause serious problems during drilling such as casing damage, uncontrolled gas release, blowouts, fires, and gas leakage outside the casing (Billy and Dick, 1974; Goodman, 1978a, 1978c; Makogon, 1974, 1981; Judge, 1982; Franklin, 1983; Collett, 1990; Yakushev and Collett, 1992; Yakushev and Chuvlin, 2000). Yakushev and Collett documented the production problems attribute to clathrate hydrates and reviewed the techniques and procedures to combat them. Generally, attempts are made to prevent hydrate decomposition (see above), or it is necessary to allow controlled hydrate dissociation, though this is difficult (Franklin, 1983).

The amount of methane and other gases trapped in clathrate hydrate is uncertain, but conservative estimates suggest that 2×10^{16} kg of carbon is trapped in oceanic sediments as methane hydrate (Kvenvolden, 1988). The actual quantity in permafrost is even more poorly known, but it represents a vast store of potential gas if it can be produced safely (Finlay and Krason, 1990). Release of the methane from clathrate hydrates may have played a role in past climate changes. Certainly they would have been far less abundant in a warm mode Earth as in the Mesozoic Era from 200–40 Ma (Harris, 2013), but is unclear whether this was an effect, a cause, or both.

Chapter 16

Mining in permafrost areas

16.1 INTRODUCTION

Mineral resources are abundant in the permafrost areas of Russia and Canada, especially on the Yakutian and Canadian Shields, as well as Central Siberia. Other important mineral resources are found in the Ural Mountains and the western Cordillera of North America. The deposits located close to where the markets are have been largely worked out, so it is now necessary to exploit those in more remote, cold areas.

Three main problems occur when mining in permafrost, *viz.,* the temperature of the permafrost, the ice content, and the restricted season for transportation. Methodology is steadily improving, but mining in permafrost is inevitably more expensive than elsewhere. As a result, the ore must be of very high grade as at Norilsk, Russia, or it must contain very valuable minerals such as gold at Dawson City (Yukon) or diamonds in the Northwest Territories and Mirny (Siberia), or it has to be needed locally as in the case of coal at Svalbard (Spitzbergen).

Proximity to low cost transportation as on Baffin Island can cause exceptions to these rules. Ice in the ore increases the problems since it is expensive to transport and the ore arrives at its destination only partly filling the container in which it is being transported. On the other hand, the cost of thawing the ice prior to transport is very large and involves the use of large quantities of energy.

For convenience, the mining projects can be divided into placer, open pit (open cast) and underground mining. The engineering involved in each of these is similar to those of other engineering work, and the placer mining has been particularly influential in developing new techniques that are now used in other engineering projects (Cysewski and Shur, 2009).

16.2 PLACER MINING

This refers to the removal of minerals from near-surface sediments. Martin Frobisher encountered ground ice during his unsuccessful attempts to mine gold on Baffin Island in the 1570's. The first Gold Rush started in Dawson City in 1896. The early gold miners in the Yukon and Alaska used fire and hot stones to thaw the ice in the permafrost. The gold bearing placer gravels in the Yukon were buried below an overburden of silty loess and peat (commonly referred to as "muck"), which was sometimes 30 m thick. In 1898, Clarence J. Berry noticed that the steam coming from the hose of a

Figure 16.1 An old gold mining dredge at Nome, Alaska. © S. A. Harris.

hoisting engine was thawing the ground beneath it. The miners quickly started to use this method, using steam points emitting a jet of steam from a narrow orifice. These were driven into the ground as the steam thawed the ice around the pipes (Wimmler, 1927). They were much more efficient than the heated stones that had been dropped down shafts, but there was a considerable loss of pressure between the boilers and the steam points (Janin, 1922), and many prospectors got scalded. The result of the use of steam was a series of conical pits about 12 m deep with frozen ground in between.

A modification by H. M. Payne of the Yukon Gold Corporation used hot water instead of steam. It could thaw four times as much gravel in two-thirds the time using half as much fuel (Wimmler, 1927). However, most miners preferred using the steaming method. By 1909, the area around Dawson City was completely deforested.

John H. Miles developed a method at Nome, Alaska, patented in 1920, using cold water delivered to the gold-bearing gravels just above the bedrock under pressure. It proved far more efficient (Table 16.1) since the water moved laterally, thawing the permafrost in a broader thawed tube than the other methods (Weeks, 1920). Unlike these earlier methods, it produced a larger cavity with vertical sides on all sediment types (Figure 16.2). Its other advantage was that the water could be used directly from adjacent streams and rivers in summer without the need for firewood. This quickly became the preferred method in Alaska and the Yukon, and prepared the way for dredging (Figure 16.1).

Starting in 1918 in the Candle Creek District, Seward Peninsula, Pearce and Johnson developed a method called the Pearce method. It uses a flow of cold surface water over frozen soil down a gentle slope to thaw the ground. It gradually thaws the underlying ground down to bedrock (Janin, 1922). This was a more efficient way of thawing the ground, and if the low areas became flooded, it was the possible to put in dredges to process the thawed ground (Figure 16.1). At Dawson City, a special canal and pipe were constructed to bring water from the river to the north, and giant dredges were brought in in pieces and were assembled in each valley. They then

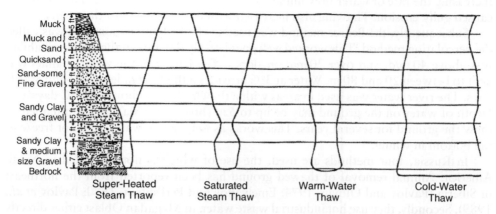

Figure 16.2 Comparison of the shape of the thawed holes in the permafrost at Nome by the thaw methods in use in 1920 by Miles at Nome (Weeks, 1920).

Table 16.1 The results of the Miles thawing experiments comparing the different methods of thawing the gravels (Weeks, 1920).

	Duration (hours)	Thermal volume (m × m × m)	Energy Spent (kW)	Energy used (kW)	Thawing Efficiency (%)
Superheated steam	156	83	231	8.9	3.8
Saturated steam	98	63	185	10.8	5.8
Hot Water	67	62	124	15.3	12.3
Cold Water	192	421	58.7	33.7	57.4

proceeded to work back and forth across the valley floors until they ran out of flood plain, after which they were abandoned. Electricity to power them was brought in from hydroelectric plants on rivers in the surrounding hills. Abandoned dredges can still be seen on the marine terraces near Nome.

Modern working involves playing water on to cliff faces using monitors, *i.e.*, spray guns with cold water. The overburden is then moved by bulldozers, and the gold-bearing sediment is then run through sluices. In this way, the gold prospectors can work claims on hillsides and terraces where the dredges could not operate. When the price of gold increases substantially, it is economic to pump water up to the higher terraces where gold also occurs.

The Miles method was first used in Russia in 1936 (Cysewski and Shur, 2009) on a gold mining operation at the Malyj Urkan creek. It was subsequently used in the Magadan area, but they used rotary-drills to speed up the installation of the points. This way, 40–50 m of overburden and gravel were thawed, opening the way for the use of dredges (Perl'shtein, 1979). The Russians experimented to determine the best spacing of the points, as well as the rate of thaw of the sediment (Gol'dtman *et al.*, 1970). These were found to be dependent on the hydraulic conductivity of the ground.

Increasing the rate of water flow initially increased the rate of thawing, but ultimately caused gully erosion. Thawing also increased with closer spacing of the points.

Subsequently, the Pearce method was used in the Kolyma region, and was called the percolation method. This worked well on relatively flat areas if the soil permeability was above 40 m/d, and over 20 m/d on slopes. The length of the water channel was kept to between 20 and 80 m. Water at 3°C provides a thawing index of around 800°C days. The river water was used in areas with a thawing index above 1000°C days/a^{-1}. Depth of water on the ground may be up to 5 m (Brown, 1970a), and is often left to thaw the ground for several years. This works as long as the water does not freeze to the bottom in winter.

In Russia, four methods are used, the use of which is not permitted in North America. Annual removal of thawed ground has been reported from the northeast of Siberia (Pavlov and Olovin, 1974; Emelyanov and Perl'shtein, 1980; Pavlov et al., 1989). Secondly, they use hot industrial waste water, in Magadan Oblast either directly or through heat exchange systems to thaw the ground. This includes waste water from atomic power stations. Thirdly, they use tubular needle electrodes in a network of equilateral triangles to thaw the ground using electric currents. Perl'shtein and Savenko (1977) developed a method for measuring the approximate amount of heat produced, and there are increasing heat losses to the atmosphere as thawing extends deeper into the ground. As a result, heating needs to be intermittent, or else the applied voltage should be reduced over time. This can only be economic where there is abundant cheap electricity available.

The fourth method involves the use of warm geothermal water. This water comes from very deep wells, or else they use hydrothermal spring water. Corrosion and blocking of pipes by minerals transported in the hot water can be a serious problem.

The Russians also use a thawing-drainage technique producing "*sushentsy*", which is a loose deposit with too little moisture to become hard on refreezing. Numerous closely-spaced, shallow drainage-ways are used to drain away the water downslope as the ice thaws. This speeds up thawing and produces a gravel deposit which can be moved by earth-moving equipment in winter (Emel'yanov, 1973). Ideally, the moisture content should be about 3% (Emel'yanov and Perl'shtein, 1980). It is only economic on large, extensive deposits.

Minerals extracted by placer mining must have a very high market value. Currently, these include gold, the platinum group of minerals, tin and Tungsten (Bundtzen, 1982). These are usually found in outwash deposits of Pleistocene age in North America.

16.3   OPEN CAST/PIT MINING

This is where the ore body lies close to the surface, so that the miners can readily remove the overlying sediment and then mine the underlying deposit (Figure 16.3). This has been used for extracting massive copper-zinc-lead ores in many places using a mine extending over extensive areas, *e.g.*, Faro, Northwest Territories. Where the ore consists of a vertical deposit, *e.g.*, in diamond pipes, the mine is not so wide but very deep (Figure 16.4). The Mirny mine is one of the deepest in the world, but when the depth reaches 550–600 m, the mines are usually converted into underground mines, assuming that enough high grade ore is present to make this an economic proposition.

Figure 16.3 View of the Faro mine developed on the Pelly Mountain slopes, Yukon Territory.

Figure 16.4 Aerial view of the Inter Diamond Mine in the Central Siberia. © A. Drozdov.

Mining at Mirny began on 1957 under extremely harsh climate conditions. The Siberian winter lasted seven months with air temperatures sometimes below −50°C, which froze the ground, making it hard to mine. During the brief summer months, the active layer would become mud, turning the entire mining operation into a sea of mud. The main processing plant had to be built on more stable ground, found 20 km away from the mine. The winter temperatures were so low that car tires and steel would shatter, and oil would freeze. During the winter, workers used jet engines to burn through the layer of permafrost or blasted it with dynamite to get access to the underlying kimberlite. The entire mine had to be covered at night to prevent the machinery from freezing. The ore had to be thawed and processed to extract the diamonds. Electricity was supplied by the electrical power plant at the Chernosevsky Dam, the electricity being brought by high voltage transmission lines south to the processing plant.

At Mirny, several problems were encountered. The ground water was highly saline (c. 50 g/l), so that the ground did not freeze properly. This also complicates rehabilitation of the site. The rock forming the slopes underwent rapid weathering, resulting in enhanced slope instability which is a major problem in a deep mine. Now the pit is a no-fly zone for helicopters that have sometimes been sucked into the pit by downdrafts.

Materials commonly extracted from open pit mines in permafrost areas include clay, gravel, coal, metal ores of copper, lead, zinc, nickel, iron, gold, silver and molybdenum, diamonds, granite, and uranium. The mines are typically enlarged until the ore is exhausted or the ratio of overburden to ore makes the operation uneconomic. Open pit mines producing building materials are commonly referred to as *quarries*.

16.3.1 Exploration

This is carried out by geologists mapping and sampling the rock. When the assay of a sample appears promising, the size and quality of the ore must be determined by drilling, usually on a grid pattern and by geophysics. This requires road access and facilities for the workers as in the case of the oil and gas industry. Often, the exploratory work is carried out in winter. The easily eroded kimberlite pipes in which diamonds are found tend to be located beneath lakes in the glaciated Canadian Shield, so winter drilling is used to prove their existence. The first such discovery was in 1991 by C. Fipke and S. Blusson. After they are located, small lakes have to be drained before mining can begin, *e.g.*, at the Ekati Diamond Mine northeast of Yellowknife. In the case of the Diavik Diamond Mine in the North Slave Lake region, the kimberlite pipes were found beneath part of the Lac de Gras. A dyke was built around the pipes with a causeway to shore, and the water pumped out. Now the mine is being converted to an underground mine (Rio Tinto, 2009).

Water must be removed continuously from open pit mines throughout their operation. This is particularly important to allow for the use of explosives in breaking up the rock (see below). The water is usually disposed of into nearby streams, or it may be used in on-site processing of the ore. The quality of the water is particularly important, as is the chemical composition of the waste water and its final disposition.

16.3.2 Extraction of the ore

The ore is usually worked by drilling and blasting back the front of a cliff or face (Figure 16.5). The idea is to break away the rock between the charge and the cliff face in Figure 6.4A. Figure 6.4B shows the results of a reasonably successful blast in contrast to Figure 6.4C. The problem is that the ice in the permafrost is elastic and it requires a stronger blast to break up the rock than where permafrost is absent. The quantity of broken-up ore must be such that it can be removed before it refreezes. This is usually operated in shifts with workers removing the broken-up ore while drillers drill the holes and set the charges to dislodge the next increment of ore at the end of the shift. In the case of a poor blast, additional drilling and blasting may be necessary to get the right gradient for the bench. For deep, vertical ore deposits, the cliff face of the bench is made to become progressively lower so that the ore trucks can remove the ore by driving round the benches that form a ramp up the sides of the pit (Figure 6.6).

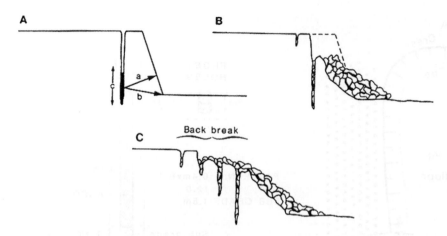

Figure 16.5 Diagram of the way blasting is used to cut back a face (A), and the results of a good blast (B) and a poor one (C) (from Harris, 1986a).

Solar thawing is impractical since it is too slow except in roadside borrow pits for road material.

Considerable experimentation has been carried out in an attempt to determine the best way to blast frozen ground in open pits (Bauer *et al.*, 1965, 1973; USSR, 1972; Morgenstern *et al.*, 1978). Part of the problem is the variability in the ice content in small distances. A good documented example is that of the Iron Ore Company of Canada mines at Schefferville, Québec. The mine was located in the discontinuous permafrost zone where ground temperatures varied from 4°C to −2.5°C, and ice content varied considerably in short distances.

Mining began in 1956 and the first problems encountered were with drilling the holes for blasting. If water was used, then the drill string and rod would often become frozen down the hole, making them difficult to retrieve. Without a fluid, drilling tended to heat up the walls of the hole, resulting in melting of the ground ice. This often refroze to the drill string, or it resulted in slumping or collapse of the walls of the hole if the hole was left open too long. This resulted in improper placement of the charges, or redrilling the hole. Ice in the wall often prevented proper placement and detonation of the explosive charge. Unexploded charges made removal of the ore dangerous. This was overcome by using a plastic liner down the hole before the charge was inserted. Dry sand and rock chips were used to backfill the hole, and the cheap explosive used was ANFO (a mixture of ammonium nitrate and fuel oil).

Ice is elastic and it requires a much larger charge to break up ice-cemented rock containing excess ice. However gravels or rock containing only pore ice can be blasted in the normal way. The pore ice increases the shear strength of the rock so that less charge is required to break up the rock than in non-frozen material. Initially, Livingstone (1956) used a varying charge, depth of placement and spacing to obtain the maximum crater volume for a given number of charges. Bauer *et al.* (1965) provide typical results of tests with placement of explosive charges in seasonally frozen ground with different materials.

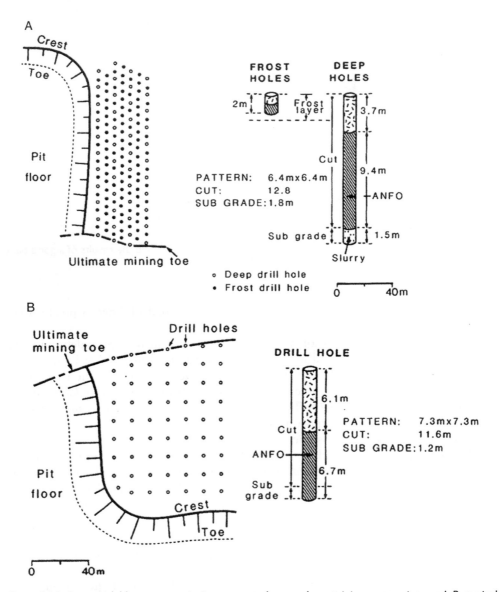

Figure 16.6 A, typical blast pattern in ice-cemented ground containing excess ice, and B, typical blast pattern in unfrozen ground containing pore ice at the Iron Ore Company Mine at Schefferville, Québec (after Garg, 1982, pp. 589–590, reproduced with the permission of the National Research Council of Canada).

In spite of this work, the Schefferville Iron Mine still encountered problems due to the high variability in temperature and ice content of the ore. These variations produced an uneven base to the bottom of the pit which made the use of mechanical shovels for loading the ore on to trucks very difficult. Any material that was broken up but not removed quickly refroze, adding to the problems. However Garg (1973,

Figure 16.7 Drillholes in the lake at Ekati Diamond Mine to establish the distribution of kimberlite beneath the water. © S.A. Harris.

1977, 1979), and King and Garg (1980) found that the ice content and unfrozen zones could be identified using geophysical techniques, and the unfrozen zones (taliks), the areas of seasonal frost with ice lenses and the zones of continuous permafrost could be mapped. The taliks were usually related to the occurrence of winter snow drifts (Nicholson, 1978b, 1979). Thus regional mapping of the temperature and ice content was carried out.

Experimental blasting established the correct pattern and depth of charges for each situation. Time lapse photography showed the best order in which to detonate the charges to produce maximum fragmentation of the ore. Computers were then used to design the placement of the charges. Only a ledge wide enough to be worked by a given shift was blasted, while the next set of charges was emplaced. At the end of each shift, the new charges were set off. The charge patterns developed by Garg (1982) were much closer in ground cemented by excess ice, and two different types of holes and charges were used (Figure 16.6A). The shallow ones broke up the seasonal ice while the deep ones broke up the main ore body. In unfrozen ground, the pattern was simpler (Figure 16.6B).

Similar techniques are used in mining diamonds on the Canadian Shield. Kimberlite is softer than granite, so the pipes occur beneath lakes. Boreholes are drilled through the frozen ice cover in winter (Figure 16.7) to establish where the kimberlite is located. Then the lake must be drained and a suitable dam built to hold the water seeping into the working (Figure 16.8). A plan is drawn up of the cross-section of the pipe and the placing of the terrace to act as a ramp for ore trucks to enter and exit the pit (Figure 16.9). Figure 16.10 shows the drilling and emplacement of the charges as well as the ramp.

Figure 16.8 Retaining dam to hold back the water from the mine operation, Panda Mine, Ekati. Note the use of thermosiphons to maintain the frozen core of the dam. © S.A. Harris.

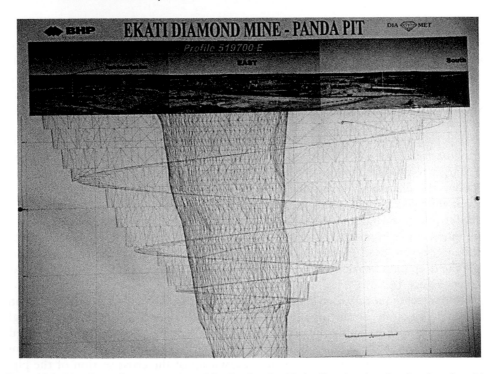

Figure 16.9 North-south cross-section of the Panda mine kimberlite pipe showing the planed position of the terrace-like ramps for ore-trucks and the proposed depth of mining, courtesy of BHP. Note the wide upper zone including much host rock that has to be removed to enable the ramp for truck traffic to be able to operate down to the planned lower limit of economic open-pit mining.

Figure 16.10 General view of the Ekati workings, showing the drilling pattern and access ramps and terracing on the margins of the kimberlite pipe. © S.A. Harris.

16.4 UNDERGROUND MINING

Many ore bodies are inclined at a steep angle and are relatively deep, discontinuous or continuous. In these cases, underground mining is the only practical way of extracting the ore economically. Fernette (1982) describes the main ways of mining underground. Key factors include the geometry, and stabilities of the host rock and ore body. Permafrost affects these strengths and stabilities, the actual modification depending on the ground temperature, the temperature of the air being circulated through the mine and the ice content of the rocks. Most deep mines in North America use ventilation at 5–10°C. This results in thawing of the ice in the rock walls, though the amount depends on the ground temperature that will normally increase with depth below the permafrost upper layers (Figure 16.11). However, if the mine is kept cold by circulating cold air, the host rock and ore body will become stronger than in the unfrozen state. This is the standard practice in Russia and it is also used by COMINCO on Little Cornwallis Island, Nunavit. At Norilsk, the main mine workings are below the permafrost at about 1,000–1,200 m below sea level, so that only the elevator shafts are refrigerated.

Most mines are designed with a main entrance through which men and supplies enter and ore is removed. There may be multiple adits called *stopes* that exploit other outcrops and provide passageways for removing ore. The actual mining is accomplished by blasting at the end of each shift when everyone is out of the mine. After ventilation of the toxic gases, the loosened ore is removed using trucks, trains, or a conveyor system, depending on the quantity and height of the gallery, *etc.* In mines with cold air ventilation, ANFO can be used as the explosive, but in warm air mines, other more reliable but expensive explosives are required. In the case of the latter, two

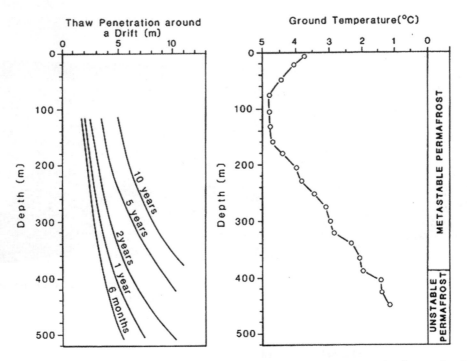

Figure 16.11 Predicted extent of thaw around a 3 m drift or decline at various depths as related to measured mean ground temperature and permafrost stability at Asbestos Hill, Québec (modified from Taylor and Judge, 1980). The thawing is based on 5°C air circulation for various periods between 5 months and 10 years.

ventilation lines are run into the mine, with one going to the lowest part (Swinzow, 1963).

Water is always a problem. Any seepage must be removed immediately by high pressure pumps. On the surface of the colder parts of the mine, any warm air that is introduced will deposit ice as icy coating on walls and tools. This must be removed regularly since the moisture can interfere with power and communication equipment. Air lines must have some methyl hydrate in them to prevent blockage by ice. However, water is used to wash down the walls to keep down the dust, and is then efficiently removed by sump pumps. Quick heating of the newly exposed walls may cause spalling of the rock (Thompson and Sayles, 1972; Pettibone, 1973). In mines ventilated with cold air, ice on the walls is left to add strength and to seal in the dust. About 70% of the energy in the cold air may be used in this way (Emyl'yanov and Perl'shtein, 1980).

The entrance to a mine is usually in the side of a hill slope, but even then, there can be problems (Linell and Lobacz, 1978). These include surface disturbance that may cause permafrost degradation (Figure 16.12A), uncontrolled drainage that can cause erosion, icings and accelerated permafrost degradation (Figure 16.12B). Creep of the slope above the portal can gradually close the portal or severely reduce its height (Figure 16.12C), falling material can seriously injure people, while landslides can block the entrance (Figure 16.12D) due to thawing of the rock in the roof resulting in its collapse

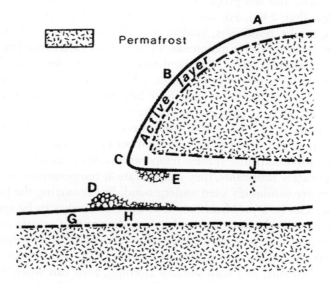

Permafrost

Figure 16.12 Problems that may occur around tunnel entrances (modified from Linell and Lobacz, 1978, p. 814, reproduced with the permission of the National Research Council of Canada). The letters correspond to the problems discussed in turn in the text.

(Figure 16.12E). Uncontrolled drainage water may enter the mine (Figure 16.12G), the portal structure may increase the depth of thaw (Figure 16.12H), seepage entering from the active layer in summer can form hanging masses of ice (Figure 16.12I), and thawing of ground ice can cause water seepage (Figure 16.12.J). In winter, the outer part of the tunnel will be subjected to freezing air temperatures.

Strength of the rock is greatly modified by the ice content and temperature. The air used in ventilation has to be humidified after warming to prevent sublimation of the ice, resulting in loss of strength of the roof and walls. Walls with temperatures of −4°C have much greater strength than those with temperatures of 0° to −2°C. The latter need many times more support in the form of props, and unsupported spans must be very short (Chaban and Gol'dtman, 1978). When the mine is warm, suitable props must be emplaced immediately to allow for the anticipated thawing, loss of strength and spalling of the walls. It is both dangerous and expensive to add extra supports later.

When a stope or adit has been mined out, mine waste from other production faces or non-toxic mill tailings may be brought in and used to backfill the space, and is then allowed to refreeze. This reduces the need for long-term supports to maintain mine stability. After the fill is thoroughly frozen, the remaining pillars of ore can be mined, increasing the tonnage from 60% to 90% of the potential total. The infilling can only be carried out in summer when it is sufficiently thawed to be moved.

16.4.1 Transport of the ore around the mine

Three basic methods are used, *viz.*, belt conveyors, trains, and load-haul-dump vehicles (Foster-Miller Associates, 1965; Aitken, 1970). In addition, slurry may be transported in pipelines. Trucks and self-propelled scrapers are used whenever the material has to be moved less than 2 km. For distances less than 900 m, screw conveyors, vibratory conveyors, trucks, bulldozers, loaders and rail cars can be used. Railless systems (trackless mining) overcome the maintenance problems of rails caused by icing, heaving and subsidence (Bakakin, 1978). Electrical machinery is preferred to diesel-powered machinery due to the fumes from the latter, which require more ventilation. Diesel exhaust also produces a significant amount of heat that must be dealt with.

Belt conveyors can be used in mines, but the whole system must be either above or below 0°C (Dubnie, 1972). Below 0°C, the materials must be dry to avoid icings and freezing of the ore to the conveyor belt. Pure rubber belts must be used since synthetic rubber becomes stiff and brittle at these temperatures. With suitable precautions, *e.g.*, using low-temperature lubricants, they will operate at temperatures as low as −40°C.

Shuttle cars are commonly used underground, but removing the build-up of fine particles and ice can be a problem. Salt or calcium chloride can be sprinkled on the cars but tends to corrode them and make them damp. Best results are obtained by keeping the ore as dry as possible and periodically cleaning the equipment. Ice must be removed periodically from all stationary switches. The advantages of keeping the mine frozen are obvious, and with suitable clothing, men can work 9 hour shifts at temperatures down to −20°C, *e.g.*, at the Polaris mine on Little Cornwallis Island. In northeastern Russia, mine temperatures of −5° to −8°C are used because they approximate the ground temperatures (Chaban and Gol'dtman, 1978). This largely avoids the development of fog and the icing-up of machinery in the mine. The air is humidified up to just below 85% relative humidity to minimize desiccation of the rock. Hydraulic and pneumatic equipment are usually impractical unless special precautions are taken.

When commodity prices fall too low, the mines may be shut down until the prices increase. In this case, water must still be removed from the mine or else it can become choked with ice. All other water and air lines must be drained until they are dry. Once a mine has become filled with ice and water, it is almost impossible to resume mining operations.

16.4.2 Support facilities

Men and supplies have to be brought in, bunk houses have to be constructed and power supplies provided. Portable housing is used where it can be brought by road. It has the advantage that it can be moved elsewhere when mining ceases. Water supply has to be available for camp and industrial use throughout the year and there has to be adequate storage for used and unwanted water. At the diamond mines on the Canadian Shield, it is usual to fly in workers for shifts of two weeks or more, rather than to build larger housing and support facilities. A similar system is often used by Gazprom in Western Siberia.

16.5 WASTE MATERIALS AND TAILINGS PONDS

When mining begins, waste material (*Gangue*) accumulates, but there is no place to put it except on the ground. Great care must be used in choosing its position in permafrost areas since it will accumulate ice in its core. The water sources include meteoric water (snow, rain, dew) and ground water. Lack of care in choosing where to place the waste has often resulted in the pile acquiring sufficient ice to become mobile, and the resulting *tecnogenic rock glacier* (Corte, 1978a) has been known to flow across the access road to the mine, or even engulf the mine entrance in various parts of Siberia (Gorbunov, 1983; Gorbunov and Severskiy, 2010).

In its natural state, oxygen mainly reaches the underlying rocks occurring in the active layer. As a result, these surficial materials have already been leached out of the products produced by chemical weathering in the local microenvironment. These materials can be stockpiled and used to reduce weathering of the surface of seasonally unfrozen layers of piles of mine waste. Placement of a layer of coarse rock at its base prevents groundwater adding to the ice in the pile. As soon as possible, good practice is to use giant rippers to break up the waste and move it underground to fill worked-out galleries and stopes during underground mining.

Below the permafrost table, or in waterlogged areas, the water and/or ice prevent much oxygen from reaching the rocks, so the component minerals remain largely unweathered. Once the rock is drained, or after it has been mined, weathering can proceed. Thus the rocks in the walls of the mine can become weathered due to thawing, just as well as any piles of rock brought to the surface of the ground. This becomes important if the ore minerals are associated with poisonous or acidic chemicals that become freed by weathering. Surface piles are often stored in piles that are also subject to surface erosion, as are the dam buttresses underground. When a mine is abandoned, the pre-existing drainage should be re-established as far as possible, and its water quality monitored. Differential settlement will occur in the tailings piles as entrapped ice piled in winter melts out, and ice collects in the tailings piled in summer. There needs to be detailed monitoring for at least 5 years, and regularly thereafter to avoid unpleasant surprises. Revegetation should be carried out as far as possible.

Many abandoned open pit mines are left with large piles of rock containing sulphide minerals. These decompose to produce sulphuric acid which then enters the hydrologic system. One such case is at the Faro Mine which occupies a hill top (Figure 16.3). Large volumes of crushed rock have been left piled on the ground and the sulphide minerals have been decomposing for over a decade. The effluent flowing down into the adjacent creek which is a tributary of the Pelly River consists of extremely acid waters that are having a negative effect of the adjacent environment. It has been left to the Yukon Territorial Government to clean it up.

Experiments have been carried out placing the tailings in layers in cold weather and then spraying them with water that is allowed to freeze before the next layer of gangue is added. The water partly thaws in summer taking much of the soluble chemical content with it (Biggar *et al.*, 2005). In Nunavut, Journeaux Associates (2012) have studied these problems of tailings in that jurisdiction. The waste is produced as a slurry with between 20% and 80% water content. In summer, a lot of the water either drains away, or evaporates before freezing. In winter, the slurry freezes quite quickly and can contain substantial quantities of ice. Retaining dams are necessary to contain the fluid waste. Again, the waste is added to the pile in increments and is allowed to freeze before another layer is added. Thus the normal sequence is a winter layer, overlain by a summer layer, and then a winter layer is added, etc. Finally the entire pile needs to be encapsulated with weathered rock.

Biggar *et al.* (2005) carried out an experiment at the Colomac Mine involving spray freezing decontamination of tailings water that was reasonably successful. 30% of the water that was sprayed on to the tailings froze, while the remainder of the water was returned to the tailings pond. Analysis of water collected from thawing ice cores obtained from the tailings indicated dissolved chemical removal of 87–99% (depending on the chemicals involved) after 39% of the spray ice column had melted. Both arsenic and cyanide concentrations were reduced considerably, but not to safe concentrations. At the study site, the authors admit that the ultimate cyanide concentration did not meet regulatory requirements. Ultimately, this process moves the more soluble chemicals from the pile into the tailings pond, where it is diluted by surface runoff from the surrounding catchment area. This assumes that there is a large catchment area above the tailings pond, or the quantity of contaminated water from the tailings is small. At Faro, this is not the case.

Journeaux Associates (2012) discussed the engineering challenges for tailings management facilities. They summarize the sub-aerial tailings disposal methods then in use in permafrost regions. When the Rankin Inlet Mine closed down, the tailings were encapsulated in weathered rock and have now become part of the permafrost. Unfortunately, encapsulation merely reduces the rate of weathering, but does not stop it. At the Greens Creek Mine at Hecla, Alaska, 80% of the rock tailings are dry stacked in cavities underground, which are then cemented with a paste backfill and left to refreeze. The Giant Mine at Yellowknife was abandoned when the mine closed down and thermosiphons are being used to refreeze the arsenic trioxide dust underground (see below).

Use of encapsulated tailings above ground is expensive and involves increased environmental hazards (Perlshtein and Pavlenkov, 2001). The usual method in Russia is to divide the containment area into sections into which are placed successive layers of waste. The next layer is not added until the previous one is completely frozen. Time taken to freeze depends on the mill output, density of the slurry, settling velocities of the solids and sediment freezing rate. Regular monitoring to ensure that the water table has not entered the tailings is essential.

16.5.1 Toxic wastes

A good example is the Giant Gold Mine (Figure 16.13) at Yellowknife (Mikes, 2013). Gold there is associated with arsenopyrite (FeAsS). When exposed to oxygen and especially when heated, it gives off arsenic vapour and sulphur dioxide, leaving behind iron oxide and gold. The arsenic vapour cools to form arsenic trioxide dust which consists of very fine particles that coat everything around. When the mine started in 1949, the arsenic vapour was released into the air, polluting the environment. It settled on everything around as a very fine, dry powder containing 60% arsenic. Dead fish appear in the bays of nearby Great Slave Lake and at least three children have died from eating snow in winter. It is estimated that 740 kg per day of arsenic trioxide was emitted into the atmosphere during this time. It dissolves in water up to 9,000 mg/l. In 1951, an electrostatic precipitator was used and modified to remove the dust. A second precipitator was used from 1963 until the mine closed in 1999. The dust was sent by pipes into large sealed chambers and stopes in permafrost where it was stored (Figure 16.14). Altogether some 237,000 tons of arsenic trioxide were buried in them.

Figure 16.13 Aerial view of the abandoned Giant Mine outside of Yellowknife, Northwest Territories in 2013. © S.A. Harris.

Yellowknife is in an area of discontinuous permafrost, and it is now known that after 1970, the permafrost was degrading, probably due to the circulation of warm air through the mine. Bulkheads were used to close off the storage bodies, but these can degrade with time. Any water that leaves the mine is treated to remove arsenic, but Yellowknife has to obtain its drinking water supply from a lake several kilometres away from the polluted area. There is too much arsenic in the adjacent arm of the Great Slave Lake to use that water safely, and that arsenic is also carried down the Mackenzie River, potentially affecting the settlements downstream as far as Inuvik.

Figure 16.14 Aerial view of the position of the chambers and stopes in old gold workings in permafrost (outlined in red). Rectangles are the chambers and irregular boxes are stopes (from Mikes, 2013).

The area is in the zone that is showing warming of the mean annual air temperatures, so there is great concern over the future prospects. Without remediation, the dust could release 12,000 kg of arsenic per year into the groundwater in a few years. Flooding of the mine by the adjacent Baker Creek could also lead to a massive escape of the arsenic into the Great Slave Lake.

Fifty-six methods were proposed to deal with the problem, and these were narrowed down to twelve (Table 16.2). Of these, freezing in place (frozen block) using thermosiphons is showing good results, and it is claimed that it will work even with a warming of the mean annual air temperature of 6°C. Using cement encapsulation is the other alternative chosen; although it is more expensive initially, it would perform better in the case of flooding by Baker Creek. An assessment of the risks resulting from using some of the proposed methods is shown in Table 16.3. Total cost of taking down the contaminated buildings, refrigerating the storage chambers, and putting in the thermosiphons capable of keeping the chambers frozen is currently over 9 million dollars. It will take another 1 million dollars a year to maintain the containment. Since the mining company no longer exists, the cost is left to the taxpayer. There are many other contaminated sites within the permafrost realm, and consideration of how to make them safe will become important in the future. Similar arsenical gold deposits occur elsewhere in the world, *e.g.*, in the Karamay area, Xinjiang Province of China (Zhou *et al.*, 2015).

Table 16.2 Diagram showing the arsenic trioxide management alternatives preferred by the community of Yellowknife and their consulting engineers (from Mikes, 2013). The two methods finally selected are B3 and G1.

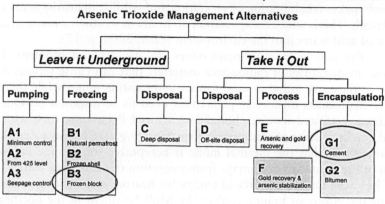

Table 16.3 Risk assessment of some of the options for dealing with arsenic trioxide at Yellowknife (from Mikes, 2013).

Alternative	Probability of Significant Arsenic Release		Worker Health & Safety Risk
	Short Term	Long Term	
A1. Water Treatment with Minimum Control	Low	High	Low
A2. Water Treatment with Drawdown	Low	Moderate	Low
A3. Water Treatment with Seepage Control	Low	Moderate	Low
B2. Frozen Shell	Very Low	Low	Low
B3. Frozen Block	Very Low	Low	Low
C. Deep Disposal	Low	Very Low	Moderate
D. Removal & Surface Disposal	High	Very Low	Moderate
F. Removal, Gold Recovery and Arsenic Stabilization	Moderate	Very Low	Moderate
G1. Removal & Cement Encapsulation	Moderate	Low	Moderate

At the Giant Mine, the sulphur dioxide emitted into the atmosphere produced sulphuric acid. Fortunately the quantity was not too great, so there was limited effect on the surrounding environment, other than to acidify the water in local water bodies such as streams and lakes. Given the size of the Great Slave Lake and its through-flow down the Mackenzie River, this was not as serious a problem as the arsenic trioxide. However, in cases of large emissions from mines smelting metal sulphide ores such as at Norilsk, the sulphur dioxide from the smelters has had a devastating effect on the surrounding vegetation. Other gold mines use cyanide to extract gold, and in that case, the tailings ponds will contain poisonous waters.

At the Rankin Inlet Mine, the waste rock weathered to produce substantial quantities of acidic and metal-rich water into Hudson Bay. The oxidized tailings dust was blown about as well. The kimberlite in diamond mines also weathers to produce acid water which makes rehabilitation of vegetation and wildlife in nearby areas and water bodies difficult. This is where encapsulation of the ore in weathered material can reduce the output of acid water into the environment (EBA, 2004, p. 17).

Radioactive tailings from uranium mines are also a major problem. They contain significant quantities of radioactive materials that can cause serious damage to life. Some of these are found around former and existing mines while there are also radioactive byproducts of nuclear plants that need to be disposed of. One solution that has been suggested involves burial in fractured Palaeozoic rock with ice-filled joints and cracks in areas of continuous permafrost, *e.g.*, the Novaya Zemlya Islands. Calculations show that the thermal mode is acceptable for storage unless there is migration of the chemicals or damage from migration of nearby sea water.

Coal has been the main source of energy for heating and cooking in rural China. A study of the effects of mining coal in the Muli Coalfield in the northeast of the Qinghai-Tibet Plateau shows that in spite of attempts at reclamation, there is serious permafrost degradation (Qin, 2009). In the mining process, the gangue sediment is warmed by the sun. This sediment is bulldozed and covered by topsoil. This surface soil is subject to marked summer temperature fluctuations, while the underlying ground is warmer than the waste gangue encapsulated below. The pile absorbs more heat in summer than undisturbed, vegetated soil, and this makes revegetation difficult at these low latitudes and relatively high altitudes. In winter, ground temperatures are virtually the same in both disturbed and undisturbed sites.

In summary, unless careful remediation is employed during actual mining, the former mines often represent major environmental challenges. In the past, the mines have tended to be abandoned with minimal remedial care. Later remediation can be extremely expensive, and represents a significant cost to the taxpayer.

Chapter 17

Provision of utilities

17.1 INTRODUCTION

Permafrost regions are characterised by extremely cold winter temperatures, large diurnal temperature variations in summer, low population densities, and usually a poor economic base. The indigenous peoples were adapted to living off the land, using only what they needed to survive. Newcomers were often people who liked solitude and whose primary concern was to minimise the effects of the cold winters. They were prepared to forgo the standard of living demanded by modern populations in big cities in warmer climates. Settlements were widely spaced out so as to provide enough resources for each settlement to survive. When their way of life became altered due to development of resources, replacing the food resources and providing modern amenities were both a tremendous challenge as well as very expensive, *e.g.*, in Nunavut.

When natural resources are discovered and exploited, the incoming workers demand more modern amenities, and are not particularly concerned about damage to the environment if it gets in the way of their making a good living. The local population soon wants a similar standard of living, partly because the development usually destroys the possibility of making a living off the land or sea. A period of upgrading of the standard of living tends to take place, partly paid for by the resources being exploited. When the resources are no longer available, the settlements fall on hard times since the cost of maintenance of facilities such as schools, water supplies, food and housing becomes unsustainable without enormous subsidies from outside. In Russia, workers usually went to the resource towns and stayed there until the resource was gone. Then they moved elsewhere. Often today, the workers needed to develop the resource are brought in from elsewhere on short-term, working visits, *e.g.*, in the oil and gas fields in Western Siberia or in the case of the Ekati mine in the Northwest Territories. This allows for concentrated settlements of minimal size that can be abandoned or moved elsewhere when the resource runs out. Where gas is used for heating in China, problems occur due to pressures on the pipe exceeding its strength (Li *et al.*, 2011).

Provision of modern services in permafrost areas is very difficult and expensive. Major factors include construction on permafrost, climate, remoteness, lack of planning for services, inadequate housing and lack of an adequate economic base to pay the costs of installing and maintaining suitable services (Gamble and Janssen, 1974; Smith and Heinke, 1981).

Figure 17.1   Utilidors on piles at Inuvik, Northwest Territories, 1983. © S. A. Harris.

In the following account, the municipal services are dealt with under three general headings, *viz.*, water supply, waste disposal and provision of electricity. With the coming of communication satellites, communication with the outside world is no major problem.

17.2   WATER SUPPLY

Water is always a key item in any settlement. It is usually needed for both housing and industry. It must be kept out of mines, but is used to generate electricity and for industrial processes. The long, cold, dark winter nights usually preclude the use of other alternative forms of energy.

17.2.1   Sources of water

Tolstikhin (1940) divided groundwater in permafrost regions into suprapermafrost, intrapermafrost and subpermafrost waters. The main characteristics of these sources are summarized in Table 17.1 (from Harris, 1986a).

Suprapermafrost water is the water occurring above the permafrost table in the active layer or in taliks or surface water bodies such as lakes and streams. It is fed by meltwater, rains, springs or seepages of groundwater, or condensation from humid air. In Central Yakutia, springs are commonly used as water sources (Anisimova *et al.*, 1973). Snow and ice can be used as a last resort, but suprapermafrost waters are particularly susceptible to pollution by natural or human means. They are only readily available in the unfrozen state for part of the year. This can be overcome by damming the waters in a suitable catchment area. The resulting lake must be large enough and deep enough to contain enough water to supply the anticipated nearby population for at least a year, in spite of the formation of thick (c. 2 m) ice in winter. Thus the water

Table 17.1   Classification and characteristics of groundwater in permafrost areas (after Tolstikhin, 1940; Church, 1974; from Harris, 1986a).

Type	Nature of Movement	Temperature	Environment	Yields	Quality
Suprapermafrost					
(a) Seasonal (freezes in winter).	Gravity flow or artesian.	Alternately >0°C and <0°C.	Organic material, alluvium, and other unconsolidated material.	Seasonal – nil in winter.	Liable to become contaminated very easily.
(b) Freezes in part during the winter.		Negative.	Alluvium and other unconsolidated material.	Can be large but liable to freeze in pipes if not heated	
(c) Does not freeze in winter.	Gravity flow.	Constantly low negative.			
Intrapermafrost					
(a) Always liquid.	Artesian or gravity flow.	Always positive or always negative.	Alluvium and unconsolidated materials; rarely in solid rock.	Can be good.	Less liable to become contaminated by human wastes. Usually good quality.
(b) Always solid (ground ice).	None.	Always negative.		Very limited and not easily available	
Subpermafrost					
(a) Shallow.	Artesian or stationary; at times sealed by saline	Low positive or negative if saline	Alluvium, fissures in bedrock, and karst solution channels.	Usually limited.	Liable to have high total dissolved solids. Not readily contaminated by human wastes.
(b) Deep.	Artesian or stationary.	Always positive.	In bedrock, e.g. porous strata, in fissures or in karst solution holes.	Usually finite.	

supply for Inuvik is the result of successfully damming and enlarging a lake as well as pumping in water from another lake. Fort MacPherson also uses a natural lake, although this is subject to seasonal flooding and pollution during the spring floods on the Mackenzie River. Deep lakes provide a plentiful supply of water which may also be relatively warm if collected from below the thermocline, *e.g.*, 2°C at Alert. At Tuktoyaktuk and Coppermine, fresh water is obtained from underneath the sea ice, having come from nearby large rivers, *e.g.*, the Mackenzie River. Surface water sources are often acidic and contamination is always a risk.

Intrapermafrost water is supplied from above and below and is often under artesian pressure. This water is not readily detected or mapped, and is usually of limited volume beneath the lower slopes of hills, even in areas of continuous permafrost. Its presence is usually indicated by icings, icing blisters and isolated hydraulic pingos which represent the places where the artesian pressure exceeds the overburden pressure. It may be highly mineralised if it is the "mother liquor" of the ice in the nearby permafrost. Pollution by humans is less likely to have occurred, but is sufficiently common that engineers are reluctant to use it as a water supply for human consumption.

Subpermafrost water is the preferred source of potable water where it is available. It requires a careful study of the water quality and volume prior to development to avoid wasting funds to drill a deep borehole into a small pocket of water. Sometimes the waters are saline (cryopegs). The relatively higher temperature of the water makes

Figure 17.2 Aerial view of the Vilyuy Dam, 75 m high, 600 m long, with a total capacity of 35.9 km³, generating 2,700 GWh of electricity. © A. Brouchkov.

it easier to pump and move compared with water at c. 0°C. The latter has a viscosity of 1.2–1.8 higher than the warmer water which is less likely to freeze in the pipes. At Faro, Yukon Territory, the subpermafrost water has temperatures of 2.8 to 4.4°C. However, high artesian pressures are common (Linell, 1973). P. Melnikov discovered the large artesian basin in central Yakutia. As the water comes to the surface, it causes thawing of the permafrost around the hole, followed by collapse of the ground or rupture of the pipe. If the water flow is interrupted for too long, the water will freeze in the pipe.

17.2.2   Dams to impound water on permafrost

Sayles (1987) provides a summary of the literature on the performance of dams built in permafrost areas up to that date. Miller *et al.* (2013) provide a summary of the history of the development of ideas for successful dam construction, including the evolution of the present methodology being used. In Russia, at least five intermediate-sized embankment dams with heights ranging from 20 to 125 m have been in operation (Biyanov, 1973, 1975; Tsytovich *et al.*, 1978; Johnson and Sayles, 1980). A typical example is the Vilyuy Dam built between 1964 and 1967 on the Vilyuy River at Chernoshevsky built to supply electricity to diamond mines including Mirny (Figure 17.2).

The Vilyuy Dam was constructed to span the valley between hard, frozen rocks with ice infilling the fissures and joints. Since it was constructed, it has been found that some water moves through these joints and escapes, though the amount varies with time. The loss is now monitored and is rather unpredictable, representing a potential threat for failure of the dam.

Smaller embankments have been operating for more than a century. As a result, the Russians have more experience of the problems involved than North Americans. The earliest dam built on permafrost was the Petrovsk Dam near the southern end of Lake Baikal in 1782. This was during the last cold Neoglacial event (c. 1650–1920 A.D.), and it performed satisfactorily until 1929, holding back a 6 m deep reservoir (Tsvetkova, 1960; Tsytovich, 1973). Soil had been placed in wooden cribs that were flooded and left to freeze. The mean annual air temperatures during this period were cold enough to allow the dam to hold back the water satisfactorily, but warming air temperatures have resulted in thawing of part of the ice.

Outside permafrost regions, earth dams are constructed with an impermeable core of clay, covered with material such as crushed rock to provide enough strength to hold back the water. Tsytovich *et al.* (1972) summarised the three main methods of construction of dams on permafrost as being those that can be maintained in a frozen state by natural environmental cooling, those that require artificial cooling, and those that are allowed to thaw. The latter are designed for an unfrozen condition with pre-thawing of the materials, after which they can be constructed as in permafrost free areas during summer (Miller *et al.*, 2013).

In areas with appreciable permafrost, the core of the dam and the foundations must be kept frozen. Their success depends on the strength and deformation characteristics of the frozen material used in their construction. No seepage of water can be permitted, or the water will thaw the surrounding material over time, resulting in settlement and failure. This is particularly critical in spillway areas. Air convection in rock can transfer heat from the water behind the dam into the dam itself, resulting in thawing and failure, as has been observed in Russia (Mukhetdinov, 1971) and Nunavut. Construction is normally carried out in winter, the dam being built using a series of layers called *lifts*. Each one is frozen before the next is emplaced, and construction can only occur in winter. Crushed rock and water can be used for the core.

Typical problems that arise include difficulty in finding inexpensive clays for the core of the dams, and cracks that appear in the upper part of the dam when the core freezes to the banks, causing leakage from these cracks that can destroy the dam. Vasiliev *et al.* (2013) have found that ice- and cryogenic-soil composites in the water-retaining elements of embankment dams can prevent these problems in cold regions with permafrost.

Between 1940 and 1972, the construction was carried out in summer with active freezing using vertical freezing pipes with forced air/brine/gas circulation to cool the lifts where the mean annual air temperature was insufficient to maintain a frozen dam for the design life of the structure. From the 1970s, passive freezing using vertical pipes has been utilized. Construction can take place in winter or summer. The Anadyr' dam in Russia (17 m high and 1,300 m long) was built using nearly saturated material in two lifts, each being emplaced in summer and left to freeze during the following winter (Kuznetsov, 1973). Passive cooling pipes were used in sandy gravelly soil with 30–80% ice over frozen clay loam. It was found that dams constructed in the thawed state do not freeze as planned, and it is essential to freeze the dam completely with passive cooling pipes prior to filling the reservoir with water (Anisimov and Sorokin, 1975).

These dams need on-going monitoring and maintenance to identify and correct any problems that develop. Typical problems include water seepage through the dam, thawing of permafrost around the spillway, changes in elevation of the top of the dam, failure of passive thermosiphons, and thawing and erosion of the banks of the lake. Zhang (2014) reports that dams in China often change from frozen to thawed and back again during the period of operation. This has also been reported from Canada (Dufour *et al.*, 1988).

Construction of dams resulting in the development of a large lake modifies the local environment. The Vilyuy impoundment has led to a warming effect of the air temperature in the river basin of up to 6°C, and has greatly reduced flooding in the downstream area. This has resulted in declines in the bird and fish populations (Cultural Survival, 2010). This modifies any regional climatic changes that may be occurring. Presently, these cannot be predicted in advance, hence the need for continuous monitoring.

17.2.3 Municipal water storage

When melted snow or ice is used as a last resort in isolated communities in Alaska, it must be stored in insulated shelters and not become contaminated. Streams used as sources of water have periods of high and low flow, with high turbidity during the latter when there is a high demand. Accordingly, an adequate quantity of water of suitable quality has to be stored until needed. A storage facility also overcomes the problem of intermittent flow from a deep well and minimizes the risk of freeze-up of the water in the pipes. Wood, steel or concrete tanks are widely used in Northern Canada and Iceland for small settlements. Above-ground tanks are usually insulated or placed inside a heated building. Partial burial may be used in ice-free ground, and special attention must be paid to heat retention to avoid freezing of the water or melting of the ground ice. On ice-rich soils, piles and insulated vented pads are necessary. The tanks must be vented to allow for expansion and contraction with temperature changes, and freezing of the water must be avoided at all costs.

In central Yakutia, the population often collects blocks of ice in winter from the surface of lakes and places it in underground storage for use in summer. This is a way of reducing pollution and salt content in drinking water.

17.2.4 Water treatment

It is essential to know the chemical composition of the salts present in the water source as well as the pH, humic acid content and composition. As elsewhere, there are specific limits for total dissolved solids (usually around 1,000 ppm), as well as for various salts. Epsom salts (sodium sulphate) is the least acceptable component. The local population can adjust to higher total dissolved solids with time, but visitors will encounter problems.

Low temperatures increase the solubility of carbon dioxide resulting in a decrease in pH of the water. pH and humus are particularly critical in permafrost regions, the humus being readily recognised by its brown colour. The organic matter corrodes copper pipes, while iron bacteria utilize the iron and manganese associated with the humus to produce brown flocs of iron hydroxide and manganese oxide. The resulting water cannot be used in laundries, nor in the textile and paper industries.

The humus also reacts with halogens to form trihalomethanes, often called THM (Rook, 1974). Thus when chlorine is added to water containing organic substances, chloroform, bromodichloromethane, dibromochloromethane, and bromoform are produced. Peters and Perry (1981) discuss the reactions and show that the yield of THM increases with increasing pH of the water to a pH of 9. The production of the THM increases with increasing time of contact of the halogen, increasing humic:fulvic acid ratio of the organic matter, increasing molecular weight of the organic matter, and increasing temperature of the water.

The importance of THM content is that these are classified as carcinogenic substances. Humus in water may promote healing of wounds (Biber and Bogolybova, 1952, in Gjessing, 1981), but it is also associated with organic and inorganic micropollutants. It often forms complexes with substances such as cadmium and albumen (Visser, 1973). Thus it may introduce undesirable micronutrients and metals into the body.

Limits for the permissible level of THM in water vary from $25 \, \mu gL^{-1}$ in Germany, $75 \, \mu gL^{-1}$ in Holland, $100 \, \mu gL^{-1}$ in the United States, and up to $350 \, \mu gL^{-1}$ in Canada. Removal of the organic matter in water to prescribed limits would be very expensive. It is generally agreed that the risks involved in ceasing chlorination of the water supply are far greater than the risks caused by the THM.

Gjessing (1981) discussed the main possible treatments to reduce the problems caused by organic matter, *viz.*, the addition of chemicals to bleach or mineralize the humus, and filtration in an attempt to remove the coloured particles and some of the "soluble" colour. The only effective way is to filter the water through activated charcoal. However, the charcoal must be changed frequently, making the process very expensive. Conventional sand filtration only removes 5–10% of the aquatic humus.

Bleaching by ozone results in colour reduction, probably by reduction in size of the molecules, but the total organic matter content scarcely changes. It increases the carbon availability to micro-organisms, resulting in increased biological growth in the water unless heavy doses of chlorine are added. However, this increases the production of THM.

Coagulation-flocculation is far more successful. Aluminium sulphate is usually used to produce a white aluminium floc. This has a large surface area which absorbs the negatively charged humus molecules. It is pH dependent.

Thallium has been found in surface waters in parts of Russian northern Europe. It is a highly toxic for humans, but the origin of the high concentrations in certain locations is not known.

Salinity is a common problem in arid and coastal areas. Desalinization techniques include distillation, reverse osmosis, and freezing, all of which have been used in the coastal districts of the Arctic. Natural freezing is the cheapest method and is still used to supplement other sources in some communities, *e.g.*, Grise Fjord, Northwest Territories and Barrow, Alaska.

17.2.5 Water requirements

Armstrong *et al.* (1981) discuss the water requirements and methods of water conservation in small remote communities in permafrost areas. Obviously, the smaller the water use, the more households can be serviced by a given system. However, the high cost per capita of providing water to a small number of households is a problem.

The minimum amount of water for subsistence in a community setting is about 5.7 L/d for a healthy person (Freedman, 1977). This basic requirement includes drinking, cooking, personal hygiene and laundry. The average household in a northern city requires additional water, which can be broken down into toilet (40%), bathing (30%),

Table 17.2 Examples of residential and community water consumption (modified from Armstrong et al., 1981).

Situation	Location	Consumption (L/person/d)	Comments	Reference
Primarily no permafrost	Canada	625	National average for 2780 municipalities, 1975. Mean residential metered (1972–4).	Environmental Protection Service, 1977; Gysi & Lamb, 1977; Howe & Linaweaver, 1967
	Calgary, Alberta	217		
	U.S.A.	240	Mean annual residential metered use, Western States, (1960s)	
Self-haul, permafrost	Chesterfield Inlet, N.W.T., Whale Cove, N.W.T.	7.5 11.5	Average, no household plumbing	Cameron, 1977
Trucked, permafrost	Yellowknife, N.W.T.	118	Mean, trucked, 2.85 persons per house	Cameron, 1977
		45–295	Range, trucked, per house	
Piped, permafrost	Inuvik, N.W.T.	480–550	Piped circulating	Smith et al., 1979
	Ft. McPherson, N.W.T.	250	Piped portion of community	
	Yellowknife,* N.W.T.	485		
	Whitehorse,* Y.T.	1680 1135–2500	Average annual Range	Stanley Associates Engineering Ltd., 1979
	Dawson City,*# Y.T.	3630–9080	Range	
	Faro,* Y.T.	790	Average annual	Cormie, in Armstrong et al., 1981
	Haines Junction,* Y.T.	570		
	Mayo,* Y.T.	2730		
	Watson Lake,* Y.T.	820		
	Fairbanks,* Alaska	650	Average annual	Smith et al., 1979
	Homer,* Alaska	1630		

*Water bleeding in winter for freeze protection.
Some leakage.

laundry (15%), kitchen (13%) and 2% for other uses. These are average figures for cases where water is provided by a system of pipes, but considerable variation occurs in individual cases.

The community uses additional water for fire-fighting, line flushing, wastage for freeze prevention (called bleeding), watering of green spaces, washing streets and in leakage. If there is any industry and commerce in the community, it can greatly increase the needs. Table 17.2 summarizes some typical values for residential and community water consumption in North America.

17.2.6 Transportation methods for water and waste water

The method of transportation of the water determines the community water consumption. In the northern areas of permafrost, three main systems are used, *viz.*, self-haul, trucking in, and pipes. The self-haul system provides subsistence supplies of water and this can be detrimental to the health of the community (White and Saviour, 1974). This system is being abandoned, but contributes to the poor living conditions in some of the native communities.

Where supplies are trucked in, conservation techniques are very important due to the high cost per litre involved. The minimum acceptable level of service is

Figure 17.3 Underpasses or overpasses must be constructed where roads and utilidors meet in Inuvik.
© S. A. Harris.

45 L/person/d in the Northwest Territories (Cameron *et al.*, 1977), although 70–80 L/person/d is desirable. This method is used in the better organized small settlements such as Teslin, Yukon Territory. It provides a full-time job for at least one person, making the rounds to every house, pumping in water and usually pumping out waste water. Costs include the truck, its maintenance, and the stipend for the operator, as well as providing a suitable water source.

Piped systems can and do supply much larger quantities of water. Inuvik and Fort Macpherson are examples of where the system is designed to prevent freezing of the water in the pipes by using insulation and heat traces. Even then, under conditions of low demand in cold weather, freezing may occur. This then disrupts the service. The water may be sufficiently cold that it must be heated in order to flow through the system in winter. The advantage of this system is that the overall use of water is lower than where the water flows constantly.

The alternative is to have the water flowing through the pipes at a constant pressure, with the excess water being bled back to the source. Even in non-permafrost areas, the quantity of water used in bleeding can be very high, *e.g.*, at Jasper, Alberta, Associated Engineering Services Ltd (1975) reported mean annual figures of 270 L/person/d for domestic use and 140 L/person/d for bleeding. The latter is higher in the winter.

The pipes may be placed in *utilidors* underground or above ground (Figures 17.1 and 17.3). The utilidor is a tunnel system enclosing the pipes in an insulated container (Smith and Heinke, 1981; Carefoot *et al.*, 1981). In Russia, the whole settlement is heated from a central steam/water heating plant, and the potable water is kept from freezing by the proximity of its pipes to the hot water. In North America, heat traces may be needed to keep the water from freezing during a disruption of service. A typical utilidor above ground consists of an insulated box carrying cold and hot water

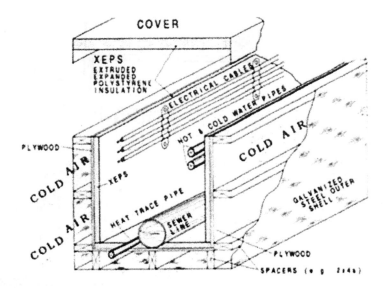

Figure 17.4 Diagram of a typical above-ground utilidor in North America.

pipes and electrical cables, sewer lines, together with a heat trace mounted on piles (Figures 17.1 and 17.4). It has to be connected to each building and greatly complicates traffic circulation in a settlement. They are used where the water table is too high to use a buried utilidor, but they were used almost exclusively to carry services until the 21st century. The problem with them is that they are very expensive to replace, *e.g.*, the 17 km system in Inuvik had a replacement cost of $140–$170 million in 2015. The wooden piles tend to rot and collapse, settlement and heaving of the ground can cause a change in direction of flow of water in the pipes, leaks and breakage of electrical cables. More details will be found in Northern News Services (2011).

In Canada, the Territorial Governments have been encouraging the small communities to upgrade their pipelines. The underground system is preferred for planning, engineering and aesthetic reasons. They can carry potable water, sewage, telephone cables, TV cables, fibre optic service and electricity. They are used in subarctic climates where at least one month has a mean monthly air temperature above 10°C, *e.g.*, in Anchorage and Fairbanks, Alaska, Whitehorse and Dawson City, Yukon Territory, Yellowknife, Northwest Territories, and Gothaab, Greenland. The pipes are protected from freezing by insulation, and by bleeding from the water line, heating, and recirculation of the water during periods of low water use to ensure continuous flow. The insulation (5–7.5 cm) is essential to avoid substantial melting of the adjacent ground ice. Often, a substantial quantity of clean gravel is placed in the trench around the pipe or utilidor to prevent thawing of ice-rich permafrost, as in Dawson City. At Barrow, Alaska, the underground utilidors are large enough to walk through. Air (9°C) is circulated through the utilidor, and is replaced every 6 hours. Elsewhere, heating is accomplished by adding water from a warm source to the water pipe or by placing a heating wire along the inside or outside of the utilidor along its length, but always inside the insulation. The heating wire uses 12 W/m in South Greenland and 20 W/m in the north. Lower wattage (8–10 W/m) is used inside the pipe itself.

All piped systems consist of one or more uninterrupted loops, originating at a central pumping station located centrally. Water is pumped into the loop at between 4°C and 7°C and returns at 1–4°C. Where electrical heating along the pipe occurs as in Greenland, the water can be pumped at 1°C and returns at 0.1°C. Interconnecting pipes between loops and isolating valves reduce the size of the area impacted by a shut-down.

Where bleeding is employed to maintain flow, there must be a series of bleeding points in the loop. This can result in water usage as high as 5,000 L/person/d. The pipes may be either single pipes to large water users, *e.g.*, apartments, or dual pipes with a large supply line and a smaller (5–15 cm) return line. The single pipe loop is preferred. Heat tracing lines are expensive to install and operate, but are the main method of avoiding freezing of the pipes. Steam and hot water thawing lines (usually using hot water from power generators) may be used to thaw the pipes.

Smith and Heinke (1981, p. 10) provide cost estimates for construction of the water and sewer lines in Canada in 1981 dollars, the cheapest being on-surface pipelines at $50–100/m. At that time, buried pipelines cost $100–300/m, on-surface utilidors cost $200–400/m, buried utilidors $400–1,200/m, while above-surface utilidors cost $600–1,700/m. The average life span of utilidors is about 50 years, after which substantial repairs or replacement is necessary. This is why most current construction uses buried pipelines if possible. However, there may be significant leakage as the underground pipes age and are subjected to recurrent or increasing heaving and subsidence stresses.

Where hot springs occur as in Iceland, the warm spring water is transported in large diameter pipes for distances over 20 km to cities and towns such as Reykjavik in Iceland with only a small loss in temperature. In Greenland, a unique phenomenon is the summer water line. These uninsulated pipes laid on the ground, only carry water in the frost-free period (Rosendahl, 1981). These pipes are drained and not used during the rest of the year. This saves money in construction costs, but this system cannot be used in colder climates.

17.3 WASTE DISPOSAL

Wastes can be divided into solid wastes and wastewater. Both can cause problems with disease transmission or damage to the environment if not dealt with properly. The methods are steadily being upgraded (Heinke, 1974; Slupsky, 1976; Rosendahl, 1981; Balmér, 1981; Smith and Heinke, 1981). Good bibliographies of the earlier methods will be found in Snodgrass (1971) and Cameron and Smith (1977).

17.3.1 Wastewater treatment and disposal

The nature of the water distribution system largely determines the characteristics of the wastewater (Smith and Heinke, 1981). Community wastewaters can be divided into four main categories, *viz.*, undiluted wastes, moderately diluted wastewater ("black water"), conventional strength wastewater ("grey water"), and very dilute waste water. Normally, a community produces wastewater falling into at least two of these categories. Each category requires a different treatment.

17.3.1.1 Undiluted wastes

These are the products of the use of bucket toilets that are commonly used in small settlements, where the cost of using pipes is prohibitive. Ideally, the "honey bags" are collected regularly and trucked to a special disposal pit as far away from the community as possible. It should be in an area not used for other purposes, and the hydrology should be checked to ensure that there is minimal risk of contaminating the water supply (Heinke and Prasad, 1977). The pit should be large enough to provide at least $0.5\,m^3$ per person per year. It must be separate from the landfill and covered over and treated with lime when full. It functions as a giant holding tank, with no decomposition taking place. Pathogens may be viable for decades. Anaerobic decomposition is possible but uneconomic. In the coastal districts of Greenland, the wastes are dumped into the sea (Rosendahl, 1981).

Incineration is coming in for sludge from other treatment processes (Eggner and Tomlinson, 1978). However, there can be problems with the operation of such a facility in a small community, and it is expensive. Another alternative used in Southern Alaska is the use of tundra ponds as wastewater treatment and disposal facilities (Schubert and Hintzman, 1994). The ponds are naturally occurring, water-filled depressions in lowland and delta areas, formed by glaciers or thermokarst. They must be deep enough not to freeze to the bottom and are usually underlain by a layer of organic matter and fine-grained frozen silts including yedoma. They provide total containment, improved structural integrity in areas of unstable soil conditions, and low operation and maintenance costs to the community (Smith and Finch, 1983; Environmental Protection Service, 1985). Their design is governed by the Wastewater Regulations of the State of Alaska 18 AAC 70 (1990).

17.3.1.2 Moderately diluted wastes

These are the products emanating from holding tanks, pressure sewers and vacuum sewers. Conventional aerobic digestion is feasible and practical at a temperature of 20°C with a 30-day minimum detention time (Prasad and Heinke, 1981). Digestion at lower temperatures requires a greater percentage of methane bacteria in the reactor, coupled with a longer detention time. An anaerobic filter treatment is often used, but both the effluent and the sludge need further treatment. Facultative methane oxidizers (Dedysh and Dunfield, 2009) and aerated lagoons, oxidation ditches, attached growth devices, extended aeration plants and physico-chemical processes are also used for medium-sized communities. The facultative lagoon is the lowest cost and most common option due to its low operating and capital costs. Both the lagoon and the outlet must not freeze in winter, so they must be suitably deep. Detention times range between 8 and 12 months.

Aerated lagoons require better maintenance which is not usually available, but oxidation ditches have been very successful (Murphy and Rangarathan, 1974). Attached growth systems require protection from freezing and are therefore not often used.

Physico-chemical methods have been used at work camps, especially along the Trans-Alaska Pipeline and at dam construction sites near James Bay. They are very costly, labour-intensive and require special chemicals and encapsulation to avoid freezing. However, they take up less space, have consistent effluent quality and minimum impact on the site. They are too expensive for normal long-term use.

17.3.1.3   Conventional strength wastewater

Hrudey and Raniga (1981) discuss the characteristics of conventional strength waste-water. These are the products returning by piped systems and are characterised by wide fluctuations in flow, organic loading, and content of both toxic substances and pharmaceutical chemicals. The content of microorganisms is high, so that disinfection must be part of the treatment. Shigellosis (bacillary dysentery) and hepatitis A were 73.9 and 21.1 times as common in the Northwest Territories as in the rest of Canada between 1975 and 1979, inclusive, indicating the consequences of lack of disinfection.

Provided the wastewater is disinfected, it can be treated by the same methods as moderately diluted wastewater. It is rich in nutrients so the effluent can provide much needed nutrients to soils. Balmér (1981) discusses the Swedish experience in treating these wastes. However, the pharmaceutical chemicals can create problems when they get into surface waters. Perhaps the worst problem is with the hormones which can cause male fish to change sex.

17.3.1.4   Very dilute waste water

This is the effluent coming from pipes where there is continuous bleeding of the water to prevent freezing-up of the sewer pipes. It is in large volume but at a very low temperature, so heat is usually added to the pipes unless the pipes are in underground utilidors, e.g., Kotzbue, Alaska. Critical factors include type of pipe, above or below ground, depth of burial, amount of insulation, nature of the outer layers of the utilidor, as well as heat loss during treatment and disposal. Back-up freeze protection devices are usually used, and affect the heat loss. At Whitehorse, Yukon Territory, a controlled blend of 1–2°C surface water and 4°C groundwater is sufficient to avoid freezing.

Smith and Given (1981) discuss the various types of treatments used in Canada. The method is usually selected according to the requirements of the regulatory agencies for purity of the effluent. Where low BOD's and suspended solids are required, a short retention lagoon and screening are used as at Hay River, Fort MacPherson, and Inuvik in the Northwest Territories, and Whitehorse, Yukon Territory. Miyamoto and Heinke (1979) describe the performance of the lagoon at Inuvik where BOD's are reduced 30% in winter and 90% in summer. Eighty % of the suspended solids are deposited as sludge, while the effluent is discharged into the east channel of the Mackenzie River.

Zero discharge of pollutants is required in some situations and involves the use of a long retention lagoon, e.g., at Watson Lake, Yukon Territory, or Imperial Oil at Tuktoyaktuk, Northwest Territories. It must have a much larger capacity, and the effluent still contains some pollutants. The lagoon must be deeper than 2 m to avoid freezing in winter, and smell can be a problem.

Where more stringent controls on the quality of the effluent are required, e.g., in Alaska, a secondary treatment plant such as a rotating biological contactor is necessary. This is very expensive because of the very large water volume involved. In these cases, bleeding of the lines is usually avoided to reduce the costs of secondary treatment. In Canada, only Churchill and Thompson, Manitoba, and Carmacks, Yukon Territory have mechanical treatment plants.

At Company villages such as Tyonek in Alaska, there are only about 199 people (200 estimate), mainly Natives. The area lies adjacent to the largest coal deposit in Alaska, and there are oil reserves nearby. The 90 homes and facilities are served by a piped water system (ACDCIS, 2008) since the residents receive good wages paid by the mines. Sewage water from each family is chemically and biologically treated in the lagoon and then released into the ground.

17.3.2 Solid waste disposal

This is one of the most neglected areas of sanitation in the Arctic. In the past, waste was stored in the house in 180 L oil drums. Average waste production was in the range of 1–1.6 L/person/day, depending on whether some of the garbage is burned prior to disposal. Refuse should be collected once or twice a week and kept separate from the liquid wastes. Unless trucks are covered, some of the garbage blows away. For most of the year, the waste material cannot easily be covered in soil, while low temperatures minimize degradation, which is essentially in cold storage (Straughn, 1972). Kent *et al.* (2003) provide recommendations on how modified solid waste sites should be planned, operated and maintained in a report to the Northwest Territorial Government. Modified landfill on slopes or in trenches is preferable where possible.

At Tyonek, the landfill is located about 4.5 km away in the forest and the residents bring their own garbage there. Combustible material is burned at high temperatures, but the rest of the unseparated garbage is then covered in soil. Burning of all combustible materials is now the norm in the Yukon Territory and Alaska. Unfortunately, low-temperature burns (300°C) produce dioxins (Kitano, 2001) and other volatile organic compounds and particulates, some of which are carcinogenic (U.S. Environmental Protection Agency, 2008).

The problem has been compounded by shopping in large centres, buying goods with great amounts of packing. Matsuura *et al.* (2008) discuss the problem in Alaska, providing examples of how the main cities and over 200 rural communities deal with the problem. At least 50% of the latter are not connected to other places by a class 1 road, so must operate their own landfills. The latter are usually an unlined open dump facility, and for much of the year, burial of the garbage is not possible. The Alaska Department of Conservation (2002) has a classification scheme; these dumps are classified as Class III Municipal Solid Waste Landfill (MSWLF) and there are no disposal fees. Class I MSWLFs have a lined disposal site of sufficient size to last for decades, *e.g.*, for Anchorage and Eagle River, the MSWLF occupies 1,112,886 m^2, and a disposal fee of US$15 per 5 m^3 is charged for loads brought in pickup trucks (Municipality of Anchorage, Alaska, 2008).

Large communities such as Fairbanks, Alaska can afford to use shredders, balers and high temperature incinerators. However, these are expensive. Without an automobile crusher, a place must be found for discarded machinery, automobiles, snowmobiles and appliances. When these are crushed, they are usually placed with the other solid waste.

17.4 ELECTRIC TRANSMISSION LINES

Present-day living is becoming ever increasingly dependent on electricity. There are three main ways of generating it, *viz.*, by burning coal, burning natural gas, and harnessing hydroelectric power. The latter creates negligible pollution and is the preferred option, and the techniques and difficulties involved have been discussed above.

In the Northern Hemisphere, Russia has 68% of its Territory underlain by permafrost. When it started to develop Siberia in the 1960s, it had to put in five transmission grids across its vast country (Moscow United Electric Grid Company, 2004). The east, central and north grids have the most permafrost that they must cross. The north grid has the most permafrost and crosses more than 86,000 km (Irkutsk Electric Grid Company, 2014).

China started construction of transmission lines in the northeast permafrost areas but has had to expand across the Qinghai-Tibet Plateau with the Qinghai-Tibet Railway Transmission Line built in 2005 (Wei, 2009). Subsequently, the Qinghai-Tibet Interconnection Project was completed in September, 2011 (Yan et al., 2012) while construction on the new Yushu-Qinghai Main Grid 330 kV transmission line began in 2012 (Cheng et al., 2013).

In North America, the Hydro-Québec transmission line carries electricity 32,000 km to the northeastern United States. It is a 735 kV AC line partly over sporadic permafrost that was put into operation in 1965 (U.S. Arctic Research Commission Permafrost Task Force, 2003). In Alaska, the Healy to Fairbanks line also crosses sporadic to discontinuous permafrost (Wyman, 2009). A second 230 kV line went into operation in 2003, and there are two lines crossing seasonally frozen ground connecting Kenai and Anchorage and Healy-Willow. Because much of the north is sparsely populated, there are relatively few major power projects, but many small local lines (Fortier and Allard, 2004).

17.4.1   Foundation problems for transmission lines built on permafrost

The major problems for transmission line construction are frost heave, thaw settlement, frost jacking (Wen et al., 2012), icings, warming of the ground, changes in vegetation cover and degree of desertification, and permafrost degradation (Yu et al., 2009). Frost heave is the biggest problem in Russia and North America. In Alaska, the 10 m long timber piles have been lifted 1–2 m by frost heave in frost susceptible soils, resulting in large repair costs (Lyazgin et al., 2004b). North of Tyumen, Russia, annual heaving can be as much as 20 cm/a^{-1}, with a 5 cm/a^{-1} differential heave in the piles on a given tower. After 20 years of operation, maximum deviation of towers can be 2.5–2.7 m (Lyazgin et al., 2003).

In northeast China, towers on the transmission lines have collapsed due to frost heave. The Halar-Yakehi 220 kV transmission line was completed in 1997, but the tops of the cast-in-place piles tilted and broke the binding beam and pile body in 2003 at the N29 tower due to differential frost heave and settlement (Jiang and Liu, 2006). Any changes in the microenvironment that alter the strength of the surrounding ground can produce failure of the towers. The major factors influencing route selection and location of towers include geological environment, harmful ecological phenomena such as icings and fault lines, different geomorphological units, thermal stability, harmful ecological conditions and construction convenience (Liu et al., 2008; Cheng et al., 2009; Qian et al., 2009).

An example of a harmful situation is the effect of shallow permafrost on flood plains among the rolling foothills of mountain ranges on the Qinghai-Tibet Plateau. Water descends through joints and taliks in the mountains which generally have

Figure 17.5 Icings along the Tuotuohe River valley adjacent to the Qinghai-Tibet High Voltage Transmission Line (You *et al.*, 2015, with permission.).

Figure 17.6 Groundwater welling up in an excavation for a pile, Tuotuohe River valley. © Cheng Guodong.

reasonably thick, continuous permafrost, and contribute to a high water table at lower altitudes, especially along river valleys. The presence of these is indicated by icings (Figure 17.5), and pingos along the stream bed and flood plain. When an attempt is made to emplace a pile, the excavation penetrates into the wet ground underneath and the

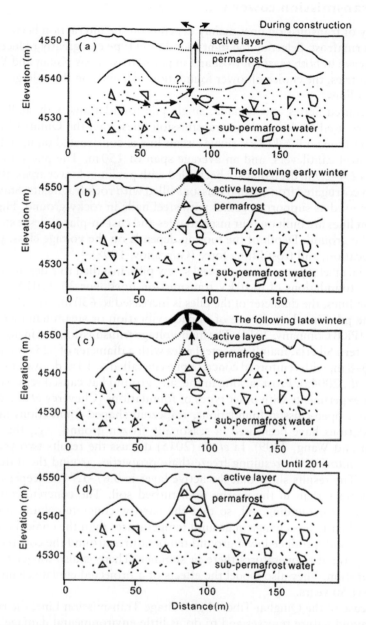

Figure 17.7 Diagram showing the evolution of the rupture resulting from passing through the permafrost base while drilling the bore hole for a tower foundation along the floodplain of the Tuotuohe River (You et al., 2016).

hole fills with water (Figure 17.6). At other times, thawing of the ground around the piles results in water coming to the surface as springs in summer and icings in winter. In these cases, the piles must be protected with additional thermosiphons or else abandoned (Figure 17.7).

17.4.2 Transmission tower foundation types

The stability of transmission lines mainly depends on the interaction between foundations and permafrost. Selecting the right foundation type can greatly reduce problems with frost heave in both seasonal frost and in permafrost areas (Guan and Wu, 2010). In most countries, the choice of tower foundations depends on the voltage class of the transmission lines.

In the United States and Canada, single pile foundations with shallow embedded depth are used for low-voltage transmission lines, *e.g.*, the Ontario Victor Mine Transmission Line with a 115 kV single circuit line is constructed on upright wooden poles with steel cantilevers, and an average span of 150 m. The posts are placed in drilled holes 2.4–2.7 m deep that are backfilled with packing. In wet areas, they may be emplaced in corrugated iron circular caissons filled with rock fragments. Stays and logs may also be used as support, together with steel nails in rocky ground. High-voltage transmission lines are built on four inserted piles and cast-in-place piles, extending 10–30 m into the ground. Depth of the footing depends on the voltage class, permafrost type, line location, and frost lifting force (Guryanov, 1998).

In Russia, steel pipes or stainless-steel screw piles installed to a depth of 5–9 m are used for the foundations of 6–10 kV transmission lines. For 35–110 kV and 220 kV transmission lines, the diameter of the piles is increased to 630 mm or 720 mm. Installation of the piles involves the use of impact, vibration or steam-hammer drilling in permafrost (PKB Company, 2014). The high-voltage transmission line in north Tyumen Oblast, Western Siberia, mainly uses steel piles with a diameter of 325 mm, installed to a depth of 6–8 m, with reinforced concrete piles installed 6–10 m deep as an alternative (Lyazgin *et al.*, 2004a). Frost jacking in the silty ground has caused serious problems, resulting in experiments with other types of foundations (Cheverev *et al.*, 2006).

In China, open-caisson foundations, trapezoid-shaped foundations and cast-in-place foundations have been used in seasonally frozen ground, *e.g.*, the Hulunbuir region (Sun and Wang, 2009). Li *et al.* (2015) discuss the results two years (2011–2012) of monitoring the resulting freeze-thaw properties around the different kinds of footings. The results show that the active layer is 0.67–1.74 m deeper around the concrete footings than in the nearby undisturbed soil. The concrete conducts heat faster than the surrounding soil, so that the narrow concrete column on a tabular base causes slightly less change in ground temperatures than the conical column on a tabular base. Both produce lower ground temperatures around the concrete in winter and higher ground temperatures there in summer. There is also longer freezing and thawing duration. They also model the effects of a projected climate change of 2.6°C over the next 50 years.

In the case of the Qinghai-Tibet High-voltage Transmission Line, the routing was chosen to avoid nature reserves and to do as little environmental damage as possible (Wang *et al.*, 2014). Construction was carried out in winter, and four foundation types (Figure 17.8) were used according to the local situation. Dug foundations were used where the hole could be dug and the pile installed in deep, stable permafrost. Precast piles were used where convenient, but where there were strong frost-heave forces with a lot of ice, the cylindrical cone foundation was used. Where there was thick ice and vulnerable permafrost, rivers, surface water and shallow groundwater, bored piles were used.

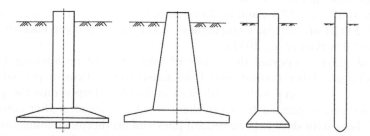

Figure 17.8 Foundation types selected by Qinghai-Tibet DC Interconnection Project, from left to right, dug foundation poured in place, precast foundation, cylindrical cone foundation, and bored piles (after Wang *et al.*, 2014).

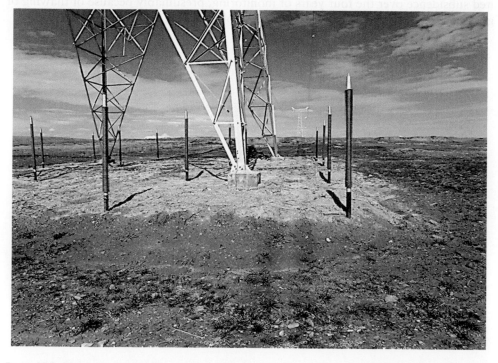

Figure 17.9 Placement of the thermosiphons used to cool the ground around the bases of the towers along the Qinghai-Tibet Transportation Corridor (from Wang *et al.*, 2014).

The ground around the footings was covered with a gravel pad from which the construction was carried out as quickly as possible in winter. Drilling was usually by rotary drills. Where there was doubt about the effects of the footings on the thermal stability of the foundations, thermosiphons were installed to combat warming (Figure 17.9). The pouring of the *in situ* cement had to be carefully controlled. The higher the moulding temperature, the better the quality of the resulting concrete (Zhu and Wang, 2011). However this significantly impacts the surrounding permafrost. As a result, moulding had to be carried out at temperatures of about 6–8°C when the air temperature is above −15°C, and could be increased to 10°C at colder temperatures (He, 2008; Ma *et al.*, 2013). Compactness of the layer-by-layer backfill had to be greater than 80% (Gar *et al.*, 2011).

Yu *et al.* (2015) reported the results of four years of monitoring 130 towers along the Qinghai-Tibet Transmission Line across the 550 km of permafrost along the route. The 10 towers with foundations embedded deeply into the permafrost using thermosiphons became frozen in the ground 2–3 months after completion of the tower, Those with shallowly embedded piles and no thermosiphons took 2–3 seasons to freeze back. The effect of construction depended on the degree of cooling of the thermosiphons versus the enhanced heat flow through the pile to depth. The latter increased the thawing below the piles by up to 2 m relative to the nearby undisturbed sites. Most subsidence occurred at the ice-rich permafrost sites. All the towers exhibited subsidence over the four years of monitoring, presumably because of thawing of ground ice, but all performed with the allowable limits of total subsidence.

Chapter 18

Agriculture and forestry

18.1 INTRODUCTION

The largest permafrost areas in the northern hemisphere stretch across both the northern part of North America and from southwestern Europe to Eastern Siberia. In the latter area, the associated vegetation ranges from alpine and meadow tundra in Scandinavia, south and east into the *Taiga*, which is the most extensive forest in the world (Figure 18.1). In eastern Siberia, it even reaches northeast China and northern Mongolia. It also underlies the northern parts of the *Boreal Forest* (as the taiga is referred to in North America) in northern Canada, extending polewards into tundra and the northern polar barrens. Both forests have a small number of tree species and a characteristic understory. When traced in the same direction, the trees become smaller

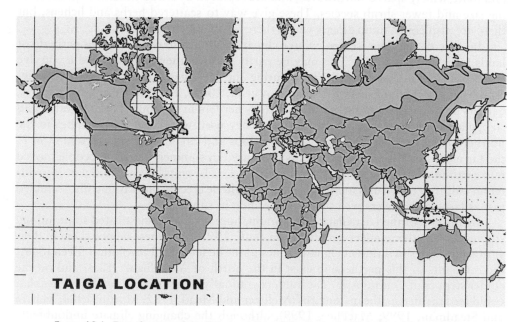

TAIGA LOCATION

Figure 18.1 Distribution of the Taiga and Boreal Forest in the Northern Hemisphere.

Figure 18.2 Typical former winter home of the Sakha people, Yakutsk. © S. A. Harris.

and more widely spaced northwards, and are replaced by shrub tundra, with decreasing size and fewer shrub species. These give way to scattered herbs and lichens, but there are also cold, Arctic deserts where there is low precipitation.

In spite of the long, severe, winter climate, humans have been using permafrost lands for a very long time. This may be partly because they had managed to survive harsh climates in Asia and Beringia during the Ice Ages, and therefore had long-established methods of survival (Mochanov *et al.*, 2008). They were primarily hunters, herders and fisherman in the north, as well as herders in the south in the forested and steppe areas in Mongolia, south to Tibet. They used wood and pit dwellings in the northern parts of the forest, together with igloos and circular tents covered in skins in the far north. In Yakutia, similar tents were used in summer, but these people moved into earth-insulated houses in winter (Figure 18.2). The tents have various names in different cultures, *e.g.*, Yurts in Mongolia, and could be erected quickly and disassembled just as fast, so they were well suited for herding in summer.

By the end of the Last Glacial Maximum, humans had domesticated some animal species such as the reindeer and dog (Vila *et al.*, 1997), and these helped them in their various lifestyles in different climatic zones until the industrial revolution. Even today, the remnants of this partnership exist in the north of Asia (Figure 18.3) and on the colder steppes of Central Asia. The hunters gradually took their toll on the species that could feed them, *e.g.*, the mammoths (Melltzer and Mead, 1982; Martin, 1984; Martin and Steadman, 1999; MacPhee, 1999), although the changing climate undoubtedly hastened the demise of many species adapted to the cold climate of Beringia and Central

Figure 18.3 Reindeer herding village camp in summer at the mouth of the River Ob, near Salekhard, Western Siberia. © S. A. Harris.

Asia. The evidence for the demise of mammoths is particularly well documented, since they survived until about 4 ka on the isolated Wrangell Island in the Arctic Ocean (Vartanyan *et al.*, 1995).

Although the overall effect of expansion of Humans was to commence a wave of extinctions of both plants and animals, they domesticated some species that they found to be useful for their survival. Along the Arctic coast of Eurasia, reindeer became domesticated as beasts of burden (Figures 18.3 and 18.4) and the herds represent a valuable resource for food and clothing. Similarly, the Inuit found Husky dogs to be important as sled dogs in North America. In the southern areas of permafrost in Russia, China and adjacent territories, camels (3,000 years B.P.) and horses (4,500 years B.P.) were the main species that were domesticated, usually being used as beasts of burden (Crandall *et al.*, 1997). Goats (Zeder and Hess, 2000) and certain types of pigs, cattle, sheep and yaks were used as livestock from about 7 ka.

The vast area of Tiaga across Siberia provided a good, extensive hunting ground for nomadic tribes, but only certain areas are suitable for cultivation, even today (Figure 18.5). Agriculture started later in Siberia than elsewhere because the inhabitants had only rudimentary cultivation skills (Naumov, 2006). Their digging sticks were inadequate to cope with the harsh environment, but the development of bronze scythes allowed them to change to a more sedentary life in the Tobol River and the Minusinsk Basin. These people of the Andronovo culture had already domesticated sheep, cows

Figure 18.4 Reindeer pulling a sleigh in winter in Siberia, followed by more reindeer. © Yakutia-
Reindeer-Tours.

Figure 18.5 Present-day land use in Asiatic Russia. Note that the mountainous areas of Eastern Siberia
are not delineated, nor are the oil and gas fields.

and horses, but could now grow wheat, which they bartered with the Chinese to the
south of their lands (Mote, 1998). During the 7th to 2nd centuries BC, the people
of the Tagar culture introduced irrigation to the region in the Minusinsk Basin of the
upper part of the Yenisei River.

The plough finally came into use during the Kirghiz Khanate in the 8th century A.D. This was the first independent Siberian state, and these people were involved in arable farming, cultivating millet, barley, wheat and hemp, as well as nomadic cattle breeding.

Simmons (1989) provides a summary of some of the earliest dates of use of cultivated crops. Wheat and barley have been grown by Asian farmers for over 9 ka, while apples and apricots originated on the northern slopes of the Tien Shan in Kazakhstan. A wide variety of herbs were used as medicine.

In the permafrost areas of North America, the Inuit had developed a culture based on seal hunting and fishing from kayaks made from driftwood and seal skin. They also hunted seals by cutting holes in the ice and killing the seals when they came up to breathe. When they got the chance, they would kill whales, which provided both meat and oil for cooking and also light during the period of winter darkness. Husky dogs were used to pull sleds, and they would build igloos to live in in winter. Southwards, they co-existed with native Indian tribes, *e.g.*, at Kujarapik, who would hunt and trap various mammals including the caribou. Fishing was practiced by both indigenous groups, based on catching salmon along the Pacific and Atlantic coasts and Arctic Char in the northern freshwater lakes.

18.2 ZONATION OF NATURAL VEGETATION ACROSS SIBERIA

Figure 18.6 shows the distribution of vegetation across Siberia (Knystautas, 1987), together with the permafrost zones. Although the taiga extends westwards into Europe, it is not underlain by permafrost there. The vegetation zonation ranges from arctic tundra in the north, through forest tundra, taiga, mixed forest, forest-steppe, steppe to semi-desert and desert in the extreme southwest. The total area of taiga covers twice the area of the Amazon rainforest, and is the largest contiguous forest left in the world. Siberia underwent only limited glaciation centered on the mountain ranges, so there are large areas underlain by sediments as well as alluvial terraces along the north-flowing rivers. This provides suitable soils for trees and potentially for agriculture.

Figure 18.7 shows the subdivision of Siberia into western, eastern and far-eastern regions. Permafrost only affects the extreme north and the mountains in the south of the Western Siberian region. The bulk of the area underlain by permafrost lies in the Eastern and Far-Eastern regions, including both mountains and plains. Permafrost extends south into northern Mongolia and northeast China. More permafrost occurs on the northern part of the Qinghai-Tibet Plateau, which is mainly covered in high-altitude tundra and steppe grasslands.

Much of the southern margins of the present-day active permafrost zone is underlain at depth by relict permafrost formed during the previous major cold events, represented elsewhere by major glacial deposits. However, a significant layer of younger permafrost also developed during the three main Neoglacial events, beginning about 6 ka and ending just over 100 years ago, and occurs in the surface layers of the ground (Figure 4.3).

The taiga in Asia is the result of the perched water table over the permafrost providing moisture for the evergreen trees in a relatively dry cold climate. The forest

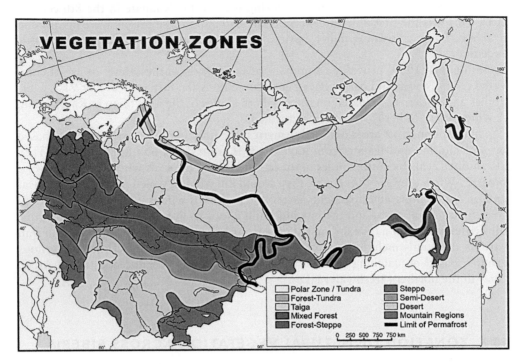

Figure 18.6 The natural vegetation zones of Asiatic Siberia compared with the southern limit of distribution of permafrost.

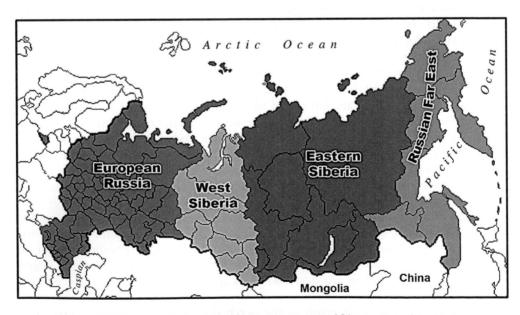

Figure 18.7 The boundaries of the three main regions of Siberia, west of the Urals.

Figure 18.8 The distribution of Boreal Forest across North America.

varies in species composition and density from north to south. The northern part has low tree density with an understory of mosses and lichens. This grades southwards into a dense forest underlain primarily by mosses while lichens mainly occur on the ground beneath the pine and larch forests. In the west, there is a transition zone with broadleaf forests forming a transition to the steppe zone, with a second similar forest on the coast near the Sea of Japan (Hays, 1993). The steppe consists of herbs, grasses and sedges, and grades into the semi-desert dominated by *Kobresia* species, chenopods, and leguminous plants.

18.3 ZONATION OF NATURAL VEGETATION IN NORTH AMERICA

Figure 18.8 shows the distribution of the Boreal Forest in North America. Black spruce is the dominant vegetation in the poorly drained areas across the region and represents 80% of the forest. Larches (*Larix laricina*) are found on base-rich fens, while Jack Pine (*Pinus banksiana*) grows on the well-drained sandy deposits. White Spruce (*Picea glauca*) is dominant on the better-drained sites in the mountains and elsewhere. There is an understorey of shrubs and herbs, the species richness decreasing with increasing with density of species and shrubs towards the North Pole. Cold, Arctic desert is present on many of the Arctic Islands. There is less floral diversity in the forested permafrost areas of North America due to repeated glaciations.

Most of the permafrost areas were covered by glaciers during the last major (Wisconsin) glaciation affecting North America. Exceptions occur in the former

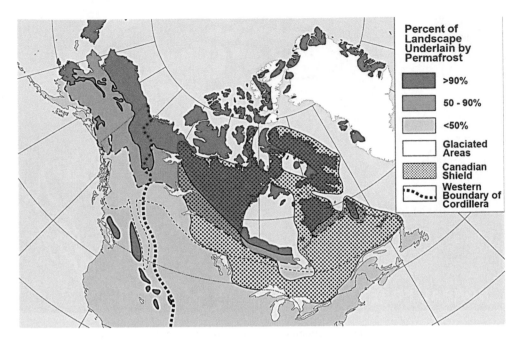

Figure 18.9 Distribution of lowlands outside the Canadian Shield that are currently underlain by permafrost.

Eastern Beringia of the Yukon Territory and Alaska, and on the tops of a few mountains along the eastern margin of the Western Cordillera, *e.g.*, Plateau Mountain, Alberta (Harris and Brown, 1978; Harris, 1997). This has resulted in the postglacial growth of permafrost occupying a smaller area in the north since the climate has been warmer during the Holocene (Figure 18.9). However, permafrost also occurs on the summits of the higher mountains along the eastern Cordillera south to New Mexico, as well as on the upper slopes of the highest volcanoes in Mexico (Figure 4.2). The two-layer arrangement of permafrost may occur in Eastern Beringia (Figure 4.3).

Another important difference is the limited area of relatively flat land underlain by alluvial or lacustrine sediments (Figure 18.9). The Western Cordillera is mountainous, with very few wide valleys suitable for agriculture. The granitic Canadian Shield underlies the central and eastern half of the mainland of Canada as well as Baffin Island. The glaciers have eroded the surface sediments, leaving large areas of bare rock with isolated sandy eskers.

Whereas the climate in Siberia consists of very cold, long winters and a reliably hot summer that lasts at least three months, the climate of northern Canada and Alaska is more varied. The high night-time temperatures affecting Siberia in summer are replaced by cool nights in North America, which also experiences significantly fluctuating summer temperatures, especially along the southern margins of the permafrost areas. These represent a significant limiting factor for expansion of agriculture.

Figure 18.10 A herd of Tibetan gazelle on wind-eroded, overgrazed meadow steppe on the Qinghai-Tibet Plateau north of Wudaoliang Station. © S. A. Harris.

18.4   SOUTHERN AND EASTERN KAZAKHSTAN, MONGOLIA AND THE QINGHAI-TIBET PLATEAU

These areas suffer from drought, resulting in the dominance of steppe grasslands, semi-deserts and deserts. Southeastern Mongolia is far from sources of precipitation, being in the rain shadow of the Altai Mountains and the Tien Shan to the west. The Qinghai-Tibet Plateau is at a sufficiently high elevation (>4500 m) that the shallower East China Monsoon cannot extend on to it under present-day conditions. Stable isotopes have shown that the limited precipitation on the northern part of the Plateau is the result of evaporation and recycling of water from the lake basins scattered across its surface.

R. Harris (2010) summarizes the evidence for rangeland degradation on the Qinghai-Tibet Plateau, together with the evidence for its magnitude and causes. Prevention and limitation of desertification is the critical factor in these areas (S. A. Harris, 2013b). The Chinese government has built small settlements outside the permafrost zone on steppe grasslands on the northeast slope of the Plateau at lower elevations and is encouraging the Tibetan herders to move there, but with limited success. The herders do not want to give up their nomadic life-style. The wind erosion along the Golmud-Lhasa Highway is widespread (Figure 18.10), but experiments placing large stones in a rectangular pattern on the bare ground is proving very successful at reducing wind erosion and allowing native plants to be planted to produce a continuous vegetation cover (Figure 18.11). Grazing is not allowed on the reclaimed ground. At lower elevations with seasonal frost, experiments using trees, shrubs and herbs indicate

Figure 18.11 The use of stones laid out in a rectangular pattern (c. 1 × 2 m) to produce sufficient surface roughness to allow herbaceous plants to develop a vegetation cover over the surface of formerly bare sand dunes on the Qinghai-Tibet Plateau. © S. A. Harris.

that the shrub cover is most effective since it traps any blowing sand or soil (Fu *et al.*, 2015).

18.5 THE EICHFELD ZONES

Eichfeld (1931) divided northern Asia into three agroclimatic zones (Figure 18.12). To provide a suitable comparison around the northern hemisphere, Harris (1968a) showed the zonation as applied to North America (Figure 18.13). Immediately, the fact stands out that there are much larger areas with a suitable climate for agriculture in Asia than in North America. This is partly due to the more extreme continental climate in northern Asia with at least three months of reliable hot summer weather. Much of the area is unglaciated lowland with podzolic soils and many rivers providing suitable drainage. The winter weather in North America is not as cold, except where there is cold air drainage, while the summers are shorter, nights are often colder, and the hot weather is more variable. As a result, crops have to endure a shorter frost-free period, colder night temperatures and unpredictable heat and moisture. A further complication is the presence of mountains in the west and in the Arctic Islands, swampy lowlands where proglacial lakes had been, a very large, rocky Canadian Shield with little soil, and many raised beaches extending far inland. The soils in the latter area are gravelly and often saline.

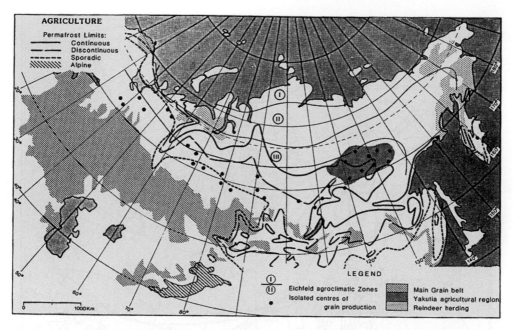

Figure 18.12 The Eichfeld agricultural zones in Eurasia compared with permafrost zonation (from Harris, 1986a).

18.5.1   Eichfeld zone I

This is the area where trees are absent and the ground cover consists of sparse tundra vegetation. Large areas of the landscape may be mountainous, while in the Arctic Islands of Northern Canada, considerable areas consist of a cold, dry desert or semi-desert. The only inhabitants are a few Inuit hunters and fishermen. Where there are settlements, vegetables can only be grown in raised hot beds or hot houses. Hydroponic greenhouses are often used, *e.g.*, using hot springs for heat at Esso in Kamchatka. The mean temperature of the vegetative period is 2.6–5.0°C. Subsistence farming includes nomadic reindeer herding in the south (Gregory, 1968) as well as fishing along the coasts.

18.5.2   Eichfeld zone II

In these subpolar areas, the summers are short, though this is partially compensated for by 24 hours of daylight at the summer solstice. Mean air temperatures during the vegetative period average 7–8°C. An important factor making a difference is the diurnal range of air and surface soil temperatures, which increase inland and towards the poles. The difference between north and south-facing slopes is low. The active layer is thin due to the short number of summer days. Field culture of cabbages and potatoes is possible. The northern stunted part of the Boreal Forest (larch, birch and balsam poplar) occurs in the warmer parts of this zone, although the forest has only 10–30% cover in Eastern Siberia. The southern part of the area in the western part of

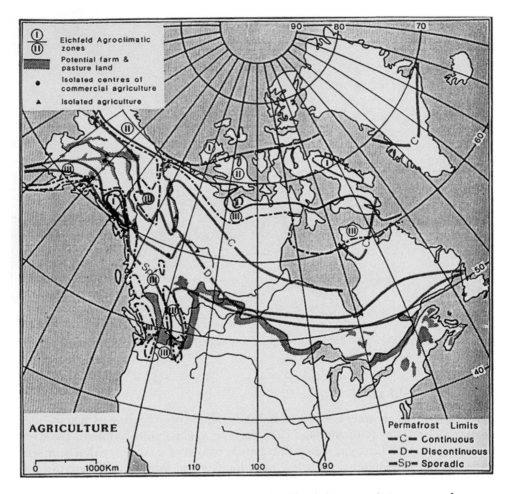

Figure 18.13 Distribution of Eichfeld zones as applied to North America relative to permafrost zones and areas of cultivatable land (from Harris, 1986a).

Eurasia is where subsistence hunting and fishing provide a living for families consisting of largely nomadic Inuit using reindeer as the main form of livestock. Figure 18.2 and 18.3 show typical winter and summer abodes for the Sakha people who live far away from towns and cities. Reindeer herding extends over a wider range of Eichfeld zones in the Far East of Siberia (Gregory, 1968).

However, in Canada, the natural vegetation of this zone is tundra, while some of the Arctic Islands have large areas of mountains coupled with little precipitation. The Inuit survived in the past by hunting seals and whales and by fishing, as noted above. The ongoing warming of the Arctic Ocean and increasing pollution of the sea are forcing the inhabitants of settlements such as Inuvik to import goods and food from the south at considerable cost. Obviously, their way of life is changing and becoming economically precarious.

18.5.3   Eichfeld zone III

18.5.3.1   The northern Taiga

This represents the zone of continuous permafrost consisting of the taiga vegetation in Siberia, but tundra in Canada, east of Hudson Bay. The difference is probably due to higher precipitation in eastern Canada. The growing season is short and vegetables are cultivated on raised beds to produce a longer growing season. Elsewhere in the Northwest Territories, a community greenhouse operates in Inuvik, producing fresh vegetables to augment the expensive and often poor quality produce trucked in from the south. Significant crops of potatoes are grown 140 km south of the Arctic Circle for local consumption (Clarke, 2013). The mean air temperature during the vegetative period is 9–10°C, though the period is shorter in western Canada than in Siberia.

Locally in Siberia, reindeer herding is carried on in the northern part of this zone, e.g., on the western side of the Urals and in the coastal areas of the Sea of Okhotsk. However, 85% of the land further south provides sustenance for sheep and cattle. The Yakutian cattle are the last remaining native Turano-Mongolian cattle breed in Siberia (Tapio et al., 2010). They provide milk and meat, and are also used as draft animals (Kantanen, 2009). There is also the rare native Yakutian horse that is noted for its adaptation to the extremely cold winter climate, and its ability to graze on vegetation that is buried under a deep snow cover (Hendricks, 1995).

In central Siberia, the longer summer permits the growth of field vegetables and early ripening cereals on the extensive gently sloping land throughout much of the southwestern area. The major objective of this agriculture is to provide local supplies of perishable foods such as milk, potatoes, vegetables and meat (Wein, 1984). These products are augmented by food brought in from designated areas in the south along the main transportation routes. Wein provides a good review of the system at that time, based on a number of Russian sources including Boyev and Gabov (1981). Since then, more grazing land has been reclaimed and drained to provide additional pasture. The area is now the heart of the Sakha Republic, which has its capital in Yakutsk.

A possible future method of farming around thermokarst lakes and alases could be sprinkler irrigation. It would need the use of electricity to pump the water uphill, but it could bring considerably greater areas into hay production. The runoff water would return to the lakes. However, this could also result in salinization unless the salinity is monitored. Tomirdiaro (1978) has previously suggested draining the lakes and the growing crops of grass on the resulting lake sediments containing organic matter.

This is also the main area of taiga in Siberia with a tall, dense stand of larch forest. It represents the main forest reserve of Russia. The trees can live up to 200 years in age, since there are few cyclones and thunderstorms causing forest fires. Summers are very hot and last for over 90 days and there is adequate precipitation in this lowland area providing water for several major north–flowing rivers. In central Siberia, the Yakut people found substantial areas of alas valleys with limited permafrost providing good agricultural land. These formed the basis for a significant agricultural zone. Sandy soils occur along the river terraces near Yakutsk. Gregory (1968, pp. 595–601) provides a concise history of the evolution of agriculture in the alas valleys along the Aldan River.

In western North America, the mountains limit the amount of potentially arable land (Figure 18.9). A limited area of agricultural land occurs in the broader valley floors in the mountains of southern Alaska and around Whitehorse and Dawson City, Yukon

Figure 18.14 Typical pasture in the grassy steppes west of Ulan Bator, Mongolia. © S. A. Harris.

Territory. In the Yukon, the land mainly produces hay for livestock. Where the summer precipitation is limited in Alaska, *e.g.*, the Matanuska River valley in Alaska, irrigation using sprinkler systems is used (Alaska Geographic, 1984, p. 69). This minimizes erosion, and the water warms up as it travels through the air. Around Fairbanks, trickle irrigation supplies water close to the plants. Many vegetables such as cabbages grow much larger at this latitude due to the long hours of daylight. However, the fact that this area is in the Western Cordillera of North America limits the area to the open river valleys and former intermontane lake-beds drained after the retreat of the glaciers.

Eastwards in the Northwest Territories, the swampy former pro-glacial lake beds along the Mackenzie valley have little agricultural potential. To the east lies the extensive Laurentian Shield, from which most of the soils were removed by the ice sheets. Granite rocks, sandy eskers and gravelly raised beaches do not provide suitable farmland. The peatlands and other wet lowlands have a forest consisting of stunted black spruce, which are poorly drained. So far, no commercial use has been found for the black spruce trees.

18.6 ASIAN STEPPE GRASSLANDS AND DESERTS

The permafrost areas of the lower Asian high-altitude mountain ranges and the southern margin of latitudinal permafrost in Mongolia have extensive steppe grasslands underlain by permafrost. The high radiation from the sun at this latitude produces a deeper active layer which permits the growth of grasses and associated herbs, but soil moisture conditions are too dry for the growth of trees. These are very productive farm lands provided that herds are moved from one pasture to another to avoid overgrazing (Figure 18.14). However, if overgrazing occurs just

Figure 18.15 A yak pulling a cart carrying part of a yurt in north-west Mongolia. © S. A. Harris.

once, the ground dries out and the grass has difficulty forming a protective cover, and thermokarst develops in the form of expanding alases (Harris, 2002b). Once started, melting of the ice in the permafrost is difficult to stop until there is no more ice, or the shallow thermokarst lake drains into a stream. The result is land lacking permafrost, but the lack of precipitation prevents the use of these areas for agriculture unless irrigation can be carried out using water from streams draining nearby mountains, *e.g.*, along the northern margin of the Hexi Corridor west of Lanzhou, China.

From northwest China to Iraq, and south to Kazakhstan, the *kareze* system of irrigation is used, consisting of underground tunnels dug into the slopes of hill slopes until they intersect the water table. These can only supply water to limited areas. Most of these are in areas lacking permafrost.

In the colder permafrost areas of the Qinghai-Tibetan Plateau, yaks replace cows as a source of milk, meat and oil. Yaks are also used as beasts of burden in Mongolia (Figure 18.15), but become ill if the air temperature rises above 10°C. Heavier loads are carried by camels which are animals that can perform well at most altitudes in the deserts and semi-deserts. However, at lower elevations such as in Mongolia and northern China, cows and sheep are the preferred animals, while donkeys and mules may also be found. However, the Mongolian horse is the predominant form of herding animal (Figure 18.16). It is very hardy and can paw through packed snow in winter to obtain food. It was the animal used by the Mongolians in their foray into Europe in the Middle Ages prior to the death of Genghis Khan, and even today, most young children in the country learn to ride before they can walk.

Figure 18.16 Mongolian horses and their traditional saddles. © S. A. Harris.

18.7 THE DEVELOPMENT OF MODERN AGRICULTURE IN PERMAFROST AREAS

Russian settlement and colonization began in 1558, when Grigory Dmitriyevich Stroganov obtained a charter from Ivan the Terrible to colonize the "empty lands" east of the Urals. He imported settlers, expropriating lands from the native tribes and ploughing up their lands. This started in the westernmost valleys and moved east. In 1620, the area around the Yenisey River in central Siberia came under cultivation. By 1660, cultivation had expanded into the upper Len and Angara River basins, growing rye, oats and barley. The Chernozems in the southwest around Krasnoyarsk were particularly fertile, whereas the poorer podzolic soils elsewhere beneath the larch forests generally did not provide enough food for the migrants, so that extra food had to be brought in from the west.

Tax benefits and free land persuaded yet more peasants to come east at the end of the 17th century. They brought with them other grains as well as vegetables including peas, carrots, turnips, onions and garlic. As the expansion east extended towards the Kamchatka Peninsula, the success rate of the new farms had been dropping, and in Kamchatka, the establishment of new farms finally failed (Forsyth, 1994). The total production of butter and grain was so great that after the completion of the Trans-Siberian Railway, a "Siberian tariff" was imposed between 1897 and 1913 to protect the European Russian farmers from very cheap Siberian agricultural products. Cost of transportation had been reduced between 5 and 6 times by construction of the railroad (Naumov, 2006).

Between 1965 and 1977, greenhouse complexes were operating near the larger population centres, and they increased vegetable production by 25% in west Siberia, 29% in East Siberia and 32% in the Far East. Of the greenhouse complexes, 61% were summer operations, the rest producing vegetables all the year round. Subsequently these have increased in number, since the new greenhouses could be located by large power plants to greatly reduce energy costs. In the north, individuals are encouraged to grow vegetables around their homes and summer cottages (dachas). The resulting produce is used to feed the immediate family, and any excess is taken to the local farmers market.

Southwards, humans almost certainly aided climate change in desertification of the grasslands and in the deforestation of the southern margins of the larch and spruce forests along both sides of the main Silk Roads across central Asia between Mongolia and China. During the last century, the Russians cut down the old larch forests in Kamchatka, leaving a poor secondary forest in its place. Unfortunately, no larches seedlings were planted, so the larch trees are now rare there. Likewise, the recent acquisition of the permafrost lands on the Qinghai-Tibet Plateau, followed by the formation of collective farms, aided the commencement of desertification there. Although the Chinese quickly reversed some of their policies, the damage had already been done (Harris, 2013b). Now, the Government is trying to persuade the Tibetans to move to new areas outside the Plateau and farm in a more sedentary way, in order to minimize the expansion of the cold high-altitude deserts.

During the early settlement along the Alaskan coast by the Russians, they started to grow their own food so that they would not have to eat fish all the time. Unfortunately, grains would not ripen in the short summer of the wet coastal climate, so they were without flour and bread. Fortunately, vegetables were more successful including tubers such as potatoes and rutabagas. Lettuce and cabbages grew well, but were watery. Turnips, radishes, beets, carrots onions and garlic did well. The traders also brought with them a few farm animals including chicken, cows, sheep pigs, horses and goats. This resulted in the meat of the chickens and pigs tasting like fish due to the fish fed to the animals. By contrast, European explorers and settlers did not usually bring seeds or animals with them when they travelled into the northern lands.

The main expansion of agriculture into Alaska came with the gold rushes. These opened up new areas away from the coasts, and it was far cheaper to grow your own food than to import it from the south. The United States set up seven agricultural stations about 1900 to carry out experiments to determine the most successful methods of agriculture there. They quickly determined that there were only two areas where this was successful, these being the Matanuska valley near Anchorage and the Tanana valley near Fairbanks. All but the Matanuska Research Station were soon abandoned. Today, it is cheaper to fly in produce from California to Anchorage and Fairbanks than to grow it locally.

18.8 FORESTRY

The European expansion into Siberia and America in the 16th century brought about tremendous changes. Although the newcomers learned a lot initially from the indigenous peoples, they greatly changed the local way of life, particularly in the Americas. In Siberia, with its vast size, the more distant peoples in extremely continental climates such as the Yakuts were able to continue largely in traditional endeavors. However, large gulags were established by Stalin in Siberia to provide lumber during the 20th century, while both there and in the permafrost areas of North America, mining was carried out, usually without any remedial action to protect the environment when the mining ceased. Since the crashing of the Russian forest industry in Siberia about 1990 and the subsequent reorganization of the Russian Federation, there have been substantial changes, especially along the southern borders of Siberia. The large areas of forest in the taiga remain an important reserve resource for the world.

Unfortunately, these reserves are being seriously damaged and exploited both in North America and Siberia (Lebedev, 2005; Newell and Simeone, 2014). In North America, the forests are being criss-crossed by cut lines for seismic exploration in areas with sedimentary rocks. Where commercially suitable resources are found, viable access roads are required to bring in drilling equipment, drilling pads and sumps must be constructed, while production wells require pipelines in which to transport the product to market. Newer methods may include fracking to break up the oil- or gas-bearing rocks to enhance production. The cut lines make the environment difficult for wildlife, and the roads and clearings open up areas for logging and cattle ranching, that had previously remained pristine. As a result of the change to Provincial control of the forestry industry in Canada during the 1970s, logging is proceeding at an unsustainable rate in many of the accessible areas south of the 60th parallel, without adequate replanting, or consideration of sustainability for the future. Summer grazing by cattle from adjacent ranches prevents the growth of young trees.

18.9   POTENTIAL EFFECTS OF CLIMATE CHANGES

While any climate warming results in problems with engineering structures and foundations, it is also liable to cause changes in the natural vegetation zones. Warming should result in desertification of steppe areas, while the latter should expand northwards into the areas currently at the southern margin of the taiga. Similarly, the taiga forests would be expected to move further north into the tundra areas. Under natural conditions, the change from forest to steppe is likely to be aided by destruction of the existing forest by a combination of fire and pests. Where deforestation occurs without replanting seedlings, the trees will tend be replaced by grasslands. The expansion of forest into tundra is likely to be rapid, as in the case of the Alpine forest rising into the tundra by at least 100 m at Marmot Basin Ski Area, Jasper, Alberta, since the end of the last Neoglacial. However, some ecologists argue that climate is less important than fire, *etc.*, in the alteration of the tundra-forest boundary in northern Canada (Trant *et al.*, 2013).

Cooling of the climate should result in a potential reversal of these changes, although the semi-desert vegetation may change via meadow-tundra into alpine tundra. Most predictions ignore possible changes in precipitation, but the latter can also determine the results. Similarly, the change from forest to tundra or grassland alters the hydrological environment and thermal regime of the ground, thus complicating the prediction of any changes. In spite of this, there are predictions that the current warming should make it possible for agriculture to extend further north in future, both in Siberia and in North America. However, any such changes would benefit Siberia far more than Canada because of the greater areas of suitable soils.

References

Aamot, H. W. C. (1966). *Dynamic Foundation measurements, Barter Island, Alaska*. U.S. Army Cold Regions Research and Engineering Laboratory Special Paper 75. 32p.

Abellán, A., Jaboyedoff, M., Oppikofer, T. & Vilaplana, J.M. (2009). Detection of millimetric deformation using a terrestrial laser scanner: experiment and application to a rockfall event. *Natural Hazards Earth System Science*, 9: 365–372. http://dx.doi.org/10.5194/nhess-9-365-2009.

Abellán, A., Oppikofer, T., Jaboyedoff, M., Rosser, N.J., Lim, M. & Lato, M.J. (2014). Terrestrial laser scanning of rock slope instabilities. *Earth Surface Processes and Landforms*, 39: 80–97.

Abraham, J. (2011). A promising tool for subsurface mapping.... USGS Fact Sheet 2011-3133.

Abyzov, S. S. (1993). Microorganisms in the Antarctic ice. In: Friedmann, E. I. (ed.). *Antarctic microbiology*. New York, N.Y., Wiley-Liss, Inc.: 265–295.

Abzhalimov, R. S. (1982). Laboratory investigations of frost heaving. *Soil Mechanics and Foundation Engineering*, 19(5): 205–207.

ACGR (1988). *Glossary of permafrost and related ground ice terms*. Permafrost Subcommittee, Associate Committee on Geotechnical Research, National Research Council of Canada Building Research Division, Ottawa. Technical Memorandum #142, 156p.

Adam, K. M. (1978). *Building and operating winter roads in Canada and Alaska*. Department of Indian and Northern Affairs, Environmental Division. Environmental Studies #4. 221p.

Adam, K. M. & Hernandez, H. (1977). Snow and Ice roads: Ability to support traffic and effects on vegetation. *Arctic* 30: 13–27.

Afanasenko, V. E., Garagulya, L. S., Ershov, E. D., Lebedenko, I. P., Maksimova, L. M. & Parmuzin, S. Y. (1991). Northern sensitivity of the ecosystem to the changes of geocryological conditions. In: Sychyov, K. I. (ed.). *Inzhenerno-Geologicheskie i Geokriologiches kie Issledocaniia v Geokriologii*. Moscow: 84–93. [In Russian].

Afanasenko, V. E., Gavroliova, A. V., Romanovsky, N. N. & Zaitsev, V. N. (1989). Geokriologiia SSSR. In: Ershov, E. D. & Romanovsly, N. N. (eds.). *Vostochnaia Sibir' I Dal'nii Vostok*. Moscow. Nedra. 515p. [In Russian].

Afanasenko, V. E. & Volkova, V. (1989). Characteristics of the formation of ground water runoff in the active water transfer zone in cryohydrological structures. In: Piguzova, V. M. (ed.). *Merzlotno-gidrogeologischeskie isle dovaniia zony vobodnogo vooobmena. Sbornik nauchnykh trudov*. Moscow. Nauka: 3–13. [In Russian].

AGI (2005). *Glossary of Geology*. 5th Edition. Neyndorf, K. K. E., Mehr, Jr., J. P. & Jackson, J. A. (Eds.). American Geological Institute.

Ahlvin, R. G. & Smoots, A. V. (1988). *Construction Guide for Soils and Foundations*. New York. John Wiley and Sons.

Åhman, R. (1976). The structure and morphology of minerogenic palsas in Norway. *Biuletyn Periglacjalny*, 26: 25–31.

Åhman, R. (1977). *Palsar I Nordnorge.* Meddelanden Fråns Lunds Universitets Geografiska Institution Avhandlinger, 78. 165p. [In Swedish].

Ahumada, A. L. (1986). Un aspect aplicado de la Geocriologia. *Acta Geocriogenica, 4th Reunion de la Subcomision Latinoamericana sobre la Importance de los Procesos Periglaciales,* 9–13. [In Spanish].

Ahumada, A. L. (1992). Periglacial climatic conditions and vertical form associations in Quebrada Benjamin Matienzo, Mendoza, Argentina. *Permafrost and Periglacial Processes,* 3: 221–224.

Aitken, G. W. (1970). Transport of frozen soil. *Proceedings of the Vermont Conference on Winter Construction (1969).* University of Vermont: 50–68.

Akagawa, S. & Fukuda, M. (1991). Frost heave mechanism in welded tuff. *Permafrost and Periglacial Processes,* 2: 301–309.

Åkerman, H. J. (1973). Preliminära resultat från undersökningar av massrörelser vid Kapp Linné, Spetsbergen. *Lunds Universtets Naturgeografiska Institution, Rapporter och Notiser,* 18: 1–13. [In Swedish].

Åkerman, H. J. (1980). *Studies on periglacial geomorphology in West Spitsbergen.* Meddelanden Fråns Lunds Universitets Geografiska Institution, Avhandlingar, 89. 298p.

Åkerman, H. J. (1982). Observations of palsas within the continuous permafrost zone in eastern Siberia and in Svalbard. *Geografisk Tiddskrift,* 82: 45–51.

Aksenov, V. I. & Brouchkov, A. V. (1993). Plastic frozen (saline) soil as bases. In: *Proceedings of the 6th International Conference on Permafrost, Beijing, China.* Wushan, Guangzhou. South China University of Technology Press, 1: 1–4.

Aksenov, V. I. & Petrukhin, I. S. (1991). Improving the structural properties of saline frozen ground by selecting optimal sandy-clay mixtures. In: Dubikov, G. I. (ed.). *Merzlye porody i kriogennye protessy; sbornik nauchnykh trudov.* Moscow. Nauka: 41–46. [In Russian].

Alaska Community Database Community Information Summaries (ACDCIS) (2008). *Tyonek.* http://www.commerse.state.ak.us/dca/commdb/CIS.cfm?Comm_Boro_Name=Tyonek.

Alaska Department of Environmental Conservation (2002). Division of Environmental Health, Solid waste Management Program. Title 18, Chapter 60 of the Alaska Administrative Code. *18 AAC 60.205, Purpose, scope and applicability classes of the MSWLF [3] and 18 AAc 60.205, Solid Waste Planning. http://www.legis.state.ak.us/foliosa.dII/aac/query*

Alaska Geographic (1984). Alaska's farms and gardens. *Alaska Geographic,* 11(2): 142p.

Alekseev, V. R., et al., Eds. (1973). *Siberian naleds.* USSR Academy of Sciences (1969). Draft translation 399. Hanover, New Hampshire. U.S. Army Cold Regions Research and Engineering Laboratory. 300p.

Alekseev, V. & Savko, N. (1975). *The theory of naled processes.* Nayka, 205p.

Alestalo, J. & Häikiö, J. (1976). Ice features and ice-thrust shore forms at Luodonselkä, Gulf of Bothnia, in winter 1972–73. *Fennia,* 144: 1–24.

Aleshinskaya, Z. V., Bondarev, L. G. & Gorbunov, A. P. (1972). Periglacial phenomena and some paleogeographical problems of central Tien Shan. *Biuletyn Peryglacyjalny,* 21: 5–13.

Allen, L. T. (1977). *The Trans-Alaska Pipeline.* Alyeska Pipeline Service Company. 2 volumes.

Allard, M., Caron, S. & Bégin, Y. (1996). Climatic and ecological controls on ice segregation and thermokarst: the case history of a permafrost plateau in Northern Québec. *Permafrost and Periglacial Processes,* 7(3): 207–227.

Allard, M. & Kasper, J. N. (1998). Temperature conditions for ice-wedge cracking: Field measurements from Salluit, Northern Québec. *Permafrost.* 7th International Conference Proceedings, Yellowknife (Canada). *Collection Nordicana,* 55: 5–12.

Allard, M. & Rousseau, L. (1999). The internal structure of a palsa and a peat plateau in the Boniface region, Québec: inferences on the formation of the ice segregation mounds. *Géographie physique et Quaternaire,* 53(3): 373–387. [In French].

Alyeska Pipeline Service Company (1977). *Summary project description of the Trans-Alaska Pipeline System*. 20p.

Amann, R., Ludwig, W. & Schleifer, K. H. (1995). Phylogenetic identification and *in situ* detection of individual microbial cells without cultivation. *Microbiological Revue*, 59: 143–169.

Amato, P., Doyle, S. M., Battista, J. R. & Christner, B. C. (2010). Implications of sub-zero metabolic activity on long-term microbial survival in terrestrial and extraterrestrial permafrost. *Astrobiology*, 10: 789–798.

AMEC Earth and Environmental (2005). *Inuvialuit settlement region drilling waste disposal sumps study*. Submitted to the Government of Canada, Environmental Studies Research Fund, ESRF 04-046. 54p.

Amitrano, D. S., Gruber, S. & Girard, L. (2012). Evidence of frost-cracking inferred from acoustic emissions in a high-alpine rock-wall. *Earth and Planetary Science and Letters*, 341–344. 86–93. Doi: 10.1016/j.epsl.2012.06.014.

An, C. B., Feng, Z. D. & Barton, L. (2006). Dry or humid? Mid Holocene humidity changes in arid and semi-arid China. *Quaternary Science Reviews*, 25: 351–361.

Ananjeva (Malkova), G. V., Melnikov, E. S. & Ponomareva, O. E. (2003). Relict permafrost in the central part of western Siberia. In: Phillips, M., Springman, S. M. and Arenson, L. U. (eds.). *Permafrost. Proceedings of the 8th International Conference on Permafrost Zurich, Switzerland*. Lisse. Swerts and Zeitlinger: 5–8.

Andersland, O. B. & Alwahhab, M. R. B. (1983). Lug behaviour for model steel piles in frozen sand. In: *Permafrost: 4th International Conference Proceedings*. Washington, D.C. National Academy Press: 16–21.

Andersland, O. B. & Ladanyi, B. (1994). *An Introduction to Frozen Ground Engineering*. New York. Chapman and Hall.

Anderson, D. M. (1971). *Remote analysis of Planetary Water*. United States Army Corps of Engineers, Cold Regions Research and Engineering Laboratory Special Report # 154. 13p.

Anderson, D. M., Chamberlin, G. L., Guymon, G. L., Kane, D. L., Kay, B. D., Mackay, J. R., O'Neill, K., Outcalt, S. I. & Williams, P. J. (1984). *Ice segregation and frost heaving*. Washington, National Academy Press, 72p.

Anderson, D. M., Reynolds R. C. & Brown, J. (1969). Bentonite debris flows in southern Alaska. *Science*, 164: 173–174.

Andersson, J. G. (1906). Solifluction: A component of subaerial denudation. *Journal of Geology*, 14: 91–112.

André, M.-F. (1990). Frequency of debris flows and slush avalanches in Spitzbergen: A tentative evaluation from lichenometry. *Polish Polar Research*, 11(3–4): 345–363.

André, M.-F. (1993). *Les versants du Spitsberg*. Nancy. Presses Universitaires de Nancy. 361p. [In French].

André, M.-F. (1994). The geomorphic impact of glaciers as indicated by tors in Northern Sweden (Aurivaara, 68°N). *Geomorphology*, 57: 403–421.

André, M.-F. (1995a). Holocene climatic fluctuations and geomorphic impact of extreme events in Svalbard. *Geografiska Annaler*, 77A: 241–250.

André, M.-F. (1995b). Post-glacial microweathering of granite *roche moutonnées* in Northern Scandinavia (Riksgränsen area, 68°N). In: Slaymaker, O. (ed.). *Steepland Geomorphology*. Chichester. John Wiley and Sons. Pp. 103–127.

André, M.-F. (1997). Holocene rockwall retreat in Svalbard: A triple-rate evolution. *Earth Surface Processes and Landforms*, 22: 423–440.

André, M.-F. (2004). The geomorphic impact of glaciers as indicated by tors in northern Sweden (Aurivaara, 68°N). *Geomorphology*, 57: 403–421.

André, M-F., Hall, K., Bertran, P. & Arocena, J. (2008). Stone runs in the Falkland Islands: Periglacial or Tropical? *Geomorphology*, 95: 524–543.

Andre'eva,Y., Larichev, O. I., Flanders, N. E. & Brown, J. (1995). Complexity and uncertainty in Arctic Resource decisions. *Polar Geography and Geology*, 19: 22–35.

Andrews, J. T. (1961). The development of scree slopes in the English Lake District and central Quebec-Labrador. *Cahiers de Géographie de Québec*, 10: 219–230.

Anisimov, O. A. & Nelson, F. E. (1996). Permafrost distribution in the Northern Hemisphere under scenarios of climate change. *Global and Planetary Change*, 14: 59–72.

Anisimov, O. & Renova, S. (2006). Permafrost and changing climate: The Russian perspective. *Ambio*, 35(4): 169–175.

Anisimov, O. A., Shiklomanov, N. I. & Nelson, F. E. (1997). Effects of global warming on permafrost and active-layer thickness: Results from transient general circulation models. *Global and Planetary Change*, 15: 61–77.

Anisimov, V. A. & Sorokin, V. A. (1975). Repair work on a frozen dam. *Hydrological Construction*, May (5): 24–25. [In Russian].

Anisimova, N. P. (1973). *Ground water in the lithosphere*. Hanover, New Hampshire. U.S. Army Cold Regions Research and Engineering Laboratory. Draft Translation 437.

Anisimova, N. P., Nikitina, N. M., Piguzova, V. M. & Shepelyev, V. V. (1973). *Water resources of Central Yakutia*. Guidebook, 2nd International Permafrost Conference, Yakutsk, USSR. 47p.

Anon. (1956). Permafrost research in Northern Canada. *Nature*, 178: 716–717.

Anon. (1982). Hydrocarbon development in the Beaufort Sea-Mackenzie Delta Region. *Environmental Impact Statement*, volume 2. Development Systems. Ottawa.

Anon. (1995). *Foundations of geocryology*. Moscow, Moscow University Press, Volume 1, 368p. ISBN 5-211-02464-8; 5-211-02637-3. [In Russian].

Antevs, E. (1932). Cenozoic climate of the Great Basin. *Geologische Ruddschen*, 40: 94–108.

Aré, F. E. (1978). The reworking of shorelines in the permafrost zone. In: *Permafrost: Proceedings of the 2nd International Conference*, USSR Contribution. Washington, D.C. National Academy of Sciences: 59–62.

Aré, F. E. (1983). Thermal abrasion of coasts. In: *Permafrost, Proceedings of the 4th International Permafrost Conference*. Washington, D.C., National Academy Press: 24–28.

Aré, F. E. (1988). Thermal abrasion of sea coast. *Polar Geography and Geology*, 12: 1–157.

Aré, F. E. (1999). The role of coastal retreat for sedimentation in the Laptev Sea. In: Kassens, H., Bauch, H. A., Eicken, H., Hubberten, H. W., Melles, M., Thiede, J. & Tomokhov, L. A. (eds). *Land-Ocean systems in the Siberian Arctic: Dynamics and History*. Berlin. Springer: 278–295.

Arenson, L. U., Azmatch, T. F. & Sego, D. C. (2008). A new hypothesis on ice lens formation in frost-susceptible soils. *Proceedings of the 9th International Conference on Permafrost, Fairbanks*: 59–64.

Arenson, L., Hoelzle, M. & Springman, S. (2002). Borehole deformation measurements and internal structures of some rock glaciers in Switzerland. *Permafrost and Periglacial Processes*, 13: 117–135.

Arkhangelov, A. A. & Novgorodova, E. V. (1991). Genesis of massive ice at "Ice Mountain", Yenesei River, Siberia, according to results of gas analyses. *Permafrost and Periglacial Processes*, 2(2): 167–170.

Armstrong, B. C., Smith, D. W. & Cameron, J. J. (1981). Water requirements and conservation alternatives for northern communities. In: Smith, D. W. and Hrudey, S. E. (eds.). *Design of water and wastewater services for cold climate communities*: 65–93.

Artières, O., Lostumbo, J., Watn, J., Bæverfjord, G. G., Delmas, P., Caquel, F., Grande, L. & Langeland, A. (2010). Geosynthetics as eco-friendly defence against erosion in arctic Regions. *GEO2*10*: 634–641.

Artyushkov, E. V. (1969). About pressing of ice-wedges by surrounding deposits. In: *Problems in Cryolithology* 1. Moscow. Moscow State University: 34–37. [In Russian].

ASARC (2015). Publications on applied snow and avalanche research. www.ucalgary.ca/asarc/publications.

Ascaso, C., Sancho, L. G. & Rodriguez-Pascual, C. (1990). The weathering action of saxicolous lichens in maritime Antarctica. *Polar Biology*, 11: 33–39.

Ashcroft, F. (2000). *Life at the Extremes*. London, U.K., HarperCollins: 326p.

Ashpiz, E. & Khrustalev, L. (2013). Evaluation of embankment foundation creep in conditions of deep bedding of roof in permafrost soils. *Sciences in Cold and Arid Regions*, 5(5): 534–539.

Ashpiz, E. S., Khrustalev, L. N., Emelyanova, L. V. & Vedernikova, M. A. (2010). Using of Synthetical Thermal Insulators for Conservation of Frozen Soil Conditions in the base of Railway Embankment. *GEO2*10*: 557–561.

Associated Engineering Services, Ltd. (1975). *Department of Environment Sewage Disposal Study, Jasper Townsite*. Report to the Environmental Protection Service, Edmonton, Alberta.

Astakhov, V. I. (1986). Geological conditions for the burial of Pleistocene glacier ice on the Yenisey. *Polar Geography and Geology*, 10(4): 286–295.

Astakhov, V. I. (1992). The last glaciations in West Siberia. *Sveriges Geologiska Undersökning*, 81: 21–30.

Astakhov, V. I. (1995). The mode of degradation of Pleistocene permafrost in West Siberia. *Quaternary International*, 28: 119–121.

Astakhov, V. I. & Isayeva, L. L. (1988). The "Ice Hill", an example of "retarded deglaciation" in Siberia. *Quaternary Science Reviews*, 7: 29–40.

Astakhov, V. I., Kaplyanskaya, F. A. & Tarnogradsky, V. D. (1996). Pleistocene permafrost of West Siberia as a deformable glacier bed. *Permafrost and Periglacial Processes*, 7: 165–191.

ASTM, Subcommittee D18.19. (2013). Standard test methods for frost heave and thaw weakening susceptibility. ASTM Book of Standards, volume 04.09.

Aubekerov, B. & Gorbunov, A. (1999). Quaternary Permafrost and Mountain Glaciation in Kazakhstan. *Permafrost and Periglacial Processes*, 10: 65–80.

Auld, R. G., Robbins, R. J., Rosenegger, L. W. & Sanster, R. H. B. (1978). Pad foundation design and performance of surface facilities in the Mackenzie Delta. *Proceedings of the 3rd International Conference on Permafrost, Edmonton*. Ottawa. National Research Council of Canada 1: 765–771.

Aulitzky, H. (1970). Der enterbach (Inzing in Tirol) am 26 Juli, 1969. *Wilbach – und Lawinenverbau*, 34: 31–66. [In German].

Aylesworth, J. M. & Egginton, P. A. (1994). Sensitivity of slopes to climate change. In: Cohen, J. (ed.). *Mackenzie Basin Impact Study (MBIS), Interim report 2*. Proceedings of the 6th Biennial AES/DIAND Meeting on Northern Climate and Mid Study Workshop of the Mackenzie Basin Impact Study, Yellowknife, Northwest Territories. Downsview, Ontario. Environment Canada.

Aylesworth, J. M., Duk-Rodkin, A., Robertson, T. & Traynor, J. A. (2000). Landslides of the Mackenzie Valley and adjacent mountainous and coastal regions. In: Dyke, L. D. and Brooks, G. R. (eds.). *The Physical Environment of the Mackenzie Valley, Northwest Territories: A Base Line for the Assessment of Environmental Change*. Ottawa. Geological Survey of Canada Bulletin #176: 167–176.

Azmatch, T. F., Sego, D. C., Arenson, L. U. & Biggar, K. W. (2012). New ice lens initiation condition for frost heave in fine-grained soils. *Cold Regions Science and Technology*, 82: 8–13.

Babb, A. L., Chow, D. M., Garlid, K. L., Popovich, R. P. & Woodruff, E. M. (1971). The thermo tube, a natural convection heat transfer device for stabilization of Arctic soils in oil producing regions. *46th Annual Meeting, New Orleans, Society of Petroleum Engineers, American Institute of Mechanical Engineering* Paper SPE 3618. 12p.

Bachéus, I. & Grab. S. W. (1995). Mires in Lesotho. *Gunneria*, 70: 243–250.

Badu, Y. B. & Trofimov, V. T. (1981). Ice content and subsidence due to melting of frozen soils of the southern part of the Western Siberian plate. *Priodnyye Usloviya Zapadnoy Sibiri*, 8: 58–63. [In Russian].

Bai, Y., Yang, D. Q., Wang, J. H., Zhang, G. S., Xu, S. J., Liu, G. X. & An, L. Z. (2005). Isolation and sreening of cold-active enzymes-producing psychrotrophic bacteria from permafrost in the Tianshan Mountains. *Journal of Glaciology and Geocryology*, 27: 615–618.

Bai, Y., Yang, Wang, J. H., Xu, S. J., Wang, X. X. & An, L. Z. (2006). Phylogenetic diversity of culturable bacteria from alpine permafrost in the Tianshan Mountains, northwestern China. *Research in Microbiology*, 157: 741–751.

Bakakin, V. P. (1978). Basic Areas of Geocryological Research in mining. *Proceedings of the 2nd International Conference on Permafrost, Yakutsk*. USSR Contribution, Washington. National Academy of Sciences: 585–586.

Bakermans, C. (2008). Psychrophiles: From Biodiversity to Biotechnology. In: Margesin, R., Schinner, F., Marx, J. C. and Gerday, C. (eds.). Berlin/Heidelberg, Germany, Springer Verlag: 17–28.

Bakulina, N. T. & Spector, V. B. (2000). *Climate and Permafrost*. In: Maksimov, G.N. and Fedorov, A.N. (eds). Yakutsk, Permafrost Institute: 21–32. [In Russian].

Balch, E. S. (1900). *Glacières and freezing caverns*. Philadelphia: Allen, Lane and Scott. 326p.

Balkwill, H. R., Roy, K. J., Hopkins, W. S. & Sliter, W. V. (1974). Glacial features and pingos, Amund Ringnes Island, Arctic Archipelago. *Canadian Journal of Earth Sciences*, 11: 1319–1325.

Ball, D. F. (1966). Late-glacial scree in Wales. *Biuletyn Peryglacjalny*, 15: 151–163.

Ballantyne, C. K. (1986). Protallus rampart development and the limits of former glaciers in the vicinity of Baosbheinn, Wester Ross. *Scottish Journal of Geology*, 22: 13–25.

Ballantyne, C. K. (1996). Formation of miniature sorted patterns by shallow ground freezing: A field experiment. *Permafrost and Periglacial Processes*, 7: 400–424.

Ballantyne, C. K. & Eckford, J. D. (1984). Characteristics and evolution of two relict talus slopes in Scotland. *Scottish Geographical Magazine*, 100: 20–33.

Ballantyne, C. K. & Harris, C. (1994). *The periglaciation of Great Britain*. Cambridge. Cambridge University Press. 330p.

Ballantyne, C. K. & Kirkbride, M. P. (1987). Rockfall activity in upland Britain during the Loch Lomond stadial. *Geographical Journal* 53(1): 86–92.

Ballantyne, C. K. & Matthews, J. A. (1982). The development of sorted circles on recently deglaciated terrain, Jotenheimen, Norway. *Arctic and Alpine Research*, 14(4): 341–354.

Ballantyne, C. K. & Matthews, J. A. (1983). Dessication cracking and sorted polygon developments, Jotenheimen, Norway. *Arctic and Alpine Research*, 15: 339–349.

Balmér, P. (1981). Swedish experiences with wastewater treatment with special reference to cold climates. In: Smith, D. W. and Hrudey, S. E. (eds.). *Design of water and wastewater services for cold climate communities*. Oxford. Pergamon Press: 125–135.

Balobayev, V. T. (1978). Reconstruction of Palaeoclimate from present-day geothermal data. *Proceedings of the 3rd International Conference on Permafrost, Edmonton*. Ottawa, National Research Council of Canada, 1: 10–14.

Balobayev, V. T. & Pavlov, A. V. (1998). Evolution of permafrost in the western and East Siberia at contemporary climate change. In: Ohata, T. and Hiyama, T. (eds.). *Proceedings of the 2nd International Workshop on Energy and Water Cycle in GAME-Siberia*. Nagoya, Kapan. Nagoya University, Institute for Hydrospheric-Atmospheric Sciences. #4: 79–82.

Baranov, I. Yu. (1949). Some glacial formations on the surface of the soil. *Piroda*, 38(10): 47–50. [In Russian].

Baranov, I. Yu. (1959). Geograficheskoye rasprostraeniye sezonnepromerzayuschchikh pochv i mnogoletnemerzlykh gornyk porod. Institut Merzlotovedeniya. In: Obrucheva, V. A. (ed.).

Osnovy geokriologii (merzlotovedeniya), Chast' pervaya, Obshchaya geokriologiya. Moscow. Akadenie Nauk S.S.S.R. 459p. [In Russian].

Baranova, U. P., Il'inskay, I. A., Nikitin, V. P., Pneva, G. P., Fradkina, A. F. & Shvareva, N. Y. (1976). *Works of Geological Institute of Russian Academy of Sciences.* Moscow, Nauka: 284p. [In Russian].

Barendregt, R. W. & Duk-Rodkin, A. (2012). Chronology and extent of Late Cenozoic ice sheets in North America: magnetostratigraphical assessment. *Studies in Geophysical Geology*, 56: 705–724. DOI: 1007/s11200-011-9019-3

Baross, J. A. & Morita, R.Y. (1978). Life at Low Temperatures: Ecological Aspects. In: Kushner, D. J. (ed.). *Microbial Life in Extreme Environments*, London, Academic Press: 9–71.

Barret, L. R. & Schaetzl, R. J. (1992). An examination of podsolization near Lake Michigan using chronofunctions. *Canadian Journal of Soil Science*, 72: 527–541.

Barry, R. G. & Gan, T. Y. (2011). *The Global Cryosphere, past, present and future.* Cambridge. Cambridge University Press. 473p.

Barsch, D. (1971). Neuere untersuchungen an blockgletschen in den Schweizer Alpen. *Verh Schweiz Naturforsch Ges*, 151: 122. [In German].

Barsch, D. (1973). Refraktionsseismische bestimmung der Obergrenze des gefrorenen schuttkörpers in versschiedenen bockgletschern Graubündens, Schweizer Alpen. *Zeitschrift für Gletscherk Glazialgeologie*, 9: 143–167. [In German].

Barsch, D. (1977). Alpiner Permafrost: ein Beitrag zur Verbreitung, zum Charakter und zur Okologie am Beispiel der Schweizer Alpen. In: Poser, H. (ed.). *Formen, Formengesellschften und Untergrenzen in den heutigen periglazialen Höhenstufen der Hochgebirge Europas und Africas zwischen Aektis und Äquator.* Academie Wissenschaften Göttingen, Mathematische/Phys. K1 Folge 3, 31: 118–141. [In German].

Barsch, D. (1987). Rock Glaciers: an approach to their systematics. In: Giardino, J. R., Shroder, J. F. and Vitek, J. D., (eds.). *Rock Glaciers.* London. Allen and Unwin: 41–44.

Barsch, D. (1996). *Rockglaciers.* Berlin. Springer-Verlag. 331p.

Barsch, D., Blümel, W. D., Flügel, W. A., Mäusbacher, R., Stäblein, G. & Zick, W. (1985). Untersuchungen zum Periglazial auf der König-Georg-Insel, Südshetlandinseln, Antarktika. *Ber Polarforschung*, 24: 75. [In German].

Barsch, D., Gude, M., Maeusbacher, R., Schukraft, G., Schulte, A. & Strauch, D. (1993). Slush stream phenomena – process and geomorphic impact. *Zeitschrift für Geomorphologie Supplementband*, 92: 39–53.

Barsch, D. & Hell, G. (1975). Photogrammetrische bewegungsmessungen am blockgletscher Murtèl I, Oberengardin, Schweitzer Alpen. *Zeitschrift für Gletscherkunde und Glaziolgeologie*, 11: 111–142. [In German].

Barsch, D. & King, L. (1989). Origin and geoelectrical resistivity of rockglaciers in semi-arid subtropical mountains(Andes of Mendoza, Argentina). *Zeitschrift für Geomorphologie* N.F., 33: 151–163.

Bates, R. L. & Jackson, J. A. (1987). Glossary of geology. *American Geological Institute*, 3rd Edition, Alexandria, Virginia.

Bauer, A., Harris, G. R., Lang, L., Prezioni, P. & Sellick, D. J. (1965). How 10c. puts Crater Research to work. *Engineering and Mining*, 166(9): 117–121.

Bauer, A., Calder, P. N., Maclachlan, R. R. & Halupka, M. (1973). Cratering and ditching with explosives in frozen soils. *Defense Research Board, Canada*. Report DREV R-699/73, 123p.

Baulig, H. (1956). Pénéplaines et pédiplaines. *Bulletin de la Sociéte Belge pour d'Etudes Géographie*, 25: 25–58. [In French].

Baulin, V. V. (1962). Osnovnye etapy istorii ratzvitya mnogoktnemerzlylch porod na territorii Zapadno-Sibirskoy nizmennosti. Akademya Nauk SSSR. Institut Merzlotovedeniya in V. A. Obrucheva *Trudy*, 19: 5–18. [In Russian].

Baulin, V. V. (1967). Ice wedges and of the upper Pleistocene of the western part of the Western Siberian plain. *Merzlotnyye Issledovaniya*, 7: 174–184. [In Russian].

Baulin, V. V., Belopukhova, Y. B., Danilova, N. S., Dubikov, G. I. & Stremyakov, A. Y. (1978a). The role of tectonics in the formation of plains. In: *Proceedings of the 3rd International Conference on Permafrost, Edmonton, Alberta*. Ottawa. National Research Council of Canada: 241–246.

Baulin, V. V., Belopukhova, Y. B., Dubikov, G. I. & Shmelev, L. M. (1967). *Geokriologicheskie (merzolotnye) usloviya Zapadno-Sibirskoy Nizmennosti*. Moscow. Izd. "Nauka". [In Russian].

Baulin, V. V., Bykov, I. Y., Sadchikov, P. B., Solovyev, V. V., Sedov, N. V., Shaposhnikova, Y. A. & Umnyakhin, A. S. (1978b). Relict permafrost in the North European part of the USSR. *Transactions (Doklady) of the Russian Academy of Sciences: Earth Science Section*, 24(1–6): 39–42. [In Russian].

Baulin, V. V., Danilova, N. S., Pavlova, O. P. & Sukhodol'skaya, L. A. (1984). Permafrost conditions of the Medvezh'ye gas field and focasting them during development of the territory. In: *Geocryological conditions and forecasting their change in regions of priority development in the north*. Moscow. Stryizdat and PNIIIS: 3–24. [In Russian].

Beardsley, G. F. & Cannon, W. A. (1930). Note on the effects of a mudflow at Mt. Shasta on the vegetation. *Ecology*, 11: 326–336.

Beaulac, I. (2006). *Impacts du pergélisol et adaptations des infrastructures de transport routier et aérien au Nunavik*. M.Sc. thesis, Départment de Génie Civil, Université Laval, Québec, Canada. 250p. [In French].

Beaulac, I. & Doré, G. (2004). *Impact du dégel du pergélisol sur les infrastuctures de transport aérien et routier au Nunavik et adaptations, état des connaissances*. Rapport GCT-04-05, Ministère des Transports du Québec. 146p. [In French].

Beaulac, I. & Doré, G. (2006a). Airfields and access roads performance assessment in Nunavik, Québec, Canada. *Proceedings of the 3rd International Conference on Cold Regions Engineering, Orono, Maine, USA*.

Beaulac, I. & Doré, G. (2006b). Development of a new heat extraction method to reduce permafrost degradation under roads and airfields. *Proceedings of the 3rd International Conference on Cold Regions Engineering, Orono, Maine, USA*.

Becher, M., Olid, C. & Klaminder, J. (2013). Buried soil organic inclusions in non-sorted circles fields in northern Sweden: Age and palaeoclimatic context. *Journal of Geophysical Research: Biogeosciences*, 118: 104–111.

Beilman, D. W., MacDonald, G. M. & Yu, Z. (2010). The northern peatland carbon pool and the Holocene carbon cycle. *PAGES news*, 18(1): 22–24.

Beke, G. J. & McKeague, J. A. (1984). Influence of tree windthrow on the properties and classification of selected forest soils. *Canadian Journal of Soil Science*, 64: 195–207.

Belopukhova, E. B. (1966). Features of permafrost relief in the western Siberia. *Materials of the 8th USSR Conference on Goecryology*. Yakutsk 6: 5–13. [In Russian].

Belopukhova, E. B. & Sukhov, A. G. (1986). Positive forms of mesorelief as the result of ground ice formation. *Polar Geography and Geology*, 10(4): 296–302.

Belotsercoskaya, G. V. (1986). Causes of deformations in buildings and structures in the Vorkuta industrial region. In: *Transactions SO NIIOSP (North Branch of Gersvanov Research Design Survey and Technological Institute for underground construction)*, Syktyvkar: 93–96. [In Russian].

Belotsercoskaya, G. V., Fedoseev, Y. G. & Yanchenko, O. M. (1989). Causes of deformation of buildings on permafrost in the railroad district of Vorkuta. *Soil Mechanics and Foundation Engineering*, 26(3): 102–106.

Benedict, J. B. (1970a). Downslope soil movement in a Colorado alpine region: rates, processes and climatic significance. *Arctic and Alpine Research*, 2: 165–226.

Benedict, J. B. (1970b). Frost cracking in the Colorado Front Range. *Geografiska Annaler*, 52A: 87–93.

Benedict, J. B. (1976). Frost creep and gelifluction features: A review. *Quaternary Research*, 6: 55–76.

Bennett, M. R., Mather, A. E. & Glasser, N. F. (1996). Earth hummocks and boulder runs at Merrivale, Dartmoor. In: Charman, D. J., Newham, R. W. and Croot, D. W. (eds). *Devon and Cornwall: Field Guide*. London. Quaternary Research Association: 81–96.

Berdnikov, V. V. (1970). Relict permafrost microrelief in the upper Volga Basin. *Geomorphology*, 4: 327–332.

Berdnikov, V. V. (1986). Late Pleistocene permafrost phenomena in the European part of the USSR and their significance for paleoclimatic reconstructions. *Biuletyn Periglacjalny*, 30: 35–43.

Berdnikov, V. V. (1970). Relict permafrost microrelief in the upper Volga Basin. *Geomorphology*, 4: 327–332.

Berezovskaya, S. L., Dmitrienko, I. A. & Kirilov, S. A. (2002). Interannual variability of the thermal regime of the bottom water layer in the Laptev Sea under different atmospheric circulation conditions. *Abstracts of the International Conference, "Extreme phenomena in the Cryosphere: Basic and applied aspects*. Puschino: 240–241. [In Russian].

Berdinnikov, V. V. (1986). Late Pleistocene permafrost phenomena in the European part of the USSR and their significance for palaeoclimatic reconstructions. *Biuletyn Peryglacjalny*, 30: 35–43.

Berg, R. L. & Aitken, G. W. (1973). Some passive methods of controlling geocryological conditions in roadway construction. *Proceedings of the 2nd International Conference on Permafrost*, Yakutsk. Washington. National Academy of Sciences. North American Contribution: 581–586.

Berg, T. E. & Black, R. F. (1966). Preliminary measurements of the growth of nonsorted polygons, Victoria Land, Antarctica. In: Tedrow, J. F. C. (ed.). *Antarctic soils and Soil Processes*. American Geophysical Union Antarctic Research Series, 8: 70–73.

Berg, T. E. & Esch, D. C. (1983). Effect of color and texture on the surface temperature of asphalt concrete pavements. *Proceedings of the 4th International Conference on Permafrost, Fairbanks, Alaska*. Washington, D.C. National Academy Press: 57–61.

Berg, T. E. & Smith, N. (1976). *Observations along the pipeline haul road between Livengood and the Yukon River*. U.S. Army Cold Regions Research and Engineering Laboratory Special Report 76-11. 83p.

Berman, L. L. (1965). Underground ice in the Northern part of the Kolyma plain. In: *Underground ice, Issue 1*. 7th International Congress on the Quaternary (INQUA), U.S.A. Moscow. Moscow University Press: 112–119. [In Russian].

Berner, R. A. (1990). Atmospheric carbon dioxide levels over Phanerozoic time. *Science*, 249: 1382–1386.

Berner, R. A. & Kothavala, Z. (2001). GEOCARB III: A revised model of atmospheric CO_2 over Phanerozoic time. *American Journal of Science*, 301: 182–204.

Berthling, I., Etzelmüller, B., Eiken, T. & Sollid, J. L. (1998). Rock glaciers on Prins Karls Forland, Svalbard. I. Internal structure, flow velocity and morphology. *Permafrost and Periglacial Processes*, 9: 135–145.

Bertran, P. (1992). Micromorphologie de grèzes litées des Charantes et du Châtillonnais (France). *Quaternaire*, 3(1): 4–15. [In French].

Bertran, P., Coutard, J.-P., Francou, B., Ozouf, J.-C. & Texier, J.-P. (1992). Données nouvelles sur l'origine du litage des grèzes: Implications paléoclimatiques. *Géographie physique et Quaternaire*, 46(1): 97–112. [In French].

Bertran, P. & Texier, J.-P. (1999). Sedimentation processes and facies on a semi-vegetated talus, Lousteau, Southwestern France. *Earth Surface Processes and Landforms*, 24: 177–187.

Beskow, G. (1935). Soil freezing and frost heaving with special applications to roads and railroads. *Swedish Geological Society, C, # 375, Year Book # 3*. (translated by J.O. Osterberg). In: *Historical Perspectives in Frost Heave Research*. Hanover, New Hampshire. U. S. Army Engineers. Cold Regions Research and Engineering Laboratory Special Report 91-23.

Betelev, N. P. (1974). Physicochemical transformations of rocks in the permafrost zone. *Lithology and Mineral Resources*, 9(5): 609–612.

Biggar, K. W., Donahue, R., Sego, D., Johnson M. & Birch, S. (2005). Spray freezing decontamination of tailings water at the Colomsac Mine. *Cold Regions Research and Technology*, 42: 106–119.

Bigoni, D. (2012). *Nonlinear Solid Mechanics: Bifurcation Theory and Material Instability*. Cambridge. Cambridge University Press.

Billings, W. D. & Mark, A. F. (1961). Interactions between alpine tundra vegetation and patterned ground in the mountains of southern New Zealand. *Ecology*, 42: 18–31.

Billings, W. D. & Peterson, K. M. (1980). Vegetational change and ice-wedge polygons through the thaw-lake cycle in arctic Alaska. *Arctic and Alpine Research*, 12: 41–432.

Billy, C. & Dick, J. W. L. (1974). Naturally occurring gas hydrates in the Mackenzie Delta, N.W.T. *Bulletin of the Canadian Petroleum Geologists*, 22: 340–352.

Bird, J. B. (1967). *The Physiography of Arctic Canada*. Baltimore. The John Hopkins Press. 336p.

Biryukov, V. Y. & Ogorodov, S. A. (2002). Bottom relief of the Pechora Sea as a result of land-ocean interactions during the Pleistocene-Holocene. In: *The 5th International Workshop on Land-Ocean Interactions in the Russian Arctic (LOIRA)*. Moscow. Shirshov Institute of Oceanology: 18–20. [In Russian].

Biryukov, V. Y. & Sovershaev, V. A. (1998). Geomorphology of the Kara Sea floor. In: *Dynamics of the Russian Arctic coasts*. Moscow. Moscow G. University: 102–115. [In Russian].

Biyanov, G. F. (1973). Experience in building dams on permafrost in Yakutia. *Permafrost: USSR Contribution to the 2nd International Permafrost Conference, Yakutsk*. Ottawa. National Research Council of Canada: 125–132. [In Russian]. Washington, D.C.

Biyanov, G. F. (1975). *Dams on Permafrost*. U.S. Army Cold Regions Research and Engineering Laboratory. Draft Translation TL 555. 234p.

Bjella, K. (2013). An investigation into a white painted airfield on permafrost: Thule Air Base, Greenland. In: Zufelt, J. E. (ed.). *10th International Symposium on Cold Regions Development, Anchorage, Alaska*. Iscord, 2013: 565–575.

Black, P. B. (1995). *Applications of the Clapeyron Equation to water and ice in porous media. Hanover, New Hampshire*. U.S. Army Corps of Engineers Cold Regions Research and Engineering Laboratory CRREL Report 95-6: 13p.

Black, R. F. (1951a). Structure in ice wedges of northern Alaska. *Bulletin, Geological Society of America*, 62(2): 1423–1424.

Black, R. F. (1951b). *Permafrost*. Smithsonian Institute Annual Report: 273–302.

Black, R. F. (1952). Growth of ice wedge polygons in permafrost near Barrow, Alaska. *Bulletin, Geological Society of America*, 63: 1235–1236.

Black, R. F. (1963). Les coins de glace et de gel permanante dans le nord de L'Alaska. *Annales de Géographie*, 72: 257–271. [In French].

Black, R. F. (1969). Thaw depressions and thaw lakes: a review. *Biuletyn Peryglacjalny*, 19: 131–150.

Black, R. F. (1974). Ice-wedge polygons of northern Alaska. In: Coates, D. R. (ed.). *Glacial Geomorphology*. 5th Annual Geomorphology Series, Binghampton, New York: 274–275.

Black, R. F. (1983). Three superposed systems of ice wedges at McLeod Point, northern Alaska, may span most of the Wisconsin stage and the Holocene. In: *Permafrost. 4th International Conference Proceedings, July 17–23rd, 1983*. Washington, D.C. National Academy Press: 68–73.

Black, R. F. & Barksdale, W. L. (1949). Oriented lakes of northern Alaska. *Journal of Geology*, 57: 105–118.

Blackwelder, E. (1931). Desert Plains. *The Journal of Geology*, 39(2): 133–140.

Blanchet, D., Skalski, J., Zhou, J., Lenstra, N. & Smith, B. (2002). Pipeline trenching in permafrost: A review. In: *Proceedings of the 4th International Pipeline Conference, Calgary*. New York. American Society of Mechanical Engineers, Paper 2002-27030.

Blatter, H. (1990). *Effect of climate on the cryosphere. Climatic conditions and the polythermal structure of glaciers*. Zürich, Federal Institute of Technology, 190p.

Blatter, H. & Hutter, K. (1991). Polythermal conditions in Arctic glaciers. *Journal of Glaciology*, 37: 261–269.

Bliss, L. C. & Wein, R. W. (1971). Changes to the active layer caused by surface disturbance. *National Research Council of Canada Associate Committee on Geotechnical Research Technical Memorandum*, 103: 37–46.

Blok, D., Heijmans, M. M. P. D., Schaepman-Strub, G., Kononov, A. V., Maximov, T. C., & Berendse, F. (2010). Shrub expansion may reduce summer permafrost thaw in Siberian tundra. *Global Change Biology*, 16: 1296–1305.

Blok, D., Heijmans, M. M. P. D., Schaepman-Strub, G., van Ruijven, J., Parmenttier, F. J. W., Maximov, T. C. & Berendse, F. (2011). The cooling capacity of mosses: Controls on water and energy fluxes in a Siberian site. *Ecosystems*, 14: 1055–1065.

Blumstengel, W. K. (1988). *Studies of an active rock glacier, east side, Slims River valley, Yukon Territory, Canada*. Unpublished M.Sc. thesis, Department of Geography, University of Calgary. 141p.

Blumstengel, W. & Harris, S. A. (1988). Observations on an active lobate rock glacier, Slims River valley, St. Elias Range, Canada. In: *Permafrost: Proceedings of the 5th International Conference*, Trondheim, Norway. Trondheim. Tapir Press 1: 689–694.

Blyakharchuk, T. A., Wright, H. E., Borodavko, P. S., van der Knaap, W. O. & Ammann, B. (2008). The role of pingos in the development of the Dzhangyskol lake-pingo complex, central Altai Mountains, southern Siberia. *Palaeogeography, Palaeoclimatology, Palaeoecology*, 257: 404–420.

Bobov, N. G. (1977). Direct indications of shrinking permafrost in the North during postglacial time. *Transactions (Doklady) of the U.S.S.R. Academy of Sciences: Earth Sciences Sections* 232(1–6): 256–259. [In Russian].

Bobov, N. G. (1999). Technogenic changes in permafrost and the stability of foundations of engineering structures. *Soil Mechanics and Foundation Engineering*, 36(2): 77–80.

Boch, S. G. & Krasnov, I. I. (1943). O nagornykh terraskh i drevnikh poverkhnostyakh vyravnivaniya na Urale I svyazannykh. *Vsesoyuznogo Geograficheskogo obshch estva Izvestiya* 75: 14–25. [In Russian].

Bockheim, J. G. (1995). Permafrost distribution in the southern circumpolar region and its relation to the environment: a review and recommendations for further research. *Permafrost and Periglacial Processes*, 6: 27–45.

Bockheim, J. G. (2007). Importance of cryoturbation in redistributing organic carbon in permafrost-affected soils. *Soil Science Society of America Journal*, 71(4): 1335–1342.

Bockheim, J. G. & Hinkel, K. M. (2012). Accumulation of excess ground ice in an age sequence of drained thermokarst lake basins, Arctic Alaska. *Permafrost and Periglacial Processes*, 23: 231–236.

Bockheim, J. G. & Tarnocai, C. (1998). Recognition of cryoturbation for classifying permafrost-affected soils. *Geoderma*, 81: 281–293.

Boelhouwers, J. (1991). Present-day periglacial activity in the Natal Drakensberg, southern Africa: A short review. *Permafrost and Periglacial Processes*, 2: 5–12.

Boelhouwers, J. (1995). Present day soil frost activity at Hexriver Mountains, western Cape, South Africa. *Zeitschrift für Geomorphologie N.F.*, 39: 237–248.

Bogushevskay, E. M. & Dimov, L. A. (2000). Pipelines on bogs: The offers in new Russian Construction Norms and Regulations 2.05.06-85. *Gas Industry*, 2000 #6: 72–73. [In Russian].

Boitsov, A. V. (1998). Underground water regime of frozen deposits in central Yakutia. In: Ohata, T. and Hiyama, T. (eds.). *Proceedings of the 2nd International Workshopon Energy and the Water Cycle in GAME – Siberia*. Nagoya, Japan, 1997. Institute for Hydrospheric-Atmospheric Sciences, Nagoya University 4: 57–63.

Bolch, T. & Gorbunov, A. P. (2014). Characteristics and origin of rock glaciers in the Northern Tien Shan (Kazakhstan/Kyrgyzstan). *Permafrost and Periglacial Processes*, 25(4): 320–332.

Bolch, T. & Marchenko, S. (2006). Significance of glaciers, rock glaciers, and ice-rich permafrost in the Northern Tien Shan as water towers under climate change conditions. *Proceedings of the workshop "Assessment of snow-glacier and water resources in Asia", 28th–30th November, 2006*. Almaty, Kazhkstan: 199–211.

Bolikhovskii, V. F. & Kyunttsel, V. V. (1990). Development of landslides in permafrost rocks of the Western Siberian tundra. *Soviet Engineering Geology*, 1: 56–60.

Bolton, W. R., Hinzman, L. D. & Yoshikawa, K. (2000). Stream flow studies in a watershed underlain by discontinuous permafrost. In: Kane, D. L., (ed.). *Proceedings of the AWRA Spring Speciality Conference, Water Resources in Extreme Environments, May 1–3, 2000*. Anchorage, Alaska. American Water Resources Association: 31–36.

Bondarenko, G. I. (1993). Prediction of stability of solifluction slopes and structures on them. *Proceedings of the 6th International Conference on Permafrost, Beijing*. Wushan, Guangzhou, China. South China University of Technology Press, 1: 851–854.

Bondarev, I. G. (1994). Subsea Permafrost, Gas Hydrates and gas pockets in Cenozoic sediments of Barents, Pechora and Kara Seas. World Petroleum Congress. New York. J. Wiley and Sons.

Bondarev, V., Okko, O. & Rokos, S. (1999). Engineering-geological investigations on permafrost in the Russian Arctic offshore and coastal areas. In: Tuhkuri, J. and Riska, K. (Eds.). *Proceedings of the International Port and Ocean Conference under Arctic Conditions*. Espo, Finland: 406–411.

Bondarev, V., Rokos, S., Kostin, D., Dlugach, A. & Polyakova, N. (2002). Under-permafrost accumulations of gas in the upper part of the sedimentary cover of the Pechora Sea. *Geology and Geophysics*, 43(7): 587–598.

Bonnaventure, P. P., Lewkowicz, A. G., Kremer, M. & Sawada, M. C. (2012). A permafrost probability model for the Southern Yukon and Northern British Columbia, Canada. *Permafrost and Periglacial Processes*, 23: 52–68.

Borisov, V. V. (1984). Hydrogeogenic soil-filtration taliks and associated water seepage. *Moscow University Geology Bulletin*, 39(2): 81–85. [In Russian].

Bosikov, N. P. (1989). The intensity of destruction of fields of inter-alas landscapes, central Yakutia. *Polar Geography and Geology*, 13(2): 149–154.

Bosikov, N. P. (1991). *Evolution of alases of the Central Yakutia*. Yakutsk. Permafrost Research Institute. 127p. [In Russian].

Bosikov, N. P. (1998). Wetness variability and dynamics of thermokarst processes in Central Yakutia. In: *Proceedings of the 7th international Conference on Permafrost, Yellowknife. Collection Nordicana*, 55: 71–76.

Bouchard, F., Francus, P., Pienitz, R., Laurion, I. & Feyte, S. (2014). Subarctic permafrost ponds: Investigating recent landscape evolution and sediment dynamics in thawed permafrost in Northern Québec (Canada). *Arctic, Antarctic and Alpine Research*, 46(1): 251–271.

Bowley, W. W. & Burghardt, M. D. (1971). Thermodynamics and stones. *EOS, Transactions of the American Geophysical Union*, 52: 407.

Bowman, J. P., McCammon, S. A., Brown, M. V., Nichols, D. S. & McMeekin, T. A. (1997). Diversity and association of psychrophilic bacteria in Antarctic sea ice. *Applied and Environmental Microbiology*, 63(8): 3068–3078.

Boyev, V. R. & Gabon, V. M. (1981). *Agriculture in the areas of industrial pioneering development in Siberia*. Moscow. [In Russian].

Bozhenova, A.P. & Bakulin, F.G. (1957). *Experimental investigations of the mechanisms of moisture migration in freezing soils*. Laboratory investigations of frozen soils. No. 3. M.: Yizdatelstvo AS USSR. [In Russian].

Brenning, A. & Trombotto, D. (2006). Logistic regression modeling of rock glacier and glacier distribution: Topographic and climatic controls in the semi-arid Andes. *Geomorphology*, 81: 141–154.

Bretz, J. H. (1929). Valley deposits immediately east of the Channeled Scablands of Washington. *Journal of Geology*, 37: 393–427.

Brewer, M. C. (1958). Some results of geothermal investigations of permafrost in northern Alaska. *Transactions of the American Geophysical Union*, 39(1): 19–26.

Brewer, R. & Haldane, A. D. (1957). Preliminary experiments in the development of clay orientation in soils. *Soil Science*, 84: 301–307.

Broll, G. & Tarnocai, C. (2002). Turf hummocks on Ellesmere Island, In: *Transactions of the 17th World Congress of Soil Science, August 14–21, 2002*. Bankok, Thailand. Paper # 1049. CD-ROM.

Brook, G. A. & Ford, D. C. (1978). The nature of labarinth karst and its implications of climaspecific models of tower karst. *Nature*, 280: 383–385.

Brook, G. A. & Ford, D. C. (1982). Hydrologic and geologic controls of carbonate water chemistry in the sub-Arctic Nahanni karst, Canada. *Earth Surface Processes and Landforms*, 7: 1–16.

Brooker, A., Fraser, R. H., Olthof, I., Kokelj, S. V. & Lacelle, D. (2014). Mapping the activity and evolution of retrogressive thaw slumps by tasselled cap trend analysis of a Landsat Satellite Image stack. *Permafrost and Periglacial Processes*, 25: 243–258.

Brooks, J., Cox, H., Bryant, W., Kennicutt, M. II, Mann, R. & McDonald, T. (1986). Association of gas hydrates and oil seepage in the Gulf of Mexico. *Organic Geochmistry*, 10: 221–234.

Brouchkov, A. V. (1998). *Frozen saline soils of the Arctic coast, their origin and properties*. Moscow. Moscow State University Press. 330p. [In Russian].

Brouchkov, A. V. (2000). Salt and water transfer in frozen soils induced by gradients of temperature and salt content. *Permafrost and Periglacial Processes*, 11(2): 153–160.

Brouchkov, A. V. (2002). Nature and distribution of frozen saline sediments on the Russian Arctic coast. *Permafrost and Periglacial Processes*, 13(2): 83–90.

Brouchkov, A. V. (2003). Frozen saline soils of the Arctic coast: their distribution and engineering properties. In: Phillips, M, Springman and Arenson Eds., *Proceedings of the 8th International Conference on Permafrost, Zurich, Switzerland*. Lisse, Swerts and Zeitlinger: 95-100.

Brouchkov, A. V., Bezrukov, V. V., Griva, G. I. & Muradyan, K. K. (2012). The effects of the relict microorganism *B*. sp. on development, gas exchange, spontaneous motor activity, stress resistance, and survival of *Drosophila melanogaster*. *Advances in Gerontology*, 25: 19–26. [In Russian].

Brouchkov, A. V. & Fukuda, M. (2002). Preliminary measurements of methane content in permafrost, Central Yakutia, and some experimental data. *Permafrost and Peryglacial Processes*, 13(3): 187–197.

Brouchkov, A.V., Fukuda, M., Federov, A., Konstantinov, P. & Iwahana, G. (2004). Thermokarst as a short-term permafrost disturbance, Central Yakutia. *Permafrost and Periglacial Processes*, 15(1): 81–87.

Brouchkov, A. V., Fukuda, M., Iwahana, G., Kobayashi, Y. & Konstantinov, P. (2005). Thermal conductivity of soils in the active layer of Eastern Siberia. *Permafrost and Periglacial Processes*, 16: 217–222.

Brouchkov, A. V., Melnikov, V. P., Sukhovei, I. G., Griva, G. I., Repin, V. E., Kalenova, L. F., Brenner, E. V., Subbotin, A. M., Trofimova, I. B., Tanaka, M., *et al.* (2009). Relict microorganisms of cryolithozone as possible objects of gerontology. *Advances in Gerontology*, 22: 253–258.

Brown, J. (1969). Soils of the Okpilak River region, Alaska. In: Péwé, T. L., (ed.). *The Periglacial Environment*. Montreal. McGill-Queen's University Press: 93-128.

Brown, J., Ferrians, Jr., O. J., Hegginbottom, J. A. & Melnikov, E. S. (1997). *Circum-Arctic map of permafrost and ground-ice conditions*. Washington, D.C. U.S. Geological Survey in cooperation with the Circum-Pacific Council for Energy, Mines and Mineral Resources, Circum-Pacific Map Series CP-45, scale 1:10,000,000.

Brown, J., Ferrians, Jr., O. J., Hegginbottom, J. A. & Melnikov, E. S. (1997). Digital Circum-Arctic map of permafrost and ground-ice conditions. In: *International Permafrost Association, Data Information Working Group, compiler Circumpolar Active-Layer Permafrost System (CAPS)*. Version 1.0. CD_ROM available from hside@kryos.colorado.edu

Brown, J. & Sellman, P. V. (1973). Permafrost and coastal plain history of Arctic Alaska. In: Britton, M. E. (ed.). *Alaskan arctic tundra*. Arctic Institute of North America Technical Paper 25. 244p.

Brown, R. J. E. (1960). The distribution of permafrost and its relation to air temperature in Canada and the USSR. *Arctic*, 13: 163–177.

Brown, R. J. E. (1963). Editor. *Proceedings of the 1st Canadian Conference on Permafrost*. Ottawa, Canadian National Research Council, Associate Committee on Soil and Snow Mechanics. Technical Memoir #76: 231p.

Brown, R. J. E. (1965). Factors influencing discontinuous permafrost in Canada. *Abstracts, 7th International Congress, International Association for Quaternary Research, Boulder, Colorado*: 47.

Brown, R. J. E. (1966a). Influence of vegetation on permafrost. In: *Permafrost: International Conference, Lafayette, Indiana*. Washington, National Academy of Sciences Publication 1287: 20–25.

Brown, R. J. E. (1966b). The relationship between mean annual air and ground temperatures in the permafrost regions of Canada. *Proceedings of the 1st Permafrost Conference, Lafayette, Indiana*. Washington. National Academy of Sciences: 241–246.

Brown, R. J. E. (1967). *Permafrost in Canada*. Ottawa. National Research Council of Canada, Division of Building Research, NRCC 9769. 2nd Edition.

Brown, R. J. E. (1968). Occurrence of permafrost in Canadian peatlands. *Proceedings of the 3rd International Peat Congress, Québec*: 174–181.

Brown, R. J. E. (1969). Factors influencing discontinuous permafrost. In: Péwé, T. L. (ed.). *The Periglacial Environment, past and present*. Monteal. McGill-Queen's University Press: 11–53.

Brown, R. J. E. (1970a). *Permafrost in Canada – Its influence on Northern Development*. Toronto. University of Toronto Press. 234p.

Brown, R. J. E. (1970b). The distribution of permafrost and its relation to air temperature in Canada and the USSR. *Arctic*, 13: 163–177.

Brown, R. J. E. (1972). Permafrost in the Canadian Archipelago. *Zeitschrift für Geomorphologie*, 13: 102–130.

Brown, R. J. E., Editor (1973a). *Proceedings of the 1st Canadian Conference on Permafrost*. Ottawa, National Research Council of Canada, Associate Committee on Soil and Snow Mechanics, Technical Memoir #76. 231p.

Brown, R. J. E. (1973b). Influence of climatic and terrain factors on ground temperature at three locations in the permafrost region of Canada. In: *North American Contribution,*

Permafrost: 2nd International Conference, Yakutsk, Washington, D. C., National Academy of Sciences: 27–34.

Brown, R. J. E. & Johnson, G. H. (1964). Permafrost and related engineering problems. *Endeavour*, 23: 66–72.

Brown, R. J. E. & Péwé, T. L. (1973). Distribution of permafrost in North America and its relationship to the environment: A review, 1963–73. *North American Contribution, Permafrost: 2nd International Conference*, Washington, D.C. National Academy of Sciences: 71–100.

Bryan, K. (1922). Erosion and sedimentation in the Papago Country, Arizona: with a sketch of the geology. *U.S. Geological Survey Bulletin # 730.*

Bryant, J. P. & Scheinberg, E. (1970). Vegetation and frost activity in an alpine fellfield on the summit of Plateau Mountain. *Canadian Journal of Botany*, 48: 751–771.

Bryant, R. B. (1989). Physical processes of Fragipan formation. In: Smeck, N. E. and Ciolkose, E. J. (ed.). *Fragipans: Their occurrence, classification and genesis.* Soil Science Society of America Special Publication #24: 141–150.

Buchenauer, H. W. (1990). Getscher-und blockgletschergeschichte der westlichen Schobergruppe (Osttirol). *Marb. Geographische Schrifter*, 117: 376. [In German].

Büdel, J. (1937). Eiszeitliche und rezente Verwitterung und Abtragung im chuncals nicht vereisten Gebeit Mitteleuropas. *Erg. Heft 229 zu Peterm. Geogr. Mitt.*, Gotha. [In German].

Budyko, M. I. (1966). Polar ice and climate. *Izvestiya AN SSSR, seriya geografiya*, 6: 3–10. [In Russian].

Budyko, M. I. (1971). *Klimat i zhizn.* Leningrad. [In Russian].

Budyko, M. I. (1980). Global ecology. Moscow. Progress Publisher. [In Russian].

Buffet, B. A. (2000). Clathrate Hydrates. *Annual Revue of Earth Planetary Science*, 28: 47–507.

Buldovich, S, N., Garagulya, L. S., Tipenko, G. S. & Seregina, N. V. (1991). Mathematical modeling of conductive and convective heat transfer in the taliks of the cryolite zone. *Moscow University Geology Bulletin*, 46(5): 55–64. [In Russian].

Buldovich, S. N., Afanasenko, L. S., Garagulya, L. S. & Ospenikov, Y. N. (2002). The effect of flooding on geocryological conditions of construction sites in West Siberia. In: *Extreme phenomena in the cryosphere: Basic and applied aspects.* Puschino. Russian Federation: Russian Academy of Sciences International Conference Abstracts: 103–104; 266. [In Russian].

Buldovich, S. N., Garagulya, L. S., Tipenko, G. S. & Seregina, N. V. (2001). Mathematical modelling of conductive heat and convective heat transfer in the taliks of the cryolite zone. *Moscow University Geology Bulletin*, 46(5): 55–64.

Bull, W. B., Schlyter, P. & Brogaard, S. (1995). Lichenomtric analysis of the Kärkerieppe slush-avalanche fan, Kärkevagge, Sweden. *Geografiska Annaler*, 77A(4): 231–240.

Bundtsen, T. K. (1982). Alaska's strategic minerals. *Alaska Geographic*, 9(4): 52–63.

Burgess, M., Judge, A., Taylor, A. & Allen, V. (1982). Ground temperature studies of permafrost growth at a drained lake site, Mackenzie Delta. In: French, H. M. (ed.). *Proceedings of the 4th Canadian Permafrost Conference, Calgary, Alberta.* Ottawa. National Research Council of Canada: 3–9.

Burgess, M. M., Grechischev, S. E., Kurfurst, P. J., Melnikov, E. S. & Moskalenko, N. G. (1993). Monitoring of engineering – geological processes along pipeline routes in permafrost terrain in Mackenzie River Valley, Canada and Nadym area, Russia. In: *Permafrost, Sixth International Conference, Proceedings*, Beijing, China. 1: 54–59.

Burgess, M. M., Robinson, S. D., Moorman, B. J., Judge, A. S. & Fridel, T. W. (1995). The application of ground penetrating radar to geotechnical investigations of insulated permafrost slopes along the Norman Wells pipeline. In: *Proceedings of the 48th Canadian Geotechnical Conference, Vacouver, B.C.* Alliston, Ontario. Canadian Geotechnical Society: 999–1006.

Burn, C. R. (1990). Frost-heave in lake-bottom sediments, Mackenzie Delta, Northwest Territories. *Proceedings of the 5th Canadian Permafrost Conference. Collection Nordicana*, 4: 103–109.

Burn, C. R. & Friele, P. A. (1989). Geomorphology, vegetation succession, soil characteristics and permafrost in retrogressive thaw slumps near Mayo, Yukon Territory. *Arctic*, 42(1): 31–40.

Burn, C. R. & Michel, F. A. (1988). Evidence for recent temperature-induced water migration into permafrost from tritium content of ground ice near Mayo, Yukon Territory, Canada. *Canadian Journal of Earth Sciences*, 25: 909–915.

Burn, C. R. & Smith, C. A. S. (1988). Observations of the "thermal offset" in near-surface mean annual temperatures at several sites near Mayo, Yukon Territory, Canada. *Arctic*, 42(2): 99–104.

Bykov, G. E. (1938). On the question of climates during the Quaternary period in the Far East and on the history of permanently frozen ground. *Doklady Akademii Nauk SSSR* 20(5): 387–390. [In Russian].

Cailleux, A. & Calkin, P. (1963). Orientation of hollows in cavernously weathered boulders in Antarctica. *Biuletyn Periglacjalny*, 12: 147–150.

Caine, T. N. (1963). Movement of low angle scree slopes in the Lake District, Northern England. *Revue de Géomorphologie Dynamique*, 14: 171–177.

Caine, N. (1968a). *The blockfields of Northeastern Tasmania*. Department of Geography Publication G/6. Canberra. Australian National University. 127p.

Caine, N. (1968b). The fabric of periglacial blockfield material on Mount Barrow, Australia. *Geografiska Annaler*, 50A(4): 193–206.

Caine, N. (1969). A model for alpine talus slope development by slush avalanching. *Journal of Geology*, 77: 92–100.

Caine, N. (1972). Air photo analysis of blockfield fabrics in Talus Valley, Tasmania. *Journal of Sedimentary Petrology*, 42(1): 33–48.

Caine, N. (1983). *The mountains of Northeastern Tasmania*. Rotterdam. Balkema.

Cairnes, D. D. (1912). Differential erosion and equiplanation in portions of Yukon and Alaska. *Bulletin of the Geological Society of America*, 23: 333–348.

Cairns, J., Overbaugh, J. & Miller, S. (1994). The origin of mutations. *Nature*, 335: 142–145.

Calkin, P. & Cailleux, A. (1962). A quantitative study of cavernous weathering (Taffoni) and its application to glacial chronology in Victoria Valley, Antarctica. *Zeitschrift für Geomorphologie*, 6: 317–324.

Callister, W. D., Jr. Rethewisch, R. G. (2015). *Fundamentals of Materials Science and Engineering*. New York. John Wiley and Sons, 2nd International edition, ISBN 0-471-66081-7, ISBN 978-0-471-66081-1.

Calmels, F. & Allard, M. (2004). Ice segregation and gas distribution in permafrost using tomodensitometric analysis. *Permafrost and Periglacial Processes*. 15(4): 367–378.

Calmels, F.C. & Allard, M. (2008). A structural interpretation of the palsa-lithalsa growth mechanism through the use of CT Scanning. *Earth Surface Processes and Landforms*, 33(2): 209–225.

Calmels, F.C., Allard, M. & Delisle, G. (2008a). Development and decay of a lithalsa in Northern Québec: a geomorphological history. *Geomorphology*, 97: 287–299.

Calmels, F. C., Delisle, G. & Allard, M. (2008b). Internal structure and the thermal and hydrological regime of a typical lithalsa: significance for permafrost growth. *Canadian Journal of Earth Sciences*, 45(1): 31–43.

Cameron, J. J. (1977). *Community water use summary*. Edmonton. Internal Report, Northern Technology Unit, Environmental Protection Service.

Cameron, J. J., Christensen, V. & Gamble, D. J. (1977). Water and Sanitation in the Northwest Territories: An overview of the setting, policies and technology. *The Northern Engineer*, 9(4): 4–12.

Cameron, J. J. & Smith, D. W. (1977). *Annotated bibliography on northern environmental engineering, 1974–75*. Environment Canada, Environmental Protection Service, Report # EPS-3-WP-77-6.

Canadian Soil Classification (1998). 3rd Edition. Agriculture Canada.

Cande, S. C. & Kent, D. W. (1995). Revised calibration of the geomagnetic polarity timescale for the Late Cretaceous and Cenozoic. *Journal of Geophysical Research*, 100: 6093–6095.

Capps, S. R. (1910). Rock glaciers in Alaska. *Journal of Geology*, 18: 359–375.

Carbiener, R. (1970). Frostmusterböden, solifluktion, pflanzengesellschafts – Mosaik und struktur, erläutert am Beispiel der Hochvogesen. In: Tuxen, R. (ed.). *Gesellschaftsmorphologie*. Berlin International Symposium Rinteln, 1966: 187–217. [In German].

Carefoot, E. I., Davies, A. L., Johnston, G. H., Lawrence, N. A., Lukomskyj, P. & Thornton, D. E. (1981). Utilities. In: Johnston, G. H. (Ed.). *Permafrost Engineering, Design and Construction*. Toronto. J. Wiley and Sons. Chapter 10: 415–472.

Carey, K. L. (1973). *Icings developed from surface water and ground water*. U.S. Army Cold Regions Research and Engineering Laboratory, Cold Regions Science and Engineering Monograph III-D3.

Carlson, L.E. & Butterwick, D.E. (1983). Testing pipelining techniques in warm permafrost. In: *Permafrost, Fourth International Conference, Proceedings*. Washington, DC. National Academy Press 97–102.

Carmack, E. C. (1986). Circulation and mixing ice-covered waters. In: Untersteiner, N. (ed.). *The geophysics of sea ice*. New York. Plenum Press 641–712. NATO AS1 Series B 146.

Carpenter, E. J., Lin, S. & Capone, D. G. (2000). Bacterial Activity in South Pole Snow. *Applied and Environmental Microbiology*, 66(10): 4514–4517.

Carr, A. (2003). *Hydrologic comparisons and model simulations of Subarctic watersheds containing continuous and discontinuous permafrost, Seward Peninsula, Alaska*. Unpublished M.Sc. thesis. Fairbanks. University of Alaska, Fairbanks.

Carson, C. E. & Hussey, K. M. (1960). Hydrodynamics in three arctic lakes. *Journal of Geology*, 68: 585–600.

Carter, L. D. (1981). A Pleistocene sand sea on the Alaskan Arctic Coastal Plain. *Science*, 211: 381–383.

Carter, L. D. (1983). Fossil sand wedges in the Alaskan arctic coastal plain and their paleoenvironmental significance. *Proceedings of the 4th International Conference on Permafrost, Fairbanks, Alaska*. Washington, D.C. National Academy Press: 109–114.

Carter, L. D. & Galloway, J. P. (1981). Earth Flows along Henry Creek, northern Alaska. *Arctic*, 34: 325–328.

Carey, K. L. (1973). Icings developed from surface water and groundwater. U.S. Army Cold Regions Research and Engineering Monograph III-D3. 65p.

Casagrande, A. (1932). A new theory of frost heaving: Discussion. *Proceeding of the U.S. Highway Research Board*, 11(1): 168–172.

Catto, N. R. (1983). Loess in the Cypress Hills, Alberta. *Canadian Journal of Earth Sciences*, 20(7): 1159–1167.

Cegla, J. & Dzulinski, S. (1970). Ukalady niestatecznie wartowane i ich wystepowanie w srodowisku periglacjalnym. *Acta Universitatis Wratislaviensis*, 124, Studia Geograficzne, 13: 17–42. [In Polish].

Cerling, T. E. & Craig, H. (1994). Geomorphology and in situ cosmogenic isotopes. *Annual Review of Earth and Planetary Sciences*, 22: 273–317.

Chaban, P. D. & Gol'dtman, V. G. (1978). Effect of geocryological conditions on the productivity and mining operations in the Northeast USSR. *Proceedings of the 2nd International Conference on Permafrost, Yakutsk*. USSR Contribution, Washington. National Academy of Sciences: 587–590.

Chalmer, W. D. (1935). Alluvial fan flooding: The Montrose, California flood of 1934. *Geographical Review*, 25: 255–263.

Chamberlain, E. J. (1981a). *Frost susceptibility criteria*. Presented at the National Transportation Research Board Meeting, January, 1981.

Chamberlain, E. J. (1981b). *Frost susceptibility of soil: Review of index tests*. Hanover, New Hampshire. U.S. Army Corps of Engineers Cold Regions Research and Engineering Laboratory. CRREL Monograph 81-2. 110p.

Chamberlain, E. J. (1983). Frost heave in saline soils. In: *Permafrost: 4th International Conference Proceedings, Fairbanks*, Washington, National Academy Press: 121–126.

Chambers, M. J. G. (1967). Investigations of patterned ground, at Signy Island, South Orkney Islands. III: Miniature patterns, frost heaving and general conclusions. *British Antarctic Survey Bulletin*, 12: 1–22.

Chan, X. (1984). Current developments in China on frost-heave processes in soil. In: Permafrost: 4th International Conference Proceedings, Fairbanks, Washington. National Academy Press: 55–60.

Chapin, J., Pernia, J. & Kjartanson, B. (2009). An approach to applying Spring Thaw Loading Restrictions for low volume roads on thermal numerical modelling. In: *Proceedings of the 14th Conference on Cold Regions Engineering*. Retson, Virginia. American Society of Civil Engineers: 469–505.

Chataignser, Y., Gosselin, L. & Doré, G. (2009). Optimization of embedded inclined open-ended channel in natural convection used as a heat drain. *International Journal of Thermal Sciences (2009)*. Doi: 101016/jiljthermasci2009.01.001

Chaudhary, V. & Gursharan, S. (2006). Structural measures for controlling avalanches in formation zone. *Defence Science Journal* 56(5): 791–799.

Chen, H., Zhu, Q., Wu, N., Wang, Y. F. & Peng, C. H. (2011). Delayed spring phenology on the Tibetan Plateau may also be attributable to other factors than winter and spring warming. *Proceedings of the National Academy of Sciences*, vol. 108, #9: E93.

Chen, J. (2000). Weathering of rocks induced by lichen colonization – a review. *Catena*, 121: 121–146.

Chen, J., Hu, Z. Y., Dou, S. & Qian, Z. Y. (2003). Yin-Yang slope problem along Qinghai-Lhasa Lines and its radiation mechanism. *Cold Regions Science and Technology*, 44: 217–224.

Chen, M., Rowland, J. C., Wilson, C. J., Altmann, G. L. & Brumby, S. P. (2013). The importance in natural variability in lake areas on the detection of permafrost degradation: A case study from the Yukon Flats, Alaska. *Permafrost and Periglacial Processes*, 24(3): 224–240.

Cheng, D. X., Liu, Z. W. & Mao, F. (2013). Research on frozen soil characteristics of 330 kV transmission line of main power grid in Yushu and Qinghai. *Electric Power Survey and Design*, 1: 20–23.

Cheng, D. X., Zhang, J. M., Liu, H. J. *et al.* (2009). The influence factor analysis for the site selection of transmission line in frozen earth area. *Journal of Engineering Geology*, 17(3): 329–334.

Cheng, G. (1982). Effect of uni-direction accumulation of unfrozen water in seasonally frozen and thawed ground. *Kexue Tongbao*, 27(9): 984–989. [In Chinese].

Cheng, G. (1983a). Vertical and horizontal zonation of high-altitude permafrost. Washington, D.C. *Proceedings of the 4th International Conference on Permafrost*, 136–141.

Cheng, G. (1983b). The mechanism of repeated-segregation for the formation of thick-layered ground ice. *Cold Regions Science and Technology*, 8: 57–66.

Cheng, G. D. (2003). The impact of local factors on permafrost distribution and its inspiring for design Qinghai-Xizang Railway. *Science in China (Series D)*, 33(6): 602–607.

Cheng, G. D. (2004). Influences of local factors on permafrost occurrence and their implications for Qinghai-Tibet railway design. *Science in China, Series D: Earth Sciences*, 47(8), 704–709.

Cheng, G. & Dramis, F. (1992). Distribution of mountain permafrost and climate. *Permafrost and Periglacial Processes* 3(2): 83–91.

Cheng, G. D., Jiang, H., Wang, K. L., *et al.* (2003). Thawing index and freezing index on the embankment surface in permafrost regions. *Journal of Glaciology and Geocryology*, 25(6): 603–607.

Cheng, G. D. & Jin, H. D. (2013). Permafrost and groundwater on the Qinghai-Tibet Plateau and in northeast China. *Hydrogeology Journal*, 21: 5–23.

Cheng, G. D. & Ma, W. (2006). Frozen ground engineering problems in construction of the Qinghai-Tibet railway. *Chinese Journal of Nature*, 28(6): 315–320.

Cheng, G. D., Sun, Z. & Nui, F. (2008). Application of the roadbed cooling approach in Qinghai-Tibet railway engineering. *Cold Regions Science and Technology*, 53(3), 241–258.

Cheng, G. D. & Tong, B. L. (1978). Experimental research on an embankment in an area with massive ground ice at the lower limit of alpine permafrost. In: *Proceedings of the 3rd International Conference on Permafrost* Ottawa. National Council of Canada 2: 199–222.

Cheng, G. D., Tong, B. L. & Luo, X. B. (1981). Two important problems of embankment construction in the section of massive ground ice. *Journal of Glaciology and Geocryology*, 3(2): 6–11. [In Chinese].

Cheng, G. & Wang, S. (1982). On the zonation of high-altitude permafrost in China. *Journal of Glaciology and Geocryology*, 4: 1–16. [In Chinese].

Cheng, G. D., Zhang, J., Sheng, Y. & Chen, J. (2004). Principle of thermal insulation for permafrost protection. *Cold Regions Science and Technology*, 40: 71–79.

Cheng, J., Jiang, M., Zan, L., Lü, X., Xu, X., Lu, P., Zhang, X. & Tian, M. (2005a). Progress in Research on the Quaternary Geology in the source area of the Yellow River. *Geoscience*, 19(2): 239–246. [In Chinese].

Cheng, J., Zhang, X., Tian, M., Yu, W. & Yu, J. (2005b). Ice-wedge casts showing climatic change since the Late Pleistocene in the source area of the Yellow River, north-east Tibet. *Journal of Mountain Science*, 2(3): 193–201.

Cheng, J., Zhang X., Tian, M., Yu, W., Tang, D. & Yue, J. (2006). Ice-wedge casts discovered in the source area of the Yellow River, Northeast Tibetan Plateau and their palaeoclimatic implications. *Quaternary Sciences*, 26 (1): 92–98. [In Chinese].

Cheng, Y. F., Ding, S. J., Lu, X. L., *et al.* (2012). Monitoring and analysis of coarse-grained frozen soil temperatures in Qinghai-Tibet DC Transmission Line engineering. *Chinese Journal of Rock Mechanics and Engineering*, 31(11): 2363–2371.

Cheng, Y. F., Lu, X. L., Liu, H. Q., *et al.* (2004). Model test study on pile foundation of 110 kV transmission line of Qinghai-Tibet Railway in frozen soils. *Chinese Journal of Rock Mechanics and Engineering*, 23 (Supplement 1): 4378–4382.

Chernov, Y. I. & Matveyeva, N. V. (1997). Arctic ecosystems in Russia. In: Wielgolaski, F. E. (Ed.). *Polar and Alpine Tundra*. Amsterdam. Elsevier. 3: 361–507.

Chernyad'yev, V. P. & Chekhovskiy, A. I. (1994). The impact of climatic warming on permafrost conditions in Russia. *Polar Geography and Geology*, 18: 121–126.

Cheverev, V. G., Pustovoit, G. P. Vidyapin, I. T., *et al.* (2006). Stabilization of tubular pile foundations in heaving soils. *Soil Mechanics and Foundation Engineering*, 43(6): 221–227.

Chigir, V. G., Khol'nov, A. P., Khrenov, N. N., Yegurtsov, S. A. & Samoylova, V. V. (1997). Many-year evolution and seasonal dynamics of thaw bulbs in perennially-frozen soil beneath gas pipelines in the far north. *Pipeline Construction*, 2: 28–31. [In Russian].

Chinn, T. J. H. (1981). Use of Rock weathering-rind thickness for Holocene absolute age-dating in New Zealand. *Arctic and Alpine Research*, 13(1): 33–45.

Chistotinov. L. V. (1973). *Moisture migration in freezing water-unsaturated soils*. M.: Nauka. [In Russian].

Chizjov, A. B., Chizhova, N. I., Morkovina, I. K. & Romanov, V. V. (1983). Tritium in permafrost and ground ice. In: *Proceedings of the 4th International Conference on Permafrost, Fairbanks, Alaska.* Washington, D. C. National Academy Press, Volume 1; 1030–1035.

Chou, Y. L. (2008). *Study on Shady-Sunny effect and the forming mechanism of longitudinal embankment crack in permafrost.* Ph.D. thesis, Graduate School of the Chinese Academy of Sciences Dissertation.

Christiansen, H. H., Åkerman, J. H. & Repelewska-Pekalowa, J. (2008). Active layer dynamics in Greenland, Svarlbad and Sweden. *8th International Conference on Permafrost. Extended Abstracts on current research and newly available information:* 19–20.

Christiansen, H. H., Etzelmüller, B., Isaksen, K., Juliussen, H., Farbrot, H., Humlum, O., Johansson, M., Ingeman-Nielsen, T., Kristensen, L., Hjort, J., Holmlund, P., Sannel, A. B. K., Sigsgaard, C., Åkerman, H. J., Foged, N., Blikra, L. H., Pernosky, M. A., & Ødegård, R. (2010). The thermal state of permafrost in the Nordic area during IPY 2007–2009. *Permafrost and Periglacial Processes,* 21: 156–181.

Christofferson, R. W. (1994). *Geosystems.* 2nd Edition. Engelwood Cliffs, N. J., MacMillan College Publishing Co. 663p.

Chudinova, S. M., Bykhovets, S. S., Sorokovikov, V. A., Gilichinsky, D. A., Zhang, T.-Z. & Barry, R. G. (2008). Could the current warming endanger the status of frozen ground regions in Asia? *8th International Conference on Permafrost. Extended Abstracts on current research and newly available data:* 21–22.

Church, M. (1974). Hydrology and permafrost with reference to northern North America. In: *Permafrost Hydrology.* Proceedings of the Workshop Seminar, Canadian National Committee, the International Hydrological Decade: 7–20.

Church, M. & Miles, M. J. (1987). Meteorological antecedents to debris flows in southwestern British Columbia. In: Costa, J. E. and Wieczoreck, G. F. (Eds.). *Debris flows/avalanches: process, recognition and mitigation.* Geological Society of America, Reviews in Engineering Geology, 7: 63–79.

Churcher, C. S. (1968). Pleistocene ungulates from the Bow River gravels at Cochrane, Alberta. *Canadian Journal of Earth Sciences,* 5: 1467–1488.

Chuvilin, L. V. & Miklyaeva, I. V. (2005). A field experiment to assess the oil pollution in the upper horizons of permafrost. *Earth Cryosphere,* 9(2): 60–66. [In Russian].

Čigir, V. G. (1977). Cryogenic slope formation. In: *Cryolithology problems.* Moscow. [In Russian].

Clague, J. J. (1981). Landslides at the south end of Kluane Lake, Yukon Territory. *Canadian Journal of Earth Sciences,* 18: 959–971.

Clark, E. F. & Simoni, O. W. (1976). *A survey of road construction and maintenance problems in Central Alaska.* U.S. Army Cold Regions Research and Engineering Laboratory, Special Report SR 76-8. 36p.

Clark, I. & Lauriol, B. (1997). Aufeis of the Firth River basin, northern Yukon, Canada: Insights into permafrost hydrology and karst. *Arctic and Alpine Research,* 29: 240–252.

Clarke, G. K. C., Leverington, D. W., Teller, J. T. & Dyke, A. S. (2004). Paleohydraulics of the last outburst flood from glacial Lake Agassiz and the 8200 BP cold event. *Quaternary Science Reviews,* 23(3): 389–407.

Clarke, K. (2013). Bud the spud moves up north: Farmers battle to make a go of it in the Northwest Territories. http://news.nationalpost.com/news/canada/bud-the-spud-moves-up-north.

Clayton, L. (1967). Stagnant glacier features of the Missouri Coteau. In: Clayton, L. and Freers, T. F. (eds.). *Glacial Geology of the Missouri Coteau and adjacent areas.* North Dakota Geological Survey Miscellaneous Papers, 30: 25–46.

Clein, J.S. & Schimel, J. P. (1995). Microbial activity of tundra and taiga soils at sub-zero temperatures. *Soil Biology and Biochemistry,* 27: 1231–1234.

Coe, M. J. (1967). *The Ecology of the alpine zone of Mount Kenya*. Monographiae Biologicae 17. The Hague. Junk. 136p.

Cogley, J. G. & McCann, S. B. (1976). An exceptional storm and its effects in the Canadian High Arctic. *Arctic and Alpine Research*, 8: 105–110.

Cohen, J. L., Furtado, J. C., Barlow, M. A., Alexeev, V. A. & Cherry, J. E. (2012). Arctic warming, increasing snow cover and widespread boreal winter cooling. *Environment Research Letters* 7(1). Doi: 10.1088/1748-9326/7/1/014007.

Collett, T. S. (1990). Potential Geological Hazards of Arctic gas hydrates. *American Association of Petroleum Geology Bulletin*, 74(5): 631–632.

Collinson, J. D. & Thompson, D. B. (1989). *Sedimentary structures*. London, Unwin Hyman.

Coltirti, M., Dramis, F. & Pambianchi, G. (1983). Stratified slope-waste deposits in the Esino River Basin, Umbria-Marche Appenines, Central Italy. *Polarforchung*, 52(2): 59–66.

Colt-KBR (2003). Conceptual geotechnical/geothermal design basis. In: *Conceptual and preliminary engineering for Mackenzie Gas Project*. Prepared for Imperial Oil Resources Ventures Limited Project # 99-C-3079.

Coltorti, M., Dramis, F. & Pambianchi, G. (1983). Stratified slope-waste deposits in the Esino River Basin, Umbria-Marche Appenines, Central Italy. *Polarforschung*, 53(2): 59–65.

Connor, B. & Harper, J. (2013). How vulnerable is Alaska's Transportation to climate change? *TR News*, 284: 23–29.

Cook, F. A. (1955). Near surface soil temperature measurements at Resolute Bay, Northwest Territories. *Arctic*, 8(4): 237–249.

Cook, F. A. (1956). Additional note on mud circles at Resolute Bay, Northwest Territories. *Canadian Geographer*, 8: 9–17.

Cook, F. A. (1966). Patterned ground research in Canada. In: *Permafrost 1st International Conference, Lafayette, Indiana*. Washington, Proceedings of the National Academy of Sciences, Publication #1287: 128–130.

Cooks, J. & Otto, E. (1990). The weathering effects of the lichen *Lecidea* aff. *sarcogynoides* (Koerb.) on Magaliesberg quartzite. *Earth Surface Processes and Landforms*. 15: 491–500.

Corner, G. D. (1980). Avalanche impact landforms in Troms, North Norway. *Geografiska Annaler*, 62A: 1–10.

Corte, A. E. (1961). *The frost behaviour of soils: Laboratory and field data for a new concept – I. Vertical sorting*. U.S. Army Corps of Engineering, Cold Regions Research and Engineering Laboratory Research Report 85(1). 22p.

Corte, A. E. (1962a). Vertical migration of particles in front of a moving freezing plane. *Journal of Geophysical Research*, 67:1085–1090.

Corte, A. E. (1962b). *Relationship between four ground patterns, structure of the active layer, and type and distribution of ice in the permafrost*. U.S. Army Corps of Engineers, Cold Regions Research and Engineering Laboratory Research Report #88. 79p.

Corte, A. E. (1962c). *The frost behavior of soils: Laboratory and field data for a new concept – II, Horizontal sorting*. U.S. Army Corps of Engineers, Cold Regions Research and Engineering Laboratory Research Report 85(2). 20p.

Corte, A. E. (1962d). The frost behavior of soils – I, Vertical sorting. In: *Soil behavior on Freezing with or without additives*. Washington, National Academy of Sciences, Highway Research Bulletin, #317: 9–14.

Corte, A. E. (1962e). The frost behavior of soils – II, Horizontal sorting. In: *Soil behavior on freezing with or without additives*. Washington, National Academy of Sciences, Highway Research Bulletin, #317: 44–66.

Corte, A. E. (1963a). *Vertical migration of particles in front of a moving freezing plane*. U. S. Army Corps of Engineers, Cold Regions research and Engineering Laboratory Report #105. 8p.

Corte, A. E. (1963b). Particle sorting by repeated freezing and thawing. *Science*, 142: 499–501.

Corte, A. E. (1966a). Experiments on sorting processes and the origin of patterned ground. In: *Permafrost: 1st International Conference, Lafayette, Indiana*. Washington, National Academy of Sciences Pulication #1287: 130–135.

Corte, A. E. (1966b). Particle sorting by repeated freezing and thawing. *Builetyn Peryglacjalny*, 15: 175–240.

Corte, A. E. (1967). Informe preliminary del progresso efectuado en el studio de las estructuras de crioturbación Pleistocenas fósiles en la Provincia de Santa Cruz. *Terceras Jornadas Geologicos Argentinas*, 2: 9–19. [In Spanish].

Corte, A. E. (1972). Laboratory formation of extrusion features by multicyclic freeze-thaw in soils. *Bulletin de Centre de Géomorphologie du CNRS*, Caen, 13, 14, 15: 157–182.

Corte, A. E. (1976a). The hydrological significance of rock glaciers. *Journal of Glaciology*, 17(75): 157–158.

Corte, A. E. (1976b). Rock Glaciers. *Biuletyn Peryglacjalny*, 26: 175–197.

Corte, A. E. (1978a). *Guide for compilation and assemblage of data for a world glacier inventory: Debris-covered glaciers, rock glaciers and ice-cored moraines*. 7p. Roneo.

Corte, A. E. (1978b). Rock glaciers as permafrost bodies with a debris cover as an active layer: a hydrological approach in the Andes of Mendoza. In: *Proceedings of the 3rd International Conference on Permafrost, Edmonton, Alberta*. Ottawa. National Research Council of Canada 1: 263–269.

Corte, A. E. (1985). Comparative study of geocryological (periglacial) conditions, features and processes in the Andes and the Himalayas: The Andes. In: *Unión Geográfica Internacional. Primera Reunión Latino-Americana sobre la importancia de los fenómenos periglaciares. Tercera Reunión Grupo Periglacial Argentino*: 35–48.

Corte, A. E. (1987a). Rock Glacier taxonomy. In: Giardino, J. R., Shroder, J. F. & Vitek, J. D. (eds.). *Rock Glaciers*. London. Allen and Unwin: 27–39.

Corte, A. E. (1987b). Central Andes rock glaciers: applied aspects. In: Giardino, J. R., Shroder, J. F. and Vitek, J. D. (eds.). *Rock Glaciers*. London. Allen and Unwin: 289–303.

Corte, A. E. (1988). Geocryology of the Central Andes and rock glaciers. In: *Proceedings of the 5th International Conference on Permafrost*. Trondheim, Tapir Press. 1: 718–723.

Corte, A. E. & Beltramone, C. (1984). Edad de las estructuras criogénicas de Pto. Madryn, Chubut, Argentina. *Actas 2da Reunión del Grupo Periglacial Argentina*, 66–72. [In Spanish].

Corte, A. E. & Espizua, L. E. (1981). *Inventario de glaciares de la Cuenca del rio Mendoza*. Instituto Argentino de Nivologia y Glaciologia, Mendoza. 64p. [In Spanish].

Costin, A. B., Thom, B. G., Wimbush, D. J. & Stuiver, M. (1967). Nonsorted steps in the Mt. Kosciusko area, Australia. *Geological Society of America Bulletin*, 78: 979–992.

Coultish, T. L. & Lewkowicz, A. G. (2003). Palsa dynamics in a subarctic mountainous environment, Wolf Creek, Yukon Territory, Canada. Chapter 30. In: Phillips, M., Springman, S. M., and Arenson, L. U. (eds.). *Permafrost, ICOP 2003*. Balkema Publishers 21: 103.

Craig, B. G. (1959). Pingo in the Thelon Valley, Northwest Territories: radiocarbon age and historical significance of the contained organic material. *Bulletin of the Geological Society of America*, 70: 509–510.

Crampton, C. B. (1965). An indurated horizon in the soils of South Wales. *Journal of Soil Science*, 16: 230–242.

Crampton, C. B. (1977). A study of the dynamics of hummocky micro-relief in the Canadian North. *Canadian Journal of Earth Sciences*, 14: 639–649.

Crandall, K. A., Lundeberg, J. & Wayne, R. K. (1997). Multiple and ancient origins of the domestic dog. *Science*, 276: 1678–1689.

Cronin, J. E. (1983). Design and performance of a liquid natural convective subgrade cooling system for construction on ice-rich permafrost. In: *Permafrost: 4th International Conference Proceedings*. Washington, D.C. National Academy Press 1524p. 198–203.

Cronin, T. M., Dwyer, G. S., Farmer, J., Bauch, H. A., Spielhagen, R. F., Jakobsson, M., Nilsson, J., Briggs, W. M., Jr. & Stepanova, A. (2012). Deep Arctic Ocean warming during the last glacial cycle. *Nature Geoscience*, 5: 631–634. Doi: 10.1038/ngeo1557.

Crory, F. E. (1965). Pile foundations in discontinuous permafrost areas. *Proceedings of the Canadian Regional Permafrost Conference, Edmonton*. Ottawa. National Research Council of Canada, Associate Committee on Snow and Ice Mechanics Technical Memoir 86: 58–76.

Crory, F. E. (1966). Pile foundations in permafrost. *Proceedings of the 1st International Permafrost Conference, Lafayette, Indiana*. Washington. National Academy of Sciences Publication 1287: 467–476.

Crory, F. E. (1968). *Bridge foundations in permafrost areas: Goldstream Creek, Fairbanks, Alaska*. U.S. Army Cold Regions Research and Engineering Laboratory Technical Report 180. 33p.

Crory, F. E. (1973). Installation of driven test piles in permafrost at Bethel Air Force Station, Alaska. *U.S. Army Cold Regions Research and Engineering Laboratory*, Technical Report 145. 27p.

Crory, F. E. (1978). The Kotzebue Hospital: A case study. *Proceedings of the Conference on applied techniques for cold environments, Anchorage, Alaska*. American Society of Civil Engineers, 1: 342–359.

Crory, F. E. (1985). *Long-term foundation studies of three bridges in the Fairbanks area*. U.S. Army Cold Regions Research and Engineering Laboratory Technical Report.

Cross, C. W. & Howe, E. (1950). Geography and general geology of the Quadrangle in the Silverton folio. *U.S. Geological Survey* Folio 120: 1–25.

Cui, Z. (1985). Comparative study of geocryological (periglacial) conditions, features and processes in the Andes and the Himalayas: The Himalayas. In: *Unión Geográfica Internacional. Primera Reunión Latino-Americana sobre la importancia de los fenómenos periglares. Tercera Reunión Grupo Periglacial Argentino*: 49–59.

Cui, Z. & Song, C. (1992). Holocene periglacial processes and environmental changes in Daqingshan Mountains, Inner Mongolia, China. *Permafrost and Periglacial Processes*, 3: 55–62.

Cui, Z., Zhao, L. & Vandenburghe, J., *et al.* (2002). Discovery of ice wedge and sand-wedge networks in Inner Mongolia and Shanxi Province and their environmental significance. *Journal of Glaciology and Geocryology*, 24(6): 708–716. [In Chinese].

Cultural Survival (2010). Silent Spring in Siberia: The plight of the Vilyuy Sakha. http://www.culturalsurvival.org/publications/cultural-survival-quarterly/Russia/silent-spring-siberia-plight-vilyuy-sakha. Retrieved 03/06/2015.

Currey, D. R. (1974). Continentallity of extratropical climates. *Annals of the Association of American Geographers* 64: 268–280.

Curry, R. P. (1966). Observations of alpine mudflows in the Tenmile Creek Range, central Colorado. *Bulletin of the Geological Society of America*, 77: 771–776.

Cysewski, M. H. & Shur, Y. (2009). Prethawing: From mining to Civil Engineering, a historical perspective. In: *Proceedings of the 14th Conference on Cold Regions Engineering*. Reston, Virginia. American Society of Civil Engineers: 22–31.

Czeppe, Z. (1961). Thermic differentiation of the active layer and its influence on frost heave in periglacial regions (Spitzbergen). *Bulletin Academie Polonnaise Sciences, Séries Science géologie et géographie*, 8: 149–152.

Czeppe, Z. (1964). Exfoliation in a periglacial climate. *Geographia Polonica*, 2: 5–10.

Czudek, T. (1990). Zum problem der kryoplanationsterrassen. *Petermanns Geographische Mitteilungen*, 134: 226–238. [In German].

Czudek, T. & Demek, J. (1970). Thermokarst in Siberia and its influence on the development of lowland relief. *Quaternary Research*, 1: 103–120.

Czudek, T. & Demek, J. (1973). The valley cryopediments in Eastern Siberia. *Biuletyn Peryglacjalny*, 22: 117–130.

Daanen, R. P., Misra, D., Epstein, H., Walker, D. & Romanovsky, V. (2008). Simulating non-sorted circle development in arctic tundra ecosystems. *Journal of Geophysical Research*, 113, G03S06, doi: 10.1029/2008JG000682.

Dahl, R. (1966). Blockfields and other weathering forms in the Narvik Mountains. *Geografiska Annaler*, 48A: 224–227.

Dallimore, S. R. & Wolfe, S. A. (1988). Massive ground ice associated with glaciofluvial sediments, Richards Island, N.W.T., Canada. *Proceedings, 5th International Conferemce on Permafrost, Trondheim*. Trondheim, Tapir Press 1: 132–137.

Dallimore, S. R., Wolfe, S. A. & Solomon, S. M. (1996). Influence of ground ice and permafrost on coastal evolution, Richards island, Beaufort Sea coast, N.W.T. *Canadian Journal of Earth Sciences*, 33: 664–675.

D'Amico, S., Collins, T., Marx, J. C., Feller, G. & Gerday, C. (2006). Psycrophilic microorganisms: Challenges for life. EMBO Report 7: 385–389.

Danilov, I. D. (1973). Subaqueous pseudomorphoses in Pleistocene deposits. *Biuletyn Peryglacjalny*, 22: 339–345.

Danilov, I. D. (1996). Formation of syncryogenic units of ice-rich deposits ("ice" or "edoma" complexes). In: *Foundations of Geocryology*, Volume 2. Lithogenous Geocryology. Moscow. Moscow University Press: 286–290. [In Russian].

Danilova, N. S. (1966). Formation of thin ice veins and resulting pseudomorphs in sandy alluvium of Lena Delta. *Vestnik Moskovskogo Universiteta (serie 4) Geologica*, 6: 108–111. [In Russian].

Darwin, C. (1839). Geology and Natural History of the various countries visited by H. M. S. Beagle. London. J. Murray, pp. 254–256.

Datsko, P. S. & Rogov, V. V. (1988). Transformation of dispersed deposits under cycles of freezing-thawing. Moscow. Moscow University Press: 132–149. [In Russian].

Davidson, D. M. & Lo, R. C. (1982). Preservation of permafrost for a fuel storage tank. *Proceedings of the 4th Canadian Permafrost Conference, Calgary*. Ottawa. National Research Council of Canada: 545–554.

Davidson, D. W. (1973). Clathrate hydrates. In: Frank (ed.). *Water: A Comprehensive Treatise*. New York, Plenum Press, vol. 2: 115–234.

Davidson, D. W., El-Defrawy, K. K., Fuglem, M. O. & Judge, A. S. (1978). Natural Gas Hydrates in Northern Canada. *Proceedings of the 3rd International Conference on Permafrost, Edmonton*. Ottawa, National Research Council, 1: 937–943.

Davidson, E. A. & Janssens, I. A. (2006). Temperature sensitivity of soil carbon decomposition and peatlands to climatic change. *Nature*, 440: 165–173.

Davidson, G. P. & Nye, J. F. (1985). *Cold Regions Science and Technology*, 11: 141.

Day, J. H. & Rice, H. M. (1964). The characteristics of some permafrost soils in the Mackenzie Valley. *Arctic* 4: 222–236.

Decaulne, A. & Saemundsson, T. (2006). Geomorphic evidence for present-day snow avalanche and debris-flow impact in the Icelandic Westfjords. *Geomorphology*, 80: 8–93.

Dedysh, S. N. & Dunfield, P. E. (2009). Facultative methane oxidizers. In: *Handbook of Hydrocarbon and Lipid Microbiology*. Heidelberg. Springer-Verlag. Chapter 27: 1968–1975.

Degenhardt, J. G. Jr. (2009). Development of tongue-shaped and multilobate rock glaciers in alpine environments – interpreted from ground penetrating radar surveys. *Geomorphology*, 109: 94–107.

de Grandpré, Fortier, D. & Stephani, E. (2010). Impact of groundwater flow on permafrost degradation: implications for transportation infrastructures. *Geo2010*, Calgary: 534–540.

De Krom, V. (1990). *Retrogressive thaw slumps and active layer slides on Herschel Island*. Unpublished M.Sc. thesis, McGill University, Montréal, Québec.

Delaloye, R., Reynard, E., Lambeil, C., Marescot, L. & Monnet, R. (2003). Thermal anomaly in a cold scree slope. *Proceedings of the 8th International Conference on Permafrost*, Lisse, Balkema, 175–180.

Delisle, G. (2000). Temporal variability of sub-sea permafrost and gas hydrate occurrences as a function of climate change in the Laptev Sea. *Siberia Polarforschung* 68: 221–225.

Delisle, G., Allard, M., Fortier, R., Calmels, F. & Larrivée, È. (2003). Umiujaq, northern Québec: innovative techniques to monitor the decay of a lithalsa in response to climatic change. *Permafrost and Periglacial Processes*, 14(4): 375–385.

De Martonne, E. (1920). Le rôle morphologique de la neige en montagne. *La Géographie*, 34: 255–267. [In French].

Demek, J. (1960). Formy zvtnváni a obnosu zuluv Krumlovkeim lese jihozápadne od Brna. *Cas. miner. a Geol.*, 5. [In Czechoslovakian].

Demek, J. (1964). Periglacial slope development in the area of the Bohemian Massif of Northern Moravia. *Biuletyn Peryglacjalny*, 14: 169–192.

Demek, J. (1968). Cryoplanation terraces in Yakutia. *Biuletyn Peryglacjalny*, 17: 91–116.

Demek, J. (1969a). Cryoplanation terraces, their geographical distribution, genesis and development. *Akadmie VED rozpravy, Rad Mathematickych A Prirodnich Ved, Rocnik*, 79(4): 80p.

Demek, J. (1969b). Cryogene processes and the development of cryoplanation terraces. *Biuletyn Peryglacjalny*, 18: 115–125.

Denny, C. S. (1936). Periglacial phenomena in southern Connecticut. *American Journal of Science*, 32: 322–342.

Denton, G. H. & Karlén, W. (1973). Holocene climatic variations – their pattern and possible cause. *Quaternary Research*, 3: 155–205.

Deryagin, B. V., Churaev N. V,, & Muller, B. N. (1985). *Surface forces*. M. [In Russian].

Desyatkin, R. V. (1993). Syngenetic soil salinization during thermokarst formation. *Eurasian Soil Science*, 25(4): 38–46.

Desyatkin, R. V. & Desyatkin, A. R. (2006). Thermokarst transformation of soil cover on cryolithozone flat territories. In: Hatano, R. and Guggenberger, G. (Eds.). *Symposium on Environmental change in Siberian Permafrost Region*. Sapporo. Hokkaido University Press: 213–223.

Dickens, H. B. & Gray, D. M. (1960). Experience with a pier-supported building over permafrost. *American Society of Civil Engineers, Journal of the Soil Mechanics and Foundation Division*, 86 (SM5): 1–14.

Dickfoss, P., Betancourt, J. L. & Thompson, L. (1997). History and paleoclimate potential of Candelaria Ice Cave, west-central New Mexico. In: Zidek, G. (Ed.). *A natural history of El Malpas*. New Mexico Bureau of Mines and Mineral Resources Bulletin 1561: 91–112.

Ding, J. (1984). Study and practice on deep foundation in permafrost areas of China. *Proceedings of the 4th International Conference on Permafrost, Fairbanks, Alaska*. Washington. National Academy of Sciences. Final Proceedings: 18–24.

Ding, J. K. & He, G. S. (2000). The critical mean value of air temperature – an important factor for designing the critical height of embankment in permafrost regions of the Tibetan Plateau. *Journal of Glaciology and Geocryology*, 22(4): 333–339.

Dionne, J.-C. (1984). Palses et limit méridionale du pergélisol dans l'hémisphère nord: le cas de Blanc-Sablon, Québec subarctique. *Géographie physique et Quaternaire*, 32(2): 165–184. [In French].

Dixon, J. C., Thorn, C. E. & Darmody, R. G. (1984). Chemical weathering processes on the Vantage Peak Nunatak, Juneau Icefield, Southern Alaska. *Physical Geography*, 5: 111–131.

Dixon, J. C., Thorn, C. E., Darmody, R. G. & Campbell, S. W. (2002). Post-glacial rock weathering processes on a roche moutonnée in the Riksgränsen area (68°N), northern Norway. *Norsk Geografisk Tidsskrift*, 56(4): 257–264.

Dobinski, W. (2012). The cryosphere and glacial permafrost as its integral component. *Central European Journal of Geosciences*, 4(4): 623–640.

Dobrovolskiy, L. A. (1974). Effects of tectonics and subsurface waters on development of Permafrost. *International Geology Review*, 16(6): 690–696.

Dokuchaev, V. V. (1879a). Abridged historical account and critical examination of the principal soil classification existing. *Transactions of the St. Petersburg Society of Nature*, 10: 64–67. [In Russian].

Dokuchaev, V. V. (1879b). Preliminary report on the investigation of the southeastern part of the chernozem of Russia. St. Petersburg. *Trudy Imp. Vol. Ekonom. Obsches*. [In Russian].

Dokuchaev, V. V. (1883). *Russian chernozem*. St. Petersburg. [In Russian].

Dokuchayev, V. V. & Gerasimov, A. S. (1988). Anchoring capacity of column foundations subjected to tangential forces induced by the frost heaving of soils. *Soil Mechanics and Foundation Engineering*, 24(6): 252–257.

Dolenko, G. I. (1913). The Lena River valley near Yakutsk. *Preliminary Report on the investigations of the soils in Asiatic Russia during 1913*. St. Petersburg, [In Russian].

Domaradzki, J. (1951). Blockströme im Kanton Graubünden. *Ergeb Wiss Untersuch Schweiz Nationparks*, 3: 177–235. [In German].

Dornbusch, U. (2005). Glacier-rock glacier relationships as climatic indicators during the late Quaternary in the Cordillera Ampato, Western Cordillera of southern Peru. *Geological Society of London Special Publications* 242: 75–82.

Dostovalov, B. N. (1952). About the physical conditions of formation of frost cracks. *Research of permafrost in Yakutia*. Volume 3. [In Russian].

Dostovalov, B. N. & Kudryavtsev, V. A. (1967). *Obshcheye merzlotovedeniya*. Moscow. Moscow State University. 403p. [In Russian].

Dostovalov, B. N. & Popov, A. I. (1966). Polygonal systems of ice wedges and conditions of their formation. In: *Proceedings of the 1st International Conference on Permafrost*, Ottawa, National Academy of Science, National Research Council of Canada, Publication #1287: 102–105.

Douglas, L. A. & Tedrow, J. F. C. (1961). Tundra soils of Arctic Alaska. In: *Proceedings of the 7th International Congress of Soil Science, 1960, Madison, Wisconsin*: 291–304.

Drachev, S. S., Savostin, L. A. & Bruni, I. E. (1995). Structural pattern and tectonic history of the Laptev Sea region. *Report on Polar Research*, 176: 348–366.

Drachev, S., Savostin, L., Groshev, V. & Bruni, I. (1998). Structure and geology of the continental shelf of the Laptev Sea, eastern Russian Arctic. *Tectonophysics*, 298: 357–393.

Draebing, D. & Eichel, J. (2016). Controlling factors of turf-banked solifluction lobe evolution on the Turtmann Glacier forefield (Switzerland). *Geophysical Research Abstracts*, 18: EGU2016-479, 2016.

Drake, J. J. & McCann, S. B. (1982). The movement of isolated boulders on tidal flats by ice floes. *Canadian Journal of Earth Sciences*, 19: 784–754.

Drew, J. F. & Tedrow, J. C. F. (1962). Arctic Soil Classification and patterned ground. *Arctic*, 15: 109–116.

Drosf-Hausen, W. (1967). The water-ice interface as seen from the liquid side. *Journal of Colloid and Interface Science*, 25(2): 243.

Duan, Q. L. (2002). The formation mechanism and treatment measure of the motive sand-flood of Cunna Lake on Qingzang Railway. *Geotechnical Engineering Technique*, 6(3): 311–314.

Duan, X. & Naterer, G. F. (2004). Heat conduction with seasonal freezing and thawing in an active layer near a tower foundation. *International Journal of Heat and Mass Transfer*, 52(7–8): 2068–2078.

Dubikov, G. I. & Ivanova, N. B. (1994). Condition and salinity of sediments at the coastal areas of the Kara Sea. *Materyaly i konferencii geokriologov Rossii*, 1. [In Russian].

Dubnie, A. (1972). *Northern mining problems with particular reference to unit operations in permafrost*. Ottawa. Department of Energy, Mines and Resources Technical Bulletin TB48. 20p.

Dubrovin, V. A. (1974). The prediction of changes in the conditions of ground freezing in connection with industrial and municipal construction in the area of the middle reaches of the Angara River. *Moscow University Geology Bulletin*, 29(3): 93–96. [In Russian].

Duchkov, A. D., Balobaev, V. T., Devyatkin, V. N., An, V. V. & Sokolova, L. S. (1995). Geothermal model of the West-Siberian permafrost. *Russian Geology and Geophysics*, 36(8): 69–79.

Duckson, D. W., Jr. (1967). Continentality. In: Oliver, J. E. & Fairbridge, R. W., (eds.). *The Encyclopedia of Climatology*. New York. Van Nostrand Reinholt Co. 365–367.

Dufour, S., Lafleche, P. & Judge, A. S. (1988). Design and monitoring of earth embankments over permafrost. *2nd International Conference on Case Histories in Geotechnical Engineering, St Louis, Missouri*. Paper #10: 1001–1010.

Duk-Rodkin, A. (1994). Tertiary-Quaternary drainage of the pre-glacial Mackenzie Basin. *Quaternary International*, 22/23: 221–241.

Dunn, J. R. & Hudec, P. P. (1965). *The influence of clays on water and ice in rock pores (2)*. New York State, Department of Public Works, Physical Research Report RR65-5. 149p.

Dunn, J. R. & Hudec, P. P. (1966). *Frost deterioration: Ice or ordered water?* Geological Society of America, Special Paper 101. 256p.

Dunn, J. R. & Hudec, P. P. (1972). Frost and sorption effects in argillaceous rocks. In: Highway Research Board, *Frost action in soils*. Washington, National Academy of Sciences – National Academy of Engineering, Highway Research Record, 393: 65–78.

Duval, P., Ashby, M. F. & Anderman, I. (1983). Rate-controlling processes in the creep of polycrystalline ice. *Journal of Physical Chemistry*, 87: 4066–4074.

Dydyshko, P. I., Kondratiev, V. G., Vasilyev, M. L., Prigoda, V. Y., Sadakova, M. N. & Valuyev, A. S. (1993). Deformed embankments on Mari and the ways of their stabilization. Permafrost; 6th International Conference Proceedings. Beijing, China. July 5–9, 1993. Wushan, Guangzhou, China: South China University of Technology Press. 1: 155–159.

Dyke, A. S. (1976). Tors and associated weathering phenomena, Somerset Island, District of Franklin. *Geological Survey of Canada*, Paper 76-1B: 209–216.

Dyke, A. S. (1978). Qualitative rates of frost heave in bedrock on Southeastern Baffin Island, District of Franklin. *Geological Survey of Canada*, Paper 78-1A: 501–502.

Dyke, A. S., (1984). Frost heaving of bedrock in permafrost regions. *Bulletin of the Association of Engineering Geologists*, XXXI(4): 389–405.

Dyke, A. S. (2004). An outline of North American deglaciation with emphasis on central and northern Canada. In: Ehlers, J. & Gibbard, P. L. (eds.). *Quaternary glaciations: Extent and chronology. Part II. North America*. Elsevier B. V.: 371–406.

Dyke, A. S. & Zoltai, S. C. (1980). Radio-carbon dated mudboils, Central Canadian Arctic. *Geological Survey of Canada*, Paper 80-1B: 271–275.

Dyke, L. D. (2000). Stability of permafrost slope in the Mackenzie Valley. In: Dyke, L. D. and Brooks, G. R. (Eds.). *The Physical Environment of the Mackenzie Valley, northwest Territories: A Base Line for the Assessment of Environmental Change*. Ottawa. Geological Survey of Canada Bulletin #547: 176–186.

Dyke, L. D. (2001). Contaminant migration through the permafrost active layer, Mackenzie Delta area, Northwest Territories, Canada. *Polar Record*, 37(202): 215–228.

Dyke, L. D. & Sladden, W. E. (2010). Permafrost and peatland evolution in the Northern Hudson Bay Lowland, Manitoba. *Arctic*, 63: 429–441.

Dylik, J. (1957). Tentative comparison of planation surfaces occurring under warm and cold semi-arid conditions. *Biuletyn Peryglacjalny*, 5: 175–186.

Dylik, J. (1966). Problems of ice wedge structure and frost fissure polygons. *Biuletyn Peryglacjalny*, 15: 241–291.

Dylik, J. (1969a). Slope development under periglacial conditions in the Łódż region. *Biuletun Peryglacjalny*, 18: 381–410.

Dylik, J. (1969b). Slope development affected by frost fissures and thermal erosion. In: Péwé, T. L. (ed.). *The Periglacial Environment*. Montreal. Mcgill-Queen's University Press: 365–386.

Dylikowa, A. (1961). Structures de pression congélistatiques et structures de gonflement par le gel près de Katarzynów près de Lódz. *Bulletin de la Société de Lettres de Lódz*, 12(9): 1–23. [In French].

Dyrness, C. T., Viereck, L. A. & Van Cleve, K. (1986). Fire in taiga communities of interior Alaska. In: Van Cleve, K., Chapin III, F. S., Flanagan, L. A., Viereck, L. A. and Dyrness, C. T. (eds.). *Forest Ecosystems in the Alaskan Tiaga*. New York, Springer-Verlag: 74–86.

Eakin, W. M. (1916). The Yukon-Koyukuk region, Alaska. *United States Geological Survey Bulletin*, 631: 67–88.

EBA (2004). *Permafrost considerations for effective mine site development in the Yukon Territory*. Mining Environment Research Group MERG report 2004-1. EBA Engineering Consultants Ltd. 26p.

Eckerstorfer, M. & Christiansen, H. H. (2012). Meteorology, topography and snowpack conditions causing two extreme mid-winter slush and wet slab avalanche periods in High Arctic maritime Svalbard. *Permafrost and Periglacial Processes*, 23: 15–25.

Edelman, C. H., Florschtz, F. & Jeswiet, J. (1936). *Über spätpleistozäne und frühholozäne kryoturbate Ablagerungen in den östlichen Niederlanden*. Verhandelingen: Geologisch-Mijnbouwkundig Genootschap voor Nederland en Kolonien, Geologisch Series 11: 301–360. [In Dutch].

Edenborn, H. M., Sama, J. I. & Kite, J. S. (2012). Thermal regime of a cold air trap in Central Pennsylvania, USA: the Trough Creel ice mine. *Permafrost and Periglacial Processes*, 23(3): 187–195.

Edlefsen, N.E. & Anderson, A.B.C. (1943). Thermodynamics of soil moisture. *Hilgardia*, 15(2).

Egginton, P. A. & Dyke, L. S. (1982). Density gradients and injection structures in mudboils in central District of Keewatin. *Geological Survey of Canada*, Paper 78-B: 203–206.

Egginton, P. A. & Shilts, W. W. (1978). Rates of movement associated with mudboils, central District of Keewatin. *Geological Survey of Canada*, Paper 78-1B: 203–206.

Eggner, C. L. & Tomlinson, B. G. (1978). Temporary wastewater treatment in remote locations. *Journal of the Water Pollution Federation*, 50: 2643–2656.

Egerov, F. G., Nikiforov, M. M. & Danilov, Y. G. (2015). Unique experience and achievements of Sakha people in development of agriculture in the north. *Modern Applied Science*, 9(5): 14–24.

Ehlers, J. & Gibbard, P. (2008). Extent and Chronology of Quaternary glaciations. *Episodes*, 31: 211–218.

Eichel, J., Krautblatter, M., Schmidtlein, S. & Dilan, R. (2013). Biogeomorphic interactions in the Turtmann Glacier forefield, Switzerland. *Geomorphology*, 201: 98–110. Doi: 10.1016/j.geomorph.2013.06.012.

Eichel, J., Corenblit, D. & Dikan, R. (2015a). Conditions for feedbacks between geomorphic and vegetation dynamics on lateral moraine slopes: A biogeomorphic feedback window. *Earth Surface Processes and Landforms*. Doi: 10.1002/esp.3859.

Eichel, J., Meyer, N., Draebing, D., Schmidtlein, S. & Dikan, R. (2015b). Controls on small-scale biogeomorphic interactions on lateral moraine slopes and their linkage to large-scale geomorphic and vegetation patterns. *Geophysical Research Abstracts*, 18: EGU2016-1860, 2016.

Eichfeld, I. G. (1931). Problema zemledelya na krayem severe. *Sovetskiy Sever*, 5. [In Russian].

Eissmann, L. (1994). Grundzüge der Quartärgeologie Mitteldeutschlands (Sachsen, Sachsen-Anhalt, Südbrandengurg, Thüringen). *Altenburger Naturwissenschaftliche Forschungen*, 7: 55–135. [In German].

Ellenburg, L. (1974). Shimobashira – kameis in Japan. *Geographica Helvetica*, 29: 1–5.

Ellenburg, L. (1976). Zur Periglazialmorphologie von Ura Nipon, der schneereichen Seite Japans. *Geographica Helvetica*, 31: 139–151. [In German].

Ellis, S. (1980). Physical and chemical characteristics of a podzolic soil formed in neoglacial, Okstindan, northern Norway. *Arctic and Alpine Research*, 12(1): 65–72.

Emel'yanov, V. I. (1973). Measures to prolong overburden removal in permafrost placers. *Kolmya*, 1: 6–7. [In Russian].

Emel'yanov, V. I. & Perl'shtein, G. Z. (1980). Water thawing of frozen ground for open pit and underground mining in the northeast part of the U.S.S.R. In: *Proceedings of the 3rd International Conference on Permafrost*. Ottawa, National Research Council of Canada. English translations, Part II: 339–354.

Encyclopedia Britanica (2007).

Englezos, P. (1993). Clathrate Hydrates. *Industrial Engineering Chemical Research*, 32: 1251–1274.

Environment Canada (1996). *Surface water and sediment data (Hydat CD-ROM)*. Boulder, Colorado. Earthinfo.

Environmental Protection Service (1977). *National Inventory of Municipal waterworks and wastewater systems in Canada, 1975*. Ottawa. Environmental Protection Services.

Environmental Protection Service (1985). *Sewage lagoons in cold climates*. Edmonton. Technical Services Branch, Environment Protection Services, Environment Canada.

Eremenko, A. (2014). Russia is running out of forest. Moscow Times, September 30th. English Edition.

Ermolaev, M. M. (1932). Geological and geomorphological description of Bol'shoi Lyakhovskii Island. Yakutsk. *Trudy SOPS AN SSSR*, ser. 7. [In Russian].

Esch, D. C. (1983). Evaluation of experimental design features for roadway construction over permafrost. *Permafrost. 4th International Conference Proceedings, Fairbanks, Alaska*. Washington, D.C., National Academy Press: 283–288.

Esch, D. (1988). Roadway embankments on warm permafrost: Problems and remedial treatments. In: *Permafrost: Proceedings of the 5th International Conference, Trondheim, Norway*. Trondheim. Tapir Press 2: 1223–1226.

Essoglou, M. E. (1957). Piling operations in Alaska. *The Military Engineer*, 49(330): 282–287.

Etzelmuller, B. (2013). Recent advances in mountain permafrost research. *Permafrost and Periglacial Processes*, 24: 99–107. DOI: 10.1002/ppp.1772.

Evans, S. G. & Clague, J. J. (1989). Rain-induced landslides in the Canadian Cordillera, July, 1988. *Geoscience Canada*, 16: 193–200.

Everett, D. H. (1961). The thermodynamics of frost action in porous solids. *Transactions of the Faraday Society*, 57: 1541–1551.

Everett, K. R. (1963). *Slope movement, Neotoma Valley Southern Ohio*. Institute of Polar Studies Report # 6. 62p.

Everett, K. R. (1966). Slope movement and related phenomena. In: Wilimovsky, N. J. & Wolfe, J. N. (eds.). Environment of the Cape Thompson region, Alaska. *U.S. Atomic Energy Commission, Division of Technical Information*: 175–220.

Everett, K. R., Hall, G. F. & Wilding, L. P. (1971). Wisconsin age cryoturbation features in Central Ohio. *Geological Society of America Bulletin*, 82: 1407–1410.

Evin, M. (1983a). Présence et signification morphoclimatic de sediments gelés à l'amont des glaciers rocheux. In: *Actes du College de l'Association des Géographes français*, Paris 8.1.83. Eboulis et environnement géographique passé et actuel: 137–142. [In French].

Evin, M. (1983b). *Structure et movement des glaciers rocheux des Alps du Sud*. Universitaire Grenoble I, Institut de Géographie Alpine. Thèse de 3e cycle, 343p. [In French].

Evin, M. (1984). Characteristiques physic-chimiques des eaux issues des glaciers rocheux des Alpes du Sud (France). *Zeitschrift für Gletscherkund und Glazialgeologie*, 20: 27–40. [In French].

Evin, M. (1987). Lithology and fracturing control of rock glaciers in southwestern Alps of France and Italy. In: Giardino, J. R., Shroder, J.F. & Vitek, J. D. (eds.). *Rock glaciers*. London. Allen and Unwin: 83–106.

Evin, M. (1993). Glaciers et glaciers rocheux dans les vallons de Mongioie et de Schiantala (Haute Stura di Demonte, Italie). *Zeitschrift für Gletscherkund und Glazialgeologie*, 27/28: 1–10. [In French].

Evin, M. & Assier, A. (1982). Mise en evidence de mouvements sur le glacier rocheux du Pic d'Asti (Queyras-Alpes du Sud-France). *Révue de Géomorphologie Dynamique*, 31(4): 127–136. [In French].

Evin, M. & Assier, A. (1983). *Relations hydrologiques entre glacier et glaciers rocheux: l'example du cirque de Marinet (Haute-Ubaye, Alpes du Sud)*. Communication, Section de Glaciologie de la Société hydrotechnique de France. Grenoble. 5p. [In French].

Evseev, V. P. (1978). Migrational Frost Mounds. In: *USSR Contribution: Permafrost, 2nd International Conference, Yakutsk, Washington, D.C. National Academy of Sciences Press*: 121–13.

Fahey, B. D. (1973). An analysis of diurnal freeze-thaw and frost heave cycles in the Indian Peaks region of the Colorado Front Range. *Arctic and Alpine Research*, 5: 269–281.

Fahey, B. D. (1974). Seasonal frost heave and frost penetration measurements in the Indian Peaks region of the Colorado Front Range. *Permafrost and Periglacial Processes*, 6: 63–70.

Fahey, B. D. (1979). Frost heaving of soils at two locations in Southern Ontario, Canada. *Geoderma*, 22: 119–126.

Fahey, B. D. (1985). Salt weathering as a mechanism of rock breakup in cold climates: an experimental approach. *Zeitschrift für Geomorphologie*, 29: 99–111.

Fahey, B. D. & Dagesse, D. F. (1984). An experimental study of the effect of humidity and temperature variations on the granular disintegration of argillaceous carbonate rocks in cold climates. *Arctic and Alpine Research*, 16(3): 291–298.

Fairchild, L. H. (1987). The importance of labor initiation processes. Review in *Engineering Geology*, 7: 51–61.

FAO (1988). *Soil map of the world: Revised legend*. Food and Agriculture Organization of the United Nations, Rome. 119p.

Farbrot, H., Hipp, T. F., Etzelmüller, B., Isaksen, K., Ødegård, R. S., Schuler, T. V. & Humlum, O. (2011). Air and ground temperature variations observed along elevation and continental gradients in Southern Norway. *Permafrost and Periglacial Processes*, 22: 343–360.

Fedosov, A. Ye. (1935). *Physico-mechanical processes in soils during their freezing and thawing*. M. Transzheldoryizdat. [In Russian].

Feldman, G. M. (1984). *Thermokarst and permafrost*. Novosibirsk. Nauka. 260p. [In Russian].

Feller, G. (2007). Life at low temperatures: is disorder the driving force? *Extremophiles*, 11: 211–216.

Feng, H. Y., Ma, X. J., Zhang, G. S., Bai, Y., Fei, G. Q., Cheng, G. D., An, L. Z. & Liu, G. X. (2004). Culturing and counting the microbial cells in permafrost on the Tibetan Plateau. *Journal of Glaciology and Geocryology*, 26: 182–187.

Feng, W. J. & Ma, W. (2006). Experimental study of the effect of awnings along the Qinghai-Tibet Railway. *Journal of Glaciology and Geocryology* 28(1): 108–115.

Feng, W. J., Sun, Z. Z., Wen, Z., Ki, G. Y., Zhang, Z. & Yu, W. B. (2012). Application of the awning measure to obstruct solar radiation in the permafrost regions of the Qinghai-Tibet Plateau. *Sciences in Cold and Arid Regions*, 4(2): 121–126.

Fernette, G. (1982). The Mining process. *Alaska Geographic*, 9(4): 120–133.

Feulner, A. J. & Williams, J. R. (1967). Development of the groundwater supply at Cape Lisborne, Alaska, by modification of the thermal regime of permafrost. *U.S. Geological Survey Professional Paper*, 575-B: 199–202.

Fife, J. A. (1960). Refrigerant piping system supports Arctic radar sites. *Heating, Piping and Air Conditioning*, 32: 112–118.

Filippov, V. M. (1981). The experimental study of the processes of dissolution of sulphate rocks in Pribaikalia. In: *Some Questions of Geomorphology of East Siberia – Irkutsk*, 131–140. [In Russian].

Finch, R. H. (1937). A tree-ring calendar for dating volcanic events at Cinder Cone, Lassen National Park, California. *American Journal of Science*, 33: 140–146.

Finlay, P. & Krason, J. (1990). Evaluation of geological relationships to gas hydrate formation and stability: Summary Report. *Gas Energy Review*, 18: 12–18.

Fisch, W. Sr, Fisch, W., Jr, and Haeberli, W. (1978). Electrical resistivity soundings with long profiles on rock glaciers and moraines in the Alps of Switzerland. *Zeitschrift für Gletscherkund und Glaciogeologie*, 13: 239–260.

Fischer, K. (1965). Mukegel, Schwemmkegel und Kegelsimse in den Alpentälern. *Mitteilungen der Geographischen Gesellschaft München*, 50: 127–159. [In German].

Fischer, L., Huggel, C., Käab, A. & Haeberli, W. (2013). Slope failures and erosion rates on a glacierized high-mountain face under climatic changes. *Earth Surface Processes and Landforms*, 38: 836–846.

FitzPatrick, E. A. (1956). An indurated horizon formed by permafrost. *Journal of Soil Science*, 7: 248–254.

FitzPatrick, E. A. (1976). Cryons and Isons. *Proceedings of the North of England Discussion Group*, 11: 31–43.

Flint, R. F. (1955). Pleistocene geology of eastern South Dakota. *U.S. Geological Survey Professional Paper*, 262. 173p.

Forbes, D. L. (1984). *Coastal geomorphology and sedimentation of Newfoundland*. Geological Survey of Canada, Paper 84-B: 11–24.

Forbes, D. L. & Taylor, R. B. (1994). Ice in the shore zone and the geomorphology of cold coasts. *Progress in Physical Geography*, 18(1): 59–89.

Forbes, D. L., Taylor, R. B. & Fobel, D. (1986). Coastal studies in the western Arctic Archipelago (Melville, McKenzie King, Lougheed and nearby islands). *Geological Survey of Canada Open File* 1409. 29p.

Ford, D. C. (1984). Karst groundwater activity and landform genesis in modern permafrost regions of Canada. In: LaFleur, R. G. (Ed.). *Groundwater as a geomorphic agent*. London, Allen and Unwin Ltd.: 340–350.

Ford, D. C. (1987). Effects of glaciations and permafrost upon the development of karst in Canada. *Earth Surface Processes and Landforms*, 12: 507–521.

Ford, D. C. & Williams, P. W. (1989). *Karst Geomorphology and Hydrology*. London. Chapman and Hall. 601p.

Forsyth, J. (1994). *A history of the peoples of Siberia: Russia's North Asian Colony, 1581–1990*. Cambridge. Cambridge University Press.

Fortier, D. & Allard, M. (2004). Late Holocene syngenetic icewedge polygons development, Bylot Island, Canadian Arctic Archipelago. *Canadian Journal of Earth Sciences*, 41(8): 997–1012.

Fortier, D. & Allard, M. (2008). Atmospheric and geothermal conditions during thermal contraction of ice wedges, Bylot Island, eastern Canadian Arctic archipelago. *8th International Conference on Permafrost. Extended Abstracts on current research and newly available information*: 37–38.

Fortier, D. & Allard, M. (2013). Late Holocene syngenetic ice-wedge polygons development, Bylot Island, Canadian Archipelago. *Canadian Journal of Earth Sciences*, 41(8): 997–1012.

Fortier, D. & Savard, C. (2010). Engineering geophysical investigation of permafrost conditions underneath airfield embankments in Northern Québec (Canada). *GEO2010*: 1307–1316.

Fortier, R. (2008). Thaw settlement of degrading permafrost; A geohazard affecting the performance of man-made infrastructures at Umiujac in Nunavik (Québec). In: Locat, J., Perret, D., Turmel, D., Demers, D., & Lerouell, S. (2008). *Comptes rendus de la 4e Conférence canadien sur les géoriques: des causes á la gestion*. Laval. Laval University Press. 594p.

Fotiev, S. M., Sheveleva, N. & Danilova, N. (1974). *Permafrost in Middle Siberia*. Moscow. Russian Academy of Sciences, Permafrost Institute. Nauka. 146p.

Fotiev, S. M. (2009). Siberian geocryological chronologies. *Earth Cryosphere*, 13(3): 3–16. [In Russian].

Foster, H. L. & Holmes, G. W. (1965). A large transitional rock glacier in the Johnson River Area, Alaska Range. *U.S. Geological Survey* Professional Paper, 525B: B112–B116.

Foster-Miller Associates (1965). *Final Phase I Report on an investigation of methods of conveying snow, ice and/or frozen ground from an excavation to a disposal area*. U.S. Army Cold Regions Research and Engineering Laboratory, Hanover, New Hampshire. Internal Report IR23. 101p.

François, R. (2002). Cool stratification. *Nature*, 428: 31–32.

Francou, B. (1988). Eboulis stratifiés dans la Hautes Andes Centrales du Pérou. *Zeitschrift für Geomorphologie*, 32: 47–76. [In French].

Franke, D., Hinz, K. & Oncken, O. (2001). The Laptev Sea Rift. *Marine and Petroleum Geology*, 18: 1083–1127.

Franklin, L. J. (1983). In situ hydrates – a potential gas source. In: Cox, J. L. (ed.). *Natural Gas Hydrates: Properties, occurrence and Recovery*. Woburn, Maine. Butterworth: 115–122.

Franzmeier, D. P., Norton, L. D. & Steinhardt, G. C. (1989). Fragipan formation in loess of the Midwestern United States. In: Smeck, N. E. and Ciolkose, E. J. (eds.). *Fragipans: Their occurrence, classification and genesis*. Soil Science Society of America Special Publication #24: 69–97.

Frauenfelder, R., Schneider, B. & Kääb, A. (2008). Using dynamic modelling to simulate the distribution of rock glaciers. *Geomorphology*, 93: 130–143.

Freden, S. (1965). Some aspects on the physics of frost heave in mineral soils. *Surface Chemistry*, 79–90.

French, H. M. (1970). Soil temperatures in the active layer, Beaufort Plain. *Arctic*, 229–239.

French, H. M. (1974). Active thermokarst processes, eastern Banks Island, western Cnadian Arctic. *Canadian Journal of Earth Sciences*, 11(6): 785–794.

French, H. M. (1975). Man-induced thermokarst, Sachs Harbour airstrip, Banks Island, N. W. T. *Canadian Journal of Earth Sciences*, 12: 132–144.

French, H. M. (1976). *The Periglacial Environment*. London and New York. Longman. 308p.

French, H. M. (1978). *Sump studies I: Terrain disturbances*. Environmental Studies #6. Ottawa. Indian and Northern Affairs. 52p.

French, H. M. (1980). Terrain, land use and waste drilling fluid disposal problems, Arctic Canada. *Arctic*, 33: 794–806.

French, H. M. (1986). Periglacial involutions and mass displacement structures, Banks Island, Canada. *Geografiska Annaler*, 68A(3): 167–174.

French, H. M. (1996). *The Periglacial Environment*. 2nd Edition. Harlow. Adison Wesley Longman, Ltd. 341p.

French, H. M. (2003). The development of Periglacial Geomorphology: 1 – up to 1965. *Permafrost and Periglacial Processes*, 14(1): 29–60.

French, H. M. (2007). *The Periglacial Environment*. 3rd Edition. New York. Wiley. 458p.

French, H. M. (2016). Do periglacial landscapes exist? A discussion of upland landscapes of Northern Interior Yukon, Canada. *Permafrost and Periglacial Processes*, 27(2): 219–218.

French, H. M. & Dutkiewcz, L. (1976). Pingos and pingo-like forms, Banks Island, western Canadian Arctic. *Biuletyn Peryglacjalny*, 26: 211–222.

French, H. M., Demitroff, M. & Forman, S. L. (2003). Evidence for Late Pleistocene permafrost in the New Jersey Pine Barrens (latitude 39°N), Eastern USA. *Permafrost and Periglacial Processes*, 14: 259–274.

French, H. M., Harris, S. A. & van Everdingen, R. O. (1983). The Klondike and Dawson. In: French, H. M. and Heginbottom, J. A. (eds.). *Northern Yukon Territory and Mackenzie Delta, Canada. Guidebook to Permafrost and related features*. 4th International Conference on Permafrost. Guidebook #3. Division of Geological and Geophysical Surveys, Fairbanks, Alaska: 35–63.

French, H. M. & Harry, D. G. (1988). Nature and origin of ground ice, Sandhills Moraine, southwest Banks Island, Western Canadian Arctic. *Journal of Quaternary Science*, 3: 19–30.

French, H. M. & Harry, D. G. (1990). Observations on buried glacier ice and massive segregated ice, western Arctic coast, Canada. *Permafrost and Periglacial Processes*, 1: 31–43.

French, H. M. & Pollard, W. H. (1986). Ground-ice investigations, Klondike District, Yukon Territory. *Canadian Journal of Earth Sciences*, 23: 550–560.

Frich, P. & Brandt, E. (1985). Holocene talus accumulation rates, and their influence on rock glacier growth. A case study from Igpik, Disco, West Greenland. *Geografisk Tidsskrift*, 85: 32–43.

Friedmann, E. I. (1994). *Viable Microorganisms in Permafrost*. In: Gilichinsky, D. (ed.). Russian Academy of Sciences, Pushchino: 21–26.

Fries, T. & Bergström, E. (1910). Några iakttagelser öfver palsar och deras förekomst i nordligaste Sverige. *Geologiska Foreningens Forhandlinger, Stockholm*, 32: 195–205.

Froelich, W. & Słupik, J. (1978). Frost mounds as indicators of water transmission zones in the active later of permafrost during the winter season (Khangay Mountains, Mongolia). *Proceedings of the 3rd International Conference on Permafrost, Edmonton*, Ottawa, National Research Council of Canada 1: 188–193.

Froelich, W. & Slupik, J. (1982). River icings and fluvial activity in extreme continental climate, Khangai Mountains, Mongolia. In: Proceedings of the 4th Canadian Permafrost Conference, Calgary.: 203–211.

Froese, D. G. (1998). Sedimentology and paleomagnetism of the Plio-Pleistocene Lower Klondike Valley terrace, Yukon Territory. Unpublished Ph.D. thesis, Department of Geography, University of Calgary. 153p.

Froese, D. G., Westgate, J. A., Retes, A. V., Enkin, R. J. & Preece, S. J. (2008). Ancient permafrost and a future, warmer Arctic. *Science*, 321: 1648.

Frost, G. V., Epstein, H. E., Walker, D. A., Matyshak, G. & Ermokhina, K. (2013). Patterned-ground facilitates shrub expansion in Low Arctic tundra. *Environmental Research Letters*, 8 015035.

Fryxell, F. M. & Horberg, L. (1943). Alpine mudflows in Grand Teton National Park, Wyoming. *Bulletin of the Geological Society of America*, 54: 457–472.

Fu, C. F., Zhao, J. B., Mei, F. M., Shao, T. J. & Zuo, J. (2015). Vertical distribution of soil moisture and surface sandy soil wind erosion for different types of sand dune on the southeastern margin of the Mu Us sandy land, China. *Sciences in Cold and Arid Regions*, 7(6): 666–674.

Fujino, K., Horiguchi, K., Shinbori, M. & Kato, K. (1983). Analysis and characteristics of cores from a massive ice body in Mackenzie Delta, N.W.T., Canada. *Proceedings of the 4th International Conference on Permafrost*. Washington, D.C., National Academy Press: 316–321.

Fujino, K., *et al.* (1988). Characteristics of the massive ground ice body in the western Canadian Arctic. *Proceedings of the 5th International Conference on Permafrost*. Trondheim, Tapir Publishers 1: 143–147.

Fukuda, M. (1982). *Experimental Studies Of Coupled Heat and Moisture Transfer in Soils During Freezing*. Contributions from the Institute of Low Temperature Science Series. No. 31.

Fukuda, M. (1983). The pore water pressure in porous rocks during freezing. In: *Proceedings of the 4th International Conference on Permafrost*. Washington, D.C., National Academy Press: 322–327.

Fukuda, M., Sento, N., Kunitky, V. V. & Nakamura, T. (1995). Radiocarbon dating results of organic materials obtained from eastern Siberian permafrost. *Proceedings of the 4th Symposium on the joint Siberian permafrost studies between Japan and Russia in 1995*. Sapporo, Kohsoku Printing Center: 155–164.

Fullard, H. & Trehaarne, R. F. (1962). *Muir's Historical Atlas*. London. George Phillip and Son, Ltd. 2nd Edition. 96p.

Fulwider, C. W. & Aitken, G. W. (1963). Effect of surface color on thaw penetration beneath a pavement in the Arctic. In: *Proceedings of the 1st International Conference on Structural Design of Asphaltic Pavements*. Ann Arbor, Michigan. Braun-Brumfield Inc.

Fürbringer, W. & Haydn, R. (1974). Zur frage der orientierung nordalaskischer seen mit hilfe des satellitenbildes. *Polarforschung*, 44: 47–53. [In German].

Furrer, G. & Bachmann, F. (1972). Solifluktionsdecken im schweizerischen Hochgebirge als speigel der postglazialen Landschaftsentwicklung. In: Cailleux, A. (Ed.). Glacial and Periglacial Morphology. *Zeitschrift für Geomorphologie Supplement-Band*, 13: 163–172. [In German].

Furrer, G., Bachmann, F. & Fitze (1971). Erdströme als Formelemente von Soliflukhousdecken im Raum Munt Chevagi/Munt Buffalora. *Ergebnisse der Wissenschaften Untersuchungen im Schwizerischen Nationalpark*, 11: 188–269. [In German].

Fursova, O., Potapov, V., Brouchkov, A., Pogorelko, G., Griva, G., Fursova, N. & Ignatov, S. (2012). Probiotic activity of bacterial strain isolated from ancient permafrost against *Salmonella* infection in mice. *Probiotics Antimicrobial Proteins*, 403: 145–153.

Gallin, W. & Erdenbat, M. (2005). Formation and organization of patterned ground in the Kharkhinaa Ull. Eighteenth Annual Keck Symposium: 140–144. http://keck.wooster.edu/publications.

Galloway, R. W., Hope, G. S., Loffler, E. & Peterson, J. A. (1973). Late Quaternary glaciations and periglacial phenomena in Australia and New Guinea. *Palaeoecology of Africa, the Surrounding Islands and Antarctica*, 8: 127–138.

Galushkin, Y. I., Sitar, K. A. & Frolov, S. V. (2013). Basin modelling of temperature and heat flow distribution and permafrost evolution, Urengoy and Kuyumbinskaya areas, Siberia. *Permafrost and Periglacial Processes*, 24(4): 268–285.

Gamble, D. J. & Janssen, C. T. L. (1974). Evaluating alternative levels of water and sanitation service for communities in the Northwest Territories. *Canadian Journal of Civil Engineering*, 1: 116–128.

Gan, X. Y., Chen, Y. & Hu, H. (2011). Frozen soil region construction quality control of direct current transmission line. *Jiangxi Electric Power*, 35(4): 37–40.

Gandahl, R. (1979). Some aspects on the design of roads with boards of plastic foam. *Proceedings of the 3rd International Conference on Permafrost*, Edmonton. Ottawa. National Research Council of Canada: 791–797.

Gao, L., Wang, X. & Wan, X. (2005). Analysis of ice cave formation in Ningwu Shanxi. *Journal of Taiyuan University of Technology*, 36(4): 455–458. [In Chinese].

Gardner, J. S. (1970). Geomorphic significance of avalanches in the Lake Louise area, Alberta, Canada. *Arctic and Alpine Research*, 2: 135–144.

Gardner, J. S. (1983). Observations of erosion by wet snow avalanches, Mount Rae, Alberta, Canada. *Arctic and Alpine Research*, 15: 271–274.

Gardner, J. S. (1989). High magnitude geomorphic events in the Canadian Rocky Mountains. *Studia Geomorphologica Carpatho-Balcanica*, 23: 39–52.

Gardner, J. S. & Bajewski, I. (1987). Hilda rock glacier stream discharge and sediment load characteristics, Sunwapta Pass area, Canadian Rocky Mountains. In: Giardino, J. R., Shroder, J. F. and Vitek, J. D., (eds.). *Rock Glaciers*. London. Allen and Unwin: 161–174.

Garg, O. P. (1973). *In situ* physicomechanical properties of permafrost using geophysical techniques. *Proceedings of the 2nd International Conference on Permafrost, Yakutsk, USSR.* Washington, D.C. National Academy of Sciences: 508–517.

Garg, O. P. (1977). Applications of geophysical techniques in permafrost studies for SubArctic mining operations. *Proceedings of the Symposium on Permafrost Geophysics, Vancouver.* Ottawa. Associate Committee on Geotechnical Research, National Research Council of Canada. Technical Memoir #119: 60–70.

Garg, O. P. (1979). Mining in frozen iron ore in Northern Québec and Labrador. *Géographie physique et Quaternaire*, 33: 369–376.

Garg, O. P. (1982). Recently developed blasting techniques in frozen iron ore at Schefferville, Québec. *Proceedings of the 4th Canadian Permafrost Conference, Calgary, Alberta, March, 1981.* Ottawa. Associate Committee on Geotechnical Research, National Research Council of Canada: 586–591.

Gavrilov, A., Romanovskii, N. & Hubbeterten, H.-W. (2006). Paleogeographical scenario of late-glacial transgression on the Laptev Sea shelf. *Earth Cryosphere*, 10(1): 39–50.

Gavrilov, A., Romanovskii, N., Romanovsky, V. & Hubberten, H. (2001). Offshore permafrost distribution and thickness in the eastern region of the Russian Arctic. In: Semiletov, I. (Ed.). *Changes in the atmosphere-land-sea system in the Amerasian Arctic*, 3: 209–218. Vladivostock, The Arctic Regional Center.

Gavrilov, A. & Tumskoy, V. (2003). Model of mean annual temperature history for the Yakutian coastal lowlands and arctic shelf during the last 4,000 years. *Proceedings of the 8th International Conference on Permafrost, Zurich, Switzerland.* Lisse, Swets and Zeitlinger: 287–290.

Gavrilova, M. K. (1967). Heat balance of larch forest in the water-parting region of the Lena and Amga. In: *Hydroclimatic studies of the Siberian forest*. Moscow. Nauka: 28–52. [In Russian].

Gavrilova, M. K. (1969). Microclimatic and thermal mode of Tungulu Lake. *Geography Matters of Yakutia*. Yakutsk: 12–23. [In Russian].

Gavrilova, M. K. (1973). Meteorological observations in Naled valley of Ulakhan-Taryn (Central Yakutia). In: Alekseyev, V. R. *et al.*, (Eds.). *Siberian naleds*. USSR Academy of Sciences (1969). Draft translation 399. Hanover, New Hampshire. U.S. Army Cold Regions Research and Engineering Laboratory: 136–157.

Gavrilova, M. K. (1978). Climatic (heat balance) factors governing formation of permafrost in the Mongolian People's Republic. In: *USSR contribution, Permafrost, 2nd International Conference*. Washington, D.C., National Academy of Sciences Translations: 35–37.

Gavrilova, M. K., Fedorov, A. N., Varlamov, S. P., *et al.* (1996). *Climate influence on permafrost landscapes of the Central Yakutia*. Yakutsk. Permafrost Institute. 149p. [In Russian].

Gell, A. (1974). Some observations on ice in the active layer and in massive ice bodies, Tuktoyaktuk Coast, N. W. T. *Geological Survey of Canada*, Paper 74-1, Part A, 387.

Gentilini, C., Gottardi, G., Govani, L., Mentani, A. & Ubertini, F. (2013). Design of falling rock protection barriers using numerical models. *Engineering Structures*, 50: 96–106.

Gerard, R., Jasek, M. & Hicks, F. (1992). *Ice-jam flood assessment, Yukon River at Dawson*. Department of Indian and Northern Affairs, Canada. Whitehorse, Yukon Territory.

Gerasimov, A. S (1975). Allowing for temporary loads when calculating creep properties of frozen ground. In: *Construction of residential buildings under Arctic conditions*. Leningrad USSR: 15–27. [In Russian].

Gerasimov, I. P. & Zimina, R. P. (1968). Recent natural landscapes and ancient glaciations of the Pamir. In: Wright H. E. & Osborn, W. H. (eds). *Arctic and Alpine Environments*. Bloomington, Indiana. Indiana University Press: 18: 267–269.

Gerber, E. (1994). *Geomorphologie und geomorphodynamik der region Lona Sassnaire (Wallis, Schweizer Alpen) unter besonderer berücksichtigung von lockersedimenten mit permafrost.* Dissertation, Université de Fribourg/Schweiz, Math Naturwiss Fakultät. 330p. [In German].

Gerrard, A. J. (1992). The nature and geomorphological relationships of earth hummocks (thufur) in Iceland. *Zeitschrift für Geomorphologie* Supplement Band, 86: 173–182.

Gersevanov, N. M. (1986). *Design Manual for Beds of Building Structures (for Construction Rule and Regulation 2.02.01–83).* Scientific-Research Institute of Foundations and Underground Structures, Stroiizdat, Moscow. [In Russian].

Gigarev, L. (1997). *Oceanic Cryolithozone.* Moscow. Moscow State University. 320p. [In Russian].

Gilichinsky, D. A. (2002). Permafrost model of extraterrestrial habitats, p. 125–142. In G. Horneck & C. Baumstark-Khan (Eds.), Astrobiology: the quest for the conditions of life. Springer Verlag, New York, NY.

Gilichinsky, D., Vishnivetskaya, T., Petrova, M., Spirina, E., Mamykin, V., & Rivkina, E., (2008). In: Margesin, R., Schinner, F., Marx, J.C., & Gerday, C. (eds.). *Psychrophiles: From Biodiversity to Biotechnology.* Berlin/Heidelberg. Springer Verlag: 83–102.

Gilichinsky, D. A., Wagener, S. & Vishnevetskaya, T. A. (1995). Permafrost microbiology. *Permafrost and Periglacial Processes,* 6: 281–291.

Gilpin, R. R. (1980). A model for the prediction of ice lensing and frost heave in soils. *Water Resources Research,* 16: 918–930.

Giovinetto, M. (1960). *Report 825–2-Part IV, IGY Project No. 4.10, U.S. National Committee for the International Geophysical Year (USNC-IGY) Antarctic Glaciological Data, Field Work (1958) and 1959 (South Pole Station).* Columbus, Ohio. Ohio State University Research Foundation.

Girard, L., Beutel, J., Gruber, S., Hunziker, J., Lin, R. & Weber, S. (2012). A custom acoustic emission monitoring system for harsh environments: Application to freezing-induced damage to alpine rock walls. *Geoscience Instrumentation and Method Data Systems,* 1: 155–167. DOI: 10.5194/gi-1-155-2012.

Girad, L., Gruber, S., Weber, S. & Beutel, J. (2013). Environmental controls of frost cracking revealed through in situ acoustic emission measurements in steep bedrock. *Geophysical Research Letters,* 40: 1–6. DOI: 10.1002/grl.50384.2013.

Gischig, V. S., Moore, J. R., Evans, K. F., Amann, F. & Loew, S. (2011). Thermomechanical forcing of deep rock slope deformation: 1. Conceptual study of a simplified slope. *Journal of Geophysical Research-Earth Surface,* 116: F04010.

Gjessing, E. T. (1981). Water treatment considerations – aquatic humus. In: Smith, D. W. & Hrudey, S. E. (eds.). *Design of water wastewater services for cold climate communities:* 95–101.

Glaznovskji, A. F. (1978). *Kurums: state of knowledge and objects of investigation.* Moscow. Unpublished manuscript. [In Russian].

Gleason, K. J., Krantz, W. B., Caine, N., George, J. H. & Gunn, R. D. (1986). Geometric aspects of sorted patterned ground in recurrently frozen soil. *Science,* 232: 216–220.

Glinka, K. D. (1963). *Treatise on Soil Science.* Jerusalem. Israel Program of Scientific Publications #558. 674p.

Global Forest Fire Assessment (1909–2000). Rome. Food and Agriculture Organization (FAO), United Nations Forestry Department. 495p.

Godin, E. & Fortier, D. (2012a). Fine-Scale Spatio-Temporal Monitoring of Multiple Thermo-Erosion Gully Development on Bylot Island, Eastern Canadian Archipelago. *Proceedings of the 10th International Conference on Permafrost.* The Northern Publisher Salekhard, Salekhard, Yamal-Nenets Autonomous District, Russia. pp. 125–130.

Godin, E. & Fortier, D. (2012b). Geomorphology of a thermo-erosion gully, Bylot Island, Nunavut, Canada. *Canadian Journal of Earth Sciences,* 49(8): 979–986.

Goering, D. J. (2003). Passively cooled railway embankments for use in permafrost areas. *Journal of Cold Regions Engineering*, 17(3): 119–133.

Goering, D. J. & Kumar, P. (2006). Winter-time convection in open-graded embankments. *Cold Regions Science and Technology*, 24(1): 57–74.

Goh, S. W., & You, Z. P. (2009). Warm mix asphalt using Sasobit® in cold region. In: *Proceedings of the 14th Conference on Cold Regions Engineering, Duluth*. Retson, Virginia. American Society of Civil Engineers: 288–298.

Gold, L. W. & Lachenbruch, A. H. (1973). Thermal conditions in Permafrost – a review of North American Literature. In: *Permafrost: 2nd* International Conference, Yakutsk, USSR. Washington, National Academy of Sciences: 3–23.

Goldshtein, M. N. (1952). *Mechanical properties of soils*. M. [In Russian].

Goldthwait, R. P. (1976). Frost sorted patterned ground. *Quaternary Research*, 6: 27–35.

Gol'dtman, V. G., Znammenskiy, V. V. & Chistopol'skiy, S. D. (1970). *Hydraulic thawing of frozen soils*. Magadan: VNII-I. 448p. [In Russian].

Golubev, V. N. (1988). Some regularities of water crystallization in fine-grained soils. In: Glaciological investigations. M. [In Russian].

Gong, Y. Y., Duan, T. Y., Chen, L. X., Li, W. L., Di, Y., Gu, C. D. & ZouTeng, W. (1997). Outline of observational study of Sino-Japan cooperative program on Asian monsoon over Tibetan Plateau. *Journal of Chengdu Institute of Meteorology*, 1: 18–27.

Gonzalez, M. A. & Corte, A. E. (1976). Pleistocene geocryogenic structures at 38°S. L., and 200 m above sea level, Gonzalez Chavez, Buenos Aires Province, Argentina. *Biuletyn Periglacyalny*, 25: 23–33.

Goodman, M. A. (1977). How permafrost affects offshore wells and structures. *World Oil*, 185 (October): 90–95.

Goodman, M. A. (1978a). Designing casings and wellheads for Arctic service. *World Oil*, 186 (February): 44–60.

Goodman, M. A. (1978b). Completion equipment fulfills special Arctic needs. *World Oil*, 186 (March): 60–70.

Goodman, M. A. (1978c). Reducing permafrost thaw around Arctic wellbores. *World Oil*, 186 (April): 71–76.

Goodman, M. A., Fischer, F. J. & Garrett, D. C. (1982). Thaw subsidence analysis for multiple wells on a gravel island. *Proceedings of the 4th Canadian Permafrost Conference, Calgary*. Ottawa. National Research Council of Canada, Associate Committee on Geotechnical Research: 497–506.

Goodman, M. A. & Franklin, L. J. (1982). Thermal model of a new concept of hydrate control during drilling. *Proceedings of the 4th Canadian Permafrost Conference, Calgary*. Ottawa. National Research Council of Canada, Associate Committee on Geotechnical Research: 349–355.

Goodman, M. A. & Mitchell, R. F. (1978). Permafrost thaw subsidence casing design. *Journal of Petroleum Design*, 20 (March): 455–460.

Goodrich, L. E. (1983). Thermal performance of a section of the Mackenzie Highway. *Proceedings of the 4th International Permafrost Conference, Fairbanks*. Washington. National Academy Press: 353–358.

Gorbunov, A. P. (1978). Permafrost investigations in high-mountain regions. *Arctic and Alpine Research*, 10(3): 283–294.

Gorbunov, A. P. (1983). Rock glaciers in the mountains of middle Asia. In: *Proceedings of the 4th International Conference on Permafrost, Fairbanks, Alaska*. Washington. National Academy Press: 359–362.

Gorbunov, A. P., Marchenko, S. & Seversky, E. V. (2004). The thermal environment of blocky materials in the mountains of Central Asia. *Permafrost and Periglacial Processes*, 15(1): 95–98.

Gorbunov, A. P. & Seversky, E. V. (1999). Solifluction in the mountains of Central Asia: Distribution, Morphology, Processes. *Permafrost and Periglacial Processes*, 10: 81–89.

Gorbunov, A. P. & Seversky, E. V. (2001). Influence of coarsely fragmental deposits on permafrost formation. *Extended Abstracts, International Symposium on Mountain and Arid Land Permafrost*. Ulanbatator. Urlah Erdam Publishing, 24–25.

Gorbunov, A. P. & Seversky, E. V. (2010). Rates of movements and deformations of rock glaciers. *Earth Cryosphere*, 14(1): 69–75. [In Russian].

Gorbunov, A. P. & Titkov, S. N. (1989). *Rock glaciers of Central Asian Mountains*. Permafrost Institute, Yakutsk. 164p. [In Russian].

Gorbunov, A. P., Titkov, S. N. & Polyakov, V. G. (1992). Dynamics of rock glaciers of the northern Tien Shan and the Djungar Ala Tau, Kazakhstan. *Permafost and Periglacial Processes*, 3: 29–39.

Gottardi, G. & Govoni, L. (2009). Full-scale Modelling of Falling Rock Protection Barriers. *Rock Mechanics Rock Engineering*, 43(3): 261–274.

Goudie, A. S. & Piggott, N. R. (1981). Quartzite tors, stone stripes and slopes at the Stiperstones, Shropshire, England. *Biuletyn Peryglacyalny*, 28: 47–56.

Grab., S. W. (1994). Thufur in the Mohlesi Valley, Lesotho, Southern Africa. *Permafrost and Periglacial Processes*, 5: 111–118.

Grab. S. W. (1997). Thermal regime for a thufur apex and its adjoining depression, Mashai Valley, Lesotho. *Permafrost and Periglacial Processes*, 8: 437–445.

Grab, S. W. (1999). A pilot study on needle ice induced stream-bank erosion in the Mashai Valley, Lesotho Highlands. *The South Afriican Geographical Journal*, 81: 126–134.

Grab, S. W. (2005a). Aspects of the geomorphology, genesis and environmental significance of earth hummocks (thufur, pounus: miniature cryogenic mounds). *Progress in Physical Geography*, 29: 139–155.

Grab, S. W. (2005b). Earth hummocks (thufur): new insights into their thermal characteristics and development in eastern Lesotho, southern Africa. *Earth Surface Processes and Landforms*, 90: 541–555.

Gradwell, M. W. (1957). Patterned ground at a High-Country Station. *New Zealand Journal of Science and Technology*, 38B: 793–806.

Graf, K. (1976). Zur mechanic von frostmusterungprozessen in Bolivia und Ecuador. *Zeitschrift für Geomorphologie N.F.*, 20: 414–467. [In German].

Graf, K. (1986). Formas glaciales y periglaciales en la Cordillera Occidental entre Bolivia y Chile. In: *Acta Geocriogenica, 4th Reunion de la Subcommission Latino Americana sobre la importancia de los procesos periglaciale*: 69–77. [In Spanish].

Granberg, H. B. (1973). Indirect mapping of snowcover for permafrost prediction at Schefferville, Québec. In: *North American Contribution, 2nd International Conference on Permafrost (Yakutsk), Proceedings*. Washington, D.C., National Academy of Sciences: 113–120.

Grave, N. N. (1944). Ground ice of Lena and Aldan watersheds. In: *Records of the Obruchev Permafrost Institute*, Moscow. Academy of Sciences Institute 4: 11–32. [In Russian].

Gravis, G. F. (1969). *Slope Sediments in Yakutia: Depositional and Freezing Environment and Cryogenic Structure*. Moscow. Nauka. 128 p. [In Russian].

Gravis, G. F., & Melnikov, E. S. (2003). Principles of classification and mapping of permafrost in Central Asia. In: Phillips, M., Springman S.M., & Arenson, L.U. (eds.). Proceedings of the 8th International Conference on Permafrost, Zurich, Switzerland. Lisse. A.A. Balkema: 297–302.

Gray, J. T. (1972). Debris accretion on talus slopes in the central Yukon Territory. In: Slaymaker, H. O. & McPherson, H. J., (eds.). *Mountain Geomorphology: Geomorphological Processes in the Canadian Cordillera*. Vancouver. Tantalus Press. British Columbia Geographical Series #14: 75–84.

Gray, J. T. (1973). Geomorphic effects of avalanches and rockfalls on steep mountain slopes in central Yukon Territory. In: Fahey, B. D. & Thompson, R. D., (eds.). *Research in Polar and Alpine Geomorphology*. Proceedings of the 3rd Guelph Symposium on Geomorphology. Norwich. Geo Abstracts: 107–117.

Gray, J., Marquette, G. C., Gosse, J., Staiger, J. & Henault-Tessier (2005). The extent, origins and age of felsenmeer surfaces of the Torngat, Kaumajet and Kiglapait Mountains of Northern Québec-Labrador, Eastern Canada. *6th International Conference on Geomorphology, Zaragoza*. Abstracts volume, p. 8.

Grebenets, V. I. (1999). Monitoring changes in geocryological conditions during the enginnering preparation, construction, and occupancy of buildings and structures. *Soil Mechanics and Foundation Engineering*, 36(5): 191–194.

Grechishchev, S. E. (1970). *Basis of method for predicting thermal stresses and deformation in frozen ground*. Moscow. Ministerstvo Geologii SSSR. Vsesoyuzni Nauchno Issledovatel'skii Institut Gidrogeolii I Inzhenernoi (VSEGINGEO). [In Russian].

Grechishchev, S. E. (1973). Basic laws of thermorheology and temperature cracking of frozen ground. *Proceedings of the 2nd International Conference on Permafrost, Yakutsk. USSR*. USSR Contribution. Washington. U.S. National Academy Press: 228–233.

Grechishchev, S. E. (1984). The principles of thermorheology of cryogenic soils. In: *Permafrost: 4th International Conference Proceedings, Fairbanks. Final Proceedings*. Washington, National Academy Press: 52–54.

Grechishchev, S. E., Instannes, A., Sheshin, Y. B., Pavlov, A. V. & Grechishcheva, O. V. (2001). Laboratory research of oil-polluted soil freezing and the model of its structure at negative temperatures. *Earth Cryosphere*, 5(2): 48–53. [In Russian].

Grechishchev, S. E., Chistotinov, L. V. & Shur, Y. L. (1980). Cryogenic processes and their forecast. *Nedra*, Moscow. [In Russian].

Grechishchev, S. Y. (1980). Thermodynamical conditions in pore water at the boundary of freezing. In: Shur, Y. L., Moskalenko, G., Gravis, G. F. and Chitotinov, L. V., (Eds.). Vsesoyuznyy Naucho-Issledovael'skiy Institut Hidrogeologii i Inzenernoy Geologii 138: 4–9. [In Russian].

Grechishcheva, E. & Motenko, R. (2015). Experimental study of freezing point and water phase composition of saline soils contaminated with hydrocarbons. *GeoQuébec 2015*, paper 379. 10p.

Greenblatt, C. L., Davis, A., Clement, B. G., Kitts, C. L., Cox, T., & Cano, R. J. (1999). Diversity of microorganisms isolated from amber. *Microbial Ecology*, 38: 58–68.

Greene, D. F. (1983). *Permafrost, fire, and the regeneration of White Spruce at treeline near Inuvik, N.W.T.* Unpublished M.Sc. thesis, Department of Geography, University of Calgary. 138p.

Greenstein, L. A. (1983). An investigation of midlatitude alpine permafrost on Niwot Ridge, Colorado, Rocky Mountains, U.S.A. *Proceedings of the 4th International Conference on Permafrost, Fairbanks*. Washington. National Academy Press: 380–383.

Gregory, J. S. (1968). Russian land, Soviet people. New York. Pegasus. 947p.

Grigoryev, M. N. (1993). *Cryomorphogenesis of the Lena Delta mouth area*. Yakutsk. Permafrost Institute, Academy of Science, USSR, Siberian Department. 176p. [In Russian].

Grigoryev, M. N. (1996). Regularities of processes of Arctic thermoerosion and thermodenudation at Laptev Sea key sites as an example. In: *Materyaly I konferentsii geokriologov Rossii*. Moscow. 1: 504–511. [In Russian].

Grigoriev, N. F. & Baranovsky, E. L. (1990). Landscape changes during geological surveys on the western coast of Yamal. In: *Geological Studies on the North of Western Siberia*, 4–8. [In Russian].

Gripp, K. (1926). Uber frost und strukturboden aus Spitzbergen. *Zeitschrift Geselschaft Erdekunde*, Berlin: 351–354. [In German].

Gruber, S. (2012). Derivation and analysis of a high-resolution estimate of global permafrost zonation. *The Cryoshere*, 6: 221–233.

Gruber, S., Burn, C. R., Arenson, L., Geertsema, M., Harris, S. A., Smith, S. L., Bonnaventure, R. & Benkert, B. (2015). Permafrost in mountainous regions of Canada. *GeoQuébec 2015*, Paper #169.

Gruber, S. & Haeberli, W. (2007). Permafrost in steep bedrock slopes and its temperature-related destabilization following climate change. *Journal of Geophysical Research*, 112, F02S18. Doi: 10.1029/2006JF000547.

Gruber, S., Hoelzle, M. & Haeberli, W. (2004). Permafrost thaw and destabilization of Alpine rock walls in the hot summer of 2003. *Geophysical Research Letters*, 31: L13504, doi: 10.1029/2004GL020051.

Gruber, S. & Hoelzle, M. (2008). The cooling effect of coarse blocks revisited: A modeling study of a purely conductive mechanism. *Proceedings of the 9th International Conference on Permafrost, Fairbanks*. 557–561.

Gruner, M. (1912). *Die Bodenkultur Islsands*. Berlin. Arch. Biontologie 3. 213p. [In German].

Grunstein, A., Todhunter, P. & Mote, T. (2005). Snowpack control over the thermal offset of air and soil temperatures in eastern North Dakota. *Geophysical Research Letters*, 32, L08503, doi: 10.1029/2005GL022532.

Gu, W., Yu, Q. H., Jin, H. J. & Zhang, J. M. (2010). Qinghai-Tibet Expressway experimental research. *Sciences in Cold and Arid Regions*, 2(5): 396–404.

Guan, S. Q. & Wu, T. (2010). Test study of foundation in high altitude permafrost area. *Engineering Journal of Wuhan University*, 43 (Supplement): 195–198.

Gudjonsson, K. A. (1992). Hummocks on the Fosheim Peninsula, Ellesmere Island, Northwest Territories. Unpublished M.Sc. thesis, Department of Geography, University of Toronto.

Guillien, Y. (1951). Les grèzes litées de Charante. *Revue Géographique des Pyrénées et du Sud-Ouest*, 22: 153–162. [In French].

Guillien, Y. & Lautridou, J.-P. (1974). *Conclusions des recherches de gélifraction expérimentale sur les calcaires des Charantes*. In: Recherches de gélifraction expérimentale du Centre de Géomorphologie de Caen Bulletin, 19. 43p. [In French].

Gurney, S. D. (1995). A reassessment of the relict Pleistocene "pingos" of west Wales: Hydraulic pingos or mineral palsas. *Quaternary Newsletter*, 77: 6–16.

Gurney, S. D. (1998). Aspects of the genesis and geomorphology of pingos: perennial permafrost mounds. *Progress in Physical Geography*, 22(3): 307–324.

Gurney, S. D. (2001). Aspects of the genesis, geomorphology and terminology of palsas: perennial permafrost mounds. *Progress in Physical Geography*, 25(2): 249–260.

Guryanov, I. E. (1998). Problems of interaction between structures and permafrost. In: *Proceedings of the 7th International Permafrost Conference, Yellowknife*. Collection Nordicana 55: 391–396.

Guy, H. P. (1971). Flood downstream from a slide. *Proceedings of the American Society of Civil Engineering, Journal of Hydraulics Division* 97: 643–646.

Guyer, S. & Keating, B. (2005). *The impact of ice roads and ice pads on tundra ecosystems*. Anchorage, Alaska. Bureau of Land Management, Open File Report #98.

Gysi, M. & Lamb, G. (1977). An example of excess urban water consumption. *Canadian Journal of Civil Engineering*, 4: 66–71.

Haag, R. W. & Bliss, L. C. (1974a). Functional effects of vegetation on the radiant heat energy budget of Boreal Forest. *Canadian Geotechnical Journal*, 11: 374–379.

Haag, R. W. & Bliss, L. C. (1974b). Energy budget changes following surface disturbance to upland tundra. *Journal of Applied Ecology*, 11: 355–374.

Hack, J. T. & Goodlet, J. C. (1960). Geomorphology and forest ecology of a mountain region of the Central Appalachians. *U.S. Geological Survey* Professional Paper 347. 65p.

Haeberli, W. (1973). Die basis-temperatur der winterlichen schnee-decke als möglicher indicator für die verbreitung von permafrost in den Alpen. *Zeitschrift für Gletscherkunde und Glacialgeologie*, 9: 221–227. [In German].

Haeberli, W. (1975). *Untersuchungen zur verbreitung von permafrost zwischen Flüelapass und Piz Grialesch (Graubünden)*. Mitteilungen der VAW/EHT Zürich 17: 1–221. [In German].

Haeberli, W. (1979). Holocene push-moraines in Alpine permafrost. *Geografiska Annaler*, 61A: 43–48.

Haeberli, W. (1985). *Creep of mountain permafrost: internal structure and flow of alpine rock glaciers*. Zürich. Mitt. Versuchsanst Wasserbau Hydrologie und Glaciologie, ETH 77. 142p.

Haeberli, W. (1989). Pilot analysis of permafrost cores from the active rockglacier Murtel I, Corvatsch, Eastern Swiss Alps. *Versuchsanst Wasserbau Hydrologie ubd Glaziologie ETH Zür Arbeitsh*, 9: 38.

Haeberli, W. (1990). Glacier and permafrost signals of 20th-century warming. *Annals of Glaciology*, 14: 99–101.

Haeberli, W., Huder, J., Keusen, H. R., Pika, J. & Röthlisberger, H. (1988). Core drilling through rock glacier permafrost. In: *Proceedings of the 5th International Conference on Permafrost, Trondheim, Norway* 2: 937–942.

Haeberli, W. & Patzelt, G. (1982). Permafrostkartierung im gebiet der Hochenbenkar-Blockgletscher, Obergurgl, Ötztaler Alpen. *Zeitschrift für Gletscherkunde und Glazialgeologie*, 18: 127–150. [In German].

Haeberli, W. & Schmidt, W. (1988). Aerophotogrammetrical monitoring of rock glaciers. In: *Permafrost. Proceedings of the 5th International Conference, Trondheim, Norway*. Trondheim. Tapir Publishers 1: 764–769.

Hagerdorn, J. (1974). Note on the occurrence of needle ice phenomena in the southern Sinai Mountains. *Zeitschrift für Geomorphologie, N. F.* Supplementary Band, 21: 35–38.

Hall, K. (1983). Observations of some periglacial features and their palaeoenvironmental implications on sub-Antarctic islands Marion and Kerguslan. *South African Journal of Antarctic Research*, 13: 35–40.

Hall, K. (2006). Sorted stripes on Sub-Antarctic Kerguelen Island. *Earth Surface Processes and Landforms*, 8(2): 115–124.

Hall, K. & André, M-F. (2001). New insights into rock weathering as deduced from high-frequency rock temperature data: an Antarctic study. *Geomorphology*, 41: 23–35.

Hall, K. & Hall, A. (1991). Thermal gradients and rock weathering at low temperatures: some simulation data. *Permafrost and Periglacial Processes*, 2: 103–112.

Hall, K. & Otte, W. (1990). A note on biological weathering on nunataks of the Juneau Icefield, Alaska. *Permafrost and Periglacial Processes*, 2: 189–196.

Hall, K., Thorn, C. E., Matsuoka, N. & Prick, A., 2002. Weathering in cold regions: some thoughts and perspectives. *Progress in Physical Geography*, 26(4): 577–603.

Halle, T. G. (1912). On the geological structure and history of the Falkland Islands. *Bulletin of the Geological Institution of the University of Uppsala*, 11: 1215–1229.

Hallet, B. (2006). Why do freezing rocks break? *Science*, 314: 1092–1093.

Hallet, B. (2013). Stone circles: form and soil kinematics. *Philosophical Transactions of the Royal Society, Section A*, 371: 20120357.

Hallet, B., Anderson, S. P., Stubs, C. W. & Gregory, E. C. (1988). Surface soil displacements in sorted circles, Western Spitzbergen. *Proceedings of the 5th International Permafrost Conference, Trondheim, Norway*. Trondheim. Tapir Press 1: 770–775.

Hallet, B. & Prestrud, S. (1986). Dynamics of periglacial sorted circles in Western Spitzbergen. *Quaternary Research*, 26(1): 81–89.

Hallet, B., Sletten, R. & Whilden, K. (2011). Micro-relief development in polygonal patterned ground in the Dry Valleys of Antarctica. *Quaternary Research*, 75: 347–355.

Hallet, B., Walder, J. S. & Stubbs, C. W. (1991). Weathering by segregation ice growth in micro-cracks at sustained sub-zero temperatures: Verification from an experimental study using acoustic emissions. *Permafrost and Periglacial Processes*, 2: 283–300.

Hallet, B. & Washington, E. D. (1991). Buoyancy forces induced by freeze-thaw in the active layer: Implications for diapirism and soil circulation. In: Dizon, J. C. and Abrhams, A. D., (Eds). *Periglacial Geomorphology*. New York. John Wiley and Sons, Ltd.: 251–279.

Halliday, M. D. (1954). Ice caves of the United States. *National Speleological Society Bulletin*, 16: 3–28.

Hallsworth, E. G., Robertson, G. K. & Gibbons, F. R. (1955). Studies in pedogenesis in New South Wales. VII. The 'gilgai' soils. *Journal of Soil Science*, 6: 1–34.

Hamberg, A. (1915). Zur Kennetnis der Vorgänge in Erdenboden beim Gefrieren und Auftauen sowie Bemerkungen über bie erste Kristallisation des Eises in Wasser. *Geologische Fören.*, Stockholm, Förh., 37: 583–619. [In German].

Hambrey, M. J. (1984). Sedimentary processes and buried ice phenomena in the pro-glacial areas of Spitzbergen glaciers. *Journal of Glaciology*, 30(104).

Hammerschmidt, E. G. (1934). Formation of gas hydrates in natural gas transmission lines. *Industrial and Engineering Chemistry*, 26: 851–855.

Han, J. & Jiang, Y. (2013). Use of geosynthetics for performance enhancement of earth structures in cold regions. *Sciences in Cold and Arid Regions*, 5(5): 517–529.

Hanna, A. J. & McRoberts, E. C. (1988). Permafrost slope design for a buried oil pipeline. In: *Proceedings of the 5th International Permafrost Conference, Trondheim, Norway*. Trondheim. Tapir Academic Press 2: 1247–1252.

Hansen, B. L. & Langway, C. C., Jr. (1966). Deep core drilling in ice core analysis at Camp Century, Greenland, 1961–1966. *Antarctic Journal*, U.S. 1: 207–208.

Hansom, J. D., Evans, D. J. A., Sanderson, D. C. W., Bingham, R. G. & Bentley, M. J. (2008). Constraining the age and formation of stone runs in the Falkland Islands using optically stimulated luminescence. *Geomorphology*, 94(1–2): 117–130.

Hao, S. C., Jiang, W. J., Ju, Q., *et al.* (2010). The features of climate changes in the Five Rivers source regions of the Tibetan Plateau. *Journal of Glaciology and Geocryology*, 32(6): 1130–1135.

Harden, D., Barnes, P. & Reimnitz, E. (1977). Distribution and character of naleds in northeastern Alaska. *Arctic*, 30(1): 28–40.

Harding, R. G. (1962). Foundation problems at Fort Macpherson, N.W.T. *Proceedings of the 1st Canadian Conference on Permafrost*. Ottawa. National Research Council of Canada, Associate Committee for Geotechnical Research Technical Memorandum 76: 159–166.

Harris, C. (1972). Processes of soil movement in turf-banked solifluction lobes, Okstindan, northern Norway. In: Price, R. J. and Sugden, D. E. (eds). *Polar Geomorphology*. Institute of British Geographers Special Publication 4: 155–174.

Harris, C. (1973). Some factors affecting the rates and processes of periglacial mass movements. *Geografiska Annaler*, 55A: 24–28.

Harris, C. (1989). Mechanisms of mass movements in periglacial environments. In: Anderson, M. G. and Richards, K. S. (Eds.). *Slope stability*. Chichester. Wiley 531–559.

Harris, C. & Lewkowicz, A. G. (1993a). Form and internal structure of active-layer detachment slides, Fosheim, Peninsula, Ellesmere Island, N.W.T., Canada. *Canadian Journal of Earth Sciences*, 30: 1708–1714.

Harris, C. & Lewkowicz, A. G. (1993b). Micromorphological investigations of active-layer detachment slides, Ellesmere Island, Canadian Arctic. *6th Intrnational Conference on Permafrost, Beijing, China*. Wushan, Guangzhou. University of Technology Press: 232–237.

Harris, C. & Lekowicz, A. G. (2000). An analysis of stability of thawing slopes, Ellesmere Island, Nunavut, Canada. *Canadian Geotechnical Journal*, 37: 449–462.

Harris, C., Davies, C. R. & Rea, B. R. (2003). Gelifluction: viscous flow or plastic creep? *Earth Surface Processes and Landforms*, 28: 1289–1301.

Harris, K. P. (1990). *The effects of low temperatures on the flammability limits of some gaseous fuels and their mixtures in air*. Unpublished M.Sc. thesis, Department of Mechanical Engineering, University of Calgary. 122p.

Harris, R. B. (2010). Rangeland degradation on the Qinghai-Tibetan Plateau: A review of the evidence of its magnitude and causes. *Journal of Arid Environments*, 74: 1–12.

Harris, S. A. (1958). The gilgaied and bad-structured soils of Central Iraq. *Journal of Soil Science*, 9: 169–185.

Harris, S. A. (1960). The Distribution of Certain Plant Species in Similar Desert and Steppe Soils of Central and Northern Iraq. *Journal of Ecology*, 48: 97–105.

Harris, S.A. (1964). Seasonal density changes in the alluvial soils of Northern Iraq. *Proceedings of the 8th International Congress of Soil Science, Bucharest* II: 291–303.

Harris, S. A. (1968). *Stratigraphy and distribution of the youngest tills near Waterloo, Ontario, and the probable bearing of these tills and parent ice sheets on the evolution of the present landscape*. Unpublished Ph.D. thesis, Department of Geology, Queen Mary College, University of London, England. 162p.

Harris, S. A. (1969). The meaning of till fabrics. *Canadian Geographer* 13: 317–337.

Harris, S. A. (1971). Preliminary observations on downslope movement of soil during the Fall in the Chinook Belt of Alberta. In: *Research methods in Pleistocene Geomorphology, 2nd Guelph Symposium on Geomorphology*: 275–285.

Harris, S. A. (1972). Three modifications to produce more accurate measurements of snowfall and evapotranspiration. *Canadian Geographer*, 16: 271–277.

Harris, S. A. (1973). Studies of soil Creep, Western Alberta, 1970–1972. *Arctic and Alpine Research*, 5(2): A171–A180.

Harris, S. A. (1975). Petrology and origin of stratified scree in New Zealand. *Quaternary Research*, 5: 199–214.

Harris, S. A. (1976). *The Vermilion Pass fire: The first seven years*. Contract report for the Parks Branch, Department of Indian Affairs and Northern Development. 176p.

Harris, S. A. (1979). Ice caves and permafrost zones in southwest Alberta. *Erdkunde* 33: 61–70.

Harris, S. A. (1981a). Climatic relationships of permafrost zones in areas of low winter snow-cover. *Arctic*, 34: 64–70.

Harris, S. A. (1981b). Distribution of active glaciers and rock glaciers compared to the distribution of permafrost landforms, based on freezing and thawing indices. *Canadian Journal of Earth Sciences*, 18(2): 376–381.

Harris, S. A. (1981c). Distribution of zonal permafrost landforms with freezing and thawing indices. *Erdkunde*, 35(2): 81–90.

Harris, S. A. (1982a). Distribution of zonal permafrost landforms with freezing and thawing indices. *Builetyn Peryglacjalny*, 29: 163–182.

Harris, S. A. (1982b). Identification of permafrost zones using selected periglacial landforms. In: French, H. M., (Ed.). *Proceedings of the 4th Canadian Permafrost Conference*: 49–58.

Harris, S. A. (1982c). Cold air drainage west of Fort Nelson, British Columbia. *Arctic*, 35: 537–541.

Harris, S. A. (1983). Infilled fissures in loess, Banks Peninsula, New Zealand. *Polarforschung*, 53: 49–58.

Harris, S. A. (1984). Climatic zonality of periglacial landforms in mountain areas. *Arctic*, 47: 184–191.

Harris, S. A. (1985). Evidence for the nature of the early Holocene climate and Palaeogeography, High Plains, Alberta, Canada. *Arctic and Alpine Research*, 17(1): 49–67.

Harris, S. A. (1986a). *The Permafrost Environment*. London, Croom Helm. 276p.

Harris, S. A. (1986b). Permafrost distribution, zonation and stability along the Eastern Ranges of the Cordillera of North America. *Arctic*, 39(1): 29–38.

Harris, S. A. (1987a). Influence of organic (Of) layer thickness on active-layer thickness at two sites in the western Canadian Arctic and Subarctic. *Erdkunde*, 41: 275–285.

Harris, S. A. (1987b). Altitude trends in permafrost active layer thickness, Kluane Lake, Yukon Territory. *Arctic*, 40: 179–183.

Harris, S. A. (1988a). Observations on the redistribution of moisture in the active layer and permafrost. *Proceedings, 5th International Conference on Permafrost, Trondheim, Norway*. Trondheim, Tapir Press 1: 364–369.

Harris, S. A. (1988b). The alpine periglacial zone. In M.J. Clark (ed.). *Advances in Periglacial Geomorphology*. Chichester. Wiley: 369–413.

Harris, S. A. (1989). Continentality Index: Its uses and limitations applied to permafrost in the Canadian Cordillera. *Physical Geography*, 10(3): 270–284.

Harris, S. A. (1990a). Dynamics and origin of saline soils on the Slims River Delta, Kluane National Park, Yukon Territory. *Arctic*, 43: 159–175.

Harris, S. A. (1990b). Long-term air and ground temperature records from the Canadian Cordillera and the probable effects of moisture changes. Proceedings of the 5th Canadian Permafrost Conference. *Nordicana*: 54: 151–157.

Harris, S. A. (1993). Palsa-like mounds in a mineral substrate, Fox Lake, Yukon Territory. In: *Proceedings of the 6th International Conference on Permafrost, Beijing, China*. 1: 238–243.

Harris, S. A. (1994a). Climatic zonality of periglacial landforms in mountain areas. *Arctic*, 47(2): 184–192.

Harris, S. A. (1994b). Chronostratigraphy of glaciations and permafrost episodes in the Cordillera of North America. *Progress in Physical Geography*, 18: 366–395.

Harris, S. A. (1996a). Lower mean annual ground temperatures beneath a block stream in the Kunlun Pass, Qinghai Province, China. *Proceedings of the 5th Chinese Permafrost Conference, Lanzhou*: 227–237.

Harris, S. A. (1996b). Permafrost distribution, zonation, and stability along the eastern ranges of the Cordillera of North America. *Arctic*, 39: 29–38.

Harris, S. A. (1997). Relict late Quaternary permafrost on a former nunatak at Plateau Mountain, S. W. Alberta. *Biuletyn Peryglacjalny*, 36: 47–63. Also in Polish, pages, 64–74.

Harris, S. A. (1998a). Nonsorted circles on Plateau Mountain, S. W. Alberta, Canada. *Proceedings of the 7th International Conference on Permafrost, Yellowknife, Collection Nordicana*: 441–448.

Harris, S. A. (1998b). Effects of vegetation cover on soil heat flux in the Southern Yukon Territory. *Erdkunde*, 52(4): 265–285.

Harris, S. A. (1998c). A Genetic Classification of the Palsa-like Mounds in Western Canada. *Biuletyn Peryglacjalny*, 37: 115–130.

Harris, S. A. (1999). Heat transfer processes at the limit of alpine permafrost, Marmot Basin, Jasper National Park, Canada. *Biuletyn Peryglacjalny*, #38: 73–93.

Harris, S. A. (2001a). Sequence of glaciations and permafrost events. In: Paape, R. and Melnikov, V. (Eds.). *Permafrost Response on Economic Development, Environmental Security and Natural Resources*. Kluwer Academic Publishers: 227–252.

Harris, S. A. (2001b). Twenty years of data on climate-permafrost-active layer variations at the lower limit of alpine permafrost, Marmot Basin, Jasper National Park, Canada. *Geografiska Annaler*, 83A: 1–13.

Harris, S. A. (2002a). Global heat budget, plate tectonics and climatic change. *Geografiska Annaler* 84A: 1–9.

Harris, S. A. (2002b). Causes and consequences of rapid thermokarst development in permafrost or glacial terrain. *Permafrost and Periglacial Processes*, 13: 237–242.

Harris. S. A. (2002c). Biodiversity of the vascular timberline flora in the Rocky Mountains of Alberta, Canada. In: Koerner, C. and Spehn, E. (Eds.). *Mountain Biodiversity: A global assessment*. Lancashire. Parthenon Publishing group, United Kingdom.: 49–57.

Harris, S. A. (2005). Evidence for multiple layers in the active layer in certain warm permafrost environments (Canada). *Earth Cryosphere*, 9(4): 3–12. [In Russian].

Harris, S. A. (2006). Reaction of continental mountain climates to the postulated "Global Warming": Evidence from Alaska and the Yukon Territory. *Proceedings, Earth Cryosphere Assessment: Theory, Applications and Prognosis of Alterations*. International Conference, Russian Academy of Sciences, Tyumen. Vol. 1: 49–54.

Harris, S. A. (2007). Reaction of continental mountain climates to the postulated "global warming": Evidence from Alaska and the Yukon Territory. *Earth Cryosphere*, 11(3): 78–84. [In Russian].

Harris, S. A. (2008). Climatic change and permafrost stability in the Eastern Canadian Cordillera. In: Kane, D. L. and Hinkel, K. M. (Eds.). Ninth International Conference on Permafrost (NICOP), Fairbanks, Alaska. Extended Abstracts: 93–94.

Harris, S. A. (2009). Climatic change and permafrost stability in the eastern Canadian Cordillera: The results of 33 years of measurement. *Sciences in Cold and Arid Regions*, 1(5): 381–403.

Harris, S. A. (2010a). Climatic change in Western North America during the last 15,000 years: The role of changes in the strengths of air masses in producing the changing climates. *Sciences in Cold and Arid Regions*, 2(5): 371–383.

Harris, S. A. (2010b). Greenhouse gases and their importance to life. In: Harris, S. A. (Ed.). *Global Warming*, Chapter 2. Rijeka, Croatia. Sciyo Publishers: 15–22.

Harris, S. A. (2010c). Evidence for increased stability of temperatures in areas of mountain permafrost in interior valleys and closed basins in wide Cordilleras in N. W. North America. In: *Cryospheric change and its influences*. Lijiang, China, 12–14th August, 2010. Program and Abstracts, pp. 52–54.

Harris, S. A. (2013a). Climatic change: Causual correlations over the last 240 Ma. *Sciences in Cold and Arid Regions*. 5(3): 259–274.

Harris, S. A. (2013b). Environmental and Land-use changes in the Tibetan Plateau section of the Upper Yangtze River Basin during the last 50 years. In: *Yangtze River: Geography, Pollution and Environmental Implications*, Rijeka, Croatia. Sciyo Publishers Chapter 2: 35–64.

Harris, S. A. (2015). Numerical simulation of formation and preservation of Ningwu ice cave, Shanxi, China: A discussion. *The Cryosphere Discus*, 9: 2367–2395.

Harris, S. A. (2016a). Identification, characteristics and classification of cryogenic block streams. *Sciences in Cold and Arid Regions*, 8(3): 00177–00186.

Harris, S. A. (2016b). Probable effects of heat advection on the adjacent environment during oil production at Prudhoe Bay, Alaska. *Sciences in Cold and Arid Regions*, 8(6): 00451-00460. DOI: 10.3724/SPJ.1226.2016.00451.

Harris, S. A., Blumstengel, W., Cook, D., Krouse, H. R. & Whitley, G. (1994). Comparison of the water drainage from an active near-slope rock glacier and a glacier, St. Elias Mountains, Yukon Territory. *Erdkunde*, 48: 81–91.

Harris, S. A. & Brown, R. J. E. (1978). Plateau Mountain: A case study of alpine permafrost in the Canadian Rocky Mountains. *Proceedings of the 3rd International Conference on Permafrost, Edmonton, Alberta*. Ottawa, National Research Council of Canada, Volume 1: 385–391.

Harris, S. A. & Brown, R. J. E. (1982). Permafrost distribution along the Rocky Mountains of Alberta. *The Roger E. Brown Memorial Volume, Proceedings of the 4th Canadian Permafrost Conference, Calgary*. Ottawa, National Research Council of Canada: 59–67.

Harris, S. A., Cheng Guodong, Zhao Xiufeng & Yongqin Ding (1998a). Nature and dynamics of an active block stream, Kunlun Pass, Qinghai Province, People's Republic of China. *Geografiska Annaler*, 80A(2): 123–133.

Harris, S. A., Cui Zhijiu & Cheng Guodong (1998b). Origin of a bouldery diamicton, Kunlun Pass, Qinghai-Xizang Plateau, People's Republic of China: Gelifluction deposit or Rock Glacier? *Earth Surface Processes and Landforms*, 23: 943–952.

Harris, S. A. & Gustafson, C. A. (1988). Retrogressive slumps, debris flows and river gully development in icy, unconsolidated sediments on hills and mountains. *Zeitschrift für Geomorphologie, N.F.* 32: 441–455.

Harris, S. A. & Gustafson, C. A. (1993). Debris flow characteristics in an area of continuous permafrost, St. Elias Range, Yukon Territory. *Zeitschrift für Geomorphologie, N.F.* 37: 41–56.

Harris, S. A. & Jin, H. J. 2012. Tessellons and "sand wedges" on the Qinghai-Tibet Plateau and their palaeoenvironmental implications. *Proceedings of the Tenth International Permafrost Conference, Salekhard* 1: 149–153.

Harris, S. A. & McDermid, G. (1998). Frequency of debris flows on the Sheep Mountain fan, Kluane Lake, Yukon Territory. *Zeitschrift für Geomorphologie, N.F.* 42(2): 159–192.

Harris, S. A. & Nyrose, D. (1992). Palsa formation in floating peat and related vegetation cover as illustrated by a fen bog in the MacMillan Pass, Yukon Territory, Canada. *Geografiska Annaler*, 74A(4): 349–362.

Harris, S. A. & Petersen, D.E. (1998). Thermal regimes beneath coarse block materials. *Permafrost and Periglacial Processes*, 9: 107–120.

Harris, S. A. & Prick, A. (1997). The periglacial environment of Plateau Mountain: Overview of current periglacial research. *Polar Geography*, 21: 113–136.

Harris, S. A. & Prick, A. (2000). Conditions of formation of stratified screes, Slims River valley, Yukon Territory: A possible analogue with some slope deposits from Belgium. *Earth Surface Processes and Landforms*, 25: 463–481.

Harris, S. A., Schmidt, I. & Krouse, H. R. (1992). Hydrogen and oxygen isotopes and the origin of the ice in peat plateaus. *Permafrost and Periglacial Processes*, 3: 19–27.

Harris, S. A. & Schmidt, I. (1994). Permafrost aggradation and peat accumulation since 1200 years B.P. in peat Plateaus at Tuchitua, Yukon Territory, Canada. *Journal of Palaeolimnology*, 12: 3–17.

Harris, S. A., van Everdingen, R. O. & Pollard, W. H. (1983). The Dempster Highway – Dawson to Eagle Plain. In: French, H. M. and Heginbottom, J. A. (Eds.). *Northern Yukon Territory and Mackenzie Delta, Canada. Guidebook to permafrost and related features.* 4th International Conference on Permafrost, Guidebook #3, Fairbanks, Alaska. Alaska Division of Geological and Geophysical Surveys: 65–86.

Harwood, T. A. (1966). Dew Line site selection and explanation. *Proceedings of the 1st International Conference on Permafrost, Lafayette, Indiana.* Washington. National Academy of Sciences Publication 1287: 359–363.

Harry, D. G., French, H. M. & Pollard, W. H. (1985). Ice wedges and permafrost conditions near King Point, Beaufort Sea Coast, Yukon Territory. *Geological Survey of Canada*, Paper 85-1A: 111–116.

Harry, D. G. & Gozdzik, J. S. (1988). Ice wedges: Growth, thaw transformation, and palaeoenvironmental significance. *Journal of Quaternary Science*, 3(1): 39–55.

Harry, D. G. & MacInnes, K. L. (1988). The effect of forest fires on permafrost terrain stability, Little Chicago-Travaillant Lake area, Mackenzie Valley, N.W.T. Ottawa. *Current Research, Part D, Geological Survey of Canada* Paper 88-1D: 91–94.

Hasler, A., Geertsema, M., Foord, V., Guber, S. & Noetzli, J. (2015). The influence of surface characteristics, topography and continentality on mountain permafrost in British Columbia. *The Cryosphere*, 9: 1025–1038.

Hasler, A., Gruber, S. & Beutel, J. (2012). Kinematics of steep bedrock. *Journal of Geophysical Research*, 117: F01016. Doi: 10.1029/2011JF001981.

Hasler, A., Gruber, S. & Haeberli, W. (2011a). Temperature variability and offset in steep alpine rock and ice faces. *The Cryosphere*, 5: 977–988.

Hasler, A., Gruber, S., Font, M. & Dubois, A. (2011b). Advective heat transport in frozen rock clefts: Conceptual model, laboratory experiments and numerical simulation. *Permafrost and Periglacial Processes*, 22: 378–389.

Hauck, C. (2013). New concepts in geophysical surveying and data interpretation for permafrost terrain. *Permafrost and Periglacial Processes*, 24(2): 131–137.

Haugland, J. E. (2004). Formation of patterned ground and fine-scale soil development within two late Holocene glacial chronosequences: Jotenheimen, Norway. *Geomorphology*, 61: 287–301.

Hay, T. (1936). Stone stripes. *Geographical Journal*, 87: 47–50.

Hays, F. W. (1993). Taiga forest and Weyerhaeuser. Case number 67. TED case study. www.american.edu/ted/TAIGA.HTML Accessed 1st September, 2015.

Hayley, D. W. (1982). Application of heat pipes to design of shallow foundations on permafrost. *Proceedings of the 4th Canadian Permafrost Conference, Calgary*. Ottawa. National Research Council of Canada: 535–544.

Hayley, D. W., Inman, D. V. & Gowan, R. J. (1984). Pipeline trench excavation in continuous permafrost using a mechanical wheel ditcher. In: *Proceedings of the 3rd International Speciality Conference on Cold Regions Engineering*. Canadian Society for Civil Engineering. Edmonton, Alberta. I: 103–118.

Hayley, D. W., Roggensack, W. D., Jubien, W. E. & Johnson, P. V. (1983). Stabilization of sinkholes on the Hudson Bay Railway. In: *Permafrost, 4th International Conference Proceedings*. Washington, D.C. National Academy Press: 468–473.

He, C. (1983). Building foundations on permafrost. *Proceedings of the 4th International Conference on Permafrost*. Washington, D.C. National Academy Press: 474–479.

He, G. L. (2008). Construction technology in iron tower foundations for electric power for QInghai-Tibet Railway in frozen area in cold seasons. *Friends of Science Amateurs*, 29: 32–33.

He, R. & Jin, H. J. (2010). Permafrost and cold-region environmental problems of the oil product pipeline from Golmud to Lhasa on the Qinghai-Tibet Plateau and their mitigation. *Cold Regions Science and Technology*, 64(3): 279–288.

Healy, T. R. (1975). Thermokarst – a mechanism for de-icing ice-cored moraines. *Boreas*, 4: 19–23.

Heginbottom, J. A. (1971). Some effects of a forest fire on the permafrost active layer at Inuvik, N.W.T. Ottawa. *National Research Council, Associate Committee on Geotechnical Research*, Technical Memorandum #103: 31–36.

Heginbottom, J. A. (1973). Some effects of surface disturbance on the permafrost active layer at Inuvik, N.W.T. *Environmental-Social Committee on Northern Pipelines*, Report 73-16: 23p.

Hegginbottom, J. A., Dubreuil, M. A. & Harker, P. T. (1995). Canada-Permafrost. In: *National Atlas of Canada*. Ottawa. Natural Resources Canada, Map 2.1.

Heine, K. (1976). Blockgletscher und Blockzungen-Generationen am Nevado de Toluca, Mexico. *Die Erde*, 107: 330–352. [In German].

Heinke, G. W. (1974). *Arctic waste disposal*. Task Force on Northern Oil Development, # 74-10. 255p.

Heinke, G. W. & Prasad, D. (1977). Disposal of human wastes in Northern Areas. Department of Civil Engineering, University of Toronto.

Hendricks, B. L. (1995). *International encyclopedia of horse breeds*. University of Oklahoma Press. http://books.google.com/?id=rhNFWOsD3jMC Retrieved 2009-04-20.

Henrikssen, M., Mangerud, J., Matiouchkov, A., Paus, A. & Svendsen, J. I. (2003). Lake stratigraphy implies an 80,000 yr delayed melting of buried dead ice in Northern Russia. *Journal of Quaternary Science*, 18(7): 663–679.

Henry, K. S. (2000). *A review of the thermodynamics of frost heave*. U.S. Army Corps of Engineers Cold Regions Research & Engineering Laboratory CRREL, ERDC/CRREL Technical Report TR-00-16. September 2000, Hanover, NH, USA, 26 p.

Hétu, B. & Gray, J. T. (2000). Effects of environmental change on scree slope development throughout the postglacial period in the Chic-Choc Mountains in the northern Gaspé Peninsula, Québec. *Geomorphology*, 32: 335–355.

Hètu, B. & Vandelac, P. (1989). La dynamique des éboulis schisteux au cours de l'hiver, Gaspésie septentrionale, Quèbec. *Géographie Physique et Quaternaire*, 43: 389–406. [In French].

Hètu, B., Van Stein, H. & Vandelac, P. (1994). Les coulees de pierres glacées: un nouveau type de coulees de pierraille sur le talus d'éboulis. *Géographie Physique et Quaternaire*, 48: 3–22. [In French].

Heyse, I. (1983). Cryoturbation types in eolian Würm Late Glacial sediments in Flanders, Belgium. *Polarforschung*, 58(2): 87–95.

Higachi, A. & Corte, A. E. (1971). Solifluction: A model experiment. *Science*, 171: 480–482.

Hinkel, K. M., Eisner, W. R., Bockheim, J. G., Nelson, F. E., Peterson, K. M. & Dai, X. (2003). Spacial extent and carbon stocks in drained thaw lake basins on the Barrow Peninsula, Alaska. *Arctic, Antarctic and Alpine Research*, 35: 291–300.

Hinkel, K. M. & Nelson, F. E. (2003). Spatial and temporal patterns of active-layer thickness at Circumpolar-Active-layer-Monitoring (CALM) sites in northern Alaska, 1995–2000. *Journal of Geophysical Research-Atmospheres*, 108: D2. Doi: 10.1029/2001JD000927

Hinkel, K. M., *et al*. (2012a). Thermokarst Lakes on the Arctic Coastal Plain of Alaska: Spatial and temporal variability in summer water temperatures. *Permafrost and Periglacial Processes*, 23: 207–217.

Hinkel, K. M., *et al*. (2012b). Thermokarst Lakes on the Arctic Coastal Plain of Alaska: Geomorphic controls on bathymetry. *Permafrost and Periglacial Processes*, 23: 218–230.

Hinzman, L. D., Boike, J., Kane, D. L. & Peltier, B. J. (2001). Regional characteristics of snowmelt across the North American Arctic. *American Geophysical Union, Fall Meeting, 2001*. Abstract #IP51A-0728. http://www.uaf.edu/water/projects/snowmelt2000/CD/CDBrowser.htm

Hinzman, L. D., Destourni, G. & Woo, M.-K. (2013). Preface: Hydrogeology of cold regions. *Hydrogeology Journal*, 20: 1–4.

Hivon, E. G. & Sego, D. C. (1993). Distribution of saline permafrost in the Northwest Territories, Canada. *Canadian Geotechnical Journal*, 30: 506–514.

Ho, P. (2000). The clash over State and Collective Property: The making of rangeland law. *The China Quarterly*, 161: 240–263.

Hodgson, D. A. (1977). A preliminary account of surficial materials, geomorphological processes, terrain sensitivity and Quaternary history of King Christian and southern Ellef Ringnes Islands, District of Franklin. *Report of Activities, Part A. Ottawa. Geological Survey of Canada*, Paper 77-1A: 485–493.

Hoekstra, P. (1969). Water movement and freezing pressures. *Soil Science Society of America Proceedings*, 33: 512–518.

Hoekstra, P. & Miller, R. D. (1965). *Movement of water in a film between glass and ice*. U.S. Cold Regions Research and Engineering Laboratory Research Report, 153. 8p.

Högbom, B. (1910). Einige illustrationen zu den geologischen wirkungen des frostes auf Spitzbergen. *Uppsala Universitet Geologiske Institutet Bulletin*, 9: 41–59. [In German].

Högbom, A. (1914). Über die geologische Bedeutung des Frostes. *Uppsala Universitet Geologiske Institutet Bulletin*, 12: 257–389. [In German].

Högström, E. (2011). *Morphometry and environment of asymmetric non-sorted stripes in the Abisko mountains, northern Sweden*. Geotryckeriet, Uppsala Universitet. 79p.

Höllermann, P. (1983). Blockgletscherstudienin europäiischen und nord-amerikanischen Gebirge. In: Poser, H. & Schunke, E., (eds.). *Mesoformen des Reliefs im heutigen Periglacialraum*. Abhandlungen der Akademie Wissenschaften Göttingen, Math Phys Kl Folge 3, 35: 116–119. [In German].

Hollingshead, G. W., Skjolingstad, L. & Rundquist, L. A. (1978). Permafrost between channels in the Mackenzie Delta, N.W.T., Canada. Proceedings of the 3rd International Conference on Permafrost, Edmonton, Alberta, National Research Council of Canada, Volume 1: 406–412.

Holmes, S. W., Hopkins, D. M. & Foster, H. L. (1968). *Pingos in Central Alaska*. U.S. Geological Survey Bulletin, 1211-H. 40p.

Holubec, I. (2008). *Flat loop thermosiphon foundations in warm permafrost*. Report to the Government of the Northwest Territories Asset Management Division, Public Works and Services, and Climate change Vulnerability Assessment, Canadian Council of Professional Engineers. 83p.

Hopfinger, E. J. (1983). Snow avalanche motion and related phenomena. *Annual Review of Fluid Mechanics*, 15: 47–76.

Hopfinger, E. J. & Tochin-Danguy, J. C. (1977). A model study of powder snow avalanches. *Journal of Glaciology*, 19: 343–356.

Hopkins, D. M. (1949). Thaw lakes and thaw sinks in the Imuruk Lake area, Seward Peninsula, Alaska. *Journal of Geology*, 57: 119–131.

Hopkins, D. M., Karlstrom, T. N., Black, R. F., Williams, J. R., Péwé, T. L., Fernold, A. T. & Muller, E. H. (1955). Permafrost and groundwater in Alaska. *U.S. Geological Survey, Professional Pape*, 264 F: 113–146.

Hopkins, D. M. & Sigafoos, R. S. (1951). Frost action and vegetation patterns on Seward Peninsula, Alaska. *U.S. Geological Survey*, Bulletin 974-C: 51–101.

Hoque, Md. A. & Pollard, W. D. (2009). Arctic coastal retreat through block failure. *Canadian Geotechnical Journal*, 46(10): 1103–1115.

Horvath, J. S. (1999). *Lessons learned from failures involving geofoam in roads and embankments*. Manhattan College, School of Engineering, Civil Engineering Department, Bronx, New York. Manhattan College Research Report #CE/GE-99-1. Downloaded on December 21st, 2014.

Hoeve, E. & Hayley, D. (2015). The Inuvik Airport runway – an evaluation of 50 years of performance. GeoQuebec 2015, #176.

Hövermann, J. (1949). Morphologische untersuchungen im Mittelharz. *Göttinger Geografisch. Abhandlungen, Heft 2*. [In German].

Howe, C. W. & Linaweaver, F. P. (1967). The impact of price on residential water demand and its relation to system design and price structure. *Water Resources Research*, 3: 13–32.

Howe, E. (1909). Landslides in the San Juan Mountains, Colorado. *United States Geological Survey, Professional Paper* 67: 31–40.

Hrudey, S. E. & Raniga, S. (1981). Greywater characteristics, health concerns and treatment technology. In: Smith, D. W. and Hrudey, S. E. (eds.). *Design of Water and Wastewater Services for Cold Climate Communities*. Oxford. Pergamon Press: 137–154.

Hu, H., Wang, G., Liu, G., Taibing, R. D., Wang, Y, Cheng, H. & Wang, J. F. (2008). Influences of alpine ecosystem degradation on soil temperature in the freezing-thawing process on Qinghai-Tibet Plateau. *Environmental Geology*, 57: 1391–1397.

Hu, W. G., Zhang, Q., Li, D. Y., Cheng, G. D., Mu, J., Wu, Q. B., Niu, F. J., An, L. Z. & Feng, H. Y. (2014). Diversity and community structure of fungi though a permafrost core profile from the Qinghai-Tibet Plateau of China. *Journal of Basic Microbiology*. Doi: 10:1002/jobm.201400232.

Hu, W. G., Zhang, Q., Tian, Cheng G. D., An, L. & Feng, H. (2015). The microbial ecology of permafrost in China: A review. *Extremophiles*. DOI: 10.1007/s00793-015-749-y.

Hu, X. & Pollard, W. (1997). The hydrologic analysis and modeling of river icing growth, North Fork Pass, Yukon Territory, Canada. *Permafrost and Periglacial Processes*, 8(2): 186–195.

Hu, Z. Y., Qian, Z. Y., Cheng, G. D. & Wang, J. M. (2002). Influence of solar radiation on embankment surface thermal regime of the Qinghai-Xizang Railway. *Journal of Glaciology and Geocryology*, 24(2): 121–128.

Huang, S. L. & Speck, R. C. (1989). An investigation into the creep behaviour of CRREL tunnel, Alaska. In: Bandopadhyay, S. and Skrudrzyk, F. J. (Eds.). *Mining in the Arctic, Proceedings of the 1st International Symposium on Mining in the Arctic, Rotterdam*: 65–73.

Hudec, P. P. (1974). Weathering of Rocks in Arctic and Sub-arctic environments. In: Aitken, J. D. and Glass, D. J. (Eds.). *Canadian Arctic Geology*. Geological Association of Canada, Canadian Society of Petroleum Geologists, Proceedings of the Symposium on the Geology of the Canadian Arctic: 313–335.

Hughes, O. L. (1969). Distribution of open-system pingos in central Yukon Territory with respect to glacial limits. *Geological Survey of Canada*, Paper 69-34. 8p.

Hughes, O. L. (1972). Surficial Geology and land classification, Mackenzie Valley Pipeline Corridor. *National Research Council of Canada, Associate Committee on Geotechnical Research, Technical Memorandum* 104: 17–24.

Hughes, O. L., Veillette, J. J., Pilon, J., Hanley, P. T. & Van Everdingen, R. O. (1973). *Terrain evaluation with respect to pipeline construction, Mackenzie Transportation Corridor, Central Part, lat. 648–688*, ESCOM Report 73-37. Environmental-Social Committee, Northern Pipelines, Task Force on Northern Oil Development. Ottawa. Information Canada. 74p.

Humlum, O. (1982). Rock glacier types on Disco, Central West Greenland. *Geogrfiska Tidsskrift*, 82: 59–66.

Humlum, O. (1998). Rock glaciers on the Faeroe Islands, the North Atlantic. *Journal of Quaternary Science*, 13(4): 293–307.

Hungr, O., Morgan, C. & Kellerhals, R. (1984). Quantitative analysis of debris torrent hazards for design or remedial measures. *Canadian Geotechnical Journal*, 21: 663–667.

Hunt, C. B. & Washburn, A. L. (1966). Patterned ground. *In: Hydrologic Basin, Death Valley, California*. U.S. Geological Survey, Professional Paper 494B: 104–133.

Huscroft, C. A., Lipovsky, P. S. & Bond, J. D. (2004). Permafrost and landslide activity: Case studies from southwestern Yukon Territory. In: Emond, D. S. and Lewis, L. L. (eds.). *Yukon Exploration and Geology, 2003*. Whitehorse. Yukon Geological Survey: 107–119.

Hyatt, J. A. (1992). Cavity development in ice-rich permafrost, Pangnirtung, Baffin Island, Northwest Territories. *Permafrost and Periglacial Processes*, 3: 293–313.

Ingólfsson, Ó. & Lokrantz, H. (2003). Massive ground ice body of glacial origin at Yugorski Peninsula, Arctic Russia. *Permafrost and Periglacial Processes*, 14: 199–215.

Irkutsk Electric Grid Company (2014). *Northern Electric Networks*. http://www.iesk.irkutsken ergo.ru/qa/1299.html accessed 2-15-14.

Isakov, A. (2013). Dangerous deformations of subgrade and methods of their calculation. *Sciences in Cold and Arid Regions*, 5(4): 353–362.

Isaksen, K., Ødegård, R. S., Eiken, T. & Sollid, J. L. (2000). Composition, flow and development of two tongue-shaped rock glaciers in the permafrost of Svalbard. *Permafrost and Periglacial Processes*, 11: 241–257.

Isarin, R. B. F. (1997). Permafrost distribution and temperatures in Europe during the Younger Dryas. *Permafrost and Periglacial Processes*, 8: 313–333.

Istomin, V. A. (1998). Overheating gas hydrates and ice: Prospect of gas, condensate, and oil field exploration and development in the Russia's offshore. Moscow. *Proceedings of VNIIGAZ, 1998*: 131–140. [In Russian].

Istomin, V. A., Yakushev, V. S., Kwon, V. G. *et al.* (2006). Self-preservation phenomenon of gas hydrates. *Gas Industry of Russia Digest*, 4: 16–27. [In Russian].

Ives, J. D. (1966). Blockfields, associated weathering forms on mountain tops, and the nunatak hypothesis. *Geografiska Annaler,* 48A: 220–223.

Iwahana, G., Fukui, K., Mikhailov, N., Ostanin, O. & Fujii, Y. (2012). Internal structure of a lithalsa in the Akkol Valley, Russian Altai Mountains. *Permafrost and Periglacial Processes,* 23: 107–118.

Jackson, L. E., Jr. & MacDonald, G. M. (1980). Movement of an ice-cored rock glacier, Tungsten, N.W.T., Canada, 1936–1980. *Arctic,* 33: 842–847.

Jackson, M. L., Tyler, S. A., Willis, A. L., Bourbeau, G. A. & Pennington, R. P. (1948). Weathering sequence of clay size minerals in soils and sediments. *Journal of Physical and Colloid Chemistry,* 52: 1234–1260.

Jaenicke, R. (1996). Stability and folding of ultrastable proteins: Eye lens crystalline and enzymes from thermophiles. *FASEB Journal,* 10: 84–92.

Jahn, A. (1948). Research on the structure and temperature of the soils in western Greenland. *Bulletin International de l'Acadéie Polonnaise des Sciences et des lêttres,* Séri A: Sciences Mathématiques, Anées 1940–46. Cracovia: 50–59.

Jahn, A. (1961). Quantitative analysis of some periglacial processes in Spitzbergen. *Universytet Wroclawski im Boleslawa Bieruta, zeszyty naukowe, nauka pryrodnice, ser. B, 5, Nauka o Ziemi,* 2: 1–54.

Jahn, A. (1975). *Problems in the periglacial zone.* Warsaw, PWN Polish Scientific Publishers, 219p.

Jahn, A. (1976). Contemporaneous geomorphological processes in Longyeardalen, Vest-Spitzbergen (Svalbard). *Biuletyn Peryglacjalny,* 26: 253–268.

Jahn, A. (1983). Soil wedges on Spitzbergen. *Proceedings 4th International Conference on Permafrost, Fairbanks, Alaska.* Washington, D.C. National Academy Press: 525–530.

Jahn, A. & Czerwinski, J. (1965). The role of impulses in the process of periglacial soil structure formation. *Acta Universitatis Wratislaviensis,* 44, Studia Geograficzne (7): 1–24. [In Polish].

Jahns, H. O., Miller, T. W., Power, L. D., Rickey, W. P., Taylor, T. P. & Wheeler, J. A. (1973). Permafrost protection for pipelines. In: *Proceedings of the 2nd International Permafrost Conference, Yakutsk.* North American Contribution. Washington, D.C. National Academy Press: 684–687.

James, F. A. (1972). The periglacial geomorphology of the Rankin Inlet area, Keewatin, N.W.T., Canada. *Biuletyn Peryglacjalny,* 21: 126–151.

Janin, C. (1922). *Recent progress in the thawing of frozen ground in placer mining.* Washington, D.C. Bureau of Mines, Department of the Interior, Technical Paper #309.

Jansson, J. K. & Taş, N. (2014). The microbial ecology of permafrost. *National Review of Microbiology,* 12: 414–425.

Jarrett, R. D. & England, J. F., Jr. (2002). Reliability of paleostage indicators for paleoflood studies: Ancient floods, modern hazards. In: *Principles and Applications of Paleoflood Hydrology.* American Geological Union, Water Science and Application Series, 5: 91– 109.

Jaworski, T. & Chutkowski, K. (2015). Genesis, morphology, age and distribution of cryogenic mounds on Kaffiøyra and Hermansenøya, Northwest Svarlbard. *Permafrost and Periglacial Processes,* 26(4): 304–320.

Jeffrey, W. W. (1967). *Forest types along the Lower Liard River, Northwest Territories.* Department of Forestry, Canada, Publication #1035. 103p.

Jenkins, R. E. L., Kanigan, J. C. N. & Kokelj, S. V. (2008). Factors contributing to long-term integrity of drilling-mud caps in permafrost terrain, MacKenzie Delta region, Northwest Territories, Canada. In: *Proceedings of the 9th International Conference on Permafrost, Fairbanks, Alaska* I: 833–838.

Jerwood, L. C., Robinson, D. A. & Williams, R. B. G. (1990a). Experimental frost and salt weathering of chalk. *Earth Surface Processes and Landforms,* 15: 699–708.

Jerwood, L. C., Robinson, D. A. & Williams, R. G. B. (1990b). Experimental frost and salt weathering of chalk – II. *Earth Surface Processes and Landforms,* 15: 699–708.

Jessop, R. W. (1960). The Stony Tableland Soils of the southeastern portion of the Australian arid zone and their evolutionary history. *Journal of Soil Science,* 11: 188–196.

Jia, M. C., Yuan, F. & Cheng., G.D. (1987). First discovery of ice wedges in Northeast China. *Journal of Glaciology and Geocryology,* 9(3): 257–260. [In Chinese].

Jiang, H. P. & Liu, Z. R. (2006). +/−500 kV direct current transmission line ground and foundation design in frozen earth area. *Inner Mongolia Electric Power,* 24(4): 1–4.

Jiang, Y. I. & Wang, M. S. (2006). Research on construction technology and plans of tunnels in the Plateau Permafrost region. *Journal of the China Railway Society,* 28(2).

Jin, H. J., Chang, X. L. & Wang, S. L. (2007). Evolution of permafrost on the Qinghai-Xizang (Tibet) Plateau since the end of the late Pleistocene. *Journal of Geophysical Research,* 112, F02S09, doi: 10.1029/2006JF000521.

Jin, H., Zhao, L., Wang, S. & Jin, R. (2006). Thermal regimes and degradation modes of permafrost along the Qinghai-Tibet Highway. *Science in China Series D: Earth Sciences,* 49(11): 1170–1183.

Jin, H., Yu, Q., Wu, J. & Wang, S. (2011). Spatiotemporal variability of permafrost degradation on the Qinghai-Tibet Plateau under a warming climate. *Sciences in Cold and Arid Regions,* 3(4): 281–305.

Joffe, J. S. (1949). *Pedology.* New Brunswick, New Jersey. Pedology Publications, 2nd Edition. 662p.

Jóhannesson, T. & Arnalds, P. (2001). Accidents and economic damage due to snow avalanches and landslides in Iceland. *Jökull,* 50: 81–94.

Johansson, M. (2009). *Changing lowland permafrost in northern Sweden: Multiple drivers of past and future trends.* Lund. Lund University. 137p.

Johansson, S. (1914). *Die festigkeit der bodenarten bei verschidenem wassergehalt nebst verschlag zu einer klassifikation.* Severiges Geologische Undersökning Årsbok 7 (arh. Och ippsatser, series c. 256). 110p. [In German].

Johnson, D. L. & Hanson, K. L. (1974). The effects of frost heave on objects in the soil. *Plains Anthropologist,* 19: 81–98.

Johnson, D. L., Muhs, D. R. & Barnhardt, M. L. (1977). The effects of frost heaving on objects in soils. II. Laboratory Experiments. *Plains Anthropologist,* 22(76): 133–147.

Johnson, E. A. & Larsen, C. P. S. (1991). Climatically induced change in fire frequency in the southern Canadian Rockies. *Ecology,* 72: 194–201.

Johnson, E. R. (1981). Buried oil pipeline design and operation in the arctic – lessons learned on the Alyeska Pipeline. *37th Petroleum and Mechanical Conference.* Dallas, Texas.

Johnson, J. P. & Nickling, W. G. (1979). Englacial temperature and deformation of a rock glacier in the Kluane Range, Yukon Territory, Canada. *Canadian Journal of Earth Sciences,* 16: 2275–2283.

Johnson, L. & Viereck, L. (1983). Recovery and active layer changes following a Tundra fire in Northwestern Alaska. *Permafrost, 4th International Conference Proceedings,* Washington, D.C., National Academy Press: 543–547.

Johnson, L. R. & Mangel, M. (2006). Life histories and the evolution of aging in bacteria and other single-celled organisms. *Mech. Ageing Dev.,* 127: 786–793.

Johnson, P. G. (1980). Glacier-rock glacier transition in the Southwest Yukon Territory, Canada. *Arctic and Alpine Research,* 12: 195–204.

Johnson, P. G. (1983). Rock Glaciers. A case for a change in nomenclature. *Geografiska Annaler,* 65A: 27–34.

Johnson, P. G. (1984a). Paraglacial conditions of instability and mass movement: a discussion. *Zeitschrift für Geomorphologie* NF, 28: 235–250.

Johnson, P. G. (1984b). Rock glacier formation by high-magnitude low-frequency slope processes in the southwest Yukon. *Annals of the Association of American Geographers*, 74: 408–419.

Johnson, P. G. & Lacasse, D. (1988). Rock glaciers of the Dalton Range, Kluane Ranges, south-west Yukon Territory, Canada. *Journal of Glaciology*, 34: 327–332.

Johnson, P. R. (1971). *Empirical heat transfer rates of small, Long and Balch thermopiles and convection loops*. University of Alaska, Institute of Arctic Environmental Engineering. Fairbanks. Report # 7102. 60p.

Johnson, S. S., Hebsgaard, M. B., Christensen, T. R., Mastepanov, M., Nielsen, R., Munch, K., Brand, T. B., Gilbert, M. T. P., Zuber, M. T., Bunce, M., *et al.* (2007). Ancient bacteria show evidence of DNA repair. *Proceedings of the National Academy of Sciences*, 104: 14401–14405.

Johnson, T. C., Berg, R. L., Carey, K. L. & Kaplar, C. W. (1974). *Roadway design in seasonal frost areas*. U.S. Army Cold Regions Research and Engineering Laboratory, Technical Report TR 259. 104p.

Johnson, T. C., McRoberts, E. C. & Nixon, J. F. (1984). Design implications of subsoil thawing. In: Berg, R. L. & Wright, E. A. (eds.). *Frost action and its control*. American Society of Civil Engineers, New York.

Johnson, T. C. & Sayles, F. H. (1980). Embankment dams on permafrost in the USSR. *U.S. Corps of Engineers Cold Regions Research and Engineering Laboratory*, Special Report 80-41.

Johnston, G. H. (1980). Permafrost and the Eagle River bridge, Yukon Territory, Canada. *Proceedings, Permafrost Engineering Workshop*. Ottawa. National Research Council of Canada. Technical Memorandum 130: 12–28.

Johnston, G. H. (1981). *Permafrost: Engineering Design and Construction*. Associate Committee on Geotechnical Research, National Research Council of Canada, Ottawa. Toronto. John Wiley and Sons. 540p.

Johnston, G. H. & Ladanyi, B. (1972). Field tests of grouted anchors in permafrost. *Canadian Geotechnical Journal*, 9: 176–194.

Johnston, J. C. (1966). *Planning and operation of hydraulic stripping plants*. Unpublished thesis, University of Alaska.

Jonasson, S. (1986). Influence of frost heaving on soil chemistry and on the distribution of plant growth forms. *Geografiska Annaler*, 68(3): 185–195.

Jones, B. M., Amundson, C. L. & Koch, J. C. (2013). *Thermokarst and thaw-related landscape dynamics – An annotated bibliography with an emphasis on potential effects on habitat and wildlife*. U.S. Department of the Interior, U.S. Geological Survey Open-File Report # 2013-1161.

Jones, B. M., Arp, C. D., Jorgenson, M. T., Hinkel, K. M., Schmutz, J. A. & Flint, P. L. (2009). Increase in the rate and uniformity of coastline erosion in Arctic Alaska. *Geophysical Research Letters*, 36: L03503.

Jørgensen, A. S. (2009). *Assessment of three mitigation techniques for permafrost protection*. Ph.D. thesis, Arctic Technology Centre, Department of Civil Engineering, Technical University of Denmark. Report R-202.

Jørgenson, A. S. (2012). Assessment of different heat drain materials for protection of permafrost under road and airfield embankments. *Proceedings of the 10th International Conference on Permafrost, Salekhard*.

Jørgenson, A. S. & Andreasen, F. (2007). Mapping permafrost surface using ground-penetrating radar at Kangerlussuaq Airport, western Greenland. *Cold Regions Science and Technology*, 48: 64–72.

Jørgenson, A. S. & Ingeman-Nielsen, T. (2008). The impact of light-colored pavements on active-layer dynamics revealed by ground-penetrating radar monitoring. In: *Proceedings of the 9th*

International Conference on Permafrost, Fairbanks. Washington, D.C. National Academy of Sciences: 865–868.

Jørgenson, M. T. (2012). Assessment of different heat drain materials for protection of permafrost under road and airfield embankments. *Proceedings of the 10th International Conference on Permafrost, Salekhard*.

Jørgensen, M. T. & Doré, G. (2009). Experimentation of several methods in Tasiulaq Airport to minimize the effects caused by the melting of permafrost. In: *Proceedings of the 14th Conference on Cold Regions Engineering, Duluth*. Reston, Virginia. ASCE: 171–182.

Jørgensen, M. T. & Grosse, G. (2016). Remote sensing of landscape change in permafost regions. *Permafrost and Periglacial Processes*, 27(4): 324–338.

Jørgenson, M. T. & Shur, Y. (2007). Evolution of lakes and basins in northern Alaska and discussion of the thaw lake cycle. *Journal of Geophysical Research*, 112: F02S17.

Journeaux Associates (2012). *Engineering challenges for tailings management facilities and associated infrastructure with regard to climate change in Nunavut*. Report # L-11-1472 to the Nunavut Government.

Judge, A. S. (1973). Deep temperature observations in the Canadian North. *North American Contribution, Permafrost*. 2nd International Conference, Yakutsk, USSR. Washington. National Academy of Sciences: 35–40.

Judge, A. S. (1982). Natural Gas Hydrates in Canada. *The Roger R.J. E. Brown Memorial Volume, Proceedings of the 4th Canadian Permafrost Conference, Calgary*. Ottawa, National Research Council of Canada: 320–328.

Jumikis, A. R. (1978). Graphs for disturbance-temperature distribution in permafrost under heated structures. *Proceedings of the 3rd International Conference on Permafrost, Edmonton*. Ottawa. National Research Council of Canada, 947p.: 589–596.

Kääb, A. & Reichmuth, T. (2005). Advance mechanisms of rock glaciers. *Permafrost and Periglacial Processes*, 16(2): 187–193.

Kachurin, S. P. (1958). Alases of Central Yakutia. Yakutsk. *Reports of the North East Division of the Permafrost Institute*, 1: 167–178. [In Russian].

Kachurin, S. P. (1959). *Principles of Geocryology. Part 1. General Geocryology*. Moscow. V. A. Obruchev Institute of Permafrost Studies. Chapter XI. Cryogenic physico-geological phenomena in permafrost regions: 365–398. [In Russian].

Kachurin, S. P. (1961). *Thermokarst on USSR Territory*. Moscow, Academy of Sciences Press. 290p. [In Russian].

Kade, A. & Walker, D. A. (2008). Experimental alteration of vegetation on nonsorted circles: Effects on cryogenic activity and implications for climatic change in the Arctic. *Arctic, Antarctic and Alpine Research*, 40: 96–113.

Kaiser, C., Meyer, H., Biasi, C., Rusalimova, O., Barsukov, P. & Richter, A. (2005). Storage and mineralization of carbon and nitrogen in soils of a frost boil tundra ecosystem in Siberia. *Applied Soil Ecology*, 29(2): 173–183.

Kalenova, L. F., Suhovey, U. G., Brouchkov, A. V., Melnikov, V. P., Fisher, T. A., Besedin, M., Novikova, M. A., Efimova, J. A. & Subbotin, A. M. (2011a). Experimental study of the effects of permafrost microorganisms on the morphofunctional activity of the immune system. *Bulletin of Experimental Biological Medicine*, 151: 201–204.

Kalenova, L. F., Sukhovei, U. G., Brouchkov, A. V., Melnikov, V. P., Fisher, T. A., Besedin, I. M., Novikova, M. A. & Efimova, J. A. (2011b). Effects of permafrost microorganisms on the quality and duration of life of laboratory animals. *Neuroscience and Behavioural Physiology*, 41: 484–490.

Kamensky, R. M. (1988). *Experimental – theoretical bases of the forecast of a thermal mode of hydraulic engineering structures and gas pipelines in cryolithozone*. Thesis of Doctor's Degree. Yakutsk. 45 p. [In Russian].

Kamensky, R. M. (1998). *Geocryological problems of construction in Eastern Russia and Northern China.* Proceedings of the International Symposium, 23–25th September, Chita, Russia. Yakutsk. Siberian Branch, Russian Academy of Sciences. Volume 1: 255p. Volume 2: 197p.

Kamensky, R. M. (2002). Geocryology – a new science in the system of earth sciences. *Yakutian Science and Technology*, 1: 12–14. [In Russian].

Kane, D. (1981). Physical mechanics of aufeis growth. *Canadian Geotechnical Journal*, 8(2): 186–195.

Kanevskiy, M., Jorgenson, T., Shur, Y., O'Donnell, J. A., Harden, J. W., Zhuang, Q. & Fortier, D. (2014). Cryostratigraphy and permafrost evolution in the lacustrine lowlands of West-Central Alaska. *Permafrost and Periglacial Processes*, 25(1): 14–34.

Kanigan, J. C. N. & Kokelj, S. V. (2010). Review of current research on drilling-mud sumps in permafrost terrain, Mackenzie Delta region, NWT, Canada. *GEO2*10*: 1473–1479.

Kanigan, J. C. N., Zajdlik, B. & Kokelj, S. V. (2010). Delineation of salt contamination in patterned ground. *GEO2*10*: 1466–1472.

Kantanen, J. (2009). Article of the month – The Yakutian cattle: the cow of permafrost. *Global-Div Newsletter*, 30(12): 3–6. http://www.globaldiv.eu/NL/GlobalDiv_Newsletter%20no%2012.pdf.

Kaplar, C. W. (1965). Stone migration by freezing of soil. *Science*, 149: 1520–1521.

Kaplar, C. W. (1969). Phenomena and mechanism of frost heaving. In: Highway Research Board 49th Annual Meeting, Washington. Preprint, 44p.

Kaplina, T. N. (1965). *Cryogenic slope processes* Moscow. Nauka. 295 p. [In Russian].

Kaplina, T. N. (2009). Alas complex of northern Yakutia. *Earth Cryosphere*, 13: 3–7. [In Russian].

Kaplyanskaya, F. A. & Tarnogradskiy, V. D. (1977). On the problem of formation of the relict glacial ice deposits and preservation of the primordially frozen tills. *Izvestiya Geog. Soc.*, 109(4): 314–319.

Kaplyanskaya, F. A. & Tarnogradskiy, V. D. (1986). Remnants of Pleistocene ice sheets in the permafrost zone as an object for paleoglaciological Research. *Polar Geography and Geology*, 10: 257–266.

Kaplar, C. W. (1965). Stone migration in freezing of soil. *Science*, 149: 1520–1521.

Kaplar, C. W. (1971). *Experiments to simplify frost susceptibility testing of soils.* U.S. Army Cold Regions Research and Engineering Laboratory, Technical Report #223. 21p.

Kaplar, C. W. (1974a). *Phenomena and mechanism of frost heaving.* Highway Research Board 49th Annual Meeting, Washington. 1970 preprint. 44p.

Kaplar, C. W. (1974b). *Freezing test for evaluating relative susceptibility of various soils.* CRREL, Technical Report #250. 36p.

Karavayeva, N. A. & Targulyan, V. O. (1960). Peculiarities of humus distribution in the tundra soils of northern Yakutia. *Soviet Soil Science*, 7: 1293–1300. [In Russian].

Karavayeva, N. A., Sokolov, I. A., Sokolova, T. A. & Targulyan, V. O. (1965). Peculiarities of the soil formation in the tundra-taiga frozen regions of Eastern Siberia and the Far East. *Soviet Soil Science*, 12: 756–766. [In Russian].

Karlén, W. (1988). Scandinavian glacial and climatic fluctuations during the Holocene. *Quaternary Science Reviews*, 7: 199–209.

Karte, J. (1987). Pleistocene periglacial conditions and geomorphology in north and central Europe. In: Boardman, J. (ed.). *Periglacial conditions in Britain and Ireland.* Cambridge Cambridge University Press: 67–75.

Kasischke, E. S., French, N. F., O'Neill, K. P., Richter, D. D., Bourgeau-Chavez, L. L. & Harrell, P. A. (2000). Influence of fire on long-term patterns of forest succession in Alaskan boreal forests. In: Kasischke, E. S. and Stocks, B. J., (eds.). *Fire, Climate Change, and carbon cycling in the Boreal Forest.* New York, Springer-Verlag, Ecological Study Series: 214–238.

Katasonov, E. M. (1954). *Lithology in the perennially frozen Quaternary deposits (cryolithology) of the Yana Lowland.* Doctoral thesis, Obruchev Permafrost Institute, Moscow. [In Russian].

Katasonov, E. M. (1969). Composition and cryogenic structure of permafrost. In: *Permafrost Investigations in the field.* National Research Council of Canada, Technical Translation #1358: 25–36.

Katasonov, E. M. (1973a). Palaeofrost (palaeocryological) studies: Objectives, methods, and some results. In: *Palaeocryology in Quaternary Stratigraphy and Palaeoecology.* Moscow, Nauka: 66–79. [In Russian].

Katosonov, E. M. (1973b). Present-day ground and ice veins in the region of the Middle Lena. *Biuletyn Peryglacialny,* 23: 81–89.

Katasonov, E. M. (1975). Frozen-ground and facial analysis of Pleistocene deposits and Paleogeography of central Yakutia. *Biuletyn Peryglacjalny,* 24: 33–40.

Katasonov, E. M. (1978). Permafrost-facies analysis as the main method of cryolithogy. In: Sange, F. J. and Hyde, P. J., (Eds.). Russian Contribution. *Proceedings of the 2nd International Conference on Permafrost*: 171–176.

Katasonov, E. M. & Ivanov, M. S. (1973). *Cryolithology of Central Yakutia.* Guidebook, 2nd International Permafrost Conference, Yakutsk. 38p.

Katayama, T., Tanaka, M., Moriizumi, J., Nakamura, T., Brouchkov, A., Douglas, T., Fukuda, M., Tomita, M. & Asano, K. (2007). Phylogenetic analysis of bacteria preserved in a permafrost ice wedge for 25,000 Years. *Applied Environmental Microbiology,* 73: 2360–2363.

Kazemian, S., Hunt, K., Prasad, A., *et al.* (2010). A review of stabilization of soft soils by injection of chemical grouting. *Australian Journal of Basic Applied Science,* 4: 5862–5968.

Kelletat, D. (1985). Patterned ground by rainstorm erosion on the Colorado Plateau, Utah. *Catena,* 12: 255–259.

Kelley, J. J. & Weaver, D. F. (1969). Physical processes at the surface of the arctic tundra. *Arctic,* 22(4): 425–437.

Kent, R., Marshall, P. & Hawke, L. (2003). *Guidelines for the planning, design, operation and maintenance of Modified Solid Waste Sites in the Northwest Territories.* Prepared for the Department Municipal and Community Affairs, Government of the Northwest Territories by Ferguson Simek Clark. FSC Project # 2001-1330.

Kerfoot, D. E. (1972). Tundra disturbance studies of the Western Canadian Arctic. *Department of Indian and Northern Affairs, Arctic Land Use Research,* 71-72-11.

Kershaw, G. P. (2003). Permafrost landform degradation over more than half a century, MacMillan Caribou Pass region, NWT/Yukon, Canada. In: Phillips, M., Springmam, S. M. and Arenson, L. U. (eds.). *Permafrost, 8th International Conference Proceedings, Zurich.* Lisse. Swets and Zeitlinger: 543–548.

Kershaw, G. P. & Gill, D. (1979). Growth and decay of palsas and peat plateaus in the MacMillan Pass – Tsichu River area, Northwest Territories, Canada. *Canadian Journal of Earth Sciences,* 16: 1362–1374.

Kershaw, K. A. & Rouse, W. R. (1971). Studies on lichen-dominated systems. I. The water relations of *Cladonia alpestris* in spruce-lichen woodland in Northern Ontario. *Canadian Journal of Botany,* 49: 1389–1399.

Kesseler, P. (1925). *Das eiszeitliche Klima und seine geologischen wirkungen im nicht vereisten Gebeit.* Stuttgart, L. Schweizerbart'sche. 204p. [In German].

Kesserli, J. E. (1941). Rock streams in the Sierra Nevada. *Geographical Revue,* 31: 203–227.

Kessler, A. (2013). *Holocene development and permafrost formation at a peat plateau and a palsa mire in Tavvavuoma, Northern Sweden.* Unpublished M.Sc. thesis, Department of Physical Geography and Quaternary Geology, Stockholm University.

Kessler, M. A. & Werner, B.T. (2003). Self-organization of sorted patterned ground. *Science*, 299: 380–383.

Kharionovsky, V. V. (1994). Distinguishing features of pipeline construction in Arctic regions. In: *Proceedings of Polartech 1994. International Conference on the Development and Commercial Utilization of Technologies in Polar Regions.* Lulea. Coldtech and Lulea University of Technology pp. 504–520.

Khrustalev, L. & Pustovoit, G. (1988). *Probability and statistical calculations of bases of buildings in the cryolithozone.* Novosibirsk. Nauka. 253 p. [In Russian].

Killingbeck, J. & Ballantyne, C. K. (2012). Earth hummocks in West Dartmoor, SW England: Characteristics, age and origin. *Permafrost and Periglacial Processes*, 23(2): 152–161.

Kim, B., Grikurov, G., & Soloviev, V. (1999). Land-Ocean System in the Siberian Arctic: Dynamics and history. In: *High resolution Seismic Studies in the Laptev Sea Shelf.* Berlin. Springer-Verlag, 683–692.

Kim, K., Zhou, W. & Huang, S. L. (2008). Frost heave predictions of buried chilled gas pipelines with the effect of permafrost. *Cold Regions Science and Technology*, 5(3): 382–396.

Kimble, J. M. (ed.) (2004). *Cryosols. Permafrost-affected soils.* Berlin. Springer-Verlag. 726p.

King, L. (1976). Permafrostuntersuchungen in (Schwedisch-Lappland) mit hilfe der hammerschlagseismik. *Zeitschrift für Gletscherkunde und Glazialgeologie*, 12: 187–204. [In German].

King, L. (1982). Qualitative und quantitative erfassung von permafrost in Tarfala (Schwedisch-Lappland) und Joten heimen (Norwegen) mit hilfe geoelektrischer sondierungen. In: Barsch, D. (Ed.). Experimente und Messungen in der Geomorphologie. *Zeitschrift für Geomorphologie* Supplement Band, 43: 139–160. [In German].

King, L. (1983). High mountain permafrost in Scandinavia. *Proceedings of the 4th International Conference on Permafrost, Fairbanks.* Washington. National Academy Press: 612–617.

King, L. (1986). Zonation and ecology of high mountain permafrost in Scandinavia. *Geografiska Annaler*, 68A: 131–139.

King, L. C. (1953). Canons of Landscape Evolution. *Geological Society of America Bulletin*, 64: 751–752.

King, L. C., Fisch, W., Haeberli, W. & Waechter, H. P. (1987). Comparison of resistivity and radio-echo soundings on rockglacier permafrost. *Zeitschrift für Gletscherkerkunde und Glazialgeologie*, 23: 77–97.

King, M. S. & Garg, O. P. (1980). Interpretation of seismic and resistivity measurements in permafrost in Northern Québec. *Proceedings of the Symposium on Geophysics (#5), Ottawa.* Associate Committee on Geotechnical Research, National Research Council of Canada Technical Memoir 128: 50–59.

King, R. A. & Reger, R. D. (1982). *Air photo analysis and summary of landform soil properties along the route of the Trans-Alaska Pipeline System.* State of Alaska, Division of Geological-Geophysical Surveys Geologic Report # 66.

Kinosita, S., *et al.* (1979). Core samplings of the uppermost layer in a tundra area. In: Kinosita, S., Ed.. *Joint Studies on Physical and Biological Environments in Permafrost, Alaska and Northern Canada.* Hokkaido, Hokkaido University Institute of Low Temperature Science, 140p.

Kitano, M. (ed.). (2001). *The book for taking good care of Global Environment.* PHP Institute, Tokyo, Japan.

Klaminder, J., Yoo, K. & Giesler, R. (2009). Soil carbon accumulation in dry tundra: the important role played by precipitation. *Journal of Geophysical Research*. G04005 (114). Doi: 10.1029/2009JG000947.

Klaminder, J., Becher, M. & Kobayashi, M. (2011). Estimating the soil mixing rate induced by cryoturbation. *American Geophysical Union*, Fall Meeting, 2011, abstract #B11A-0474.

Klatka, T. (1962). Geneza i wiek goloborozy kysogórskich. *Acta Geographica Lodziensia*, 12. [In Polish].

Kleman, J. & Borgström, I. (1990). The boulder fields of Mount Fulufjället, West-Central Sweden. *Geografiska Annaler*, 72A: 63–78.

Kling, J. (1997). Observations on sorted circle development, Abisko, Northern Sweden. *Permafrost and Periglacial Processes*, 8: 447–435.

Knystautas, A. (1987). *The Natural History of the USSR*. New York. McGraw-Hill Book Company. 219p.

Knutson, A. (1993). *Frost action in soils*. Oslo. Norwegian Road Research Laboratory. 40p.

Kojima, S. (1994). Relationships of vegetation, earth hummocks and topography in the high arctic environment of Canada. *Polar Biology*, 7: 256–269.

Kokelj, S. V. & Burn, C. R. (2004). Tilt of spruce trees near ice wedges, Mackenzie Delta, Northwest Territories, Canada. *Arctic, Antarctic and Alpine Research*, 36(4): 615–623.

Kokelj, S. V., Burn, C. R. & Tarnocai, C. (2007). The structure and dynamics of earth hummocks in the Subarctic Forest near Inuvik, Northwest Territories, Canada. *Arctic, Antarctic and Alpine Research*, 39(1): 99–109.

Kokelj, S. V. & Lewkowicz, A. G. (1998). Long-term influence of active-layer detachment sliding on permafrost slope hydrology, Hot Weather Creek, Ellesmere Island, Canada. *Proceedings of the 7th International Conference on Permafrost, Yellowknife*. Québec City. Centre d'Etudes Nordiques, Université Laval. Collection Nordicana # 57: 583–589.

Kokelj, S. V. & GeoNorth Limited (2002). *Drilling mud sumps in the Mackenzie Delta region: Construction, abandonment and past performance*. Submitted to the Water Resources Division, Indian and Northern Affairs Canada, Yellowknife, Northwest Territories. 55p.

Kokelj, S. V., Lacelle, D., Lantze, T. C., Malone, L., Clark, I. D. & Chin, K. S. (2013). Thawing of massive ground ice in mega slumps drives increases in stream sediment and solute flux across a range of watershed scales. *Journal of Geophysical Research, Earth Surface* 118: 681–692.

Kokelj, S. V., Pisaric, M. F. J. & Burn, C. R. (2007). Cessation of ice-wedge development during the 20th century in spruce forests of eastern Mackenzie Delta, Northwest Territories, Canada. *Canadian Journal of Earth Sciences*, 44: 1503–1515.

Kokelj, S. V., Riseborough, D., Coutts, R. & Kanigan, J. N. C. (2010). Permafrost and terrain conditions at northern drilling-mud sumps: Impacts of vegetation and climate change and the management implications. *Cold Regions Science and Technology*, 64(1): 46–56.

Kokelj, S. V., Tunnicliffe, J., Lacelle, D., Lantze, T. C., Fraser, R. H. & Chin, K. S. (2015a). Increased precipitation drives megaslump development and destabilization of ice-rich permafrost terrain, northwest Canada. *Global and Planetary Change*, 129: 56–68.

Kokelj, S. V., Tunnicliffe, J., Lacelle, D., Lantze, T. C. & Fraser, R. H. (2015b). Retrogressive thaw slumps: From slope process to the landscape sensitivity of Northwest Canada. *GeoQuebec, 2015*.

Kolomyts, E. G. (1976). *Snow Structure and Landscape Indications*. Moscow. Nauka. 206p. [In Russian].

Kondratiev, V. G. (1996). Strengthening railroad base constructed on icy permafrost soil. *Proceedings of the 8th International Conference on Cold regions Engineering*. Fairbanks. University of Alaska, Fairbanks: 688–699.

Kondratiev, V. G. (2004). Strengthening of supports for contact systems and overhead transmission lines erected on heaving seasonal thawing soils. *Soil Mechanics and Foundation Engineering*, 41(5): 185–190.

Kondratiev, V. (2010). Some geocryological problems of railways and highways on permafrost of Transbaikal and Tibet. *GEO2*10*: 541–548.

Kondratiev, V. G. (2013). Roadbed, embankment, tower support and culvert stability problems on permafrost. *Sciences in Cold and Arid Regions*, 5(4): 377–386.

Konishchev, V. N. (1973). Kriogennoye vyretrivaniye. In: Akedemiya Nauk SSSR, Sektsiya Nauk o zemle, Sibirskoye Otdeleniye,II. Mezhdunarodnaya Konferentsiya po Merzlotovedeniyu, Doklady; soobshcheniya 3. Yakutsk, Yakutskoye Knizhnoye Izdatel'stovo. 100p. [In Russian].

Konishchev, V. N. (1978). Frost weathering. USSR contribution, Permafrost 2nd International Conference on Permafrost, Yakutsk. Washington, D.C., National Academy of Sciences: 176–181.

Konishchev, V. N. (1982). Characteristics of cryogenic weathering in European USSR. *Arctic and Alpine Research*, 14: 261–265.

Konishchev, V. N. (1998). Relationship between the lithology of active-layer materials and mean annual ground temperature in the former USSR. *Proceedings of the 7th International Conference on Permafrost, Yellowknife*. Québec City. Centre d'Etudes Nordiques, Université Laval. Collection Nordicana # 57: 591–594.

Konishchev, V. N., Fanstova, M. A. and Rogov, V. V. (1963). Cryogenic processes as reflected in ground microstructures. *Builetyn Peryglacjalny*, 22: 213–219.

Konishchev, V. N. & Kartashova, G. G. (1972). The main periods of deposition and vegetation development of the southern part of the Yana-Indigirka Plain during the Cenozoic Era. *Viestnik Moskovskogo Universiteta*. 2: 67–73. [In Russian].

Konishchev, V. N. & Maslov, A. D. (1969). Physical reasons of frontal growth of syngenetic ice-wedges. In: *Problems in Cryolithology, Volume 1*. Moscow. Moscow State University Press. [In Russian].

Konishchev, V. N. & Rogov, V. V. (1993). Investigations of cryogenic weathering in Europe and Northern Asia. *Permafrost and Periglacial Processes*, 4: 49–64.

Konishchev, V. N., Rogov, V. V. & Shurina, G. N. (1975). Cryogenic transformation of clayey sediment rocks. Fondation Française d'Etudes Nordiques, 6th Congrés International, *Les problems poses par le gèlifraction*. Recherches fondamentales et appliqués. Report 104.

Konishchev, V. N., Rogov, V. V. & Shurina, G. N. (1976). Cryogenic factor influence on primary minerals (results of experimental investigation). *Problems in Cryolithology*, 5: 50–61. [In Russian].

Konstantinov, I. P. & Gurianov, I. E. (2001). Numerical estimation of loads and stresses on deformed sites of the pipeline Mastakh-Yakutsk. *Earth Cryosphere*, 5: 68–75. [In Russian].

Konyakhin, M. (1988). *Isotopic oxygen content of ice wedges on the Kolyma plain*. Unpublished Ph.D. thesis, Moscow. Moscow State University. 27p. [In Russian].

Konrad, J.-M. (1999). Frost susceptibility related to soil index properties. *Canadian Geotechnical Journal*, 36: 403–417.

Konrad, J.-M. & Morgenstern, J. R. (1982). Effects of applied pressure on freezing soils. *Canadian Geotechnical Journal*, 19: 494–505.

Konrad, J.-M. & Morgenstern, N. R. (1983). Frost-susceptibility of soils in terms of their segregation potential. *Proceedings of the 4th International Conference on Permafrost*; 660–665.

Konrad, J.-M. & Morgenstern, J. R. (1984). Frost-heave prediction of chilled pipelines buried in unfrozen soils. *Canadian Geotechnical Journal*, 21: 100–115.

Košták, B., Dobrev, N., Zika, P. & Ivanov, P. (1998). Joint monitoring on a rock face bearing a historical bas-relief. *Quarterly Journal of Engineering Geology*, 31: 37–45.

Kostyaev, A. G. (1965). Polygonal ice-wedges in the Amguma River basin. In: *Underground ice, Issue 1*. For the 7th International Congress on the Quaternary (INQUA), U.S.A. Moscow. Moscow University Press: 133–140. [In Russian].

Kotlyakov, V. M., Editor in Chief (1997). *World Atlas of Snow and Ice Resources*. Moscow. Institute of Geography, Russian Academy of Sciences, Vol. 1, Atlas, 392p. Vol. 2. Snow and Ice phenomena and processes. 372p. Vol. 3. Legends and explanations of all the maps in English, 144p.

Kotlyakov, V. & Khromova, T. (2002). Maps of permafrost and ground ice. In: Stlbovoi, V. and McCallum, I. (Eds.). *Land resources of Russia*. Laxenburg, Austria: International Institute for Applied Systems Analysis and the Russian Academy of Sciences. CD_ROM. Distributed by the National Snow and Ice Data center, Boulder, Colorado, U.S.A. http://nsidc.org/data/docs/fgdc/ggd600_russia_pf_maps/russian_perm.

Kotlyakov, V. M., Rzheevskii, B. N. & Samoilov, V. A. (1977). The dynamics of avalanches in the Khibins. *Journal of Glaciology*, 19: 431–439.

Kotov, A. N. (1998). Alas and ice complex deposits of North-West Chukotka (East Siberian coast). *Earth Cryosphere*, 2(1): 11–18. [In Russian].

Kovacs, A. (1983). *Shore ice ride-up and pile-up features. Part 1: Alaska's Beaufort Sea coast*. Hanover, New Hampshire. U.S. Corps of Engineers Cold Regions Research and Engineering Laboratory. Report 83-9.

Kovacs, A. & Sodhi, D. S. (1980). Shore ice pile-up and ride-up: field observations, models, theoretical analyses. *Cold Regions Science and Technology*, 2: 209–288.

Kovda, V. A. (1946). *Origin and regime of saline soils*. Moscow-Leningrad. Russian Academy of Sciences, volume 1. 382p. [In Russian].

Kowalewski, D. E., Marchant, D. R., Head, J. W. III & Jackson, D. W. (2013). A 2D model for characterising first-order variability in sublimation of buried glacier ice, Antarctica: Assessing the influence of polygon troughs, desert pavements and shallow subsurface salts. *Permafrost and Periglacial Processes*, 23: 1–14.

Kowalkowski, A. (1978). The catena of permafrost soils in the Bayen-Naurin-Khotnor Basin, Khangai Mountains, Mongolia. *Proceedings of the 3rd International Conference on Permafrost, Edmonton*, Ottawa, National Research Council of Canada 1: 413–418.

Kristiansen, K. J. & Sollid, J. L. (1985). *Börselvfjellet-Lille Porsangen, Nord Norge*. Kvartägeolisk og geomorfologist kart 1:75,000. Geografisk Institut, Universitetet Oslo. [In Norwegian].

Križek, M. & Uxa, T. (2013). Morphology, sorting and microclimates of relict sorted polygons, Krkonoše Mountains, Czech Republic. *Permafrost and Periglacial Processes*, 24(4): 313–321. DOI: 10.1002/ppp.1789.

Kromer, R. A., Hutchinson, D. J., Lato, M. J., Gauthier, D. & Edwards, T. (2015). Identifying rock slope failure precursors using LiDAR for transportation corridor hazard management. *Engineering Geology*, 195: 93–103.

Krumbein, W. C. (1941). Measurement and geological significance of shape and roundness in sedimentary particles. *Journal of Sedimentary Petrology*, 11: 164–172.

Krumme, O. (1935). *Frost und schnee in ihrer wirking auf den boden in Hochtaunus*. Rhein-Mainische Forschungen, 13. 73p. [In German].

Kubota, Z., Shimizu, K., Tanaka, Y. & Makita, T. (1984). Thermodynamic properties of R13 ($CClF_3$), R33 (CHF_3), R152a ($C_2H_4F_2$) and propane hydrates for desalinization of sea water. *Journal of Chemical Engineering of Japan*, 17(4): 423–429.

Kudryavtsev, V. A. (1954). (ed.). *Obshcheye Merslotovedeniya (Geokriologiya)*. Moscow, Izdatel'stovo Moskovskogo Universiteta. [In Russian].

Kudryavtsev, V. A. (1978). (ed.). *Obshcheye Merslotovedeniya (Geokriologiya)*, Izd. 2, (edu 2). Moscow, Izdatel'stovo Moskovskogo Universiteta. [In Russian].

Kudryavtsev, V. A., Garagula, L. S., Kondrat'yeva, K. A. & Melamed, V. G. (1974). *Osnovy merzlotnogo prognoza*. MGU. 431p. [In Russian].

Kudryavtsev, V. A., Garagula, L. S., Kondrat'yeva, K. A. & Melamed, V. G. (1977). Fundamentals of frost forcasting in geocryological engineering investigations. US Army CRREL draft translation 606, Hanover, New Hampshire. 489p.

Kudryavtsev, V. A., Garagula, L. S., Kondrat'yeva, K. A., Buldovich, S. N., Brouchkov, A. V., Koshurnikov, A. V. & Motenko R. G. (2016). Fundamentals of frost forcasting in geocryological engineering investigations. Moscow University Press, 512p [In Russian].

Kuenen, P. H. (1958). Experiments in Geology. *Geological Society of Glasgow, Transactions*, 23: 1–28.

Kuhry, P., Grosse, G., Harden, J. W., Hugelius, G., Koven, C. D., Ping, C.-L., Schirrmeister, L. & Tarnocai, C. (2013). Characterisation of the permafrost carbon pool. *Permafrost and Periglacial Processes*, 24(2): 146–155.

Kunský, J. (1954). *Homes of Primeval Man*. Prague. Artia.

Kurdyakov, A. G. (1965). Polygonal ice-wedges in Amguema River basin. In: *Underground ice, Issue 1*. 7th International Congress on Quaternary (INQUA), USA. Moscow. Moscow University Press: 87–103. [In Russian].

Kurfurst, P. J. & Van Dine, D. F. (1973). Terrain sensitivity and mapping, Mackenzie Valley Transportation Corridor. *Geological Survey of Canada*, Paper 73-1, Part B: 155–159.

Kurylyk, B. L. (2015). Discussion of "A simple thaw-freeze algorithm for a multi-layered soil using the Stefan equation" by Xie and Gough (2013). *Permafrost and Periglacial Processes*, 26(2): 200–206.

Kuznetsov, A. L. (1973). Dam of the Anadyr' heat and electric power plant. *Trudy Gidroproekta*, 34: 88–100. [In Russian].

Kvenvolden, K. A. (1988). Methane hydrate – a major reservoir of carbon in the shallow geosphere? *Chemical Geology*, 71: 41–51.

Kvenvolden, K. A., Ginsburg, G. & Solovyev, V. (1993). Worldwide distribution of subaquatic hydrates. *Geo-Marine Letters*, 13: 32–40.

Laberge, M.-J. & Payette, S. (1995). Long-term monitoring of permafrost change in a palsa peatland in Northern Québec, Canada: 1983–1993. *Arctic and Alpine Research*, 27(2): 167–171.

Lacelle, D., Lapalme C., Davila, A. F., Pollard, W., Marinova, M., Heldmann, J. & McKay, C. P. (2016). Solar radiation and air and ground temperature relations in the Cold and Hyper-Arid Quartermain Mountains, McMurdo Dry Valleys, Antarctica. *Permafrost and Periglacial Processes*, 27(2): 163–176.

LaChapelle, E. R. (1969). *Field Guide to ice crystals*. Washington. University of Washington Press. 101p.

Lachenbruch, A. H. (1962). *Mechanics of thermal contraction cracks and ice wedge polygons in permafrost*. Geological Society of America, Special Paper #78. 69p.

Lachenbruch, H. A. (1968). Permafrost. In R. W. Fairbridge (ed.), *Encyclopedia of Geomorphology*. New York, Reinhold Book Company: 833–838.

Lachenbruch, A. H. (1970). Thermal considerations in Permafrost. In: Adkinson, W. L. and Borage, M. M., (Eds.). Geological Seminar on the North Slope of Alaska. *Proceedings American Association of Petroleum Geologists*, Pacific Section, A1-Rio, J1–2 and Discussion J2–5.

Lachenbruch, A. H. & Marshall, B. V. (1969). Heat Flow in the Arctic. *Arctic*, 22: 300–311.

Ladanyi, B. (1972). An engineering theory of creep of frozen soils. *Canadian Geotechnical Journal*, 9(1): 63–80.

Ladanyi, B. (1981). Mechanical behaviour of frozen soils. In: Selvadurai, A. P. S., (Ed.). *Mechanics of structured media*. Proceedings of a Symposium on the mechanical behaviour of Structured Media, Ottawa, Canada. Amsterdam. Elsevier. Part B: 205–245.

Ladanyi, B. (1984). Design and construction of deep foundations in permafrost: North American practice. *Permafrost, 4th International Conference, Fairbanks, Alaska*. Washington, D.C. National Academy Press. Final Proceedings: 43–50.

Lafortune, M., Filion, L. & Hètu, B. (1997). Dynamique d'un front forestier sur un talus d'éboulis actif en climat tempéré froid (Gaspésie, Québec). *Gèographie physique et Quaternaire*, 51(1): 67–80. [In French].

Lagov, P. A. & Parmuzina, O. Y. (1978). Ice formation in the seasonally thawing layer. In: *General Geocryology*. Novosibirsk, Nauka, USSR: 56–59. [In Russian].

Lai, Y., Zhang, S. & Yu, W. *et al.* (2006a). Laboratory study of particle size for optmal cooling effect of closed crushed-rock layer. *Journal of Glaciology and Glaciology*, 45(2): 114–121.

Lai, Y., Ma, W., Zhang, M., Yu, W. & Gao, Z. (2006b). Experimental investigation on influence of boundary conditions on cooling effect and mechanism of crushed-rock layers. *Cold Regions Science and Technology*, 45: 114–121.

Lai, Y., Zhang, S. & Yu, W. (2012). A new structure to control frost boiling and frost heave of embankments in cold regions. Doi: 10.1016/j.coldregions.2012.04.002.

Lambert, J. D. H. (1972a). Plant succession on tundra mudflows: preliminary observations. *Arctic*, 25(2): 99–106.

Lambert, J. D. H. (1972b). Botanical changes resulting from seismic and drilling operations, Mackenzie Delta Area. Department Of Indian and Northern Affairs, Ottawa. ALUR 71-72-148.

Lambiel, C. & Pieracci, K. (2008). Permafrost distribution in Talus Slopes located within the alpine periglacial belt, Swiss Alps. *Permafrost and Periglacial Processes, 19*(3): 293–304.

Lamontagne, V., Périer, L., Lemieux, C., Doré, G., Allard, M., Roger, J. & Guinond, A. (2015). Suivi du compartement thermique et mécanique de l'adaptation de la route d'access à l'aérort de Salluit au Nunavik, Canada. *GEOQuébec*. [In French].

Lamonthe, C. & St.-Onge, D. (1961). A note on a periglacial erosion process in the Isachsen area, N. W. T. *Geographical Bulletin*, 16: 104–113.

Landschützer, P., Gruber, N. & Bakker, D. C. E. (2015). A 30 years observation-based global monthly gridded sea surface pCO2 product from 1982 through 2011. http://cdiac.ornl.gov/ftp/oceans/SPCO2_1982_2011_ETH_SOM_FFN. Carbon Dioxide Information Analysis Center, Oak Ridge National Laboratory, US Department of Energy, Oak Ridge, Tennessee. doi: 10.3334/CDIAC/OTG.SPCO2_1982_2011_ETH_SOM-FFN.

Langway, C. C. Jr. (1967). Stratigraphic analysis of a deep ice core from Greenland. *U.S. Army, CRREL Research Department*, 77: 132p.

Lantuit, H., Overduin, P. P., *et al.* (2012). The Arctic coast dynamics database: A new classification scheme and statistics on Arctic permafrost coastlines. *Estuaries and Coasts*, 35: 383–400.

Lantuit, H. & Pollard, W. H. (2008). Fifty years of coastal erosion and retrogressive thaw slump activity on Herschel Island, southern Beaufort Sea, Yukon Territory, Canada. *Geomorphology* 95: 84–102.

Lantuit, H., Pollard, W. H., Couture, N., Fritz, M., Schirrmeister, L., Meyer, H. & Hubberten, H.-W. (2012b). Modern and Late Holocene retrogressive thaw slump activity on the Yukon coastal plain and Herschel Island, Yukon Territory, Canada. *Permafrost and Periglacial Processes*, 23: 39–51.

Laprise, D. & Payette, S. (1988). Evolution recente d'une tourbiere a palses (Québec subactique): une analyse cartographique et dendrochronologique. *Canadian Journal of Botany* 66(11): 2217–2227. [In French].

Laroque, S. J., Hétu, B. & Filion, L. (2001). Geomorphic and dendrochronological impacts of slush-flows in Central Gaspé Peninsula (Québec, Canada). *Geografiska Annaler*, 83A: 191–201.

Larsen, D. E. (1983). *Erosion of perennially frozen streambanks*. Hanover, New Hampshire. U.S. Army Cold Regions Research and Engineering Laboratory, CRREL Report 83-29. 26p.

Lato, M.J., Diederichs, M.S., Hutchinson, D.J., & Harrap, R. (2012). Evaluating roadside rockmasses for rockfall hazards using LiDAR data: optimizing data collection and processing protocols. *Natural Hazards*, 60(3): 831–864.

Lato, M., Hutchinson, J., Diederichs, M., Ball, D. & Harrap, R. (2009). Engineering monitoring of rockfall hazards along transportation corridors: using mobile terrestrial LiDAR. *Natural Hazards Earth System Science*, 9: 935–946. http://dx.doi.org/10.5194/nhess-9-935-2009.

Lato, M. J., Gauthier, D. & Hutchinson, D. J. (2015). Selecting the Optimal 3D Remote Sensing Technology for the Mapping, Monitoring and Management of Steep Rock Slopes Along Transportation Corridors. In: *Transportation Research Board, 94th Annual Meeting No. 15-3055.*

Lato, M. J., Hutchinson, D. J., Gauthier, D., Edwards, T. & Ondercin, M. (2014). Comparison of airborne laser scanning, terrestrial laser scanning, and terrestrial photogrammetry for mapping differential slope change in mountainous terrain. *Canadian Geotechnical Journal*, 52: 1–12.

Lauriol. B. & Gray, J. T. (1990). Drainage karstique en milieu de pergélisol: le cas de l'île d'Akpatok, T. N. O., Canada. *Permafrost and Periglacial Processes*, 1: 129–144.

Laursen, L. (2010). Climate scientists shine light on cave ice. *Science*, 239: 746–747.

Lautala, P., Harris, D., Ahtborn, T., Alkire, B. & Hodel, R. (2012). *Synthesis of railroad engineering best practices in areas of deep seasonal permafrost and permafrost*. Final report. Michigan Tech. Transportation Institute. 255p.

Lawler, D. M. (1988a). A bibliography of needle ice. *Cold Regions Science and Technology*, 15: 295–310.

Lawler, D. M. (1988b). Environmental limits of needle ice: a global survey. *Arctic and Alpine Research*, 20: 137–159.

Lawler, D. M. (1993). Needle ice processes and sediment mobilization on river banks: the River Ilston, West Glamorgan, UK. *Journal of Hydrology*, 150: 81–114.

Lawrence, D. M. & Swenson, S. C. (2011). Permafrost response to increasing Arctic Shrub abundance depends on the relative influence of shrubs on local soil cooling versus large-scale climate warming. *Environment Research Letters*, 6: 045504. Doi:10.1099/1748-9326/6/4/044504.

Lawson, D. (1983). Ground ice in perennially frozen sediments, Northern Alaska. In: *Permafrost: Proceedings of the 4th International Conference, Fairbanks, Alaska*. Washington, D.C. National Academy Press: 695–700.

Leaf, C. F. & Martinelli, Jr. M. (1977). *Avalanche dynamics engineering applications for land use planning*. Port Collins, Colorado. Rocky Mountain Forest and Range Experimental Station. U.S. Forest Service Research Paper RM-183. 51p.

Lebarge, M.-J. & Payette, S. (1995). Long-term monitoring of permafrost chamge in a palsa peatland in northern Québec, Canada: 1983–1993. *Arctic and Alpine Research*, 27(2): 167–171.

Lee, I. W., Lee, S. J., Lee, S. H. & Shin, H. Y. (2013). Innovative restoration method for concrete track settlement. *Sciences in Cold and Arid Regions*, 5(4): 461–467.

Leffingwell, E. de K. (1915). Ground-ice wedges; The dominant form of ground-ice on the north coast of Alaska. *Journal of Geology*, 23: 635–654.

Leffingwell, E. de K. (1919). *The Canning River Region, Northern Alaska*. United States Geological Survey, Professional Paper 109. 251p.

Legget, R. F., Brown, R. J. E. & Johnston, G. H. (1966). Alluvial fan formation near Aklavik, Northwest Territories, Canada. *Bulletin of the Geological Society of America*, 77: 15–29.

Leibman, M. O. (1995). Cryogenic landslides on the Yamal Peninsula, Russia: Preliminary observations. *Permafrost and Periglacial Processes*, 6: 259–264.

Leibman, M. O. & Egorov, I. P. (1996). Climate and environmental controls of cryogenic landslides, Yamal, Russia. In: Senneset, K. (ed.). *Proceedings of the 7th International Symposium on Landslides, Trondheim.* Rotterdam. A. A. Balkema: 1941–1946.

Leibman, M. O., Kizakov, A. I., Sulerzhitsky, L. D. & Zaretskaia, N. E. (2003). Dynamics of landslide slopes and their development on Yamal Peninsula. In: Phillips, M., Springman, S. M. and Arenson, L. U. (eds.). *Permafrost, Proceedings of the 8th International Conference on Permafrost. Lisse.* Swets and Zeitlinger: 651–656.

Leibman, M. O., Khomutov, A. & Kizyakov, A. (2014). Cryogenic Landslides in the West-Siberian Plain of Russia: Classification, Mechanisms and Landforms. In: Shan, W., Guo, Y., Wang, F., Marui, H. and Strom, A., (eds.). *Landslide in Cold Regions in the context of climate change, Environmental Science and Engineering.* Switzerland. Springer International Publishing. 310p. DOI: 10.1007/978-3-319-00867-7_11.

Leibman, M. O., Rivkin, F. M. & Saveliev, V. S. (1993). Hydrogeological aspects of cryogenic slides on the Yamal Peninsula. *Proceedings of the 6th International Conference on Permafrost, Beijing.* Wushan, Guangzhou, China. South China University of Technology Press 1: 380–382.

Lemmon, D. S., Duk-Rodkin, A. & Bednarski, J. M. (1994). Late glacial drainage systems along the northwest margin of the Laurentide ice sheet. *Quaternary Science Reviews,* 13: 805–828.

Leshenhikov, F. N. & Ryashchenko, T. G. (1973). *Izmeneniye sostava i svoystv glinistykh gruntov pri promerzanii.* In: Akademiya Nauk SSSR, Sektsiya Nauk o Zemle, Sibirskoye Otdeleniye. II. Mezhdunarodnaya Konferentsiya po Merzlotovedeniyu, Doklady I soobshcheniya, 3. Yakutsk, Yakutskoye Knizhnoye Izdatel'stovo. 102p. [In Russian].

Leshenhikov, F. N. & Ryashchenko, T. G. (1978). Changes in composition and properties of clay soils during freezing. In: Sanger, F. J. (ed.). *USSR Contribution,* Permafrost 2nd International Conference, Yakutsk, USSR. Washington, National Academy of Sciences: 201–213.

Letavernier, G. & Ozouf, J.-C. (1987). La gélifraction des roches et des parois calcaires. *Bulletin de l'Association Française pour l'Etude du Quaternaire,* 3: 139–145. [In French].

Levy, J. S., Fountain, A. G., Gooseff, M. N., Welch, K. A. & Lyons, W. B. (2011). Water tracks and permafrost in Taylor Valley, Antarctica: Extensive and shallow groundwater connectivity in a cold desert ecosystem. *Geological Society of America Bulletin,* 123(11/12): 2259–2311.

Levy, M. & Miller, S. L. (1998). The stability of the RNA bases: Implications for the origin of life. *Biochemistry,* 95: 7933–7938.

Lewis, C. A. (1994). Protalus ramparts and the altitude of the local equilibrium line during the last glacial stage in Bokspruit, East Cape Drackensberg, South Africa. *Geografiska Annaler,* 76A: 37–48.

Lewis, C. A. & Hanvey, P. M. (1993). The remains of rock glaciers in Bottleneck, East Cape Drakensberg, South Africa. *Transactions of the Royal Society of Africa,* 48: 265–289.

Lewkowicz, A. G. (1987). Headwall retreat of ground-ice slumps, Banks Island, Northwest Territories. *Canadian Journal of Earth Sciences,* 24: 1077–1085.

Lewkowicz, A. G. (1990). Morphology, frequency and magnitude of active-layer detachment slides, Fosheim Peninsula, Ellesmere Island, N.W.T. Proceedings of the 5th Canadian Permafrost Conference. *Nordicana:* 54: 111–118.

Lewkowicz, A. G. (1992). Factors influencing the distribution and initiation of active-layer detachment slides on Ellesmere Island, Arctic Canada. In: Dixon, J. C. and Abrahams, A. D. (eds.). *Periglacial Geomorphology.* Proceedings of a Symposium in Geomorphology. Chichester. John Wiley and Sons: 223–250.

Lewkowicz, A. G. (2011). Slope hummock development, Fosheim Peninsula, Ellesmere Island, Nunavit, Canada. *Quaternary Research,* 75: 334–346.

Lewkowicz, A. G. & Coultish, T. L. (2004). Beaver damming and palsa dynamics in a Subarctic mountainous environment, Wolf Creek, Yukon Territory, Canada. *Arctic, Antarctic and Alpine Research*, 36(2): 208–218.

Lewkowicz, A. G. & Ednie, M. (2004). Probability mapping of mountain permafrost using the BTS method., Wolf Creek, Yukon Territory, Canada. *Permafrost and Periglacial Processes*, 15: 67–80.

Lewkowicz, A. G., Etzelmüller, B. & Smith, S. (2011). Characteristics of discontinuous permafrost based on ground temperature measurements and electrical resistivity tomography, Southern Yukon Territory. *Permafrost and Periglacial Processes*, 22: 320–342.

Lewkowicz, A. G. & French, H. M. (1982a). The hydrology of small runoff plots in an area of continuous permafrost, Banks Island. *In: Proceedings of the 4th Canadian Permafrost Conference, Calgary.* Ottawa, National Research Council of Canada: 151–162.

Lewkowicz, A. G. & Gudjonsson, K. A. (1992). Slope hummocks on Fosheim Peninsula Northwest Territories. *Geological Survey of Canada, Current Research*, Part B. Paper 92-1B: 97–102.

Lewkowicz, A. G. & Harris, C. (2005a). Morphology and geotechnique of active-layer detachment failures in discontinuous and continuous permafrost, Northern Canada. *Geomorphology*, 69: 275–297.

Lewkowicz, A. G. & Harris, C. (2005b). Frequency and magnitude of active-layer detachment failures in discontinuous and continuous permafrost, Northern Canada. *Permafrost and Periglacial Processes*, 16: 115–130.

Li, C. M., Zhang, X. F., Zhao, L., Cheng, G. D. & Xu, S. J. (2012). Phylogenetic diversity of bacterial isolates and community function in permafrost-affected soil along different vegetation types in the Qinghai-Tibet Plateau. *Journal of Glaciology and Geocryology*, 34: 713–725.

Li, G. Y., Sheng, Y., Jin, H. J., Ma, W. & Wen, Z. (2009). Thermal Characteristics of oil and permafrost along the proposed China-Russia Crude Oil Pipeline. In: Moores, H. D. and Hinzman, J., Jr, (Eds.). Cold Regions Engineering 2009. Proceedings of the 14th Conference on Cold Regions Engineering, Duluth. Reston, Virginia. American Society of Civil Engineers: 226–241.

Li, G. H., Yu, G. H., Ma, W., Chen, Z., Mu, Y. H., Guo, L. & Wang, F. (2015). Freeze-thaw properties and long-term thermal stability of the unprotected tower foundation soils in permafrost regions along the Qinghai-Tibet power transmission line. *Cold Regions Research and Technology*. http://doi: org/10.1016/j.cold regions.2015.05.004

Li, H. J. (2009). Study on construction technology for tunnelling in the Plateau permafrost region. *Proceedings of the 14th Conference on Cold Regions Engineering. Cold Regions Impact on Research, Design and Construction, Duluth, Minnesota.* American Society of Civil Engineers: 16–21.

Li, J., Sheng, Y., Jiao, S. & Yang, G. (2009). Analysis of factors affecting the development of alpine permafrost in central-eastern Quilanshan Mountains, Northwest China. 978-1-4244-3395-7/09/$25.00©2009IEEE.

Li, S. (1996). Characteristics of existing glacial development in the Hoh Xil Region, Qinghai-Tibet Plateau. *Scientia Geographica Sinica*, 16(1): 10–17. [In Chinese].

Li, S. (2005). Discussion on the anti-melting techniques for the construction of the tunnel embedded in long-term frozen ground. *Modern Tunnelling Technology*, 42(6).

Li, Shude & Cheng Guodong (1996). *Map of Frozen Ground on the Qinghai-Xizang (Tibet) Plateau.* Lanzhou. Ganzu Culture Press. [In Chinese].

Li Shuzun, Cheng Guodong & Guo Dongxin (1996). The future thermal regime of numerically simulated permafrost on the Qinghai-Xizang (Tibet) Plateau under climate warming. *Science in China (Series D)*, 39: 434–441. [In Chinese].

Li. Y. M., Li, H. S., Liu, Z. L. & Chen, L. S. (2011). Incident analyses of frost heaving failure of municipal underground gas pipelines in cold regions in northern China. *Sciences in Cold and Arid Regions*, 3(6): 473–477.

Lide, D. R. (ed.) (2005). *CRC Handbook of Chemistry and Physics*. (86th edition). Boca Raton Florida. CRC Press. ISBN 0-8493-0486-5.

Liestøl, O. (1977). Pingos, springs and permafrost in Spitzbergen. *Norsk. Polarinstitutt Årbok*, 1975: 7–29.

Linder, L. & Marks, L. (1985). Types of debris slope accumulations and rock glaciers in South Spitzbergen. *Boreas*, 14: 139–153.

Lindsay, J. D. & Odynsky, W. (1965). Permafrost in organic soils of Northern Alberta. *Canadian Journal of Soil Science*, 45: 265–269.

Linell, K. A. (1973). Risk of uncontrolled flow from wells through permafrost. *Proceedings of the 2nd International Conference on Permafrost, Yakutsk, USSR*. Washington. North American Contribution: 462–468.

Linell, K. A. & Johnston, G. H. (1973). Principles of engineering design and construction in Permafrost regions. In: *Permafrost, North American Contribution, 2nd International Conference on Permafrost, Yakutsk*. Washington, National Academy of Sciences: 553–575.

Linell, K. A. & Kaplar, C. W. (1959). The factor of soil and material type in frost action. *U.S. Highway Research Board Bulletin*, 225: 81–126.

Linnell, K. A. & Kaplar, C. W. (1963). Description and classification of frozen soils. *Proceedings of the First International Conference on Permafrost, Lafayette, Indiana, USA*. Washington DC: National Academy of Sciences, National Research Council Publication 1287: 481–486.

Linell, K. A. & Lobacz, E. F. (1978). Some experiences with tunnel entrances in permafrost. *Proceedings of the 3rd International Conference on Permafrost, Edmonton*. Ottawa. National Research Council of Canada 1: 813–819.

Ling, F., Wu, Q., Zhang, T. & Niu, F. (2012). Modelling open-talik formation and permafrost lateral thaw under a thermokarst lake, Beiluhe Basin, Qinghai-Tibet Plateau. *Permafrost and Periglacial Processes*, 23(4): 312–321.

Liston, G. E. & Hall, D. K. (1995). Sensitivity of lake freeze-up and break-up to climate change: a physically based modeling study. *Annals of Glaciology*, 21: 387–393.

Liu, B., Jin, H. L., Sun, Z., Su, Z. Z. & Zhang, C. X. (2012). Geochemical evidences of dry climate in the Mid-Holocene in Gonghe Basin, northeastern Qinghai-Tibet Plateau. *Sciences in Cold and Arid Regions*, 4(6): 472–483.

Liu, H. J., Fan, C. B., Cheng, D. X., *et al.* (2008). The route, site and foundation selection in frozen ground region of Qinghai-Tibet Direct Current Transmission Line. *Electric Power Survey and Design*, 2: 12–16.

Livingston, J. M. (2004). *Floodbed Sedimentology: A new method to reconstruct paleo-ice-jam flood frequency*. Unpublished M. Sc. Thesis, Department of Geography, University of Calgary. 159p.

Livingston, H. & Johnson, E. (1978). Insulated roadway subdrains in the Subarctic for the prevention of spring icings. *Proceedings of the Conference on Applied Techniques for Cold Environments, Anchorage. American Society of Civil Engineers*, 1: 474–487.

Livingstone, C. W. (1956). *Excavations in frozen ground, Part I. Explosion tests in Keweenaw silt*. U.S. Army, Cold Regions Research and Engineering Laboratory (SIPRE). Hanover, New Hampshire. Technical Report TR 30. 97p.

Livingston, D. A. (1954). On the orientation of lake basins. *American Journal of Science*, 252: 547–554.

Lobacz, E. F. & Quinn, W. F. (1966). Thermal regime beneath buildings constructed on permafrost. *Proceedings of the 1st Conference on Permafrost, Lafayette, Indiana*. Washington, D.C. National Academy of Sciences Publication #1287: 247–252.

Lokrantz, H., Ingólfsson, Ó. & Forman, S. L. (2003). Glaciotectonised Quaternary sediments at Cape Shpindler, Yugorski Peninsula, Arctic Russia: implications for glacial history, ice movements and Kara Sea Ice configuration. *Journal of Quaternary Science*, 18(6): 527–543.

Long, E. L. (1966). The Long thermopile. *Proceedings of the 1st International Conference on Permafrost, Lafayette, Indiana*. Washington, D.C. National Academy Press Pulication # 1287: 487–491.

Lopez, C. M. L., Brouchkov, A., Nakayama, H., Takakai, F., Federov, A. N. & Fukuda, M., 2006. Epigenetic salt accumulation and water movement in the active layer of Central Yakutia in Eastern Siberia. *Hydrological Processes*, 21(1): 103–109.

Lorrain, R. D. & Demeur, P. (1985). Isotopic evidence for relic Pleistocene glacier ice on Victoria Island, Canadian Arctic Archipelago. *Arctic and Alpine Research*, 17: 89–98.

Lötschert, W. (1972). Über die vegetation frostgeformter Böden auf Island. *Natur. und Muscum*, 102: 1–12. [In German].

Low., W. W. (1925). Instability of viscous fluid motion. *Nature*, 115(2887): 299–300.

Loziński, W. (1909). Über die mechanische Verwitterung der Sandsteine im germässigten Klima. *Bulletin de l'International Science de Cracovie*. [In German].

Loziński, W. (1912). Die periglaciale fazies der mechanischen Verwitterung. *Compte Rendu, XI International Géological Congress, Stockholm, 1910*. [In German].

Lu, Z.Q., Zhu, Y.H., Zhang, Y.Q., Wen, H.J., Li, Y.H., & Liu, C.L. (2011), Gas hydrate occurrences in the Qilian Mountain permafrost, Qinghai Province, China, Cold Regions Science and Technology, 66: 93–104. DOI:10.1016/j.coldregions2011.01,008.

Lubomirov, A. S. (1987). Origin and development of some lakes of Andir Lowlands. In: *Natural conditions of digested regions of Siberia*. Yakutsk, Permafrost Institute, 89–99. [In Russian].

Luckman, B. H. (1977). The geomorphic activity of snow avalanches. *Geographiska Annaler*, 59A: 31–46.

Luckman, B. H. (1978a). Geomorphic work of snow avalanches in the Canadian Rocky Mountains. *Arctic and Alpine Research*, 10: 261–276.

Luckman, B. H. (1978b). Debris accumulation patterns on talus slopes in Surprise valley, Alberta. *Géographie physique et Quaternaire*, 42(3): 247–278.

Luckman, B. H. (1992). Debris flows and snow avalanche landforms in the Lairig Ghru, Cairngorm Mountains, Scotland. *Geografiska Annaler*, 74A: 109–121.

Luckman, B. H. & Crockett, K. J. (1978). Distribution and characteristics of rock glaciers in the southern part of Jasper National Park, Alberta. *Canadian Journal of Earth Sciences*, 15: 540–550.

Liu, M.-H., Niu, F.-J., Fang, J.-H., Lin, Z.-J., Luo, J. & Yin, G.-A. (2014). In-situ testing study on convection and temperature characteristics of a new crushed-rock slope embankment design in a permafrost region. *Sciences in Cold and Arid Regions*, 6(4): 378–387.

Lundqvist, J. (1962). *Patterned ground and related frost phenomena in Sweden*. Sveriges Geologiska Undersönkning Årsbok, 55(7): 101p.

Lundqvist, J. (1969). Earth and ice mounds: a terminological discussion. In: Péwé, T. L., (Ed.). *The periglacial environment*. Montreal. McGill-Queen's University Press: 203–215.

Luoto, M. & Seppälä, M. (2002). Characteristics of earth hummocks (pounus) with and without permafrost in Finnish Lapland. *Geografiska Annaler*, 84A(2): 127–136.

Luscher, U., Black, W. T. & Nair, K. (1975). Geotechnical aspects of Trans-Alaska pipeline. *American Society of Civil Engineers, Journal of the Transportation Engineering Division*, 101 (TE4): 669–680.

Lüthi, M. P. & Funk, M. (2001). Modelling heat flow in a cold, high altitude glacier: Interpretation of measurements from Colle Gnifetti, Swiss Alps. *Journal of Glaciology*, 47(157): 314–324.

Lyazgin, A. L., Baysasan, R. M., Chisnik, S. A., *et al*. (2003). Stabilization of pile foundation subjected to frost heave and in thawing permafrost. *Proceedings of the 8th International Conference on Permafrost*. Lisse. Swets and Zeitlinger: 707–711.

Lyazgin, A. L., Lyashenko, V. S., Ostroborodov, S. V., *et al*. (2004a). Experience in the prevention of frost heave of pile foundations of transmission towers under northern conditions. *Power Technology and Engineering*, 38(2): 124–126.

Lyazgin, A. L., Ostroborodov, S. V., Pustovoit, G. P., *et al*. (2004b). Leveling of pile foundations supporting electric transmission lines by temperature control of bed soils. *Soil Mechanics and Foundation Engineering*, 41(1): 23–26.

Lyov, A. V. (1916). *Search and exploration for water sources along the western part of the Amur Railroad in permafrost conditions*. Irkutsk. Irkutsk Books Publishers. [In Russian].

Ma, H. T., Song, X. J. & Wang, Q. L. (2013). Foundation construction techniques of steel tower cast-in-situ concrete in permafrost regions. *Qinghai Electric Power*, 32(1): 25–29.

Ma, W., Wu, Q.B., Liu, Y. & Bing, H. (2008). Analysis of the cooling mechanism of a crushed rock embankment in warm and lower temperature permafrost regions along the Qinghai-Tibet Railway. *Sciences in Cold and Arid Regions, Initial Issue*: 14–25.

Mabliudov, B. R. (1985). Regularities of ice cave spreading. *Data of Glaciological Studies*, Academy of Science, USSR 54: 193–200.

MacFarlane, I. C. (ed.). (1969). *Muskeg Engineering Handbook*. Toronto. University of Toronto Press. 320p.

Mackay, J. R. (1953). Fissures and mud circles on Cornwallis Island, N.W.T. *Canadian Geographer*, 3: 31–37.

Mackay, J. R. (1956). Notes on oriented lakes in the Liverpool Bay area, N.W. Territories. *Revue Canadienne de Géographie* 10: 169–173.

Mackay, J. R. (1958). *A subsurface organic layer associated with permafrost in the western Arctic*. Canadian Department of Mines and Technical Surveys, Geographical Branch, Geographical Paper 18: 21p.

Mackay, J. R. (1963). *The Mackenzie Delta Area*. Geographical Branch Memoir #8, 202p.

Mackay, J. R. (1965). Gas-domed mounds in permafrost, Kendall Island, N.W.T. *Geographical Bulletin*, 7: 105–115.

Mackay, J. R. (1966). Segregated epigenetic ice and slumps in permafrost, Mackenzie Delta area, N.W.T. *Geographical Bulletin*. 8: 59–80.

Mackay, J. R. (1970). Disturbances to the Tundra and Forest Tundra environment in the Western Arctic. *Canadian Geotechnical Journal*, 7: 420–432.

Mackay, J. R. (1971a). Ground ice in the active layer and the top portion of the permafrost. *Proceedings of a seminar on the permafrost active layer*. Ottawa, National Research Council, Technical Memoir 103: 26–30.

Mackay, J. R. (1971b). Origin of massive icy beds in permafrost, western Arctic coast. *Canadian Journal of Earth Sciences*, 8: 397–422.

Mackay, J. R. (1972a). Offshore permafrost and ground ice, Southern Beaufort Sea, Canada. *Canadian Journal of Earth Sciences*, 9: 1550–1561.

Mackay, J. R. (1972b). The world of underground ice. *Annals of the Association of American Geographers*, 62: 1–22.

Mackay, J. R. (1973a). Problems in the origin of massive icy beds, western Arctic, Canada. *North American Contribution, Permafrost*. Second International conference, Yakutsk, USSR. Washington, National Academy of Sciences: 223–228.

Mackay, J. R. (1973b). A frost tube for the determination of freezing in the active layer above permafrost. *Canadian Geotechnical Journal*, 10: 392–396.

Mackay, J. R. (1973c). The growth of pingos, western Arctic coast, Canada. *Canadian Journal of Earth Sciences*, 10: 979–1004.

Mackay, J. R. (1974a). Ice-wedge cracks, Garry Island, North-West Territories. *Canadian Journal of Earth Sciences*, 11: 1366–1383.

Mackay, J. R. (1974b). Measurement of upward freezing above permafrost with a self-positioning thermistor probe. *Geological Survey of Canada*, Paper 74-1B: 250–254.

Mackay, J. R. (1974c). Reticulate ice veins in permafrost, Northern Canada. *Canadian Geotechnical Journal*, 11: 230–237.

Mackay, J. R. (1975a). The closing of ice-wedge cracks in permafrost, Garry Island, Northwest Territories. *Canadian Journal of Earth Sciences*, 12: 1668–1674.

Mackay, J. R. (1975b). The stability of permafrost and recent climatic change in the Mackenzie Valley, N.W.T. *Geological Survey of Canada, Report of Activities Part B*, Paper 75-1B: 173–176.

Mackay, J. R. (1977). Changes in the active layer from 1968–1976 as a result of the Inuvik fire. *Geological Survey of Canada*, Paper 77-1B: 273–275.

Mackay, J. R. (1978). The use of snow fences to reduce ice-wedge cracking, Garry Island, Northwest Territories. *Geological Survey of Canada*, Paper 78-1A: 523–524.

Mackay, J. R. (1979a). Pingos of the Tuktoyaktuk Peninsula area, Northwest Territories. *Géographie physique et Quaternaire*, 33: 3–61.

Mackay, J. R. (1979b). An equilibrium model for hummocks (non-sorted circles), Garry Island, Northwest Territories. *Geological Survey of Canada*, Paper 79-1A: 165–167.

Mackay, J. R. (1980a). Illisarvik: An experiment in lake drainage. In: *Proceedings of the Symposium on Permafrost Geophysics*. National Research Council of Canada, Associate Committee on Geotechnical Research, Technical Memorandum #128: 1–4.

Mackay, J. R. (1980b). The origin of hummocks, western Arctic coast. *Canadian Journal of Earth Sciences*, 17: 996–1006.

Mackay, J. R. (1981a). Active layer slope movement in a continuous permafrost environment, Garry Island, Northwest Territories, Canada. *Canadian Journal of Earth Sciences*, 19: 1666–1680.

Mackay, J.R. (1981b). An experiment in lake drainage, Richards Island, Northwest Territories: a progress report. *Geological Survey of Canada*, Paper 81-1A: 63–68.

Mackay, J. R. (1982). Active layer growth, Illisavik experimental drained lake site, Richards island, Northwest Territories. *Geological Survey of Canada*, Paper 82-1A: 123–126.

Mackay, J. R. (1983). Downward water movement into frozen ground, western Arctic coast, Canada. *Canadian Journal of Earth Sciences*, 20: 120–134.

Mackay, J. R. (1984a). The frost heave of stones in the active layer above permafrost with downward and upward freezing. *Arctic and Alpine Research*, 16: 413–417.

Mackay, J. R. (1984b). The direction of ice-wedge cracking in permafrost: downward or upward? *Canadian Journal of Earth Sciences*, 21: 516–524.

Mackay, J. R. (1986a). Growth of Ibyuk pingo, Western Canadian Arctic, Canada, and some implications for environmental reconstructions. *Quaternary Research*, 26: 68–80.

Mackay, J. R. (1986b). The first seven years (1978–1985) of ice wedge growth, Illisarvik experimental drained lake site, western Arctic coast. *Canadian Journal of Earth Sciences*, 23(11): 1782–1795.

Mackay, J. R. (1988). Ice wedge growth in newly aggrading permafrost, western Arctic coast, Canada. In: *Permafrost, 5th International Conference, Proceedings, Trondheim, Norway*. Trondheim. Tapir Press 1: 809–814.

Mackay, J. R. (1989a). Massive ice: some field criteria for the identification of ice types. *Geological Survey of Canada*, Paper 89-1G: 5–11.

Mackay, J. R. (1989b). Ice-wedge cracks, western Arctic coast. *Canadian Geographer*, 33(4): 365–368.

Mackay, J. R. (1990a). Seasonal growth bands in pingo ice. *Canadian Journal of Earth Sciences*, 27(8): 1115–1125.

Mackay, J. R. (1990b). Some observations on the growth and deformation of epigenetic, syngenetic and anti-syngenetic ice wedges. *Permafrost and Periglacial Processes*, 1: 15–29.

Mackay, J. R. (1992). The frequency of ice-wedge cracking (1967–1987) at Garry Island, western Arctic coast, Canada. *Canadian Journal of Earth Sciences*, 29(2): 236–248.

Mackay, J. R. (1993a). The sound and speed of ice-wedge cracking, Arctic Canada. *Canadian Journal of Earth Sciences*, 30: 509–518.

Mackay, J. R. (1993b). Air temperature, snow cover, creep of frozen ground, and the time of ice-wedge cracking, western Arctic Coast. *Canadian Journal of Earth Sciences*, 30: 1720–1729.

Mackay, J. R. (1995). Active-layer changes (1968–1993) following the forest-tundra fire near Inuvik, NWT, Canada. *Arctic and Alpine Research*, 27: 323–336.

Mackay, J. R. (1997). A full-scale field experiment (1978–1995) on the growth of permafrost by means of lake drainage, western Arctic coast; a discussion of the method and some results. *Canadian Journal of Earth Sciences*, 34(1): 17–33.

Mackay, J. R. (1998). Pingo growth and collapse, Tuktoyaktuk Peninsula area, Western Arctic coast, Canada; A long-term field study. *Géographie physique et Quaternaire*, 52(3): 271–323.

Mackay, J. R. (1999). Cold-climate shattering (1974–1993) of 200 glacial erratics on the exposed bottom of a recently drained Arctic lake, western Arctic coast, Canada. *Permafrost and Periglacial Processes*, 10: 125–136.

Mackay, J. R. (2000). Thermally induced movements in ice-wedge polygons, western Arctic coast: A long-term study. *Géographie physique et Quaternaire*, 54(1): 41–68.

Mackay, J. R. & Black, R. F. (1973). Origin, composition and structure of perennially frozen ground and ground ice: A review. *Permafrost: North American Contribution, 2nd International Conference*, Washington, D.C., National Academy of Sciences: 185–192.

Mackay, J. R. & Burn, C. R. (2002). The first 20 years (1978–1979 to 1998–1999), experimental drained lake site, western Arctic coast, Canada. *Canadian Journal of Earth Sciences*, 39(1): 95–111.

Mackay, J. R. & Burrows, C. (1979). Uplift of objects by an upfreezing surface. *Canadian Geotechnical Journal*, 17: 609–613.

Mackay, J. R. & Dalimore, S. R. (1992). Massive ice of the Tuktoyaktuk area, western Arctic coast, Canada. *Canadian Journal of Earth Sciences*, 29: 1235–1249.

Mackay, J. R., Konishchev, V. N. & Popov, A. I. (1978). Geological controls of the origin, characteristics, and distribution of ground ice: A review. In: *3rd International Conference on Permafrost, Proceedings*. Ottawa. National Research Council of Canada 2: 1–18.

Mackay, J. R. & Leslie, R. V. (1987). A simple probe for the measurement of frost heave within frozen ground in a permafrost environment. *Geological Survey of Canada*, Paper 87-1A: 37–41.

Mackay, J. R. & MacKay, D. K. (1974). Snow cover and ground temperatures, Garry Island, N.W.T. *Arctic*, 27: 288–296.

Mackay, J. R. & MacKay, D. K. (1976). Cryostatic pressures in nonsorted circles (mud hummocks), Inuvik, Northwest Territories. *Canadian Journal of Earth Sciences*, 13: 889–897.

Mackay, J. R. & Mathews, W. H. (1973). Geomorphology and Quaternary history of the Mackenzie River Valley near Fort Good Hope, N.W.T., Canada. *Canadian Journal of Earth Sciences*, 19: 26–41.

Mackay, J. R. & Mathews, W. H. (1974a). Needle ice striped ground. *Arctic and Alpine Research*, 6(1): 79–84.

Mackay, J. R. & Mathews, W. H. (1974b). Movement of sorted stripes, the Cinder Cone, Garibaldi Park, B. C., Canada. *Arctic and Alpine Research*, 6(4): 347–359.

Mackay, J. R., Ostrick, J., Lewis, C. P. & MacKay, D. K. (1979). Frost heave at ground temperatures below 0°C, Inuvik, Northwest Territories. *Geological Survey of Canada*, Paper 79-1A: 403–406.

Mackay, J. R., Rampton, V. N. & Fyles, J. G. (1972). Relic Pleistocene permafrost, Western Arctic, Canada. *Science*, 176: 1321–1323.

Mackay, J. R. & Stager, J. K. (1966a). The structure of some pingos in the Mackenzie delta area, N.W.T. *Geographical Bulletin*, 8: 360–368.

Mackay, J. R. & Stager, J. K. (1966b). Thick tilted beds of segregated ice, Mackenzie Delta area, N.W.T. *Biuletyn Peryglacjalny*, 15: 39–43.

Mackay, J. R. & Terasmae, J. (1963). Pollen diagrams in the Mackenzie Delta area, N.W.T. *Arctic*, 16: 229–238.

MacDonald, G. J. (1990). Role of methane clathrates in past and future climates. *Climate Change*, 16: 247–281.

MacPhee, R. D. E. (1999). *Extinctions in near time*. New York. Kluwer Academic/Plenum.

Maejima, J. (1977). Global pattern of temperature lapse rate in the lower troposphere with special reference to the altitude of snow line. *Geographical Reports of Tokyo Metropolitan University*, 12: 117–126.

Maggioni, M., Freppaz, M., Piccini, P., Williams, M. W. & Zanini, E. (2009). Snow cover effects on Glacier Ice Surface Temperature. *Arctic, Antarctic, and Alpine Research*, 41(3): 323–329.

Mahaney, W. C. (1980). Late Quaternary rock glaciers, Mount Kenya, Kenya *Glaciology*, 25: 492–497.

Makogon, Yu. F. (1982). Perspectives for the development of gas hydrate deposits. *The Roger J. E. Brown Memorial Volume, Proceedings of the 4th Canadian Permafrost Conference, Calgary, Alberta*. Ottawa, National Research Council of Canada: 299–304.

Makoto, K. & Klaminder, J. (2012). The influence of non-sorted circles on species diversity of vascular plants, bryophytes and lichens in Sub-Arctic Tundra. *Polar Biology*. Doi: 10.1007/s00300-012-1206-3.

Malde, H. E. (1964). Patterned ground in the western Snake River Plain, Idaho, and its possible cold-climate origin. *Geological Society of America Bulletin*, 75: 191–208.

Manikian, V. (1983). Pile driving and load tests in permafrost for the Kuparuk pipeline system. *Permafrost, Proceedings of the 4th International Conference, Fairbanks, Alaska*. Washington, D.C. National Academy Press: 804–810.

Marangunic, C. (1976). El glaciar de roca Pedregosa, Rio Colorado, V region. *Congresso Geologica Chili Actas*, 1: D71–80. [In Spanish].

Marchant, D. R., Lewis, A. R., Phillips, W. M., Moore, E. J., Souchez, R. A., Denton, G. H., Sugden, D. E., Potter, N., Jr. & Landis, G. P. (2002). Formation of patterned ground and sublimation till over Miocene glacier ice in Beacon Valley, southern Victoria Land, Antarctica. *Geological Society of America Bulletin*, 114(6): 718–730.

Marchenko, S. S. & Gorbunov, A. P. (1997). Permafrost changes in the Northern Tien Shan during the Holocene. *Permafrost and Periglacial Processes*, 8: 427–435.

Margesin, R. (2009). *Permafrost Soils*. Berlin/Heidelberg, Springer Verlag, Germany. 348p.

Mark, A. F. (1994). Patterned ground activity in a southern New Zealand high-alpine cushion field. *Arctic and Alpine Research*, 26: 270–280.

Marker, M. E. & Whittington, G. (1971). Observation on some valley forms and deposits in Sani Pass area, Lesotho, South Africa. *South African Geographical Journal*, 53: 96–99.

Markov, K. K. (1973). *Cross-Section of the Newest Sediments*. Moscow. Moscow University Press. 198p.

Markvart, T. & Luis CastaŁżer, L. (2003). Practical Handbook of Photovoltaics: Fundamentals and Applications. Elsevier. (ISBN 1-85617-390-9).

Marquette, G. C., Gray, J. T., Courchesne, F. Stockli, L., Macpherson, G. & Finkel, R. (2004). Felsenmeer persistence under non-erosive ice in the Torngat and Kaumajet mountains, Quebec and Labrador, as determined by soil weathering and cosmogenic nuclide exposure. *Canadian Journal of Earth Sciences*, 41(1): 19–38.

Marshall, P. W. (1981). The formation and age of ice in caves. *Biuletyn Peryglacyalny*, 28: 79–84.

Martin, H. E. & Whalley, W. B. (1978). A glacier ice-cored rock glacier, Tröllaskahi, Iceland. *Jökull*, 37: 45–55.

Martin, P. S. (1984). Catastrophic extinctions and Late Pleistocene blitzkrieg: two radiocarbon tests. In: Nitecki, M. H. (Ed.). *Extinctions*. Chicago. University of Chicago Press: 153–189.

Martin, P. S. & Steadman, D. W. (1999). Prehistoric extinctions on islands and continents. In: MacPhee, R. D. E. (Ed.). *Extinctions in near time*. New York. Kluwer Academic/Plenum: 17–55.

Masters, A. M. (1990). Temporal and special change in forest fire history of Kootenay National Park, Canadian Rockies. *Canadian Journal of Botany*, 68: 1763–1767.

Mathewson, C. C. & Mayer-Cole, T. A. (1984). Development and runout of a detachment slide, Bracebridge Inlet, Bathhurst Island, Northwest Territories, Canada. *Bulletin of the Association of Engineering Geologists*, 21: 407–424.

Matsumoto, S. (1970). Block streams in the Kitakami Mountains – with special reference to Hitekamidake Area. *Scientific Reports of Tokoka University, 7th Series (Geography)*, 26: 221–235.

Matsuoka, N. (1990). The rate of bedrock weathering by frost action: field measurements and a predictive model. *Earth Surface Processes and Landforms*, 15: 73–90.

Matsuoka, N. (1996). Soil moisture variability in relation to diurnal frost heaving on Japanese High Mountain slopes. *Permafrost and Periglacial Processes*, 7: 139–151.

Matsuoka, N. (1998). The relationship between frost heave and downslope soil movement: Field measurements in the Japanese Alps. *Permafrost and Periglacial Processes*, 9: 121–133.

Matsuoka, N. (2001a). Direct observation of frost wedging in alpine bedrock. *Earth Surface Processes and Landforms*, 26: 601–614.

Matsuoka, N. (2001b). Solifluction rates, processes and landforms: a global review. *Earth-Science Reviews*, 55: 107–134.

Matsuoka, N. (2008). Frost weathering and rockwall erosion in the southeastern Swiss Alps: Long-term (1994–2006) observations. *Geomorphology*, 99: 353–368.

Matsuoka, N., Abe, M. & Ijiri, M. (2003). Differential frost heave and sorted patterned ground: Field measurement and a laboratory experiment. *Geomorphology*, 52: 73–85.

Matsuoka, N. & Hirakawa, K. (2000). Solifluction resulting from one-sided and two-sided freezing: Field data from Svalbard. *Polar Geoscience*, 13: 187–201.

Matsuoka, N., Hirakawa, K., Watanabe, T. & Moriwaki, K. (1997). Monitoring of periglacial slope processes in the Swiss Alps: the first two years of frost shattering, heave and creep. *Permafrost and Periglacial Processes*, 8: 155–177.

Matsuoka, N., Ikeda, A. & Date, T. (2005). Morphometric analysis of solifluction lobes and rock glaciers in the Swiss Alps. *Permafrost and Periglacial Processes*, 16(1): 99–113.

Matsuoka, N. & Morikawa, K. (1992). Frost heave and creep in the Sør Rondane Mountains, Antarctica. *Arctic and Alpine Research*, 24: 271–280.

Matsuoka, N. & Murton, J. (2008). Frost weathering: Recent advances and future directions. *Permafrost and Periglacial Processes*, 19: 195–210.

Matsuura, H., Lung, D. E. & Nakazawa, A. (2008). Commentary: Solid waste as it impacts community sustainability in Alaska. *Journal of Rural and Community Development*, 3(3): 108–122.

Matthews III, V. (1999). Origin of horizontal needle ice at Charit Creek Station, Tennessee. *Permafrost and Periglacial Processes*, 10: 205–207.

Matthews, J. A. (1974). Families of lichenometric dating curves from Storbreen gletschervorfeld, Jotenheimen, Norway. *Norsk Geografisk Tidsskrift*, 28: 215–235.

Matthews, J. A. (1975). Experiments on the reproducibility and reliability of lichenometric dates, Storbreen gltschervorfeld, Jotenheimen, Norway. *Norsk Geografisk Tidsskrift*, 29: 97–109.

Matthews, J. (1977). A lichenometric test of the 1750 hypothesis: Storbreen gltschervorfeld, southern Norway. *Norsk Geografisk Tidsskrift*, 31: 129–136.

Matthews, J. A., Dahl, S.-O., Berrisford, M. S. & Nesje, A. (1997). Cyclic development and thermokarstic degradation of palsas in the mid-Alpine zone at Leirpullan, Dovrefjell, southern Norway. *Permafrost and Periglacial Processes*, 8: 107–122.

Matveev, A. P. (1963). The dynamics and age of screes and stone rivers of the North Urals goletz zone in the Denezhkin Kamen massif. In: *Problems of the North*. Moscow. [In Russian].

Mavlyudov, B. (2008). Geography of caves glaciation. In: Kadebskaya, O. and Mavlyudov, B. (eds.). International Workshop on Ice Caves IWIC-III. Kunger Ice Cave, Perm Region, Russia: 35–44.

Mayewski, P. A. & Hassinger, J. (1980). Characteristics and significance of rock glaciers in southern Victoria Land, Antarctic. *Antarctic Journal of the U.S.*, 15: 68–69.

McCarroll, D. (1990). Differential weathering of feldspar and pyroxene in an Arctic-Alpine environment. *Earth Surface Processes and Landforms*, 15: 641–651.

McCarroll, D. & Vines, H. (1995). Rock-weathering by the lichen *Lecidea auriculata* in an arctic alpine environment. *Earth Surface Processes and Landforms*, 20: 199–2.

McColl, S. T. & Davies, T. R. H. (2013). Large ice-contact slope movements: glacial buttressing, deformation and erosion. *Earth Surface Processes and Landforms*, 8: 1102–1115.

McHattie, R. L. & Esch, D. C. (1983). Benefits of a peat underlay used in road construction on permafrost. *Permafrost; 4th International Conference*. Washington. National Academy Press: 826–831.

McNamara, J. P., Kane, D. L. & Hinzman, L. D. (1999). An analysis of an arctic channel network using a digital elevation model. *Geomorphology*, 29: 339–353.

McRoberts, E. C. & Morgenstern, N. R. (1974a). The stability of thawing slopes. *Canadian Geotechnical Journal*, 11: 447–469.

McRoberts, E. C. & Morgenstern, N. R. (1974b). Stability of slopes in frozen soil, Mackenzie Valley, N.W.T. *Canadian Geotechnical Journal*, 11: 554–573.

McRoberts, E. C. & Nixen, J. F. (1975). Reticulate ice veins in permafrost, Northern Canada: Discussion. *Canadian Geotechnical Journal*, 12: 159–162.

Mears, A. I. (1976). *Guidelines and methods for detailed snow avalanche hazard investigations in Colorado*. Denver. Colorado Geological Survey Bulletin #38. 125p.

Mears, B. (1981). Periglacial wedges in the Late Pleistocene environment of Wyoming's intermontane basins. *Quaternary Research*, 15: 171–198.

Meentemeyer, V. & Zippin, J. (1981). Soil moisture and texture controls of selected parameters of needle ice growth. *Earth Surface Processes and Landforms*, 6: 113–125.

Melnikov, P. I. (1962). About soil temperature changes for the last century in the Shargin Well in Yakutsk, and the longevity and persistence of the heat processes and recovery of disturbed permafrost temperatures. In: Melnikov, P. I. (ed.). *Permafrost and Permafrost Processes on the Territory of Yakutia SSR*. Moscow. Academy of Sciences of USSR: 54–67. [In Russian].

Melnikov, P. I. & Tolstikhin, N. I. (1974). Editors. *Obschcheye Merzlotovediye*. Izdatel'stvo "Nauka", Novosibirsk, Sibirskoye Otdeleniye. [In Russian].

Melnikov, V. P. & Spesiivtev, V. I. (2000). *Cryogenic formations in the Earth's Lithosphere*. Novosibirsk, Russia. Novosibirsk Scientific Publishing Center UIGGM, Siberian Branch, Russian Academy of Sciences. 343p. [In Russian and English].

Meltzer, D. J. & Mead, J. I. (1982). The timing of late Pleistocene mammalian extinctions in North America. *Quaternary Research*, 19: 130–135.

Meng, X.-G., Zhu, D.-G., Shao, Z.-G., Tu, J., Ha, J.-R. & Meng, Q.-W. (2004). The discovery and sense of the Quaternary period glacial traces on the northern part of Lüliang Mountain of Ningwu, Shanxi Province. *Journal of Geomechanics*, 10(4): 327–336. [In Chinese].

Meng, X.-G., Zhu, D.-G., Shao, Z.-G., Tu, J., Ha, J.-R., & Meng, Q.-W. (2006). A discussion of the formation mechanism of the "Ten-Thousand-Year-Old Ice Cave" in Shanxi Province. *Acta Geoscientica Sinica*, 27(2): 163–168. [In Chinese].

Messerli, B. (1972). Formen und Formungsprozesse in Hochgebirgsregion des Tibesti. *Hochgebirgforschung*, 2: 22–86. [In German].

Messerli, B. & Zurbuchen, M. (1968). Blockgletscher im Weissmies und Aletsch und ihre photogrammetrische Kartierung. *Sondersbruck aus dem Quartalschaft, 3, Die Alpen. 13p.* [In German].

Metz, M. C. (1984). Pipeline workpads in Alaska. *Permafrost. 4th International Conference, Fairbanks. Final Proceedings.* Washington, D.C. National Academy Press 106–108.

Metz, M. C., Krzewinski, T. G. & Clarke, E. S. (1982). The TransAlaska Pipeline System workpad – an evaluation of the present evidence. *Proceedings of the 4th Canadian Permafrost Conference, Calgary.* Ottawa. National Research Council of Canada: 523–534.

Meyer, K. (1999). *Horizontal directional drilling (HDD) Colvile River Crossing.* Presentation for the Alaska Arctic Pipeline Workshop, Anchorage, Alaska, 8–9th November, 1999. www.boemre.gov/tarworks/WorkshopPages/PipelineWorkshops/workshop%2025/Presentations/meyer1.pdf.

Michaelson, G. J., Ping, C. L., Epstein, H., Kimble, J. M. & Walker, D. A. (2008). Soils and frost boil systems across the North American Arctic Transect. Journal of Geophysical Research, 113, G03S11. Doi: 1029/2007JG000672.

Michel, B. (1971). The winter regimes of rivers and lakes. US Army Corps of Engineers, *Cold Regions Research and Engineering Laboratory*, Monograph III-B1a. 139p.

Middendorf, A. (1867–1978). A trip to the north and east of Siberia. Part 1 (1867) 242p; Part 2 (1878). 331p. St. Petersburg. Imperial Academy of Sciences. [In Russian].

Mikes, P. (2013). Giant Mine remediation project – Arsenic trioxide management. In: *Pan-Territorial Permafrost Workshop, November 5–7th, 2013.* Presentation for SRK Consulting.

Mikhailov, G. P. (1971). Temperature regime of embankment consisting of coarse rock on permafrost. *Transportation Construction*, 12: 32–33. [In Russian].

Militzer, B. & Wilson, H. F. (2010). New phases of water ice predicted at Megabar pressures. *Physical Review Letters*, 105: 195701.

Millar, C. I. & Westfall, R. D. (2008). Rock glaciers and related periglacial landforms in the Sierra Nevada, CA, USA; inventory, distribution and climatic relationships. *Quaternary International*, 188: 90–104.

Miller, D. (1998). Tibetan pastoralism: Hard times on the Plateau. *Chinabrief*, 1(2): 17–22.

Miller, G. H., Brigham-Grette, J., Alley, R. B., *et al.* (2010). Temperature and precipitation history of the Arctic. *Quaternary Science Reviews*, 29: 1679–1715.

Miller, J. M. (1971). Pile foundations in thermally fragile frozen soils. *Proceedings of the Symposium on Cold Regions Engineering, 1970, University of Alaska.* American Society of Civil Engineers 1: 34–72.

Miller, M., Kurylo, J. B. & Rykaart, M. (2013). Frozen Dams in permafrost regions. Lima, Peru. http://www.minewatersolutions.com

Miller, R. D. (1972). *Freezing and heaving of saturated and unsaturated soils.* Highway Research Record, 393: 1–11.

Miller, R. D. (1978). Frost heaving in non-colloidal soils. *Proceedings of the 3rd International Conference on Permafrost, Edmonton, Alberta.* Ottawa. National Research Council of Canada: 707–713.

Miller, R. D. (1984). Thermally induced vegetation: A qualitative discussion. *Proceedings of the 4th International Conference on Permafrost, Fairbanks.* Washington. National Academy Press: 61–63.

Miller, S. L. & Smythe, W. D. (1970). Carbon dioxide clathrate on the Martian ice cap. *Science*, 170: 531–533.

Minervin, A. V. (1982). The role of cryogenic processes in forming of loess deposits. *Problems in Cryolithology*, 10: 41–61. [In Russian].

Miyamoto, H. K. & Heinke, G. W. (1979). Performance evaluation of an Arctic sewage lagoon. *Canadian Journal of Civil Engineering*, 6: 324–328.

Mjagkov, S. M. (1980). Kammene gletcery transantarkticeskichgor. *Antarctica*, 20: 89–92. [In Russian].

Mochanov, Y. A., Fedoseev, S. A., Bland, R. L. & Carlson, R. L. (2008). Archaeology, the Paleolithic of Northeast Asia, a non-tropical origin for humanity, and the earliest stages of the settlement of America. *Canadian Journal of Archaeology*, 32(2): 285–288.

Moore, J. R., Egloff, J., Nagelisen, J., Hunziker, M., Aerne, U. & Christen, M. (2013). Sediment transport and bedrock erosion by wet snow avalanches in the guggigraben Matter Valley, Switzerland. *Arctic, Antarctic and Alpine Research*, 45(3): 350–362.

Moorman, B. J. (2003). Glacier-permafrost hydrology interactions, Bylot Island, Canada. In: Phillips, M., Springman, S. M. & Arenson, L. U. (eds). *8th International Permafrost Conference Proceedings*. Lisse. Swets and Zeitlinger: 783–788.

Moorman, B. J. & Michel, F. A. (2000). The burial of ice in a periglacial environment on Bylot Island, Arctic Canada. *Permafrost and Periglacial Processes*, 11: 161–175.

Morard, S., Delaloye, R. & Dorthe, J. (2008). Seasonal thermal regime of a mid-latitude ventilated debris accumulation. In: Kane, D. L. and Hinkel, K. M. (eds.). *Proceedings of the 9th International Conference on Permafrost*. Fairbanks. Institute of Northern Engineering, University of Alaska, 2: 1233–1238.

Morgan, A. V. (1972). Late Wiscosinan ice-wedge polygons near Kitchener, Ontario, Canada. *Canadian Journal of Earth Sciences*, 9: 607–617.

Morganstern, A., Ulrich, M., Günther, F., Roessler, S., Federova, I. V., Rudaya, N. A., Wetterich, S., Boike, J. & Schirrmeister, L. (2013). Evolution of thermokarst in East Siberian ice-rich permafrost: A case study. *Geomorphology*, 201: 363–379.

Morgenstern, N. R. & Nixon, J, F. (1971). One-dimensional consolidation of thawing soils. *Canadian Geotechnical Journal*, 8(4): 58–565.

Morgenstern, N. R., Thomson, S. & Mageau, D. (1978). Explosive cratering in permafrost: State of the Art. *Department of National Defence, Defence Research Establishment, Suffield, Ralston, Alberta*. Contract 8 SU 77-00015. 420p.

Morohashi, R., Anma, S. & Hanaoka, M. (2007). Slush Avalanche, which had attacked Fujisan Skyline Road on March 25, 2007 at Mt. Fuji. *Journal of the Japan Society of Erosion Control Engineering*, 60(2).

Morse, P. D. & Burn, C. R. (2013). Field observations of syngenetic ice wedge polygons, outer Mackenzie Delta, western Canadian Arctic coast, Canada. *Journal of Geophysical Research, Earth Surface*, 118. doi:10.1002/jgrf.20086

Morse, P. D. & Burn, C. R. (2014). Perennial frost blisters of the outer Mackenzie Delta, western Arctic coast, Canada. *Earth Surface Processes and Landforms*, 39: 200–213.

Morse, P. D., Burn, C. & Kokelj, S. V. (2009). Near-surface ground-ice distribution, Kendall Island Bird Sanctuary, western Arctic coast. *Permafrost and Periglacial Processes*, 20: 155–171.

Morse, P. D., Wolfe, S. A., Kokelj, S. V. & Gaanderse, A. J. R. (2016). The occurrence and thermal disequilibrium state of permafrost in forest ecotypes of the Great Slave Region, Northwest Territories, Canada. *Permafrost and Periglacial Processes*, 27(2): 145–162.

Mortensen, H. (1932). Uber die Physikalische Möglickeit der "Brodel" hypothese. *Centralblatt Mineralogie, Geologie und Paläontologie*, Abhandlung B: 417–422. [In German].

Moscow United Electric Grid Company (2014). JSC "MOESK". http://www.moesk.ru/ Accessed 15th February, 2015.

Moses, C. A. & Smith, B. J. (1993). A note on the role of the lichen *Collema auriforma* in (Koerb.) on Magaliesberg quartzite. *Earth Surface Processes and Landforms*, 15: 491–500.

Moskalenko, N. G. (1998). Impact of vegetation removal and its recovery after disturbance on permafrost. *Proceedings of the 7th International Conference on Permafrost, Yellowknife.* Québec City. Centre d'Etudes Nordiques, Université Laval. Collection Nordicana # 57: 763–769.

Motenko, R. G., Kolesnikova, A. A. & Juravlev, I. I. (2001). Experimental study of water phase composition in frozen soils polluted with oil or oil products. *International Conference on the Conservation and transformation of material and energy in Earth Cryosphere.* Pushchino, Russia. Abstracts: 101–102. [In Russian].

Moulton, K. L. & Berner, R. A. (1988). Quantification of the effect of plants on weathering: studies in Iceland. *Geology*, 26(10): 895–898.

MRPRC (2009). Code for design of High Speed Railway (TB10621–2009). [In Chinese].

Mukhetdinov, N. A. (1971). Effect of non-linear air penetration on the thermal regime of rock-filled dams. Vsesoiuznyi Nauchno Isseldovatel'skii Institut Gidrotekhniki. *Izvestiia*, 96: 205–217. [In Russian]. USA Cold Regions Research and Engineering Laboratory Translation 586.

Muir, D. & Ford, D. (1985). *Castleguard.* Ministry of Supply and Services, Canada.

Müller, F. (1943). *Permafrost or permanently frozen ground and related engineering problems.* Strategic Engineering Study Special Report Studies #62. United States Army.

Muller, F. (1947). *Permafrost or permanently frozen ground and related engineering problems.* Ann Arbor, Michigan, J. W. Edwards. 231p.

Muller, F. (1959). Beobachtung uber pingos. *Meddelelser om Grønland*, 153(3): 127p.

Müller, F. (1963). *Englacial temperature measurements on Axel Heiberg Island, Canadian Arctic Archipelago.* I. S. A. H. Publication #61. Commission on Snow and Ice: 168–180.

Muller, S. W. (1946). *Permafrost.* Ann Arbor, Michigan. J. W. Edwards, Inc. 231p.

Municipality of Anchorage, Alaska (2008). *Anchorage Regional Landfill.* Anchorage, AK: Solid Waste Services. http://www.muni.org/sws/disposalARL.cfm

Murphy, R. E. & Rangarathan, K. R. (1974). Bioprocesses of the oxidation ditch in a subarctic climate. In: *Symposium on wastewater treatment in cold climates.* Ottawa. Environmental Protection Service Report #EPS 3-WP-74-3.: 332–357.

Murray, B. J., Knopf, D. A. & Bertram, A. K. (2005). The formation of cubic ice under conditions relevant to Earth's atmosphere. *Nature*, 434(7030): 202–205.

Murton, J. B. (1993). *Thermokarst Sedimentology of the Tuktoyaktuk coastlands, NWT.* Unpublished Ph.D. thesis, University of Ottawa.

Murton, J. B. & French, H. M. (1993a). Thaw modification of frost-fissure wedges, Richards Island, Pleistocene Mackenzie Delta, western Canadian Arctic. *Journal of Quaternary Science*, 8: 185–196.

Murton, J. B. & French, H. M. (1993b). Thermokarst involutions, Summer Island, Pleistocene Mackenzie Delta, western Arctic Canada. *Permafrost and Periglacial Processes*, 4: 217–229.

Murton, J. B. & French, H. M. (1994). Cryostructures in permafrost, Tuktoyaktuk coastlands, western Arctic Canada. *Canadian Journal of Earth Sciences*, 31: 737–747.

Murton, J. B., Goslar, T., Edwards. M. E., Baterman, M. D., Danilov, P. P., Savvinov, G. N., Gubin, B., Haile, J., Kanevskiy, M., Loxhkin, A. V., Murton, D. K., Shur, Y., Tikhonov, A., Vasil'chuk, A. C., Vasil'chuk, Y. K. & Wolfe, S. A. (2015). Palaeoenvironmental interpretation of Yedoma Silt (Ice Complex) deposition as cold-climate loess, Duvanny Yar, Northeast Siberia. *Permafrost and Periglacial Processes*, 26(3): 208–288.

Murton, J. B. & Kolstrup, E. (2003). Ice-wedge casts as indicators of palaeotemperatures: precise proxy or wishful thinking? *Progress in Physical Geography*, 27(2): 155–170.

Murton, J. B., Peterson, R. & Ozouf, J.-C. (2006). Rock fracture by ice segregation in cold regions. *Science*, 314: 1127–1129.

Murton, J. B., Whiteman, C. A., Waller, R. I., Pollard, W. H., Clark, I. D. & Dallimore, S. R. (2005). Basal ice facies and supraglacial melt-out till of the Laurentide Ice Sheet, Tuktoyaktuk Coastlands, western Arctic Canada. *Quaternary Science Reviews*, 24: 681–708.

Muschell, F. E. (1970). Pile tips and barbs prevent ice uplift. *Civil Engineering*, 40: 41–43.

Mutter, E. Z. & Phillips, M. (2012). Active layer characteristics at ten borehole sites in alpine permafrost terrain, Switzerland. *Permafrost and Periglacial Processes*, 23(3): 138–151.

Nan Zhoutang, Li Shuxun & Li Yongzhi (2002). Mean annual ground temperature distribution on the Tibetan Plateau: Permafrost distribution mapping and further applications. *Journal of Glaciology and Geocryology*, 24(2): 142–148. [In Chinese].

Nansen, F. (1922). *Spitzbergen*. Leipzig, F. A. Brockhaus, 3rd Edition. 327p. [In German].

Nasmith, H. W. & Mercer, A. G. (1979). Design of dykes to protect against debris flows at Port Alice, British Columbia. *Canadian Geotechnical Journal*, 16: 748–757.

Naumov, I. V. (2006). In: Collins, D. N. (ed.). *The history of Siberia*. Norfolk. Routledge.

Nebogina, N. A. (2009). *The influence of the composition of oil and the extent of water content on the structural and mechanical properties of emulsions*. Ph.D. thesis, Tomsk University. Russia. [In Russian].

Nelson, F. E. (1986). Permafrost distribution in central Canada: Applications of a climate beased predictive model. *Association of American Geographers, Annals*, 76: 550–569.

Nelson, F. E., Anisimov, O. A. & Shiklomanov, N. I. (2002). Climatic change and hazard zonation in the circum-Arctic permafrost regions. *Natural Hazards*, 26: 203–225.

Nelson, F. E. & Outcalt, S. I. (1987). A computational method of prediction and regionalization of permafrost. *Arctic and Alpine Research*, 19(3): 279–288.

Nesje, A. (1989). The geographical and altitudinal distribution of blockfields in southern Norway, and its significance to the Pleistocene ice sheets. *Zeitschrift für Geomorphologie, Supplementband*, 72: 41–53.

Newberry, R. W., Beaty, K. G. & McCullough, G. R. (1978). Initial shoreline erosion in a permafrost affected reservoir, southern Indian Lake, Canada. *Proceedings of the 3rd International Conference on Permafrost, Edmonton*. Ottawa. National Research Council of Canada 1: 427–433.

Ng, H.-J. & Robinson, D.B. (1985). Equilibrium phase compositions and hydrating conditions in systems containing methanol, light hydrocarbons, carbon dioxide and hydrogen sulfide. *Gas Processors Association Research Report*, RR-66. Tulsa, Oklahoma.

Nichols, R. L. (1953). Geomorphologic observations at Thule, Greenland and Resolute Bay, Cornwallis Island, N.W.T. *American Journal of Science*, 251: 268–275.

Nicholson, F. H. (1976). Patterned ground formation and description as suggested by Low Arctic and Subarctic examples. *Arctic and Alpine Research*, 8(4): 329–342.

Nicholson, F. H. (1978a). Permafrost modifications by changing the natural energy budget. *Proceedings of the 3rd International Conference on Permafrost, Edmonton* 1: 427–433.

Nicholson, F. H. (1978b). Permafrost distribution and characteristics near Schefferville, Québec: Recent studies. *Proceedings of the 3rd International Conference on Permafrost, Edmonton* 1: 427–434.

Nicholson, F. H. (1979). Permafrost spatial and temporal variations near Schefferville, Nouveau-Québec. *Géographie physique et Quaternaire*, 33: 265–278.

Nicholson, F. H. & Granberg, H. B. (1973). Permafrost and snow-cover relationships near Scheffervile, Nouveau-Québec. *Permafrost. North American Contribution*. 2nd International Permafrost conference, Yakutsk, USSR. Washington, National Academy of Sciences, Publication 2115: 151–158.

Nicholson, W.L., Munakata, N., Horneck, G., Melosh, H.J. & Setlow, P. (2000). Resistance of *Bacillus* endospores to extreme terrestrial and extraterrestrial environments. *Microbiological Molecular Biology*, 64: 548–572.

Nicolsky, D. S., Romanovsky, V. E., Tipenko, G. S. & Walker, D. A. (2008). Modeling biogeophysical interactions in nonsorted circles in the Low Arctic. *Journal of Geophysical Research*, 113, G03305. Doi: 10.2029/2007JG000565.

Nikiforoff, C. (1928). The perpetually frozen subsoil of Siberia. *Soil Science*, 26: 61.

Nishu, R. & Matsuoka, N. (2012). Kinematics of an alpine retrogressive rockslide in the Japanese Alps. *Earth Surface Processes and Landforms*, 37: 1641–1650.

Niu F., Cheng, G. D., Luo, J. & Lin, Z. J. (2014). Advances in thermokarst lake research in permafrost regions. *Sciences in Cold and Arid Regions*, 6(4): 388–397.

Niu, F. J., Liu, M. H., Cheng, G. D., Lin, Z. J. & Yin, G. A. (2015). Long-term thermal regimes of the Qinghai-Tibet Railway embankments in plateau permafrost regions. *Science China: Earth Sciences*. Doi: 10.1007/s11430-015-5063-0

Niu, Wenyuan (1980). Theoretical analysis of physico-geographical zonation. *Acta Geographica Sinica*, 35(4): 288–298.

Nixon, J. F. (1973a). *The consolidation of thawing soil*. Unpublished Ph.D. thesis, Department of Civil Engineering, University of Alberta, Edmonton.

Nixon, J. F. (1973b). Thaw consolidation of some layered systems. *Canadian Geotechnical Journal*, 10(4): 617–631.

Nixon, J. F. (1978). Geothermal aspects of ventilated pad design. *Proceedings of the 3rd International Conference on Permafrost, Edmonton*. Ottawa. National Research Council of Canada 1: 840–846.

Nixon, J. F. (1982). Seasonal and climatic warming effects on pile creep in permafrost. *The Roger E. Brown Memorial Volume, Proceedings of the 4th Canadian Permafrost Conference, Calgary*. Ottawa, National Research Council of Canada: 335–340.

Nixon, J. F. (1986). Thermal simulation of subsea permafrost. *Canadian Journal of Earth Sciences*, 23: 2039–2046.

Nixon, J. F. & Hazen, B. (1993). Uplift resistance of pipelines buried in frozen ground. In: *Proceedings of the 6th International Conference on Permafrost, Beijing, China*. Chinese Society of Glaciology and Geocryology: 494–499.

Nixon, J. F. & Lem, G. (1984). Creep and strength testing of fine-grained frozen soil. *Canadian Geotechnical Journal*, 21(3): 518–529.

Nixon, J. F. & Morgenstern N. R. (1973). Practical extensions to a theory of consolidation for thawing soils. In: *Proceedings of the 2nd International Permafrost Conference, Yakutsk, U.S.S.R.* Washington, D.C. National Academy of Sciences, North American Contribution: 369–377.

Nixon, J.F., Morgenstern, N. & Reesor, S.M. (1983). Frost heave – pipeline interaction using continuum mechanics. *Canadian Geotechnical Journal*, 20(2): 251–261.

Nixon, J. F., Saunders, R. & Smith, J. (1991). Permafrost and thermal interfaces from Norman Wells pipeline ditchwall logs. *Canadian Geotechnical Journal*, 28(5): 738–745.

Nixon, J. F. & Vebo, Å. L. (2005). Discussion of "Frost heave and pipeline buckling". *Canadian Geotechnical Journal*, 42(1): 321–322.

Nixon, M., Liu, B., Zhou, J. & Lawrence, K. (2010). Probabilistic Estimation of Uplift Resistance for Chilled Gas Pipelines. *GEO2*10*: 587–594.

Niyazov, B. S. & Degovets, A. S. (1975). Estimation of the parameters of catastrophic mudflows in the basins of the lesser and greater Almatinka Rivers. *Soviet Hydrology, Selected Papers*, 2: 75–92.

Noetzli, J. & Gruber, S. (2009). Transient effects in alpine permafrost. *The Cryosphere*, 3: 85–99.

Nogami, M. (1980). Periglacial environment in Japan: Present and past. *Geojournal*, 4(2): 125–132.

Noguchi, Y., Tabuchi, H. & Hasegawa, H. (1967). Physical factors controlling the formation of patterned ground on Haleakala, Maui. *Geografiska Annaler.* 69 A: 329–342.

Nordenskjord, O. (1909). *Die Polarwelt und thre Nachbarlander.* Leipzig, B. Teubner. [In German].

Northern News Services (2011). Utilidor woes continue. http://www.nnsl.com/frames/newspapers/2011-11/nov17_11uti.html Accessed June 6th, 2015.

Nurmikolu, S. (2010). *Fouling and frost susceptibility of railway ballast and subballast, field and laboratory study.* ISBN 978-3-639-23623-1. VDM Publishing House. p. 235, app. 65.

Nurmikolu, S. & Kolisoja, P. (2008). The effect of fines content and quality on frost heave susceptibility of crushed rock aggregates used in railway track structure. In: *Proceedings of the 9th International Conference on Permafrost, Fairbanks* 2: 1299–1305.

Nurmikolu, S. & Silvast, M. (2013). Causes, effects and control of seasonal frost action in railways. *Sciences in Cold and Arid Regions,* 5(4): 363–367.

Nyberg, R. (1985). *Debris flows and slush avalanches in northern Swedish Lappland: Distribution and geomorphological significance.* Unpublished Thesis, University of Lund. 222p.

Nyman, K. J. (1983). Thaw settlement analysis for buried pipelines in permafrost. *Proceedings of the Conference on Pipelines in Adverse Environments, San Diego.* Pipeline Division of the ASCE. II: 300–325.

Oberman, N. G. (1974). Regional 'nyye osobennosti merzloy zony Timano-Ural'skoy oblasti. Vysshihk uchcbn. Zavedenni, *Geologiya i razveddka Izvestia,* 11: 98–103. [In Russian].

Odum, W. B. (1983). Practical application of underslab ventilation system: Prudhoe Bay case study. In: *Permafrost. 4th International Conference Proceedings, Fairbanks.* Washington. National Academy Press: 940–944.

Ohata, T., Furukawa, T. & Higuchi, K. (1994a). Glacioclimatological study of perennial ice in the Fuji Ice Cave, Japan. Part 1: Seasonal variation and mechanism of maintenance. *Arctic and Alpine Research,* 26(3): 227–237.

Ohata, T., Furukawa, T. & Osada, K. (1994b). Glacioclimatological study of perennial ice in the Fuji Ice Cave, Japan. Part 2. Interannual variation and relation to climate. *Arctic and Alpine Research,* 26(3): 238–244.

Oksanen, P. O. (2005). Development of palsa mires on the northern European continent in relation to Holocene climatic and environmental changes. *Acta Universtatis Ouluensis, Scientiae Rerum Naturalium,* A 446: 1–50.

Oksanen, P. O. & Väliranta, M. (2006). Palsa mires in a changing climate. *Suoseura,* 57(2): 33–43. [In Finnish].

Ollier, C. (2010). Glaciers – Science. *Geoscientist,* 20(3): 16–21.

Ollivier, J., Yang, S. Z., Dörfer, C., Welzl, G., Kühn, P., Scholten, T., Wagner, D. & Schloter, M. (2013). Bacterial community structure in soils of the Tibetan Plateau affected by discontinuous permafrost or seasonal thawing. *Biological Fertility of Soils,* 50: 555–559.

O'Neill, B. & Burn, C. R. (2015). Subdivision of ice-wedge polygons, western Arctic coast. *GeoQuébec,* Québec City, September, 2015.

Onikienko, T. C. (1995). Refinement of relation between the reservoir volume and level of the Ust-Khantaiskya hydropower plant. *Hydroengineering Construction,* 3: 19–23.

Orakogla, M. E. & Liu, J. L. (2014). Thermal conductivity of reinforced soils: a literature review. *Sciences in Cold and Arid Regions,* 6(4): 409–414.

Orlov, V. O. (1962). *Cryogenic frost heave of fine-grained soils.* Academy of Sciences of the USSR. 186p. [In Russian].

Osokin, I. M. (1973). Zonation and regime of naleds in Trans-Baikal region. *Proceedings of the 2nd International Conference on Permafrost. USSR Contribution.* Washington, D.C. 391–396.

Ospennikov, Y. N. (1979). Results of observations on rates of movement of stone flows of the Chulman plato. *Merzlotnyye Issledovaniya,* 18: 129–133. [In Russian].

Osterkamp, T. E. & Harrison, W. (1985). *Sub-sea permafrost: Probing, thermal regime and data analyses, 1975–1981*. Summary Report, 1985. Fairbanks. Fairbanks Geophysical Institute, University of Alaska: 108p.

Osterkamp, T. E., Jorgenson, M. T., Schuur, E. A. G., Shur, Y. L., Kanevskiy, M. Z., Vogel, J. G. & Tumskoy, V. E. (2009). Physical and ecological changes associated with warming permafrost and thermokarst in interior Alaska. *Permafrost and Periglacial Processes*, 20: 235–256.

Oswell, J. M. (2011). Pipelines in permafrost: Geotechnical issues and lessons. *Canadian Geotechnical Journal*, 48: 1412–1431.

Oswell, J. M., Skibinsky, D. & Cavanagh, P. C. (2005). Discussion of "frost heave and pipeline upheaval buckling". *Canadian Geotechncal Journal*, 42(1): 323–324.

Outcalt, S. I. (1971a). An algorithm for needle ice growth. *Water Resources Research*, 7: 394–400.

Outcalt, S. I. (1971b). The climatology of a needle ice event: an experiment in simulation climatology. *Archives of Meteorological Geophysics and Bioklimatology.*, Series B, 19: 325–338.

Outcalt, S. E. & Benedict, J. B. (1965). Photointerpretation of two types of rock glaciers in the Colorado Front Range, USA. *Journal of Glaciology*, 5: 849–856.

Overeem, I., Anderson, R. S., Wobus, C. W., Clow, G. D., Erban, F. E. & Matell, N. (2011). Sea ice loss enhances wave action at the Arctic coast. *Geophysical Research Letters*, 38: L17503.

Owen, L. A., Richards, B., Rhodes, E. J., Cunningham, W. D., Windley, B. F., Badamgarav, J. & Dorjnamjaa, D. (1998). Relic permafrost structures in the Gobi of Mongolia: age and significance. *Journal of Quaternary Science*, 13(6): 539–547.

Owens, I. F. (1972). Morphological characteristics of Alpine mudflows in the Nigel Pass area. In: Slaymaker, H. O. and MacPherson, H. J., (eds). *Mountain Geomorphology*. B.C. Geographical Series #14, Vancouver. Tantalus Press: 93–100.

Pagani *et al.* (2005). Marked decline in carbon dioxide concentrations during the Paleocene. *Science*, 309(5734): 600–603.

Palmer, A. C. (1977). *Settlement of a pipeline on thawing permafrost*. Division of Engineering, Brown University, Providence, Rhode Island. Department of Defense, Advanced Research Projects Agency, Contract SD-86. Materials Research Division.

Palmer, A. C. & Williams, P. J. (2003). Frost heave and pipeline upheaval buckling. *Canadian Geotechnical Journal*, 40(5): 1033–1038.

Pan, B. T. & Chen, F. H. (1997). Permafrost evolution in the northeastern Qinghai-Tibetan Plateau during the last 150,000 years. *Journal of Glaciology and Geocryology*, 19(2): 124–132. [In Chinese].

Panikov, N.S. & Sizova, M.V. (2007). Growth kinetics of microorganisms isolated from Alaskan soil and permafrost in solid media frozen down to −5°C. *FEMS Microbiological Ecology*, 54: 500–512.

Parameswaran, V. R. (1978). Adfreeze strength of frozen sand to model piles. *Canadian Geotechnical Journal*, 15: 494–500.

Parmuzin, C. Y. (2008). *Land Use in the Cryolithozone*. Moscow. Moscow University Press. 171p. [In Russian].

Parsekian, A. D., Jones, B. M., Jones, M., Grosse, G., Walter Anthony, K. M. & Slater, L., (2011). Expansion rate and geometry of floating vegetation mats on the margins of thermokarst lakes, northern Seward Peninsula, Alaska, USA. *Earth Surface Processes and Landforms*, 36: 1889–1897.

Parson, C. G. (1987). Rock glaciers and site characteristics on the Blanca Massif, Colorado, U.S.A. In: Giardino, J. R., Shroder, J. F. and Vitek, J. D. (eds.). *Rock glaciers*. London. Allen and Unwin: 127–144.

Pavlov, A. V. (1984). *Thermal exchange in the earth landscape.* Novosibirsk. Nauka. 254p. [In Russian].

Pavlov, A. V. (1994). Current changes in climate and permafrost in the Arctic and Sub-Arctic of Russia. *Permafrost and Periglacial Processes,* 5(2): 101–110.

Pavlov, A. V. (1996). Permafrost-climatic monitoring of Russia: Analysis of field data and forcast. *Polar Geography,* 20(1): 44–66.

Pavlov, A. V. (1999). The thermal regime of lakes in Northern Plains Regions. *Earth Cryosphere,* 3(3): 59–70. [In Russian].

Pavlov, A. V., Dubrovin, V. A. & Kotlov, S. B. (1989). Studying the thermal regime of natural complexes of tundra zones in Western Siberia. In: Pavlov, A. V. (ed.). *Methods of studying the thermal regime of soils in the cryolithozone. Collected scientific papers.* Moscow. VSEGINGEO, USSR: 6–20. [In Russian].

Pavlov, A. V. & Olovin, B. A. (1974). *Artificial thawing of frozen ground by heat from solar radiation in placer working.* Novosibirsk. Nauka Press: 141–153. [In Russian].

Payette, S., Delwide, A., Caccianiga, M. & Beauchemin, M. (2004). Accelerated thawing of subarctic peatland permafrost over the last 50 years. *Geophysical Research Letters,* 31: L18208.

Payette, S., Gauthier, L. & Grenier, I. (1986). Dating ice-wedge growth in subarctic peatlands following deforestation. *Nature,* 322: 724–727.

Payette, S., Samson, H. & Lagarec, D. (1976). The evolution of permafrost in the taiga and in the forest-tundra, western Quebec-Labrador Peninsula. *Canadian Journal of Forest Research,* 6: 203–220.

Payton, R. W. (1992). Fragipan formation in argillic brown earths (Fragiudalfs) of the Milfield Plain, north-east England. I. Evidence for a periglacial stage of development. *Journal of Soil Science,* 43: 621–644.

PBK Company (2014). Company profile. http://www.moesk.ru/, accessed 2–15–2014.

Pelletier, J. D. (2005). Formation of oriented thaw lakes by thaw slumping. *Journal of Geophysical Research,* 110: F02018. DOI: 10.1029/2004JF000158

Peng, H., Mayer, B., Harris, S. & Krouse. H. R. (2007). The influence of below-cloud secondary effects on the stable isotope composition of hydrogen and oxygen in precipitation at Calgary, Alberta, Canada. *Tellus,* 59(4): 698–704.

Peng, H. Y. & Cheng, G. D. (1990). Ice-wedge networks in Da Hinggan Mountains and their palaeoclimatic significance. In: *Proceedings of the 4th Conference on Glaciology and Geocryology.* Beijing. Science Press: 9–16.

Penner, E. (1960). The importance of freezing rate in frost action in soils. *Proceedings of the American Society for Testing and Materials,* 60: 1151–1165.

Penner, E. & Goodrich, L. E. (1980). Location of segregated ice in frost susceptible soil. In: *Proceedings of the 2nd International Symposium on Ground Freezing, Trondheim, Norway:* 626–639.

Pèrez, F. L. (1987). Needle-ice activity and the distribution of stem-rosette species in a Venezuelan páramo. *Arctic and Alpine Research,* 19(2): 135–153.

Périer, L., Doré, G. & Burn, C. R. (2014). The effects of water flow and temperature on thermal regime around a culvert built on permafrost. *Sciences in Cold and Arid Regions,* 6(5): 415–422.

Perla, R. (1978). Artificial release of avalanches in North America. *Arctic and Alpine Research,* 10(2): 235–240.

Perl'shtein, G. Z. (1979). *Water and thermal impact on frozen soils in North-East of USSR.* Novosibirsk: Nauka. 232p. [In Russian].

Perl'shtein G. Z. & Pavlenkov D. A. (2001). Natural freezing of tailings. Materials of the Second Conference of Russian Geocryologists. July 6–8, 2001, volume 4. Moscow University, Moscow: 215–221. [In Russian].

Perl'shtein, G. Z. & Savenko, L. P. (1977). Approximate calculation of distribution of heat source intensity in electric thawing of frozen ground. *Proceedings of the All-Union Research Institute of gold and rare metals, Magadan*, 37: 156–160. [In Russian].

Perov, V. F. (1969). Block fields of the Khibiny Mountains. *Biuletyn Peryglacjalny*, 19: 381–387.

Perşoiu, A., Onac, B. P., Wynn, J., Bojar, A.-V. & Holmgren, K. (2011). Stable isotope behaviour during cave ice formation by water freezing in Scarisoara ice cave, Romania. *Journal of Geophysical Research*, 116(D2).

Peters, C. J. & Perry, R. (1981). The formation and control of trihalomethanes in water treatment processes. In: Smith, D. W. & Hrudey, S. E. (eds.). *Design of water wastewater services for cold climate communities*. Oxford. Pergamon Press: 103–121.

Peterson, R. A. & Krantz, W. B. (2003). A mechanism for differential frost heave and its implications for patterned ground formation. *Journal of Geology*, 49: 69–80.

Peterson, R. A., Walker, D. A., Romanovsky, V. E., Knudson, J. A., Raynolds, M. K. & Krantz, W. B. (2003). A differential frost heave model: Cryoturbation-vegetation interactions. In: Phillips, M., Springman, S. M. & Arenson, L. U. (eds.). *Permafrost. Proceedings of the 8th International Conference on Permafrost*. Lisse. A. A. Balkema Publishers: 885–890.

PetroChina Daqing Oilfield Engineering Company (2009). *Report on analyses and numerical simulations of thermal and stress field of the buried pipeline in permafrost regions*. Daqing. Petrochina Daqing Oilfield Engineering Company, Ltd.

Petrone, K. C., Hinzman, L. D. & Boone, R. D. (2000). Nitrogen and carbon dynamics of storm runoff in three sub-arctic streams. In: Kane, D. L., (ed.). *Proceedings of the AWRA Spring Speciality Conference, Water Resources in Extreme Environments, May 1–3, 2000*. Anchorage, Alaska. American Water Resources Association: 167–172.

Petryaev, A. & Morozova, A. (2013). Railroad bearing strength in the period of thawing and methods of its enhancement. *Sciences in Cold and Arid Regions*, 5(5): 548–553.

Pettapiece, W. W. (1974). A hummocky permafrost soil from the Subarctic of northwestern Canada and some influences of fire. *Canadian Journal of Soil Science*, 54(4): 343–355.

Pettapiece, W. W. (1975). Soils of the Subarctic in the lower Mackenzie Basin. *Arctic*, 28: 35–53.

Pettibone, H. C. (1973). Stability of an underground room in frozen gravel. *Proceedings of the 2nd International Conference on Permafrost, Yakutsk, USSR. North American Contribution*. Washington. National Academy of Sciences. Publication 1287: 699–706.

Pettijohn, F. J. (1949). *Sedimentary Rocks*. New York, Harper and Brothers, Publishers. 526p.

Petzold, D. E. & Rencz, A. N. (1975). The albedo of selected subarctic substrates. *Arctic and Alpine Research*, 7: 393–398.

Péwé, T. L. (1954). *Effect of permafrost on cultivated fields, Fairbanks area, Alaska*. US Geological Survey, Bulletin 989-f. 351p.

Péwé, T. L. (1959). Sand-wedge polgyons (tesselations) in the McMurdo Sound Region, Antarctica – A progress report. *American Journal of Science*, 257: 545–552.

Péwé, T. L. (1962). Ice wedges in permafrost, Lower Yukon River near Galena, Alaska. *Biuletyn Peryglacjalny*, 11: 65–76.

Péwé, T. L. (1966). Ice wedges in Alaska – classification, distribution and significance. In: *Proceedings of the 1st International Conference on Permafrost*, Ottawa. National Academy of Science, National Research Council of Canada, Publication #1287: 76–81.

Péwé, T. L. (1970a). Permafrost and vegetation on flood-plains of Sub-Arctic rivers (Alaska); A summary. In: *Ecology of the Subarctic Regions*, Paris, UNESCO: 141–142.

Péwé, T. L. (1970b). Altiplanation terraces of early Quaternary age near Fairbanks, Alaska. *Acta Geographica Łódziensia*, 24: 357–363.

Péwé, T. L. (1973a). Ice-wedge casts and past permafrost distribution in North America, *Geoforum*, 15: 15–26.

Péwé, T. L. (1973b). *Permafrost Conference in Siberia. Geotimes*, 18(12): 23–26.

Péwé, T.L. (1983a). Alpine permafrost in the contiguous United States: A review. *Arctic and Alpine Research*, 15(2):145–156.

Péwé, T. L. (1983b). *Geological hazards of the Fairbanks area, Alaska*. Division of Geological and Geophysical Surveys, Fairbanks, Alaska, Special Report #15, 109p.

Péwé, T. L., Schmidt, R. A. M. & Sloan, C. E. (1990). Permafrost and thermokarst: Geomorphic effects of subsurface water on landforms in cold regions. In: Higgins, C. G. and Coates, D. R. (eds.). *Groundwater geomorphology: The role of subsurface water in earth-surface processes and landforms*. Geological Society of America Special Paper, 252: 211–218.

Pierson, T. C. (1980). Erosion and deposition by debris flows at Mt. Thomas, North Canterbury, New Zealand. *Earth Surface Processes and Landforms*, 5: 227–247.

Pihlainen, J. A. (1959). Pile construction in permafrost. *American Society of Civil Engineers, Journal of Soil Mechanics, Foundation Division* 85 (SM6) Part 1: 75–95.

Pikylevich, L. D. (1963). On stages of frost processes and moisture variation in seasonally thawing ground. In: *Frozen Ground Study*. Moscow. Moscow University, Publication 3: 158–167. [In Russian].

Ping, C. L., Bockheim, J. G., Kimble, J. M., Michaelson, G. J. & Walker, D. A. (1998). Characteristics of cryogenic soils along a latitudinal transect in Arctic Alaska. *Journal of Geophysical Research*, 103, #D22: 917–928.

Pissart, A. (1963). Les trace de "pingos" du Pays du Galles (Grande Bretagne et du Plateau des Hautes Fagnes (Belgique). *Zeitschrift für Geomorphologie*, 10: 226–236. [In French].

Pissart, A. (1967a). Les pingos de l'ile Prince Patrick (76°N–120°W). *Annales, Société Géologique de Belgique*, 9: 189–217. [In French].

Pissart, A. (1967b). Les modalités de l'écoulement de l'eau sur l'isle Prince Patrick. *Biuletyn Peryglacjalny*, 16: 217–224. [In French].

Pissart, A. (1968). Les polygones des fente de gel de l'ile Prince Patrick (Arctique Canadien – 76°lat. N.). *Biuletyn Periglacjalny*, 17: 171–180. [In French].

Pissart, A. (1969). La mechanism périglaciaire dressant les pierres dans le sol. Resultats d'expériences. *Academie de Science, Paris*, Comptes Rendus 268: 3015–3017. [In French].

Pissart, A. (1970). Vitesse des mouvements de pierres dans des sols et sur les versants périglaciaires au Chambeyron, (Basse Alpes). In: Macar, P. & Pissart, A. (eds.). *Processus périglaciaires*. Liège. Les Congrès et Colloques de l'Université De Liège 67: 295–321. [In French].

Pissart, A. (1972). Variations de volume de sols gelés submissant des fluctuations du temperature sous 0°C. *Centre de Géomorphologie de Caen*, Bulletin 13-14-15: 17–33. [In French].

Pissart, A. (1973a). Résultats d'expériences sur l'action du gel dans le sol. *Builetyn Peryglacjalny*, 23: 101–113. [In French].

Pissart, A. (1973b). L'origin des sols polygonaux et striés du Chamberon *Bulletin de la Société Géographique de Liège*, 9: 33–53. [In French].

Pissart, A. (1976). Sols à buttes cercles non-triés et sols striés non-triés de l'île de Banks (Canada, N.W.T.). *Builetyn Peryglacjalny*, 26: 275–285. [In French].

Pissart, A. (1977). Apparition et évolution des sola structuraux périglaciaires de haute montagne. Expériences de terrain au Chambeyron (Alpes, France). In: Poser, H. (ed.). *Formen, formengesellschaften und Untergrenzen in den heutigen periglazialen Höhenstufen der Hochgebirge Europas und Afrikas zwischen Arctis und Äquator. Berich über en Symposium*. Göttingen. Abhandlungen der Akademie der Wissenschaften in Göttingen, Mathematisch-Physik Klasse, 31: 142–156. [In French].

Pissart, A. (1982). Déformations de cylinders de limon entourés de graviers sous l'action d' alternances gel-dégel. Expéperiences sur l'origine des cryoturbations. *Builetyn Periglacjalny*, 29: 219–229. [In French].

Pissart, A. (1987). Les traces de pingos et de palses en Belgique et dans le monde. *Géomorphologie pèriglaiaire. Textes des leçons de la Chaire Francqui belge*. Université de Liège,

Belgique: Rijksuniversiteit Ghent. Edition Laboratoire de Géomorphologie et de Gèologie du Quaternaire: 45–53. [In French].

Pissart, A. (1990). Advances in periglacial geomorphology. *Zeitschrift für Geomorphologie*, 79: 119–131.

Pissart, A. (1992). Vertical movements of boulders in a subnival boulder pavement at 2800 m a.s.l. in the Alps (France). *Permafrost and Periglacial Processes*, 3: 203–208.

Pissart, A. (2000a). Remnants of lithalsas of the Hautes Fagnes, Belgium: A summary of present-day knowledge. *Permafrost and Periglacial Processes*, 11(4): 327–355.

Pissart, A. (2000b). Les traces de lithalses et de pingos connues dans le monde. *Hautes Fagnes*, 237: 16–25. [In French].

Pissart, A. (2002). Palsas, lithalsas and remnants of these periglacial mounds. A progress report. *Progress in Physical Geography*, 26(4): 605–621.

Pissart, A. (2003). The remnants of Younger Dryas lithalsas on the Hautes Fagnes Plateau in Belgium and elsewhere in the world. *Geomorphology*, 52: 5–38.

Pissart, A. (2010). Remnants of lithalsas on the Hautes Fagnes plateau (Belgium) are on weathered quartzitic rocks. *Zeitschrift für Geomorphologie*, 54(5): 1–15.

Pissart, A., Calmels, F. & Wastians, C. (2011). The potential lateral growth of lithalsas. *Quaternary Research*, 75(2): 371–377.

Pissart, A. & French, H. M. (1976). Pingo investigations, north-central Banks Island, Canadian Arctic. *Canadian Journal of Earth Sciences*, 13: 937–946.

Pissart, A. & Gangloff, P. (1984). Les palses minerals et organiques de la vallé de l'Aveneau, près de Kuujjuaq, Québec. *Géographie physique et Quaternaire*, 38(3): 217–228.

Pissart, A., Harris, S., Prick, A. & Van Vliet-Lanoë, B. (1998). La signification paléoclimatique des lithalses (palses minerals). *Biuletyn Peryglacjalny*, 37: 141–145. [In French].

Pissart, A. & Lautridou, J. P. (1984). Variations de longueur de cylindres de pierre de Caen (calcaire bathonian) sous l'effect de sechage et d'humidification. *Zeitschrift für Geomorphologie*, 49: 111–116. [In French].

Platzer, K., Bartelt, P. & Jaedicke, C. (2007). Basal shear and normal stresses of dry and wet snow avalanches after a slope deviation. *Cold Regions Science and Technology*, 49: 11–25.

Plug, L. J. & Werner, B. T. (2002). Non linear dynamics of ice-wedge networks and resulting sensitivity to severe cooling events. *Nature*, 417: 929–933.

Pollard, W. (1998). The nature and origin of ground ice in the Herschel Island area, Yukon Territory. In: *Permafrost, 7th International Conference Proceedings, Yellowknife*. Collection Nordicana 55: 23–30.

Pollard, W. & French, H. M. (1983). Seasonal frost mound occurrence, North Fork Pass, Ogilve Mountains, northern Yukon, Canada. In: *Proceedings of the 4th International Conference on Permafrost, Fairbanks*. Washington. National Academy Press: 1000–1004.

Pollard, W. & French, H. M. (1984). The groundwater hydraulics of seasonal frost mounds, northern Yukon. *Canadian Journal of Earth Sciences*, 21: 1073–1081.

Pollard, W. & French, H. M. (1985). The internal structure and ice crystallography of seasonal frost mounds. *Journal of Glaciology*, 31: 157–162.

Ponomarev, V. D. & Khrustalev, L. N. (1980). Construction by the method of stabilzing perennially frozen foundation soils. *Proceedings of the 3rd International Conference on Permafrost. English Translations, Part II*. Ottawa. National Research Council of Canada: 283–296.

Popescu, R., Onaca, A., Urdea, P. & Vaspremeanu-Stroe (2017). Spatial distribution and main characteristics of alpine permafrost from Southern Carpathians, Romania. In: Rădoane, M. & Vespremeanu (eds.). *Landform dynamics and evolution in Romania*. Springer Geography. Chapter 6. DOI: 10.1007/978-3-319-32589-7_6.

Popov, A. I. (1953). Specific features of lithogenesis in the alluvial plains under cold conditions. *Izvestiya AN ESSR* Series, Geography 2. [In Russian].

Popov, A. I. (1962). *The origin and development of massive fossil ice.* Issue II. Academy of Sciences of the USSR, V. A. Obruchev Institute of Permafrost Studies, Moscow. National Research Council of Canada, Technical Translation #1006: 5–24.

Popov, A. I. (1965). Underground Ice. In: *Underground Ice.* Issue 1. 7th International Congress on Quaternary (INQUA), USA. Moscow. Moscow University Press: 8: 7–39. [In Russian].

Popov, A. I. (1978a). Cryolithogenesis, the composition and structure of frozen rocks, and ground ice (the current state of the problem). *Biuletyn Peryglacjalny,* 27: 155–169.

Popov, A. I. (1978b). Cryolithogenesis. *Proceedings, U.S.S.R. Contribution, 2nd International Conference on Permafrost.* Washington, D.C. National Academy of Sciences, 181–184.

Popov, A. I., Gvozdetskiy, N. A., Chiksishev, A. G. & Kudelin, B. I. (1972). Karst in the USSR. In: Herak, M. and Stingfield, V. T. (eds.). *Karst.* Amsterdam, Elsevier: 355–416.

Popov, A. I., Kachurin, S. P. & Grave, N. A. (1966). Features of the development of frozen geomorphology in northern Eurasia. *Proceedings of the 1st International Conference on Permafrost.* Ottawa. National Academy of Sciences-National Research Council of Canada, publication 1287: 181–185.

Popov, A. I., Rosenbaum, G. E. & Tumel, N. V. (1985). *Cryolithology.* Moscow. Moscow State University Press. 238p. [In Russian].

Porkhaev, G. V., Valershtein, R. L., Eroshenko, A. L., Mindich, A. L., Mirenburg, Y. S., Ponomurev, V. D. & Khrustalev, L. N. (1980). Construction by the method of stabilzing perennially frozen foundation soils. *Proceedings of the 3rd International Conference on Permafrost. English Translations, Part II.* Ottawa. National Research Council of Canada: 283–296.

Porsild, A. E. (1945). The alpine flora of the east slope of the Mackenzie Mountains, Northwest Territories. *National Museums of Canada* Bulletin 101 (Biological series #30). Ottawa. Edmond Cloutier.

Porsild, A. E. (1951). Botany of southeastern Yukon adjacent to the Canol Road. *National Museums of Canada* Bulletin 21 (Biological series #41). Ottawa. Queens Printer.

Porter, S. C. (1966). *Pleistocene geology of Anakturak Pak, Central Brooks Range, Alaska.* Arctic Institute of North America, Technical Paper 18. 100p.

Porter, S. C., Ashol Singhvi, Zhisheng, A. & Zhongping, L. (2001). Luminescent age and palaoenvironmental implications of a Late Pleistocene ground wedge on the Northeastern Tibetan Plateau. *Permafrost and Periglacial Processes,* 12(2): 203–210.

Portnov, A. Smith, A. J., Mienert, J., Cherkashov, G., Rekant, P., Semenov, P., Serov, P. & Vanshtein, B. (2013). Offshore permafrost decay and massive seabed methane escape in water depths >20 m at the south Kara Sea shelf. *Geophysical Research Letters,* 40: 1–6. Doi: 10.1002/grl.50735.

Poser, H. (1948). Boden-und klimaverhältnisse in Mittel-und Westeuropa während der Wümeiszeit. *Erdkunde,* 2: 53–68. [In German].

Potter, N. (1969). *Rock glaciers and mass-wastage in the Galena Creek area, northern Absaroka Mountains.* Unpublished Ph.D. thesis, University of Minnesota. 150p.

Potter, N. (1972). Ice-cored rock glacier, Galena Creek, northern Absaroka Moutains, Wyoming *Bulletin of the Geological Society of America,* 83: 3025–3058.

Potts, A. S. (1970). Frost action in Rocks: Some experimental data. *Transactions of the Institute of British Geographers* 49: 109–124.

Prasad, D. & Heinke, G. W. (1981). Disposal and treatment of concentrated human wastes. In: Smith, D. W. and Hrudey, S. E. (eds.). *Design of water and wastewater services for cold climate communities.* Oxford. Pergamon Press: 125–135.

Prest, V., Grant, D. R. & Rampton, V. N. (1968). *Glacial map of Canada.* Geological Survey of Canada Map 1253A.

Price, L. W. (1970). Up-heaved blocks: A curious feature of instability in the tundra. *Proceedings of the American Association of Geographers,* 2: 106–110.

Price, L. W. (1971). Vegetation, microtopography, and depth of active layer on different exposures in Subarctic Tundra. *Ecology*, 52: 638–647.

Price, L. W. (1973). Rates of mass wasting in the Ruby Range, Yukon Territory. *Permafrost, 2nd International Conference, Yakutsk, USSR*. North American Contribution. Washington, D.C. National Academy of Sciences: 235–245.

Price, L. W. (1991). Rates of mass wasting in the Ruby Range, Yukon Territory, Canada: a 20 year study. *Arctic and Alpine Research*, 23: 200–205.

Price, L. W. (1968). Oriented lakes. In: Fairbridge, R. W., (Ed.). *Encyclopedia of Geomorphology*. New York. Reinholt Book Co.: 784–796.

Price, P. B., Nagornov, O. V., Bay, R., Chirkin, D., He, Y., Miocinovic, P., Richards, A., Woschnagg, K., Koci, B. & Zagorodnov, V. (2002). Temperature profile for glacial ice at the South Pole: Implications for life in a nearby subglacial lake. *Proceedings of the National Academy of Sciences*, 99(12): 7844–7847.

Prick, A. (2003). Rock weathering and rock fall in an arctic environment, Longyearbyen, Svalbard. In: Phillips, M., Springman, S. M. and Arenson, L. U. (eds.). *Permafrost,Proceedings of the 8th International Conference on Permafrost. Lisse*. Swets and Zeitlinger: 907–912.

Prick, A., Pissart, A. & Ozouf, J.-C. (1993). Variations dilatométriques de cylinders de roche calcaires sibissant des cycles de gel-dégel. *Permafrost and Periglacial Processes*, 4: 1–15. [In French].

Priesnitz, K. (1988). Cryoplanation. In: Clark, M. J., (Ed.). *Advances in Periglacial Geomorphology*. Chichester. Wiley and Sons: 49–67.

Priesnitz, K. & Schunke, E. (1983a). Meeting and Field Trip of the IGU Commission "The significance of periglacial phenomena" in Iceland, 22nd August–2nd September, 1982. Reykjavik. 39p.

Priesnitz, K. & Schunke, E. (1983b). Periglaciale pediplanation in der Kanadischen Kordillere. In: Poser, H. and Schunke, E. (eds.). *Mesoformen des reliefs im heutigen Periglacialraum*. Abhandlungen Akademie Wissenschaften in Göttingen, Math.-Phys. Klasse 35: 266–280. [In German].

Priesnitz, K. & Schunke, E. (2002). The fluvial morphodynamics of two small permafrost drainage basins, northwestern Canada. *Permafrost and Periglacial Processes*, 13(3): 207–217.

Protz, R., Ross, G. J., Martini, I. P. & Terasmae, J. (1984). Rate of podzolic soil formation near Hudson Bay, Ontario. *Canadian Journal of Earth Science*, 64(1): 31–49.

Psenner R. & Sattler B. (1998). Life at the freezing point. *Science*, 280: 2073–2074.

Pufahl, D. E. (1976). *The behaviour of thawing slopes in permafrost*. Unpublished Ph.D. thesis, Department of Civil Engineering, University of Alberta, 345p.

Pufahl, D. E. & Morgenstern, N. R. (1979). Stabilization of planar landslides in permafrost. *Canadian Geotechnical Journal*, 16: 734–747.

Pufahl, D. E., Morgenstern, N. R. & Roggensack, W. D. (1974). *Observations on recent highway cuts in permafrost*. Environmental-Social Program, Northern Pipelines, Task Force on Northern Oil Development, Report 74-32. 53p.

Pullman, E. R., Jorgenson, M. T. & Shur, Y. (2007). Thaw settlement in soils of the Arctic Coastal Plain, Alaska. *Arctic, Antarctic and Alpine Research*, 39: 468–476.

Puskeppeleit M., Quintern L. E., El Naggar, S., Schott, J. U., Eschweiler, U., Horneck, G. & Bücker, H. (1992). Long-term dosimetry of solar UV radiation in Antarctica with spores of *Bacillus subtilis*. *Applied Environmental Microbiology*, 58: 2355–2359.

Pylkkänen, K. & Nurmikulu, A. (2011). Frost susceptibility of railway subballast materials. *Proceedings of the International Heavy Haul Association Conference, IHHA 2011, Calgary*.

Pylkkänen, K., Luomala, H., Guthrie, W. S. & Nurmikulu, A. (2012). Real-time in-situ monitoring of frost depth, seasonal frost heave, and moisture in track structures. *Proceedings of the 15th International Conference on Cold Regions Engineering. Québec, Canada*.

Puzakov, N. A. (1960). *Water-heat regime of roadway beds*. M.: Avtotransizdat. [In Russian].

Qian, J., Liu, H. J., Yu, Q.H., *et al.* (2009). Permafrost engineering geological characteristic and discussion of route selection in Qinghai-Tibet Plateau. *Journal of Engineering Geology*, 17(4): 508–515.

Qin, Y. (2009). Estimate of the permafrost degradation at Muli coalfield, Qinghai-Tibet Plateau. *Cold Regions Engineering*, 2009: 162–171. Doi: 10.1061/41072(359)19

Rabassa, J., Coronato, A. & Salemme, M. (2005). Chronology of the Late Cenozoic Patagonian Glaciations and their correlation with biostratigraphic units in the Pampean region (Argentina). *Journal of South American Earth Sciences*, 20(1–2): 363–379.

Rabassa, J. (2008). The Late Cenozoic of Patagonia and Tierra del Fuego. *Developments in Quaternary Science*, 11: 151–204.

Rachold, V., Bolshiyanov, D., Grigorev, M., Hubberten, H.-W., Junker, R., Kunitsky, V., Merker, F., Overdum, P. & Schneider, W. (2007). Near-shore Arctic subsea permafrost in transition. *Eos, Transactions of the American Geophysical Union*, 88(13): 149. Doi: 10.1029/2007EO130001

Radd, F. J. & Oertle, D. H. (1973). Experimental pressure studies of frost heaving mechanisms and the growth-fusion behavior of ice. *Proceedings of the 2nd International Conference on Permafrost, Yakutsk, USSR*. North American Contribution. Washington, National Academy of Sciences: 377–384.

Radforth, J. R. (1972). Analysis of disturbance effects of operation of off-road vehicles on tundra. *Department of Indian Affairs and Northern Development. Arctic Land Use Research Program 72-73-12*.

Railton, J. B. & Sparling, J. H. (1973). Preliminary studies on the ecology of palsa mounds in northern Ontario. *Canadian Journal of Botany*, 51: 1037–1044.

Rampton, V. N. (1974). The influence of ground ice and thermokarst upon the geomorphology of the Mackenzie-Beaufort region. In Fahey, B. D. and Thompson, R. D., (eds.). *Research in Polar and Alpine Geomorphology*. Proceedings of the 3rd Guelph Symposium on Geomorphology. Norwich. Geobooks, 43–59.

Rampton, V. N. & Walcott, R. I. (1974). Gravity profiles across ice-cored topography. *Canadian Journal of Earth Sciences*, 11: 110–122.

Ran, Y., Li, X., Cheng, G., Zhang, T., Wu, Q., Jin, H. & Jin, R. (2012). Distribution of permafrost in China: An overview of existing permafrost maps. *Permafrost and Periglacial Processes*, 23: 322–333.

Rangecroft, S., Harrison, S., Anderson, K., Magrath, J., Castel, A. P. & Pacheco, P. (2014). A first rock glacier inventory for the Bolivian Andes. *Permafrost and Periglacial Processes*, 25(4): 333–342.

Rapp, A. (1959). Avalanche boulder tongues in Lappland. *Geografiska Annaler*, 41: 34–38.

Rapp, A. (1960a). Recent development of mountain slopes in Kärkevagge and surroundings, northern Sweden. *Geografiska Annaler*, 42: 71–200.

Rapp, A. (1960b). *Talus slopes and mountain walls at Tempelfjorden, Spitzbergen*. Norsk Polarinstitutt Skrifter, 119. 96p.

Rapp. A. (1960c). *Talus slopes and mountain walls at Tempelfjorden, Spitzbergen*. Thesis, Meddelanden Uppsala Universitets Geografiska Institution, 158A: 71–200.

Rapp, A. (1967). Pleistocene activity and Holocene stability of hillslopes, with examples from Scandinavia and Pennsylvania. *Les Congrès et Colloques de L'Université de Liège* 40: 229–244.

Rapp, A. (1985). Extreme rainfall and rapid snowmelt as causes of mass movements in high latitude mountains. In: Church, M. and Slaymaker, O., (eds.). *Field and theory: Lectures in Geocryology*. Vancouver. University of British Columbia Press: 35–56.

Rapp, A. (1995). Slush-avalanches in the Abisko area in May to June, 1995. From the Vale of Tears to Illumination Mountain. *Svensk Geografisk Arsbok*, 71: 100–107.

Rapp, A. & Åkermann, H. J. (1993). Slope processes and climate in the Abisko Mountains, northern Sweden. In: Frenzel, B. (ed). *Solifluction and climate variation in the Holocene.* Stuttgart. Gustav Fisher Verlag: 163–177.

Rapp, A. & Rudberg, S. (1964). Studies on periglacial phenomena in Scandinavia 1960–1963. *Biuletyn Peryglacjalny,* 14: 79–81.

Rauser, C. L., Mueller, L. D. & Rose, M.R. (2005). Evolution of late life. *Ageing Research Revue,* 5: 14–32.

Ray, R. J., Krantz, W. B., Caine, T. N. & Gunn, R. D. (1983a). A mathematical model for patterned ground sorted polygons and stripes and underwater polygons. *Proceedings of the 4th International Conference on Permafrost, Fairbanks, Alaska.* Washington, D. C., National Academy Press, 1036–1041.

Ray, R. J., Krantz, W. B., Caine, T. N. & Gunn, R. D. (1983b). A model for sorted patterned ground regularity. *Journal of Glaciology,* 29: 317–337.

Raymo, M. E. (1992). Global climate change: a three million year perspective. In: Kukla, G. J. and Went, E. (eds.). *Start of a glacial.* NATO ASI Series, Series 1. Global Environmental Change, 3: 207–223.

Raynolds, M. K., Walker, D. A., Munger, C. A., Vonlanthen, C. M., & Kade, A. N. (2008). A map analysis of patterned-ground along a North American Arctic Transect. *Journal of Geophysical Research,* 113., G03S03. doi: 10.1029/2007JG000512.

Razaqpur, A. G. & Wang, D. (1996). Frost-induced deformations and stresses in pipelines. *International Journal of Pressure Ves. and Piping,* 69: 105–118.

Rea, B. R., Whalley, W. B., Rainey, M. M. & Gordon, J. E. (1996). Blockfields, old or new? Evidence and implications from some plateaus in northern Norway. *Geomorphology,* 15: 109–121.

Red, M., Pulina, J. B., & Trzcinski, Ju. (1996). *Guide des terrains Karstiques choisis de la Siberie Orientale et de l'Oural.* Sosnowjec, Universitete de Silisie. 126p. [In French].

Reed, R. E. (1966). Refrigeration of a pipe pile by air circulation. *U.S. Army, Cold Regions Research and Engineering, Laboratory Technical Report* #156. 19p.

Reedyk, S., Woo, M.-K. & Prowse, T. (1995). Contribution of ice ablation to streamflow in a discontinuous permafrost area. *Canadian Journal of Earth Sciences,* 32(1): 13–20.

Refsdal, G. (1987). Frost protection of road pavements. Oslo. Norwegian Committee on Permafrost. *Frost action and Soils,* 26: 3–19.

Reger, R. D. & Péwé, T. L. (1976). Cryoplanation terraces; indicators of a permafrost environment. *Quaternary Research,* 6: 99–109.

Reid, J. R. & Nesje, A (1988). A giant ploughing block, Finse, Southern Norway. *Geografiska Annaler,* 70A: 27–33.

Reid, R. L., Tennant, J. S. & Childs, K. W. (1975). The modelling of a thermosyphon type permafrost protection device. *American Society of Mechanical Engineering, Journal of Heat Transfer,* August: 382–386.

Reimnitz, E., Graves, S. M. & Barnes, P. W. (1988). Beaufort Sea coastal erosion, sediment flux, shoreline evolution, and the erosional shelf profile, *Map 1-1881-G,* p. 22. Reston, Virginia. U.S. Geological Survey.

Reimnitz, E., Barnes, P. W. & Harper, J. R. (1990). A review of beach nourishment from ice transport of shoreface material, Beaufort Sea, Alaska. *Journal of Coastal Research,* 6: 439–470.

Rekant, P. & Vasilev, A. (2011). Distribution of subsea permafrost at the Kara Sea shelf. *The Earth Cryosphere,* 15(4): 69–72.

Remizov, V.V., Lanchakov, G.A., Chigir, V.G., Khrenov, N.N., Yegurtsov, S.A. & Samoylova, V.V. (1997). Technogenic frost heave and ground settlement on the rights-of-way of northern pipelines: practical results. *Gas Industry,* 11: 17–20. [In Russian].

Rempel, A. W., Wettlaufer, J. S. & Worster, M. G. (2004). Premelting dynamics in a continuum model of frost heave. *Journal of Fluid Mechanics,* 498: 227–244.

Retallack, G. J. (2001). A 300-million year record of atmospheric carbon dioxide from fossil plant cuticles. *Nature*, 411: 287–290.

RICGDR (1975). *Permafrost*. Research Institute of Glaciology, Cryopedology and Desert Research, Acadaemica Sinica, Lanzhou, China. Technical Translation #2006, The Canada Institute for Scientific and Technical Information, National Research Council of Canada, 1981. 146p.

Richter, H., Haase, G. & Barthel, H. (1963). Die Goletzterrassen. *Petermanns Geographische Mitteilungen*, 107: 183–192. [In German].

Rieger, S. (1983). *Soils of the Cold Regions*. New York, Academic Press. 230p.

Rieke, R., Vinson, T. S. & Mageau, D. W. (1983). The role of specific surface area and related index properties in the frost heave susceptibility of soils. In: *Proceedings of the 4th International Conference on Permafrost, Fairbanks, Alaska* 1066–1071.

Rilo, I. P., Dolgikh, G. M., & Blasov, V. F. (2013). New soil thermal stabilization systems for building fundamentals in permafrost regions. *Sciences in Cold and Arid Regions*, 5(4): 387–392.

Rio Tinto (2009). *Diavik Diamond Mine Factbook*. Pdf from diavik.ca website.

Ripmeester, J. A., Tse, J. S., Ratcliffe, C. I. & Powell, B. M. (1987). A new clathrate structure. *Nature*, 325: 135–136.

Riseborough, D.W., Williams, P.J. & Smith, M.W. (1993). Pipelines buried in freezing ground: A comparison of two ground-thermal conditions. In: Yoon, M., Murray, A. and Thygesen, J. (eds.). *Proceedings of the 12th International Conference on Offshore Mechanics and Arctic Engineering*. Book G00681. American Society of Mechanical Engineers: 187–193.

Robin, G. de Q. (1972). Polar ice sheets: A review. *Polar Record*, 100: 5–22.

Robinson, R. S. & Johnsson, M. J. (1997). Chemical and physical weathering of fluvial sands in an arctic environment: Sands of the Sagavanirktok River, North Slope, Alaska. *Journal of Sedimentary Research*, 67(3): 560–570.

Robinson, S. D. & Pollard, W. H. (1998). Massive ground ice within Eureka Sound bedrock, Ellesmere Island, Canada. In: *Permafrost, 7th International Conference Proceedings, Yellowknife*. Collection Nordicana, 55: 949–954.

Robjent, L. & Dosh, W. (2009). Warm-mix asphalt for rural county roads. In: *Proceedings of the 14th Conference on Cold Regions Engineering*. Reston, Virginia. American Society of Civil Engineers: 438–454.

Rockie, W. A. (1942). Pitting on Alaskan farms: a new erosion problem. *Geographical Review*, 32: 128–134.

Roggensack, W. D. & Morgenstern, N. R. (1978). Direct shear tests on natural fine-grained permafrost soils. *Proceedings of the 3rd International Conference on Permafrost, Edmonton, Alberta*. Ottawa. National Research Council of Canada: 728–735.

Rogers, J., Allard, M., Sarrazin, D., L'Hérault, E., Doré, G. & Guimond, A. (2015). Evaluating the use of distributed temperature sensing for permafrost monitoring in Salluit, Nunavik. In: *Compendium of selected papers, GEOQuébec 2015*. Transport Canada: 93–98.

Rognon, P. (1967). *Le massif de l'Atouko et ses bordures*. Paris. Etude géomorphologique. Etude du Centre National de la Recherche Scientifique. 559p. [In French].

Rogov, V. V. (1987). The role of gas-liquid inclusions in mechanism of cryogenic disintegration of quartz. *Vestnik, Moscow State University, Geography*, 3: 81–85. [In Russian].

Rohne, R. J. & Lebens, M. A. (2009). Subgrade temperature and freezing cycle in pervious pavements. In: *Proceedings of the 14th in shallow on Cold Regions Engineering*. Reston, Virginia. American Society of Civil Engineers: 429–437.

Rokos S. I., Kostin, D. A. & Dlugach, A. G. (2001). Free gas and permafrost sediments of the Pechora and Kara inner shelves. In: *Sedimentological processes an the evolution of marine systems in oceanic periglacial conditions*. KNC RAN. Apatity Book 1: 40–51. [In Russian].

Roland, J. C., Jones, C. E., Altmann, G., *et al.* (2010). Arctic landscapes in transition: Responses to thawing permafrost. *Eos*, 91(26): 229–331.

Romanovsky, N. N. (1973). Regularities in formation of frost-fissures and development of frost-fissure polygons. *Biuletyn Peryglacjalny*, 23: 237–277.

Romanovsky, N. N. (1974). Development principles of the polygonal-wedge microrelief. *Moscow University Geology Bulletin*, 29(5): 67–78. [In Russian].

Romanovsky, N. N. (1976). The scheme of correlation of polygonal wedge structures. *Biuletyn Peryglacjalny*, 26: 287–294.

Romanovsky, N. N. (1977). Formation of polygon wedge structures. *Novosibirsk Izdatelstov 'Nauka'*. Akademie Nauk SSR, Sibirskoye Otdeliniye: 215p. [In Russian].

Romanovsky, N. N. (1985). Distribution of recently active ice and soil wedges in the USSR. In: Church, M. and Slaymaker, O. (eds). *Field and theory: Lectures in geocryology*. Vancouver. University of British Columbia Press: 154–165.

Romanovsky, N. N. (1993). *Fundamentals of the cryogenesis of the Lithosphere*. Moscow. Moscow University Press. 336p. [In Russian].

Romanovsky, N. N., Afanasenko, V. E. & Koreisha, M. M. (1978). Long term dynamics of groundwater icings. Proceedings of the *3rd International Conference on Permafrost, Edmonton, Alberta*. Vol.1, Part 1., English translations of 26 of the Soviet papers. Ottawa. National Research Council of Canada: 195–207.

Romanovsky, N. N. & Hubberton, H. W. (2001a). Permafrost formation and evolution on shelf and lowlands (on the example of Laptev Sea). *Izvestiya Academii Nauk Seriya Geograficheskaya*, 3: 15–28.

Romanovsky, N. N. & Hubberton, H. W. (2001). Results of permafrost modeling of the lowlands and shelf of the Laptev Sea region, Russia. *Permafrost and Periglacial Processes*, 12: 191–202.

Romanovsky, N. N., Hubberton, H. W., Gavrilov, A. V., Eliseeva, A. & Tipenko, G. (2005). Offshore permafrost and gas hydrate stability zone on the shelf of East Siberian Seas. *Geo-Marine Letters* 25: 167–182.

Romanovsky, N. N., Hubberton, H. W., Gavrilov, A. V., Tumskoy, V. E. & Kholodov, A. I. (2004). Permafrost of the east Siberian shelf and coastal lowlands. *Quaternary Science Reviews*, 23: 1359–1369.

Romanovsky, N. N., Hubberton, H. W., Gavrilov, A. V., Tumskoy, V. E., Tipenko, G., Grigoreiev, M. & Siegert, C. (2000). Thermokarst and land-ocean interactions, Laptev Sea region, Russia. *Permafrost and Periglacial Processes*, 11: 137–152.

Romanovsky, N. N. & Tyurin, A. I. (1983). Rock stream deserption. *Proceedings of the 4th International Permafrost Conference*. Washington, D.C. National Academy Press: 1078–1082.

Romanovsky, N. N. & Tyurin, A. I. (1986). Kurums. *Biuletyn Peryglacjalny*, 31: 249–259.

Romanovsky, N. N., Tyurin, A. I. & Sergeev, D. O. (1989). *Kurums of Bald-Mountain belt*. Novosobirsk. Nauka. [In Russian].

Romanovsky, V. & Cailleux, A. (1942). Sols polygonaux et fentes de desiccation. *Bulletin de la Société Géologique Française, 5th série*, 12(7-8-9): 321–327. [In French].

Romanovsky, V., Jafarov, E., Genet, H., McGuire, D. & Marchenko, S. S. (2011). *Fire and Permafrost*. FRI-0850_VRomanovsky.pdf (accessed on 30th March, 2013).

Romanovsky, V.E., Cable, W. L., Kholohov, A.L., *et al.* (2014). Changes in permafrost and active layer thickness due to climate in Prudhoe Bay region and North Slope, Alaska. www.geobotany.uaf.sdu/library/posters/Romanovsky/2014_OttawaAC2014_pas20141205pdf. Accesses on May 5, 2016.

Römkens, M. J. M. (1969). Migration of mineral particles in ice with a temperature gradient. Unpublished Ph.D. thesis, Cornell University. 109p.

Römkens, M. J. M. & Miller, R. D. (1973). Migration of mineral particles in ice with a temperature gradient. *Journal of Colloid and Interface Science*, 42: 103–111.

Rook, J. J. (1974). Formation of haloforms during chlorination of natural waters. *Journal of Water Treatment Exam.*, 23: 234–243.

Roscow, J. (1977). *800 miles to Valdez*. New York. Prentice-Hall Inc. 277p.

Roseen, R. M., Ballestero, T. P., Houle, K. M., Briggs, J. F. & Houle, J. J. (2009). Pervious concrete and porous asphaltic pavements performance for stormwater management in Northern Climates. In: *Proceedings of the 14th Conference on Cold Regions Engineering, Duluth*. Reston, Virginia. American Society of Civil Engineers: 311–327.

Rosen, Y. (2014). Experiment finds wicking fabric battles "frost boils" on Dalton Highway. *Alaska Despatch*, April 9th, 2014.

Rosendahl, G. P. (1981). Alternative strategies used in Greenland. In: Smith, D. W. & Hrudey, S. E. (eds.). *Design of waste water services for cold climate communities*. Oxford. Pergamon Press: 3–16.

Ross, N. (2006). *A re-evaluation of the origins of Late Quaternary ramparted depressions in Wales*. Unpublished Ph.D. thesis, School of Earth, Ocean and Planetary Sciences, Cardiff University. 424p. UMI U584973, published on microfilm, 2013.

Rothschild, L. J. & Mancinelli, R. L. (2001). Life in extreme environments. *Nature*, 409: 1092–1101.

Rouse, W. R. (1976). Microclimatic changes accompanying burning in subarctic woodland. *Arctic and Alpine Research*, 8(4): 291–304.

Rouse, W. R. & Kershaw, K. A. (1971). The effects of burning on heat and water regimes of lichen-dominated Subarctic surfaces. *Journal of Arctic and Alpine Research*, 3: 291–304.

Rowley, R. K., Watson, G. H. & Ladanyi, B. (1973). Vertical and lateral pile load tests in permafrost. *Proceedings of the 2nd International Permafrost conference, Yakutsk, USSR. North American Contribution*. Washington, D.C. National Academy of Sciences: 712–721.

Rowley, R.K., Watson, G.H. & Ladanyi, B. (1975). Prediction of pipe performance in permafrost under lateral load. *Canadian Geotechnical Journal*, 12: 510–523.

Roy-Leveillee, P. & Burn, C. R. (2015). Geometry of oriented lakes in Old Crow Flats, northern Yukon. GeoQuébec 2015 #284.

Royer, D. L., Berner, R. A., Montañez, I. P., Tabor, N. J. & Beerling, D. J. (2004). CO_2 as a primary driver of Phanerozoic climate. *GSA Today*, 14(3): 4–10.

Rozenbaum, G. Z. (1987). Mechanism of syngenetic growth of ice-wedges in alluvial deposits. In: *Cryogenic processes*. Moscow. Moscow University Press: 4–28. [In Russian].

Ruddiman, W. F., Raymo, M. E. & McIntyre, A. (1986). Matuyma 41,000-year cycles. *Earth and Planetary Science Letters*, 80: 117–129.

Ruddiman, W. F. (2001). *Earth's climate: Past and future*. New York. W. H. Freeman and Sons.

Rudram, A. K. (1994). *The influence of periglacial processes on the stability of shallow slopes*. Unpublished M.Sc. dissertation, School of Engineering, University of Wales, Cardiff.

Ruuhijärvi, R. (1960). Über die regionaleinteilung der nord-finnischen Moore. *Annales Botanici Societatisoologicae Botanicae Fennicae "Vanamo"*, 31:1. [In German].

Rybkin, V. V., Nastchik, N. P. & Marcul, R. V. (2013). Stability of the continuous weld rail track on the concrete sleepers on the curves with radius $R < l = 300$ m. *Sciences in Cold and Arid Regions*, 5(5): 654–658.

Sadovsky, A. V. & Dorman, Y. A. (1981). The artificial freezing and cooling of soils at engineering sites. *Engineering Geology*, 18(1–4): 327–331.

Saemundsson, T., Arnalds, O., Kneisel, C., Johnsson, H. P. & Decaulne, A. (2012). The Orravatnsrustir palsa site in Central Iceland – Palsas in an Aeolian sedimentation environment. *Geomorphologie*, 167–168: 13–20.

Salt, K. E. & Ballantyne, C. K. (1997). The structure and sedimentology of relict talus, Knockan, Assynt, N. W. Scotland. *Scottish Geographical Magazine*, 113: 82–89.

Salukov, V. V., Kolotovsky, A. H., Teplinsky, Y. A. & Kuzbojev, A. S. (2000). Predisposition of pipes of the large diameter to stress – corrosion destructions. *Gas Industry*, 12: 44–46. [In Russian].

Salvigsen, O. & Elgersma, A. (1985). Large-scale karst features and open taliks at Varde-borsletta, Outer Isfjorden, Svalbard. *Polar Research*, 3: 145–153.

Samson, H. (1974). *Évolution di pergelisol en milieu tourbeux en relation avec le dynamism de la vegetation, golfe de Richmond, Nouveau-Québec*. Unpublished M.Sc. thesis. Univertsité Laval, Québec. 158p. [In French].

Sanders, D., Widera, L. & Osterman, M. (2014). Two-layer scree/snow-avalanche triggered by rockfall (Eastern Alps): Significance for sedimentology of scree slopes. *Sedimentology*, 61: 996–1030.

Sanger, F. J. (1969). Foundation of structures in cold regions. *U.S. Army, Cold Regions Research and Engineering Laboratory*. Monograph #M111-C4. 91p.

Sannel, A. B. K., Hugelius, G., Jansson, P. & Kuhry, P. (2016). Permafrost warming in a Subarctic Peatland – which meteorological controls are most important? *Permafrost and Periglacial Processes*, 27(2): 177–188.

Sass, O. (2006). Determination of the internal structure of alpine talus deposits using different geophysical methods (Lechtaler Alps, Austria). *Geomorphology*, 80: 45–58.

Sass, O. & Krautblatter, M. (2007). Debris flow-dominated and rockfall dominated talus slopes: Genetic models derived from GPR measurements. *Geomorphology*, 86: 176–192.

Sassen, R. & MacDonald, I. R. (1994). Evidence for structure H hydrate, Gulf of Mexico continental slope. *Organic Geochemistry*, 22: 1029–1032.

Satake, K. (1977). Disappearance of puddle water during night with growth of ice needles and its reappearance by day. *Nature*, 265: 519–520.

Saunders, R. J. (1989). Relationships between soil parameters and wheel ditching production rates in permafrost. M. Eng. Thesis, University of Alberta, Edmonton.

Saveliev, B. A. (1963). *Structure, content and properties of ice cover of sea and fresh water bodies*. Moscow. Moscow State University Press. 541 p. [In Russian].

Saveliev, V. S. (1962a). *Solifluktsa. Vechnaya Merzlota Chukotka*. Trudy Sev-vost. Kompl. Nauch. Issled. Instituta Volume 10. [In Russian].

Saveliev, V. S. (1962b). *Sklonovye processy na Chukotke I ikh vozdeistviye na inzhenernye sooruzheniya*. Fondy Instituta Merzlotovedeniya, Akademiya Nauk SSSR. [In Russian].

Sayles, F. H. (1984). Design and performance of water-retaining embankments in permafrost. *Proceedings of the 4th International Conference On Permafrost, Fairbanks, Alaska*. Final Proceedings: 31–42.

Sayles, F. H. (1987). *Embankment dams on permafrost*. U.S. Army Corps of Engineers Special Report 87-11. 109p.

Scapozza, C., Lambiel, C., Baron, L., Marescot, L. & Reunard, E. (2011). Internal structure and permafrost distribution in two alpine periglacial talus slopes, Valais, Swiss Alps. *Geomorphology*, 132: 208–221.

Schaefer, J. M., Shlubchter, C., Wieler, R., Ivy-Ochs, S., Kubik, P, Marchant, D., Korschinek, G., Kine, K., Herzog, G. & Serefiddin, F. (1995). News of the oldest ice on Earth buried in Antarctica, and a new cosmogenic tool. *Geochimica et Cosmochimica Acta Supplement, 69*, Supplement 1 (Goldschmidt Conference Abstracts 2005, A164).

Schaerer, P. A. (1962). Planning avalanche defence works for the Trans-Canada Highway at Rogers Pass, B.C. *Engineering Journal*, 45(3): 31–38.

Schaerer, P. A. (1972). Terrain and vegetation of snow avalanche sites at Rogers Pass, British Columbia. In: Slaymaker, O. & McPherson, H. J., (eds.). *Mountain Geomorphology*. Vancouver. Tantalus Research Press. 274p. Chapter 5-3: 215–222.

Schirrmeister, L., Siegert, K. & Kuntsky, V. (2002). Quaternary ice-rich permafrost sequences as a paleoenvironmental archive for the Laptev Sea Region in northern Siberia. *International Journal of Earth Sciences*, 91(1): 154–167.

Schmertmann, J. H. & Taylor, R. S. (1965). *Quantitative data from a patterned ground site over permafrost*. U.S. Army Cold Regions Research and Engineering Laboratory Research Report #96.

Schmid, J. (1955). *Der bodenfrost als morphologischer factor*. Hiedelburg. Dr. Alfred Hüthig Verlag. 144p. [In German].

Schofield, R. K. (1935). The pf of water in soil. *Transactions of the Third International Congress on Soil Science*, Oxford, Great Britain 2: 37–48.

Schubert, D. H. & Heintzman, T. (1994). Tundra ponds as natural wastewater treatment and disposal facilities in rural Alaska. *Proceedings of the 7th National Symposium on Individual and small community sewage systems, Atlanta, Georgia*.

Schumacher, B. A. (2002). *Methods for the determination of total organic carbon (TOC) in soils and sediments*. Ecological Risk Assessment Support center, Office of Research and Development, US. Environmental Protection Agency, NCEA-C-1282. EMASC-001. 23p.

Schunke, E. (1973). Palsen und kryokarst in Zentral Island. *Nachr. Wissenschaften Göttingen*, Matematische-Phys. Klasse, 2: 65–102. [In German].

Schunke, E. (1975). *Die periglazialerscheinungen Islands in abhängigkeit von Klima und Substrat*. Academie Wiss. Göttingen Abhandlungen, Mathematische-Phys. Kl. Folge, 30(3): 273p. [In German].

Schunke, E. (1977a). Zur Okologie der Thufur Islands. *Ber. A. d Forschungsstelle "Nedri As"*, 26. Hveragerdi, Iceland. 69p. [In German].

Schunke, E. (1977b). Zur genese der Thufur Islands und Ost-Grönlands. *Erdkunde*, 31: 279–287. [In German].

Schunke, E. (1977c). The ecology of thufurs in Iceland. *Berichte ans der forschurgestelle Nedri As, Hveragerdi (Iceland)*, 26: 39–69.

Schunke. E. (1981). Zur kryogenen Bodendynamik der arktischen Tundren Nordamerikas und Nordeuropas. *Polarforschung*, 51: 161–174. [In German].

Schunke, E. & Tarnocai, S. C. (1988). Earth hummocks (Thufur). In: Clark, M. J. (ed.). *Advances in Periglacial Geomorphology*. New York. Wiley and Sons. Chapter 10: 231–245.

Scotter, G. W. & Zoltai, S. C. (1982). Earth hummocks in the Sunshine Area of the Rocky Mountains, Alberta and British Columbia. *Arctic*, 35(3): 411–416.

Sedih, A. D. (1993). *Losses of gas on objects of main transport*. Manuscript, IRC Gasprom. 47p. [In Russian].

Sedykh, A. D. & Khrenov, N. N. (1999). Current safety problems on operational gas pipelines in northwest Siberia. *Oil and Gas Construction*, 1: 58–61. [In Russian].

Sekretov, S. (1999). Euasian Basin – Laptev Sea geodynamic system: Tectonic and structural evolution. *Polarforschung*, 69: 51–54.

Selby, M. J. (1971). Salt weathering of landforms, and an Antarctic example. *Proceedings, 6th Geography Conference*, New Zealand Geographical Society, Christchurch: 30–35.

Seligman, B. J. (1998). *Key factors influencing the reliability of trunk gas pipelines in the Western Siberian North*. Unpublished Ph.D. thesis, University of Cambridge, England.

Seligman, B. J. (2000). Long-term variability of pipeline-permafrost interactions in north-west Siberia. *Permafrost and Periglacial Processes*, 11: 5–22.

Selker, J. S., van de Giessen, N., Westhoff, M., Lexenburg, W. & Parlange, M. B. (2006). Fibre optics opens window on stream dynamics. *Geophysical Research Letters*, 33 (L24401): 1–4.

Sellman, P. V., Brown, J., Lewellen, R. I., McKim, H. & Merry, C. (1975). The classification and geomorphic implications of thaw lakes on the Arctic Coastal Plain, Alaska. US Army Cold Regions Research and Engineering Laboratory Research, Hanover, New Hampshire. Report 334.

Semmel, A. (1969). *Verwitterungs-und abtrgungserscheinungen in rezentum periglazialgebieten (Lappland und Spitzbergen).* Würzburger Geographische Arbeiten, 26. 82p. [In German].

Senneset, K. (2000). *Proceedings, International Workshop on Permafrost Engineering, Longyearbyen, Svalbard, Norway, 18–21 June, 2000.* Norwegian University of Science and Technology (NTNU)/The University Courses on Svalbard (UNIS). 327p.

Seppälä, M. (1971). Evolution of eolian relief of the Kaamasjoki-Kiellajoki river basin in Finnish Lapland. *Fennia*, 104: 88p.

Seppälä, M. (1972). The term "palsa". *Zeitschrift für Geomorphologie* N.F., 16: 463.

Seppälä, M. (1982). An experimental study of the formation of palsas. In: *Proceedings of the 4th Canadian Permafrost Conference, Calgary.* Ottawa, National Research Council: 36–42.

Seppälä, M. (1986). On the origin of palsas. *Geografiska Annaler*, 68A: 141–147.

Seppälä, M. (1988). Palsas and related forms. In: Clark, M. J., (ed.). *Advances in Periglacial Geomorphology.* Chichester. Wiley and Sons: 247–278.

Seppälä, M. (1993). Surface abrasion of palsas by wind action in Finnish Lapland. *Geomorphology*, 52: 141–148.

Seppälä, M. (1994). Snow depth controls palsa growth. *Permafrost and Perglacial Processes*, 5(4): 283–288.

Seppälä, M. (2004). *Wind as a geomorphic agent in cold climates.* Cambridge. Cambridge University Press.

Seppälä, M. (2005a). Frost heave on earth hummocks (pounus) in Finnish Lapland. *Norsk Geografisk Tidsskrift*, 59(2): 171–176.

Seppälä, M. (2005b). Dating palsas. In: Ojala, A. K. (Ed.). *Quaternary studies in the northern and Arctic regions of Finland.* Geological Survey of Finland, Special Paper 40: 79–84.

Seppälä, M. (2006). Palsa mires in Finland. *The Finnish Environment*, 23: 155–162.

Sergeev, G. B. & Batyuk, B. A. (1978). *Cryochemistry.* Moscow. [In Russian].

Serrano, E. & López-Martinez, J. (1998). Caracterización y distribucion de las formas y los procesos periglaciares en las Islas Shetland del Sur (Antártida). In: Órtiz, A. G., Franch, F. S., Schulte, L. & Navarro, A. G. (eds.). *Procesos biofisicos actuals en melos frios. Estudios recientes*, 181–204. Barcelona. Publicacion Universitat de Barcelona. [In Spanish].

Serreze, M. C. & Barry, R. G. (2009). *The Arctic Climate System.* Cambridge. Cambridge University Press.

Shackleton, N. J., Hall, M. A. & Pate, D. (1995). Pliocene stable isotope stratigraphy of site 846. In: Pisas, N. G., Nayer, C. A., Janecek, T. R., Palmer-Jackson, A. & van Andel, T.H. (eds.). *Proceedings of the Ocean Drilling Program*, Scientific Results, 138: 337–353.

Shakhova, N. & Semiletov, I. (2007). Methane release and coastal environment in the east Siberian Arctic Shelf. *Journal of Marine Systematics*, 66: 227–243.

Shakhova, N., Semiletov, I. & Panteleev, G. (2005). The distribution of methane on the Siberian Arctic Shelves: Implications for the marine methane cycle. *Geophysical Research Letters*, 32 Lo9601.

Shakhova, N. Semiletov, I., Salyuk, A., Yusupov, V., Kosmach, D, & Gustafsson, O. (2010). Extensive methane venting to the atmosphere from sediments of the East Siberian Arctic shelf. *Science*, 327: 1246–1250.

Shamanova, I. I. & Uvarkin, Y. T. (1973). Particularities of thermokarst in the northern taiaga of the Western Siberia. *Izvestiya Akademii Nauk SSSR Seriya Geograficheskaya*, 1: 89–94. [In Russian].

Sharkhuu, N. & Luvsandagva, D. (1975). *Basic features of permafrost in Mongolia*. Ullanbaator. Mongolian Academy of Sciences Press. 108p. [In Mongolian].

Sharp, R. P. (1942a). Ground ice mounds in tundra. *Geographical Review*, 32: 417–423.

Sharp, R. P. (1942b). Periglacial involutions in Northeastern Illinois. *Journal of Geology*, 50: 113–133.

Sharp, R. P. (1942c). Mudflow Levees. *Journal of Geomorphology*, 5: 222–227.

Sharp, R. P. & Noble, L. H. (1953). Mudflow of 1941 at Wrightwood, Southern California. *Bulletin of the Geological Society of America*, 64: 547–560.

Sharpe, C. F. S. (1938). *Landslides and related phenomena*. New York. Columbia University.

Shavrina, E. V. & Guk, E. V. (2005). Modern dynamics of ice formations in Pinega caves. In: Mavlyudov, B. R. (ed.). *Glacier caves and glacial karst in high mountains and polar regions*. Moscow. Institute of Geography of the Russian Academy of Sciences: 113–117.

Shavrina, E. V., Maalov, V. N. & Gurkalo, E. I. (2005). Role of continental glaciations in karst development of Russian European North. In: Mavlyudov, B. R. (Ed.). *Glacier caves and glacial karst in High Mountains and Polar Regions*. Moscow, Institute of Geography of the Russian Academy of Sciences: 118–122.

Sheng, D., Zhang, S., Yu, Z. & Zhang, J. (2013). Assessing frost susceptibility of soils using PC heave. *Cold Regions Research and Engineering*, 95: 27–38.

Sher, A. V., Kaplina, T. N. & Oveander, M. G. (1987). Unified Regional Stratigraphic Chart for the Quaternary deposits in the Yana-Kolyma Lowland and its mountainous surroundings. Explanatory Note. In: *Decisions of Interdepartmental Stratigraphic Conference on the Quaternary of the East USSR, Magadan, 1982*. Russian Academy of Sciences, Far-Eastern Branch, North-eastern Complex Research Institute, Magadan, USSR: 29–69. [In Russian].

Sheridan, P. P., Miteva, V. I. & Brenchley, J. E. (2003). Phylogenetic analysis of anaerobic psychrophilic enrichment cultures obtained from a Greenland glacier ice core. *Applied Environmental Microbiology*, 69(4): 2153–2160.

Sheskin, Yu., Gorbunov, V. P. & Kulagin, B. A. (1992). The impact of oil on the strength properties of frozen soil. In: *Digest methods of study of cryogenic, physical and geological processes*: 30–34.

Shevchenko, L. Y. & Shitsova, I. V. (2008). Mechanical properties of frozen clay soils. *Geoecology*, 1: 78–84.

Shi, Y. & Yang, S. (2014). Numerical modeling of formation of a static ice cave – Ningwu ice cave, Shanxi, China. In: *International Workshop on Ice Caves, NICKRI Symposium* 4: 7–11.

Shilts, W. W. (1978). Nature and genesis of mudboils, central Keewatin, Canada. *Canadian Journal of Earth Sciences*, 15: 1053–1068.

Shroder, J. F. (1978). Dendrogeomorphological analysis of mass movement on Table Cliffs Plateau, Utah. *Quaternary Research*, 9: 168–185.

Shu, D. & Huang, X. (1983). Design and construction of a cutting at sections of thick-layer ground ice. In: *Permafrost: 4th International Conference Proceedings*. Washington. National Academy Press #1524: 1152–1156.

Shumskiy, P. A. (1955). *Basis of structural ice studies*. Akademiya Nauk, SSSR. Izdatel'stavo Academiya Nauk, SSSR. [In Russian].

Shumskiy, P. A. (1964). *Ground (subsurface) ice*. Ottawa, National Research Council of Canada, Technical Translation #1130.

Shumskiy, P. A. & Vtyurin, B. I. (1966). Underground Ice. In: Proceedings of the 1st International Conference on Permafrost. Ottawa. National Academy of Sciences, National Research Council of Canada Publication #1287: 108–113.

Shur, Y. (1988a). *The upper horizon of permafrost and thermokarst.* Akademia Moscow Nauka. 210p. [In Russian].

Shur, Y. (1988b). The upper horizon of permafrost soils. In: Senneset, K., (Ed.). *Permafrost, Proceedings of the 5th International Conference on Permafrost, 2–5th August, 1988. Trondheim. Tapir Press.* 1: 867–871.

Shur, Y., Hinkel, K. M. & Nelson, F. E. (2005). The transient layer: Implications for geocryology and climate-change science. *Permafrost and Periglacial Processes,* 16: 5–18.

Shur, Y., Jorgenson, T., Kanevskiy, Y. & Ping, C.-L. (2008). Formation of frost boils and earth hummocks. In: Kane, D. L. & Hinkel, K. M. (Eds.). *9th International Conference on Permafrost, Extended Abstracts*: 287–288.

Shur, Y., Kanevskiy, M. Z., Jorgenson, M. T., Fortier, D., Dillon, M., Stephani, E. & Bray, M. (2009). Yedoma and thermokarst in the northern part of Seward Peninsula, Alaska. *American Geophysical Union,* Fall Meeting, 2009. Abstract #C41A-0443.

Shur, Y. & Ping, C. (2003). The driving force of frost boils and hummocks formation. *American Geophysical Union,* Fall Meeting, 2003. Abstract #C21B-0823.

Schuur, E. A. G., *et al.* (2008). Vulnerability of permafrost carbon to climate change: Implications for the global carbon cycle. *Bioscience,* 58: 701–714.

Sibirtsev, N. M. (1895). Genetic classification of soils. *Zap. Novo-Aleksasandr. Agricultural Institute*: 1–23. [In Russian].

Siewert, M. B., Krautblatter, M., Chriatiansen, H. H. & Eckerstorfer, M. (2012). Arctic rock-wall retreat rates estimated using laboratory-calibrated ERT measurements of talus cones in Longyeardalen, Svalbard. *Earth Surface Processes and Landforms,* 37: 1542–1555.

Silvast, M., Nurmikolu, A., Wiljanen, B. & Levonmäki, M. (2010). An inspection of railway ballast quality using ground penetrating radar in Finland. *Proceedings of the Institute of Mechanical Engineers, Part F: Journal of Rail and Rapid Transit,* 224(F5): 345–351.

Silvast, M., Nurmikolu, A., Wiljanen, B. & Levonmäki, M. (2012). Identifying frost-suseptible areas on Finnish railways using GPR technique. *Proceedings of the Institute of Mechanical Engineers, Part F: Journal of Rail and Rapid Transit,* June 27. Doi: 10.1177/0954409712452076.

Silvast, M., Nurmikolu, A., Wiljanen, B. & Mäkelä, E. (2013). Efficient track rehabilitation planning by integrating track geometry and GPR data. *Proceedings of the 10th Heavy Haul Conference, New Delhi, India.*

Simmons, I. G. (1989). *The changing face of the Earth.* Oxford. Blackwell.

Sims, R. A. (1978). The use of "muskeg caps" to deter frost penetration under transmission line tower bases. In: Williams, G. F. and Curran, J. (Eds.). *National Research Council of Canada, Associate Committee on Geotchnical Research,* Technical Memorandum #122: 116–131.

Singhroy, V., Murnaghan, K. & Couture, R. (2010). InSAR monitoring of a retrogressive thaw flow at Thunder River, lower Mackenzie. Proceedings of GEO2*10, Calgary: 1317–1322.

Sjöberg, Y., Hugelius, G. & Khury, P. (2013). Thermokarst lake morphometry and erosion features in two peat plateau areas of northeast European Russia. *Permafrost and Periglacial Processes,* 24(1): 75–81.

Skaven-Haug, S. (1959). Protection against frost heaving on the Norwegian railways. *Geotechnique,* 9(3).

Skidmore, M. L., Foght, J. M. & Martin, J. (2000). Microbial Life beneath a High Arctic Glacier. *Applied and Environmental Microbiology,* 66(8): 3214–3220.

Skyles, E. & Vanchig, G. (2007). Palsa fields and cryoplanation terraces, Hangay Nuruu, Central Mongolia. *20th Annual Keck Symposium*: 49–53. http://keck.wooster.edu/publications

Slaughter, C. W. (1982). Occurrence and recurrence of aufeis in an upland Taiga catchment. *Proceedings of the 4th Canadian Conference on Permafrost, Calgary.* Ottawa. National Research Council of Canada: 182–188.

Sletten, R. S., Hallet, B. & Fletcher, R. C. (2003). Resurfacing time of terrestrial surfaces in the formation and maturation of polygonal patterned ground. *Journal of Geophysical Research*, 108(E4): 8044. Doi: 10.1029/2002JE001914.

Slupsky, J. W. (ed.) (1976). *Some problems of solid and liquid waste disposal in the northern environment*. Environmental Protection Series Report, EPS-4-NW-76-2. 230p.

Smith, A. P. (1974). Population dynamics and life forms of *Espeletia* in the Venezuelan Andes. Unpublished Ph.D. dissertation, Duke University. 254p.

Smith, D. I. (1972). The solution of limestone in an arctic environment. In: *Polar Geomorphology*. Institute of British Geographers Special Publication, #4: 187–200.

Smith, D. J. (1987). Late Holocene solifluction lobe activity in the Mount Rae area, Canadian Rocky Mountains. *Canadian Journal of Earth Sciences*, 24: 1634–1642.

Smith, D. J. (1988). Rates and controls of soil movement on a solifluction slope in the Mount Rae area, Canadian Rocky Mountains. *Zeitschrift für Geomorphologie N.F. Supplement Band*, 71: 25–44.

Smith, D. J. (1992). Long-term rates of contemporary solifluction in the Canadian Rocky Mountains. In: Dixon, J. C. & Abrahams, A. D. (eds). *Periglacial Geomorphology*. Chichester. Wiley: 203–221.

Smith, D. W. & Finch, G. R. (1983). *A critical evaluation of the operation and performance of lagoons in cold climates*. Department of Civil Engineering, University of Alberta, Edmonton, Alberta.

Smith, D. W. & Given, P. W. (1981). Treatment alternatives for dilute, low temperature wastewater. In: Smith, D. W. & Hrudy, S. E. (eds.). *Design of water and wastewater services for cold climate communities*. Oxford. Pergamon Press: 165–179.

Smith, D. W. & Heinke, G. W. (1981). Cold climate environmental engineering – an overview. In: Smith, D. W. & Hrudy, S. E. (eds.). *Design of water and wastewater services for cold climate communities*. Oxford. Pergamon Press: 3–16.

Smith, D. W., Reed, S., Cameron, J. J., Heinke, G. W., James, F., Reid, B., Ryan, W. L. & Scribner, J. (1979). *Cold climate utility manual*. Ottawa. Environmental Protection Service, Report EPS 3-WP-79-2. 650p.

Smith, D. W. & Sego, D. C. (1994). *Cold Regions Science and Engineering: A global Perspective*. Proceedings of the 7th International Cold Regions Engineering Speciality Conference, March 7–9th, 1994, Edmonton. Montreal. Canadian Society for Civil Engineering. 869p.

Smith, H. T. U. (1953). The Hickory Run boulder field, Carbon County, Pennsylvania. *American Journal of Science*, 251: 625–642.

Smith, M. W. (1975). Microclimatic influences on ground temperatures and permafrost distribution, Mackenzie Delta, Northwest Territories. *Canadian Journal of Earth Sciences*, 12: 1421–1438.

Smith, M. W. (1985). Observations of soil freezing and frost heave at Inuvik, Northwest Territories, Canada. *Canadian Journal of Earth Sciences*, 22: 283–290.

Smith, M. W. & Riseborough, R. W. (2002). Climate and Limits of Permafrost: A Zonal Analysis. *Permafrost and Periglacial Processes*, 13: 1–15.

Smith, N. & Berg, R. L. (1973). Encountering massive ground ice during road construction in Central Alaska. *Proceedings of the 2nd International Conference on Permafrost, Yakutsk, USSR. North American Contribution*. Washington. National Academy of Sciences: 736–745.

Smith, S. L. & Burgess, M. M. (1999). Mapping the sensitivity of Canadian permafrost to climatic warming. In: *Interactions between the cryosphere, climate and greenhouse gases*. Proceedings of the International Union of Geodesy and Geophysics, 99, Symposium HS2, Birmingham. IAHS #A256: 71–80.

Smith, S. L., Riseborough, D. W. & Bonnaventure, P. P. (2015). Eighteen year record of forest fire effects on ground thermal regimes and permafrost in the central Mackenzie valley, NWT, Canada. *Permafrost and Periglacial Processes*, 26(4): 289–303.

Sneath, D. (1998). State Policy and pasture degradation in Inner Asia. *Science*, 281: 1147–1148.

Snodgrass, M. P. (1971). *Waste disposal and treatment in permafrost areas; A bibliography.* Washington, D.C. U.S. Department of the Interior, Office of Library Services, Bibliography Series 22. 29p.

Sobol, I. S. & Sobol, S. V. (2014). Assessment and forecast of changes in reservoir volumes due to thermal settling in permafrost areas of Russia. *Sciences in Cold and Arid Regions*, 6(5): 428–431.

Söderman, G. (1980). *Slope processes in cold environments of northern Finland.* Fennia, 152(2): 1–86.

Soil Classification Working Group (Canada) (1998). *The Canadian system of soil classification.* 3rd Edition. Ottawa. Agriculture and Agri-Food, Canada.

Soil Survey Staff (2014). *Soil Taxomomy, a basic system of soil classification for making and interpreting soil surveys.* 2nd Edition. Agriculture Handbook # 436. Lincoln, Nebraska. Natural Resources Conservation Center.

Sokolov, B. L. (1973). Regime of naleds. *Proceedings of the 2nd International Conference on Permafrost, Yakutsk. USSR Contribution.* Washington. National Academy of Sciences Press: 408–411.

Sokolov, B. L. (1978). Regime of naleds. In: *Permafrost: the USSR contribution to the 2nd International Conference.* Washington. National Academy of Sciences: 408–411.

Sokolov, I. A. (1980). Nongley hydromorphic soil formation. *Soviet Soil Science*, 10: 17–28. [In Russian].

Sokolova, O. V. & Gorkovenko, N. B. (1997). Assessing the frost susceptibility of coarse-fragment soils with a silty-clayey filler. *Soil Mechanics and Foundation Engineering*, 34(2): 46–51.

Solbraa, K. (1971). The durabilty of bark in road construction. *Royal Norwegian Council on Science and Industry*, December, 1971.

Sollid, J. L. & Sørbel, L. (1974). Palsa bogs at Haugtjørnin, Dovrefjell, South Norway. *Norsk Geografisk Tidsskrift*, 28: 53–60.

Sollid, J. L. & Sørbel, L. (1992). Rock glaciers in Svalbard and Norway. *Permafrost and Periglacial Processes*, 9: 215–220. [In Russian].

Solntsava, N. V. (1998). *Oil extraction and the geochemistry of the natural landscape.* Moscow. Moscow State University. [In Russian].

Solomatin, V. I. (1986). *Petrology of underground Ice.* Novosibirsk. Academy Nauka. 215p. [In Russian].

Solomonov, N. G., Dasyatkin, R. V., Larionov, V. P., Ivanov, B. I., Isaev, A. P. & Borisov, Z. Z. (1998). Environmental problems of oil and gas resources in Yakutia. *Earth Cryosphere*, 5(4): 30–35. [In Russian].

Solonenko, V. P. (1960). *Essays on the engineering geology of Eastern Siberia.* Irkutsk. Irkutsk Publishing Books. [In Russian].

Soloviev, P. A. (1972). Alanyy rel'yef Centralnoy Takutii i ego proiskhozhdeniye. Mnogoletne-merzlyye porody i soputstvuyushchiye im yavlenia na territorri Yakutskoy, SSR. Moscow. Izdatel'stvo AN SSSR: 38–53. [In Russian].

Soloviev, P. A. (1973a). *Alas thermokarst relief of Central Yakutia.* Guidebook, 2nd International Permafrost Conference, Yakutsk, USSR. 48p.

Soloviev, P. A. (1973b). Thermokarst phenomena and landforms due to frost heaving in Central Yakutia. *Biuletyn Peryglacjalny*, 23: 135–155. [In Russian].

Soloviev, V., Ginzburg, G., Telepnev, E. & Mikhaluk, Y. (1987). Cryothermia and Gas Hydrates in the Arctic Ocean. Leningrad. Sevmorgelogia. 150p. [In Russian].

Soons, J. M. (1968). Erosion by needle ice in the Southern Alps, New Zealand. In: Osburn, W. H. & Wright, H. E. (eds.). *Arctic and Alpine Environments.* Proceedings of the 7th INQUA Congress, Colorado. Bloomington. Indiana University Press: 217–227.

Soons, J. M. & Greenland, D. K. (1970). Observations on the growth of needle ice. *Water Resources Research*, 6(2): 579–593.

Souchez, R. (1971). Rate of frost-shattering and slope development in dolomitic limestone, southwest Ellesmere Island (Arctic Canada). *Quaternaria*, 14: 21.

Souchez, R. A. (1984). On the isotopic composition of δD and $\delta^{18}O$ of water and ice ring freezing. *Journal of Glaciology*, 30: 369–372.

Souchez, R. A. & Lorrain, R. D. (1991). *Ice composition and glacier dynamics*. Berlin. Springer Series in Physical Environment #8. 207p.

Spector, V. B. (2002). Karst phenomena in the cryolithozone of the platform part of Yakutia. *Science and Education*, 3: 69–73. [In Russian].

Spector V. B. & Spector V. V. (2009). Karst processes and phenomena in the perennially frozen carbonate rocks of the middle Lena basin. *Permafrost and Periglacial Processes*, 20: 71–78.

Spesivtsev, V. I. (2001). The cryolithozone of the shelf of the Barents and Kara Sea. In: Paepe, R., Melnikov, V. P., Van Overloop, E. & Gorokhov, V. D. (eds.). *Permafrost response on economic development, environmental security and natural resources*. Dordrecht, Netherlands. Kluwer Academic Publishers 76: 105–134.

Spolanskaya, N. A. & Evseyev, V. P. (1973). Domed-hummocky peat bogs of the Northern Taiga in Western Siberia. *Biuletyn Peryglacjalny*, 22: 271–283.

Springer, M. E. (1958). Desert pavement and vesicular layer of some soils in the desert of the Lahontian Basin, Nevada. *Proceedings of the Soil Science Society of America*, 22: 63–66.

Stalker, A. M. (1973). *Surficial Geology of the Drumheller area, Alberta*. Geological Survey of Canada, Memoir 370: 122p.

Stangl, K. O., Roggensack, D. W. & Hayley, D. W. (1982). Engineering Geology of surficial soils, eastern Melville Island. *Proceedings of the 4th Canadian Permafrost Conference, Calgary, Alberta*. Ottawa. National Research Council of Canada: 136–147.

Stanley Associates Engineering Ltd. (1979). *Dawson City, Yukon, Utilities pre-design report*. Prepared for the Government of the Yukon Territory, Whitehorse, Yukon Territory.

Stanley, J. M. & Cronin, J. E. (1983). Investigations and implications of subsurface conditions beneath the TransAlaska Pipeline at Aitigun Pass. *Permafrost: 4th International Conference Proceedings*. Washington, D.C. National Academy Press: 1188–1193.

Stark, T. R., Arellano, D. & Horvath, J. S. (2004). *Guideline and recommended standard for Geofoam applications in highway embankments*. Washington, D.C. U.S. Transportation Research Board NCHRP report # 529.

State Council (2015). Fantastic Luyashan ice cave in China's Shanxi Province. http://english. gov.cn/news/photos/2015/04/18/content_2814750912799. Retrieved on 5th May, 2015.

Steenstrup, K. J. V. (1883). Bidrag til Kjenskab til Braeerne og Brae-Isen I Nord-Grönland. *Meddeleser om Grönland*, 4: 69–112. [In Danish].

Steever, B. (2001). *Tundra revegetation in Alaska's Arctic*. BP technical brief, BP Exploration (Alaska).

Stefan, J. (1889). On the theory of ice formation, particularly ice formation in the Arctic Ocean. *S-B Wien Akad.*, 98: 17.

Steven, B., Leveille, R., Pollard, W. H. & Whyte, L. G. (2006). Microbial ecology and biodiversity in permafrost. *Extremophiles*, 10: 259–267.

Steven, B., Niederberger, T. D. and Whyte, L. G. (2009). In: Margesin, R. (ed.). *Permafrost Soils*. Berlin/Heidelberg, Springer Verlag: 59–72.

Stevens, C. W., Gaanderse, A. J. R. & Wolfe, S. A. (2013). Lithalsa distribution, morphology and landscape associations in the Slave Lake Lowland, Northwest Territories. *Geological Survey of Canada*, Open File 6531, issue 7255.

Stewart, E.J., Madden, R., Paul, G. & Taddei, F. (2005). Ageing and death in an organism that reproduces by morphologically symmetric division. *PLOS Biology:* 3: e45.

Stone, J. O., Ballantyne, C. K. & Fifield, L. K. (1998). Exposure dating and validation of periglacial weathering limits, northwest Scotland. *Geology*, 26(7): 587–590.

Streletskiy, D. A., Tananaer, N. I., Opel, T., Shiklomanov, N. I., Nyland, K. E., Streletskaya, I. D., Tokarev, I. & Shiklomanov, A. I. (2015). Permafrost hydrology in changing climatic conditions: Seasonal variability of stable isotope composition in rivers in discontinuous permafrost. *Environmental Research Letters*, 10 095003. http://iopscience.iop.org/1748–9326/10/9/095003.

St-Onge, D. A. (1969). Nivation landforms. *Geological Survey of Canada*, Paper 69–30, 12p.

St-Onge, D. A. & Pissart, A. (1990). Un pingo en système fermé dans des dolomites paléozoïques de l' Arctique canadien. *Permafrost and Periglacial Processes*, 1: 275–282. [In French].

Stoy, P. C., Street, L. E., Johnson, A. V., Prieto-Blanco, A. & Ewing, S. (2012). Temperature, heat flux, and reflectance of common Subarctic mosses and lichens under field conditions might change the community composition and impact climate-relevant surface fluxes? *Arctic, Antarctic and Alpine Research*, 44: 500–508.

Straughn, R. G. (1972). The sanitary landfill in the Subarctic. *Arctic*, 25: 40–48.

Street, R. B. & Melnikov, P. I. (1990). Seasonal snow cover, ice and permafrost. In: Tegart, W. J. McG., Sheldon, G. W. & Griffiths, D. C. (eds.). *Climate change*, the IPPC Impacts Assessment. Chapter 7. Report prepared for the IPPC by Working Group II, Canberra. WMO-UNEP, Australian Government Publishing Service, 7-1 to 7.33.

Strelietskaya, I. D. & Vasiliev, A. A. (2009). Isotopic composition of ice wedges of West Tamyr. *Earth Cryosphere*, 13(3): 59–69. [In Russian].

Strunk, H. (1983). Pleistocene diapiric upturnings of lignites and clayey sediments as periglacial phenomena in central Europe. In: *Proceedings of the 4th International Conference on Permafrost, Fairbanks*. Washington. National Academy Press: 1200–1204.

Sturm, M. & Liston, G. E. (2003). The snow cover on lakes on the Arctic Coastal Plain of Alaska. *Journal of Glaciology*, 49: 370–380.

Sturm, M., Racine, C. & Tae, K. (2001). Climate change: increasing shrub abundance in the Arctic. *Nature*, 411: 546–547.

Strömquist, L. (1973). *Geomorfologiska studier av blockfält I narra Skamdinavien*. Uppssala Universitet, Naturgeografiska Institut, Advelningen für Naturgeografi, UNGI rapport #22. 161p. [In Swedish].

Sturzenegger, M. & Stead, D. (2009). Close-range terrestrial digital photogrammetry and terrestrial laser scanning for discontinuity characterization on rock cuts. *Engineering Geology*, 106: 163–182.

Su, K., Zhang, J. M., Feng, W. J., *et al.* (2013). Model tests on initial freezing process of column foundation on slope in permafrost regions. *Chinese Journal of Geotechnical Engineering*, 35(4): 794–799.

Su, Q. & Huang, J. J. (2013). Deformation and failure modes of composite foundation with sub-embankment plain concrete piles. *Sciences in Cold and Arid Regions*, 5(5): 624–625.

Sukhodrovsky, V. L. (1979). *Exogenous Relief Forming in the Cryolithic Zone*. Moscow. Nauka. 280 p. [In Russian].

Sukhodrovsky, V. L. (2002). About the genesis of ice complex and alas relief. *Earth Cryosphere*, 6(1): 56–61. [In Russian].

Sukha, (2013). An underground Million years ice cavern – Ningwu. *Sukha, Sunday, March 10th, 2013.* http://sukhasights.blogspot.ca/2014/03/an-underground-million-years-ice-cavern-Ningwu. Retrieved on 8th May, 2015.

Sumgin, M. I. (1927). *Vechnaya merzlota pochv v pradelakh S.S.S.R.* Vladivostok. [In Russian].

Sumgin, M. I. (1931). Conditions of soil formation in permafrost regions. *Pochvovedenie*, 26(3): 5–17. [In Russian].

Sumgin, M. I., Kachurin, S. P., Tolstukhin, N. I. & Tumel, V. F. (1940). *Obshchee Merzlotovedeniye*, Moscow., Russian Academy of Sciences. 240p. [In Russian].

Sun, H., Ge, X.-R., Niu, F.-J., Liu, G. & Zhang, J.-Z. (2014). Cooling effect of convection-intensifying composite embankment with air doors on permafrost. *Sciences in Cold and Arid Regions*, 6(4): 372–377.

Sun, H. B. & Wang, X. L. (2009). Design of tower foundation of electric power lines in permafrost region. *Inner Mongolia Electric Power*, 27(6): 32–35.

Sutherland Brown, A. (1969). The Aiyansh lave flow, British Columbia. *Canadian Journal of Earth Sciences*, 6: 1460–1468.

Svensson, H. (1962). Några iakttagelser från palsområden. *Norsk geografisk Tidsskift*, 18: 212–227. [In Norwegian].

Svensson, H. (1964). Structural observations in the minerogenic core of a pals. *Geografiska Annaler*, 17: 138–142.

Svensson, H. (1977). Observations on polygonal fissuring in non-permafrost areas of the Norden countries. *Abhandlungen der Akademie der Wissenschaften in Göttingen, Mathematishe Physikalische Klasse*, 31: 63–76.

Svensson, H. (1988a). Ice-wedge casts and relict polygonal patterns in Scandinavia. *Journal of Quaternary Science*, 3: 57–67.

Svensson, H. (1988b). Recent frost fissuring in a coastal area of southwest Sweden. *Norsk Geografisk Tidsskift*, 42: 271–277.

Sverdrup, O. (1904). *The new land – four years in Arctic regions*. London. Longmans Green. 2 volumes.

Swanson, D. K. (1996). Susceptibility of permafrost soils to deep thaw after forest fires in Interior Alaska, U.S.A., and some ecological implications. *Arctic and Alpine Research*, 28(2): 217–227.

Swanson, D. K., Ping, C. L., & Michaelson, G. J. (1999). Diapirism in soils due to thaw of ice-rich materials near the permafrost table. *Permafrost and Periglacial Processes*, 10: 349–367.

Sweet, L. (1982). Solar assisted culvert thawing device, research notes. Fairbanks. *State of Alaska, Department of Transportation and Public Facilities*, 2(1).

Swinzow, G. K. (1963). *Tunneling in permafrost*. U.S. Army Cold Regions Research and Engineering Laboratory, Hanover New Hampshire. Technical Report, 93: 14–27.

Sykes, D. J. (1971). Effects of fire and fire control on soil and water relations in Northern Forest Regions – A preliminary Review. *Proceedings, Fire in the Northern Environment – A Symposium, College, Fairbanks, Alaska*: 37–44.

Taber, S. (1918). Ice forming in clay soils will lift weights. *Engineering News-Record*, 80: 262–263.

Taber, S. (1929). Frost heaving. *Journal of Geology*, 37: 428–461.

Taber, S. (1930a). The mechanics of frost heaving. *Journal of Geology*, 38: 303–317.

Taber, S. (1930b). Freezing and thawing of soils as factors in destruction of road pavements. *Public Roads*, 11: 113–132.

Tai, X. S., Yang, X. L., Liu, G. X., Xue, L. G., Zhang, Y., Chen, T., Zhang, W., Wu, X. K. & Mao, W. L. (2014). Variations of N-and P-cucling bacterial populaions along with altitude in the Qilian Mountains. *Journal of Glaciology and Geocryology*, 36: 214–221.

Takahashi, T. (1981). Debris flow. *Annual Review of Fluid Mechanics*, 13: 57–77.

Tallman, A. M. (1973). Resistivity methodology for permafrost delineation. In: Fahey, B. D. & Thompson, R. D. (eds.). *Research in Polar and Alpine Geomorphology*. Department of Geography, University of Guelph. Geographical Publication 3: 73–83.

Tallman, A. M. (1975). Glacial and periglacial geomorphology of the Fourth of July Creek valley, Atlin region, Cassiar District, northwestern British Columbia. Unpublished Ph.D. thesis, Michigan State University. 150p.

Tang, G. H. & Wang, X. H. (2007). Effect of temperature control on tunnel construction in permafrost. *Rock and Soil Mechanics*, 28(3).

Tape, K., Sturm, M. and Racine, C. (2006). The evidence for shrub expansion in Alaska and the pan-Arctic. *Global Change Biology*, 12: 686–702.

Tapio, I., Tapio, M., Li, M. H., Popov, R., Ivanova, Z., & Kantanen, J. (2010). Estimation of the relatedness among non-pedigreed Takutian cryo-bank bulls using molecular data: Implications for conservation and breed management. *Genetics Selection Evolution*, 42(1): 28.

Tarnocai, C. (1973). *Soils of the Mackenzie River area*. Environmental Social Committee, Task Force on Northern Oil Development, Report 73–26. 136p.

Tarnocai, C. & Zoltai, S. C. (1978). Earth hummocks of the Canadian Arctic and Subarctic. *Arctic and Alpine Research*, 10: 581–594.

Tarnocai, C., Candell, J. G., Schuur, E.A.G., Kuhry, P., Mazhitova, G. & Zimov, S. (2009). Soil organic carbon pools in the northern circumpolar permafrost region. *Global Biogeochemical Cycles*, 23: GB2023. Doi: 10.1029/2008GB03327.

Taylor, A., Dallimore, S. & Outcalt, S. (1996). Late Quaternary history of the Mackenzie-Beaufort region, Arctic Canada, from modeling of permafrost temperatures. I. The onshore-offshore transition. *Canadian Journal of Earth Sciences*, 33: 52–61.

Taylor, A. & Judge, A. S. (1980). Permafrost studies in northern Québec. Scott, W. J. & Brown, R. J. E. (eds.). *Proceedings of a Symposium on Permafrost Geophysics, Calgary*. Ottawa. Associate Committee on Geotechnical Research, National Research Council of Canada. Technical Memorandum, 128: 94–102.

Taylor, R. B. (1978). The occurrence of grounded ice ridges and shore ice piling along the northern coast of Somerset Island, NWT. *Arctic*, 31: 133–149.

Tchebakova, N. M., Parfenova, E. & Soja, A. J. (2009). The effects of climate, permafrost and fire on vegetation change in Siberia in a changing climate. *Environmental Research Letters*, 4 (2009)045013. doi:10.1088/1748-9326/4/4/045013

Tedrow, J. C. (1963). Arctic Soils. *Proceedings of the 1st Permafrost International Conference*. National Academy of Sciences, National Research Council Publication 1287: 50–55.

Tedrow, J. C. F. (1966). Polar Desert Soils. *Soil Science Society of America Proceedings*, 30: 381–387.

Tedrow, J. F. C. (1969). Thaw lakes, thaw sinks and soils in northern Alaska. *Biuletyn Peryglacjalny*, 20: 337–345.

Tedrow, J. F. C. & Douglas, L. A. (1958). *Carbon 14 dating of some Arctic soils*. Department of Soils, Rutgers University. Mimeograph note. 6p.

Tedrow, J. F. C. & Ugolini, F. C. (1966). Antarctic soils. In: Tedrow, J. F. C. (Ed.). *Antarctic soils and soil processes*. Washington. American Geophysical Union Antarctic Series, #8: 161–177.

Tenthorey, G. (1992). Perennial névés and the hydrology of rock glaciers. *Permafrost and Periglacial Processes*, 3: 247–252.

Terzaghi, K. (1952). Permafrost. *Boston Society of Civil Engineers Journal*, 39: 1–50.

Tews, J. (2004). Hummock vegetation at the Arctic tree-line near Churchill, Manitoba. *The Canadian Field-Naturalist*, 118(4): 590–594.

Thaler, K. (2008). *Analyse der Temperaturverhältnisse in der Eisriesenwelt-Höhle im Tennengebirge anhand einer 12 jährigen Messreihe*. M.Sc. thesis, Institut für Meteorologie und Geophysik, Leopold-Franzens Universität, Innsbruck. 101p. [In German].

Thie, J. (1974). Distribution and thawing of permafrost in the southern part of the discontinuous permafrost zone in Manitoba. *Arctic*, 27: 187–200.

Thomas, H. P. & Ferrell, J. E. (1983). Thermokarst features associated with the buried sections of the Trans-Alaska Pipeline. *Proceedings of the 4th International Permafrost Conference, Fairbanks*. Washington, D. C. National Academy of Sciences Press: 1245–1250.

Thomas, H. P., Johnson, E. R., Stanley, J. M., Shuster, J. A. & Pearson, S. W. (1982). Pipeline stabilization project at Aitigun Pass. *Proceedings of the 3rd International Symposium on Ground Freezing, Hanover, New Hampshire*. U.S. Cold Regions Research and Engineering Laboratory.

Thompson, E. G. & Sayles, F. H. (1972). *In situ* creep analysis of a room in frozen soil. *American Society of Civil Engineers. Journal of Soil Mechanics and Foundation Division*, #SM6: 899–915.

Thompson, P. (1976). *Cave Exploration in Canada*. Edmonton. The Canadian Caver. 183p.

Thomson, S. (1963). Icings on the Alaska Highway. *Proceedings of the 1st Intrnational Conference on Permafrost, Lafayette, Indiana*. Washington, National Academy of Sciences Publication 1287: 526–529.

Thoraninsson, S. (1951). Notes on patterned ground in Iceland, with particular reference to the Icelandic "flás". *Geografiska Annaler*, 33: 144–156.

Thorn, C. E., Darmody, R. G. and Dixon, J. C. & Schlyter, P. (2001). The chemical weathering regime at Karkevagge, arctic-alpine Sweden. *Geomorphology*, 41: 37–52.

Thorn, C. E., Darmody, R. G. & Dixon, J. C. (2014). Rethinking weathering and pedogenesis in alpine periglacial regions: some Scandinavian evidence. In: Martini, I. P., French, H. M. & Pérez, A. A., (eds.). *Ice-marginal and Periglacial Processes and Sediments*. London. Geological Society Special Publications, #354: 183–193.

Thoroddsen, Th. (1913). Polygonböden und "thufur" auf Island. *Petermanns Geogaphische Mittelungen*, 59: 253–255. [In German].

Thoroddsen, Th. (1914). An account of the physical geography of Iceland with special reference to plant life. In: Kolderup-Rosenvinge, L. & Warming, E. (eds.). *The botany of Iceland*, volume 1, 1912–1918: 187–343. Copenhagen, J. Frimodt; London, John Weldon. 675p.

Tian, Y. H., Fang, J. H. & Shen, Y. P. (2013). Research on EPS application to very wide highway embankments in permafrost regions. *Sciences in Cold and Arid Regions*, 5(4): 503–508.

Tiggelaar, J. W. & Lisi, R. D. (2009). Porous asphalt proven successful in Cold Region: WBUUC porous parking lot. In: *Proceedings of the 14th Cold Regions Engineering Conference, Duluth*. Reston, Virginia. American Society of Engineers: 374–383.

Titkov, S. N. & Grebenets, V. I. (2009). Geocryological conditions of the railway bridge construction site, South Yamal Peninsula (Western Siberia, Russia). *Proceedings of the 14th Conference on Cold Regions Engineering. Cold Regions Impact on Research, Design and Construction, Duluth, Minnesota*. American Society of Civil Engineers: 1–7.

Tobiasson, W. (1973). Performance of the Thule Hanger cooling systems. *Proceedings of the 2nd International Conference on Permafrost, Yakutsk, USSR*. North American Contribution. Washington. National Academy of Sciences; 752–758.

Todhunter, P. E. & Popham, J. L. (2005). Relationship between snow cover and thermal offset of soil and air temperatures in the Great Plains of the United States. *American Geophysical Union*, Fall Meeting, 2005, abstract PP52A-0651.

Tolman, C. F. (1909). Erosion and deposition in Southern Arizona Bolson Rgion. *The Journal of Geology*, 17: 136–163.

Tolstikhin, N. I. (1940). The regime of ground and surface waters in the region of permafrost distribution. In: Shumgin *et al.* (eds.). *General Geocryology*. Moscow. Academie Nauka, SSSR. 340p. [In Russian].

Tomirdiaro, S. V. (1978). *Natural processes of the ice permafrost zone*. Moscow. Nedra. 145p. [In Russian].

Tomirdiaro, S. V. (1983). Loess-like yedoma-complex deposits in northeastern USSR: Stages and interruptions in their accumulation and their cryotextures. Proceedings of the 4th International Conference on Permafrost, Fairbanks. Washington, D.C. National Academy Press: 1263–1266.

Tomirdiaro, S. V., Arslanov, K. A., Chernenkiy, B. I., Tertychnaya, T. V. & Prokhorova, T. N. (1984). New data on the formation of loess-ice sequences in northern Yakutia and ecological conditions of the mammoth fauna in the Arctic during the Late Pleistocene. *Doklady, AN SSSR* 278: 1446–1449. [In Russian].

Tomirdiaro, S. V. & Chernenkiy, B. I. (1987). *Cryogenic deposits of the East Arctic and Subarctic.* AN SSSR Far-East-Science Centre. 196p. [In Russian].

Trans-Alaska Pipeline System (2013). The Facts. www.alyeska.pipe.com. Accessed on May 8, 2016.

Trant, A., Brown, C. D., Cairns, D. M., Danby, R. K., Lloyd, A. H., Marnet, S. D., Matheson, I. E., Tremblay, G. D., Walker, X., Wilmking, M., Boudreau, S., Harper, K., Henry, G. H. R., Hermanutz, L., Hik, D., Hofgaard, A., Johnstone, J. F., Kershaw, P., Laroque, C. & Weir, J. (2013). Ecological factors, not climate warming, explain variability in treeline patterns. *98th Ecological Society of America Annual Meeting, August, 2013.* COS 84-9.

Trautman, M. A. (1963). Isotopes Incorporated radiocarbon measurements III. *Radiocarbon,* 5: 62–79.

Tripati, A.K, Backman, J., Elderfield, H. & Ferretti, P. (2008). Eocene bipolar glaciations associated with global carbon cycle changes. *Nature,* 463: 341–346.

Trofimov, V. T., Krasilova, N. S., Afanasenko, V. E., Buldovich, S. N., Gerasimova, A. S. & Ospennikov, E. N. (2000). Map of rock mass stability under man-made impacts in the permafrost zone. *Moscow University Geology Bulletin, Geologiya,* 52(2): 65–77. [In Russian].

Trofimov, V. T. & Vasil'chuk, Y. K. (1983). Syncryogenic ice wedges and massive ice in Pleistocene deposits on the north of Western Siberia. *Byulleten Moskovskogo Obshchestva Ispytateley Prirody, Otdel Geologicheskiy,* 58(4): 113–121. [In Russian].

Trofimova, E. V. (2005). Cave ice of Priolhonie (Eastern Siberia, Russia). In: Mavlyudov, B. R., (ed.). *Glacier caves and glacier karst in high mountains and Polar Regions.* Moscow, Institute of Geography of the Russian Academy of Sciences: 127–131.

Troll, C. (1944). Strukturböden, solifluktion und frostklimate der Erde. *Geologische Rundschau,* 34: 545–694. [In German].

Troll, C. (1958). *Structure soils, solifluction and frost climates of the Earth.* U.S. Army Corps of Engineers, Snow, Ice and Permafrost Research Establishment, Translation #43. 121p.

Troll, C. (ed.). (1972). *Geoecology of the High-Mountain regions of Eurasia.* Weisbaden. Franz Steiner Verlag GMBH. 299p.

Trzhtsinsky, Yu. B. (1996). Gypsum karst in the south of the Siberian Platform, Russia. *International Journal of Speleology,* 25(3–4): 293–295.

Trzhtsinsky, Yu. B. (2002). Human-induced activation of gypsum karst in the Southern Priangaria (East Siberia, Russia). *Carbonates and Evaporites,* 17(2): 154–158.

Tsevetkova, S. G. (1960). Experience in dam construction in permafrost regions. *Materialy k osnovam ucheniia o merzlykh zonahk zemnoi kory,* 6: 87–112. [In Russian]. USA Cold Regions Research and Engineering Laboratory, Translation 161.

Tsytovich, N. A. (1957). The fundamentals of frozen ground dynamics. Proceedings of the 4th International Conference on Soil Mechanics and Foundation Engineering, London: 1: 116–119.

Tsytovich, N. A. (1959). Osnovy geokriologii (merzlotovedeniia). *Inst. Merzlotovedeniia, Akad. nauk SSSR, Moscow, Akad. nauk SSSR,* Pt. 2, p. 28–79, 1959. See also: CRREL Acc. No: 14017836. [In Russian].

Tsytovich, N. A. (1966). Permafrost problems. In: *Proceedings of the 1st Permafrost Conference, Lafayette, Indiana, 1963.* Washington, D.C. National Academy of Sciences and National Research Council: 7.

Tsytovich, N. A. (1973). *The mechanics of frozen ground*. Moscow. Vysshaya Shkola Press. 446p. [In Russian].

Tsytovich, N. A. (1975). *The mechanics of frozen ground*. McGraw-Hill Book Company. 426p.

Tsytovich, N. A., Ukhova, N. V. & Ukhov, S. B. (1972). Prediction of the temperature stability of dams built of local materials on pemafrost. Leningrad. Stroiizdat: p. 143. [In Russian]. USA Cold Regions Research and Engineering Translation 435.

Tsytovich, N. A., Kronik, Ya. A. & Biyanov, G. F. (1978). Design. Construction, and performance of earth dams on pemafrost in the Far North. *Proceedings of the 3rd International Conference on Permafrost*. Ottawa. National Research Council of Canada: 137–149.

Tufnell, L. (1966). Some little-studied British Landforms. *Proceedings of the Cumberlkand Geological Society*, 2(1): 50–56.

Tumel, V. F. (1946). History of permafrost in the USSR. In: *Problems of the palaeogeography during the Quaternary Period*. Transactions of the Institute of Geography, Russian Academy of Sciences #37. Izd-vo AN SSSR. [In Russian].

Twiss, R. Y. & Moores, E. M. (2006). *Structural Geology*. 2nd Edition. New York. W. H. Freeman.

Tyrrell, J. B. (1897). Report on the Doobaunt, Kazan and Ferguson Rivers and the North-west coast of Hudson Bay, and on two overland routes from Hudson Bay to Lake Winnepeg. *Geological Survey of Canada Annual Report, 1896 (new series)* IX: 1F-218F.

Tyurin, A. I. (1979). *Kurums as a form of cryogenic slope process*. Unpublished Ph.D. thesis, Moscow. Moscow State University. 16p. [In Russian].

Tyurin, A. I. (1985). Formation of rock streams under conditions of road construction. *Moscow State University Geology Bulletin*, 40(4): 56–62. [In Russian].

U.S. Arctic Research Commission Permafrost Task Force (2003). *Climate change, permafrost, and impacts on civil infrastructure*. Special Report 01–03. U.S. Arctic Research Commission, Arlington, Virginia.

U.S. Army/U.S. Air Force (1966). Arctic and Subarctic construction: Calculation methods for determination of depths of freeze-thaw in soils. *Technical Manual*, TM5–852–6/AFM88-19.

U.S. Environmental Protection Agency (EPA) (2008). *Backyard burning: Human health*. http://www.epa.gov./osw/nonhaz/municipal/bacyard/health.htm

USSR (1972). Guide for the application of the drilling and blasting methods of loosening frozen and perennially frozen ground and moraines. Ottawa, *Technical Translation #1877*, (1976). 27p.

Vallee, S. & Payette, S. (2007). Collapse of permafrost mounds along a subarctic river over the last 100 years (Northern Québec). *Geomorphology*, 90(1–2): 162–170.

Van Cleve, K. & Viereck, L. A. (1983). A comparison of successional sequences following fire on permafrost-dominated and permafrost-free sites in Interior Alaska. *Permafrost: 4th International Conference Proceedings*, Washington. National Academy Press: 1286–1291.

Vandenberghe, J. (1983). Some periglacial phenomena and their stratigraphic position in the Weichselian deposits in the Netherlands. *Polarforschung*, 53: 97–107.

Vandenberghe, J. (1988). Cryoturbations. In: Clark, M. J. (Ed.). *Advances in Periglacial Geomorphology*. Chichester, John Wiley and Sons, Ltd., 179–198.

Vandenberghe, J. (1992). Cryoturbations: A sediment structural analysis. *Permafrost and Periglacial Processes*, 3: 343–352.

Vandenburghe, J. & Van Den Broek, P. (1982). Weichselian convolution phenomena and processes in fine sediments. *Boreas*, 11: 299–315.

Vandenburghe, J., Zhijiu, C., Liang, Z. & Wei, Z., 2004 Thermal contraction crack networks as evidence for Late-Pleistocene permafrost in Inner Mongolia, China. *Permafrost and Periglacial Processes*, 15(1): 21–30.

Van Everdingen, R. O. (1976). Geocryological terminology. *Canadian Journal of Earth Sciences*, 13: 862–867.

Van Everdingen, R. O. (1978). Frost mounds near Bear Rock near Fort Norman, N.W.T., 1975–1976. *Canadian Journal of Earth Sciences*, 15: 263–276.

Van Everdingen, R. O. (1981). *Morphology, hydrology and hydrochemistry of karst in permafrost terrain near Great Bear lake, Northwest Territories*. National Hydrology Research Institute, Paper #11, Inland Waters Directorate Scientific Series, 114. 53p.

Van Everdingen, R. O. (1982). The management of groundwater discharge for the solution of icing problems in the Yukon. *Proceedings of the 4th Canadian Permafrost Conference, Calgary*. Ottawa. National Research Council of Canada: 212–226.

van Everdingen, R. O. (1987). The importance of permafrost in the hydrological regime. In: Healey, M. C. & Wallace, R. R. (eds.). *Canadian Aquatic Resources. Canadian Bulletin of Fisheries and Aquatic Sciences*, 215: 243–276

Van Everdingen, R. O. (1990). Groundwater Hydrology. In: Prouse, T. D. and Ommaney, C. S. I., (eds.). *Northern Hydrology: Canadian perspectives*. Saskatoon, National Hydrology Research Institute Report #1: 77–101.

Van Everdingen, R. O. (1998). *Multilanguage glossary of permafrost and related ground-ice terms*. Version #2. Calgary. The Arctic Institute of North America.

Van Everdingen, R. O. (2005). *Multilanguage glossary of permafrost and related ground-ice terms*. Calgary. The Arctic Institute of North America.

Van Everdingen, R. O. & Allen, H. D. (1983). Ground movements and dendrochronology in a small icing area on the Alaska Highway, Yukon, Canada. *Permafrost; Proceedings of the 4th Internatinal Conference, Fairbanks, Alaska*. Washington, D.C. National Academy Press: 1292–1297.

Van Stejin, H., Bertran, P., Francou, B., Hétu, B. & Texier, J. P. (1995). Models for the genetic and environmental interpretation of stratified slope deposits: a review. *Permafrost and Periglacial Processes*, 6: 125–146.

Van Steijn, H., Boelhouwers, J., Harris, S. & Hétu, B. (2002). Recent research on the nature, origin and climatic relations of blocky and stratified slope deposits. *Progress in Physical Geography*, 26: 551–575.

Van Steijn, H. & Hètu, B. (1997). Rain-generated overland flow as a factor in the development of some stratified slope deposits: A case study from the Pays du Buëch (Préalpes, France). *Gèographie physique et Quaternaire*, 51(1): 1–14.

Van Steijn, H., de Ruig, J. & Hoozemans, F. (1988). Morphological and mechanical aspects of debris flows in parts of the French Alps. *Zeitschrift für Geomorphologie*, 32: 143–161.

Van Vleit-Lanöe, B. (1985). Frost effects in soils. In: Boardman, J., (ed.), *Soil and Quaternary Landscape Evolution*. Chichester, Wiley: 117–158.

Van Vleit-Lanöe, B. (1988). The significance of cryoturbation phenomena in environmental reconstruction. *Journal of Quaternary Science*, 3(1): 85–96.

Van Vleit Lanöe, B. (1991). Differential frost heave, load casting and convection: converging mechanisms; a discussion of the origin of cryoturbations. *Permafrost and Periglacial Processes*, 2: 123–129.

Van Vleit, B. & Langhor, R. (1981). Correlation between fragipans and permafrost with special reference to Weischel silty deposits in Belgium and North France. *Catena*, 8: 137–154.

Varnes, D. J. (1958). Landslide types and processes. In: Eckel, E. (ed.). *Landslides and Engineering Practice*. Washington, D.C. Highway Research Board Special Report #29: 20–47.

Vartanyan, S. L., Arslanov, Kh. A., Tertychnaya, T. V. & Chernov, S. B. (1995). Radiocarbon dating evidence for Mammoths on Wrangel Island. *Radiocarbon*, 37(1): 1–6.

Vasil'chuk, A. C. & Vasil'chuk, Yu. K. (2012). Pollen and spores as indicators of the origin of Massive Ice. *Proceedings of the 10th International Conference on Permafrost, Salekhard, Russia*. 1: 487–491.

Vasil'chuk, Yu. K. (1992). *Oxygen isotope composition of ground ice. (Application to paleocryological reconstructions)*. Moscow. Theoretical Problems Department, The Russian Academy of Sciences, Geological Faculty of Moscow State University. Volumes 1 and 2, 418 and 262 pp. respectively. [In Russian].

Vasil'chuk, Yu. K. (2012). The Pleistocene-Holocene transition (at 10 ka B.P.) as the time of radical changes of typical geocryological formations. *Earth Cryosphere*, 16(3): 29–38. [In Russian].

Vasil'chuk, Yu. K., van der Plicht, J., & Jungner, H. (2000). AMS-dating of Late Pleistocene and Holocene syngenetic ice-wedges. *Nuclear Instruments and methods in Physics Research*, B, 172: 637–641.

Vasil'chuk, Yu. K. & Vasil'chuk, A. C. (1995). Ice-wedge formation in northern Asia during the Holocene. *Permafrost and Periglacial Processes*, 6(3): 273–279.

Vasil'chuk, Yu. K., Vasil'chuk, A. C., Budantseva, N. A. & Chizhova, J. N. (2008). *Palsa of frozen peat mires*. Moscow. Moscow University Press: 559p. [In Russian].

Vasil'chuk, Yu. K., Vasil'chuk, A. C., Budantseva, N. A. & Chizhova, J. N. (2012). Palsas in the north of Western Siberia. *Engineering Geology*, June, 2012: 18–32. [In Russian].

Vasil'chuk, Yu. K., Vasil'chuk, A. C., Jungner, H. & van der Plich, J. (1999). The syngenetic ice wedge formation during Holocene Optimum in fast accumulated peat in Central Yamal Peninsula. *Earth Cryosphere*, 3(10): 11–22. [In Russian].

Vasilev, N. K., Ivanov, A. A., Shatalina, I. N. & Sokurov, V. V. (2013). Ice- and cryogel composites in water-retaining elements in embankment dams constructed in cold regions. *Sciences in Cold and Arid Regions*, 5(4): 444–450.

Vasilyev, A. A. (2002). Assessment of factors affecting the destruction of sea coasts in the western sector of the Russian Arctic. In: *Extreme Cryospheric Phenomena: Fundamental and Applied Aspects, Pushchino*. Publication PSC of the Russian Academy of Sciences: 73–74. [In Russian].

Vedernikov, A. E. (1959). Investigation of frozen coarse skeleton grounds. Magadan. *Trudy VNII-I* 13. [In Russian].

Veillette, J. & Thomas, R. (1979). Icings and seepage in frozen glaciofluvial deposits, District of Keewatin, NWT. *Canadian Geotechnical Journal*, 16(4): 789–798.

Velichko, A. A., Andreev, A.A & Klimanov, V. A. (1997). Climate and vegetation dynamics in the tundra and forest zone during the late Glacial and Holocene. *Quaternary International*, 41/42: 71–96.

Velli, Y. Y. (1973). *Stability of buildings and engineering construction in the Arctic*. Leningrad, Stroiizdat. [In Russian].

Velli, Y. Y. (1977). Studies, projecting and construction on frozen saline soils. In: *Fundamentals of icy and frozen saline soils*. Leningrad: 35–45. [In Russian].

Velli, Y. Y. (1980). Foundations on complex permafrost soils. In: *U.S. Army Cold Regions Research and Engineering Laboratory, SR 90-40 and building under cold climates and on Permafrost: Collection of papers from a U.S.-Soviet joint seminar*. Leningrad, USSR, December, 1980: 204–217.

Venkateswaran, K., Kempf, M., Chen, F., Satomi, M., Nicholson, W. & Kern, R. (2003). *Bacillus nealsonii* sp. nov., isolated from a spacecraft-assembly facility, whose spores are gamma-radiation resistant. *International Journal of Systematic Evolutional Microbiology*, 53(1): 165–172.

Veuille, S., Fortier, D., Verpaelst, M., Grandmont, K. & Charbonneau, S. (2015). Heat advection in the active layer of permafrost: Physical modelling to quantify the impact of subsurface flow on soil thawing. In: *Compendium of selected papers*. Transport Canada: 63–70.

Vieira, G. & Ramos, M. (2003). Geographic factors and geocryological activity in Livingston Island, Antarctic. Preliminary results. In: Phillips, M., Springman, S. M. & Arebson, L. U.

(eds.). *Proceedings of the 8th International Conference on Permafrost, Zurich*. Lisse. Swets and Zeitlinger: 1183–1188.

Viereck, L. A. (1970). Soil temperatures in river bottom stands in Interior Alaska. In: *Ecology of Subarctic Regions*, Paris, UNESCO: 223.

Viereck, L. A. (1973). Ecological effects of river flooding and forest fires on permafrost in the Taiga of Alaska. In: *North American Contribution. 2nd International Conference on Permafrost, Yakutsk*. Washington, D.C., National Academy of Sciences: 60–87.

Viereck, L. A. (1975). Forest Ecology of the Alaskan Taiga. *Proceedings of the Circumpolar Conference on Northern Ecology:* 1: 1–22.

Viereck, L. A. (1982). Effects of fire and firelines on active layer thickness and soil temperatures in Interior Alaska. *Proceedings of the 4th Canadian Permafrost Conference*, Ottawa, National Research Council of Canada: 123–125.

Vinson, T. S., Ahmad, F. & Rieke, R. (1986). *Factors important to the Development of Frost Heave Susceptibility Criteria for coarse-grained soils*. Transportation Research Record 1089. Washington, D.C. Transportation Research Board.

Vinson, T. S. & Lofgren, D. (2003). Denali Park access road icing problems and mitigation options. In: Phillips, M., Springman, S. M. and Arenson, L. U. (eds.). *Permafrost, Proceedings of the 8th International Conference on Permafrost. Lisse*. Swets and Zeitlinger: 1189–1194.

Visser, S. A. (1973). Some biological effects of humic acids in the rat. *Acta Biologica Med. Germ.*, 31: 569–581.

Vitt, D. H., Halsey, L. A. & Zoltai, S. C. (1994). The bog landforms of continental western Canada in relation to climate and permafrost patterns. *Arctic and Alpine Research*, 26(1): 1–13.

Voeikov, A. I. (1899). Permafrost in Siberia along the prospective railroad route. *Journal of the Ministry of Rail Communications*, 13: 14–18. [In Russian].

Voellmy, A. (1955). Uber die Zerstorungskraft von Lawinen. *Bauzeeitung, Jahrg. 73*, S. 159–165, 212–217, 246–249, 280–285. [In German].

Voellmy, A. (1964). On the destructive force of avalanches. *Alberta Avalanche Study Center Translation #2*.

Vogt, T. & Larqué, P. (1998). Transformations and neotransformations of clay in the cryogenic environment: Examples from Transbaikalia (Siberia) and Patagonia (Argentina). *European Journal of Soil Science*, 49: 367–376.

Voitkovsky, K. F. (1960). *Mechanical properties of ice*. M.:Yizdatelstvo. AS USSR. [In Russian].

Volkwein, A., Schellenberg, K., Labiouse, V., Agliardi, F., Berger, F., Bourrier, F., Dorren, L.K.A., Gerber, W., & Jaboyedoff, M. (2011). Rockfall characterisation and structural protection – a review. *Natural Hazards and Earth System Science*, 11(9): 2617–2651.

Vologodsky, G. P. (1975). *Karst of the Irkutsk amphitheatre*. Moscow: Nauka. 123p. [In Russian].

Von Wakonigg, H. (1996). Unterkühlte Schutthalden (Undercooled talus). *Arbeiten aus dem Institut für Geographie der Karl-Franzens Universität, Graz*, 33: 209–233. [In German].

Vonder Mühll, D. (1992). Evidence of intrapermafrost groundwater flow beneath an active rock glacier in the Swiss Alps. *Permafrost and Periglacial Processes*, 3: 169–173.

Vonder Mühll, D. (1993). *Geophysikalische untersuchungen im permafrost des Oberengardins*. Mittelungen Versuchsanst Wasserbau Hydrologie Edigenöss Tahnische Hochschule Zürich, 122: 222. [In German].

Vonder Mühll, D. & Haeberli, W. (1990). Thermal characteristics of the permafrost within an active rock glacier (Murtèl/Corvatsch, Grisons, Swiss Alps). *Glaciology*, 36: 151–158.

Vorndrang, G. (1972). *Kriopedologische untersuchungen mit Hilfe von bodentemperaturmessungen (an einen zonalen strukturboden-vorkommen in der Silvrettagruppe)*. Münchener Geografer Abhandlungen, 6. 70p. [In German].

Voroshilov, G. D. (1978). Effect of coagulators on the magnitude of frost heave of Far Eastern Supesses and Suglinoks. *Proceedings of the 2nd International Conference on Permafrost,*

Yakutsk, USSR. Washington, D.C. USSR Contribution, National Academy of Sciences: 261–254.

Vorren, K-D. (1979). Recent palsa datings, a brief survey. *Norsk geografisk Tidsskrift*, 33: 217–219.

Vorren, K.-D. & Vorren, B. (1975). The problem of dating a palsa: two attempts involving pollen diagrams, determination of moss subfossils, and C^{14} datings. *Astarte*, 8: 73–81.

Votyakov, Yi. N. & Grechishchev, S. Ye. (1969). Temporary after-effect of temperature strains and stresses in frozen soils. In "*Construction in East Siberia and Far North*", No 14. [In Russian].

Vtyurin, B. I. (1975). *Underground ice in the USSR*. Moscow. Nauka. 212p. [In Russian].

Vtyurina, E. (1962). Basic features of formation of cryogenic structure of active layer soils, and cryotexture method of estimation of the depth of the active layer. *Permafrost of the USSR*. Moscow. Russian Academy of Sciences. [In Russian].

Vyalov, S. S. (1959). *Rheological properties and bearing capacity of frozen soils*. Hanover, New Hampshire. U.S. Army Cold Regions Research and Engineering Laboratory Translation #74. 219p.

Vyalov, S. S. (1978). Rheological basis of soil mechanics. In: Yizdatelstvo "Vysshaya Shkola". [In Russian].

Vyalov, S. S. (1984). Placing of deep pile foundations in permafrost in the USSR. *Permafrost: 4th International Conference on Permafrost*, Washington, D.C. National Academy Press Final Proceedings: 16–17.

Vyalov, S. S., Fotiev, S. M., Gerazimov, A. S. & Zolotar, A. I. (1997). Change in the boundaries of geotemperature zones in Western Siberia during global warming. *Hydrotechnical Construction*, 31(11): 655–659. [In Russian].

Vyalov, S. S., Gerasimov, A. S., Zolotar, A. J. & Fotiev, S. M. (1993). Ensuring structural stability and durability in permafrost ground areas at global warming of the Earth's climate. In: *Proceedings of the 6th International Conference on Permafrost*. Wushan, Guangzhou, China. South China University of Technology Press 1: 955–960.

Wagner, S. (1992). Creep of alpine permafrost, investigated on the Murtel rock glacier. *Permafrost and Periglacial Processes*, 3: 157–162.

Wahrhaftig, C. & Cox, A. (1959). Rock glaciers in the Alaska Range. *Geological Society of America Bulletin*, 70: 383–436.

Walder, J. S. & Hallet, B. (1985). A theoretical model of the fracture of rock during freezing. *Geological Society of America, Bulletin*, 96(3): 336–346.

Walker, D. A., *et al.* (2003). Vegetation-soil-thaw-depth relationships along a low-arctic bioclimatic gradient, Alaska: Synthesis of information and ATLAS studies. *Permafrost and Periglacial Processes*, 14: 103–123.

Walker, D. A., *et al.* (2004). Frost boil ecosystems: Complex interaction between landforms, soils, vegetation and climate. *Permafrost and Periglacial Processes*, 15: 171–188.

Walker, D. A., *et al.* (2008). Arctic patterned-ground ecosustems: a synthesis of field studies and models along a North American Arctic Transect. *Journal of Geophysical Research: Biogeosciences*, 113. Doi: 10.1029/2007JG000504

Walker, D. A., Webber, P. J., Binnian, E. F., Everett, K. R., Lederer, N. D., Nordstrand, E. A. & Walker, M. D. (1987). Cumulative impacts of oil fields on northern Alaskan landscapes. *Science*, 238: 757–761.

Walker, H. J. & Arnborg, L. (1966). Permafrost ice wedge effect on riverbank erosion. *Proceedings of the 1st International Permafrost Conference*. Washington. National Academy of Science, National Research Council Publication 1287: 164–171.

Wallace, R. E. (1948). Cave-in lakes in the Nebesna, Chisana and Tanana river valleys, eastern Alaska. *Journal of Geology*, 56: 171–181.

Walter, K. M., Zimov, S. A., Chanton, J. P., Verbyla, D. & Chapin, F. S. III (2006). Methane bubbling from Siberian thaw lakes as a positive feedback to climate warming. *Nature*, 443: 71–75.

Walsh, J. E. (1991). The Arctic as a bellweather. *Nature*, 352: 19–20.

Walton, D. W. H. (1985). A preliminary study of the action of crustose lichens on rock sufaces in Antarctica. In: Siegfried, W. R., Condy, P. R. & Laws, R. M. (eds.). *Antarctic nutrient cycles and food webs*. Berlin, Springer-Verlag: 180–185.

Wan, X. H., Lai, Y. & Wang, C. (2015). Experimental study on the freezing temperatures of saline silty soils. *Permafrost and Periglacial Processes*, 26: 175–187.

Wan, Y. L. (2008). Transmission line pole foundations design in permafrost regions on Qinghai-Tibet Plateau. *Electric Railway*, 6: 9–12.

Wang, A. G., Ma, W. & Wu, Z. J. (2005). Thickness of overlying sand and gravel layer and cooling effect of crushed rocks on roadbeds. *Chinese Journal of Rock Mechanics and Engineering*, 24(3): 2333–2341. [In Chinese with English Abstract].

Wang, B. (2011). Retrogression rate of thaw slumps in permafrost – an update from the latest monitoring data, *Proceedings of the 2011 Pan-Am CGS Geotechnical Conference*. 6p.

Wang, B. & French, H. M. (1994). Climate controls and high altitude permafrost, Qinghai-Xizang (Tibet) Plateau, China. *Permafrost and Periglacial Processes*, 5: 269–282.

Wang, B. & French, H. M. (1995). Implications of frost heave for patterned ground, Tibet Plateau, China. *Arctic and Alpine Research*, 27(4): 377–344.

Wang, B. L. (1990). Massive ground ice within bedrock. *Journal of Glaciology and Geocryology*, 12(3): 209–218. [In Chinese].

Wang, G.-S., Yu, Q.-H., You, Y.-H., Zhang, Z., Guo, L. Wang, S.-J. & Yu, Y. (2014). Problems and counter measures in construction of transmission line projects in permafrost regions. *Sciences in Cold and Arid Regions*, 6(5): 432–439.

Wang, J., Hua, J., Sui, J., Wu, P., Lui, T., & Chen, P. P. (2016). The impact of bridge pier on ice jam evolution – an experimental study. *Journal of Hydrology and Hydromechanics*, 64(1): 75–82.

Wang, K., Sun, J., Cheng, G. & Jiang, K. (2011a). On the influence of altitude and latitude on mean surface air temperature across the Qinghai-Tibet Plateau. *Journal of Mountain Science*, 8: 808–816.

Wang, L., Dong, X. P., Zhang, W., Zhang, G. S., Liu, G. X. & Feng, H. Y. (2011b). Quantitative characters of microorganisms in permafrost at different depths and their relation to soil physicochemical properties. *Journal of Glaciology and Geocryology*, 33: 436–441.

Wang, L. Y. & Xu, X. Y. (2010). Analysis on the ground temperature around transmission line tower foundation in permafrost regions. *Construction Technology in Cold Regions*, 3: 81–83.

Wang, N., Zhao, Q., Li, J., Hu, G. & Cheng, H. (2003). The sand wedges of the last ice age in the Hexi Corridor, China: paleoclimatic interpretation. *Geomorphology*, 51: 313–320.

Wang, S. J., Chen, J. B. & Qi, J. L. (2009). Study on the technology for highway construction and engineering practices in permafrost regions. *Sciences in Cold and Arid Regions*, 1(5): 412–422.

Wang, S. L. (1989). Formation and evolution of permafrost on the Qinghai-Xizang Plateau since the Late Pleistocene. *Journal of Glaciology and Geocryology*, 11(1): 69–75. [In Chinese].

Wang, X. L. & Zhang, H. Y. (2004). Pole and tower foundation design and their construction measures in frozen ground. *Inner Mongolia Electic Power*, 22(3): 65–66.

Wang, X. S., Fan, T. T., Gao, Y. X. & Hu, Y. (2013). Research on China's high-grade highway design index of asphalt pavement with granular base.

Wang, Z. Y., Ling, X. Z., Zhang, F., Wang, L. N., Chen, S. J. & Zhu, Z. Y. (2013). Field monitoring of railroad embankment vibration responses in seasonally frozen regions. *Sciences in Cold and Arid Regions*, 5(4): 393–398.

Washburn, A. L. (1947). Reconnaisance geology of portions of Victoria Island and adjacent regions, Arctic Canada. *Geological Society of America*, Memoir 22. 142p.

Washburn, A. L. (1952). Patterned ground. *Revue Canadienne de Géographie*, IV: 5–9.

Washburn, A. L. (1956). Classification of patterned ground and review of suggested origins. *Geological Society of America Bulletin*, 67: 823–865.

Washburn, A. L. (1967). Instrumental observations of mass wasting in the Mesters Vig District, Northeast Greenland. *Meddeleser om Grønland*, 166: 297p.

Washburn, A. L. (1969). Weathering, frost action and patterned ground in the Mesters Vig District, Northeast Greenland. *Meddeleser om Grønland*, 176: 303p.

Washburn, A. L. (1973). *Periglacial Processes and Environments*. London, Edward Arnold. 320p.

Washburn, A. L. (1979). *Geocryology – A survey of periglacial processes and environmemts*. London, Edward Arnold. 406p.

Washburn, A. L. (1985). Periglacial problems. In: Church, M. & Slaymaker, O. (eds.). *Field and theory: Lectures in Geocryology*. Vancouver. University of British Columbia Press: 166–202.

Washburn, A. L. (1997). *Plugs and plug circles: A basic form of patterned ground, Cornwallis Island, Arctic Canada – Origin and Implications*. Geological Society of America, Memoir #190. 87p.

Washburn, A. L. & Goldthwait R. P. (1958). Slushflows (Abstract). *Bulletin of the Geological Society of America*, 69: 1657–1658.

Washburn, A. L., Smith, D. D. & Goddard, R. H. (1963). Frost cracking in a middle-latitude climate. *Builetyn Peryglacjalny*, 12: 175–183.

Waters, R. S. (1962). Altiplanation terraces and slope development in Vest-Spitzbergen and southwest England. *Biuletyn Peryglacjalny*, 11: 89–101.

Watts, S. H. (1983). Weathering pit formation in bedrock near Cory Glacier, southeastern Ellesmere Island, Northwest Territories. *Geological Survey of Canada*, Paper 83-1A: 487–491.

Weaver, J. S. & Stewart, J. M. (1982). *In-situ* Hydrates under the Beaufort Sea Shelf. *Proceedings of the 4th Canadian Permafrost Conference, Calgary, Alberta*. Ottawa, Associate Committee on Geotechnical Research, National Research Council of Canada: 312–319.

Weed, R. & Norton, S. A. (1991). Siliceous crusts, quartz rinds and biotic weathering of sandstones in the cold desert of Antarctica. In: Berthelin, J. (ed.). *Diversity and environmental biochemistry*. Amsterdam. Elsevier: 327–329.

Weeks, W. S. (1920). Thawing of frozen ground with cold water. *Mining and Scientific Press* 120 (March 13th): 367–370.

Wei, Z. Y. (2009). Analysis of the reason of subsidence and displacement of cone cylindrical foundation for Qinghai-Tibet Railway 110 kV transmission project and treatment measure. *Qinghai Electric Power*, 28(3): 22–24.

Wein, N. (1976). Frequency and characteristics of Arctic tundra fires. *Arctic*, 29: 213–222.

Wein, N. (1984). Agriculture in the pioneering regions of Siberia and the Far East: Present status, problems and prospects. *Soviet Geography*, 25: 592–620.

Wen, Z., Yu, Q. H., Wang, D. Y., *et al.* (2012). The risk evaluation of the tower foundation frost jacking along Qinghai-Tibetan transmission line and its countermeasures. *Proceedings of the 15th International Speciality Conference on Cold Regions Engineering*. ASCE2012. Québec City, Canada. American Society of Civil Engineers, Reston, Virginia: 573–582.

Whalley, W. B., Rea, B. R., Rainey, M. M. & McAlister. (1997). Rock weathering in blockfields: Some preliminary data from mountain plateaus in Northern Norway. In: Widdowson, M. (ed.). *Palaeosurfaces: Recognition, reconstruction and interpretation*. Geological Society of London, Special Publication 129: 133–145.

Whalley, W. B., Rea, B. R., & Rainey, M. M. (2001). Weathering, blockfields and fracture systems and the implications for long-term landscape formation: Some evidence from Lygen and Øksfjodjøken areas in North Norway. *Polar Geography*, 28(2): 93–119.

White, A. U. & Seviour, C. (1974). *Rural water supply and sanitation in less developed countries – a selected annotated bibliography*. Ottawa. International Development Research Centre Report. IDRC-028e.

White, B. (2013). Buried Alaska gas pipeline could face powerful bending forces. 5p. http://www. acrticgas.gov/printmail/buried-alaska-gas-pipeline-could-face-powerful-bending-forces

White, P. G. (1979). Rock glacier morphometry, San Juan Mountains, Colorado. *Geological Society of America, Bulletin* 90(6): 1515–1518; II924-II952.

White, S. E. (1971a). Rock glacier studies in the Colorado Front Range, 1961 to 1968. *Arctic and Alpine Research*, 3: 43–64.

White, S. E. (1971b). Debris falls at the front of the Arapaho rock glacier, Colorado Front Range, U.S.A. *Geografiska Annaler*, 53A: 86–91.

White, S. E. (1974). Rock glaciers and blockfields. *Geological Society of America, Abstracts with Programs* 6: 1005.

White, S. E. (1975). Additional data on Arapaho rock glacier in Colorado Front Range, U.S.A. *Journal of Glaciology*, 14: 529–530.

White, S. E. (1976a). Is frost action really only hydration shattering? A review. *Arctic and Alpine Research*, 8: 1–6.

White, S. E. (1976b). Rock glaciers and blockfields. Review and new data. *Quaternary Research*, 6: 77–97.

White, S. E. (1981). Alpine mass movement forms (noncatastrophic): Classification, description and significance. *Arctic and Alpine Research*, 13: 127–137.

White, S. E. (1987). Differential movement across transverse ridges on Arapahoe rock glaciers, Colorado Front Range, U.S.A. In: Giardino, J. R, Shroder, J. F. and Vitek, J. D. (eds.). *Rock glaciers*. London. Allen and Unwin: 145–149.

Whittacker, B. N. & Reddish, D. J. (1989). *Subsidence occurrence, prediction and control*. Amsterdam, Oxford, New York and Tokyo. Elsevier. 528p.

Wielicki, B. A., Barkstrom, B. R., Harrison, E. F., Lee, R. B., Smith, G. L. & Cooper, J. E. (1996). Clouds and the Earth's Radiant Energy System (CERES): An Earth Observing System Experiment. Bulletin of the American Meteorological Society, 77, 853–868). (http://earthobservatory.nasa.gov/GlobalMaps/view.php?d1=CERES_NETFLUX_M).

Wild, G. I. (1882). *On the air temperature in the Russian Empire*. St. Petersburg. Russian Geological Society. 359p. [In Russian].

Willerslev, E. & Cooper, A. (2005). Ancient DNA. *Proceedings of the Royal Society*, B 272: 3–16.

Wilkerson, A. S. (1932). Some frozen deposits in gold fields of interior Alaska. *American Museum Novitat*, 5.

Willams, D. M. (1996). The barbed walls of China: A contemporary grassland drama. *The Journal of Asian Studies*, 55(3): 665–691.

Williams, J. R. (1970). Groundwater in permafrost regions of Alaska. *U.S. Geological Survey, Professional Paper* 696.

Williams, J. R. & van Everdingen, R. O. (1973). Groundwater investigations in permafrost regions of North America. In: *Permafrost, North American Contribution to the 2nd International Conference on Permafrost, Yakutsk*. Washington, National Academy of Sciences: 435–446.

Williams, M. A. J. (2000). Desertification: General debates explored through local studies. *Progress in Environmental Science*, 2(3): 229–251.

Williams, M. A. J. & Balling, Jr., R. C. (1996). *Interactions between desert and climate*. London. Edward Arnold.

Williams, P. J. (1967). *Properties and behavior of freezing soils*. Norwegian Technical Institute, Publication #72. 119p.

Williams, P. J. (1979). *Pipelines and Permafrost*. London. Longman. 98p.

Williams, P. J. (1984). Moisture migration in frozen soils. *Final Proceedings of the 4th International Conference on Permafrost*. Washington. National Academy Press: 64–66.

Williams, P. J. (1986). *Pipelines and Permafrost. Science in a cold climate*. Ottawa. Carleton University Press. 137p.

Williams, P. J. & Smith, M. W. (1989). *The frozen earth. Fundamentals of Geocryology*. Cambridge. Cambridge University Press. 306p.

Williams, R. G. B. & Robinson, D. A. (1981). Weathering of sandstone by the combined action of frost and salt. *Earth Surface Processes and Landforms*, 6: 1–9.

Williams, R. H. (1959). Ventilated building foundations in Greenland. *American Society of Civil Engineers, Journal of the Construction Division*, 85(602): 23–36.

Willman, H. B., Glass, K. D. & Frye, J. C. (1963). *Mineralogy of glacial tills and their weathering profiles in Illinois. Part 1. Glacial Tills*. Illinois State Geological Survey, Urbana. Circular #347. 55p.

Wilson, P. (1990). Clast size variations on talus: some observations from Northwest Ireland. *Earth Surface Processes and Landforms*, 15: 183–188.

Wilson, P. (2007). Periglacial landforms, rock forms, and block/rock streams. In: Elias, S. (ed.). *Encyclopedia of Quaternary Science*. Oxford. Elsevier: 2217–2225.

Wilson, P. (2014). Rockfall talus slopes and associated talus-foot features in the glaciated Great uplands of Britain and Ireland: periglacial, paraglacial or composite landforms? *Geological Society of London, Special Publications* 320: 133–144.

Wilson, P., Bentley, M. J., Schnabel, C., Clark, R. & Xu, S. (2008). Stone run (block stream) formation in the Falkland Islands over several cold stages, deduced from cosmogenic isotope (^{10}Be and ^{26}Al) surface exposure dating. *Journal of Quaternary Science*, 23: 461–473. DOI:10.1002/jqs.1156

Wimmer, M. (2007). Eis-und temperaturmessungen im Schönberg System (Totes Gebirge, Öbersterreich/Steiermark). *Alpin Untertagr, Berchesgarden 9–11 Proceedings*. Munich. Deutsches Höhlen- und Karstforscher: 93. [In German].

Wimmler, N. L. (1927). *Placer-mining methods and costs in Alaska*. Washington, D.C. U.S Government Printing Office.

Winterbottom, K. M. (1974). *The effects of slope angle, aspect and fire on snow avalanching in the Field, Lake Louise, and Marble Canyon region of the Canadian Rocky Mountains*. Unpublished M.Sc. thesis, Department of Geography, University of Calgary. 148p.

Wisshak, M., Straub, R. & Lopez-Correa, M. (2005). Das Eisrohrhöle-Bammelschacht-Systm (1337/118) im Kleinen Weitschartenkopf (Reiteralm). *Berschtesgadener Alpen Karst und Höhle 2004/2005*. Munich: Verband Deutschen-Höhlen-und Karstforscher: 68–81. [In German].

Wolfe, S. A., Kotler, E. & Dallimore, S. R. (2001). Surficial characteristics and the distribution of thaw landforms (1970 to 1999), Shingle point to Kay Point, Yukon Territory. *Geological Survey of Canada*. Open file 4088.

Wolfe, S. A., Stevens, C. W., Gaanderse, A. J. & Oldenborger, G. A. (2014). Lithalsa distribution, morphology and landscape associations in the Great Slave Lowland, Northwest Territories, Canada. *Geomorpholog,y* 204: 302–313.

Woo, M.-K. (1986). Permafrost hydrology in North America. *Atmosphere-Ocean*, 24(3): 201–234.

Woo, M.-K. (2012). *Permafrost Hydrology*. Heidelburg, Sprnger-Verlag. 575p.

Woo, M.-K. & Heron, R. (1981). Occurrence of ice layers at the base of High Arctic snowpacks. *Arctic and Alpine Research*, 13: 225–230.

Woo, M.-K., Yang, Z., Xia, Z. & Yang, D. (1994). Streamflow processes in an alpine permafrost catchment, Tianshan, China. *Permafrost and Periglacial Processes*, 5: 71–85.

Woo, M.-K. & Young, K. L. (1990). Thermal and hydrological effects of slope disturbances in a continuous permafrost environment. Proceedings of the 5th Canadian Permafrost Conference. *Nordicana*: 54: 175–180.

Woods, C. B. (1977). *Distribution and selected characteristics of high altitude patterned ground in the summit area of Plateau Mountain, Alberta*. Unpublished M.Sc. thesis, Department of Geography, University of Calgary. 171p.

Worsley, P. & Gurney, S. D. (1996). Geomorphology and hydrogeological significance of the Holocene pingos in the Karup Valley area, Trail Island, northern East Greenland. *Journal of Quaternary Science*, 11: 249–262.

Worsley, P., Gurney, S. D. & Collins, P. (1995). Late Holocene 'Mineral palsas' and associated vegetation patterns: A case study from Lac Hendry, Northern Québec, Canada and significance for European Pleistocene themokarst. *Quaternary Science Reviews*, 14: 179–192.

Wrangel, F. P. (1841). *A journey to the northern shores of Siberia and along the Arctic Ocean made in 1820–1824*. St. Petersburg. [In Russian].

WRB (2014). *World reference base for soil resources*. IUSS working group. Rome. FAO. World Soil Resources Report # 106. 181p.

Wu, Q. B., Jiang, G. L., Zhang, P., Deng, Y. S., Hou, Y. D. & Zhang, B. G. (2015). Evidence of natural gas in Kunlun Pass Basin, Qinghai-Tibet Plateau, China. *Chinese Science Bulletin*, 60(1): 68–74.

Wunnemann, B., Reinhardt, C., Kotlia, B. S. & Riedel, F. (2008). Observations on the relationship between lake formation, permafrost activity and lithalsa development during the last 20,000 years in the Tso Kar Basin, Ladakh, India. *Permafrost and Periglacial Processes*, 19: 341–358.

Wyman, G., (2009). *Transmission line construction in Sub-Arctic Alaska case study: Golden Valley Electric Association's 230 kV Northern Intertie. Electrical transmission and substation structures, 2009*. American Society of Civil Enineers, Reston, Virginia: 1–13.

Xie, C. & Gough, W. A. (2013). A simple thaw-freeze algorithm for a multi-layered soil using the Stefan Equation. *Permafrost and Periglacial Processes*, 24: 252–260.

Xie, S. B., Qu, J. Q., Zu, R. F., Zhang, K. C. & Han, Q. J. (2012). New discoveries on the effects of desertification on the ground temperature of permafrost and its significance to the Qinghai-Tibet Railway. *Chinese Science Bulletin*, 57(8): 838–842.

Xie, S. B., Qu, J. Q., Lai, Y. & Pang, J. (2015). Formation mechanisms and suitable controlling pattern of sand hazards at Honglianghe River section of Qinghai-Tibet Railway. *Natural Hazards*, 76(2): 855–871. Doi: 10.1007/s11069-014-1523-7

Xu, A. H. (2011). Discussion on the relationship between longitudinal cracks and alignment of subgrade in permafrost regions. *Sciences in Cold and Arid Regions*, 3(2): 132–136.

Xu, S., Zhang, D., Xu, Q. & Shi, S. (1984). Periglacial development in the northeast marginal region of Qinghai-Xizang Plateau. *Journal of Glaciology and Geocryology*, 19: 280–283.

Xu, X. Z., Cheng, G. D., & Yu, Q. H. (1999). Research prospect and suggestions of gas hydrates in permafrost regions on the Qinghai Tibet Plateau. *Advance in Earth Sciences*, 14(2): 201–204. [In Chinese].

Yachevskiy, L. A. (1989). Permafrost soils in Siberia. *Proceedings of the Russian Geographical Society*, 25: 341–355. [In Russian].

Yakushev, V. S. & Collett, T. S. (1992). Gas hydrates in Arctic Regions: Risk to drilling and production. In: Chung, J. S., Natvig, B. J., Li, Y.-C. & Das, B. M. (eds.). *Proceedings, 2nd International Offshore and Polar Engineering Conference, San Francisco, June 14–19th, 1992*: 317–324.

Yakushev, V. S. & Chuvilin, E. M. (2000). Natural gas and hydrate accumulations within Permafrost in Russia. *Cold Regions Science and Technology*, 31: 187–197.

Yan, F. Z., Li, P. & Cheng, G. D. (2012). Principal problems and solutions of the foundation engineering in the high-altitude permafost region. *Electric Power*, 45(12): 34–41.

Yan, W.-J., Niu, F. J., Zhang, X.-J., Luo, J. & Yin, G.-A. (2014). Advances in studies on concrete durability and countermeasures against freezing-thawing effects. *Sciences in Cold and Arid Regions*, 6(4): 398–408.

Yang, D., Kane, D. L., Hinzman, L. D., Zhang, X., Zhang, T. & Ye, H. (2002). Siberian Lena River hydrologic regime and recent change. *Journal of Geophysical Research – Atmospheres*, 107: 4694. DOI: 10.1029/2002JD002542.

Yang, S. & Shi, Y. (2015a). Numerical simulation of formation and preservation of Ningwu Ice Cave, Shanxi, China. *The Cryosphere Discuss*, 9: 2367–2395.

Yang, S. & Shi, Y. (2015b). Numerical simulation of formation and preservation of Ningwu Ice Cave, Shanxi, China. *The Cryosphere*, 9, 1983–1993. doi:10.5194/tc-9-1983-2015.

Yang, S. Z., Cao, X. & Jin, H. J. (2015). Validation of wedge ice isotopes at Yitul'he, Northeast China as climatic proxy. *Boreas*, 44(3): 502–510.

Yang, S. Z. & Jin, H. (2011). $\delta^{18}O$ and δD records of inactive ice-wedges in Yituli'he, Northeastern China and their paleoclimatic implications. *Science China, Earth Sciences*, 54: 119–126.

Yang, Y. H., Zhu, B. Z., Jiang, F. Q., Wang, X. L. & Li, Y. (2012). Prevention and management of wind-blown sand damage along Qinghai-Tibet Railway in Cuonahu Lake area. *Sciences in Cold and Arid Regions*, 4(2): 132–139.

Yarie, J. (1981). Forest fire cycles and life tables: A case study from interior Alaska. *Canadian Journal of Forest Research*, 11: 554–562.

Yarnal, B. M. (1979). *The sequential development of a rock glacier-like landform, Mount Assiniboine Provincial Park, British Columbia*. Unpublished M. Sc. thesis, Department of Geography, University of Calgary. 141p.

Yershov, E. D. (1979). *Moisture transfer and cryogenic textures in fine grained soils*. Moscow. Moscow University Publications, USSR. 214p.

Yershov, E. D. (1984). Transformation of dispersed deposits under repeated freezing-thawing. *Engineering Geology*, 3: 59–66. [In Russian].

Yershov, E. D. (ed.). (1985). Strains and stresses in freezing and thawing soils. Moscow. Moscow State University. M.: Yizdatelstvo [In Russian].

Yershov, E. D. (1986). *Physical chemistry and mechanics of frozen soils*. M.: Yizdatelstvo, Moscow. Moscow State University. [In Russian].

Yershov, E. D. (1990). *Obshchaya Geokriologiya*. Nedra. [In Russian].

Yershov, E. D. (ed.). (1998a). *Foundations of Geocryology*. Moscow. Moscow State University. 575p. [In Russian].

Yershov, E. D. (1998b). *General Geocryology*. Studies in Polar Research, Cambridge, Cambridge University Press. 580p.

Yonge, C. J. & Macdonald, W. D. (2014). Stable isotope composition of perennial ice in caves as an aid to characterizing ice cave types. In: Land, L., Kern, Z., Maggi, V. and Turri, S. (Eds.). *Proceedings of the 6th International Workshop on ice caves*: 41–49.

Yoshikawa, K. (1993). Notes on open system pingo ice, Adventdalen, Spitzbergen. *Permafrost and Periglacial Processes*, 4: 327–334.

Yoshikawa, K. (1998). The groundwater hydraulics of an open system pingo. *Proceedings, 7th International Conference, Yellowknife*. Collection Nordicana, 55: 1179–1184.

Yoshikawa, K., Bolton, W. R., Romanovsky, V. E., Fukuda, M. & Hinzman, L. D. (2003). Impacts of wildfire on the permafrost in the boreal forests of Interior Alaska. *Journal of Geophysical Research*, 108, NO DI, 8148. doi:10.1029/2001JD000438

Yoshikawa, K. & Harada, K. (1995). Observations on nearshore pingo growth, Adventdalen, Spitzbergen. *Permafrost and Periglacial Processes*, 6: 361–372.

Yoshikawa, K. & Hinzman, L. D. (2003). Shrinking thermokarst ponds and groundwater dynamics in discontinuous permafrost near Council, Alaska. *Permafrost and Periglacial Processes*, 14: 151–160.

Yoshikawa, K., Sharkhuu, N. & Sharkhuu, A. (2013). Groundwater hydrology and stable iso-tope analysis of an open-system pingo in northwestern Mongolia. *Permafrost and Periglacial Processes*, 24: 175–183.

You, Y. H., Yang, M. B., Yu, Q. H., Wang, X. B., Li, X. & Yue, Y. (2016). Investigation of an icing near a tower foundation along the Qinghai-Tibet Power Transmission Line. *Cold Regions Science and Technology*, 126: 250–259. http://dx.doi.org/10.1016/j.coldregions.2015.05.005

Young, A. (1972). *Slopes*. Edinburgh. Allen and Unwin. 288p.

Yount, J., Decker, R., Rice, R. & Wells, L. (2004). Reducing avalanche hazard to US Route 89/191 in Jackson, Wyoming using snow sails. *Proceedings of the International Snow Science Workshop, ISSW, Jackson Hole, Wyoming, September 19–24th, 2004*.

Yu, Q. H., Liu, H. J., Qian, J., *et al.* (2009). Research on frozen engineering of Qinghai-Tibet 500 kV DC Power Transmission Line. *Chinese Journal of Engineering Geophysics*, 6(6): 806–812.

Yu, Q. H., Zheng, G. S., Guo, L., Wang, X. B., Wang, P. F. & Bao, Z. H. (2015). Analysis of tower foundation stability along the Qinghai-Tibet Power Transmission Line and impact of the route on permafrost. *Cold Regions Research and Engineering*. Doi: 10.1016/j.coldregions.2015.06.15

Yu, T. L., Lei, J. G. & Li, C. Y. (2009). Compression strength of floating ice and calculation of ice force on bridge piers. *Proceedings of the International Conference on Modeling and Simulation, Cape Town*, South Africa: 29–31.

Yu, X., Yu, X.Y., Zhang, B. & Li, N. (2009). An innovative sensor for assisting spring load restrictions: Results from a field demonstration study. In: *Proceedings of the 14th Conference on Cold Regions Engineering*. Reston, Virginia. American Society of Civil Engineers: 417–428.

Yue, Z. R., Wang, T. L., Ma, C. & Sun, T. C. (2013). Frost control of fine round gravel fillings in deep seasonal frost frozen regions. *Sciences in Cold and Arid Regions*, 5(4): 425–432.

You, Y., Yang, M., Yu, Q., Wang, X., Li, X. and Yoe, Y. 2015. Investigating the genesis of ground icing near a Qinghai-Tibet Power Transmission Line tower foundation by electrical resistivity tomography. *Cold Regions Science and Technology*.

Yu, H., Luedeling, E. & Xu, J. (2010). Winter and spring warming result in delayed spring phenology on the Tibetan Plateau. *Proceedings of the National Academy of Sciences*, 107: 22151–22156.

Yu, J. L., Lei, J. Q., Li, C., Yu, H. & Shan, S. (2009). Compressive strength of floating ice and calculation of ice force on bride piers during ice collision. In: *Proceedings of the 14th Conference on Cold Regions Engineering*. Reston, Virginia. American Society of Civil Engineers: 609–617.

Yu, J. L. & Shi, H. X. (2011). Changes of microbes population in the different degraded alpine meadows on the Qinghai-Tibetan Plateau. *Acta Agriculture Boreali-occidentalis Sinica*, 20: 77–81.

Yu, Q. H., Niu, F. J., Pan, X, Bai, Y. & Zhang, M. (2008). Investigation of embankment with temperature-controlled ventilation along the Qinghai-Tibet Railway. *Cold Regions Science and Technology*, 53: 193–199.

Yu, Q. H., Wen, Z., Ding, Y. S., *et al.* (2009). Monitoring the tower foundations in the permafrost regions along the Qinghai-Tibet DC Transmission Line from Qinghai Province to Tibetan Autonomous Region. *Journal of Glaciology and Geocryology*, 34(5): 1165–1172.

Yu, R. F. (2002). Research on frozen foundation around pole of transmission line. *Jilin Electric Power*, 2: 9–12.

Yu, Z., Loisel, J., Brosseau, D., Beilman, D. & Hunt, S. (2010). Global peatland dynamics since the last glacial maximum. *Geophysical Research Letters*, 37, L13402. 5p.

Žák, K., Onac, B. P. & Persoiu, A. (2008). Cryogenic carbonates in cave environments: A review. *Quaternary International*, 187: 84–96.

Žák, K., Orvošá, M., Filippi, M., Vlček, L., Onac, B. P., Perşoiu, A., Rohovec, J. & Svętlik, I., (2013). Cryogenic cave pearls in the periglacial zones of ice caves. *Journal of Sedimentary Research*, 83: 207–220.

Zarling, J. P., Conner, B. & Goering, D. J. (1983). Air duct systems for road stabilization over permafrost areas. In: *Permafrost. 4th International Conference*. Washington. National Academy Press Publication: 1463–1468.

Zeder, M. A. & Hesse, B. (2000). The initial domestication of goats in the Zagros Mountains 10,000 years ago. *Science*, 287: 2254–2257.

Zemtsov, A. A. (1959). Relict permafrost in the West-Siberian depression. In: *Ice Age on the European part of the USSR*. Moscow. Moscow State University: 331–334. [In Russian].

Zenin Xing, Xiaoling Wu & Hongkang Qu. (1980). Determination of the ancient permafrost table, based on the variation in the content of clay minerals. *Chinese Journal of Glaciology and Cryopedology (Bingchuan Dongtu)*, 2, Special Issue 29-46. [In Chinese].

Zenin Xing, Xiaoling Wu & Hongkang Qu. (1984). Determination of the ancient permafrost table, based on the variation in the content of clay minerals. Ottawa. *National Research Council of Canada Technical Translation* NRU/CNR TT-2082.

Zhan, Z. S. & Kuang, C. M. (2006). Research on blasting technology for construction of Fenghoushan tunnel in Permafrost. *Journal of Rock Mechanics and Engineering*, 25(5).

Zhang, E. M. & Wu, Z. W., (1999). *The degradation of permafrost and highway engineering*. Lanzhou. Lanzhou University Press: 105–115.

Zhang, G. S., Nui, F. J., Ma, X. J., Liu, W., Dong, M. X., Feng, H. Y., An, L. Z. & Cheng, G. D. (2007a). Phylogenetic diversity of bacteria isolates from the Qinghai-Tibet Plateau permafrost region. *Canadian Journal of Microbiology*, 53: 1000–1010.

Zhang, G. S., Ma, X. J., Niu, F. J., Dong, M. X., Feng, H. Y., An, L. Z. & Cheng, G. D. (2007b). Diversity and distribution of alkaliphilic psychrotolerant bacteria in the Qinghai-Tibet Plateau permafrost region. *Extremophiles*, 11: 415–424.

Zhang, G. S., Jiang, N., Liu, X. L. & Dong, X. Z. (2008a). Methanogenesis from methanol at low temperatures by a novel psychrophilic methanogen, "*Methanolobus psychrophilus*" sp. nov., prevalent in Zoige wetland of the Tibetan Plateau. *Applied Environmental Biology*, 74: 6114–6120.

Zhang, G. S., Tian, J. Q., Jiang, N., Guo, X. P., Wang, Y. F. & Dong, X. Z. (2008b). Methanogen community in Zoige wetland of Tibetan plateau and phenotypic characterization of a dominant uncultured methanogen cluster ZC-1. *Environmental Microbiology*, 10: 1850–1860.

Zhang, J., Zhang, M. & Lui, Y. (2009). Reasonable height of roadway embankment in permafrost regions. In: *Proceedings of the 14th Conference on Cold Regions Engineering*. Reston, Virginia. American Society of Civil Engineers: 486–495.

Zhang, K. C., Qu, J. J., K. T., Niu, G. H. & Han, Q. J. (2010). Damage by wind-blown sand and its control along the Qinghai-Tibet Railway in China. *Aeolian Research*, 1: 143–146.

Zhang, K. C., Niu, Q. H., Qu, Liao, J. J., Yao, Z. Y. & Han, Q. J. (2011). Charateristics of sand damages and dynamic environment along the Tuotuohe section of the Qinghai-Tibet Railway. *Sciences in Cold and Arid Regions*, 3(2): 137–142.

Zhang, L. M., Wang, M., Prosser, J. I., Zheng, Y. M. & He, J. Z. (2009). Alititude ammonia-oxidizing bacteria and archaea in soils on Mount Everest. *FEMS Microbial Ecology*, 70: 52–61.

Zhang, L. N. (1988). Study of the adherent layer on different types of ground in permafrost region on the Qinghai-Tibet Plateau. *Journal of Glaciology and Geocryology*, 10(1): 8–14.

Zhang, L. X., Yuan, S. C. & Yang, Y. P. (2003). Mechanism and prevention of deformation cracks of embankments in the permafrost regions along the Qinghai-Xizang Railway. *Quaternary Sciences*, 23(6): 604–610.

Zhang, M. & Niu, F. J. (2009). Numerical study on the cooling effect of closed crushed-rock embankment along Qinghai-Tibet Railway in China. In: *Proceedings of the 14th Conference on Cold Regions Engineering.* Reston, Virginia. American Society of Civil Engineers: 467–495.

Zhang, T., Barry, R. G., Knowles, K., Heginbottom, J. A. & Brown, J. (1999). Statistics and characteristics of permafost and ground-ice distribution in the northern hemisphere. *Polar Geography*, 23(2): 132–154.

Zhang, X., Presler, W., Li, L., Jones, D. & Odgers, B. (2014). Use of wicking fabric to help prevent frost boils in Alaskan pavements. *Journal of Materials in Civil Engineering*, 26(4): 728–740. Doi: 10.1061/(ASCE)MT.1943-5533.0000828.

Zhang, R. V. (2014). Geocryolody principles of earth dams for low and medium pressures in permafrost with a changing climate. [In Russian]. http://www.rae.ru/fs/572-r34841/?Ing=en

Zhang, X. F. & Sun, X. M. (2003). Analyzing the effect of tunnelling on temperature of permafrost in the Kunlun Mountains, Tibetan Plateau. *Journal of Glaciology and Geocryology*, 25(6).

Zhang, Y. & Lai, Y. (2003). Analysis of hydration heat effects of cast-in-situ concrete foundations of cold regions culverts. *Highway*, 2: 50–56. [In Chinese].

Zhao, L., Cheng, G. D., Yu, Q. H., *et al.* (2010a). Permafrost injury and its control countermeasures along the key section of Qinghai-Xizang Highway. *Chinese Journal of Nature*, 32(1): 9–13.

Zhao, L., Ding, Y. J., Liu, G. Y., Wang, S. L. & Jin, H. J. (2010b). Estimates of the reserves of ground ice in permafrost regions on the Qinghai-Tibetan Plateau. *Journal of Glaciology and Geocryology*, 32(1): 1–9. [In Chinese].

Zhao, Y. & Wang, J. (1983). Calculation of the thaw depth beneath heated buildings in permafrost regions. In: *Proceedings of the 4th International Permafrost Conference, Fairbanks.* Washington, D.C. National Academy Press: 1490–1495.

Zhou, S., Zhou, K., Yao, C., Wang, J. L. & Ding, J. L. (2015). Exploratory data analysis and singularity mapping in geochemical anomaly identification in Karamay, Xinjiang, China. *Journal of Geochemical Exploration*, 154: 171–179.

Zhou, S. Z., Yi, C. L., Shi, Y. F. *et al.* (2001). Study on the ice age MIS 12 in western China. *Journal of Geomechanics*, 7(4): 321–327. [In Chinese].

Zhou, Y.W. & Guo, D.X. (1983). Some features of permafrost in China. *Proceedings of 4th International Conference on Permafrost.* Vol. 1. Washington, D. C. National Academy Press: 496–501.

Zhu Cheng, Cui Zhijiu & Yao Zeng (1992). Research on the feature rock glaciers on the Central Tian Shan Mountains. *Acta Geographica Sinica* 1992(3): 233–241.

Zhu, L.-P., Wang, J.-C. & Li, B.-Y. (2003). The impact of solar radiation upon rock weathering at low temperatues: A laboratory study. *Permafrost and Periglacial Processes*, 14(1): 61–67.

Zhu, L. N., Wang, G. R. & Guo, X. M. (1989). Calculation of the critical height of asphalt-paved embankment of the Qinghai-Xizang Highway. *Proceedings of the 3rd Chinese National Conference on Permafrost, Harbin, China, Aug. 19–24, 1986.* Beijing. Kexue chubanshe (Science Press). China: 339–346.

Zhu, Y. H., Zhang, Y. Q., Wen, H. J., Lu, Z. Q., Jia, Z. Y., Li, Y. H., Li, Q. H., Liu, C. L., Wang, P. K. & Guo, X. W. (2010). Hydrates in the Qilian Mountain permafrost, Qinghai, northwest China. *Acta Geological Sinica (English Edition)*, 84(1): 1–10.

Zhu, Y. K. & Wang, B. C. (2011). Cause analysis to foundation freeze damage of transmission power line towers in severe cold district and its prevention measures. *Inner Mongolia Electric Power*, 29(6): 90–93.

Zolotarev, G. S. (ed.) (1990). *Textbook on engineering geology.* Moscow. Moscow University Press. 294 p. [In Russian].

Zoltai, S. C. (1971). Southern limit of permafrost features in peat landforms, Manitoba and Saskatchewan. *Geological Association of Canada*, Special Paper 4: 305–310.

Zoltai, S. C. (1972). Palsas and peat plateaus in Central Manitoba and Saskatchewan. *Canadian Journal of Forest Research*, 2: 291–302.

Zoltai, S. C. (1975). Tree ring record of soil movements on permafrost. *Arctic and Alpine Research*, 7: 331–340.

Zoltai, S. C. & Tarnocai, C. (1971). Properties of a wooded palsa in Northern Manitoba. *Arctic and Alpine Research*, 3(2): 115–129.

Zoltai, S. C. & Tarnocai, C. (1974). *Soils and vegetation of hummocky terrain*. Environmental Task Force on Northern Oil Development. Report 74-5. 86p.

Zoltai, S. C. & Tarnocai, C. (1975). Perennially frozen peatlands in the Western Arctic and Subarctic of Canada. *Canadian Journal of Earth Sciences*, 12(1): 28–43.

Zoltai, S. C., Tarnocai, C. & Pettapiece, W. W. (1978). Age of cryoturbated organic materials in earth hummocks from the Canadian Arctic. *Proceedings of the 3rd International Conference on Permafrost, Edmonton, Alberta*. Ottawa. National Research Council of Canada: 326–331.

Zotov, V. D. (1940). Certain types of soil erosion and resultant relief features on the higher mountains of New Zealand. *New Zealand Journal of Science and Technology*, 21B: 256–262.

Zubov, N. N. (1945). *Arctic ice*. Springfield, Virginia. National Technical Information Service, A 0426082. Translated from the Russian.

Zuidhoff, E. S. & Kolstrup, E. (2000). Changes in palsa distribution in relation to climate change in Laivadalen, Northern Sweden, especially 1960–1997. *Permafrost and Periglacial Processes*, 11(1): 55–69.

Zuidhoff, E. S. & Kolstrup, E. (2005). Palsa development and associated vegetation in Northern Sweden. *Arctic, Antarctic and Alpine Research*, 37(1): 49–60.

Subject index

Acoustic emissions 76
Active layer 1, 3, 6, 14, 16–17, 18, 20, 27, 32, 48, 51–53, 62, 65, 68, 69, 74, 77, 78, 103, 119, 244, 255, 283, 337, 369, 379, 388, 402, 627
Active layer detachment slides 297, 298–300, 304, 425, 435
Adfreeze strength 480
  short-term 481, 482
  sustained 481, 482
Advection 9–10
Aggregation 78, 445
A horizon offset 94, 95
Air exchange 12, 494, 495
Airfield construction 538–541
Air masses 10, 86, 89, 96–99, 106, 129, 132
Alas 214, 417, 418–424
Alaskan coastal plain 186, 189
Albedo 10, 85, 86, 89, 99, 111, 118, 119, 232, 252, 402, 404, 470, 502, 508
Aldan River 162, 301, 629
Altiplanation terraces 276, 285
Amderma 151, 465, 469, 483, 492, 540
Andes 16, 59, 103, 200, 345, 347
Angle of slope of embankment sides 492
Antarctica 3, 16, 30, 31, 43, 78, 89, 95, 97, 112, 121, 145, 151, 167, 169, 274, 306, 319, 321, 345, 362, 377, 381, 391, 399
Antarctic permafrost 30
Anti-syngenetic ice-wedges 158
Arctic Canada 19, 196, 392, 443
Arctic coastal erosion rates 429, 430
Arctic mudboils 375, 376–378, 379
Argentina 16, 20, 73, 136, 347, 365
Artificial refrigeration 489–490, 563, 567
Asian steppe grassland 630–632
Aufeis 180, 195–198, 215, 366

Australia 38, 250, 280, 354, 362
Azonal soils 145

Baffin Island 38, 60, 185, 403, 416, 577, 624
Bahada/Bajada 284
Balch effect 12, 203, 204, 490
Ball-and-pillow structure 66, 70
Ballast railway tracks 524–527
Banded ice 200, 201, 206, 207
Barents Sea 41
Baromechanical processes 445
Barrow Alaska 11, 13, 62, 484, 603, 606
Baydjarakhs 172, 174, 407
Beaded streams 411–412
Bearing capacity of soil 193, 448, 459, 551
Beaufort Sea 38, 434, 435
Beaver Creek Yukon Territory 502
Beihu'he 490, 493
Belt conveyors 589, 590
Belts 145, 154, 156, 356
Big Lyakovsky Island 40
Bilar 418, 419
Biogenic weathering 319–320, 320–321, 374
Blind taliks 4, 7
Block fields 12, 323, 350, 355, 361, 363, 365, 366, 373
Blockgletscher 334
Block slope 12, 77, 96, 146, 317, 322, 326, 332, 355, 366, 372, 493
Block streams
  active dynamic block streams 353, 355–357, 359
  inactive block streams 359–365
  lag block streams 353, 354, 359
  relict block streams 353, 354, 359–365
  Siberian type block streams 359
  Tibetan type rock streams 359

Blokströme 350
Bottom snow temperature 91
Braking block 280
Bridges 427, 530–532, 534
British Columbia 58, 59, 96, 98, 107, 110, 183, 200, 271, 288, 306, 326
British Isles 58
Brodelböden 65
Buffering of air temperatures 108–109
Bugor pucheniya 214
Bugry 196
Building materials 471–472, 582
Building stability 465–500
Bulgannyakh 214, 215
Buoyancy of ice-wedges 193–194
Buried glacier ice 18, 184, 186, 187

Capillary water 43, 45, 51
Causes of climatic change 217
Cavity development 402–403
  subcritical 403
  supracritical 403
Central Siberia 17, 37, 75, 145, 151, 235, 306, 346, 351, 353, 417, 418, 577, 581, 629
Chemical weathering 74, 77, 112, 315, 317, 319, 320, 363, 591
Chimney effect 12, 329
China 8, 9, 16, 26, 32, 34, 38, 70, 80, 104, 110, 129, 137, 169, 170, 204, 207, 208, 210, 226, 256, 280, 281, 284, 321, 346, 350, 351, 399, 406, 409, 414, 443, 471, 480, 493, 501, 511, 514, 521, 522, 527, 528, 538, 559, 594, 597, 601, 614, 633
Chugach range 16
Clathrate hydrates 35, 571, 572, 573, 575
Clay mineralogy 53, 463
Climate change 34–35, 86, 108, 140–142, 143, 171, 206, 214, 341, 415, 418, 441, 527, 550, 575, 633, 634
Climatic factors 85–109
Closed talik 4, 5
Coagulation 445, 446
Cold air drainage 48, 107–108, 109, 129, 131, 230, 239, 471, 626
Cold-based glaciers 120, 284
Colloid plucking 74, 77
Columns (ice), 200
Composite slope failures

active layer detachment slides 297, 298–300
  retrogressive thaw failures 300–304
  retrogressive thaw flow slides 300
  retrogressive thaw slumps 300, 302, 303, 304
  skin flows 299
  slush avalanches 310–312
  snow avalanches 304–310
Compressor stations 559, 560, 566–569
Concrete railway tracks 524–527
Conduction 9–10, 12, 401, 413, 414
Confluent permafrost 7, 123
Continentality index 15–16
Continental type salinity 26, 405
Continuous permafrost 7, 16, 78, 80, 85, 106, 119, 122, 126, 129, 132, 142, 186, 187, 198, 203, 207, 214, 218, 234, 247, 284, 332, 336, 373, 381, 386, 417, 422, 471, 476, 478, 484, 501, 509, 530, 552, 553, 557, 585, 596, 599, 612, 629
Convection 9–10, 67, 115, 401
Cosmic-ray dating 78
Cracking of ground 62, 99, 100, 150, 155, 157, 368, 408
Creep deformation 270
Cryofacial analysis 61
Cryofacies 60–61
Cryogenic block fields 322–325
Cryogenic creep 269–270, 275, 284
Cryogenic debris flows 287–296
Cryogenic fans 326–329
Cryogenic fast flows 287–312
Cryogenic patterned ground
  macroforms 373–387
  microforms 394–395
  nonsorted patterned ground 324, 386
  plug circles 386–387
  sorted patterned ground 387–393
    sorted circles and nets 388
    sorted polygons 391–393
    sorted stripes 391–392
    stone pits 392–393
Cryogenic weathering 315
Cryogenic weathering index (CWI), 75
Cryolithology 61, 398
Cryolithozone 1, 2, 123
Cryology 3
Cryopedolith 182
Cryopegs 4, 5, 24, 111, 540, 599

Cryoplanation
  Cryoplanation terraces 285, 287
Cryosols 22
Cryosols 22
Cryostatic pressures 67, 261, 376
Cryostratigraphy 60–61
Cryostructure 20, 38, 60, 61
Cryosuction 45, 180, 186–187, 189, 214,
      229, 230, 238, 239, 240, 243, 244
Cryotexture 60–61, 446
Cryotic 1, 3, 207, 306
Cryoturbation 65–71, 73, 77, 102, 145,
      253, 254, 255, 261, 263, 265, 384

Dalton Highway 48, 499, 500, 531
Dams to impound water 600–602
Data loggers 9, 14, 52
Dawson City 20, 409, 427, 429, 506, 577,
      578, 606, 629
Debris flow fans 287, 289
Debris flows 250, 287, 290, 292, 294, 295,
      296, 310, 311, 313, 331, 337, 351
Defluction 269
Delli 276
Densification 60
Depth of zero amplitude 15, 52, 54, 338,
      403
Deserption 269
Desiccation zone 446, 447
Differential settlement 468, 469, 470, 525,
      591
Dilation cracking 63–64, 149, 226
Diode effect 92, 203, 230, 264, 322, 323,
      329, 332, 337, 351, 357, 402,
      493–496
Discontinuous permafrost 16, 79, 92, 118,
      123, 126, 129, 142, 210, 218, 220,
      222, 225, 233, 259, 265, 284, 299,
      373, 381, 399, 498, 511, 552, 553,
      583, 593, 611
Dispersion 445
Distributed temperature sensing 9
Distribution of permafrost 85, 92, 106, 122,
      124, 125, 131, 132, 145, 332, 400
Drakensberg 59
Drilling rigs 543, 545, 546–547
Druza 58
Dry permafrost 3, 16, 112, 225, 472
Dry Valleys of Antarctica 78, 95
Dujoda 419, 420
Duricrust 22

Earth hummocks
  niveo-aeolian hummocks 261–263
  oscillating hummocks 252–256
  silt-cycling hummocks 260–261
  thufurs 256–260
East Greenland type pingos 218
East Slims rock glacier Kluane National Park
      340
Effective radiating temperature 89
Eichfeld agricultural zones 627
Eisfilamente 58
Ekati 586, 587, 597
Elastic strain 275
Electric transmission lines 610
Embankment consolidation 513
Embankment heights 508–509
Epigenetic ice-wedges 419
Epigenetic permafrost 61
Epigenetic salinity 24
Equilibrium permafrost 403
Europe 8, 38, 89, 104, 207, 214, 228, 231,
      256, 354, 441, 501, 621, 631
Evaporation 26, 48, 81, 90, 109, 112, 118,
      243, 301, 414, 417, 423, 625
Experimental embankments 504–505
Exposure thermal offset 95

False taliks 4, 7
Faro 443, 580, 581, 591, 592, 600
Felsenmeer 316, 322
Fenghoushan 516
Filter dams 295
Fire 30, 85, 110, 118–119, 140, 248, 255,
      256, 306, 380, 400, 404, 418
Forced preservation (clathrate hydrates), 36
Forecast permafrost maps 128
Fossil cracking 63, 177
Foundations 465–500, 614, 615, 634
Fragipans 22–24, 403
Freezing fringe 447
Freezing of moisture 16, 20–21
Frost boils 20, 196, 374, 375, 376, 381,
      385–386, 500
Frost comminution 74–78
Frost cracking 62, 65, 69, 75, 149–177, 394,
      540
Frost creep 270, 272–274, 280, 322, 356,
      357, 373
Frost design classification 65, 460
Frost heaving 45, 48, 59–60, 274, 279, 316,
      318, 356, 445–455, 460, 502

Frost jacking 71, 315, 351, 352, 357, 393, 476, 478, 479, 480, 490, 568, 611, 614
Frost-pull theory 72
Frost susceptibility classifications 460, 461
Frost-susceptible soils 465
Frost wedging 60, 75, 316, 317
Frozen fringe 246, 446

Gabions 332, 538
Gas hydrate forced preservation 36
Gas hydrates 35–37, 83, 138, 139, 222, 443, 543, 547, 566, 571, 572, 573, 574
Gas hydrate self-preservation 36
Gelifluction 146, 268, 269, 273, 274–276, 278, 279, 280, 281, 284, 285, 286, 317, 322, 373
Gelisols 22
Geocryological zones 16
Geocryology 3, 441
Geophysical methods 128, 131, 228, 238, 243, 337, 557
Geotextiles 494, 498–500, 511
Geothermal gradient 11, 15, 54, 55
Geothermal heat flow 10, 15, 54, 85, 118, 120
Giant Gold Mine Yellowknife 592
Glacial grass 58
Glacier rocheux 334
Glaciers 1, 119–122, 347, 349, 624
Global climatic change 89
Gold dredge 578
Goletz terraces 285
Golmud-Lhasa highway 503, 518, 625
Golmud-Lhasa railway 494, 496, 501, 504
Gorodetsky rock glacier 335, 341, 342, 344
Gravitational water 43, 44, 47, 137
Gravity processes 65–71
Great Bear Lake region 79
Greenhouse effect 89, 90
Greenhouse gases 34, 89
Ground cracking 61–63, 167, 390
Ground heat flux 90, 95
Ground vegetation offset 95
Ground veins 173
Ground wedges 153
Growth of thick ice-wedges 190–192

Hazard zonation 140–142
Heat balance on the Earth surface 85–91
Herschel Island 436

Hexi Corridor China 170, 207
High altitude permafrost 32
High-centre polygons 171, 174
High-latitude permafrost 31–32
Himalayas 16, 38, 106
Hoar frost 58, 202, 207, 498
Hydration shattering 74, 76, 77, 315, 319
Hydrology 85, 106, 111–115, 144, 214, 236, 268, 347, 369, 375, 400, 402, 465, 470, 471, 608
Hygroscopic water 43, 44, 45

Ice caves 9, 80, 95, 180, 202, 203, 204, 207–210
Ice complex 39, 153, 161, 180, 189–190, 192, 193, 194, 214, 420, 422
Ice jams 425–429, 506, 530, 531
Iceland 2, 8, 59, 99, 145, 175, 177, 200, 231, 233, 244, 248, 256, 259, 261, 305, 306, 310, 345, 392, 410, 411, 602, 607
Ice-push ridges 432, 433, 438
Ice segregation in microcracks 74
Ice stagnation deposits 183, 184
Ice tesselon 153
Ice-wedge casts 20, 21, 24, 38, 67, 153, 159, 161, 167, 169, 170, 171–173, 176, 210, 399, 406
Ice-wedge polygons 112, 154, 160, 161, 164, 225, 415, 416, 417, 433, 436, 439
Ice-wedges 149–177, 186, 189, 190, 192, 193, 240, 404, 407, 411, 413
Icing blisters 63, 226–228
Icings 111, 112, 114, 180, 195, 198, 446, 506, 532–537, 559, 588, 590, 599, 611, 612, 613
Icy string bogs 264–265
Incoming solar radiation 10, 103, 111, 405, 422
Injection ice 186, 187, 189, 214, 215, 216, 222, 225, 338
Injection structures 65–71
Inner Mongolia 170
Insulation 62, 109, 252, 255, 475, 476, 477, 486, 498, 500, 503, 507, 509, 513, 537, 538, 547, 555, 556, 605, 606, 607, 609
Intra-permafrost taliks 4
Intra-sedimental ice 186
Intrazonal soils 145

Involutions 65
Isolated patches 105, 126, 559
Isothermal 47, 52, 116, 122, 202, 267, 414

Japan 58, 145, 200, 271, 295, 297, 394, 443, 623
Jordtuva 252

Kamennye gletcery 334
Kammeis 58
Kara Sea 151, 187, 469, 483, 492
Karkevagge Sweden 77, 311, 312, 313, 321
Karst 78–80
Kazakhstan 89, 177, 335, 342, 347, 621, 631
Khatanga Russia 8
Kunlun Pass 27, 138, 221, 285, 493
Kurums 12, 146, 315, 350, 355–357

Lake ice 31, 115, 117, 189
Lakes 7, 46, 62, 115–117, 413, 414, 417, 595, 598, 602
Lapse rate 48, 94, 106, 107, 209
Laptev Sea 38, 40, 81, 83, 397, 435, 437, 492
Laramie Wyoming 167, 360
Latent heat 10, 12, 20, 47, 48, 51, 59, 90, 92, 93, 95
Latitudinal permafrost 98, 129, 130, 630
Layered permafrost 7, 123
Leptosols 22, 145
Lithalsa plateaus 240, 244
Lithalsas 26, 63, 214, 215, 228, 229, 230, 235, 239–243, 244, 248, 402
Lithalsa scars 409, 410
Lithology 61, 251, 252, 313, 355
Little Cornwallis Island Nunavut 587
Load casting 65, 66, 67, 69–71, 169, 380
Load-haul-dump vehicles 589
Loess tessellons 170, 409
Low-centre polygons 164
Lower limit of permafrost 4, 16, 103, 104, 121, 364, 395

Mackenzie Delta 37, 38, 51, 60, 62, 68, 69, 70, 110, 129, 161, 164, 170, 187, 188, 189, 216, 217, 218, 226, 276, 549, 550
Mackenzie mountains 161, 471
Mackenzie pingos 216
Mackenzie valley 302, 304, 400, 479, 630

Main oil and gas producing fields in the Northern Hemisphere 544
Marine type salinity 193
Marmot Basin Jasper National Park 51, 114, 135, 140
Massive ground ice 179–211
Massive ice blocks 169, 211
Mass wasting 77, 189, 259, 261, 267–313, 363, 392
Maximum embankment heights 509
Mean annual air temperature (MAAT), 34, 54, 55, 91, 95, 96, 98, 100, 103, 108, 109, 111, 118, 123, 126, 131, 143, 150, 151, 157, 160, 198, 202, 203, 204, 209, 259, 322, 354, 359, 400, 508, 509
Mean annual freezing index 15, 91
Mean annual ground surface temperature (MAGST), 54, 56, 62, 108
Mean annual ground temperature 12, 91, 96, 111, 128, 130, 142–144, 150, 159, 359, 515, 530
Mean annual thawing index 15, 91
Methods of construction
  active method 470
  passive method 470–471
Micro-landscapes 128
Micro-organisms 30, 31, 32, 33, 34, 45, 603
Mineral soil offset 94
Mining
  miles method 579
  mine tailings 12
  open cast/open pit 580–587
  pearce method 578, 580
  placer mining 577–580
  steam points 578
Mirny 577, 580, 581, 582, 600
Modeling of permafrost 130–131
Moisture movement 16–17
Moisture regimes 1, 9, 17, 375, 418
Moisture wicking material 499
Mongolia 8, 70, 96, 111, 196, 213, 221, 226, 256, 284, 393, 417, 423, 618, 625, 630, 631, 633
Monitoring 562, 566
Moose warmer 534, 535, 537
Morenny rock glacier 335
Mountain permafrost 48, 129, 130, 136, 142, 143, 144
Mount Garibaldi 59, 278
Muli coalfield China 596

Multilayered permafrost 7, 123
Multi-stage ice-wedges 157
Municipal water storage 602
Mushfrost 58

Nahanni National Park Northwest
    Territories 7
Naleds 180, 195, 198
Needle ice 48, 58–59, 73, 270, 373, 391,
    394
Needle ice creep 270–272, 273, 279
Nelson frost number 92
Net solar radiation 86, 90, 95
New Zealand 38, 58, 78, 98, 145, 189, 256,
    260, 261, 271, 280, 295, 297
Non-confluent permafrost 123
Non-conventional permafrost 198
Non-sorted circles 20, 21, 69, 261, 371
Nonsorted circles in maritime climates
    384–385
Norilsk 443, 466, 501, 551, 559, 577, 587,
    595
Northern Russia 19, 103, 512
North Rockies 16
Norway 6, 78, 90, 145, 230, 231, 233,
    235, 264, 362, 369, 394, 461,
    498, 543

Offshore permafrost 38–41, 56
Oil and gas exploration 543–545
Organic layer offset 94
Organic matter 27–30, 34, 35, 43, 118, 235,
    239, 248, 254, 255, 273, 376, 403,
    415, 571, 602, 603, 608, 629
Oriented lakes 415–417
Orographic precipitation 99, 104
Over-compaction of soil 403

Pads 473–474, 475, 498
Palsa plateaus 235, 237, 410
Palsas
    continental 230, 233, 234, 235, 236, 248,
        510
    maritime 231, 233, 234, 235, 399, 410
Palsa scars 410
Paving roads and airfields 527–529
Peatlands 29, 96, 110, 230, 248, 415, 630
Peat plateaus 29, 118, 160, 214, 230, 235,
    244–249
Pechora Sea 41, 83
Pediments 268, 284, 285
Perched water table 3, 19, 379, 401, 538,
    621

Pereletok 7, 8, 123
Perennial frost blisters 225
Permafrost 3
Permafrost base 4, 6, 8, 54, 55, 222, 230,
    613
Permafrost carbon content 139
Permafrost distribution 123–144, 264, 394,
    400
Permafrost ice content 398
Permafrost key sites 128
Permafrost landforms 18, 89, 145–147, 250,
    405
Permafrost landscapes 304
Permafrost mapping 127–128
Permafrost probability maps 130, 132, 265
Permafrost sensitivity to climate change 140
Permafrost southern boundary 7, 123, 125,
    207
Permafrost stability 109, 142–144,
    468–470, 588
Permafrost table 3, 4, 7, 14, 17, 18, 19, 20,
    51, 53, 54, 68, 77, 111, 112, 119, 138,
    152, 157, 238, 250, 253, 255, 256,
    267, 276, 297, 299, 323, 356, 376,
    379, 382, 389, 399, 403, 479, 492,
    501, 504, 512, 518, 520, 529, 534,
    538, 539, 540, 541, 549, 591, 598
Permafrost thickness 1, 103, 123, 131, 135,
    137
Pile foundations 614
Pingos
    closed system/hydrostatic/Mackenzie type
        216
    open system/hydraulic/East Greenland type
        218
    pingo plateaus 48, 222–225
    pingo scar 216, 217, 409–411
Pipelines
    above ground pipelines 551
    buried pipelines 552, 607
    cold pipelines 553, 554
    crossing rivers 552
    monitoring 566
    pipeline crossings 569–570
    temperature profiles 56, 136
    warm pipelines 552, 553
Pipkrake 58, 270
Piston effect 446
Plastic strain 445, 456
Plateau Mountain Alberta 21, 52, 100, 146,
    202, 324, 325, 382, 383, 390, 624
Plateau permafrost 129, 135

Pleistocene Mackenzie Delta 170, 188
Ploughing block 281, 283
Plug-like flow 276
Polar jet stream 98
Pollution effects on soil freezing 50
Polycyclic retrogressive thaw slumps 301
Polygonal beaded relief 154
Polynias 431
Polythermal glaciers 122
Pounus 265–266
Precipitation 10, 26, 48, 54, 78, 81, 99–102,
    146, 284, 293, 306
Primary heaving 59
Primary sandy wedges 153
Primary wedges 153–170, 409
Prince Patrick Island 170, 431
Production wells 83, 546, 548, 634
Prudhoe Bay 10, 11, 56, 77, 174, 377, 479,
    500, 546, 570, 571
Pseudomorphs of repeated wedges 153

Qilian Mountains China 36, 70
Qinghai-Tibet Highway 130, 139, 493, 527,
    533
Qinghai-Tibet Plateau 16, 32, 36, 37, 38, 48,
    53, 54, 92, 101, 102, 103, 104, 106,
    107, 112, 120, 129, 130, 135, 146,
    170, 210, 213, 219, 220, 221, 226,
    230, 235, 242, 243, 284, 306, 321,
    350, 351, 353, 357, 369, 399, 406,
    472, 493, 494, 496, 501, 502, 504,
    514, 515, 518, 523, 532, 537, 558,
    572, 596, 611, 621, 625–626, 633

Rafts 474–475
Regelation 73, 186
Regional climatic changes 602
Reliability of mapping 131–135
Relict permafrost 1, 6, 7, 15, 41, 54, 105,
    106, 123, 621
Repeated ice-wedges 153, 157
Resolute Bay Cornwallis Island 14, 69
Retrogressive thaw flow slides 300
Rheological properties of soils 50
River ice 187, 189
Rockfalls 20, 268, 313, 328, 333, 337,
    341
Rock glaciers
   active rock glaciers 136, 332, 335,
       337–339, 340–345, 345–347
   complex rock glaciers 336, 338, 347
   fossil rock glaciers 347–348

inactive rock glaciers 136, 336, 347,
    348
lobate rock glaciers 334, 335, 337,
    346
near slope rock glaciers 337, 338
piedmont rock glaciers 334
spatulate rock glaciers 331, 334
tongue-shaped rock glaciers 334, 335,
    337, 341, 342
Rock streams 350
Rock tessellons 38, 62, 149–177, 399
Roc Noir rock glacier 348
Rossby waves 25, 97, 98
Rotational slump 425
Rotten porous ice 5
Rubble streams 350

Saline soils 15, 24, 25, 26, 49, 60, 161, 193,
    195, 405, 406, 477, 502, 532
Sandar China 172
Sand fences 517
Sand tessellons 167–170, 210
Sand wedge 169, 210, 399, 406, 409
Schefferville 100, 111, 583, 584
Screes 12, 326
Sea density 80–83
Seasonal frost cracks 173
Seasonal frost mounds 63, 127, 215,
    225–226
Seasonally-frozen soil 1
Secondary heaving 60
Secondary wedges 153, 170–177
Segregated ice 19, 158, 186, 188, 210, 229,
    256, 298, 413, 435, 451, 454, 460
Sensible heat flux 90, 95, 300
Shading 111, 493, 496–498
Shargin's well 13, 133
Shield layer 4, 17
Shrinkage limit 446
Shrub offset 95
Siberia 7, 8, 29, 31, 37, 56, 62, 80, 97, 108,
    115, 142, 150, 154, 157, 158, 159,
    164, 165, 166, 180, 182, 189, 198,
    215, 218, 222, 226, 227, 231, 235,
    283, 285, 336, 350, 353, 354, 355,
    357, 381, 398, 400, 402, 406, 414,
    422, 441, 466, 475, 511, 530, 540,
    568, 580, 591, 611, 619, 620,
    621–623, 629, 633, 634, 635
Silcrete 22
Sills 186, 187, 189, 475–476
Slabs 309, 474–475

Slides
    landslides 296, 297, 317, 322, 341, 424,
        435, 502, 504
    rotational 296, 425
Slope thermal offset 95
Slow flows 269–287
Slumps
    active slumps 301, 304
    relict slumps 304
    stable slumps 304
Slush flows 304
Snow avalanches
    dry 310
    gully 308
    Lee cliff 307
    wet 310
Snow depth 100, 130, 507, 550
Snow pack offset 94, 95, 99
Snow removal 128, 492
Soil creep 192, 195
Soil wedges 150, 171, 173–177
Solar radiation 10, 74, 81, 85, 89, 97, 103,
        105, 109, 111, 300, 401, 434, 514,
        518
Solifluction
    fast solifuction 275
    slow solifluction 275
Solonchaks 24, 25, 26
Solonetz soils 24
Sorted polygons 368, 374, 388, 391
Sorted stone circles 21, 367
Sorting of sediments 369, 371, 386
South America 38, 89, 146, 200, 345, 399
Speleothems 200, 201, 203, 205, 207
Sporadic permafrost 16, 92, 98, 108, 126,
        132, 198, 200, 250, 264, 332, 385,
        415, 530, 565, 611
Spread footings 476–477
Stabilizing cut slopes 538, 539
Stagnant glacial ice 185
Stalactites 200, 206
Stalagmites 200, 206
Stefan-Boltzmann constant 86
Stefan-Boltzmann law 86
Stone-banked terraces/lobes 279
Stratified screes 326, 331, 334
String bog 252, 264–265, 266, 410
String fen 237
Structure of the water molecule 45
Subarctic mudboils 375, 377, 379–381
Sublimation 12, 93, 99, 102, 589

Sub-permafrost waters 4
Sub-sea permafrost 81
Summit Lake British Columbia 287, 288
Sump problems 549
Super cooling 47
Supra-permafrost taliks 4
Syngenetic/epigenetic permafrost 61
Syngenetic ice-wedges 157, 164
Syngenetic permafrost 61
Syngenetic salinity 24

Taffoni 321
Taliks 4, 7, 24, 78, 83, 106, 111, 502, 552,
        585, 598, 611
Talus 12, 322, 333–349
Talus slopes 12, 316, 326, 327, 328, 329,
        332, 333, 334, 346, 359
Tanggula Shan 54, 120, 139
Taryn 180, 195, 198
Temperature shift 10
Tessellations 20
Tessellons 152, 153, 169
Thaw bulb 469, 470, 473, 474, 553
Thawing fringe 3, 214, 244–249
Thawing front 3, 53, 267, 357
Thaw lakes 56, 401, 403, 415
Thaw settlement 60, 565, 611
Thaw weakening 460
Thermal abrasion 38
Thermal conductivity 6, 7, 9, 10, 12, 13, 51,
        53, 54, 59, 72, 94, 95, 99, 101, 102,
        112, 118, 122, 135, 357, 401, 403,
        406, 471, 472, 475, 478, 498, 508,
        555, 561
Thermal contraction cracking 149, 152,
        238
Thermal creep 270, 277
Thermal diode 10, 12, 27, 29, 109, 110,
        117, 160
Thermal erosion 119, 397–439
Thermal offset 30, 91, 93–96, 99, 100, 112,
        115, 117, 252, 402
Thermal regimes 10–15, 143
Thermal semi-conductors 493
Thermistors 9, 52, 247
Thermocouples 9, 52
Thermo-erosion 398, 403, 404
Thermokarst
    thermokarst lakes 30, 96, 140, 284,
        412–415, 416, 417, 629

thermokarst mounds 112, 172, 174, 407–409
thermokarst pits 406–407
Thermo-mechanical forcing 75
Thermo-mechanical processes 445
Thermopiles
  flat loop 488
  sloped 488
  vertical piles 488
Thermosiphons
  single phase 484, 487
  two phase 484, 487
Thermo-terraces 436
Through taliks 4, 7, 24
Thufurs 214, 249, 251, 252, 256–260, 262
Tibetan Plateau 12, 20, 24, 25, 34, 48, 57, 80, 96, 159, 161, 250, 251, 274, 354, 483
Tien Shan 16, 120, 177, 276, 322, 338, 343, 345, 354, 621, 625
Timpi 419, 420
Toe berms 494, 510, 511
Tors 78, 79, 285, 286
Total thermal offset 94, 95
Trains 518, 521, 533, 587, 589
Transient layer 3, 17, 123, 240, 401, 418, 422
Transmission tower foundations 614–616
Tree offset 94
Tubular heat drains 490, 492
Tuktoyaktuk 61, 549, 599, 609
Turf-banked terraces/lobes 268, 278, 279, 280, 283

Ulan Bator Mongolia 8, 630
Unpaved embankments 509–518
Up-heaving of objects 71–72
Up-turning of sediments 158
Upward injection of sediments 69, 70
Utilidors 598, 605, 606, 607, 609

Vegetation 2, 85, 106, 109–111, 118, 128, 134, 145, 146, 252, 253, 255, 263, 353, 400, 617, 621–623
Ventilation ducts 490–492, 494
Verkhoyansk 16
Vertical foliation 155, 158, 193
Vertical support members 562, 563
Viscous strain 445, 456

Warm-based glaciers 122
Waste materials
  conventional strength wastewater 607, 609
  incineration 608
  moderately diluted wastes 608
  solid waste disposal 610
  undiluted wastes 607, 608
  very dilute waste water 607, 609–610
  wastewater treatment 607
Waterproof plastics 498–500
Water requirements 603–604
Water supply
  intrapermafrost water 599
  subpermafrost water 218, 598, 599, 600
  suprapermafrost water 111, 598
Water treatment 602–603
Wave-cut notch 437, 438
Weathering 53, 74–78, 315, 318, 319, 320, 321, 324, 326, 347, 365, 502, 591
Wedges
  primary 153–170
  secondary 170–177
Winter roads 505–507
  ice bridges 506
  ice roads 506
  manufactured snow roads 506
  processed snow roads 506
  snow roads 506
  whole river ice roads 506
  winter trails 505

Xeric nonsorted circles 381–384

Yakutia 24, 72, 151, 192, 218, 222, 401, 405, 618
Yakutsk 8, 13, 55, 133, 180, 404, 418, 441, 465, 468, 470, 481, 488, 629
Yedoma 150, 153, 161, 180, 181, 182, 189, 190, 193, 195, 400, 414, 608
Yellow River 170, 171, 225
Yitulihe 156, 158, 559

Zailijskiy Alatau Kazahkstan 335
Zero curtain effect 13, 14, 47, 48
Zonal soils 145
Zonation of permafrost 8, 35, 126–127
Zone of minimum amplitude 14

Printed and bound by CPI Group (UK) Ltd, Croydon, CR0 4YY

24/10/2024

01778286-0010